FOURTH EDITION

DIGITAL SYSTEMS

PRINCIPLES AND APPLICATIONS

RONALD J. TOCCI
Monroe Community College

Prentice Hall, Englewood Cliffs, New Jersey 07632

LIBRARY OF CONGRESS
Library of Congress Cataloging-in-Publication Data

Tocci, Ronald J.
    Digital systems : principles and applications / Ronald J. Tocci. -
- 4th ed.
        p.   cm.
    Includes index.
    ISBN 0-13-213034-3
    1. Digital electronics.   I. Title.
TK7868.D5T62 1988
    621.395--dc19   88-473
                                                                     CIP

Editorial/production supervision: *Ed Jones*
Interior and cover design: *Lorraine Mullaney*
Manufacturing buyer: *Pete Havens and Bob Anderson*

© 1988, 1985, 1980, 1977 by Prentice-Hall, Inc.
A Division of Simon & Schuster
Englewood Cliffs, New Jersey 07632

Printed in the United States of America

10 9 8 7 6 5 4 3 2 1

ISBN   0-13-213034-3

Prentice-Hall International (UK) Limited, *London*
Prentice-Hall of Australia Pty. Limited, *Sydney*
Prentice-Hall Canada Inc., *Toronto*
Prentice-Hall Hispanoamericana, S.A., *Mexico*
Prentice-Hall of India Private Limited, *New Delhi*
Prentice-Hall of Japan, Inc., *Tokyo*
Simon & Schuster Asia Pte. Ltd., *Singapore*
Editora Prentice-Hall do Brasil, Ltda., *Rio de Janeiro*

CONTENTS

6 DIGITAL ARITHMETIC: OPERATIONS AND CIRCUITS   256

7 COUNTERS AND REGISTERS   310

8 INTEGRATED-CIRCUIT LOGIC FAMILIES   394

PREFACE

This book is a comprehensive up-to-date study of the principles and techniques of modern digital systems. It is intended for use in two-year and four-year programs in technology, engineering, and computer science. Although a background in basic electronic devices is helpful, a major portion of the material requires no electronics training. Those portions of the text that utilize electronic concepts can be skipped over without adversely affecting the comprehension of the logic principles.

GENERAL IMPROVEMENTS

This fourth edition contains many general improvements over the previous edition. All the material has been updated to reflect the latest developments in the digital field. A good portion of the material has been rewritten with greater clarity, more extensive illustrations, and additional applications. Each chapter now begins with a list of learning objectives and ends with an extensive glossary of terms that also serves as a chapter summary. A number of section review questions has been added to provide more reinforcement of the text material. A significant number of in-depth end-of-chapter problems has been added, bringing the total to well over 400. Many more illustrative examples have been added so that there are now 200 of them.

In addition to these pedagogical improvements, two major topic areas have been incorporated throughout the text: (1) the new IEEE/ANSI standard for logic symbols, and (2) troubleshooting. The inclusion of the IEEE/ANSI symbols is done gradually, with minimum disruption of the topic flow, and, if desired, the instructor can omit all or part of this material. The extensive troubleshooting coverage is spread over Chapters 4 through 11 and includes presentation of troubleshooting principles and techniques, case studies, 23 troubleshooting examples, and 56 *real* troubleshooting problems (these are printed in blue in the problem sections). When supplemented with proper laboratory exercises, this material should help foster the development of good troubleshooting skills.

SPECIFIC CHANGES

The major topical changes are:

Chapter 2:
Added number systems conversions using an intermediate number system.

Chapter 3:
More timing diagrams; added introduction of IEEE/ANSI symbols.

Chapter 4:
Added introduction of digital IC families and their basic characteristics; new material on common digital IC faults, circuit faults, and troubleshooting techniques.

Chapter 5:
Improved coverage of edge-triggered FFs and transparent latch; added coverage of data lockout FF, Schmitt trigger, astable multivibrators, 555 timer IC; new material on problems caused by clock skew.

Chapter 6:
Improved coverage of 2's-complement representation and hexadecimal arithmetic.

Chapter 8:
Added material on speed–power product, 74AS, 74ALS, 74HC and 74HCT families; increased emphasis on obtaining information from data sheets; introduction of logic pulser and current tracer as troubleshooting tools.

Chapter 9:
New material on digital magnitude comparator; more applications; more on bus signals.

Chapter 10:
Added R/2R ladder, bipolar DACs, and the ADC 0801 A/D converter; more on D/A converter specifications, data acquisition, and digitizing.

Chapter 11:
Added material on PLDs and PLAs, dynamic RAM controllers, sequential memories, FIFOs, CPU–memory interface, troubleshooting RAM systems, and ROM testing; expanded coverage of masked ROM and dynamic RAM.

RETAINED FEATURES

This edition retains all of the features that made the previous editions so widely accepted. It utilizes a block diagram approach to teach the basic logic operations without confusing the reader with the details of internal operation. All but the most basic electrical characteristics of the logic ICs are withheld until the reader has a firm understanding of logic principles. In Chapter 8 the reader is introduced to the internal IC circuitry. At that point, the reader can interpret a logic block's input and output characteristics and "fit" it properly into a complete system.

The treatment of each new topic or device typically follows these steps: the principle of operation is introduced; thoroughly explained examples and applications are presented often using actual ICs; short review questions are posed at the end of the section; and finally, in-depth problems are available at the end of the chapter. Ranging from simple to complex, these problems provide instructors with a wide choice of student assignments. These problems are often intended to reinforce the material without simple repetition of the principles. They require the student to demonstrate comprehension of the principles by applying them to different situations. This also helps the student develop confidence and expand his/her knowledge of the material.

SEQUENCING

It is a rare instructor who uses the chapters of a textbook in the sequence in which they are presented. In fact, I must admit that, for many different reasons, I do not use my own books in that way. This book was written so that, for the most part, each chapter builds on previous material, but it is possible to alter the chapter sequence somewhat. The first part of Chapter 6 (arithmetic operations) can be covered right after Chapter 2 (number systems), although this would produce a long interval before the arithmetic circuits of Chapter 6 are encountered. Much of the material in Chapter 8 (IC characteristics) can be covered earlier (e.g., after Chapter 4 or 5) without causing any serious problems.

This book can be used in either a one-term course or in a two-term sequence. When used in one term, it may be necessary, depending on available class hours, to omit some topics. Here is a list of sections and chapters that can be deleted with minimum disruption. Obviously, the choice of deletions will depend on factors such as program/course objectives and student background.

1. *Chapter 1:* all
2. *Chapter 2:* Sections 6 and 7
3. *Chapter 4:* Sections 7 and 8; Sections 10–13 if troubleshooting is not to be covered
4. *Chapter 5:* Sections 3 and 24
5. *Chapter 6:* Sections 5, 7, 11, 13, 16–18, 20
6. *Chapter 7:* Sections 10, 14–23
7. *Chapter 8:* Sections 9, 17–21
8. *Chapter 9:* Sections 6, 10, 11
9. *Chapter 10:* Sections 6, 13, 14
10. *Chapter 11:* Sections 10, 19–24
11. *Chapter 12:* all

ACKNOWLEDGMENTS

Prior to starting work on this edition, we sent an extensive questionnaire to many users and former users of the third edition. I am grateful to all of those who responded with their comments, critiques, and suggestions. Their input was invaluable as I went through the process of deciding what changes to incorporate in the new edition. I am also very grateful to the people at Prentice-Hall for all of their expertise and professionalism. A special fond thank-you to Alice Barr for her patience and flexibility in dealing with a procrastinating author. She was a constant source of encouragement, and her enthusiasm for the project was, at times, all that kept me going.

Finally, I would like to express my deepest gratitude to my colleague, Frank Ambrosio, who graciously agreed to assist me in preparing this revision. His excellent work on the chapter objectives and glossaries, index, and Instructors' Solutions Manual made it possible for me to meet the publication deadline.

Ron Tocci
Monroe Community College

INTRODUCTORY CONCEPTS

1

Upon completion of this chapter, you will be able to:

- Distinguish between analog and digital representations.
- Name the advantages, disadvantages, and major differences among analog, digital, and hybrid systems.
- Understand the need for analog-to-digital converters (ADCs) and digital-to analog converters (DACs).
- Convert between decimal and binary numbers.
- Identify typical digital signals.
- Cite several integrated-circuit fabrication technologies.
- Identify a timing diagram.
- State the differences between parallel and serial transmission.
- Name various memory elements.
- Describe the major parts of a digital computer and understand their functions.

INTRODUCTION

When most of us hear the term "digital," we immediately think of "digital calculator" or "digital computer." This can probably be attributed to the dramatic way that low-cost, powerful calculators and computers have become accessible to the average person. It is important to realize that calculators and computers represent only one of the many applications of digital circuits and principles. Digital circuits are used in electronic products such as video games, microwave ovens, and automobile control systems, and in test equipment such as meters, generators, and oscilloscopes. Digital techniques have also replaced a lot of the older "analog circuits" used in consumer products such as radios, TV sets, and high-fidelity sound recording and playback equipment.

In this book we are going to study the principles and techniques that are common to all digital systems from the simplest on/off switch to the most complex computer. If this book is successful, you should gain a deep understanding of how all digital systems work, and you should be able to apply this understanding to the analysis and troubleshooting of any digital system.

We start by introducing some underlying concepts that are a vital part of digital technology; these concepts will be expanded on as they are needed later in the text. We will also introduce some of the terminology that is so necessary when embarking on a new field of study, and will add to it in every chapter.

1-1 NUMERICAL REPRESENTATIONS

In science, technology, business, and, in fact, in most other fields of endeavor, we are constantly dealing with *quantities*. These quantities are measured, monitored, recorded, manipulated arithmetically, observed, or in some other way utilized in most physical systems. It is important when dealing with various quantities that we be able to represent their values efficiently and accurately. There are basically two ways of representing the numerical value of quantities: *analog* and *digital*.

Analog Representations In *analog representation* one quantity is represented by another which is directly proportional to the first. An example is an automobile speedometer, in which the deflection of the needle is proportional to the speed of the auto. The angular position of the needle represents the value of the auto's speed, and the needle follows any changes that occur as the auto speeds up or slows down.

Another example is the common room thermostat, in which the bending of the bimetallic strip is proportional to the room temperature. As the temperature changes gradually, the curvature of the strip changes proportionally.

Still another example of an analog quantity is found in the familiar audio microphone. In this device an output voltage is generated in proportion to the amplitude of the sound waves that impinge on the microphone. The variations in the output voltage follow the same variations as the input sound.

Analog quantities such as those cited above have an important characteristic: *they can vary over a continuous range of values*. The automobile speed can have *any* value between zero and, say, 100 mph. Similarly, the microphone output might be anywhere within a range of zero to 10 mV (for example, 1 mV, 2.3724 mV, 9.9999mV).

Digital Representations In *digital representation* the quantities are represented not by proportional quantities but by symbols called *digits*. As an example, consider the digital watch, which provides the time of day in the form of decimal digits which represent hours and minutes (and sometimes seconds). As we know, the time of day continuously changes, but the digital watch reading does not change continuously; rather, it changes in steps of one per minute (or per second). In other words, this digital representation of the time of day changes in *discrete* steps, as compared to the representation of time provided by an analog watch, where the dial reading changes continuously.

The major difference between analog and digital quantities, then, can be simply stated as follows:

$$analog \equiv continuous$$

$$digital \equiv discrete \ (step \ by \ step)$$

Because of the discrete nature of digital representations, there is no ambiguity when reading the value of a digital quantity, whereas the value of an analog quantity is often open to interpretation.

EXAMPLE 1-1

Which of the following involve analog quantities and which involve digital quantities?

(a) Resistor substitution box.

(b) Current meter.

(c) Temperature.

(d) Sand grains on a beach.

(e) Radio volume control.

SOLUTION

(a) Digital.

(b) Analog.

(c) Analog.

(d) Digital, since the number of grains can be only certain discrete (integer) values and not any value over a continuous range.

(e) Analog.

REVIEW QUESTION

1. Concisely describe the major difference between analog and digital quantities.

1-2 DIGITAL AND ANALOG SYSTEMS

A *digital system* is a combination of devices designed to manipulate physical quantities that are represented in digital form; that is, they can take on only discrete values. These devices are most often electronic, but they can also be mechanical, magnetic, or pneumatic. Some of the more familiar digital systems include digital computers and calculators, digital watches, traffic-signal controllers, and typewriters.

An *analog system* contains devices that manipulate physical quantities that are represented in analog form. In an analog system, the quantities can vary over a continuous range of values. For example, the amplitude of the output signal to the speaker in a radio receiver can have any value between zero and its maximum limit. Other common analog systems are the telephone system, magnetic tape recording and playback equipment, and the automobile odometer.

Advantages of Digital Techniques   An increasing majority of applications in electronics, as well as in most other technologies, use digital techniques to perform operations that were once performed using analog methods. The chief reasons for the shift to digital technology are:

1. *Digital systems are easier to design.* This is because the circuits that are used are *switching circuits,* where *exact* values of voltage or current are not important, only the range (HIGH or LOW) in which they fall.

2. *Information storage is easy.* This is accomplished by special switching circuits that can latch onto information and hold it for as long as necessary.

3. *Accuracy and precision are greater.* Digital systems can handle as many digits of precision as you need simply by adding more switching circuits. In analog systems, precision is usually limited to three or four digits because the values of voltage and current are directly dependent on the circuit component values.

4. *Operation can be easily programmed.* It is very easy to design digital systems whose operation is controlled by a set of stored instructions called a *program.* Analog systems can also be *programmed,* but the variety and complexity of the available operations is severely limited.

5. *Digital circuits are less affected by noise.* Spurious fluctuations in voltage (noise) are not as critical in digital systems because the exact value of a voltage is not important, as long as the noise is not large enough to prevent us from distinguishing a HIGH from a LOW.

6. *More digital circuitry can be fabricated on IC chips.* It is true that analog circuitry has also benefited from the tremendous development of IC technology, but its relative complexity, and its use of devices that cannot be economically integrated (high-value capacitors, precision resistors, inductors, transformers), have prevented analog systems from achieving the same high degree of integration.

Limitations of Digital Techniques
There is really only one major drawback when using digital techniques:

The real world is mainly analog

Most physical quantities are analog in nature, and it is these quantities that are often the inputs and outputs that are being monitored, operated on, and controlled by a system. Some examples are temperature, pressure, position, velocity, liquid level, flow rate, and so on. We are in the habit of expressing these quantities *digitally,* such as when we say that the temperature is 64° (63.8° when we want to be more accurate); but we are really making a digital approximation to an inherently analog quantity.

To take advantage of digital techniques when dealing with analog inputs and outputs, three steps must be followed:

1. Convert the ''real-world'' analog inputs to digital form.
2. Process (operate on) the digital information.
3. Convert the digital outputs back to real-world analog form.

Figure 1-1 shows a block diagram of this for a typical temperature control system. As the diagram shows, the analog temperature is measured and the measured value is then converted to a digital quantity by an *analog-to-digital converter (ADC).* The digital quantity is then processed by the digital circuitry, which may or may not include a digital computer. Its digital output is converted back to an analog quantity by a *digital-to-analog converter (DAC).* This analog output is fed to a controller which takes some kind of action to adjust the temperature.

The need for conversion between analog and digital forms of information can be considered a drawback because of the added complexity and expense. Another factor that is often important is the extra time required to perform these conversions. In many applications, these factors are outweighed by the numerous advantages of using digital techniques, so the conversion between analog and digital quantities has become quite commonplace in the current technology.

There are situations, however, where using only analog techniques is simpler

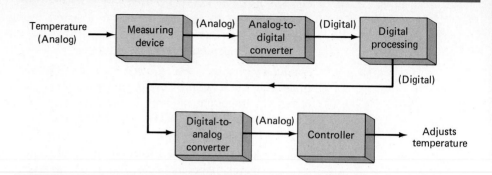

FIGURE 1-1   **Block diagram of a temperature control system that requires analog/digital conversions in order to allow the use of digital processing techniques.**

and more economical. For example, the process of signal amplification is most easily accomplished using analog circuitry.

It is becoming more and more common to see both digital and analog techniques employed within the same system in order to profit from the advantages of each. In these *hybrid* systems, one of the most important parts of the design phase involves determining what parts of the system are to be analog and what parts are to be digital.

Finally, it is safe to predict that as time goes on, we should see a greater and greater shift to digital techniques as the economic benefits of integration become even more overriding than they are now.

REVIEW QUESTIONS

1.   What are the advantages of digital techniques over analog?
2.   What is the chief limitation to the use of digital techniques?

1-3   DIGITAL NUMBER SYSTEMS

Many number systems are in use in digital technology. The most common are the decimal, binary, octal, and hexadecimal systems. The decimal system is clearly the most familiar to us because it is a tool that we use every day. Examining some of its characteristics will help us to better understand the other systems.

Decimal System   The *decimal system* is composed of *10* numerals or symbols. These 10 symbols are 0, 1, 2, 3, 4, 5, 6, 7, 8, 9; using these symbols as *digits* of a number, we can express any quantity. The decimal system, also called the *base-*

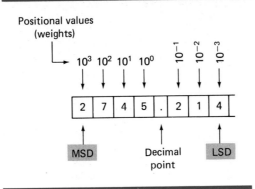

FIGURE 1-2 Decimal position values as powers of 10.

10 system, because it has 10 digits, has evolved naturally as a result of the fact that man has 10 fingers. In fact, the word "digit" is the Latin word for "finger."

The decimal system is a *positional-value system* in which the value of a digit depends on its position. For example, consider the decimal number 453. We know that the digit 4 actually represents 4 *hundreds,* the 5 represents 5 *tens,* and the 3 represents 3 *units*. In essence, the 4 carries the most weight of the three digits; it is referred to as the *most significant digit (MSD)*. The 3 carries the least weight and is called the *least significant digit (LSD)*.

Consider another example, 27.35. This number is actually equal to two tens plus seven units plus three tenths plus five hundredths, or $2 \times 10 + 7 \times 1 + 3 \times 0.1 + 5 \times 0.01$. The decimal point is used to separate the integer and fractional parts of the number.

More rigorously, the various positions relative to the decimal point carry weights that can be expressed as powers of 10. This is illustrated in Figure 1.2, where the number 2745.214 is represented. The decimal point separates the positive powers of 10 from the negative powers. The number 2745.214 is thus equal to

$$(2 \times 10^{+3}) + (7 \times 10^{+2}) + (4 \times 10^{1}) + (5 \times 10^{0})$$
$$+ (2 \times 10^{-1}) + (1 \times 10^{-2}) + (4 \times 10^{-3})$$

In general, any number is simply the sum of the products of each digit value times its positional value.

Decimal Counting

When counting in the decimal system, we start with 0 in the units position and take each symbol (digit) in progression until we reach 9. Then we add a 1 to the next higher position and start over with zero in the first position (see Figure 1.3). This process continues until the count of 99 is reached. Then we add a 1 to the third position and start over with zeros in the first two positions. The same pattern is followed continuously as high as we wish to count.

It is important to note that in decimal counting the units position (LSD) changes upward with each step in the count, the tens position changes upward every 10 steps in the count, the hundreds position changes upward every 100 steps in the count, and so on.

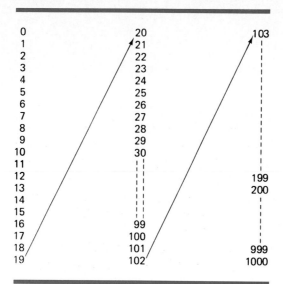

0	20	103
1	21	
2	22	
3	23	
4	24	
5	25	
6	26	
7	27	
8	28	
9	29	
10	30	
11		
12		199
13		200
14		
15		
16	99	
17	100	
18	101	999
19	102	1000

FIGURE 1-3 **Decimal counting.**

Another characteristic of the decimal system is that using only two decimal places we can count through $10^2 = 100$ different numbers (0 to 99).* With three places we can count through 1000 numbers (0 to 999); and so on. In general, with N places or digits we can count through 10^N different numbers, starting with and including zero. The largest number will always be $10^N - 1$.

Binary System   Unfortunately, the decimal number system does not lend itself to convenient implementation in digital systems. For example, it is very difficult to design electronic equipment so that it can work with 10 different voltage levels (each one representing one decimal character, 0-9). On the other hand, it is very easy to design simple, accurate electronic circuits that operate with only two voltage levels. For this reason, almost every digital system uses the binary number system (base 2) as the basic number system of its operations, although other systems are often used in conjunction with binary.

In the *binary system* there are only two symbols or possible digit values, 0 and 1. Even so, this base-2 system can be used to represent any quantity that can be represented in decimal or other number systems. In general though, it will take a greater number of binary digits to express a given quantity.

All the statements made earlier concerning the decimal system are equally applicable to the binary system. The binary system is also a positional-value system, wherein each binary digit has its own value or weight expressed as powers of 2. This is illustrated in Figure 1-4. Here, places to the left of the *binary point* (counterpart of the decimal point) are positive powers of 2 and places to the right are negative powers of 2. The number 1011.101 is shown represented in the figure. To find its equivalent in the decimal system we simply take the sum of the products of each digit value (0 or 1) times its positional value.

*Zero is counted as a number.

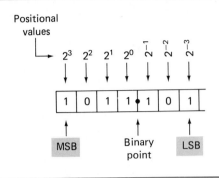

Positional values

MSB

Binary point

LSB

FIGURE 1-4 **Binary position values as powers of 2.**

$$1011.101_2 = (1 \times 2^3) + (0 \times 2^2) + (1 \times 2^1) + (1 \times 2^0)$$
$$+ (1 \times 2^{-1}) + (0 \times 2^{-2}) + (1 \times 2^{-3})$$
$$= 8 + 0 + 2 + 1 + 0.5 + 0 + 0.125$$
$$= 11.625_{10}$$

Notice in the preceding operation that subscripts (2 and 10) were used to indicate the base in which the particular number is expressed. This convention is used to avoid confusion whenever more than one number system is being employed.

In the binary system, the term *binary digit* is often abbreviated to the term *bit* which we will use henceforth. Thus, in the number expressed in Figure 1.4 there are 4 bits to the left of the binary point, representing the integer part of the number, and 3 bits to the right of the binary point, representing the fractional part. The most significant bit (MSB) is the leftmost bit (largest weight). The least significant bit (LSB) is the rightmost bit (smallest weight). These are indicated in Figure 1.4.

Binary Counting   When we deal with binary numbers, we will usually be restricted to a specific number of bits. This restriction is based on the circuitry that is used to represent these binary numbers. Let's use 4-bit binary numbers to illustrate the method for counting in binary.

The sequence (shown in Figure 1-5) begins with all bits at 0; this is called the *zero count*. For each successive count, the units (2^0) position *toggles;* that is, it changes from one binary value to the other. Each time it changes from a 1 to a 0, the twos (2^1) position will toggle. Each time the twos position changes from 1 to 0, the fours (2^2) position will toggle. Likewise, each time the fours position goes from 1 to 0, the eights (2^3) position toggles. This same process would be continued for the higher-order bit positions if the binary number had more than 4 bits.

The binary counting sequence has an important characteristic, as shown in Figure 1-5. The units bit (LSB) changes either from 0 to 1 or 1 to 0 with *each* count. The second bit (twos position) stays at 0 for two counts, then 1 for two counts, then 0 for two counts, and so on. The third bit (fours position) stays at 0 for four counts, then 1 for four counts; and so on. The fourth bit (eights position) stays at 0 for eight counts, then at 1 for eight counts. If we wanted to count further we would add more places, and this pattern would continue with 0s and 1s alternating in groups of 2^{N-1}. For example, using a fifth binary place, the fifth bit would alternate sixteen 0s, then sixteen 1s, and so on.

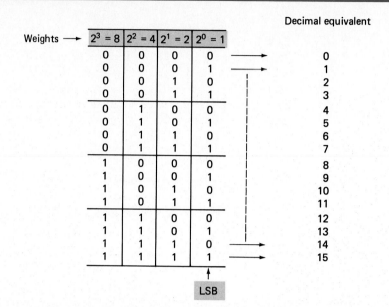

Weights →	$2^3 = 8$	$2^2 = 4$	$2^1 = 2$	$2^0 = 1$		Decimal equivalent
	0	0	0	0	→	0
	0	0	0	1	→	1
	0	0	1	0		2
	0	0	1	1		3
	0	1	0	0		4
	0	1	0	1		5
	0	1	1	0		6
	0	1	1	1		7
	1	0	0	0		8
	1	0	0	1		9
	1	0	1	0		10
	1	0	1	1		11
	1	1	0	0		12
	1	1	0	1		13
	1	1	1	0	→	14
	1	1	1	1	→	15

LSB

FIGURE 1-5   Binary counting sequence.

As we saw for the decimal system, it is also true of the binary system that by using N bits or places we can go through 2^N counts. For example, with 2 bits we can go through $2^2 = 4$ counts (00_2 through 11_2); with 4 bits we can go through $2^4 = 16$ counts (0000_2 through 1111_2); and so on. The last count will always be all 1s and is equal to $2^N - 1$ in the decimal system. For example, using 4 bits, the last count is $1111_2 = 2^4 - 1 = 15_{10}$.

EXAMPLE 1-2

What is the largest number that can be represented using 8 bits?

SOLUTION
$2^N - 1 = 2^8 - 1 = 255_{10} = 11111111_2$.

REVIEW QUESTIONS

1. What is the decimal equivalent of 1101011_2? (*Ans.* 107)
2. What is the next binary number following 10111_2 in the counting sequence? (*Ans.* 11000)
3. What is the largest decimal value that can be represented using 12 bits? (*Ans.* 4095)

1-4 REPRESENTING BINARY QUANTITIES

In digital systems the information that is being processed is usually present in binary form. Binary quantities can be represented by any device that has only two operating states or possible conditions. For example, a switch has only two states: open or closed. We can arbitrarily let an open switch represent binary 0 and a closed switch represent binary 1. With this assignment we can now represent any binary number as illustrated in Figure 1-6(a), where the states of the various switches represent 10010_2.

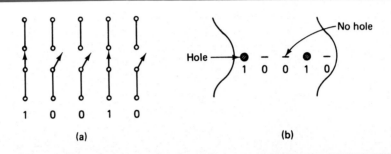

(a)   (b)

FIGURE 1-6   Using (a) switches and (b) punched paper tape to represent binary numbers.

Another example is shown in Figure 1-6(b), where holes punched in paper are used to represent binary numbers. A punched hole is a binary 1 and absence of a hole is a binary 0.

There are numerous other devices which have only two operating states or can be operated in two extreme conditions. Among these are: light bulb (bright or dark), diode (conducting or nonconducting), relay (energized or deenergized), transistor (cut off or saturated), photocell (illuminated or dark), thermostat (open or closed), mechanical clutch (engaged or disengaged), and magnetic tape (magnetized or demagnetized).

In electronic digital systems, binary information is represented by voltages (or currents) that are present at the inputs and outputs of the various circuits. Typically, the binary 0 and 1 are represented by two nominal voltage levels. For example, zero volts (0 V) might represent binary 0, and +5 V might represent binary 1. In actuality, because of circuit variations, the 0 and 1 would be represented by voltage ranges. This is illustrated in Figure 1-7(a), where any voltage between 0 and 0.8 V represents a 0 and any voltage between 2 and 5 V represents a 1. All input and output signals will normally fall within one of these ranges except during transitions from one level to another. Figure 1-7(b) shows a typical digital signal as it sequences through the binary value 01010.

We can now see another significant difference between digital and analog systems. In digital systems, the exact value of a voltage *is not* important; for example, a voltage of 3.6 V means the same as a voltage of 4.3 V. In analog systems, the exact value of a voltage *is* important. For instance, if the analog voltage is propor-

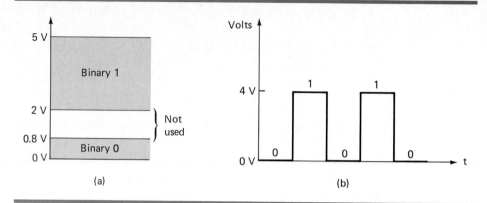

FIGURE 1-7 (a) Typical voltage assignments in digital system; (b) typical digital signal.

tional to the temperature measured by a transducer, 3.6 V would represent a different temperature than would 4.3 V. In other words, the voltage value carries significant information. This characteristic means that the design of accurate analog circuitry is generally more difficult than digital circuitry because of the way in which exact voltage values will fluctuate due to variations in component values, temperature, and noise.

1-5 DIGITAL CIRCUITS

As explained in the last section, digital circuits are designed to produce output voltages that fall within the prescribed 0 and 1 voltage ranges such as those defined in Figure 1-7. Likewise, digital circuits are designed to respond predictably to input voltages that are within the defined 0 and 1 ranges. What this means is that a digital circuit will respond in the same way to all input voltages that fall within the allowed 0 range; similarly, it will not distinguish between input voltages that lie within the allowed 1 range.

To illustrate, Figure 1-8 represents a typical digital circuit with input v_i and output v_o. The output is shown for two different input signal waveforms. Note that v_o is the same for both cases because the two input waveforms, while they differ in their exact voltage levels, are at the same binary levels.

Logic Circuits   The manner in which a digital circuit responds to an input is referred to as the circuit's *logic*. Each type of digital circuit obeys a certain set of logic rules. For this reason, digital circuits are also called *logic circuits*. We will use both terms interchangeably throughout the text. In Chapter 3 we will see more clearly what is meant by a circuit's ''logic.''

We will be studying all the types of logic circuits that are currently used in digital systems. Initially, our attention will be focused only on the logical operation which these circuits perform—that is, the relationship between the circuit inputs and outputs. We will defer any discussion of the internal circuit operation of these logic circuits until after we have developed an understanding of their logical operation.

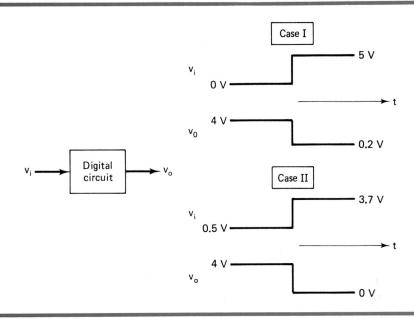

FIGURE 1-8 **A digital circuit responds to an input's binary level (0 or 1) and not to its actual voltage.**

Digital Integrated Circuits Almost all the digital circuits used in modern digital systems are integrated circuits (ICs). The wide variety of available logic ICs has made it possible to construct complex digital systems that are smaller and more reliable than the discrete-component counterparts.

Several integrated-circuit fabrication technologies are used to produce digital ICs, the most common being TTL, CMOS, NMOS, and ECL. Each differs in the type of circuitry used to provide the desired logic operation. For example, TTL (transistor-transistor logic) uses the bipolar transistor as its main circuit element, while CMOS (complementary metal-oxide semiconductor) uses the enhancement MOSFET as its principal circuit element. We will learn about the various IC technologies, their characteristics, and their relative advantages and disadvantages after we master the basic logic circuit types.

REVIEW QUESTIONS

1. *True or false:* The exact value of an input voltage is critical for a digital circuit.
2. Can a digital circuit produce the same output voltage for different input voltage values?
3. A digital circuit is also referred to as a _____ circuit.

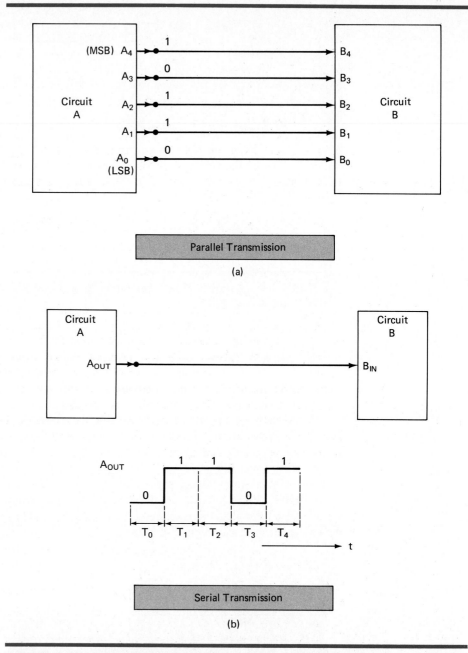

FIGURE 1-9   (a) Parallel transmission uses one connecting line per bit, and all bits are transmitted simultaneously; (b) serial transmission uses only one signal line, and the individual bits are transmitted serially (one at a time).

1-6 PARALLEL AND SERIAL TRANSMISSION

One of the most common operations that occurs in any digital system is the transmission of information from one place to another. The information can be transmitted over a distance as small as a fraction of an inch on the same circuit board, or over a distance of many miles when an operator at a computer terminal is communicating with a computer in another city. The information that is transmitted is in binary form and is generally represented as voltages at the outputs of a sending circuit that are connected to the inputs of a receiving circuit. Figure 1-9 illustrates the two basic methods for digital information transmission: *parallel* and *serial*.

Figure 1-9(a) shows how the binary number 10110 is transmitted from circuit A to circuit B using parallel transmission. Each bit of the binary number is represented by one of the circuit A outputs, where output A_4 is the MSB and A_0 is the LSB. Each of the circuit A outputs is connected to a corresponding input of circuit B so that all 5 bits of information are transmitted simultaneously (in parallel).

In Figure 1-9(b) there is only one connection from circuit A to circuit B when serial transmission is used. Here the output of circuit A will produce a digital signal whose voltage level will change at regular intervals in accordance with the binary number being transmitted. In this way the information is being transmitted a bit at a time (serially) over the one signal line. The *timing diagram* in Figure 1-9(b) shows how the signal level varies with time. During the first time interval, T_0, the signal is at the 0 level; during interval T_1 the signal is at the 1 level; and so forth.

The principal trade-off between parallel and serial representations is one of speed versus circuit simplicity. The transmission of binary data from one part of a digital system to another can be done more quickly using parallel representation because all the bits are transmitted simultaneously, while serial representation transmits one bit at a time. On the other hand, parallel requires more signal lines connected between the sender and receiver of the binary data than does serial. In other words, parallel is faster and serial requires fewer signal lines. This comparison between parallel and serial methods for representing binary information will be encountered many times in discussions throughout the text.

REVIEW QUESTION

1. Describe the relative advantages of parallel and serial transmission of binary data.

1-7 MEMORY

When an input signal is applied to most devices or circuits, the output somehow changes in response to the input, and when the input signal is removed, the output returns to its original state. These circuits do not exhibit the property of *memory*, since their outputs revert back to normal. In digital circuitry certain types of devices

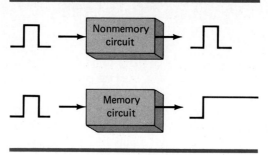

FIGURE 1-10   Comparison of nonmemory and memory operation.

and circuits do have memory. When an input is applied to such a circuit, the output will change its state, but it will remain in the new state even after the input is removed. This property of retaining its response to a momentary input is called memory. Figure 1-10 illustrates nonmemory and memory operations.

Memory devices and circuits play an important role in digital systems because they provide means for storing binary numbers either temporarily or permanently, with the ability to change the stored information at any time. As we shall see, the various memory elements include magnetic types and those which utilize electronic latching circuits (called *flip-flops*).

1-8   DIGITAL COMPUTERS

Digital techniques have found their way into innumerable areas of technology, but the area of automatic *digital computers* is by far the most notable and most extensive. Although digital computers affect some part of all of our lives, it is doubtful that many of us know exactly what a computer does. In simplest terms, *a computer is a system of hardware that performs arithmetic operations, manipulates data (usually in binary form), and makes decisions.*

For the most part, human beings can do whatever computers can do, but computers can do it with much greater speed and accuracy. This is in spite of the fact that computers perform all their calculations and operations one step at a time. For example, a human being can take a list of 10 numbers and find their sum all in one operation by listing the numbers one over the other and adding them column by column. A computer, on the other hand, can add numbers only two at a time, so that adding this same list of numbers will take nine actual addition steps. Of course, the fact that the computer requires only a microsecond or less per step makes up for this apparent inefficiency.

A computer is faster and more accurate than people are, but unlike most people it has to be given a complete set of instructions that tell it *exactly* what to do at each step of its operation. This set of instructions, called a *program,* is prepared by one or more persons for each job the computer is to do. Programs are placed in the computer's memory unit in binary-coded form, with each instruction having a unique code. The computer takes these instruction codes from memory *one at a time* and performs the operation called for by the code. Much more will be said about this later.

Major Parts of a Computer   There are several types of computer systems, but each can be broken down into the same functional units. Each unit performs specific functions, and all units function together to carry out the instructions given in the program. Figure 1-11 shows the five major functional parts of a digital computer and their interaction. The solid lines with arrows represent the flow of information. The dashed lines with arrows represent the flow of timing and control signals.

The major functions of each unit are:

1. *Input unit.* Through this unit a complete set of instructions and data are fed into the computer system and into the memory unit, to be stored until needed. The information typically enters the input unit by means of punched cards, magnetic tape, magnetic disk, or a keyboard.

2. *Memory unit.* The memory stores the instructions and data received from the input unit. It stores the results of arithmetic operations received from the arithmetic unit. It also supplies information to the output unit.

3. *Control unit.* This unit takes instructions from the memory unit one at a time and interprets them. It then sends appropriate signals to all the other units to cause the specific instruction to be executed.

4. *Arithmetic Unit.* All arithmetic calculations and logical decisions are performed in this unit, which can then send results to the memory unit to be stored.

5. *Output unit.* This unit takes data from the memory unit and prints out, displays, or otherwise presents the information to the operator (or process, in the case of a process control computer).

How Many Types of Computers Are There?   The answer depends on the criteria used to classify them. Computers are often classified according to physical size, which usually, but not always, is also indicative of their relative capabilities. The *microcomputer* is the smallest and newest member of the computer family. It

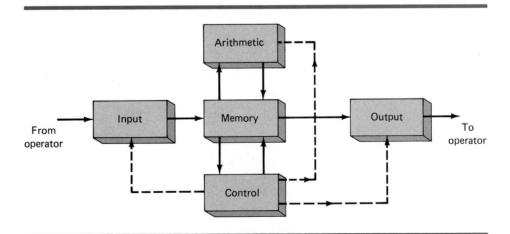

FIGURE 1-11   Functional diagram of a digital computer.

generally consists of several IC chips, including a *microprocessor* chip, *memory* chips, and *input/output interface* chips. These chips are a result of tremendous advances in IC fabrication technology, which have also served to bring prices down so that it is possible to buy a fully assembled microcomputer now for less than $100.

Minicomputers are larger than microcomputers and have prices that go into the tens of thousands of dollars (including input/output peripheral equipment). Minis are widely used in industrial control systems, scientific applications for schools and research laboratories, and in business applications for small businesses. Although more expensive than microcomputers, minicomputers continue to be widely used because they are generally faster and possess more capabilities. These differences in speed and capability, however, are rapidly shrinking.

The largest computers are those found in large corporations, banks, universities, and scientific laboratories. These "maxicomputers" can cost as much as several million dollars and include complete systems of peripheral equipment such as magnetic tape units, magnetic disk units, punched card punchers and readers, keyboards, printers, and many more. Applications of maxicomputers range from computationally oriented science and engineering problem solving to the data-oriented business applications, where emphasis is on maintaining and updating large quantities of data and information.

REVIEW QUESTIONS

1. Explain how a digital circuit that has memory differs from one that does not.
2. Name the five major functional units of a computer.
3. Name the three types of computers and briefly describe how they differ.

GLOSSARY **Analog Representation** Representation of a quantity that varies over a continuous range of values. [Sec. 1-1]

Analog System Combination of devices designed to manipulate physical quantities that are represented in analog form. [Sec. 1-2]

Analog-to-Digital Converter: Circuit that converts an analog quantity to a corresponding digital quantity. [Sec. 1-2]

Arithmetic-Logic Unit Part of the computer dedicated to all of the arithmetic and logical operations. [Sec. 1-8]

Binary Digit Bit. [Sec. 1-3]

Binary Point Mark that separates the integer from the fractional portion of a binary quantity. [Sec. 1-3]

Binary System Number system in which there are only two possible digit values, 0 and 1. [Sec. 1-3]

Bit A digit in the binary system. [Sec. 1-3]

CMOS (Complementary Metal-Oxide Semiconductor) Integrated-circuit technology that uses MOSFETS as the principal circuit element. [Sec. 1-5]

Control Unit Provides decoding of program instructions and the necessary timing and control signals for the execution of such instructions. [Sec. 1-8]

Decimal system Number system that uses 10 different digits or symbols to represent a quantity. [Sec. 1-3]

Digital Computer System of hardware that performs arithmetic and logic operations, manipulates data, and makes decisions. [Sec. 1-8]

Digital Integrated Circuits: Self-contained circuits that have been made by using one of several integrated-circuit fabrication technologies. [Sec. 1-5]

Digital Representation Representation of a quantity that varies in discrete steps over a range of values. [Sec. 1-1]

Digital System Combination of devices designed to manipulate physical quantities that are represented in digital form. [Sec. 1-2]

Digital-to-Analog Converter Circuit that converts a digital quantity to a corresponding analog quantity. [Sec. 1-2]

Flip-flop Memory device capable of storing a logic level. [Sec. 1-7]

Hybrid System System that employs both analog and digital techniques. [Sec. 1-2]

Input Unit Facilitates the feeding of information into the computer's memory unit. [Sec. 1-8]

Least Significant Bit (LSB) Rightmost bit (smallest weight) of a binary expressed quantity. [Sec. 1-3]

Least Significant Digit (LSD) Digit that carries the least weight in a particular number. [Sec. 1-3]

Logic Circuits Any circuit that behaves according to a set of logic rules. [Sec. 1-5]

Maxicomputer Largest computers available, used to maintain and update large quantities of data and information. [Sec. 1-8]

Memory Ability of a circuit's output to remain at one state even after the input condition that caused that state is removed. [Sec. 1-7]

Memory Unit Stores instructions and data received from the Input unit, as well as results from the arithmetic-logic unit. [Sec. 1-8]

Microcomputer Newest member of the computer family consisting of microprocessor chip, memory chips, and I/O interface chips. In some cases all of the aforementioned are in one single IC. [Sec. 1-8]

Minicomputers Computers that are generally larger, faster, and possess more capabilities than microcomputers. [Sec. 1-8]

Most Significant Bit (MSB) Leftmost binary bit (largest weight) of a binary expressed quantity. [Sec. 1-3]

Most Significant Digit (MSD) Digit that carries the most weight in a particular number. [Sec. 1-3]

NMOS (N-channel Metal-Oxide Semiconductor) Integrated-circuit technology that uses N-channel MOSFETS as the principal circuit element. [Sec. 1-5]

Output Unit Receives data from the memory unit and presents it to the operator. [Sec. 1-8]

Parallel Transmission Simultaneous transfer of all bits of a binary number from one place to another. [Sec. 1-6]

Positional-Value System System in which the value of a digit is dependent on its relative position. [Sec. 1-3]

Program Sequence of binary-coded instructions designed to accomplish a particular task by a computer. [Secs. 1-2 and 1-8]

Serial Transmission Transfer of binary information from one place to another a bit at a time. [Sec. 1-6]

Timing Diagram Depiction of logic levels as related to time. [Sec. 1-6]

Toggles Process of changing from one binary state to the other. [Sec. 1-3]

TTL (Transistor-Transistor Logic) Integrated-circuit technology that uses the bipolar transistor as the principal circuit element. [Sec. 1-5]

PROBLEMS

Section 1-2

1-1. Which of the following are analog quantities, and which are digital?
(a) The number of atoms in a sample of material.
(b) The altitude of an aircraft.
(c) The pressure in a bicycle tire.
(d) Current through a speaker.
(e) Timer setting on a microwave oven.

Section 1-3

1-2. Convert the following binary numbers to their equivalent decimal values.
(a) 11001
(b) $1001.1001_2 = \underline{\hspace{1cm}}_{10}$
(c) $10011011001.10110 = \underline{\hspace{1cm}}_{10}$

1-3. Using 6 bits, show the binary counting sequence from 000000 to 111111.

1-4. What is the maximum number that we can count up to using 10 bits?

1-5. How many bits are needed to count up to a maximum of 511?

1-6. Suppose that the decimal integer values from 0 to 15 are to be transmitted.
(a) How many lines will be needed if parallel representation is used?
(b) How many will be needed if serial representation is used?

NUMBER SYSTEMS AND CODES

2

OUTLINE

Upon completion of this chapter, you will be able to:

- Use two different methods to perform decimal-to-binary conversions.
- Cite several advantages of the octal number system.
- Convert from the hexadecimal or octal number systems to either the decimal or binary number systems.
- Express decimal numbers using the BCD code.
- Understand the difference between the BCD code and the straight binary code.
- Cite the major differences between the Gray code and the binary code.
- Understand the need for alphanumeric codes, especially the ASCII code.
- Describe the parity method for error detection.
- Determine parity (odd or even) of digital data.

INTRODUCTION

The binary number system is the most important one in digital systems, but several others also are important. The decimal system is important because it is universally used to represent quantities outside a digital system. This means that there will be situations where decimal values have to be converted to binary values before they are entered into the digital system. For example, when you punch a decimal number into your hand calculator (or computer), the circuitry inside the device converts the decimal number to a binary value.

Likewise, there will be situations where the binary values at the outputs of a digital circuit have to be converted to decimal values for presentation to the outside world. For example, your calculator (or computer) uses binary numbers to calculate answers to a problem, then converts the answers to a decimal value before displaying them.

In addition to binary and decimal, two other number systems find widespread applications in digital systems. The *octal (base-8)* and *hexadecimal (base-16)* number systems are both used for the same purpose—to provide an efficient means for representing large binary numbers. As we shall see, both these number systems have the advantage that they can be easily converted to and from binary.

In a digital system, three or four of these number systems may be in use at the same time, so that an understanding of the system operation requires the ability to convert from one number system to another. This chapter will show you how to perform these conversions. Although some of them will not be of immediate use in our study of digital systems, you will need them when you begin to study microprocessors.

This chapter will also introduce some of the *binary codes* that are used to represent various kinds of information. These binary codes will use 1s and 0s, but in a way that differs somewhat from the binary number system.

2-1 BINARY-TO-DECIMAL CONVERSIONS

As explained in Chapter 1, the binary number system is a positional system where each binary digit (bit) carries a certain weight based on its position relative to the

LSB. Any binary number can be converted to its decimal equivalent simply by summing together the weights of the various positions in the binary number which contain a 1. To illustrate:

$$1 \quad 1 \quad 0 \quad 1 \quad 1_2 \quad (binary)$$

$$2^4 + 2^3 + 0 + 2^1 + 2^0 = 16 + 8 + 2 + 1$$

$$= 27_{10} \quad (decimal)$$

Let's try another example with a greater number of bits.

$$1 \; 0 \; 1 \quad 1 \; 0 \; 1 \; 0 \quad 1_2 =$$

$$2^7 \; + \; 2^5 \; + \; 2^4 \; + \; 2^2 \; + \; 2^0 \; = \; 181_{10}$$

Note that the procedure is to find the weights (i.e., powers of 2) for each bit position that contains a 1, then add them up. Also note that the MSB has a weight of 2^7 even though it is the 8th bit; this is because the LSB is the first bit and it always has a weight of 2^0.

REVIEW QUESTIONS

1. Convert 100011011011_2 to its decimal equivalent. (*Ans.* 2267)
2. What is the weight of the MSB of a 16-bit number? (*Ans.* 32768)

2-2 DECIMAL-TO-BINARY CONVERSIONS

There are two ways to convert a decimal number to its equivalent binary-system representation. The first method is the reverse of the process described in Section 2.1. The decimal number is simply expressed as a sum of powers of 2 and then 1s and 0s are written in the appropriate bit positions. To illustrate:

$$45_{10} = 32 + 8 + 4 + 1 = 2^5 + 0 + 2^3 + 2^2 + 0 + 2^0$$

$$= 1 \quad 0 \quad 1 \quad 1 \quad 0 \quad 1_2$$

Note that a 0 is placed in the 2^1 and 2^4 positions, since all positions must be accounted for. Another example is shown below.

$$76_{10} = 64 + 8 + 4 = 2^6 + 0 + 0 + 2^3 + 2^2 + 0 + 0$$

$$= 1 \quad 0 \quad 0 \quad 1 \quad 1 \quad 0 \quad 0_2$$

Repeated Division Another method uses repeated division by 2. The conversion, illustrated below for 25_{10}, requires repeatedly dividing the decimal number by 2 and writing down the remainders after each division until a quotient of 0 is obtained. Note that the binary result is obtained by writing the first remainder as the LSB and the last remainder as the MSB.

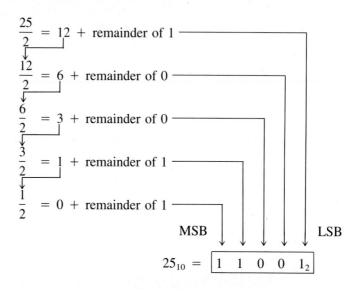

$$\frac{25}{2} = 12 + \text{remainder of } 1$$

$$\frac{12}{2} = 6 + \text{remainder of } 0$$

$$\frac{6}{2} = 3 + \text{remainder of } 0$$

$$\frac{3}{2} = 1 + \text{remainder of } 1$$

$$\frac{1}{2} = 0 + \text{remainder of } 1$$

MSB   LSB

$$25_{10} = \boxed{1 \quad 1 \quad 0 \quad 0 \quad 1}_2$$

If a calculator is being used to perform the divisions by 2, the remainders can be determined by noting whether the quotient has a fractional part or not. For example, the calculator would produce $25/2 = 12.5$. The .5 indicates that there is a remainder of 1. The calculator would also produce $12/2 = 6.0$, which indicates a remainder of 0. Another example is shown below as it would occur using a calculator.

$$\frac{37}{2} = 18.5 \longrightarrow \text{remainder of } 1 \text{ (LSB)}$$

$$\frac{18}{2} = 9.0 \longrightarrow \qquad\qquad 0$$

$$\frac{9}{2} = 4.5 \longrightarrow \qquad\qquad 1$$

$$\frac{4}{2} = 2.0 \longrightarrow \qquad\qquad 0$$

$$\frac{2}{2} = 1.0 \longrightarrow \qquad\qquad 0$$

$$\frac{1}{2} = 0.5 \longrightarrow \qquad\qquad 1 \text{ (MSB)}$$

Thus, $37_{10} = 100101_2$.

Review Questions

2-3 OCTAL NUMBER SYSTEM

The octal number system is very important in digital computer work. The octal number system has a base of *eight,* meaning that it has eight possible digits: 0, 1, 2, 3, 4, 5, 6, and 7. Thus, each digit of an octal number can have any value from 0 to 7. The digit positions in an octal number have weights as follows:

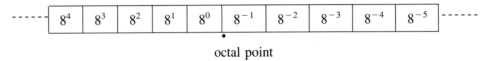

$$\begin{array}{|c|c|c|c|c|c|c|c|c|c|} \hline 8^4 & 8^3 & 8^2 & 8^1 & 8^0 & 8^{-1} & 8^{-2} & 8^{-3} & 8^{-4} & 8^{-5} \\ \hline \end{array}$$

octal point

Octal-to-Decimal Conversion   An octal number, then, can be easily converted to its decimal equivalent by multiplying each octal digit by its positional weight. For example,

$$372_8 = 3 \times (8^2) + 7 \times (8^1) + 2 \times (8^0)$$
$$= 3 \times 64 + 7 \times 8 + 2 \times 1$$
$$= 250_{10}$$

Another example:

$$24.6_8 = 2 \times (8^1) + 4 \times (8^0) + 6 \times (8^{-1})$$
$$= 20.75_{10}$$

Decimal-to-Octal Conversion   A decimal integer can be converted to octal by using the same repeated-division method that we used in the decimal-to-binary conversion, but with a division factor of 8 instead of 2. An example is shown below.

$$\frac{266}{8} = 33 + \text{remainder of } 2$$
$$\frac{33}{8} = 4 + \text{remainder of } 1$$
$$\frac{4}{8} = 0 + \text{remainder of } 4$$

$$266_{10} = \boxed{412_8}$$

Note that the first remainder becomes the least significant digit (LSD) of the octal number, and the last remainder becomes the most significant digit (MSD).

If a calculator is used to perform the divisions in the above process, the result will include a decimal fraction instead of a remainder. The remainder can be obtained, however, by multiplying the decimal fraction by 8. For example, 266/8 produces 33.25. The remainder becomes $0.25 \times 8 = 2$. Similarly, 33/8 will be 4.125 and the remainder becomes $0.125 \times 8 = 1$.

Octal-to-Binary Conversion

The primary advantage of the octal number system is the ease with which conversion can be made between binary and octal numbers. The conversion from octal to binary is performed by converting *each* octal digit to its 3-bit binary equivalent. The eight possible digits are converted as indicated in Table 2-1.

TABLE 2-1

OCTAL DIGIT	0	1	2	3	4	5	6	7
BINARY EQUIVALENT	000	001	010	011	100	101	110	111

Using these conversions, any octal number is converted to binary by individually converting each digit. For example, we can convert 472_8 to binary as follows:

$$
\begin{array}{ccc}
4 & 7 & 2 \\
\downarrow & \downarrow & \downarrow \\
100 & 111 & 010
\end{array}
$$

Hence, octal 472 is equivalent to binary 100111010. As another example, consider converting 5431 to binary:

$$
\begin{array}{cccc}
5 & 4 & 3 & 1 \\
\downarrow & \downarrow & \downarrow & \downarrow \\
101 & 100 & 011 & 001
\end{array}
$$

Thus, $5431_8 = 101100011001_2$.

Binary-to-Octal Conversion

Converting from binary integers to octal integers is simply the reverse of the foregoing process. The bits of the binary number are grouped into groups of *three* bits starting at the LSB. Then each group is converted to its octal equivalent (Table 2-1). To illustrate, consider the conversion of 100111010_2 to octal.

$$
\begin{array}{ccc}
\underbrace{1\ 0\ 0} & \underbrace{1\ 1\ 1} & \underbrace{0\ 1\ 0} \\
\downarrow & \downarrow & \downarrow \\
4 & 7 & 2_8
\end{array}
$$

Sometimes the binary number will not have even groups of 3 bits. For those cases, we can add one or two 0s to the left of the MSB of the binary number to fill out the last group. This is illustrated below for the binary number 11010110.

$$
\begin{array}{ccc}
\underbrace{0\ 1\ 1} & \underbrace{0\ 1\ 0} & \underbrace{1\ 1\ 0} \\
\downarrow & \downarrow & \downarrow \\
3 & 2 & 6_8
\end{array}
$$

Note that a 0 was placed to the left of the MSB in order to produce even groups of 3 bits.

Counting in Octal   The largest octal digit is 7, so that when counting in octal, a digit is incremented upward from 0 to 7. Once it reaches 7, it recycles to 0 on the next count and causes the next-higher digit to be incremented. This is illustrated in the following sequences of octal counting: (a) 65, 66, 67, 70, 71; (b) 275, 276, 277, 300.

With N octal digits, we can count from zero up to $8^N - 1$, for a total of 8^N different counts. For example, with three octal digits we can count from 000_8 to 777_8, which is a total of $8^3 = 512_{10}$ different octal numbers.

Usefulness of Octal System   The ease with which conversions can be made between octal and binary make the octal system attractive as a "shorthand" means of expressing large binary numbers. In computer work, binary numbers with up to 64 bits are not uncommon. These binary numbers, as we shall see, do not always represent a numerical quantity but are often some type of code which conveys non-numerical information. In computers, binary numbers might represent (1) actual numerical data; (2) numbers corresponding to a location (address) in memory; (3) an instruction code; (4) a code representing alphabetic and other nonnumerical characters; or (5) a group of bits representing the status of devices internal or external to the computer.

When dealing with a large quantity of binary numbers of many bits, it is convenient and more efficient for us to write the numbers in octal rather than binary. Keep in mind, however, that the digital circuits and systems work strictly in binary; we are using octal only as a convenience for the operators of the system.

EXAMPLE 2-1

Convert 177_{10} to its 8-bit binary equivalent by first converting to octal.

SOLUTION

$$\frac{177}{8} = 22 + \text{remainder of 1}$$

$$\frac{22}{8} = 2 + \text{remainder of 6}$$

$$\frac{2}{8} = 0 + \text{remainder of 2}$$

Thus, $177_{10} = 261_8$. Now we can convert this octal number to its binary equivalent 010110001_2, so we finally have

$$177_{10} = 10110001_2$$

Note that we chop off the leading 0 to express the result as 8 bits.

This method of decimal-to-octal-to-binary conversion is often quicker than going directly from decimal to binary, especially for large numbers. Likewise, it is often quicker to convert binary to decimal by first converting to octal.

REVIEW QUESTIONS

1. Convert 614_8 to decimal. (*Ans.* 396)
2. Convert 146_{10} to octal, then from octal to binary. (*Ans.* 222 and 010010010)
3. Convert 10011101_2 to octal. (*Ans.* 235)
4. Write the next three numbers in this octal counting sequence: 624, 625, 626, _ _ _, _ _ _, _ _ _. (*Ans.* Last number is 631)
5. Convert 975_{10} to binary by first converting to octal. (*Ans.* 1111001111_2)
6. Convert binary 1010111011 to decimal by first converting to octal. (*Ans.* 699)

2-4 HEXADECIMAL NUMBER SYSTEM

The hexadecimal system uses base 16. Thus, it has 16 possible digit symbols. It uses the digits 0 through 9 plus the letters A, B, C, D, E, and F as the 16 digit symbols. Table 2-2 shows the relationships among hexadecimal, decimal, and binary. Note that each hexadecimal digit represents a group of four binary digits. It is important to remember that hex (abbreviation for hexadecimal) digits A through F are equivalent to the decimal values 10 through 15.

TABLE 2-2

HEXADECIMAL	DECIMAL	BINARY
0	0	0000
1	1	0001
2	2	0010
3	3	0011
4	4	0100
5	5	0101
6	6	0110
7	7	0111
8	8	1000
9	9	1001
A	10	1010
B	11	1011
C	12	1100
D	13	1101
E	14	1110
F	15	1111

Hex-to-Decimal Conversion   A hex number can be converted to its decimal equivalent by using the fact that each hex digit position has a weight that is a power of 16. The LSD has a weight of $16^0 = 1$, the next higher digit has a weight of 16^1

= 16, the next higher digit has a weight of $16^2 = 256$, and so on. The conversion process is demonstrated in the examples below.

$$356_{16} = 3 \times 16^2 + 5 \times 16^1 + 6 \times 16^0$$

$$= 768 + 80 + 6$$

$$= 854_{10}$$

$$2AF_{16} = 2 \times 16^2 + 10 \times 16^1 + 15 \times 16^0$$

$$= 512 + 160 + 15$$

$$= 687_{10}$$

Note that in the second example the value 10 was substituted for A and the value 15 for F in the conversion to decimal.

Decimal-to-Hex Conversion   Recall that we did decimal-to-binary conversion using repeated division by 2, and decimal-to-octal using repeated division by 8. Likewise, decimal-to-hex conversion can be done using repeated division by 16. The examples below will illustrate. Again note how the remainders of the division

EXAMPLE 2-2

Convert 423_{10} to hex.

SOLUTION

$$\frac{423}{16} = 26 + \text{remainder of } 7$$

$$\frac{26}{16} = 1 + \text{remainder of } 10$$

$$\frac{1}{16} = 0 + \text{remainder of } 1$$

$$423_{10} = \boxed{1A7_{16}}$$

EXAMPLE 2-3

Convert 214_{10} to hex.

SOLUTION

$$\frac{214}{16} = 13 + \text{remainder of } 6$$

$$\frac{13}{16} = 0 + \text{remainder of } 13$$

$$214_{10} = \boxed{D6_{16}}$$

processes form the digits of the hex number. Also note that any remainders that are greater than 9 are represented by the letters A–F.

If a calculator is being used to perform the divisions in the conversion process, the results will include a decimal fraction instead of a remainder. The remainder can be obtained by multiplying the fraction by 16. To illustrate, in Example 2.2 the calculator would have produced

$$\frac{214}{16} = 13.375$$

The remainder becomes $(.375) \times 16 = 6$.

Hex-to-Binary Conversion   Like the octal number systetm, the hexadecimal number system is used primarily as a "shorthand" method for representing binary numbers. It is a relatively simple matter to convert a hex number to binary. *Each hex digit is converted to its 4-bit binary equivalent* (Table 2.2). This is illustrated below for $9F2_{16}$.

$$9F2_{16} = \quad 9 \qquad\qquad F \qquad\qquad 2$$
$$\downarrow \qquad\quad\;\; \downarrow \qquad\quad\;\; \downarrow$$
$$1\;0\;0\;1 \quad 1\;1\;1\;1 \quad 0\;0\;1\;0$$
$$= 100111110010_2$$

Binary-to-Hex Conversion   This conversion is just the reverse of the process above. The binary number is grouped into groups of *four* bits, and each group is converted to its equivalent hex digit.

$$1\;0\;1\;1\;1\;0\;1\;0\;0\;1\;1\;0_2 = \underbrace{1\;0\;1\;1}_{B} \quad \underbrace{1\;0\;1\;0}_{A} \quad \underbrace{0\;1\;1\;0}_{6}$$

$$= BA6_{16}$$

In order to perform these conversions between hex and binary, it is necessary to know the 4-bit binary numbers (0000–1111) and their equivalent hex digits. Once these are mastered, the conversions can be performed quickly without the need for any calculations. This is why hex (and octal) are so useful in representing large binary numbers.

Counting in Hexadecimal   When counting in hex, each digit position can be incremented (increased by 1) from 0 to F. Once a digit position reaches the value F, it is reset to 0 and the next digit position is incremented. This is illustrated in the hex counting sequences below.

(a)   38, 39, 3A, 3B, 3C, 3D, 3E, 3F, 40, 41, 42

(b)   6F8, 6F9, 6FA, 6FB, 6FC, 6FD, 6FE, 6FF, 700

Note that when there is a 9 in a digit position, it becomes an A when it is incremented.

EXAMPLE 2-4

Convert decimal 378 to a 16-bit binary number by first converting to hex-adecimal.

SOLUTION

$$\frac{378}{16} = 23 + \text{remainder of } 10$$

$$\frac{23}{16} = 1 + \text{remainder of } 7$$

$$\frac{1}{16} = 0 + \text{remainder of } 1$$

Thus, $378_{10} = 17A_{16}$. This hex value can easily be converted to binary 000101111010. Finally, we can express 378_{10} as a 16-bit number by adding four leading 0s.

$$378_{10} = 0000 \quad 0001 \quad 0111 \quad 1010_2$$

EXAMPLE 2-5

Convert $B2F_{16}$ to octal.

SOLUTION

It's easiest to first convert hex to binary, then to octal.

$$B2F_{16} = 1011 \quad 0010 \quad 1111$$
$$= 101 \quad 100 \quad 101 \quad 111$$
$$= 5 \quad 4 \quad 5 \quad 7 \quad _8$$

REVIEW QUESTIONS

1. Convert $24CE_{16}$ to decimal. (*Ans.* 9422)
2. Convert 3117_{10} to hex, then from hex to binary. (*Ans.* C2D and 110000101101)
3. Convert 1001011110110101_2 to hex. (*Ans.* 97B5)
4. Write the next four numbers in this hex counting sequence: E9A, E9B, E9C, E9D, _ _ _, _ _ _, _ _ _, _ _ _. (*Ans.* Last number is EA1)
5. Convert 3527_8 to hex. (*Ans.* 757_{16})

2-5 BCD CODE

When numbers, letters, or words are represented by a special group of symbols, this is called *encoding,* and the group of symbols is called a *code*. Probably one of the most familiar codes is the Morse code, where series of dots and dashes represent letters of the alphabet.

We have seen that any decimal number can be represented by an equivalent binary number. The group of 0s, and 1s in the binary number can be thought of as a code representing the decimal number. When a decimal number is represented by its equivalent binary number, we call it *straight binary coding*.

Digital systems all use some form of binary numbers for their internal operation, but the external world is decimal in nature. This means that conversions between the decimal and binary systems are being performed often. We have seen that the conversions between decimal and binary can become long and complicated for large numbers. For this reason, a means of encoding decimal numbers that combines some features of both the decimal and binary systems is used in certain situations.

Binary-Coded-Decimal Code   If *each* digit of a decimal number is represented by its binary equivalent, this produces a code called *binary-coded-decimal* (hereafter abbreviated BCD). Since a decimal digit can be as large as 9, 4 bits are required to code each digit (the binary code for 9 is 1001).

To illustrate the BCD code, take a decimal number such as 874. Each digit is changed to its binary equivalent as follows:

$$
\begin{array}{ccc}
8 & 7 & 4 \quad \text{(decimal)} \\
\downarrow & \downarrow & \downarrow \\
1000 & 0111 & 0100 \quad \text{(BCD)}
\end{array}
$$

As another example, let us change 943 to its BCD-code representation:

$$
\begin{array}{ccc}
9 & 4 & 3 \quad \text{(decimal)} \\
\downarrow & \downarrow & \downarrow \\
1001 & 0100 & 0011 \quad \text{(BCD)}
\end{array}
$$

Once again, each decimal digit is changed to its straight binary equivalent. Note that 4 bits are *always* used for each digit.

The BCD code, then, represents each digit of the decimal number by a 4-bit binary number. Clearly, only the 4-bit binary numbers from 0000 through 1001 are used. The BCD code does not use the numbers 1010, 1011, 1100, 1101, 1110, and 1111. In other words, only 10 of the 16 possible 4-bit binary code groups are used. If any of these "forbidden" 4-bit numbers ever occurs in a machine using the BCD code, it is usually an indication that an error has occurred. (See Examples 2-6 and 2-7, page 37.)

Comparison of BCD and Binary   It is important to realize that BCD is not another number system like binary, octal, decimal, and hexadecimal. It is, in fact, the decimal system with each digit encoded in its binary equivalent. It is also important to understand that a BCD number is *not* the same as a straight binary number. A straight binary code takes the *complete* decimal number and represents it in

EXAMPLE 2-6

Convert 0110100000111001 (BCD) to its decimal equivalent.

SOLUTION
Divide the BCD number into 4-bit groups and convert each to decimal.

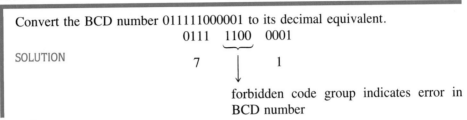

$$0110 \quad 1000 \quad 0011 \quad 1001$$
$$6 \qquad 8 \qquad 3 \qquad 9$$

EXAMPLE 2-7

Convert the BCD number 011111000001 to its decimal equivalent.

$$0111 \quad 1100 \quad 0001$$

SOLUTION

$$7 \qquad \qquad 1$$

forbidden code group indicates error in BCD number

binary; the BCD code converts *each* decimal *digit* to binary individually. To illustrate, take the number 137 and compare its straight binary and BCD codes:

$$137_{10} = 10001001_2 \qquad \text{(binary)}$$

$$137_{10} = 0001 \quad 0011 \quad 0111 \quad \text{(BCD)}$$

The BCD code requires 12 bits while the straight binary code requires only 8 bits to represent 137. BCD requires more bits than straight binary to represent decimal numbers of more than one digit. This is because BCD does not use all possible 4-bit groups, as pointed out earlier, and is therefore somewhat inefficient.

The main advantage of the BCD code is the relative ease of converting to and from decimal. Only the 4-bit code groups for the decimal digits 0–9 need be remembered. This ease of conversion is especially important from a hardware standpoint because in a digital system it is the logic circuits that perform the conversions to and from decimal.

BCD is used in digital machines whenever decimal information is either applied as inputs or displayed as outputs. Digital voltmeters, frequency counters, and digital clocks all use BCD because they display output information in decimal. Electronic calculators use BCD because the input numbers are entered in decimal via the keyboard and the output numbers are displayed in decimal.

BCD is not often used in modern high-speed digital computers for two good reasons. First, as was already pointed out, the BCD code for a given decimal number requires more bits than the straight binary code and is therefore less efficient. This is important in digital computers because the number of places in memory where these bits can be stored is limited. Second, the arithmetic processes for numbers represented in BCD code are more complicated than straight binary and thus require more complex circuitry. The more complex circuitry contributes to a decrease in the speed at which arithmetic operations take place. Calculators that use BCD are therefore considerably slower in their operation than computers.

REVIEW QUESTIONS

1. Represent the decimal value 178 by its straight binary equivalent. Then encode the same decimal number using BCD. (*Ans.* 10110010_2 and 000101111000 [BCD])
2. How many bits are required to represent an eight-digit decimal number in BCD?
3. What is an advantage of encoding a decimal number in BCD as compared to straight binary? What is a disadvantage?

2-6 EXCESS-3 CODE

The *excess-3 code* is related to the BCD code and is sometimes used instead because it possesses advantages in certain arithmetic operations. The excess-3 code for a decimal number is performed in the same manner as BCD except that *3* is added to *each* decimal digit before encoding it in binary. For example, to encode the decimal digit 4 into excess-3 code, we must first add 3 to obtain 7. Then the 7 is encoded in its equivalent 4-bit binary code 0111.

As another example, let us convert 48 into its excess-3 code representation.

$$
\begin{array}{cc}
4 & 8 \\
+3 & +\ 3 \\
\hline
7 & 11
\end{array}
\quad \text{add 3 to each digit}
$$

$$
\downarrow \qquad \downarrow
$$

$$
0111 \quad 1011 \qquad \text{convert to 4-bit binary code}
$$

Table 2-3 lists the BCD and excess-3-code representations for the decimal digits. Note that both codes use only 10 of the 16 possible 4-bit code groups. The excess-3 code, however, does not use the same code groups. For excess-3, the invalid code groups are 0000, 0001, 0010, 1101, 1110, and 1111.

TABLE 2-3

DECIMAL	BCD	EXCESS-3
0	0000	0011
1	0001	0100
2	0010	0101
3	0011	0110
4	0100	0111
5	0101	1000
6	0110	1001
7	0111	1010
8	1000	1011
9	1001	1100

2-7 GRAY CODE

The *Gray code* belongs to a class of codes called *minimum-change codes,* in which only *one* bit in the code group changes when going from one step to the next. The Gray code is an *unweighted* code, meaning that the bit positions in the code groups do not have any specific weight assigned to them. Because of this, the Gray code is not suited for arithmetic operations but finds application in input/output devices and some types of analog-to-digital converters.

Table 2-4 shows the Gray-code representation for the decimal numbers 0 through 15, together with the straight binary code. If we examine the Gray-code groups for each decimal number, it can be seen that in going from *any* one decimal number to the next, only *one* bit of the Gray code changes. For example, in going from 3 to 4, the Gray code changes from 0010 to 0110, with only the second bit from the left changing. Going from 14 to 15 the Gray-code bits change from 1001 to 1000, with only the last bit changing. This is the principal characteristic of the Gray code. Compare this with the binary code, where anywhere from one to all of the bits change in going from one step to the next.

TABLE 2-4

DECIMAL	BINARY CODE	GRAY CODE	DECIMAL	BINARY CODE	GRAY CODE
0	0000	0000	8	1000	1100
1	0001	0001	9	1001	1101
2	0010	0011	10	1010	1111
3	0011	0010	11	1011	1110
4	0100	0110	12	1100	1010
5	0101	0111	13	1101	1011
6	0110	0101	14	1110	1001
7	0111	0100	15	1111	1000

The Gray code is often used in situations where other codes, such as binary, might produce erroneous or ambiguous results during those transitions in which more than one bit of the code is changing. For instance, using binary code and going from 0111 to 1000 requires that all four bits change simultaneously. Depending on the device or circuit that is generating the bits, there may be a significant difference in the transition times of the different bits. If so, the transition from 0111 to 1000 could produce one or more intermediate states. For example, if the most significant bit changes faster than the rest, the following transitions will occur:

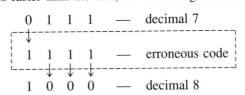

```
0  1  1  1   —   decimal 7
1  1  1  1   —   erroneous code
1  0  0  0   —   decimal 8
```

The occurrence of 1111 is only momentary but it could conceivably produce erroneous operation of the elements that are being controlled by the bits. Obviously, using the Gray code would eliminate this problem, since only one bit change occurs per transition and no "race" between bits can occur.

> **1.** What is the principal advantage of the Gray code over straight binary?

2-8 ALPHANUMERIC CODES

In addition to numerical data, a computer must be able to handle nonnumerical information. In other words, a computer should recognize codes that represent letters of the alphabet, punctuation marks, and other special characters as well as numbers. These codes are called *alphanumeric codes*. A complete alphanumeric code would include the 26 lowercase letters, 26 uppercase letters, 10 numeric digits, 7 punctuation marks, and anywhere from 20 to 40 other characters, such as + , /, #, %, *, and so on. We can say that an alphanumeric code represents all of the various characters and functions that are found on a standard typewriter (or computer) keyboard.

ASCII Code The most widely used alphanumeric code, the American Standard Code for Information Interchange (ASCII), is used in most microcomputers and minicomputers, and in many mainframes. The ASCII code (pronounced ''askee'') is a 7-bit code, so it has $2^7 = 128$ possible code groups. This is more than enough to represent all of the standard keyboard characters as well as control functions such as the ⟨RETURN⟩ and ⟨LINEFEED⟩ functions. Table 2-5 shows a partial listing of the ASCII code. In addition to the binary code group for each character, the table gives the octal and hexadecimal equivalents.

EXAMPLE 2-8

The following is a message encoded in ASCII code. What is the message?

1001000   1000101   1001100   1010000

SOLUTION

Convert each 7-bit code to its hex equivalent. The results are

48   45   4C   50

Now locate these hex values in Table 2-5 and determine the character represented by each. The results are

H E L P

The ASCII code is used for the transfer of alphanumeric information between a computer and input/output devices such as video terminals or printers. A computer also uses it internally to store the information that an operator types in at the computer's keyboard. The following examples illustrates this.

TABLE 2-5 Partial Listing of ASCII Code

CHARACTER	7-BIT ASCII	OCTAL	HEX	CHARACTER	7-BIT ASCII	OCTAL	HEX
A	100 0001	101	41	Y	101 1001	131	59
B	100 0010	102	42	Z	101 1010	132	5A
C	100 0011	103	43	0	011 0000	060	30
D	100 0100	104	44	1	011 0001	061	31
E	100 0101	105	45	2	011 0010	062	32
F	100 0110	106	46	3	011 0011	063	33
G	100 0111	107	47	4	011 0100	064	34
H	100 1000	110	48	5	011 0101	065	35
I	100 1001	111	49	6	011 0110	066	36
J	100 1010	112	4A	7	011 0111	067	37
K	100 1011	113	4B	8	011 1000	070	38
L	100 1100	114	4C	9	011 1001	071	39
M	100 1101	115	4D	blank	010 0000	040	20
N	100 1110	116	4E	.	010 1110	056	2E
O	100 1111	117	4F	(	010 1000	050	28
P	101 0000	120	50	+	010 1011	053	2B
Q	101 0001	121	51	$	010 0100	044	24
R	101 0010	122	52	*	010 1010	052	2A
S	101 0011	123	53	)	010 1001	051	29
T	101 0100	124	54	−	010 1101	055	2D
U	101 0101	125	55	/	010 1111	057	2F
V	101 0110	126	56	,	010 1100	054	2C
W	101 0111	127	57	=	011 1101	075	3D
X	101 1000	130	58	⟨RETURN⟩	000 1101	015	0D
				⟨LINEFEED⟩	000 1010	012	0A

EXAMPLE 2-9

An operator is typing in a BASIC program at the keyboard of a certain micro-computer. The computer converts each keystroke into its ASCII code and stores the code in memory. Determine the codes that will be entered into memory when the operator types in the following BASIC statement:

GOTO 25

SOLUTION
Locate each character (including the space) in Table 2.5 and record its ASCII code.

G	1000111
O	1001111
T	1010100
O	1001111
(space)	0100000
2	0110010
5	0110101

REVIEW QUESTIONS

1. Encode the following message in ASCII code using the hex representation: "COST = $72." (*Ans.* 43, 4F, 53, 54, 20, 3D, 20, 24, 37, 32)
2. The following ASCII-coded message is stored in successive memory locations in a computer:

    1010011 1010100 1001111 1010000

    What is the message? (*Ans.* STOP)

2-9 PARITY METHOD FOR ERROR DETECTION

The movement of binary data and codes from one location to another is the most frequent operation performed in digital systems. Here are some examples:

- The reading of instruction codes and data from internal memory as a computer executes a program
- The storage and retrieval of data from external memory devices such as magnetic tape and disk
- The transmission of information from a computer to a remote user terminal or another computer

Whenever information is transmitted from one device (the transmitter) to another device (the receiver), there is a possibility that errors can occur such that the receiver does not receive the identical information that was sent by the transmitter. The major cause of any transmission errors is *electrical noise,* which consists of spurious fluctuations in voltage or current that are present in all electronic systems to varying degrees. Figure 2-1 is a simple illustration of a type of transmission error.

The transmitter sends a relatively noise-free serial digital signal over a signal line to a receiver. However, by the time the signal reaches the receiver, it contains a certain degree of noise superimposed on the original signal. Occasionally, the noise is large enough in amplitude so that it will alter the logic level of the signal

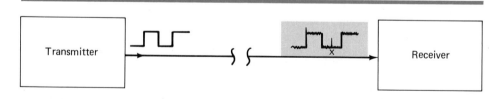

FIGURE 2-1 **Example of noise causing an error in the transmission of digital data.**

as it does at point *x*. When this occurs, the receiver may incorrectly interpret that bit as a logic 1, which is not what the transmitter has sent.

Most modern digital equipment is designed to be relatively error-free, and the probability of occurrence of errors such as that shown in Figure 2-1 is very low. However, we must realize that digital systems often transmit thousands, even millions of bits per second, so that even a very low rate of occurrence of errors can produce an occasional error that might prove to be bothersome, if not disastrous. For this reason, many digital systems employ some method for detection (and sometimes correction) of errors. One of the simplest and most widely used schemes for error detection is the *parity method*.

Parity Bit A *parity bit* is an extra bit that is attached to a code group that is being transferred from one location to another. The parity bit is made either 0 or 1, depending on the number of 1s that are contained in the code group. Two different methods are used.

In the *even parity* method, the value of the parity bit is chosen so that the total number of 1s in the code group (including the parity bit) is an *even* number. For example, suppose that the code group is 1000011. This is the ASCII character C. The code group has *three* 1s. Therefore, we will add a parity bit of 1 to make the total number of 1s an even number. The *new* code group, including the parity bit, thus becomes

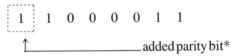

If the code group contains an even number of 1s to begin with, the parity bit is given a value of 0. For example, if the code group were 1000001 (the ASCII code for "A"), the assigned parity bit would be 0, so the new code, including the parity bit, would be 01000001.

The *odd-parity* method is used in exactly the same way except that the parity bit is chosen so the total number of 1s (including the parity bit) is an *odd* number. For example, for the code group 1000001, the assigned parity bit would be a 1. For the code group 1000011, the parity bit would be a 0.

Regardless of whether even parity or odd parity is used, the parity bit becomes an actual part of the code word. For example, adding a parity bit to the 7-bit ASCII code produces an 8-bit code. Thus, the parity bit is treated just like any other bit in the code.

The parity bit is used to detect any *single-bit* errors that occur during the transmission of a code from one location to another (e.g., from a computer to a video terminal). For example, suppose the character "A" is being transmitted and *odd* parity is being used. The transmitted code would be

$$\boxed{1} \quad 1 \quad 0 \quad 0 \quad 0 \quad 0 \quad 0 \quad 1$$

When the receiver circuit receives this code, it will check to see that the code contains an odd number of 1s (including the parity bit). If so, the receiver will

---

*The parity bit can be placed at either end of the code word but is usually placed to the left of the MSB.

assume that the code has been correctly received. Now, suppose that because of some noise or malfunction, the receiver actually receives the following code:

$$\boxed{1} \quad 1 \quad 0 \quad 0 \quad 0 \quad 0 \quad 0 \quad 0$$

The receiver will find that this code has an *even* number of 1s. This tells the receiver that there must be an error in the code, since presumably the transmitter and receiver have agreed to use odd parity. There is no way, however, that the receiver can tell which bit is in error, since it does not know what the code is supposed to be.

It should be apparent that this parity method would not work if *two* bits were in error, because two errors would not change the ''oddness'' or ''evenness'' of the number of 1s in the code. In practice, the parity method is used only in situations where the probability of a single error is very low, and the probability of double errors is essentially zero.

When the parity method is being used, the transmitter and receiver must have agreement, in advance, as to whether odd or even parity is being used. There is no advantage of one over the other, although even parity seems to be used more often. The transmitter must attach an appropriate parity bit to each unit of information that it transmits. For example, if the transmitter is sending ASCII-coded data, it will attach a parity bit to each 7-bit ASCII code group. When the receiver examines the data that it has received from the transmitter, it checks each code group to see that the total number of 1s (including the parity bit) is consistent with the agreed-upon type of parity. This is often called *checking the parity* of the data. In the event that it detects an error, the receiver may send a message to the transmitter asking it to retransmit the last set of data. The exact procedure that is followed when an error is detected will depend on the particular system design.

EXAMPLE 2-10

A transmitter is sending ASCII-coded data to a receiver with an even-parity bit. Show the actual codes when the transmitter is sending the message ''HELLO.''

SOLUTION

First look up the ASCII codes for each character in the message. Then for each code, count the number of 1s. If it is an even number, attach a 0 as the MSB. If it is an odd number, attach a 1. Thus the resulting 8-bit codes will all have an even number of 1s (including parity).

attached even-parity bits

H-	0	1	0	0	1	0	0	0
E-	1	1	0	0	0	1	0	1
L-	1	1	0	0	1	1	0	0
L-	1	1	0	0	1	1	0	0
O-	1	1	0	0	1	1	1	1

REVIEW QUESTIONS

1. Attach an odd-parity bit to the ASCII code for the $ symbol, and express the result in hexadecimal. (*Ans*. A4)
2. Attach an even-parity bit to the BCD code for decimal 69. (*Ans*. 001101001)
3. Why can't the parity method detect a double error in transmitted data?

GLOSSARY

Alphanumeric Codes Codes that represent numbers, letters, punctuation marks, and special characters. [Sec. 2-8]

ASCII Code (American Standard Code for Information Interchange) Seven-bit alphanumeric code used by most computer manufacturers. [Sec. 2-8]

Binary-Coded-Decimal Code (BCD Code) Four-bit code used to represent each digit of a decimal number. [Sec. 2-5]

Electrical noise Spurious fluctuations in voltage and/or current in any electronic system. [Sec. 2-9]

Encoding When a group of symbols is used to represent numbers, letters, or words. [Sec. 2-5]

Even Parity Total number of 1s (including the parity bit) that are contained in the code group is an even number. [Sec. 2-9]

Excess-3 Code Four-bit code in which 3 is added to the decimal digit before encoding it in binary. [Sec. 2-6]

Gray Code Code that never has more than one bit changing when going from one step to the next. [Sec. 2-7]

Hexadecimal Number System Number system that has a base of 16; digits 0 through 9 plus letters A through F are used to express a hexadecimal number. [Sec. 2-4]

Minimum-Change Codes Codes in which only one bit in the code group changes when going from one step to the next. [Sec. 2-7]

Octal Number System Number system that has a base of eight; digits from 0 to 7 are used to express an octal number. [Sec. 2-3]

Odd Parity Total number of 1s (including the parity bit) that are contained in the code group is an odd number. [Sec. 2-9]

Parity Bit Additional bit that is attached to each code group before being transferred from one location to another. [Sec. 2-9]

Parity Method Scheme used for error detection during the transmission of data. [Sec. 2-9]

Repeated Division Method used to convert a decimal number to its equivalent binary-system representation. [Sec. 2-2]

Straight Binary Coding Decimal number represented by its equivalent binary number. [Sec. 2-5]

Unweighted Code Code in which bit positions in the code groups do not have any specific weight assigned to them. [Sec. 2-7]

PROBLEMS

Sections 2-1 and 2-2

2-1. Convert these binary numbers to decimal.
(a) 10110   (b) 10001101   (c) 100100001001
(d) 1111010111   (e) 10111111

2-2. Convert the following decimal values to binary.
(a) 37   (b) 14   (c) 189   (d) 205   (e) 2313   (f) 511

2-3. What is the largest decimal value that can be represented by an 8-bit binary number? A 16-bit number?

Section 2-3

2-4. Convert each octal number to its decimal equivalent.
(a) 743   (b) 36   (c) 3777   (d) 257   (e) 1204

2-5. Convert each of the following decimal numbers to octal.
(a) 59   (b) 372   (c) 919   (d) 65536   (e) 255

2-6. Convert each of the octal values from Problem 2-4 to binary.

2-7. Convert the binary numbers in Problem 2-1 to octal.

2-8. List the octal numbers in sequence from 165_8 to 200_8.

2-9. When a large decimal number is to be converted to binary, it is sometimes easier to convert it first to octal, and then from octal to binary. Try this procedure for 2313_{10} and compare it to the procedure used in Problem 2-2(e).

Section 2-4

2-10. Convert these hex values to decimal.
(a) 92   (b) 1A6   (c) 37FD   (d) 2CO   (e) 7FF

2-11. Convert these decimal values to hex.
(a) 75   (b) 314   (c) 2048   (d) 25619   (e) 4095

2-12. Convert the binary numbers in Problem 2-1 to hexadecimal.

2-13. Convert the hex values in Problem 2-10 to binary.

2-14. In most microcomputers the *addresses* of memory locations are specified in hexadecimal. These addresses are sequential numbers that identify each memory circuit

(a) A particular microcomputer can store an 8-bit number in each memory location. If the memory addresses range from 0000_{16} to $FFFF_{16}$, how many memory locations are there?

(b) Another microcomputer is specified to have 4096 memory locations. What range of hex addresses does this computer use?

2-15. List the hex numbers in sequence from 280 to 2A0.

Section 2-5

2-16. Encode these decimal numbers in BCD.
(a) 47   (b) 962   (c) 187   (d) 42,689,627   (e) 1204

2-17. How many bits are required to represent the decimal numbers in the range from 0 to 999 using straight binary code? Using BCD code?

2-18. The following numbers are in BCD. Convert them to decimal.
(a) 1001011101010010   (b) 000110000100
(c) 0111011101110101   (d) 010010010010

Section 2-8

2-19. Represent the statement "X = 25/Y" in ASCII code.

2-20. Attach an *even* parity bit to each of the ASCII codes for Problem 2-19 and given the results in hex.

2-21. The following code groups are being transmitted. Attach an *even*-parity bit to each group.
(a) 10110110   (b) 00101000   (c) 11110111

Section 2-9

2-22. Convert the following decimal numbers to BCD code and then attach an *odd*-parity bit.
(a) 74   (b) 38   (c) 165   (d) 9201

2-23. In a certain digital system, the decimal numbers from 000 through 999 are represented in BCD code. An *odd*-parity bit is also included at the end of each code group. Examine each of the code groups below and assume that each one has just been tansferred from one location to another. Some of the groups contain errors. Assume that *no more than* two errors have occurred for each group. Determine which of the code groups has a single error and which of them *definitely* has a double error. (*Hint:* Remember that this is a BCD code.)
(a) 1001010110000
        ⌐__ parity bit
(b) 0100011101100
(c) 0111110000011
(d) 1000011000101

2-24. Suppose that the receiver received the following data from the transmitter of Example 2-10:

```
0 1 0 0 1 0 0 0
1 1 0 0 0 1 0 1
1 1 0 0 1 1 0 0
1 1 0 0 1 0 0 0
1 1 0 0 1 1 0 0
```

What can the receiver determine as far as errors in this received data?

Sections 2-1–2-8

2-25. Perform each of the following conversions. For some of them, you may want to try several methods to see which one works best for you. For example, a binary-to-decimal conversion may be done directly, or it may be done as a binary-to-octal conversion followed by an octal-to-decimal conversion.

(a) $1417_{10} = $ _____$_2$

(b) $255_{10} = $ _____$_2$

(c) 11010001_2 _____$_{10}$

(d) $1110101000100111_2 = $ _____$_{10}$

(e) $2497_{10} = $ _____$_8$

(f) $511_{10} = $ _____$_8$

(g) $235_8 = $ _____$_{10}$

(h) $4316_8 = $ _____$_{10}$

(i) $7A9_{16} = $ _____$_{10}$

(j) $3E1C_{16} = $ _____$_{10}$

(k) $1600_{10} = $ _____$_{16}$

(l) $38,187_{10} = $ _____$_{16}$

(m) $865_{10} = $ _____ (BCD)

(n) $100101000111(BCD) = $ _____$_{10}$

(o) $465_8 = $ _____$_{16}$

(p) $B34_{16} = $ _____$_8$

(q) $01110100(BCD) = $ _____$_2$

(r) $111010_2 = $ _____(BCD)

2-26. The circuits inside a computer must deal in the binary number system because the transistors are designed to be either fully conducting (ON) or nonconducting (OFF). All the programs and data that are used by a computer are stored in its memory as groups of ones and zeros. In many microcomputers, the information is stored in groups of 8 bits called *bytes*. The information represented by each byte can be of many different types, and in many cases the type is known only to the programmer. For example, consider the following byte:

01000101

It could represent the binary equivalent of the decimal value 69_{10}. It could also be the BCD representation of 45_{10}. It might even be the ASCII code for the character "E" with an odd-parity bit. Actually, it could be several other types of information with which you are not yet familiar. Examine the three sets of bytes below. For each set, try to determine what type or types of information they could be representing.

(a) 00100000
    00101001
    01000110

(b) 00100000
    01010010
    11000100

(c) 11000011
    00110001
    01000100

LOGIC GATES AND BOOLEAN ALGEBRA

Upon completion of this chapter, you will be able to:

- Analyze the INVERTER circuit.
- Describe the operation of and construct the truth tables for the AND, NAND, OR, and NOR gates.
- Draw timing diagrams for the various logic-circuit gates.
- Simplify complex logic circuits by applying the various Boolean algebra laws and rules.
- Simplify intricate Boolean equations by applying DeMorgan's theorems.
- Use either of the universal gates (NAND or NOR) to implement the circuit represented by a Boolean expression.
- Explain the advantages of constructing a logic circuit diagram using the alternate gate symbols, versus the standard logic-gate symbols.
- Describe the concept of active-LOW and active-HIGH logic signals.
- Draw and interpret logic circuits that use the new IEEE/ANSI standard symbols.

As pointed out in Chapter 1, digital (logic) circuits operate in the binary mode where each input and output voltage is either a 0 or 1; the 0 and 1 designations represent predefined voltage ranges. This characteristic of logic circuits allows us to use *Boolean algebra* as a tool for the analysis and design of digital systems. In this chapter we will study *logic gates*, which are the most fundamental logic circuits, and we will see how their operation can be described using Boolean algebra. We will also see how logic gates can be combined to produce logic circuits, and how these circuits can be described and analyzed using Boolean algebra.

3-1 BOOLEAN CONSTANTS AND VARIABLES

Boolean algebra differs in a major way from ordinary algebra in that Boolean constants and variables are allowed to have only two possible values, 0 or 1. A Boolean variable is a quantity that may, at different times, be equal to either 0 or 1. The Boolean variables are often used to represent the voltage level present on a wire or at the input/output terminals of a circuit. For example, in a certain digital system the Boolean value of 0 might be assigned to any voltage in the range from 0 to 0.8 V while the Boolean value of 1 might be assigned to any voltage in the range 2 to 5 V.*

Thus, Boolean 0 and 1 do not represent actual numbers but instead represent the state of a voltage variable or what is called its *logic level*. A voltage in a digital circuit is said to be at the logic level 0 or the logic level 1, depending on its actual numerical value. In the digital logic field several other terms are used synonomously with 0 and 1. Some of the more common ones are shown in Table 3-1. We will use the 0/1 and LOW/HIGH designations most of the time.

*Voltages between 0.8 and 2V are undefined (neither 0 nor 1) and under normal circumstances should not occur.

TABLE 3-1

LOGIC 0	LOGIC 1
False	True
Off	On
Low	High
No	Yes
Open switch	Closed switch

Boolean algebra is used to express the effects that various digital circuits have on logic inputs, and to manipulate logic variables for the purpose of determining the best method for performing a given circuit function. In all our work to follow we shall use letter symbols to represent logic variables. For example, A might represent a certain digital circuit input or output, and at any time we must have either $A = 0$ or $A = 1$: if not one, then the other.

Because only two values are possible, Boolean algebra is relatively easy to work with as compared to ordinary algebra. In Boolean algebra there are no fractions, decimals, negative numbers, square roots, cube roots, logarithms, imaginary numbers, and so on. In fact, in Boolean algebra there are only *three* basic operations:

1. *Logical addition,* also called *OR addition* or simply the *OR operation.* The common symbol for this operation is the plus sign (+).

2. *Logical multiplication*, also called *AND multiplication* or simply the *AND operation.* The common symbol for this operation is the multiplication sign (·).

3. *Logical complementation or inversion*, also called the *NOT operation.* The common symbol for this operation is the overbar ($^-$).

3-2 TRUTH TABLES

Many logic circuits have more than one input and only one output. A *truth table* shows how the logic circuit's output responds to the various combinations of logic levels at the inputs. The format for two-, three-, and four-input truth tables is shown in Figure 3-1.

In each truth table the possible combinations of 0 and 1 logic levels for the inputs (A, B, C, D) are listed on the left, and the resultant logic level for output x is listed on the right. The entries for x are shown as "?" for now, because these values will be different for each type of logic circuit.

Note that there are 4 table entries for the two-input truth table, 8 entries for a three-input truth table, and 16 entries for the four-input truth table. The number of input combinations will equal 2^N for an N-input truth table. Also note that the list of all possible input combinations follows the binary counting sequence, so it is an easy matter to write down all the combinations without missing any.

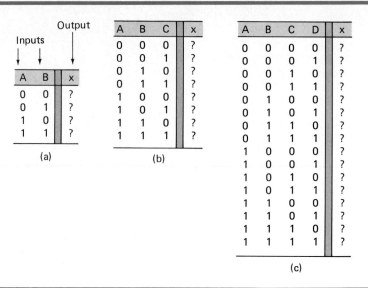

A	B	x
0	0	?
0	1	?
1	0	?
1	1	?

(a)

A	B	C	x
0	0	0	?
0	0	1	?
0	1	0	?
0	1	1	?
1	0	0	?
1	0	1	?
1	1	0	?
1	1	1	?

(b)

A	B	C	D	x
0	0	0	0	?
0	0	0	1	?
0	0	1	0	?
0	0	1	1	?
0	1	0	0	?
0	1	0	1	?
0	1	1	0	?
0	1	1	1	?
1	0	0	0	?
1	0	0	1	?
1	0	1	0	?
1	0	1	1	?
1	1	0	0	?
1	1	0	1	?
1	1	1	0	?
1	1	1	1	?

(c)

FIGURE 3-1   Truth tables for (a) two-input, (b) three-input, and (c) four-input cases.

3-3   OR OPERATION

Let A and B represent two independent logic variables. When A and B are combined using the OR operation the result, x, can be expressed as

$$x = A + B$$

In this expression the $+$ sign does not stand for ordinary addition; it stands for the OR operation, whose rules are given in the truth table shown in Figure 3.2(a).

It should be apparent from the truth table that except for the case where $A = B = 1$, the OR operation is the same as ordinary addition. However, for $A = B = 1$ the OR sum is 1 (not 2 as in ordinary addition). This is easy to remember if we recall that only 0 and 1 are possible values in Boolean algebra, so that the largest

FIGURE 3-2   (a) Truth table defining the OR operation; (b) circuit symbol for a two-input OR gate.

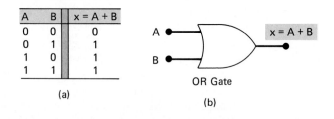

A	B	x = A + B
0	0	0
0	1	1
1	0	1
1	1	1

(a)

OR Gate

(b)

value we can get is 1. This same result is true if we have $x = A + B + C$, for the case where $A = B = C = 1$. That is,

$$x = 1 + 1 + 1 = 1$$

We can therefore say that the OR operation result will be 1 if any one *or* more variables is a 1. This is also clear in the table in Figure 3.2(a).

The expression $x = A + B$ can either be read as "*x* equals *A* plus *B*" or as "*x* equals *A* OR *B*." Both expressions are in common use. The key thing to remember is that the + sign stands for the OR operation as defined by the truth table in Figure 3.2(a), and not for ordinary addition.

OR Gate In digital circuitry an *OR gate* is a circuit that has two or more inputs and whose output is equal to the OR sum of the inputs. Figure 3.2(b) shows the symbol for a two-input OR gate. The inputs *A* and *B* are logic voltage levels and the output *x* is a logic voltage level whose value is the result of the OR operation on *A* and *B*; that is, $x = A + B$. In other words, the OR gate operates such that its output is high (logic 1) if either input *A* or *B* or *both* are at a logic 1 level. The OR gate output will be low (logic 0) only if all its inputs are at logic 0.

This same idea can be extended to more than two inputs. Figure 3.3 shows a three-input OR gate and its truth table. Examination of this truth table shows again that the output will be 1 for every case where one or more inputs is 1. This general principle is the same for OR gates with any number of inputs.

Using the language of Boolean algebra, the output *x* can be expressed as $x = A + B + C$, where again it must be emphasized that the + represents the OR operation. The output of any OR gate, then, can be expressed as the OR of its various inputs. We will put this to use when we subsequently analyze logic circuits.

Summary of the OR Operation The important points to remember concerning the OR operation and OR gates are:

1. The OR operation produces a result of 1 when *any* of the input variables is 1.
2. The OR operation produces a result of 0 *only* when all the input variables are 0.
3. In OR addition, $1 + 1 = 1$, $1 + 1 + 1 = 1$, and so on.

FIGURE 3-3 **Symbol and truth table for a three-input OR gate.**

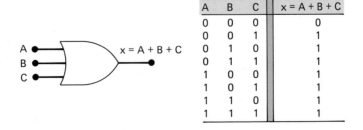

A	B	C	x = A + B + C
0	0	0	0
0	0	1	1
0	1	0	1
0	1	1	1
1	0	0	1
1	0	1	1
1	1	0	1
1	1	1	1

EXAMPLE 3-1

In many industrial control systems it is required to activate an output function whenever any one of several inputs is activated. For example, in a chemical process it may be desired that an alarm be activated whenever the process temperature exceeds a maximum value *or* whenever the pressure goes above a certain limit. Figure 3-4 is a block diagram of this situation. The temperature transducer circuit produces an output voltage proportional to the process temperature. This voltage, V_T, is compared to a temperature reference voltage, V_{TR}, in a voltage comparator circuit. The comparator output is normally a low voltage (logical 0), but it switches to a high voltage (logical 1) whenever V_T exceeds V_{TR}, indicating that process temperature is excessive. A similar arrangement is used for the pressure measurement, so its associated comparator output goes from low to high when the pressure is excessive.

Since we want the alarm to be activated when either temperature *or* pressure is too high, it should be apparent that the two comparator outputs can be fed to a two-input OR gate. The OR gate output thus goes HIGH (1) for either alarm condition and will activate the alarm. This same idea can be obviously extended to situations with more than two process variables.

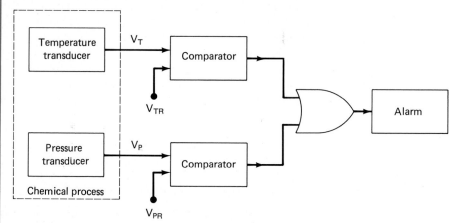

FIGURE 3-4 **Example of the use of OR gate in alarm system.**

EXAMPLE 3-2

Determine the OR gate output in Figure 3-5. The OR gate inputs *A* and *B* are varying according to the timing diagrams shown. For example, *A* starts out LOW at time t_0, goes HIGH at t_1, back LOW at t_3, and so on.

SOLUTION

The OR gate output is determined by realizing that it will be HIGH whenever *any* of the inputs are at the high level. When *A* goes HIGH at t_1, OUTPUT will go HIGH. OUTPUT will stay HIGH until t_4, when both inputs are LOW. Note that the changes in the input levels that occur at t_2 and t_3 have no effect on OUTPUT since one of the inputs remains at the HIGH level while the other is changing. As long as one input to an OR gate is HIGH, the output will stay HIGH regardless of what is happening at the other inputs. The same reasoning is used to determine the rest of the timing diagram for OUTPUT.

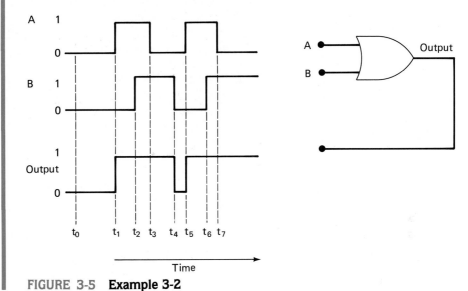

FIGURE 3-5 **Example 3-2**

EXAMPLE 3-3

For the situation depicted in Figure 3-6, determine the waveform at the OR gate output.

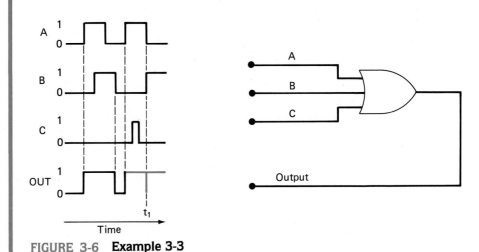

FIGURE 3-6 **Example 3-3**

SOLUTION

The three OR gate inputs A, B, and C are varying, as shown by their waveform diagrams. The OR gate output is determined by realizing that it will be high whenever *any* of the three inputs is at a high level. Using this reasoning, the OR output waveform is as shown in the figure. Particular attention should be paid to what occurs at time t_1. The diagram shows that at that instant of time, input A is going from high to low while input B is going from low to high. Since these inputs are making their transitions at approximately the same time and since these transitions take a certain amount of time, there is a short interval when these OR gate inputs are both in the undefined range between 0 and 1. When this occurs, the OR gate output also becomes a value in this range, as evidenced by the glitch or spike on the output waveform at t_1. The occurrence of this glitch and its size (amplitude and width) depends on the speed with which the input transitions occur. It should be noted that if the C input had been sitting high while A and B were changing, the glitch would not occur because the high at C would keep the OR output high.

REVIEW QUESTIONS

1. What is the only set of input conditions that will produce a LOW output for any OR gate?
2. Write the Boolean expression for a six-input OR gate.
3. If the A input in Figure 3-6 is permanently kept at the 1 level, sketch the resultant output waveform. (*Ans.* Output will be a constant HIGH)

3-4 AND OPERATION

If two logic variables A and B are combined using AND multiplication, the result, x, can be expressed as

$$x = A \cdot B$$

In this expression the · sign stands for the Boolean operation of AND multiplication, whose rules are given in the truth table shown in Figure 3-7(a).

FIGURE 3-7 **(a) Truth table for the AND operation; (b) AND gate symbol.**

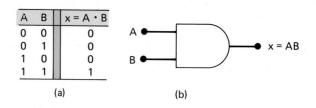

A	B	x = A · B
0	0	0
0	1	0
1	0	0
1	1	1

(a)   (b)

It should be apparent from the table that AND multiplication is *exactly* the same as ordinary multiplication. Whenever *A* or *B* are 0, their product is zero; when both *A and B* are 1, their product is 1. We can therefore say that in the AND operation the result will be 1 *only if* all the inputs are 1; for all other cases the result is 0.

The expression $x = A \cdot B$ is read "*x* equals *A and B*." The multiplication sign is generally omitted as in ordinary algebra, so the expression becomes $x = AB$. The key thing to remember is that the AND operation is the same as ordinary multiplication, where the variables can be either 0 or 1.

AND Gate   A two-input AND gate is shown symbolically in Figure 3-7(b). The AND gate output is equal to the AND product of the logic inputs; that is, $x = AB$. In other words, the AND gate is a circuit that operates such that its output is high only when all its inputs are high. For all other cases the AND gate output is low.

This same operation is characteristic of AND gates with more than two inputs. For example, a three-input AND gate and its accompanying truth table are shown in Figure 3-8. Once again, note that the gate output is 1 only for the case where $A = B = C = 1$. The expression for the output is $x = ABC$. For a four-input AND gate, the output is $x = ABCD$, and so on.

Note the difference between the symbols for the AND gate and the OR gate. Whenever you see the AND symbol on a logic circuit diagram, it tells you that the output will go HIGH *only* when *all* inputs are HIGH. Whenever you see the OR symbol, it means that the output will go HIGH when *any* input is HIGH.

Summary of the AND Operation

1. The AND operation is performed exactly like ordinary multiplication of 1s and 0s.

2. An output equal to 1 occurs only for the single case where all inputs are 1.

3. The output is 0 for any case where one or more inputs are 0.

FIGURE 3-8   **Truth table and symbol for a three-input AND gate.**

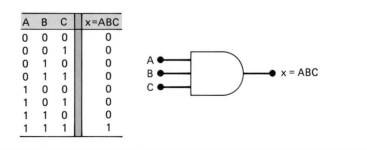

A	B	C	x=ABC
0	0	0	0
0	0	1	0
0	1	0	0
0	1	1	0
1	0	0	0
1	0	1	0
1	1	0	0
1	1	1	1

EXAMPLE 3-4

Determine the output x from the AND gate in Figure 3-9 for the given input waveforms.

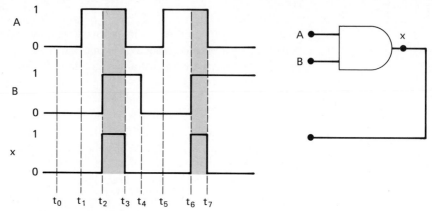

FIGURE 3-9 Example 3-4

SOLUTION

The output of an AND gate is determined by realizing that it will be high only when all inputs are high at the same time. For the input waveforms given, this condition is met only during intervals t_2–t_3 and t_6–t_7. At all other times, one or more of the inputs is a 0, thereby producing a low output. Note that input level changes that occur while the other input is low have no effect on the output.

EXAMPLE 3-5

Determine the output waveform for the AND gate shown in Figure 3-10.

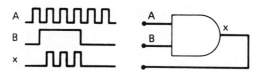

FIGURE 3-10 Example 3-5

SOLUTION

The output x will be at 1 only when A and B are both high at the same time. Using this fact, the x waveform can be determined as shown in the figure.

Notice that the x waveform is 0 whenever B is 0, regardless of the signal at A. Also notice that the x waveform is the same as A whenever B is 1. Thus, we can think of the B input as a *control* input whose logic level determines whether the A waveform gets through to the x output or not. In this situation, the AND gate is used as an *inhibit circuit*. We can say that $B = 0$ is the inhibit condition producing a 0 output. Conversely, $B = 1$ is the *enable* condition, which enables A to reach the output. This inhibit operation is an important application of AND gates.

EXAMPLE 3-6

What will happen to the *x* output waveform in Figure 3-10 if the *B* input is kept at the 0 level?

SOLUTION

With *B* kept LOW, the *x* output will also stay LOW. This can be reasoned in two different ways. First, with $B = 0$ we have $x = A \cdot B = A \cdot 0 = 0$, since anything multiplied (ANDed) by 0 will be 0. Another way to look at it is that an AND gate requires that all inputs be HIGH in order for the output to be HIGH, and this cannot happen if *B* is kept LOW.

REVIEW QUESTIONS

1. What is the only input combination that will produce a HIGH at the output of a five-input AND gate?
2. What logic level should be applied to the second input of a two-input AND gate if the logic signal at the first input is to be inhibited (prevented) from reaching the ouput?
3. *True or false:* An AND gate output will always differ from an OR gate output for the same input conditions.

3-5 NOT OPERATION

The NOT operation is unlike the OR or AND operations in that it can be performed on a single input variable. For example, if the variable *A* is subjected to the NOT operation, the result *x* can be expressed as

$$x = \overline{A}$$

where the overbar represents the NOT operation. This expression is read "*x* equals NOT *A*" or "*x* equals the *inverse* of *A*" or "*x* equals the *complement* of *A*." Each of these are in common usage and they all indicate that the logic value of $x = \overline{A}$ *is opposite to* the logic value of *A*. The truth table in Figure 3-11(a) clarifies this for the two cases $A = 0$ and $A = 1$. That is,

$$\overline{1} = 0 \qquad \text{because NOT 1 is 0}$$

and

$$\overline{0} = 1 \qquad \text{because NOT 0 is 1}$$

The NOT operation is also referred to as *inversion* or *complementation,* and these terms will be used interchangeably throughout the text. Although we will always

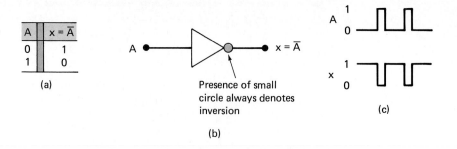

(a)

(b)

Presence of small
circle always denotes
inversion

(c)

FIGURE 3-11   (a) Truth table and (b) symbol for the NOT circuit.

use the overbar indicator to represent inversion, it is important to mention that another indicator for inversion is the prime symbol ('). That is,

$$A' = \overline{A}$$

Both should be recognized as indicating the inversion operation.

NOT Circuit (INVERTER)   Figure 3-11(b) shows the symbol for a NOT circuit, which is more commonly called an INVERTER. This circuit *always* has only a single input, and its output logic level is always opposite to the logic level of this input. Figure 3-11(c) shows how the INVERTER affects an input signal. It inverts (complements) the input signal at all points on the waveform.

Summary of Boolean Operations   The rules for the OR, AND, and NOT operations may be summarized as follows:

OR	*AND*	*NOT*
$0 + 0 = 0$	$0 \cdot 0 = 0$	$\overline{0} = 1$
$0 + 1 = 1$	$0 \cdot 1 = 0$	$\overline{1} = 0$
$1 + 0 = 1$	$1 \cdot 0 = 0$	
$1 + 1 = 1$	$1 \cdot 1 = 1$	

REVIEW QUESTIONS

1.   The output of the INVERTER of Figure 3-11 is connected to the input of a second INVERTER. Determine the output level of the second INVERTER for each level of input A. (*Ans.* In each case it is the same as A)

2.   The output of the AND gate in Figure 3-7 is connected to the input of an INVERTER. Write the truth table showing the INVERTER output, y, for each combination of inputs A and B.

3-6 DESCRIBING LOGIC CIRCUITS ALGEBRAICALLY

Any logic circuit, no matter how complex, may be completely described using the Boolean operations previously defined, because the OR gate, AND gate, and NOT circuit are the basic building blocks of digital systems. For example, consider the circuit in Figure 3-12. This circuit has three inputs, *A, B,* and *C,* and a single output, *x.* Utilizing the Boolean expression for each gate, we can easily determine the expression for the output.

The expression for the AND gate output is written $A \cdot B$. This AND output is connected as an input to the OR gate along with *C,* another input. The OR gate operates on its inputs such that its output is the OR sum of the inputs. Thus, we can express the OR output as $x = A \cdot B + C$. (This final expression could also be written as $x = C + A \cdot B$, since it does not matter which term of the OR sum is written first.)

Occasionally, there may be confusion as to which operation in an expression is performed first. The expression $A \cdot B + C$ can be interpreted in two different ways: (1) $A \cdot B$ is ORed with *C,* or (2) *A* is ANDed with the term $B + C$. To avoid this confusion, it will be understood that if an expression contains both AND and OR operations, the AND operations are performed first, unless there are *parentheses* in the expression, in which case the operation inside the parentheses is to be performed first. This is the same rule that is used in ordinary algebra to determine the order of operations.

To illustrate further, consider the circuit in Figure 3-13. The expression for the OR-gate output is simply $A + B$. This output serves as an input to the AND gate along with another input, *C.* Thus, we express the output of the AND gate as $x = (A + B) \cdot C$. Note the use of parentheses here to indicate that *A* and *B* are ORed *first,* before their OR sum is ANDed with *C.* Without the parentheses it would be interpreted *incorrectly,* since $A + B \cdot C$ means *A* is ORed with the product $B \cdot C$.

FIGURE 3-12 **Logic circuit with its Boolean expression.**

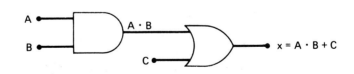

FIGURE 3-13 **Logic circuit whose expression requires parentheses.**

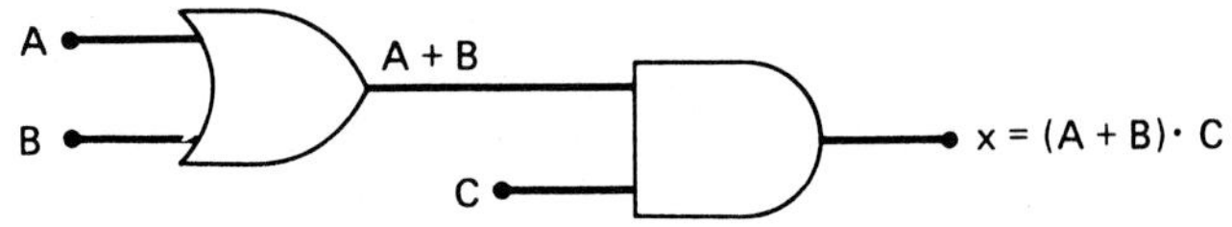

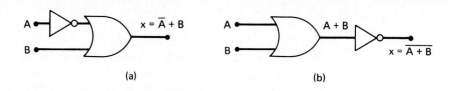

(a)   (b)

FIGURE 3-14   Circuits using INVERTERs.

Circuits Containing INVERTERs   Whenever an INVERTER is present in a logic-circuit diagram, its output expression is simply equal to the input expression with a bar over it. Figure 3-14 shows two examples using INVERTERs. In Figure 3-14(a) the input A is fed through an INVERTER, whose output is therefore $\bar{A}$. The INVERTER output is fed to an OR gate together with B, so the OR output is equal to $\bar{A} + B$. Note that the bar is only over the A, indicating that A is first inverted and then ORed with B.

In Figure 3.-14(b) the output of the OR gate is equal to $A + B$ and is fed through an INVERTER. The INVERTER output is therefore equal to $\overline{(A + B)}$, since it inverts the *complete* input expression. Note that the bar covers the entire expression $(A + B)$. This is important because, as will be shown later, the expressions $\overline{(A + B)}$ and $(\bar{A} + \bar{B})$ are *not* equivalent. The expression $\overline{(A + B)}$ means that A is ORed with B and then their OR sum is inverted, whereas the expression $(\bar{A} + \bar{B})$ indicates that A is inverted and B is inverted and the results are then ORed together.

FIGURE 3-15   More examples.

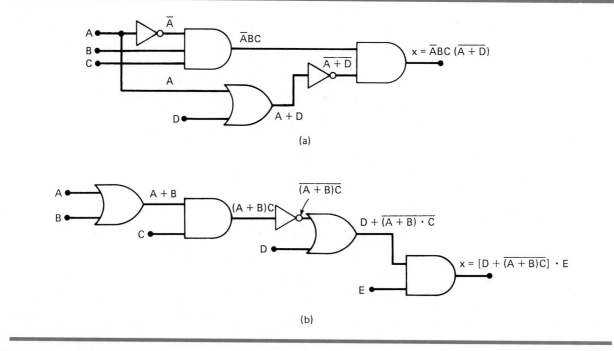

(a)

(b)

Figure 3-15 shows two more examples, which should be studied carefully. Note especially the use of *two* separate sets of parentheses in Figure 3-15(b). Also notice in Figure 3-15(a) that the input variable A is connected as input to two different gates.

REVIEW QUESTION

1. In Figure 3-15(a) change each AND gate to an OR gate, and change the OR gate to an AND gate. Then write the expression for output x. (*Ans.* $x = \overline{A} + B + C + \overline{A \cdot D}$)

3-7 EVALUATING LOGIC-CIRCUIT OUTPUTS

Once the Boolean expression for a circuit output is obtained, the logic level of the output can be determined for any values of circuit inputs. For example, suppose that we want to know the logic level of the output x for the circuit in Figure 3-15(a) for the case where $A = 0$, $B = 1$, $C = 1$, and $D = 1$. As in ordinary algebra, the value of x can be found by plugging the values of the variables into the expression and performing the indicated operations as follows:

$$x = \overline{A}BC\,\overline{(A + D)}$$
$$= \overline{0} \cdot 1 \cdot 1 \cdot \overline{(0 + 1)}$$
$$= 1 \cdot 1 \cdot 1 \cdot \overline{(0 + 1)}$$
$$= 1 \cdot 1 \cdot 1 \cdot \overline{(1)}$$
$$= 1 \cdot 1 \cdot 1 \cdot 0$$
$$= 0$$

As another illustration, let us evaluate the output of the circuit in Figure 3-15(b) for $A = 0$, $B = 0$, $C = 1$, $D = 1$, and $E = 1$.

$$x = [D + \overline{(A + B)C}] \cdot E$$
$$= [1 + \overline{(0 + 0) \cdot 1}] \cdot 1$$
$$= [1 + \overline{0 \cdot 1}] \cdot 1$$
$$= [1 + \overline{0}] \cdot 1$$
$$= [1 + 1] \cdot 1$$
$$= 1 \cdot 1$$
$$= 1$$

As a general rule, the following rules must always be followed when evaluating a Boolean expression:

1. First, perform all inversions of single terms; that is, $\overline{0} = 1$ or $\overline{1} = 0$.

2. Then perform all operations within parentheses.

3. Perform an AND operation before an OR operation unless parentheses indicate otherwise.

4. If an expression has a bar over it, perform the operations of the expression first and then invert the result.

Determining Output Level from a Diagram The output logic level for given input levels can also be determined directly from the circuit diagram *without* using the Boolean expression. This technique is often used by technicians during the troubleshooting or testing of a logic system since it also tells them what each gate output is supposed to be as well as the final output. To illustrate, the circuit of Figure 3-15(a) is redrawn in Figure 3-16 with the input levels $A = 0$, $B = 1$, $C = 1$, and $D = 1$. The procedure is to start from the inputs and to proceed through each INVERTER and gate, writing down each of their outputs in the process until the final output is reached.

In Figure 3-16, AND gate 1 has *all* three inputs at the 1 level because the INVERTER changes the $A = 0$ to a 1. This condition produces a 1 at the AND output since $1 \cdot 1 \cdot 1 = 1$. The OR gate has inputs of 1 and 0, which produces a 1 output since $1 + 0 = 1$. This 1 is inverted to 0 and applied to AND gate 2 along with the 1 from the first AND output. The 0 and 1 inputs to AND gate 2 produce an output of 0, because $0 \cdot 1 = 0$.

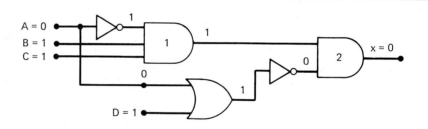

FIGURE 3-16 **Determining output level from circuit diagram.**

REVIEW QUESTIONS

1. Use the expression for x to determine the output of the circuit in Figure 3-15(a) for the conditions $A = 0$, $B = 1$, $C = 1$, and $D = 0$. (*Ans. x = 1*)

2. Use the expression for x to determine the output of the circuit in Figure 3-15(b) for the conditions $A = B = E = 1$, $C = D = 0$. (*Ans. x = 1*)

3. Determine the answers to questions 1 and 2 by finding the logic levels present at each gate input and output as was done in Figure 3-16.

3-8 IMPLEMENTING CIRCUITS FROM BOOLEAN EXPRESSIONS

If the operation of a circuit is defined by a Boolean expression, a logic-circuit diagram can be implemented directly from that expression. For example, if we needed a circuit that was defined by $x = A \cdot B \cdot C$, we would immediately know that all that was needed was a three-input AND gate. If we needed a circuit that was defined by $x = A + \overline{B}$, we would use a two-input OR gate with an INVERTER on one of the inputs. The same reasoning used for these simple cases can be extended to more complex circuits.

Suppose that we wanted to construct a circuit whose output is $y = AC + B\overline{C} + \overline{A}BC$. This Boolean expression contains three terms (AC, $B\overline{C}$, $\overline{A}BC$), which are ORed together. This tells us that a three-input OR gate is required with inputs that are equal to AC, $B\overline{C}$, and $\overline{A}BC$, respectively. This is illustrated in Figure 3-17(a), where a three-input OR gate is drawn with inputs labeled as AC, $B\overline{C}$, and $\overline{A}BC$.

Each OR-gate input is an AND product term, which means that an AND gate with appropriate inputs can be used to generate each of these terms. This is shown in Figure 3-17(b), which is the final circuit diagram. Note the use of INVERTERs to produce the $\overline{A}$ and $\overline{C}$ terms required in the expression.

This same general approach can always be followed, although we shall find that there are some clever, more efficient techniques that can be employed. For

FIGURE 3-17 Constructing a logic circuit from a Boolean expression.

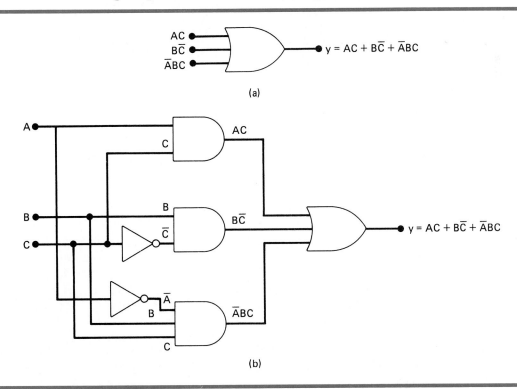

(a)

(b)

now, however, this straightforward method will be used to minimize the number of new things that are to be learned.

EXAMPLE 3-7

Draw the circuit diagram that implements the expression $x = AB + \overline{B}C$.

SOLUTION
This expression indicates that the terms AB and $\overline{B}C$ are inputs to an OR gate, and each of these two terms is generated from a separate AND gate. The result is shown in Figure 3-18.

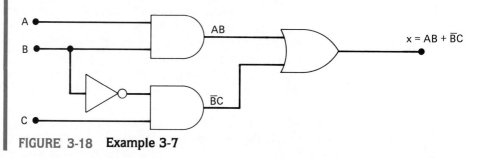

FIGURE 3-18 **Example 3-7**

REVIEW QUESTION

1. Draw the circuit diagram that implements the expression $x = \overline{ABC}(\overline{A} + D)$ and compare it to Figure 3-15(a).

3-9 NOR GATES AND NAND GATES

Two other types of logic gates, NOR gates and NAND gates, are used extensively in digital circuitry. These gates actually combine the basic operations AND, OR, and NOT, which makes it relatively easy to describe them using the Boolean algebra operations learned previously.

NOR Gate   The symbol for a two-input NOR gate is shown in Figure 3-19(a). It is the same as the OR gate symbol except that it has a small circle on the output. This small circle represents the inversion operation. Thus, the NOR gate operates like an OR gate followed by an INVERTER, so that the circuits in Figure 3-19(a) and (b) are equivalent, and the output expression for the NOR gate is $x = \overline{A + B}$.

   The truth table in Figure 3-19(c) shows that the NOR gate output is the exact inverse of the OR-gate output for all possible input conditions. Whereas an OR-gate output goes HIGH when any input is HIGH, the NOR gate output goes LOW when any input is HIGH. This same operation can be extended to NOR gates with more than two inputs.

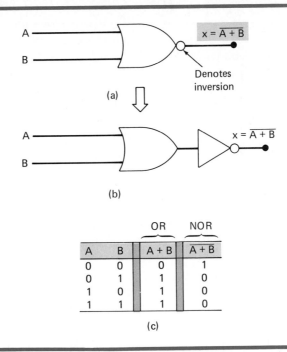

(a)

(b)

		OR	NOR
A	B	A + B	$\overline{A + B}$
0	0	0	1
0	1	1	0
1	0	1	0
1	1	1	0

(c)

FIGURE 3-19   (a) NOR symbol; (b) equivalent circuit; (c) truth table.

EXAMPLE 3-8

Determine the waveform at the output of a NOR gate for the input waveforms shown in Figure 3.20.

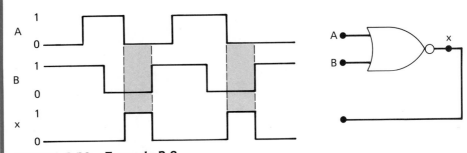

FIGURE 3-20   Example 3-8

SOLUTION
There are several ways to determine the NOR output waveform. One way is to first find the OR output waveform and then invert it (change all 1s to 0s, and vice versa). Another way utilizes the fact that a NOR-gate output will be HIGH *only* when all inputs are LOW. Thus you can examine the input waveforms, find those time intervals where they are all LOW, and make the NOR output HIGH for those intervals. The NOR output will be LOW for all other time intervals. The resultant output waveform is shown in the figure.

NOR GATES AND NAND GATES

EXAMPLE 3-9

Determine the Boolean expression for a three-input NOR gate followed by an inverter.

SOLUTION

Refer to Figure 3.21 where the circuit diagram is shown. The expression at the NOR output is $\overline{(A + B + C)}$, which is then fed through an INVERTER to produce

$$x = \overline{\overline{(A + B + C)}}$$

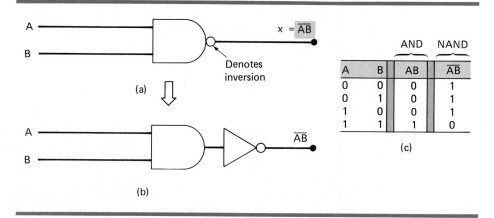

FIGURE 3-21   Example 3-9

The presence of the double inversion signs indicates that the quantity $(A + B + C)$ has been inverted and then inverted again. It should be clear that this simply results in the expression $(A + B + C)$ being unchanged. That is,

$$x = \overline{\overline{(A + B + C)}} = A + B + C$$

Whenever two inversion bars are over the same variable or quantity, they cancel each other out, as in the example above. However, in cases such as $\overline{\overline{A}} + \overline{B}$ the inversion bars do not cancel. This is because the smaller inversion bars invert the single variables A and B, respectively, while the wide bar inverts the quantity $(\overline{A} + \overline{B})$. Thus, $\overline{\overline{A} + \overline{B}} \neq A + B$. Similarly, $\overline{\overline{A} \, \overline{B}} \neq AB$.

NAND Gate   The symbol for a two-input NAND gate is shown in Figure 3-22(a). It is the same as the AND-gate symbol except for the small circle on its output. Once again this small circle denotes the inversion operation. Thus, the NAND operates like an AND gate followed by an INVERTER, so that the circuits

FIGURE 3-22   (a) NAND symbol; (b) equivalent circuit; (c) truth table.

A	B	AND AB	NAND $\overline{AB}$
0	0	0	1
0	1	0	1
1	0	0	1
1	1	1	0

(c)

of Figure 3-22(a) and (b) are equivalent, and the output expression for the NAND gate is $x = \overline{AB}$.

The truth table in Figure 3-22(c) shows that the NAND-gate output is the exact inverse of the AND gate for all possible input conditions. The AND output goes HIGH only when all inputs are HIGH, while the NAND output goes LOW only when all inputs are HIGH. This same characteristic is true of NAND gates having more than two inputs.

EXAMPLE 3-10

Determine the output waveform of a NAND gate having the inputs shown in Figure 3-23.

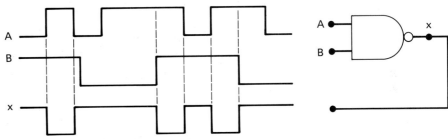

FIGURE 3-23   Example 3-10

SOLUTION

The output can be determined in several ways. One way is to first draw the output for an AND gate, and then invert it. Another way utilizes the fact that a NAND output will be LOW only when all inputs are HIGH. Thus, you can find those time intervals during which the inputs are all HIGH, and make the NAND output LOW for those intervals. The output will be HIGH at all other times.

EXAMPLE 3-11

Implement the logic circuit that has the expression $x = \overline{AB \cdot (\overline{C + D})}$ using only NOR and NAND gates.

SOLUTION

The $(\overline{C + D})$ term is the expression for the output of a NOR gate. This term is ANDed with A and B and the result is inverted; this, of course, is the NAND operation. Thus, the circuit is implemented as shown in Figure 3-24. Note that the NAND gate first ANDs the A, B, and $(\overline{C + D})$ terms, and then inverts the *complete* result.

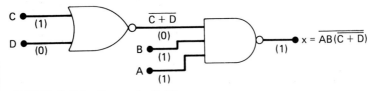

FIGURE 3-24   Example 3-11

EXAMPLE 3-12

Determine the output level in Figure 3-24 for $A = B = C = 1$ and $D = 0$.

SOLUTION

In the first method we use the expression for x.

$$
\begin{aligned}
x &= \overline{AB(\overline{C + D})} \\
&= \overline{1 \cdot 1 \cdot (\overline{1 + 0})} \\
&= \overline{1 \cdot 1 \cdot (\overline{1})} \\
&= \overline{1 \cdot 1 \cdot 0} \\
&= \overline{0} = 1
\end{aligned}
$$

In the second method we write down the input logic levels on the circuit diagram (shown in parentheses) and follow these levels through each gate to the final output. The NOR gate has inputs of 1 and 0 to produce an output of 0 (an OR would have produced an output of 1). The NAND gate thus has input levels of 0, 1, and 1 to produce an output of 1 (an AND would have produced an output of 0).

REVIEW QUESTIONS

1. What is the only set of input conditions that will produce a HIGH output from a three-input NOR gate?
2. What type of gate is equivalent to a NAND gate followed by an IN-VERTER? (*Ans.* An AND gate)
3. Change the NOR gate of Figure 3-24 to a NAND gate, and change the NAND to a NOR. What is the new expression for x? (*Ans.* $x = A + B + \overline{CD}$)

3-10 BOOLEAN THEOREMS

We have seen how Boolean algebra can be used to help analyze a logic circuit and express its operation mathematically. We will continue our study of Boolean algebra by investigating the various Boolean theorems (rules) that can help us to simplify logic expressions and logic circuits. The first group of theorems is given in Figure 3-25. In each theorem, x is a logic variable that can be either a 0 or 1. Each theorem is accompanied by a logic-circuit diagram that demonstrates its validity.

Theorem (1) states that if any variable is ANDed with 0, the result has to be 0. This is easy to remember because the AND operation is just like ordinary multiplication, where we know that anything multiplied by 0 is 0. We also know that

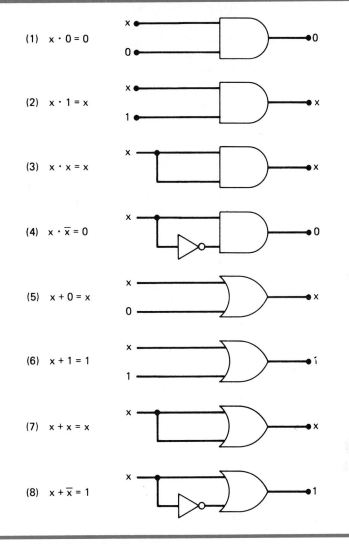

FIGURE 3-25 Single-variable theorems.

the output of an AND gate will be 0 whenever any input is 0, regardless of the level on the other input.

Theorem (2) is also obvious by comparison with ordinary multiplication.

Theorem (3) can be proved by trying each case. If $x = 0$, then $0 \cdot 0 = 0$; if $x = 1$, then $1 \cdot 1 = 1$. Thus, $x \cdot x = x$.

Theorem (4) can be proved in the same manner. However, it can also be reasoned that at any time either x or its inverse $\bar{x}$ has to be at the 0 level, so their AND product always has to be 0.

Theorem (5) is straightforward since 0 *added* to anything does not affect its value, either in regular addition or in OR addition.

Theorem (6) states that if any variable is ORed with 1, the result will always be 1. Checking this for both values of x: $0 + 1 = 1$ and $1 + 1 = 1$. Equivalently, we can remember that an OR-gate output will be 1 when *any* input is 1, regardless of the value of the other input.

Theorem (7) can be proved by checking for both values of x: $0 + 0 = 0$ and $1 + 1 = 1$.

Theorem (8) can be proved similarly, or we can just reason that at any time either x or $\bar{x}$ has to be at the 1 level so that we are always ORing a 0 and a 1, which always results in 1.

Before introducing any more theorems, it should be pointed out that in applying theorems (1)–(8) the variable x may actually represent an expression containing more than one variable. For example, if we have $A\bar{B}(\overline{A\bar{B}})$, we can invoke theorem (4) by letting $x = A\bar{B}$. Thus, we can say that $A\bar{B}(\overline{A\bar{B}}) = 0$. The same idea can be applied to the use of any of these theorems.

Multivariable Theorems

The theorems presented below involve more than one variable:

(9) $x + y = y + x$

(10) $x \cdot y = y \cdot x$

(11) $x + (y + z) = (x + y) + z = x + y + z$

(12) $x(yz) = (xy)z = xyz$

(13a) $x(y + z) = xy + xz$

(13b) $(w + x)(y + z) = wy + xy + wz + xz$

(14) $x + xy = x$

(15) $x + \bar{x}y = x + y$

Theorems (9) and (10) are called the *commutative laws*. These laws indicate that the order in which we OR or AND two variables is unimportant; the result is the same.

Theorems (11) and (12) are the *associative laws*, which state that we can group the variables in an AND expression or OR expression any way we want.

Theorem (13) is the *distributive law*, which states that an expression can be expanded by multiplying term by term just the same as in ordinary algebra. This theorem also indicates that we can factor an expression. That is, if we have a sum of two (or more) terms, each of which contains a common variable, the common variable can be factored out just like in ordinary algebra. For example, if we have the expression $A\bar{B}C + \bar{A}\,\bar{B}\,\bar{C}$, we can factor out the $\bar{B}$ variable:

$$A\bar{B}C + \overline{AB}\,\overline{C} = \bar{B}(AC + \overline{AC})$$

As another example, consider the expression $ABC + ABD$. Here the two terms have the variables A and B in common, so $A \cdot B$ can be factored out of both terms. That is,

$$ABC + ABD = AB(C + D)$$

Theorems (9)–(13) are easy to remember and use since they are identical to those of ordinary algebra. Theorems (14) and (15), on the other hand, do not have any counterparts in ordinary algebra. Each can be proved by trying all possible cases for x and y. This is illustrated for theorem (14) as follows:

Case 1: For $x = 0$, $y = 0$,

$$x + xy = x$$
$$0 + 0 \cdot 0 = 0$$
$$0 = 0$$

Case 2: For $x = 0$, $y = 1$,

$$x + xy = x$$
$$0 + 0 \cdot 1 = 0$$
$$0 + 0 = 0$$
$$0 = 0$$

Case 3: For $x = 1$, $y = 0$,

$$x + xy = x$$
$$1 + 1 \cdot 0 = 1$$
$$1 + 0 = 1$$
$$1 = 1$$

Case 4: For $x = 1$, $y = 1$,

$$x + xy = x$$
$$1 + 1 \cdot 1 = 1$$
$$1 + 1 = 1$$
$$1 = 1$$

Theorem (14) can also be proved by factoring and using theorems (6) and (2) as follows:

$$x + xy = x(1 + y)$$
$$= x \cdot 1 \quad \text{[using theorem (6)]}$$
$$= x \quad \text{[using theorem (2)]}$$

All of these Boolean theorems can be useful in simplifying a logic expression—that is, in reducing the number of terms in the expression. When this is done, the reduced expression will produce a circuit that is less complex than the one which the original expression would have produced. A good portion of the next chapter will be devoted to the process of circuit simplification. For now, the following examples will serve to illustrate how the Boolean theorems can be applied.

EXAMPLE 3-13

Simplify the expression $y = A\overline{B}D + A\overline{B}\,\overline{D}$.

SOLUTION
Factor out the common variables $A\overline{B}$ using theorem (13):
$$y = A\overline{B}(D + \overline{D})$$
Using theorem (8), the term in parentheses is equivalent to 1. Thus,
$$y = A\overline{B} \cdot 1$$
$$y = A\overline{B} \quad \text{[using theorem (2)]}$$

EXAMPLE 3-14

Simplify $z = (\overline{A} + B)(A + B)$.

SOLUTION
The expression can be expanded by multiplying out the terms [theorem (13)].
$$z = \overline{A} \cdot A + \overline{A} \cdot B + B \cdot A + B \cdot B$$
Invoking theorem (4), the term $\overline{A} \cdot A = 0$. Also, $B \cdot B = B$ [theorem (3)].
$$z = 0 + \overline{A} \cdot B + B \cdot A + B = \overline{A}B + AB + B$$
Factoring out the variable B [theorem (13)], we have
$$z = B(\overline{A} + A + 1)$$
Finally, using theorem (6),
$$z = B$$

EXAMPLE 3-15

Simplify $x = ACD + \overline{A}BCD$.

SOLUTION
Factoring out the common variables CD, we have
$$x = CD(A + \overline{A}B)$$
Utilizing theorem (15), we can replace $A + \overline{A}B$ by $A + B$, so
$$x = CD(A + B)$$
$$= ACD + BCD$$

REVIEW QUESTIONS

1. Use theorems (13) and (14) to simplify $y = A\overline{C} + AB\overline{C}$. (*Ans.* $y = A\overline{C}$)
2. Use theorems (13) and (8) to simplify $y = \overline{A}\,\overline{B}CD + \overline{A}\,\overline{B}\,\overline{C}\,\overline{D}$. (*Ans.* $y = \overline{A}\,\overline{B}\,\overline{D}$)

3-11 DEMORGAN'S THEOREMS

Two of the most important theorems of Boolean algebra were contributed by a great mathematician named DeMorgan. *DeMorgan's theorems* are extremely useful in simplifying expressions in which a product or sum of variables is inverted. The two theorems are:

$$(16)\ \overline{(x + y]} = \bar{x} \cdot \bar{y}$$

$$(17)\ \overline{(x \cdot y)} = \bar{x} + \bar{y}$$

Theorem (16) says that when the OR sum of two variables is inverted, this is the same as inverting each variable individually and then ANDing these inverted variables. Theorem (17) says that when the AND product of two variables is inverted, this is the same as inverting each variable individually and then ORing them. Each of DeMorgan's theorems can be readily proven by checking for all possible combinations of x and y. This will be left as an end-of-chapter exercise.

Although these theorems have been stated in terms of single variables x and y, they are equally valid for situations where x and/or y are expressions that contain more than one variable. For example, let's apply them to the expression $\overline{(A\bar{B} + C)}$ as shown below:

$$\overline{(A\bar{B} + C)} = \overline{(A\bar{B})} \cdot \bar{C}$$

Note that here we treated $A\bar{B}$ as x and C as y. The result can be further simplified since we have a product $A\bar{B}$ that is inverted. Using theorem (17), the expression becomes

$$\overline{A\bar{B}} \cdot \bar{C} = (\bar{A} + \bar{\bar{B}}) \cdot \bar{C}$$

Notice that we can replace $\bar{\bar{B}}$ by B, so we finally have

$$(\bar{A} + B) \cdot \bar{C} = \bar{A}\,\bar{C} + B\bar{C}$$

This final result contains only inverter signs that invert a single variable.

EXAMPLE 3-16

Simplify the expression $z = \overline{(\bar{A} + C) \cdot (B + \bar{D})}$.

SOLUTION
Using theorem (17), we can rewrite this as

$$z = \overline{(\bar{A} + C)} + \overline{(B + \bar{D})}$$

We can think of this as breaking the large inverter sign down the middle and changing the AND sign ($\cdot$) to an OR sign ($+$). Now the term $\overline{(\bar{A} + C)}$ can be simplified by applying theorem (16). Likewise, $\overline{(B + \bar{D})}$ can be simplified.

$$z = \overline{(\bar{A} + C)} + \overline{(B + \bar{D})}$$
$$= (\bar{\bar{A}} \cdot \bar{C}) + \bar{B} \cdot \bar{\bar{D}}$$

Here we have broken the larger inverter signs down the middle and replaced the ($+$) with a ($\cdot$). Canceling out the double inversions, we have finally

$$z = A\bar{C} + \bar{B}D$$

Example 3-16 points out that when using DeMorgan's theorems to reduce an expression, an inverter sign may be broken at any point in the expression and the operator at that point in the expression is changed to its opposite ($+$ is changed to $\cdot$, and vice versa). This procedure is continued until the expression is reduced to one in which only single variables are inverted. Two more examples are given below.

1.
$$
\begin{aligned}
z &= \overline{A + \overline{B} \cdot C} \\
&= \overline{A} \cdot \overline{(\overline{B} \cdot C)} \\
&= \overline{A} \cdot (\overline{\overline{B}} + \overline{C}) \\
&= \overline{A} \cdot (B + \overline{C})
\end{aligned}
$$

2.
$$
\begin{aligned}
q &= \overline{(A + BC) \cdot (D + EF)} \\
&= \overline{(A + BC)} + \overline{(D + EF)} \\
&= (\overline{A} \cdot \overline{BC}) + (\overline{D} \cdot \overline{EF}) \\
&= [\overline{A} \cdot (\overline{B} + \overline{C})] + [\overline{D} \cdot (\overline{E} + \overline{F})] \\
&= \overline{A}\,\overline{B} + \overline{A}\,\overline{C} + \overline{D}\,\overline{E} + \overline{D}\,\overline{F}
\end{aligned}
$$

DeMorgan's theorems are easily extended to more than two variables. For example, it can be proved that

$$\overline{x + y + z} = \overline{x} \cdot \overline{y} \cdot \overline{z}$$

$$\overline{x \cdot y \cdot z} = \overline{x} + \overline{y} + \overline{z}$$

and so on for more variables. Again, realize that any one of these variables can be an expression rather than a single variable.

Implications of DeMorgan's Theorems
Let us examine these theorems (16) and (17) from the standpoint of logic circuits. First, consider theorem (16),

$$\overline{x + y} = \overline{x} \cdot \overline{y}$$

The left-hand side of the equation can be viewed as the output of a NOR gate whose inputs are x and y. The right-hand side of the equation, on the other hand, is the result of first inverting both x and y and then putting them through an AND gate. These two representations are equivalent and are illustrated in Figure 3-26(a). What this means is that an AND gate with INVERTERs on each of its inputs is equivalent to a NOR gate. In fact, both representations are used to represent the NOR function. When the AND gate with inverted inputs is used to represent the NOR function, it

FIGURE 3-26 **(a) Equivalent circuits implied by theorem (16); (b) alternative symbol for the NOR function.**

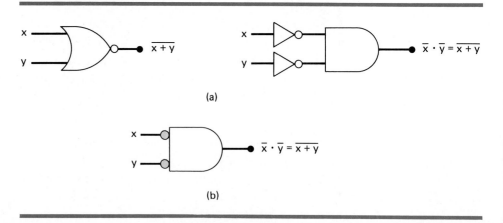

(a)

(b)

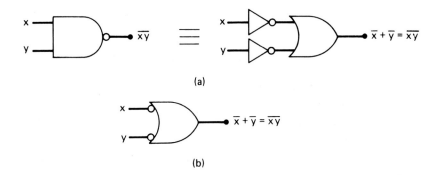

(a)

$$\overline{x} + \overline{y} = \overline{xy}$$

(b)

FIGURE 3-27 (a) Equivalent circuits implied by theorem (17); (b) alternative symbol for the NAND function.

is usually drawn as shown in Figure 3-26(b), where the small circles on the inputs represent the INVERTERs.

Now consider theorem (17),

$$\overline{x \cdot y} = \overline{x} + \overline{y}$$

The left side of the equation can be implemented by a NAND gate with inputs x and y. The right side can be implemented by first inverting inputs x and y and then putting them through an OR gate. These two equivalent representations are shown in Figure 3-27(a). The OR gate with INVERTERs on each of its inputs is equivalent to the NAND gate. In fact, both representations are used to represent the NAND function. When the OR gate with inverted inputs is used to represent the NAND function, it is usually drawn as shown in Figure 3-27(b), where the circles again represent inversion.

EXAMPLE 3-17

Implement a circuit having the output expression $Z = \overline{A} + \overline{B} + C$ using a NAND gate and an INVERTER.

SOLUTION
$z = \overline{A} + \overline{B} + C$ can be written as

$$z = \overline{\overline{\overline{A}} \cdot \overline{\overline{B}} \cdot \overline{C}} = \overline{A \cdot B \cdot \overline{C}}$$

by invoking theorem (17) and by canceling the double inversions over A and B. In this new form, it is easy to see how a NAND gate can be used to produce z. The circuit is drawn in Figure 3-28.

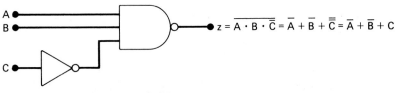

$z = \overline{A \cdot B \cdot \overline{C}} = \overline{A} + \overline{B} + \overline{\overline{C}} = \overline{A} + \overline{B} + C$

FIGURE 3-28 Example 3-17

REVIEW QUESTIONS

3-12 UNIVERSALITY OF NAND GATES AND NOR GATES

All Boolean expressions consist of various combinations of the basic operations of OR, AND, and INVERT. Therefore, any expression can be implemented using OR gates, AND gates, and INVERTERs. It is possible, however, to implement any logic expression using *only* NAND gates and no other type of gate. This is because NAND gates, in the proper combination, can be used to perform each of the Boolean operations OR, AND, and INVERT. This is demonstrated in Figure 3-29.

First, in Figure 3-29(a) we have a two-input NAND gate whose inputs are purposely connected together so that the variable A is applied to both. In this configuration, the NAND simply acts as an INVERTER, since its output is $x = \overline{A \cdot A} = \overline{A}$.

In Figure 3-29(b) we have two NAND gates connected so that the AND operation is performed. NAND gate 2 is used as an INVERTER to change $\overline{AB}$ to $\overline{\overline{AB}} = AB$, which is the desired AND function.

The OR operation can be implemented using NAND gates connected as shown in Figure 3-29(c). Here NAND gates 1 and 2 are used as INVERTERs to invert the inputs, so the final output is $x = \overline{\overline{A} \cdot \overline{B}}$, which can be simplified to $x = A + B$ using DeMorgan's theorem.

In a similar manner, it can be shown that NOR gates can be arranged to implement any of the Boolean operations. This is illustrated in Figure 3-30. Part (a) shows that a NOR gate with its inputs connected together behaves as an INVERTER, since the output is $x = \overline{A + A} = \overline{A}$.

In Figure 3-30(b) two NOR gates are arranged so that the OR operation is performed. NOR gate 2 is used as an INVERTER to change $\overline{A + B}$ to $\overline{\overline{A + B}} = A + B$, which is the desired OR function.

The AND operation can be implemented with NOR gates as shown in Figure 3-30(c). Here NOR gates 1 and 2 are used as INVERTERs to invert the inputs, so the final output is $x = \overline{\overline{A} + \overline{B}}$, which can be simplified to $x = A \cdot B$ by use of DeMorgan's theorem.

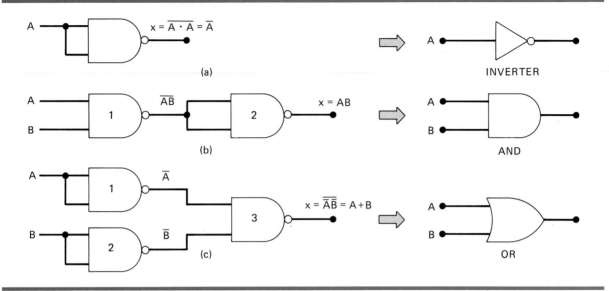

(a)

(b)

(c)

INVERTER

AND

OR

FIGURE 3-29 NAND gates can be used to implement any Boolean function.

Since any of the Boolean operations can be implemented using only NAND gates, then any logic circuit can be constructed using only NAND gates. The same is true for NOR gates. This characteristic of NAND and NOR gates can be very useful in logic-circuit design as the following example illustrates.

FIGURE 3-30 NOR gates can be used to implement any Boolean operation.

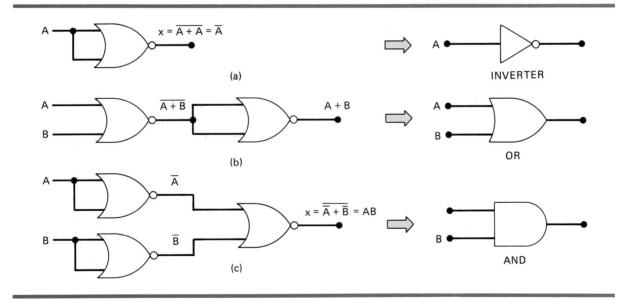

(a)

(b)

(c)

INVERTER

OR

AND

EXAMPLE 3-18

A logic-circuit designer has to implement a circuit that satisfies the expression $x = AB + CD$ using the minimum number of ICs. He has the TTL ICs shown in Figure 3-31 available for his use. Each of these ICs is a *quad*, which means that it contains *four* identical two-input gates on one chip.

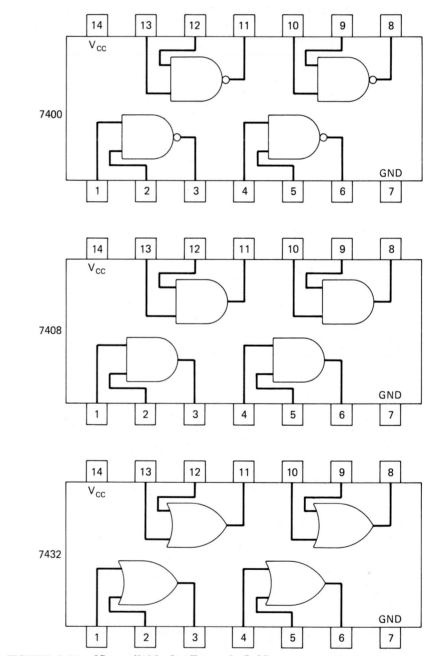

FIGURE 3-31   ICs available for Example 3-18.

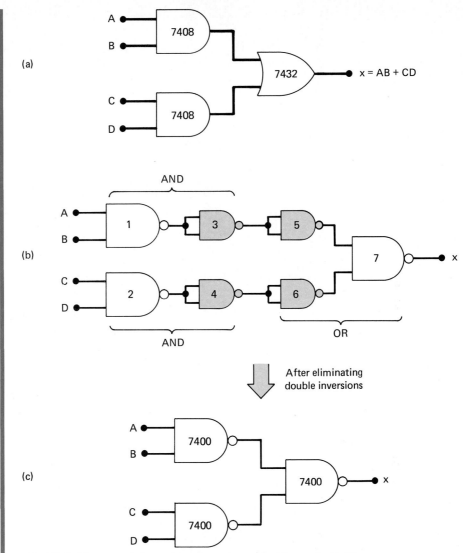

FIGURE 3-32 **Possible implementations for Example 3-18.**

The straightforward method for implementing the given expression uses two AND gates and an OR gate as shown in Figure 3-32(a). This implementation requires the 7408 and 7432 ICs, and some of the gates on each IC are wasted since they are not used.

Another implementation can be accomplished by taking the circuit of Figure 3-32(a) and replacing each AND gate and OR gate by its equivalent NAND-gate implementation from Figure 3.29. The result is shown in Figure 3-32(b).

At first glance this new circuit looks like it requires seven NAND gates. However, NAND gates 3 and 5 are connected as INVERTERs in series and can be eliminated from the circuit since they perform a double inversion of the signal out of NAND gate 1. Similarly, NAND gates 4 and 6 can be eliminated. The final circuit, after eliminating the double INVERTERs, is drawn in Figure 3-32(c).

This final circuit is more efficient than the one in Figure 3-32(a) because it uses three two-input NAND gates that can be implemented from one IC, the 7400.

REVIEW QUESTIONS

1. How many different ways do we now have to implement the inversion operation in a logic circuit? (*Ans. Three*)

2. Implement the expression $x = (A + B)(C + D)$ using OR and AND gates. Then implement the expression using only NOR gates by converting each OR and AND gate to its NOR implementation from Figure 3-30. Which circuit is more efficient? (*Ans.* The NOR circuit is more efficient because it can be implemented from a single IC, the 7402, which is a QUAD two-input NOR chip)

3. Write the output expression for the circuit of Figure 3-32(c) and use DeMorgan's theorems to show that it is equivalent to the expression for the circuit of Figure 3-32(a).

3-13 ALTERNATE LOGIC-GATE REPRESENTATIONS

We have introduced the five basic logic gates (AND, OR, INVERTER, NAND, and NOR) and the standard symbols used to represent them on logic-circuit diagrams. Although you may find that many circuit diagrams still use these standard symbols exclusively, it has become increasingly more common to find circuit diagrams that utilize an alternate set of symbols *in addition* to the standard symbols.

Before discussing the reasons for using an alternate symbol for a logic gate, we will present the alternate symbols for each gate and show that they are equivalent to the standard symbols. Refer to Figure 3-33; the left side of the illustration shows the standard symbol for each logic gate, and the right side shows the alternate symbol. The alternate symbol for each gate is obtained from the standard symbol by doing the following:

1. Invert each input and output of the standard symbol. This is done by adding bubbles (small circles) on input and output lines that do not have bubbles, and by removing bubbles that are already there.

2. Change the operation symbol from AND to OR, or from OR to AND. (In the special case of the INVERTER, the operation symbol is not changed.)

For example, the standard NAND symbol is an AND symbol with a bubble on its output. Following the steps outlined above, remove the bubble from the out-

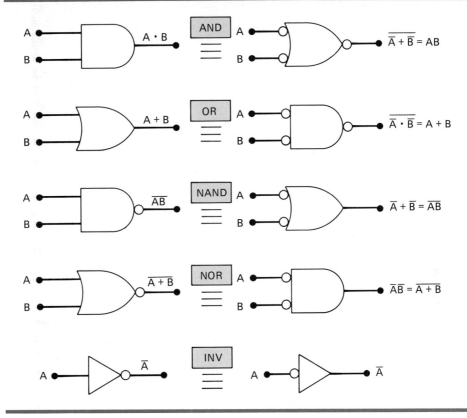

FIGURE 3-33 Standard and alternate symbols for various logic gates and inverter.

put, and add a bubble to each input. Then change the AND symbol to an OR symbol. The result is an OR symbol with bubbles on its inputs.

We can easily prove that this alternate symbol is equivalent to the standard symbol by using DeMorgan's theorems and recalling that the bubble represents an inversion operation. The output expression from the standard NAND symbol is $\overline{AB} = \overline{A} + \overline{B}$, which is the same as the output expression for the alternate symbol. This same procedure can be followed for each pair of symbols in Figure 3-33.

Several points should be stressed regarding the logic symbol equivalences:

1. The equivalences are valid for gates with *any* number of inputs.

2. None of the standard symbols have bubbles on their inputs, and all the alternate symbols do.

3. The standard and alternate symbols for each gate represent the same physical circuit; there is no difference in the circuits represented by the two symbols.

4. NAND and NOR gates are inverting gates, so both the standard and alternate symbols for each will have a bubble on *either* the input or the output. AND and OR gates are *noninverting* gates, so the alternate symbols for each will have bubbles on *both* inputs and output.

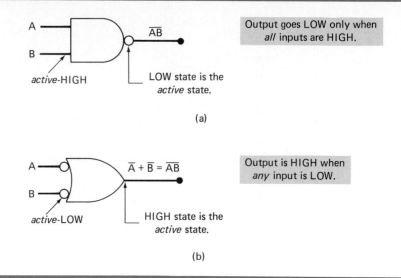

(a)

(b)

FIGURE 3-34 Interpretation of the two NAND gate symbols.

Logic-Symbol Interpretation   Each of the logic-gate symbols of Figure 3-33 provides a unique interpretation of how the gate operates. Before we can demonstrate these interpretations, we must first establish the concept of *active* logic levels.

When an input or output line on a logic gate symbol has *no bubble* on it, that line is said to be active-HIGH. When an input or output line *does* have a *bubble* on it, that line is said to be active-LOW. The presence or absence of a bubble, then, determines the active-HIGH/active-LOW status of a gate's inputs and output, and is used to interpret the gate operation.

To illustrate, Figure 3-34(a) shows the standard symbol for a NAND gate. The standard symbol has a bubble on its output and no bubbles on its inputs. Thus, it has an active-LOW output and active-HIGH inputs. The logic operation represented by this symbol can therefore be interpreted as follows:

The output goes LOW only when all the inputs are HIGH

Note that this says that the output will go to its active state only when *all* the inputs are in their active states. The word "all" is used because of the AND symbol.

The alternate symbol for a NAND gate shown in Figure 3-34(b) has an active-HIGH output and active-LOW inputs so its operation can be stated as

The output goes HIGH only when any input is LOW

Again, this says that the output will be in its active state whenever *any* of the inputs are in their active states. The word "any" is used because of the OR symbol.

With a little thought, it can be seen that the two interpretations for the NAND symbols in Figure 3-34 are different ways of saying the same thing.

Summary   At this point you are probably wondering why there is a need to have two different symbols and interpretations for each logic gate. Hopefully, the reasons will become clear after reading the next section. For now, let us summarize the important points concerning the logic-gate representations.

1. To obtain the alternate symbol for a logic gate, take the standard symbol and change its operation symbol (OR to AND, or AND to OR) and change the bubbles on both inputs and output (that is, delete bubbles that are present, and add bubbles where there are none).

2. To interpret the logic-gate operation, first note which logic state, 0 or 1, is the active state for the inputs, and which is the active state for the output. Then, realize that the output's active state is produced by having *all* the inputs in their active state (if an AND symbol is used), or by having *any* of the inputs in their active state (if an OR symbol is used).

EXAMPLE 3-19

Give the interpretation of the two OR gate symbols.

SOLUTION
The results are shown in Figure 3-35. Note again how the word *any* is used when the gate symbol includes an OR symbol and the word *all* is used when it includes an AND symbol.

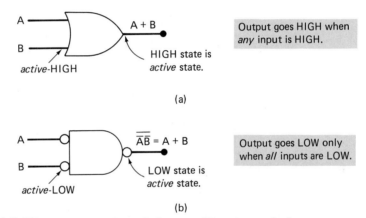

FIGURE 3-35 **Interpretation of the two OR gate symbols.**

REVIEW QUESTIONS

1. Write the interpretation of the operation performed by the standard NOR gate symbol in Figure 3-33. (*Ans*. The output goes LOW only when any input is HIGH)

2. Repeat question 1 for the alternate NOR gate symbol. (*Ans*. The output goes HIGH only when all inputs are LOW)

3. Repeat question 1 for the alternate AND gate symbol. (*Ans*. The output goes LOW when any input is LOW)

4. Repeat question 1 for the standard AND gate symbol. (*Ans*. The output goes HIGH only when all inputs are HIGH)

3-14 WHICH GATE REPRESENTATION TO USE

Some logic-circuit designers and many textbooks use only the standard logic-gate symbols in their circuit schematics. While this practice is not incorrect, it does nothing to make the circuit operation easier to follow. Proper use of the alternate gate symbols in the circuit diagram can make the circuit operation much clearer. This can be illustrated by considering the example shown in Figure 3-36.

The circuit in Figure 3-36(a) contains three NAND gates connected to produce an output Z that depends on inputs A, B, C, D. The circuit diagram uses the standard symbol for each of the NAND gates. While this diagram is logically correct,

FIGURE 3-36 (a) Original circuit using standard NAND symbols; (b) equivalent representation where output z is active-HIGH: (c) equivalent representation where output z is active-LOW; (d) truth table.

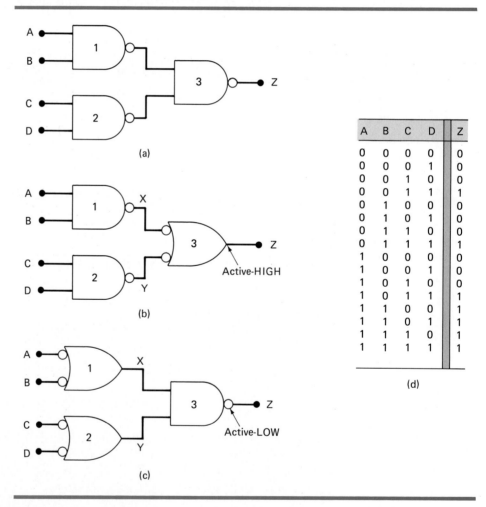

A	B	C	D	Z
0	0	0	0	0
0	0	0	1	0
0	0	1	0	0
0	0	1	1	1
0	1	0	0	0
0	1	0	1	0
0	1	1	0	0
0	1	1	1	1
1	0	0	0	0
1	0	0	1	0
1	0	1	0	0
1	0	1	1	1
1	1	0	0	1
1	1	0	1	1
1	1	1	0	1
1	1	1	1	1

(d)

it does not facilitate any understanding of how the circuit functions. The improved circuit representations given in Figure 3-36(b) and (c), however, can be analyzed more easily to determine the circuit operation.

The representation of Figure 3-36(b) is obtained from the original circuit diagram by replacing NAND gate 3 with its alternate symbol. In this diagram output Z is taken from a NAND gate symbol that has an active-HIGH output. Thus, we can say that Z will go HIGH when either X or Y is LOW. Now, since X and Y appear at the output of NAND symbols having active-LOW outputs, we can say that X will go LOW only if $A = B = 1$, and Y will go LOW only if $C = D = 1$. Putting this all together, we can describe the circuit operation as follows:

Output Z will go HIGH whenever either A = B = 1 or C = D = 1

This description can be translated to truth-table form by setting $Z = 1$ for those cases where $A = B = 1$, and for those cases where $C = D = 1$. For all other cases, Z is made a 0. The resultant truth table is shown in Figure 3-36(d).

The representation of Figure 3-36(c) is obtained from the original circuit diagram by replacing NAND gates 1 and 2 by their alternate symbols. In this equivalent representation the Z output is taken from a NAND gate that has an active-LOW output. Thus, we can say that Z will go LOW only when $X = Y = 1$. Since X and Y are active-HIGH outputs, we can say that X will be HIGH when either A or B is LOW, and Y will be HIGH when either C or D is LOW. Putting this all together, we can describe the circuit operation as follows:

Output Z will go LOW only when A or B is LOW and C or D is LOW

This description can be translated to truth-table form by making $Z = 0$ for all cases where at least one of the A or B inputs is LOW at the same time that at least one of the C or D inputs is LOW. For all other cases, Z is made a 1. The resultant truth table is the same as that obtained for the circuit diagram of Figure 3-36(b).

Which Circuit Diagram Should Be Used?

The answer to this question depends on the particular function being performed by the circuit output. If the circuit is being used to cause some action (e.g., turn ON a device, activate another logic circuit) when output Z goes to the 1 state, then we say that Z is active-HIGH, and the circuit diagram of Figure 3-36(b) should be used. On the other hand, if the circuit is being used to cause some action when Z goes to the 0 state, then Z is active-LOW and the diagram of Figure 3-36(c) should be used.

Of course, there will be situations where *both* output states are used to produce different actions and either one can be considered to be the active state. For these cases, either circuit representation can be used.

Bubble Placement

Refer to the circuit representatiion of Figure 3-36(b) and note that the symbols for NAND gates 1 and 2 were chosen to have active-LOW outputs to match the active-LOW inputs of NAND gate 3. Refer to the circuit representation of Figure 3-36(c) and note that the symbols for NAND gates 1 and 2 were chosen to have active-HIGH outputs to match the active-HIGH inputs of

NAND gate 3. This leads to the following general rule for preparing logic-circuit schematics:

Whenever possible, connect bubble outputs to bubble inputs, and nonbubble outputs to nonbubble inputs

The following examples will show how this rule can be applied.

EXAMPLE 3-20

The logic circuit in Figure 3-37(a) is being used to activate an ALARM when its output Z goes HIGH. Modify the circuit diagram so that it more effectively represents the circuit operation.

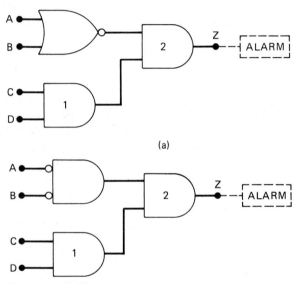

(a)

FIGURE 3-37 **Example 3-20**

SOLUTION

Since $Z = 1$ will activate the ALARM, Z is active-HIGH. Thus, the AND gate 2 symbol does not have to be changed. The NOR gate symbol should be changed to the alternate symbol with a nonbubble (active-HIGH) output to match the nonbubble input of AND gate 2. The result is shown in Figure 3-37(b).

EXAMPLE 3-21

When the output of the logic circuit in Figure 3-38(a) goes LOW, it activates another logic circuit. Modify the circuit diagram to more effectively represent the circuit operation.

SOLUTION

Since Z is active-LOW, the symbol for OR gate 2 has to be changed to its alternate symbol as shown in Figure 3-38(b). The new OR-gate 2 symbol has bubble inputs, so the AND gate and OR gate 1 symbols have to be changed to

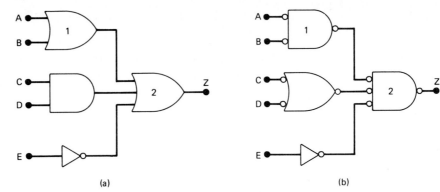

FIGURE 3-38 **Example 3-21**

bubbled outputs as shown in Figure 3-38(b). The INVERTER already has a bubble output.

Analyzing Circuits   When a logic-circuit schematic is drawn using the rules we followed in these examples, it is much easier for an engineer or technician (or student) to follow the signal flow through the circuit and to determine the input conditions that are needed to activate the output. This will be illustrated in the following examples—which, incidentally, use circuit diagrams taken from the logic schematics of an actual microcomputer.

EXAMPLE 3-22

The logic circuit in Figure 3-39 generates an output, *MEM*, that is used to activate the memory ICs in a particular microcomputer. Determine the input conditions necessary to activate *MEM*.

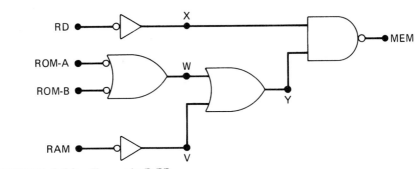

FIGURE 3-39 **Example 3-22**

SOLUTION
One way to do this would be to write the expression for *MEM* in terms of the inputs *RD, ROM-A, ROM-B,* and *RAM,* and to evaluate it for the 16 possible combinations of these inputs. While this method would work, it would require a lot more work than is necessary.

A more efficient method is to interpet the circuit diagram using the ideas we have been developing in the last two sections. These are the steps:

1. *MEM* is active-LOW, and it will go LOW only when X and Y are HIGH.
2. X will be HIGH only when $RD = 0$.
3. Y will be HIGH when either W or V is HIGH.
4. V will be HIGH when $RAM = 0$.
5. W will be HIGH when either $ROM\text{-}A$ or $ROM\text{-}B = 0$.
6. Putting this all together, *MEM* will go LOW only when $RD = 0$ *and* at least one of the three inputs *ROM-A, ROM-B,* or *RAM* is LOW.

EXAMPLE 3-23

The logic circuit in Figure 3-40 is used to control the motor on a cassette recorder when the microcomputer is sending data to or receiving data from the recorder. The circuit will turn on the motor when $CASS = 1$. Determine the input conditions necessary to turn on the motor.

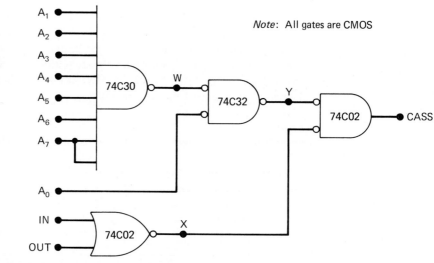

FIGURE 3-40   **Example 3-23**

SOLUTION

Once again we will interpret the diagram in a step-by-step fashion:

1. *CASS* is active-HIGH, and it will go HIGH only when $X = Y = 0$.
2. X will be LOW when either *IN* or *OUT* is HIGH.
3. Y will be LOW only when $W = 0$ and $A_0 = 0$.
4. W will be LOW only when A_1–A_7 are all HIGH.
5. Putting this all together, *CASS* will be HIGH when $A_1 = A_2 = A_3 = A_4 = A_5 = A_6 = A_7 = 1$ and $A_0 = 0$, and either IN or OUT or both are 1.

Note the strange symbol for the 8-input CMOS NAND gate (74C00); also note that the signal A_7 is connected to two of the NAND inputs.

Asserted Levels We have been describing logic signals as being active-LOW or active-HIGH. For example, the output *MEM* in Figure 3.39 is active-LOW, and the output CASS in Figure 3-40 is active-HIGH, since these are the output states that cause something to happen. Likewise, Figure 3-40 has active HIGH inputs A_1 to A_7, and active-LOW input A_0.

When a logic signal is in its active state, it can be said to be *asserted*. For example, when we say that input A_0 is asserted, we are saying that it is in its active-LOW state. When a logic signal is not in its active state, it is said to be *unasserted*. Thus, when we say that *CASS* is unasserted, we mean that it is in its inactive state (LOW).

Clearly, the terms "asserted" and "unasserted" are synonomous with "active" and "inactive," respectively.

$$asserted = active$$

$$unasserted = inactive$$

Both sets of terms are in common use in the digital field, so you should recognize both ways of describing a logic signal's active state.

Labeling Active-LOW Logic Signals It has become common practice to use an overbar to label active-LOW signals. The overbar serves as another indication that the signal is active-LOW; of course, the absence of an overbar means that the signal is active-HIGH.

To illustrate, all of the signals in Figure 3-39 are active-LOW, so they can be labeled as follows:

$$\overline{RD}, \quad \overline{ROM\text{-}A}, \quad \overline{ROM\text{-}B}, \quad \overline{RAM}, \quad \overline{MEM}$$

Remember, the overbar is simply a way to emphasize that these are active-LOW signals. We will employ this convention for labeling logic signals whenever appropriate.

Summary The examples in this section have shown how to draw a logic-circuit diagram so that the circuit operation may be easily determined. By choosing the proper symbol for each logic gate, we may quickly analyze the circuit diagram to determine the input conditions required to activate the output.

We will use this method for many of the circuit diagrams presented throughout the text. In some cases, however, for various reasons we will use only standard gate symbols in the circuit diagram.

REVIEW QUESTIONS

1. Use the method of Examples 3-22 and 3-23 to determine the input conditions needed to activate the output of the circuit in Figure 3-37(b). (*Ans.* Z will go HIGH when $A = B = 0$ and $C = D = 1$)
2. Repeat question 1 for the circuit of Figure 3-38(b). (*Ans.* Z will go LOW when $A = B = 0$, $E = 1$, and either C or D or both are LOW)
3. How many NAND gates are in Figure 3-39? (*Ans.* 2)

3-15 NEW IEEE/ANSI STANDARD LOGIC SYMBOLS

The logic symbols that we have used throughout this chapter are the well-known standard symbols that have been widely used in the digital industry for many years. These symbols work well for the basic logic gates because each gate symbol has a distinctive shape, and each gate input has the same function. They do not provide enough useful information, however, for the more complex logic devices such as flip-flops, counters, decoders, multiplexers, memories, and microprocessor interface ICs. These complex ICs often have several inputs and outputs with different functions and modes of operation.

In order to provide more useful information about these complex logic devices, a new set of standard symbols was introduced in 1984 under IEEE/ANSI Standard 91–1984. These newer symbols are gradually being accepted by more and more electronics companies and IC manufacturers, and have begun to appear in their published literature. In addition, U.S. military contracts now require the use of these new symbols. It is therefore important to become familiar with the new standard symbols since eventually they will completely replace the traditional ones, although the process may take a long time.

The principal difference in the new standard is that it uses rectangular symbols for all devices instead of different symbol shapes for each device. A special notation system is used to indicate how the inputs and outputs are related. Figure 3-41 shows the traditional symbols with the newer rectangular symbols for the basic logic gates. Study them carefully and note the following points:

1. The newer symbols use a small right triangle (◁) in place of the small bubble of the traditional symbols. Like the bubble, this triangle indicates an inversion of the logic level. The presence or absence of the triangle also signifies whether an input or output is active-LOW or active-HIGH.

2. A special notation inside each rectangular symbol describes the logic relation between inputs and output. The ''1'' inside the INVERTER symbol denotes a device with only *one* input; the triangle on the output indicates that the output will go to its active-LOW state when that one input is in its active-HIGH state.

   The ''&'' inside the AND symbol means that the output will go to its active-HIGH state when all the inputs are in their active-HIGH state. The ''≥'' inside the OR gate means that the output will go to its active state (HIGH) whenever *one or more* inputs are in their active state (HIGH).

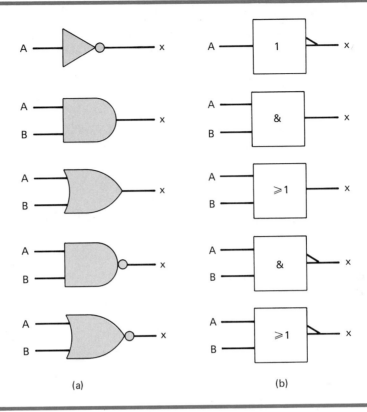

(a)   (b)

FIGURE 3-41   Standard logic symbols: (a) traditional; (b) rectangular.

3.   The rectangular symbols for the NAND and NOR are the same as those for
     the AND and OR, respectively, with the addition of the small inversion tri-
     angle on the output.

New IEEE/ANSI Symbols for Logic Gate ICs

The rectangular symbols
can also be used to represent the complete logic of an IC package that contains a
number of independent gates. This is illustrated in Figure 3-42 for the 7404 TTL
hex INVERTER IC, and in Figure 3-43 for the 7420 dual four-input NAND IC.
Each logic gate is represented as a separate rectangular block. Note how the rectangular symbols indicate the logic operation notation only in the top block; it is understood to apply to all the other blocks representing the other gates on the chip.
    It is important to understand the difference between the two alternative ways
to represent a logic gate in a circuit, and the two different standards for logic gate
symbols. You choose which set of standard logic symbols to use—either the traditional symbols (distinctive shapes for each type of gate) or the new standard rectangular symbols. Regardless of which symbols you choose to use, there are two possible ways to represent a gate in a circuit diagram depending on its active output
state. This is illustrated in the example on pages 96-97.

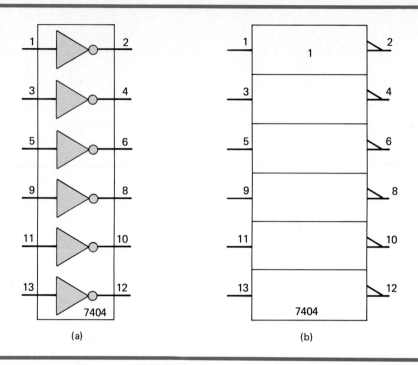

(a)   (b)

FIGURE 3-42   7404 hex INVERTER IC: (a) traditional logic symbol; (b) rectangular logic symbol. The notation "1" appears only in the top rectangle, but applies to all blocks below.

FIGURE 3-43   7420 dual four-input NAND IC: (a) traditional symbol; (b) rectangular symbol.

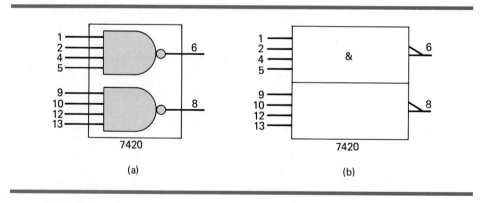

(a)   (b)

EXAMPLE 3-24

Figure 3-44(a) shows the two representations for a NOR gate using the traditional logic symbols. Remember, the choice of which representation to use in a circuit diagram is determined by the desired active state of the output. Redraw the two representations using the new IEEE/ANSI symbols.

SOLUTION
Figure 3.44(b) shows the results.

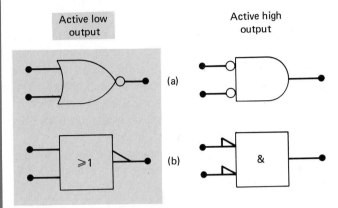

FIGURE 3-44 **Both representations of a NOR gate using the two types of symbols: (a) traditional; (b) rectangular.**

New IEEE/ANSI Symbols for Complex ICs   There would be no real advantage to the new symbols if all we had to deal with were the basic logic gates. For the more complex logic devices, however, the new standard symbols with their *dependency notation* will specify the complete logic operation of the device. This makes it virtually unnecessary to refer to the manufacturer's data manual to find out how a particular logic IC is functioning in a circuit. We will see examples of this when we encounter the more complex logic circuits in subsequent chapters.

Throughout this text we will continue to use mainly the traditional logic symbols in our circuit diagrams. There are two reasons for this: (1) the logic circuit schematics for most of the digital equipment already in the field will still use the traditional symbology, and (2) it will be some time before the many thousands of engineers and technicians make the complete switch to the new standard. We will, however, present and describe the new IEEE/ANSI symbols for most of the logic devices that are covered in the text. In this way, you can become familiar with the various aspects of the dependency notation that is the major benefit provided by the new standard.

REVIEW QUESTIONS

1. What is the major advantage of the new IEEE/ANSI symbols?
2. Draw all of the basic logic gates using both the traditional symbols and the new standard symbols.
3. Repeat question 2 for the alternative representation of each gate.

Active Logic Levels When an input or output line of a logic circuit symbol has a bubble, that line is active-LOW. On the other hand, if it does not have a bubble, the line is active-HIGH. [Sec. 3-13]

AND Gate Digital circuit that implements the AND operation. The output of this circuit is HIGH (logic level 1) only if all its inputs are HIGH. [Sec. 3-4]

AND Operation Boolean algebra operation in which the symbol ($\cdot$) is used to indicate the ANDing of two or more logic variables. The result of the AND operation will be HIGH (logic level 1) only if all variables are HIGH. [Secs. 3-1 and 3-4]

Asserted Term used to describe the state of a logic signal. The term ''asserted'' is synonymous with ''active.'' [Sec. 3-14]

Associative Laws Laws which state that the manner in which variables in an AND expression or OR expression are grouped together does not affect the final result. [Sec. 3-10]

Boolean Algebra Algebraic process used as a tool in the design and analysis of digital systems. In Boolean algebra only two values are possible, ''0'' and ''1.'' [Sec. 3-1]

Boolean Theorems Rules that can be applied to Boolean algebra to simplify logic expressions. [Sec. 3-10]

Bubbles Small circles on the input or output lines of logic circuit symbols which represent inversion of that particular signal. If a bubble is present, that input or output is said to be active-LOW. [Sec. 3-13]

Commutative Laws Laws which state that the order in which two variables are ORed or ANDed is unimportant. [Sec. 3-10]

Complement *See* Invert.

DeMorgan's Theorems The first theorem states that the complement of a sum (OR operation) equals the product (AND operation) of the complements. The second theorem states that the complement of a product (AND operation) equals the sum (OR operation) of the complements. [Sec. 3-11]

Dependency Notation Method used to represent pictorially the relationship between inputs and outputs of logic circuits. This method employs the use of qualifying symbols embedded near the top center or geometric center of a symbol element. [Sec. 3-15]

Distributive Law Law which states that an expression can be expanded by multiplying term by term. This law also states that if we have a sum of two or more terms, each of which contains a common variable, the common variable can be factored out. [Sec. 3-10]

IEEE/ANSI Institute of Electrical and Electronics Engineers/American National Standards Institute. [Sec. 3-15]

Invert Cause a logic level to go to the opposite state.

INVERTER Also referred to as the NOT circuit, this logic circuit implements the

NOT operation. An INVERTER has only one input, and its output logic level is always the opposite of this input's logic level. [Sec. 3-5]

Logic Level State of a voltage variable. The range of such voltage is expressed either by a 1 (HIGH) or a 0 (LOW). [Sec.3-1]

NAND Gate Logic circuit which operates like an AND gate followed by an INVERTER. The output of a NAND gate is LOW (logic level 0) only if all inputs are HIGH (logic level 1). [Sec. 3-9]

NOR Gate Logic circuit which operates like an OR gate followed by an INVERTER. The output of a NOR gate is LOW (logic level 0) when any or all inputs are HIGH (logic level 1). [Sec. 3-9]

NOT Circuit *See* INVERTER. [Sec. 3-5]

NOT Operation Boolean algebra operation in which the overbar ($\frown$) or the prime (') symbol is used to indicate the inversion of one or more logic variables. The result of a NOT operation is always the complement of the expression being NOTed. [Secs. 3-1 and 3-5]

OR Gate Digital circuit that implements the OR operation. The output of this circuit is HIGH (logic level 1) if any or all of its inputs are HIGH. [Sec. 3-3]

OR Operation Boolean algebra operation in which the symbol (+) is used to indicate the ORing of two or more logic variables. The result of the OR operation will be HIGH (logic level 1) if one or more variables is HIGH. [Secs. 3-1 and 3-3]

Truth Table Logic table that depicts a circuit's output response to the various combinations of the logic levels at its inputs. [Sec. 3-2]

Unasserted Term used to describe the state of a logic signal. The term "unasserted" is synonymous with "inactive." [Sec. 3-14]

PROBLEMS   Section 3-3

3-1.   Draw the output waveform for the circuit of Figure 3-45.

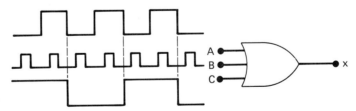

FIGURE 3-45

3-2.   Suppose that the *A* input in Figure 3-45 is unintentionally shorted to ground (i.e., *A* = 0). Draw the resulting output waveform.

3-3.   Suppose that the *A* input in Figure 3-45 is unintentionally shorted to the +5-V supply line (i.e., *A* = 1). Draw the resulting output waveform.

3-4.   Read the statements below concerning an OR gate. At first they may appear to be valid, but after some thought you should realize that neither one is *always* true. Prove this by showing a specific example to refute each statement.

(a) If the output waveform from an OR gate is the same as the waveform at one of its inputs, then the other input is being held permanently LOW.

(b) If the output waveform from an OR gate is always HIGH, then one of its inputs is being held permanently HIGH.

3-5.   How many different sets of input conditions will produce a HIGH output from a five-input OR gate?

Section 3-4

3-6.   Change the OR gate in Figure 3-45 to an AND gate.
(a) Draw the output waveform.
(b) Draw the output waveform if the A input is permanently shorted to ground.
(c) Draw the output waveform if A is permanently shorted to $+5$ V.

3-7.   Refer to Figure 3-4. Modify the circuit so that the alarm is to be activated only when the pressure and temperature exceed their maximum limits at the same time.

3-8.   Change the OR gate in Figure 3-6 to an AND gate and draw the output waveform.

Sections 3-5–3-7

3-9.   Add an INVERTER to the output of the OR gate from Figure 3-45. Draw the waveform at the INVERTER output.

3-10.   Write the Boolean expression for the output x in Figure 3-46(a). Determine the value of x for all possible input conditions and list them in a truth table.

3-11.   Repeat Problem 3-10 for the circuit in Figure 3-46(b).

3-12.   Change each OR to an AND, and each AND to an OR in Figure 3-15(b). Then write the expression for the output.

3-13.   Determine the complete truth table for the circuit of Figure 3-16 by finding the logic levels present at each gate output for each of the 16 possible combinations of input levels.

Section 3-8

3-14.   For each of the following expressions, construct the corresponding logic circuit, using AND and OR gates and INVERTERs.
(a) $x = \overline{AB(C + D)}$
(b) $z = (A + B + \overline{CD\overline{E}}) + \overline{BCD}$
(c) $y = (\overline{M + N} + \overline{P}Q)$

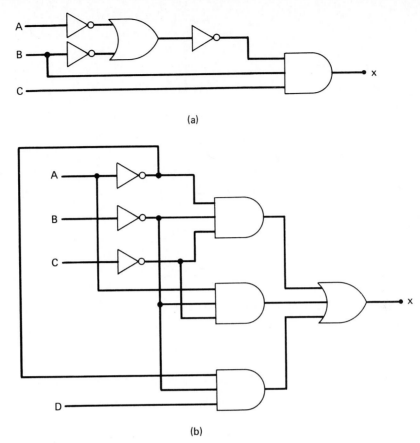

(a)

(b)

FIGURE 3-46

Section 3-9

3-15. Apply the input waveforms of Figure 3-47 to a NOR gate and draw the output waveform.

3-16. Repeat Problem 3-15 with the *C* input held permanently LOW. Then repeat for *C* held permanently HIGH.

3-17. Repeat Problem 3-15 for a NAND gate.

3-18. Repeat Problem 3-16 for a NAND gate.

3-19. Write the output expression for the circuit of Figure 3-48.

3-20. Determine the complete truth table for the circuit of Figure 3-48.

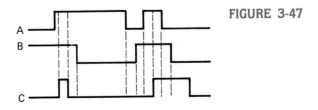

FIGURE 3-47

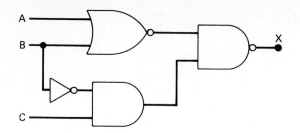

FIGURE 3-48

3-21. Modify the circuits that were constructed in Problem 3-14 so that NAND gates and NOR gates are used wherever appropriate.

Section 3-10

3-22. Prove theorem (15) by trying all possible cases.

3-23. Simplify the following expression using theorems (13b), (3), and (4).
$$x = (M + N)(\overline{M} + P)(\overline{N} + \overline{P})$$

3-24. Simplify the following expression using theorems (13a), (8), and (6).
$$z = \overline{A}B\overline{C} + AB\overline{C} + B\overline{C}D$$

Sections 3-11 and 3-12

3-25. Prove DeMorgan's theorems by trying all possible cases.

3-26. Simplify each of the following expressions using DeMorgan's theorems.
(a) $\overline{\overline{AB\overline{C}}}$ (b) $\overline{\overline{A} + \overline{BC}}$ (c) $\overline{\overline{AB\overline{CD}}}$ (d) $\overline{A(B + \overline{C})D}$

3-27. Use DeMorgan's theorems to simplify the expression for the output of Figure 3-48.

3-28. Convert the circuit of Figure 3-46(b) to one using only NAND gates. Then write the output expression for the new circuit, simplify it using De-Morgan's theorems, and compare it to the expression for the original circuit.

3-29. Convert the circuit of Figure 3-46(a) to one using only NOR gates. Then write the expression for the new circuit, simplify it using DeMorgan's theorems, and compare it to the expression for the original circuit.

3-30. Show how a two-input NAND gate can be constructed from two-input NOR gates.

3-31. Show how a two-input NOR gate can be constructed from two-input NAND gates.

Sections 3-13 and 3-14

3-32. Draw the standard representations for each of the basic logic gates. Then draw the alternative representations.

3-33. For each statement below, draw the appropriate logic gate representation and indicate the type of gate.
(a) A HIGH output occurs only when all three inputs are LOW.

(b) A LOW output occurs when any of the four inputs is LOW.

(c) A LOW output occurs only when all eight inputs are HIGH.

3-34. The output of the circuit of Figure 3-48 is supposed to turn on an indicator lamp when it goes LOW.

(a) Modify the circuit diagram so that it more effectively represents the circuit operation.

(b) Use the new circuit diagram to determine the input conditions that will activate the output. Do this by working back from the output using the information given by the gate symbols as was done in Examples 3-22 and 3-23. Compare the results to the truth table obtained in Problem 3-20.

3-35. (a) Determine the input conditions needed to activate output Z in Figure 3-37(b). Do this by working back from the output as was done in Examples 3-22 and 3-23.

(b) Assume that it is the LOW state of Z that is to activate the alarm. Change the circuit diagram to reflect this, and then use the revised diagram to determine the input conditions needed to activate the alarm.

3-36. Modify the circuit of Figure 3-40 so that $A_1 = 0$ is needed to produce $CASS = 1$ instead of $A_1 = 1$.

3-37. Determine the input conditions needed to cause the output in Figure 3-49 to go to its active state.

3-38. Use the results of Problem 3-37 to obtain the complete truth table for the circuit of Figure 3-49.

3-39. What is the asserted state for the output of Figure 3-49? For the output of Figure 3-36(c)?

3-40. Figure 3-50 shows an application of logic gates that simulates a two-way switch like the ones used in our homes to turn a light on or off from two different switches. Here the light is an LED which will be ON (conducting) when the NOR gate output is LOW. Note that this output is labeled $\overline{LIGHT}$ to indicate that it is active-LOW. Determine the input conditions needed to turn on the LED. Then verify that the circuit operates as a two-way switch

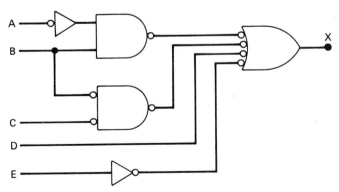

FIGURE 3-49

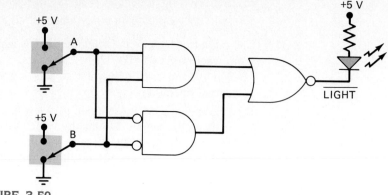

FIGURE 3-50

using switches A and B. In Chapter 4 you will learn how to design circuits like this one to produce a given relationship between inputs and outputs.

Section 3-15

3-41. Redraw the circuit of Figure 3-49 using the IEEE/ANSI symbology.

3-42. Determine the Boolean expression for the output Z in Figure 3-51.

3-43. The output of the circuit of Figure 3-51 is supposed to be active-LOW. Redraw it to more effectively represent the circuit operation.

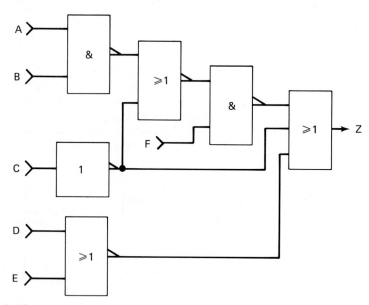

FIGURE 3-51

3-44. Use the redrawn version of the circuit of Figure 3-51 and do each of the following:

(a) Determine the various input conditions that will produce the active-LOW output state. Do this using only the circuit diagram without writing the expression for Z and without generating the complete truth table. The results should be

A	B	C	D	E	F
1	1	1	1	1	1
1	1	1	1	0	1
1	1	1	0	1	1

(b) Verify that the simplified expression for output Z is given by:

$$Z = \overline{ABCF\,(D + E)}$$

(c) Plug each of the sets of conditions from (a) into the expression you determined in (b), and verify that each one produces Z = 0.

COMBINATORIAL LOGIC CIRCUITS

4

Upon completion of this chapter, you will be able to:

- Modify a logic expression into a sum-of-products expression.
- Perform the necessary steps to derive a sum-of-products expression in order to design a combinatorial logic circuit in its simplest form.
- Use the Karnaugh map as a tool to simplify and design logic circuits.
- Explain the operation of both exclusive-OR and exclusive-NOR circuits.
- Design logic circuits with and without the help of a truth table.
- Identify and understand inhibit circuits.
- Cite the basic characteristics of digital ICs.
- Understand the inherent operative differences between TTL and CMOS.
- Use the basic troubleshooting rules of digital systems.
- Deduce from measured results the faults of malfunctioning combinatorial logic circuits.

In Chapter 3 we studied the operation of all the basic logic gates, and we used Boolean algebra to describe and analyze circuits that were made up of combinations of logic gates. These circuits can be classified as *combinatorial* logic circuits because, at any time, the logic level at the output depends on the combination of logic levels present at the inputs. A combinatorial circuit has no *memory* characteristic, and so its output depends *only* on the current value of its inputs.

In this chapter we will continue our study of combinatorial logic circuits. To start, we will go further into the simplification of logic circuits. Two methods will be used: one will use Boolean algebra theorems, the other a *mapping* technique. In addition, we will study simple techniques for designing logic circuits to satisfy a given set of requirements. A complete study of logic-circuit design is not one of our objectives, but the methods we introduce are more than sufficient for the types of design situations that a technician will encounter.

The last portion of the chapter is devoted to the troubleshooting of combinatorial circuits. This first exposure to troubleshooting should begin to develop the type of analytical skills needed for successful troubleshooting. To make this material as practical as possible, we will first present some of the basic characteristics of logic gate ICs in the TTL and CMOS logic families along with a description of the most common types of faults encountered in digital IC circuits.

4-1 SUM-OF-PRODUCTS FORM

The methods of logic-circuit simplification and design that we will study require the logic expression to be in a *sum-of-products* form. Some examples of this form are

1. $ABC + \overline{A}B\overline{C}$
2. $AB + \overline{A}B\overline{C} + \overline{C}D + D$
3. $\overline{A}B + C\overline{D} + EF + GK + H\overline{L}$

Each of these sum-of-products expressions consists of two or more AND terms (products) that are ORed together. Each AND terms consists of one or more variables appearing in either complemented or uncomplemented form. For example, in the sum-of-products expression $ABC + \overline{A}B\overline{C}$, the first AND product contains the variables A, B, and C in their uncomplemented (not inverted) form. The second AND term contains A and C in their complemented (inverted) form. Note that in a sum-of-products expression, an inversion sign *cannot* appear over more than one variable in a term (e.g., we cannot have $\overline{ABC}$ or $\overline{RST}$).

4-2 SIMPLIFYING LOGIC CIRCUITS

Once the expression for a logic circuit has been obtained, we may be able to reduce it to a simpler form containing fewer terms or fewer variables in one or more terms. The new expression can then be used to implement a circuit that is equivalent to the original circuit but that contains fewer gates and connections.

To illustrate, the circuit of Figure 4-1(a) can be simplified to produce the circuit of Figure 4-1(b). Since both circuits perform the same logic, it should be obvious that the simpler circuit is more desirable because it contains fewer gates and will therefore be smaller and cheaper than the original. Furthermore, the circuit reliability will improve because there are fewer interconnections that can be potential circuit faults.

In subsequent sections we will study two methods for simplifying logic circuits. One method will utilize the Boolean algebra theorems and, as we shall see, is greatly dependent on inspiration and experience. The other method (Karnaugh mapping) is a systematic, cookbook approach. Some instructors may wish to skip over this latter method because it is somewhat mechanical and does not contribute to a better understanding of Boolean algebra. This can be done without affecting the continuity or clarity of the rest of the text.

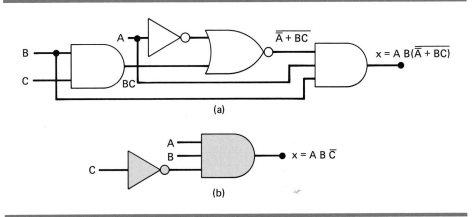

(a)

(b)

FIGURE 4-1

4-3 ALGEBRAIC SIMPLIFICATION

The Boolean algebra theorems that we studied in Chapter 3 can be used to help us simplify the expression for a logic circuit. Unfortunately, it is not always obvious which theorems should be applied in order to produce the simplest result. Furthermore, there is no easy way to tell whether the simplifed expression is in its simplest form or whether it could have been simplified further. Thus, algebraic simplification often becomes a process of trial and error. With experience, however, one can become adept at obtaining reasonably good results.

The examples that follow will illustrate many of the ways in which the Boolean theorems can be applied in trying to simplify an expression. You should notice that these examples contain two essential steps:

1. The expression is put into the sum-of-products form.
2. Once it is in this form, the product terms are checked for common factors, and factoring is performed wherever possible. Hopefully, the factoring results in the elimination of one or more terms.

EXAMPLE 4-1

Simplify the logic circuit shown in Figure 4-2(a).

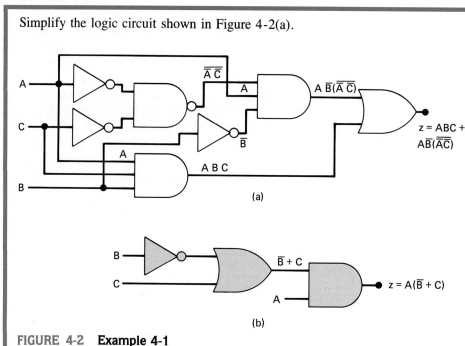

FIGURE 4-2 **Example 4-1**

SOLUTION
The first step is to determine the expression for the output. The result is

$$z = ABC + A\overline{B} \cdot (\overline{\overline{A}\,\overline{C}})$$

Once the expression is determined, it is usually a good idea to break down all large inverter signs using DeMorgan's theorems and then multiply out all terms.

$$\begin{aligned} z &= ABC + A\overline{B}(\overline{\overline{A}} + \overline{\overline{C}}) && \text{[theorem (17)]} \\ &= ABC + A\overline{B}(A + C) && \text{[cancel double inversions]} \\ &= ABC + A\overline{B}A + A\overline{B}C && \text{[multiply out]} \\ &= ABC + A\overline{B} + A\overline{B}C && [A \cdot A = A] \end{aligned}$$

With the expression in sum-of-products form, we should look for common variables among the various terms with the intention of factoring. The first and third terms above have AC in common, which can be factored out:

$$z = AC(B + \overline{B}) + A\overline{B}$$

Since $B + \overline{B} = 1$, then

$$\begin{aligned} z &= AC(1) + A\overline{B} \\ &= AC + A\overline{B} \end{aligned}$$

We can now factor out A, which results in

$$z = A(C + \overline{B})$$

This result can be simplified no further. Its circuit implementation is shown in Figure 4-2(b). It is obvious that the circuit in (b) is a great deal simpler than the original circuit in (a).

EXAMPLE 4-2

Simplify the expression $z = ABC + AB\overline{C} + A\overline{B}C$.

SOLUTION
We will look at two different ways to arrive at the same result.

Method 1: The first two terms in the expression have the variables AB in common. Thus,

$$\begin{aligned} z &= AB(C + \overline{C}) + A\overline{B}C \\ &= AB(1) + A\overline{B}C \\ &= AB + A\overline{B}C \end{aligned}$$

We can factor the variable A from both terms:

$$z = A(B + \overline{B}C)$$

Invoking theorem (15),

$$z = A(B + C)$$

Method 2: The original expression is $z = ABC + AB\overline{C} + A\overline{B}C$. The first two terms have the variables AB in common. The first and last terms have the variables AC in common. How do we know whether to factor AB from the first two terms or AC from the two end terms? Actually, we can do both by using the ABC term *twice*. In other words, we can rewrite the expression as

$$z = ABC + AB\overline{C} + A\overline{B}C + ABC$$

where we have added an extra term ABC. This is valid and will not change the value of the expression since $ABC + ABC = ABC$ [theorem (7)]. Now we can factor AB from the first two terms and AC from the last two terms:

$$z = AB(C + \overline{C}) + AC(\overline{B} + B)$$
$$= AB \cdot 1 + AC \cdot 1$$
$$= AB + AC = A(B + C)$$

This is, of course, the same result as method 1. This trick of using the same term twice can always be used. In fact, the same term can be used more than twice if necessary.

EXAMPLE 4-3

Simplify $z = \overline{AC} \, (\overline{\overline{ABD}}) + \overline{AB}\overline{C}\,\overline{D} + A\overline{B}C$.

SOLUTION
First, use DeMorgan's theorem on the first term.

$$z = \overline{AC}(A + \overline{B} + \overline{D}) + \overline{AB}\overline{CD} + A\overline{B}C$$

Multiplying out,

$$z = \overline{ACA} + \overline{AC}\overline{B} + \overline{AC}\overline{D} + \overline{AB}\overline{CD} + A\overline{B}C$$

Since $\overline{A} \cdot A = 0$, the first term is eliminated.

$$z = \overline{AB}C + \overline{AC}\overline{D} + \overline{AB}\overline{CD} + A\overline{B}C$$

This is the desired sum-of-products form. Now we have to look for common factors among the various product terms. The idea is to check for the largest common factor between any two or more product terms. For example, the first and last terms have the common factor $\overline{B}C$, and the second and third terms share the common factor $\overline{A}\,\overline{D}$. We can factor these out as follows:

$$z = \overline{B}C(\overline{A} + A) + \overline{A}\overline{D}(C + B\overline{C})$$

Now, since $\overline{A} + A = 1$, and $C + B\overline{C} = C + B$ [theorem (15)], we have

$$z = \overline{B}C + \overline{A}\overline{D}(B + C)$$

This same result could have been reached with other choices for the factoring. For example, we could have factored C from the first, second, and fourth product terms to obtain

$$z = C(\overline{AB} + \overline{A}\overline{D} + A\overline{B}) + \overline{AB}\overline{CD}$$

The expression inside the parentheses can be factored further:

$$z = C(\overline{B}[\overline{A} + A] + \overline{A}\overline{D}) + \overline{AB}\overline{CD}$$

Since $\overline{A} + A = 1$, this becomes

$$z = C(\overline{B} + \overline{A}\overline{D}) + \overline{AB}\overline{CD}$$

Multiplying out yields

$$z = \overline{B}C + \overline{AC} \cdot \overline{AB}\overline{CD}$$

Now we can factor $\overline{A}\,\overline{D}$ from the second   .erms to get

$$z = \overline{B}C + \text{A.}\qquad \text{3C)}$$

Using theorem (15) the expression in parentheses becomes $B + C$. Thus we finally have

$$z = \overline{B}C + \overline{A}\overline{D}(B + C)$$

This is the same result as we obtained earlier, but it took us many more steps.

This illustrates why you should look for the largest common factors: it will generally lead to the final expression in the fewest number of steps.

EXAMPLE 4-4

Simplify the expression $x = (\overline{A} + B)(A + B + D)\overline{D}$.

SOLUTION

The expression can be put into sum-of-products form by multiplying out all the terms. The result is

$$x = \overline{A}A\overline{D} + \overline{A}B\overline{D} + \overline{A}D\overline{D} + BA\overline{D} + BB\overline{D} + BD\overline{D}$$

The first term can be eliminated, since $\overline{A}A = 0$. Likewise, the third and sixth terms can be eliminated, since $D\overline{D} = 0$. The fifth term can be simplified to $B\overline{D}$, since $BB = B$. This gives us

$$x = \overline{A}B\overline{D} + AB\overline{D} + B\overline{D}$$

We can factor $B\overline{D}$ from each term to obtain

$$x = B\overline{D}(\overline{A} + A + 1)$$

Clearly, the term inside the parentheses is always 1, so we finally have

$$x = B\overline{D}$$

EXAMPLE 4-5

Simplify the circuit of Figure 4-3(a).

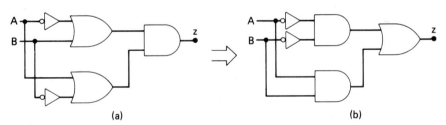

(a)   (b)

FIGURE 4-3   **Example 4-5**

SOLUTION

The expression for output z is

$$z = (\overline{A} + B)(A + \overline{B})$$

Multiplying out to get the sum-of-products form, we obtain

$$z = \overline{A}A + \overline{A}\,\overline{B} + BA + B\overline{B}$$

We can eliminate $\overline{A}A = 0$ and $B\overline{B} = 0$ to end up with

$$z = \overline{A}\,\overline{B} + AB$$

This expression is implemented in Figure 4-3(b), and if we compare it to the original circuit, we see that both circuits contain the same number of gates and connections. In this case the simplification process produced an equivalent, but not simpler, circuit.

EXAMPLE 4-6

Simplify $x = A\overline{B}C + \overline{A}BD + \overline{C}\overline{D}$.

SOLUTION
You can try, but you will not be able to simplify this expression any further.

REVIEW QUESTIONS

1. State which of the following expressions are *not* in the sum-of-products form: (a) $RS\overline{T} + \overline{R}S\overline{T} + \overline{T}$, (b) $A\overline{D}C + \overline{A}DC$, (c) $MN\overline{P} + (M + \overline{N})P$, (d) $AB + \overline{A}B\overline{C} + A\overline{B}\overline{C}D$. (*Ans.* b and c)
2. Simplify the circuit in Figure 4-1(a) to arrive at the circuit of Figure 4-1(b).
3. Change each AND gate in Figure 4-1(a) to a NAND gate. Determine the new expression for x and simplify it. (*Ans.* $x = \overline{A} + \overline{B} + \overline{C} = \overline{ABC}$)

4-4 DESIGNING COMBINATORIAL LOGIC CIRCUITS

When the desired output level of a logic circuit is given for all possible input conditions, the results can be conveniently displayed in a truth table. The Boolean expression for the required circuit can then be derived from the truth table. For example, consider Figure 4-4(a), where a truth table is shown for a circuit that has two inputs, A and B, and output x. The table shows that output x is to be at the 1 level *only* for the case where $A = 0$ and $B = 1$. It now remains to determine what logic circuit will produce this desired operation. It should be apparent that one possible solution is that shown in Figure 4-4(b). Here an AND gate is used with inputs $\overline{A}$ and B, so $x = \overline{A} \cdot B$. Obviously x will be 1 *only if* both inputs to the AND gate are 1, namely $\overline{A} = 1$ (which means $A = 0$) and $B = 1$. For all other values of A and B, the output x has to be 0.

A similar approach can be used for the other input conditions. For instance, if

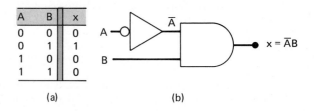

A	B	x
0	0	0
0	1	1
1	0	0
1	1	0

(a)   (b)

FIGURE 4-4

x were to be high only for the $A = 1$, $B = 0$ condition, the resulting circuit would be an AND gate with inputs A and $\overline{B}$. In other words, for any of the four possible input conditions we can generate a high *x* output by using an AND gate with appropriate inputs. The four different cases are shown in Figure 4-5. Each of the AND gates shown generates an output that is 1 *only* for the one given input condition and is 0 for all other conditions. It should be noted that the AND inputs are inverted or not inverted depending on the values that the variables have for the given condition. If the variable is 0 for the given condition, it is inverted before entering the AND gate.

Let us now consider the case shown in Figure 4-6(a), where we have a truth table which indicates that the output *x* is to be 1 for two different cases: $A = 0$, $B = 1$ and $A = 1$, $B = 0$. How can this be implemented? We know that the AND term $\overline{A} \cdot B$ will generate a 1 only for the $A = 0$, $B = 1$ condition, and the AND term $A \cdot \overline{B}$ will generate a 1 for the $A = 1$, $B = 0$ condition. Since *x* has to be

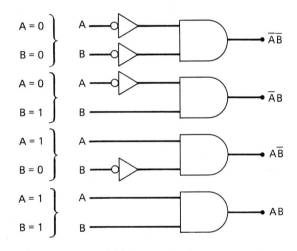

FIGURE 4-5 **AND gates used to generate high outputs for each of the possible input conditions.**

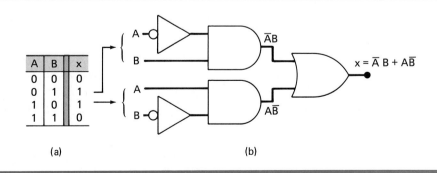

A	B	x
0	0	0
0	1	1
1	0	1
1	1	0

(a)

$x = \overline{A} \, B + A \overline{B}$

(b)

FIGURE 4-6

high for *either* condition, it should be clear that these terms should be ORed together to produce the desired output, x. This implementation is shown in Figure 4-6(b), where the resulting expression for the output is $x = \overline{A}B + A\overline{B}$.

In this example, an AND term is generated for each case in the table where the output x is to be a 1. The AND-gate outputs are then ORed together to produce the total output x, which will be high when either AND term is high. This same procedure can be extended to examples with more than two inputs. Consider the following truth table for a three-input circuit:

A	B	C	x	
0	0	0	0	
0	0	1	0	
0	1	0	1	$\rightarrow \overline{A}B\overline{C}$
0	1	1	1	$\rightarrow \overline{A}BC$
1	0	0	0	
1	0	1	0	
1	1	0	0	
1	1	1	1	$\rightarrow ABC$

Here there are three cases where the output x is to be 1. The required AND term for each of these cases is shown. Again, note that for each case where a variable is 0, it appears complemented in the AND term. The final expression for x is obtained by ORing the three AND terms. Thus.

$$x = \overline{A}B\overline{C} + \overline{A}BC + ABC$$

This expression can be implemented with three AND gates feeding an OR gate.

In all the preceding examples the expression for the output x was derived from the truth table in sum-of-products form. The general procedure for obtaining the output expression from a truth table can be summarized as follows:

1. Write an AND term for each case in the table where the output is 1.

2. Each AND term contains each input variable in either inverted or noninverted form. If the variable is 0 for that particular case in the table, it is inverted in the AND term.

3. All the AND terms are then ORed together to produce the final expression for the output.

Complete Design Problem   Once the output expression has been determined from the truth table in sum-of-products form, it can easily be implemented using AND and OR gates. There will be one AND gate for each term in the expression and one OR gate, which is fed by the outputs of each AND gate. Usually, however, the expression can be simplified, thereby resulting in a more efficient circuit. The following example illustrates the complete design procedure.

EXAMPLE 4-7

Design a logic circuit that has three inputs, *A, B,* and *C,* and whose output will be high only when a majority of the inputs is high.

SOLUTION

The first step is to set up the truth table based on the problem statement. The eight possible input cases are shown in Figure 4-7(a). Based on the problem statement, the output x should be a 1 whenever two or more inputs are 1. For all other cases the output should be 0. The next step is to write the AND terms for each case where $x = 1$, as shown in the figure. The expression for x can then be written as

$$x = \overline{A}BC + A\overline{B}C + AB\overline{C} + ABC$$

This expression can be simplified in several ways. Perhaps the quickest way is to realize that the last term ABC has two variables in common with each of the other terms. Thus, we can use the ABC term to factor with each of the other terms. The expression is rewritten with the ABC term occurring three times (recall that this is legal in Boolean algebra).

$$x = \overline{A}BC + ABC + A\overline{B}C + ABC + AB\overline{C} + ABC$$

Factoring the appropriate pairs of terms, we have

$$x = BC(\overline{A} + A) + AC(\overline{B} + B) + AB(\overline{C} + C)$$

Since each term in parentheses is equal to 1, we have

$$x = BC + AC + AB$$

This expression is implemented in Figure 4-7(b). Since the expression is in sum-of-products form, the circuit consists of a group of AND gates working into a single OR gate.

This expression can be factored further, but it will not result in the elimi-

A	B	C	x	
0	0	0	0	
0	0	1	0	
0	1	0	0	
0	1	1	1	$\overline{A}BC$
1	0	0	0	
1	0	1	1	$A\overline{B}C$
1	1	0	1	$AB\overline{C}$
1	1	1	1	ABC

(a)

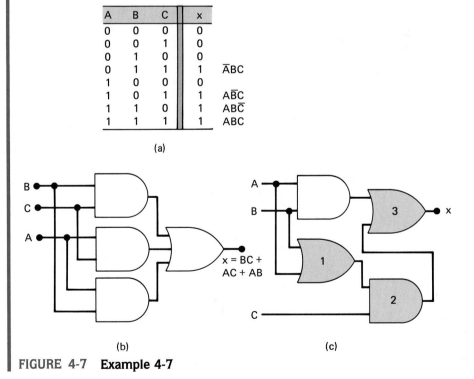

(b)   (c)

FIGURE 4-7   Example 4-7

nation of any terms. It might, however, produce a slightly simpler circuit. For example, let's factor C from the first two terms to obtain

$$x = C(B + A) + AB$$

This expression is implemented in Figure 4-7(c). The resulting circuit requires the same number of gates as the circuit in Figure 4-7(b), but it requires one less connection because the final OR gate has only two inputs. In practice, a logic-circuit designer may still select the implementation in Figure 4-7(b) because of some other considerations, one of which is the fact that the A and B input signals in Figure 4-7(c) have to propagate through three gates (1, 2, and 3) before reaching the output. This might be an important consideration in a high-speed digital system.

EXAMPLE 4-8

Refer to Figure 4-8(a), where four logic-signal lines A, B, C, D are being used to represent a 4-bit binary number with A as the MSB and D as the LSB. The binary inputs are fed to a logic circuit that produces a HIGH output only when the binary number is greater than $0110_2 = 6_{10}$. Design this circuit.

SOLUTION

The truth table is shown in Figure 4-8(b). For each case in the truth table we have indicated the decimal equivalent of the binary number represented by the $ABCD$ combination.

The output z is set equal to 1 for all those cases where the binary number is greater than 0110. For all other cases, z is set equal to 0. This truth table gives us the following sum-of-products expression

$$z = \overline{A}BCD + A\overline{B}\,\overline{C}\,\overline{D} + A\overline{B}\,\overline{C}D + A\overline{B}C\overline{D} + A\overline{B}CD + AB\overline{C}\,\overline{D}$$
$$+ AB\overline{C}D + ABC\overline{D} + ABCD$$

Simplification of this expression will be a formidable task, but with a little care

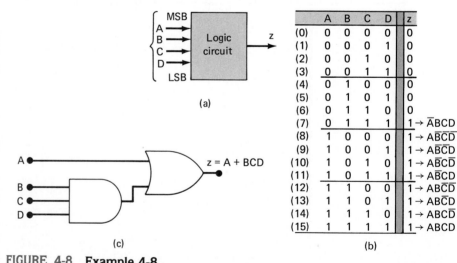

		A	B	C	D	z	
(0)		0	0	0	0	0	
(1)		0	0	0	1	0	
(2)		0	0	1	0	0	
(3)		0	0	1	1	0	
(4)		0	1	0	0	0	
(5)		0	1	0	1	0	
(6)		0	1	1	0	0	
(7)		0	1	1	1	1	$\rightarrow \overline{A}BCD$
(8)		1	0	0	0	1	$\rightarrow A\overline{B}\,\overline{C}\,\overline{D}$
(9)		1	0	0	1	1	$\rightarrow A\overline{B}\,\overline{C}D$
(10)		1	0	1	0	1	$\rightarrow A\overline{B}C\overline{D}$
(11)		1	0	1	1	1	$\rightarrow A\overline{B}CD$
(12)		1	1	0	0	1	$\rightarrow AB\overline{C}\,\overline{D}$
(13)		1	1	0	1	1	$\rightarrow AB\overline{C}D$
(14)		1	1	1	0	1	$\rightarrow ABC\overline{D}$
(15)		1	1	1	1	1	$\rightarrow ABCD$

(a)

(c)

(b)

FIGURE 4-8 Example 4-8

it can be accomplished. The step-by-step process involves factoring and eliminating terms of the form $A + \overline{A}$.

$$
\begin{aligned}
z &= \overline{A}BCD + A\overline{B}\overline{C}(\overline{D} + D) + A\overline{B}C(\overline{D} + D) + AB\overline{C}(\overline{D} + D) \\
&\quad + ABC(\overline{D} + D) \\
&= \overline{A}BCD + A\overline{B}\overline{C} + A\overline{B}C + AB\overline{C} + ABC \\
&= \overline{A}BCD + A\overline{B}(\overline{C} + C) + AB(\overline{C} + C) \\
&= \overline{A}BCD + A\overline{B} + AB \\
&= \overline{A}BCD + A(\overline{B} + B) \\
&= \overline{A}BCD + A
\end{aligned}
$$

This can be reduced further by invoking theorem (15), which says that $x + \overline{x}y = x + y$. In this case $x = A$ and $y = BCD$. Thus,

$$
z = \overline{A}BCD + A = BCD + A
$$

This final expression is implemented in Figure 4-8(c).

As this example demonstrates, the algebraic simplification method can be quite lengthy when the original expression contains a large number of terms. This is a limitation that is not shared by the Karnaugh mapping method, as we will see later.

EXAMPLE 4-9

Suppose the design problem of Example 4-8 had been stated as follows: "The logic circuit output is to go LOW whenever the binary number is *less than* $0111_2 = 7_{10}$." Design the circuit.

SOLUTION

Clearly, the statement above is a different, but equivalent, way of stating the same design problem as Example 4-8. Thus, the truth table will be exactly the same as in Figure 4-8(b). Likewise, the expression for z will be exactly the same, and it will produce the same circuit as in Figure 4-8(c).

The only difference is that the statement above implies that z is to be an active-LOW output. If we wish to convey this information, then the circuit diagram should be converted accordingly. This is done by changing the OR-gate symbol to its alternate symbol, as shown in Figure 4-9. This in turn requires changing the AND symbol to its alternate symbol, so that a bubble output is connected to a bubble input.

Remember, both circuits are physically the same AND and OR gates, but the circuit diagram in Figure 4-9(b) more accurately represents the circuit's operation as given in the statement of the design problem.

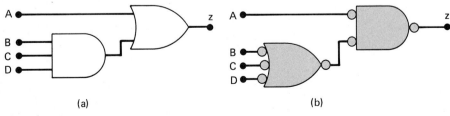

(a)   (b)

FIGURE 4-9   Example 4-9

120

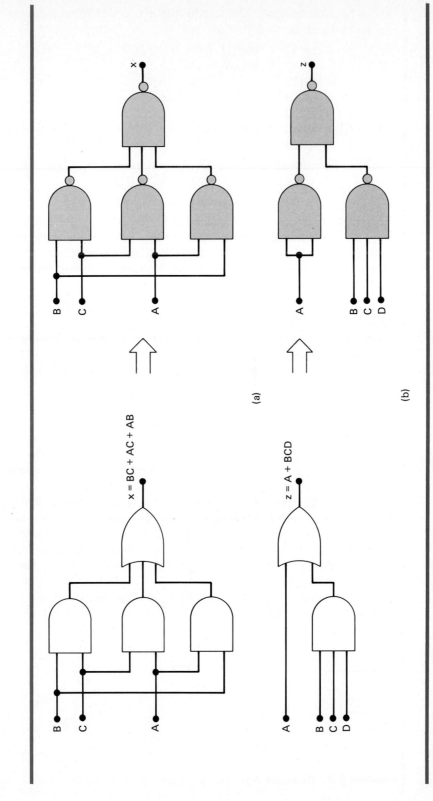

$x = BC + AC + AB$

$z = A + BCD$

(a)

(b)

FIGURE 4-10 (a) Converting circuit of Figure 4-7(b) to NANDS; (b) converting circuit of Figure 4-8(c) to NANDs.

Implementing the Final Design   In the design examples we have shown, the final circuit was implemented using AND and OR gates. In fact, the sum-of-products form always yields a circuit containing one or more AND gates driving a single OR gate. One of the reasons for using the sum-of-products form is that it can be implemented using all NAND gates with little, if any, increase in circuit complexity over the AND/OR implementation. Since NAND gates are the most widely available logic gates in the TTL logic family, this is an important characteristic.

To illustrate, Figure 4-10 shows the equivalent NAND implementation for the circuits of Figures 4-7(b) and 4-8(c). You may wish to work out these conversions yourself as a review of the procedure covered in Chapter 3.

Comparing the NAND implementation with the circuit in Figure 4-10(a), we see that they are identical in structure; that is, *each* gate of the original circuit has been replaced by a single NAND gate. This characteristic is true only if the original circuit is in sum-of-products form. The only exception to this is when the sum-of-products form contains a single-variable term such as $z = A + BCD$ in Figure 4-10(b). Here the NAND implementation requires an extra NAND gate used as an INVERTER on the A input.

We can streamline the process of converting a *sum-of-products* circuit from AND/OR to NAND gates as follows:

1.   Replace each AND gate, OR gate, and INVERTER by a *single* NAND gate.
2.   Use a NAND gate to invert any single variable that is feeding the final OR gate.

Check out this process for the circuits in Figure 4-10.

REVIEW QUESTIONS

1.   Write the sum-of-products expression for a circuit with four inputs and an output that is to be HIGH only when input A is LOW at the same time that exactly two other inputs are LOW. (*Ans.* $x = \overline{A}\,\overline{B}\,\overline{C}D + \overline{A}\,\overline{B}CD + \overline{A}BC\,\overline{D}$)
2.   Implement the expression of question 1 using all NAND gates. How many are required? (*Ans.* Eight)

4-5   KARNAUGH MAP METHOD*

The Karnaugh map is a graphical device used to simplify a logic equation or to convert a truth table to its corresponding logic circuit in a simple, orderly process. Although a Karnaugh map (henceforth abbreviated *K map*) can be used for problems involving any number of input variables, its practical usefulness is limited to six

*This topic may be omitted without affecting the continuity of the remainder of the text.

variables. The following discussion will be limited to problems with up to four inputs, since even five- and six-input problems are too involved and are best done by a computer program.

Karnaugh Map Format   The K map, like a truth table, is a means for showing the relationship between logic inputs and the desired output. Figure 4-11 shows three examples of K maps for two, three, and four variables, together with the corresponding truth tables. These examples illustrate the following important points:

1.   The truth table gives the value of output X for each combination of input values. The K map gives the same information in a different format. Each

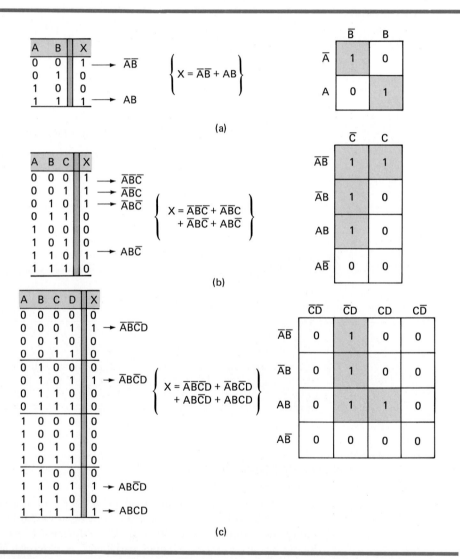

FIGURE 4-11   Karnaugh maps and truth tables for two, three, and four variables.

case in the truth table corresponds to a square in the K map. For example, in Figure 4-11(a), the $A = 0$, $B = 0$ condition in the truth table corresponds to the $\overline{A}\,\overline{B}$ square in the K map. Since the truth table shows $X = 1$ for this case, a 1 is placed in the $\overline{A}\,\overline{B}$ square in the K map. Similarly, the $A = 1$, $B = 1$ condition in the truth table corresponds to the AB square of the K map. Since $X = 1$ for this case, a 1 is placed in the AB square. All other squares are filled with 0s. This same idea is used in the three- and four-variable maps shown in the figure.

2. The K-map squares are labeled so that horizontally adjacent squares differ only in one variable. For example, the upper-left-hand square in the four-variable map is $\overline{A}\,\overline{B}\,\overline{C}\,\overline{D}$, while the square immediately to its right is $\overline{A}\,\overline{B}\,CD$ (only the D variable is different). Similarly, vertically adjacent squares differ only in one variable. For example, the upper-left-hand square is $\overline{A}\,\overline{B}\,\overline{C}\,\overline{D}$ while the square directly below it is $\overline{A}B\overline{C}\,\overline{D}$ (only the B variable is different).

3. Once a K map has been filled with 0s and 1s, the sum-of-products expression for the output X can be obtained by ORing together those squares that contain a 1. In the three-variable map of Figure 4-11(b), the $\overline{A}\,\overline{B}\,\overline{C}$, $\overline{A}\,\overline{B}C$, $\overline{A}BC$, and $AB\overline{C}$ squares contain a 1, so $X = \overline{A}\,\overline{B}\,\overline{C} + \overline{A}\,\overline{B}C + \overline{A}BC + AB\overline{C}$.

Looping   The expression for output X can be simplified by properly combining those squares in the K map which contain 1s. The process for combining these 1s is called *looping*.

Looping Groups of Two (Pairs)   Figure 4-12(a) is the K map for a particular three-variable truth table. This map contains a pair of 1s that are vertically adjacent to each other; the first one represents $\overline{A}B\overline{C}$ and the second one represents $AB\overline{C}$. Note that in these two terms only the A variable appears in both normal and complemented form (B and $\overline{C}$ remain unchanged). These two terms can be looped (combined) to give a resultant that elimates the A variable since it appears in both uncomplemented and complemented forms. This is easily proved as follows:

$$X = \overline{A}B\overline{C} + AB\overline{C}$$
$$= B\overline{C}(\overline{A} + A)$$
$$= B\overline{C}(1) = B\overline{C}$$

This same principle holds true for any pair of vertically or horizontally adjacent 1s. Figure 4-12(b) shows an example of two horizontally adjacent 1s. These two can be looped and the C variable eliminated since it appears in both its uncomplemented and complemented forms to give a resultant of $X = \overline{A}B$.

Another example is shown in Figure 4-12(c). In a K map the top row and bottom row of squares are considered to be adjacent. Thus, the two 1s in this map can be looped to provide a resultant of $\overline{A}\,\overline{B}\,\overline{C} + A\overline{B}\,\overline{C} = \overline{B}\,\overline{C}$.

Figure 4-12(d) shows a K map that has two pairs of 1s which can be looped. The two 1s in the top row are horizontally adjacent. The two 1s in the bottom row are also adjacent since in a K map the leftmost column and rightmost column of squares are considered to be adjacent. When the top pair of 1s is looped, the D variable is eliminated (since it appears as both D and $\overline{D}$) to give the term $\overline{A}B C$. Looping the bottom pair eliminates the C variable to give the term $A\overline{B}D$. These two terms are ORed to give the final result for X.

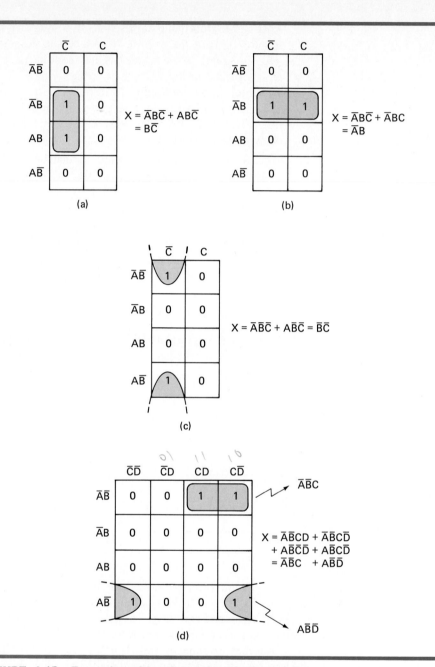

FIGURE 4-12 **Examples of looping pairs of adjacent 1s.**

To summarize: *Looping a pair of adjacent 1s in a K map eliminates the variable that appears in complemented and uncomplemented form.*

Looping Groups of Four (Quads)   A K map may contain a group of four 1s that are adjacent to each other. This group is called a *quad*. Figure 4-13 shows several examples of quads. In (a) the four 1s are vertically adjacent and in (b) they

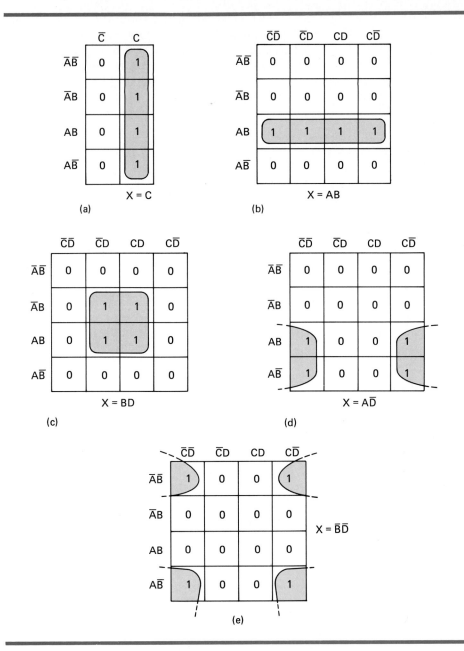

FIGURE 4-13 **Examples of looping groups of four 1s (quads).**

are horizontally adjacent. The K map in (c) contains four 1s in a square and they are considered adjacent to each other. The four 1s in (d) are also adjacent, as are those in (e) because, as pointed out earlier, the top and bottom rows and the leftmost and rightmost columns are considered to be adjacent to each other.

When a quad is looped, the resultant term will contain only the variables that do not change form for all the squares in the quad. For example, in (a) the four squares which contain a 1 are $\overline{A}\overline{B}C$, $\overline{A}BC$, ABC, and $A\overline{B}C$. Examination of these terms reveals that only the variable C remains unchanged (both A and B appear in complemented and uncomplemented form). Thus, the resultant expression for X is simply $X = C$. This can be proved as follows:

$$X = \overline{A}\overline{B}C + \overline{A}BC + ABC + A\overline{B}C$$
$$= \overline{A}C(\overline{B} + B) + AC(B + \overline{B})$$
$$= \overline{A}C + AC$$
$$= C(\overline{A} + A) = C$$

As another example, consider Figure 4-13(d), where the four squares containing 1s are $AB\overline{C}D$, $A\overline{B}\overline{C}D$, $ABC\overline{D}$, and $A\overline{B}C\overline{D}$. Examination of these terms indicates that only the variables A and $\overline{D}$ remain unchanged, so the simplified expression for X is

$$X = A\overline{D}$$

This can be proved in the same manner that was used above.

The reader should check each of the other cases in Figure 4-13 to verify the indicated expressions for X. To summarize: *Looping a quad of 1s eliminates the two variables that appear in both complemented and uncomplemented form.*

Looping Groups of Eight (Octets)

A group of eight 1s that are adjacent to each other is called an *octet*. Several examples of octets are shown in Figure 4-14. When an octet is looped in a four-variable map, three of the four variables are eliminated because only one variable remains unchanged. For example, examination of the eight looped squares in (a) shows that only the variable B is in the same form for all eight squares; the other variables appear in complemented and uncomplemented form. Thus, for this map, $X = B$. The reader can verify the results for the other examples in Figure 4-14.

To summarize: *Looping an octet of 1s eliminates the three variables that appear in both complemented and uncomplemented form.*

Complete Simplification Process

We have seen how looping of pairs, quads, and octets on a K map can be used to obtain a simplified expression. We can summarize the rule for loops of *any* size: *When a variable appears in both complemented and uncomplemented form within a loop, that variable is eliminated from the expression. Variables that are the same for all squares of the loop must appear in the final expression.*

It should be clear that a larger loop of 1s eliminates more variables. To be exact, a loop of two eliminates one variable, a loop of four eliminates two, and a loop of eight eliminates three. This principle will now be used to obtain a simplified logic expression from a K map that contains any combination 1s and 0s.

The procedure will first be outlined and then applied to several examples. The steps below are followed in using the K-map method for simplifying a Boolean expression:

1. Construct the K map and place 1s in those squares corresponding to the 1s in the truth table. Place 0s in the other squares.

2. Examine the map for adjacent 1s and loop those 1s which are *not* adjacent to any other 1s. These are called *isolated* 1s.

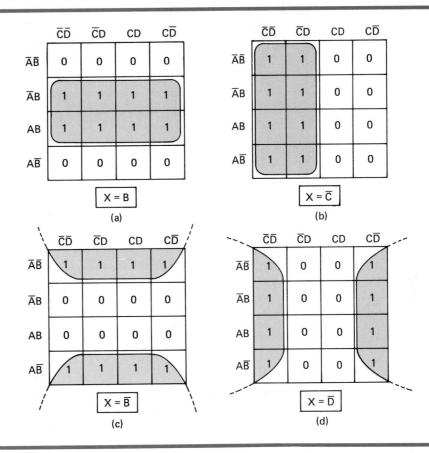

FIGURE 4-14   Examples of looping groups of eight 1s (octets).

3.   Next, look for those 1s which are adjacent to only one other 1. Loop *any* pair containing such a 1.

4.   Loop any octet even if some of the 1s have already been looped.

5.   Loop any quad that contains one or more 1s which have not already been looped.

6.   Loop any pairs necessary to include any 1s that have not yet been looped, making sure to use the minimum number of loops.

7.   Form the OR sum of all the terms generated by each loop.

These steps will be followed exactly and referred to in the following examples. In each case, the resulting logic expression will be in its simplest sum-of-products form.

EXAMPLE 4-10

Figure 4-15(a) shows the K map for a four-variable problem. We will assume that the map was obtained from the problem truth table (step 1). The squares are numbered for convenience in identifying each loop.

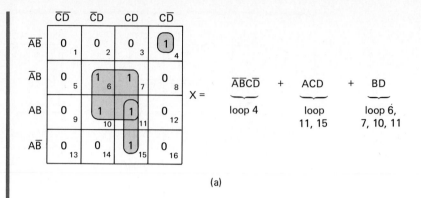

(a)

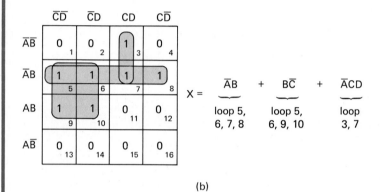

(b)

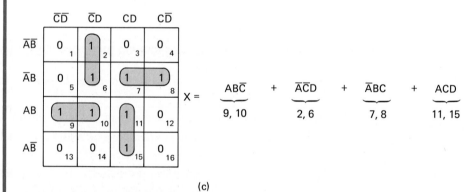

(c)

FIGURE 4-15 Examples 4-10–4-12

Step 2: Square 4 is the only square containing a 1 that is not adjacent to any other 1. It is looped and is referred to as loop 4.

Step 3: Square 15 is adjacent *only* to square 11. This pair is looped and referred to as loop 11, 15.

Step 4: There are no octets.

Step 5: Squares 6, 7, 10, and 11 form a quad. This quad is looped (loop 6, 7, 10, 11). Note that square 11 is used again, even though it was part of loop 11, 15.

Step 6: All 1s have already been looped.

Step 7: Each loop generates a term in the expression for X. Loop 4 is simply $\overline{A}\overline{B}C\overline{D}$. Loop 11, 15 is ACD (the B variable is eliminated). Loop 6, 7, 10, 11, is BD (A and C are eliminated).

EXAMPLE 4-11

Consider the K map in Figure 4-15(b). Once again we can assume that step 1 has already been performed.

Step 2: There are no isolated 1s.

Step 3: The 1 in square 3 is adjacent *only* to the 1 in square 7. Looping this pair (loop 3, 7) produces the term $\overline{A}CD$.

Step 4: There are no octets.

Step 4: There are two quads. Squares 5, 6, 7, and 8 form one quad. Looping this quad produces the term $\overline{A}B$. The second quad is made up of squares 5, 6, 9, and 10. This quad is looped because it contains two squares that have not been looped previously. Looping this quad produces $B\overline{C}$.

Step 6: All 1s have already been looped.

Step 7: The terms generated by the three loops are ORed together to obtain the expression for X.

EXAMPLE 4-12

Consider the K map in Figure 4-15(c).

Step 2: There are no isolated 1s.

Step 3: The 1 in square 2 is adjacent only to the 1 in square 6. This pair is looped to produce $\overline{A}\,\overline{C}D$. Similarly, square 9 is adjacent only to square 10. Looping this pair produces $AB\overline{C}$. Likewise, loop 7, 8, and loop 11, 15 produce the terms $\overline{A}BC$ and ACD, respectively.

Step 4: There are no octets.

Step 5: There is one quad formed by squares 6, 7, 10, and 11. This quad, however, is *not* looped because all the 1s in the quad have been included in other loops.

Step 6: All 1s have already been looped.

Step 7: The expression for X is shown in the figure.

EXAMPLE 4-13

Consider the K map in Figure 4-16(a).

Step 2: There are no isolated 1s.

Step 3: There are no 1s that are adjacent to only one other 1.

Step 4: There are no octets.

Step 5: There are no quads.

Steps 6 and 7: There are many possible pairs. The looping must use the minimum number of loops to account for all the 1s. For this map there are *two*

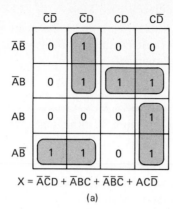

$$X = \overline{A}\overline{C}D + \overline{A}BC + \overline{A}B\overline{C} + AC\overline{D}$$

(a)

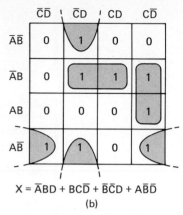

$$X = \overline{A}BD + BC\overline{D} + \overline{B}\overline{C}D + A\overline{B}\overline{D}$$

(b)

FIGURE 4-16   The same K map with two equally good solutions.

possible loopings, which require only four looped pairs. Figure 4-16(a) shows one solution and its resultant expression. Figure 4-16(b) shows the other. Note that both expressions are of the same complexity, so neither is better than the other.

EXAMPLE 4-14

Use the K map to simplify the expression $x = \overline{A}\overline{B}\overline{C} + \overline{B}C + \overline{A}B$.

SOLUTION
In this problem we are not given the truth table from which to fill in the K map. Instead, we must fill in the K map by taking each of the product terms in the expression and placing 1s in the corresponding squares.

The first term, $\overline{A}\overline{B}\overline{C}$, tells us to enter a 1 in the $\overline{A}\overline{B}\overline{C}$ square of the map (see Figure 4-17). The second term, $\overline{B}C$, tells us to enter a 1 in each square that contains a $\overline{B}C$ in its label. In Figure 4-17 this would be the $A\overline{B}C$ and $\overline{A}\overline{B}C$ squares. Likewise, the $\overline{A}B$ term tells us to place a 1 in the $\overline{A}BC$ and $\overline{A}B\overline{C}$ squares. All other squares will be filled with 0s.

Now the K map can be looped for simplification. The result is $x = \overline{A} + \overline{B}C$, as shown in the figure.

FIGURE 4-17   Example 4-14

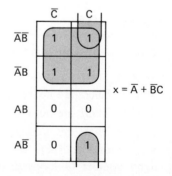

$$x = \overline{A} + \overline{B}C$$

"Don't Care" Conditions Some logic circuits can be designed so that there are certain input conditions for which there are no specified output levels, usually because these input conditions will never occur. In other words, there will be certain combinations of input levels where we "don't care" whether the output is HIGH or LOW. This is illustrated in the truth table of Figure 4-18(a).

Here the output z is not specified as either 0 or 1 for the conditions A, B, C = 1, 0, 0 and A, B, C = 0, 1, 1. Instead, an x is shown for these conditions. The x represents the "don't care" condition. A "don't care" condition can come about for several reasons, the most common being that in some situations certain input combinations can never occur, and so there is no specified output for these conditions.

A circuit designer is free to make the output for any "don't care" condition either a 0 or a 1 in order to produce the simplest output expression. For example, the K map for this truth table is shown in Figure 4-18(b) with an x placed in the $A\overline{B}\overline{C}$ and $\overline{A}BC$ squares. The designer here would be wise to change the x in the $A\overline{B}\overline{C}$ square to a 1 and the x in the $\overline{A}BC$ square to a 0, since this would produce a quad that can be looped to produce $z = A$, as shown in Figure 4-18(c).

Whenever "don't care" conditions occur, we have to decide which ones to change to 0 and which to 1 to produce the best K-map looping (i.e., the simplest expression). This decision is not always an easy one. Several end-of-chapter problems will provide practice in dealing with "don't care" cases.

Summary The K-map process has several advantages over the algebraic method. K mapping is a more orderly process with well-defined steps as compared to the trial-and-error process sometimes used in algebraic simplification. K mapping usually requires fewer steps, especially for expressions containing many terms, and it always produces a minimum expression.

Nevertheless, some instructors prefer the algebraic method because it requires a thorough knowledge of Boolean algebra and is not simply a mechanical procedure. Each method has its advantages, and though most logic designers are adept at both, being proficient in one method is all that is necessary to produce acceptable results.

FIGURE 4-18 "Don't care" conditions should be changed to 0 or 1 to produce K-map looping that yields the simplest expression.

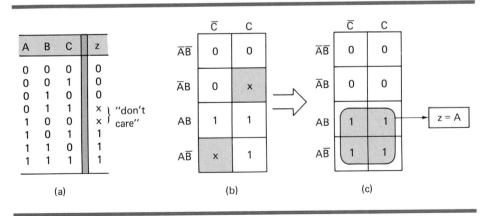

1. Use K mapping to simplify the expression of Example 4-7.
2. Use K mapping to simplify the expression of Example 4-8. This should emphasize the advantage of K mapping for expressions containing many terms.
3. What is meant by a "don't care" condition?

4-6 EXCLUSIVE-OR AND EXCLUSIVE-NOR CIRCUITS

Two special logic circuits that occur quite often in digital systems are the *exclusive-OR* and *exclusive-NOR* circuits.

Exclusive-OR Consider the logic circuit of Figure 4-19(a). The output expression of this circuit is

$$x = \overline{A}B + A\overline{B}$$

The accompanying truth table shows that $x = 1$ for two cases: $A = 0$, $B = 1$ (the $\overline{A}B$ term) and $A = 1$, $B = 0$ (the $A\overline{B}$ term). In other words, *this circuit produces*

FIGURE 4-19   (a) Exclusive-OR circuit and truth table; (b) traditional EX-OR gate symbol; (c) IEEE/ANSI symbol for EX-OR gate.

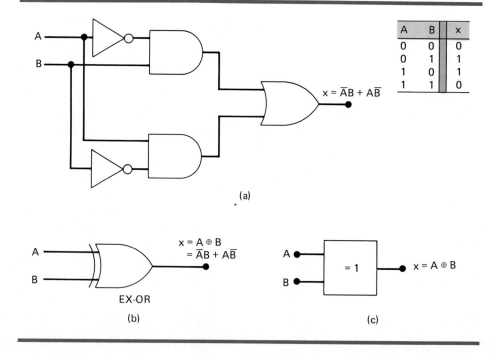

A	B	x
0	0	0
0	1	1
1	0	1
1	1	0

$x = \overline{A}B + A\overline{B}$

(a)

$x = A \oplus B$
$= \overline{A}B + A\overline{B}$

EX-OR

(b)

$= 1$

$x = A \oplus B$

(c)

a HIGH output whenever the two inputs are at opposite levels. This is the exclusive-OR circuit, which will hereafter be abbreviated EX-OR.

This particular combination of logic gates occurs quite often and is very useful in certain applications. In fact, the EX-OR circuit has been given a symbol of its own, shown in Figure 4-19(b). This symbol is assumed to contain all the logic contained in the EX-OR circuit and therefore has the same logic expression and truth table. This EX-OR circuit is commonly referred to as an EX-OR *gate,* and we consider it as another type of logic gate. The new IEEE/ANSI symbol for an EX-OR gate is shown in Figure 4-19(c). It indicates that the output will be active-HIGH when *only* one input is HIGH.

An EX-OR gate has only *two* inputs; there are no three-input or four-input EX-OR gates. The two inputs are combined such that $x = \overline{A}B + A\overline{B}$. A shorthand way that is sometimes used to indicate the EX-OR output expression is

$$x = A \oplus B$$

where the symbol $\oplus$ represents the EX-OR gate operation.

The characteristics of an EX-OR gate are summarized as follows:

1. It has only two inputs and its output is
$$x = \overline{A}B + A\overline{B} = A \oplus B$$

2. Its output is *HIGH* only when the two inputs are at *different* levels.

Several ICs are available that contain EX-OR gates. Those listed below are *quad* EX-OR chips containing four EX-OR gates.

- **7486**   QUAD EX-OR (TTL family)
- **74C86**   QUAD EX-OR (CMOS family)
- **74HC86**   QUAD EX-OR (high-speed CMOS)

Exclusive-NOR   The exclusive-NOR circuit (abbreviated EX-NOR) operates completely opposite to the EX-OR circuit. Figure 4-20(a) shows an EX-NOR circuit and its accompanying truth table. The output expression is

$$x = AB + \overline{AB}$$

which indicates along with the truth table that x will be 1 for two cases: $A = B = 1$ (the AB term) and $A = B = 0$ (the $\overline{AB}$ term). In other words, *this circuit produces a high output whenever the two inputs are at the same level.*

It should be apparent that the output of the EX-NOR circuit is the exact inverse of the output of the EX-OR circuit. The traditional symbol for an EX-NOR gate is obtained by simply adding a small circle at the output of the EX-OR symbol [Figure 4-20(b)]. The IEEE/ANSI symbol adds the small triangle on the output of the EX-OR symbol. Both symbols indicate an output that goes to its active-LOW state when *only* one input is HIGH.

The EX-NOR gate also has only *two* inputs, and it combines them such that its output is

$$x = AB + \overline{AB}$$

A shorthand way to indicate the output expression of the EX-NOR is

$$x = \overline{A \oplus B}$$

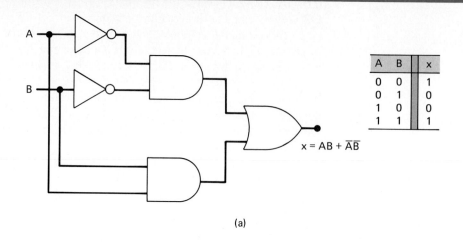

A	B	x
0	0	1
0	1	0
1	0	0
1	1	1

$x = AB + \overline{A}\overline{B}$

(a)

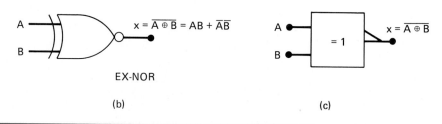

$x = \overline{A \oplus B} = AB + \overline{A}\overline{B}$

EX-NOR

(b)

$x = \overline{A \oplus B}$

(c)

FIGURE 4-20 (a) Exclusive-NOR circuit; (b) traditional symbol for EXNOR gate; (c)IEEE/ANSI symbol.

which is simply the inverse of the EX-OR operation. The EX-NOR gate is summarized as follows:

1. It has only two inputs and its output is

$$x = AB + \overline{A}\overline{B} = \overline{A \oplus B}$$

2. Its output is HIGH only when the two inputs are at the *same* level.

Several ICs are available that contain EX-NOR gates. Those listed below are quad EX-NOR chips containing four EX-NOR gates.

- **74LS8266** QUAD EX-NOR (TTL family)
- **74C266** QUAD EX-NOR (CMOS)
- **74HC266** QUAD EX-NOR (high-speed CMOS)

Each of these EX-NOR chips, however, has special output circuitry that limits its use to special types of applications. Very often, a logic designer will obtain the EX-NOR function simply by connecting the output of an EX-OR to an INVERTER.

EXAMPLE 4-15

Determine the output waveform for the input waveforms given in Figure 4-21.

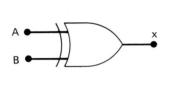

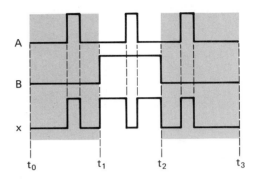

FIGURE 4-21 **Example 4-15**

SOLUTION

The output waveform is obtained using the fact that the EX-OR output will go HIGH only when its inputs are at different levels. The resulting output waveform reveals several interesting points:

1. The x waveform matches the A input waveform during those time intervals when $B = 0$. This occurs during the $t_0 - t_1$ and $t_2 - t_3$ time intervals.

2. The x waveform is the *inverse* of the A input waveform during those time intervals when $B = 1$. This occurs during the $t_1 - t_2$ interval.

3. These observations show that an EX-OR gate can be used as a *controlled INVERTER;* that is, one of its inputs can be used to control whether the signal at the other input will be inverted or not. This property will be useful in certain applications.

EXAMPLE 4-16

$x_1 x_0$ represents a 2-bit binary number that can have any value (00, 01, 10, or 11); for example, when $x_1 = 1$ and $x_0 = 0$, the binary number is 10, and so on. Similarly, $y_1 y_0$ represents another 2-bit binary number. Design a logic circuit, using x_1, x_0, y_1, and y_0 inputs, whose output will be HIGH only when the two binary numbers $x_1 x_0$ and $y_1 y_0$ are *equal*.

SOLUTION

The first step is to construct a truth table for the 16 input conditions.

The output z has to be high whenever the $x_1 x_0$ values match the $y_1 y_0$ values; that is, whenever $x_1 = y_1$ and $x_0 = y_0$. The table shows that there are four such cases. We could now continue with the normal procedure, which would be to obtain a sum-of-products expression for z, attempt to simplify it, and then implement the result. However, the nature of this problem makes it ideally suited for implementation using EX-NOR gates, and a little thought will produce a simple solution with minimum work. Refer to Figure 4-22; in this logic diagram x_1 and

x_1	x_0	y_1	y_0	z (OUTPUT)
0	0	0	0	1
0	0	0	1	0
0	0	1	0	0
0	0	1	1	0
0	1	0	0	0
0	1	0	1	1
0	1	1	0	0
0	1	1	1	0
1	0	0	0	0
1	0	0	1	0
1	0	1	0	1
1	0	1	1	0
1	1	0	0	0
1	1	0	1	0
1	1	1	0	0
1	1	1	1	1

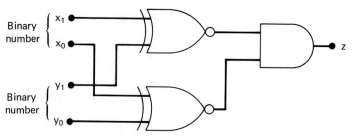

FIGURE 4-22 **Circuit for detecting equality of two 2-bit binary numbers.**

y_1 are fed to one EX-NOR gate and x_0 and y_0 are fed to another EX-NOR gate. The output of each EX-NOR will ge HIGH only when its inputs are equal. Thus, for $x_0 = y_0$ and $x_1 = y_1$ both EX-NOR outputs will be high. This is the condition we are looking for, because it means that the two 2-bit numbers are equal. The AND gate output will be HIGH only for this case, thereby producing the desired output.

EXAMPLE 4-17

When simplifying the expression for the output of a combinatorial logic circuit, you may encounter the EX-OR or EX-NOR operations as you are factoring. This will often lead to the use of EX-OR or EX-NOR gates in the implementation of the final circuit. To illustrate, simplify the circuit of Figure 4-23(a).

SOLUTION
The unsimplified expression for the circuit is obtained as

$$Z = ABCD + A\overline{B}\,\overline{C}D + \overline{A}\,\overline{D}$$

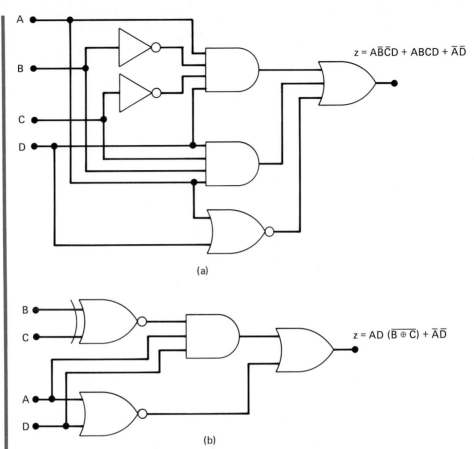

$$z = A\bar{B}\bar{C}D + ABCD + \bar{A}\bar{D}$$

(a)

$$z = AD\,(\overline{B \oplus C}) + \bar{A}\bar{D}$$

(b)

FIGURE 4-23 Example 4-17 showing how an EX-NOR gate may used to simplify circuit implementation.

We can factor AD from the first two terms:

$$Z = AD(BC + \bar{B}\bar{C}) + \bar{A}\bar{D}$$

At first glance you might think that the expression in parentheses can be replaced by 1. But that would be true only if it were $BC + \overline{BC}$. You should recognize the expression in parentheses as the EX-NOR combination of B and C. This fact can be used to reimplement the circuit as shown in Figure 4-23(b). This circuit is much simpler than the original since it uses gates with fewer inputs, and two INVERTERs have been eliminated.

REVIEW QUESTIONS

1. Use Boolean algebra to prove that the EX-NOR output expression is the exact inverse of the EX-OR output expression.
2. What is the output of an EX-NOR gate when a logic signal and its exact inverse are connected to its inputs? (*Ans*. A constant LOW)

3. A logic designer needs an INVERTER, and all that is available is one EX-OR gate from a 7486 chip. Does he need another chip? (*Ans.* No, he can use the EX-OR gate as an INVERTER by connecting one of its inputs permanently HIGH)

4-7 DESIGNING WITHOUT A TRUTH TABLE

Some logic design problems can be solved without going through the process of constructing a truth table, writing the sum-of-products expression, and simplifying it. For these simple situations, you can arrive at the circuit implementation directly by using your understanding of the various logic operations and the logic gates. The following examples will illustrate.

EXAMPLE 4-18

Design a logic circuit with an output Z that goes HIGH *only* when inputs C and D are *both* LOW at the same time that either A or B or both are HIGH.

SOLUTION

As the circuit requirements are stated, two conditions have to be met simultaneously for output Z to go HIGH. This makes it clear that Z is an active-HIGH output that is generated from an AND operation, so we can start by drawing an AND gate with active-HIGH output Z as shown in Figure 4-24(a). The inputs to this AND gate can be made either active-LOW or active-HIGH; we'll use active-HIGH. These inputs are derived from the two stated conditions:

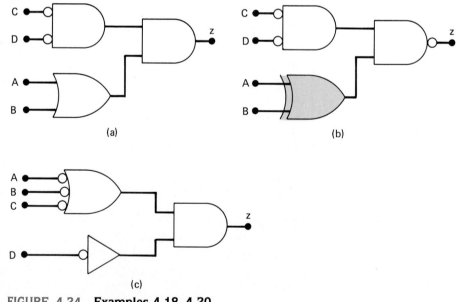

FIGURE 4-24 **Examples 4-18–4-20.**

1. *C and D both LOW.* This has to produce a HIGH at one of the AND inputs. This can be accomplished by putting *C* and *D* into active-LOW inputs of another AND gate and connecting the active-HIGH output of this new AND gate to one input of the output AND gate.

2. *A or B or both HIGH.* This has to produce a HIGH at the other input of the output AND gate. This can be done by combining *A* and *B* in an OR gate with active-HIGH inputs and connecting the active-HIGH output of this OR gate to the output AND gate.

The final circuit shown in Figure 4-24(a) satisfies the requirements stated. You can check it out by working backwards from output *Z* to find the input conditions needed to activate the output.

EXAMPLE 4-19

Design a logic circuit with output *Z* that goes LOW only when inputs *C* and *D* are both LOW at the same time that either *A* or *B*, but not both, are HIGH.

SOLUTION
This is the same as Example 4-18 except for two differences. First, output *Z* is active-LOW. Second, only one of the inputs *A*, *B* can be HIGH to activate *Z*. Thus, the required circuit can be obtained from the circuit of Figure 4-24(a) by putting a bubble on output *Z*, and changing the OR gate to an EX-OR gate. The result is shown in Figure 4-24(b).

EXAMPLE 4-20

Draw a circuit with an output that goes HIGH only when any of the inputs *A, B, C* are LOW at the same time that *D* is LOW.

SOLUTION
The result is drawn in Figure 4-24(c).

4-8 INHIBIT CIRCUITS

Each of the basic logic gates can be used to control the passage of an input logic signal through to the output. This is depicted in Figure 4-25, where a logic signal, *A*, is applied to one input of each of the basic logic gates. The other input of each gate is the control input, *B*. The logic level at this control input will determine whether the input signal is *enabled* to reach the output, or *inhibted* from reaching the output.

Examine Figure 4-25 and you should notice that when the noninverting gates (AND, OR) are enabled, the output will follow the *A* signal exactly. Conversely, when the inverting gates (NAND, NOR) are enabled, the output will be the exact inverse of the *A* signal.

Also notice that AND and NOR gates produce a constant LOW output when

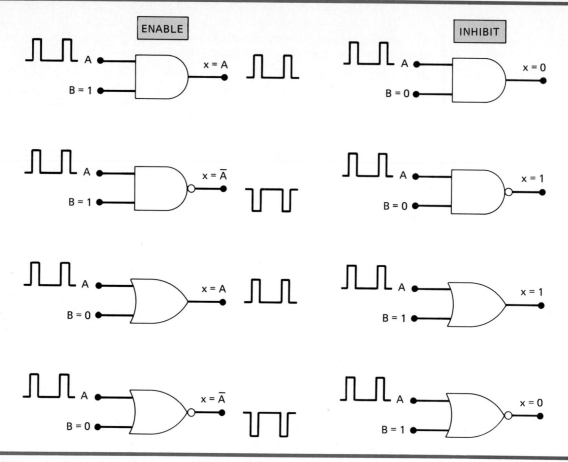

FIGURE 4-25 **Four basic gates can either enable or inhibit the passage of an input signal, _A_, under control of the logic level at control input, _B_.**

they are in the inhibited condition. Conversely, the NAND and OR gates produce a constant HIGH output in the inhibited condition.

There will be many situations in digital-circuit design where the passage of a logic signal is to be enabled or inhibited, depending on conditions present at one or more control inputs. Several are shown in the following examples.

EXAMPLE 4-21

Design a logic circuit that will allow a signal to pass to the output only when control inputs B and C are both HIGH; otherwise, the output will stay LOW.

SOLUTION
An AND gate should be used because the signal is to be passed without inversion, and the inhibit output condition is a LOW. Since the enable condition has to occur only when $B = C = 1$, a three-input AND gate is used, as shown in Figure 4-26(a).

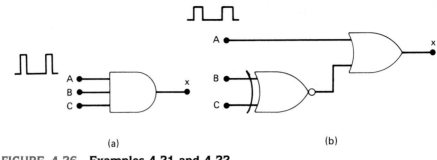

(a)   (b)

FIGURE 4-26   Examples 4-21 and 4-22

EXAMPLE 4-22

Design a logic circuit that allows a signal to pass to the output only when one, but not both, of the control inputs are HIGH; otherwise, the output will stay HIGH.

SOLUTION

The result is drawn in Figure 4-26(b). An OR gate is used because we want the output inhibit condition to be a HIGH, and we do not want to invert the signal. Control inputs B and C are combined in an EX-NOR gate. When B and C are different, the EX-NOR sends a LOW to enable the OR gate. When B and C are the same, the EX-NOR sends a HIGH to inhibit the OR gate.

EXAMPLE 4-23

Design a logic circuit with input signal A, control input B, and outputs X and Y to operate as follows:

1. When $B = 1$, output X will follow input A, and output Y will be 0.
2. When $B = 0$, output X will be 0, and output Y will follow input A.

SOLUTION

The two outputs will be 0 when inhibited and will follow the input signal when enabled. Thus, an AND gate should be used for each output. Since X is to be enabled when $B = 1$, its AND gate must be controlled by B, as shown in Figure 4-27. Since Y is to be enabled when B = 0, its AND gate is controlled by $\bar{B}$.

   This circuit is called a *pulse-steering* circuit because it steers the input pulse to one output or the other depending on B.

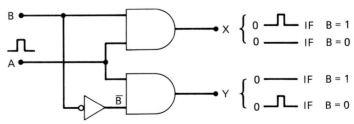

FIGURE 4-27   Example 4-23

1. Design a logic circuit with three inputs A, B, C and an output that goes LOW only when A is HIGH while B and C are different. (*Ans.* $x = A[B \oplus C]$)
2. Which logic gates produce a 1 output when they are inhibited?
3. Which logic gates pass the inverse of the input signal when they are enabled?

4-9 BASIC CHARACTERISTICS OF DIGITAL ICS

Digital ICs are a collection of resistors, diodes, and transistors fabricated on a single piece of semiconductor material (usually silicon) called a *substrate* which is commonly referred to as a "chip." The chip is enclosed in a protective plastic or ceramic package from which pins extend for connecting the IC to other devices. The most common type of package is a *dual-in-line package (DIP)* shown in Figure 4-28. It is so-called because it consists of two parallel rows of pins. The pins are numbered counterclockwise when viewed from the top of the package with respect to an identifying notch or dot at one end of the chip. The DIP shown is a 14-pin package; 16-, 20-, 24-, 28-, 40-, and 64-pin packages are also used.

Digital ICs are often categorized according to their circuit complexity as measured by the number of equivalent logic gates on the substrate. There are currently four standard levels of complexity that are defined as follows:

COMPLEXITY	NUMBER OF GATES
Small-scale integration (SSI)	Fewer than 12
Medium-scale integration (MSI)	12 to 99
Large-scale integration (LSI)	100 to 9999
Very large-scale integration (VLSI)	10,000 or more

FIGURE 4-28 (a) Dual-in-line package; (b) top view showing pin numbers.

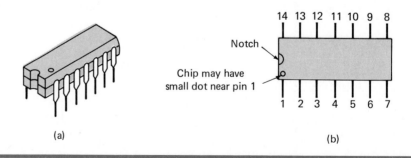

A fifth level called ultralarge-scale integration (ULSI) is currently undergoing development.

Chips that are classified as SSI are those containing a small number of gates. All of the specific ICs that we have referred to in Chapter 3 and this chapter are SSI chips. In later chapters we will encounter ICs that fall into the other categories. For now, our interest will concentrate on SSI.

Bipolar and Unipolar Digital ICs Digital ICs can also be categorized according to the principal type of electronic component used in their circuitry. *Bipolar ICs* are those that are made using the bipolar junction transistor (NPN and PNP) as its main circuit element. *Unipolar ICs* are those that use the unipolar field-effect transistors (P-Channel and N-channel MOSFETs) as their main element.

The most widely used family of bipolar digital ICs is the TTL (transistor-transistor logic) family. Figure 4-29(a) shows a standard TTL INVERTER circuit. Notice that it contains several bipolar transistors. The TTL family is especially prominent in small-scale and medium-scale integration (SSI and MSI) and has been a front-runner in these categories for quite some time. Its leading position in SSI and MSI, however, is being challenged by the CMOS (complementary metal-oxide semiconductor) logic family, which belongs to the category of unipolar digital ICs. Figure 4-29(b) shows a standard CMOS INVERTER circuit that uses enhancement-type MOSFETs as its main circuit element. Later in the text we will take a detailed look at these TTL and CMOS circuits.

FIGURE 4-29 **(a) TTL INVERTER circuit; (b) CMOS INVERTER circuit. Pin numbers are given in parentheses.**

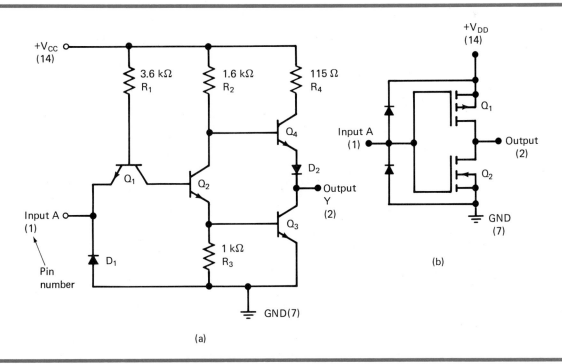

TTL Family   This logic family actually consists of several subfamilies or series. Table 4-1 lists the name of each TTL series together with the prefix designation used to identify different ICs as being part of that series. For example, ICs that are part of the standard TTL series have an identification number that starts with 74. The 7402, 7438, and 74123 are all ICs in this series. Likewise ICs that are part of the low-power Schottky TTL series will have an identification number that starts with 74LS. The 74LS02, 74LS38, and 74LS123 are examples of devices in the 74LS series.

TABLE 4-1   Various Series within the TTL Logic Family

TTL SERIES	PREFIX	EXAMPLE IC
Standard TTL	74	7404 (hex INVERTER)
High-speed TTL	74H	74H04 (hex INVERTER
Low-power TTL	74L	74L04 (hex INVERTER)
Schottky TTL	74S	74S04 (hex INVERTER)
Low-power Schottky TTL	74LS	74LS04 (hex INVERTER)
Advanced Schottky	74AS	74AS04 (hex INVERTER)
Advanced low-power Schottky TTL	74ALS	74ALS04 (hex INVERTER)

The differences between the various TTL series are in their electrical characteristics, such as power dissipation, delay times, and switching speed. They do not differ in the pin layout or logic operations performed by the internal circuitry. For example, the 7402, 74H02, 74L02, 74S02, 74LS02, 74ALS02, and 74AS02 are all quad two-input NOR gates. We will compare the electrical characteristics of the different TTL series in a later chapter.

CMOS Family   There are several CMOS series available. They are listed in Table 4-2.

TABLE 4-2   Various Series within CMOS Logic Family

CMOS SERIES	PREFIX	EXAMPLE IC
Metal-gate CMOS	40 or 140	4001 or 14001 (quad NOR gates)
Metal-gate, pin-compatible with TTL	74C	74C02 (quad NOR gates)
Silicon-gate, pin-compatible with TTL, high-speed	74HC	74HC02 (quad NOR gates)
Silicon-gate, high-speed, electrically compatible with TTL	74HCT	74HCT02 (quad NOR gates)

The 4000 and 14000 series are the oldest CMOS series. These series contain many of the same logic functions as the TTL family, but they were not designed to be pin-compatible with TTL devices. For example, the 4001 quad NOR chip contains four two-input NOR gates as does the TTL 7402 chip, but the gate inputs and outputs on the CMOS chip will not have the same pin numbers as the corresponding signals on the TTL chip.

The 74C, 74HC, and 74HCT series are newer CMOS series, the latter two being the most recent. All of them are pin-compatible with correspondingly numbered TTL devices. For example, the 74C02, 74HC02, and 74HCT02 are all pin-compatible with the 7402, 74H02, 74LS02, and so on. The 74HC and 74HCT series operate at a higher speed than the 74C devices. The 74HCT series is designed to be electrically compatible with TTL devices. This means that 74HCT ICs, unlike 74C and 74HC ICs, can be connected directly to TTL devices without any interfacing circuitry.

Power and Ground   To use digital ICs, it is necessary to make the proper connections to the IC pins. The most important connections are *dc power* and *ground*. These are required for the circuits on the chip to operate correctly. Referring to Figure 4-29, you can see that both the TTL and CMOS circuits have a dc power supply voltage connected to one of their pins, and ground to another. The power supply pin is labeled V_{CC} for the TTL circuit, and V_{DD} for the CMOS circuit. Many of the newer CMOS ICs that are designed to be compatible with TTL ICs also use V_{CC} as their power pin.

If either the power or ground connection is not made to the IC, the logic gates on the chip will not respond properly to the logic inputs and will not produce the expected output logic levels.

Logic-Level Voltage Ranges   For TTL devices, V_{CC} is nominally + 5 V. For CMOS ICs, V_{DD} can range from +3 to +18 V, although +5 V is most often used when CMOS ICs are used in the same circuit with TTL ICs.

For standard TTL devices the acceptable voltage ranges for the logic 0 and logic 1 levels are defined as shown in Figure 4-30(a). A logic 0 is any voltage in the range 0 to 0.8 V; a logic 1 is any voltage between 2 V and 5 V. Voltages that are not in either of these ranges are said to be *indeterminate* and should not be used as inputs to any TTL device. The IC manufacturers cannot guarantee how a TTL circuit will respond to input levels that are in the indeterminate range (0.8 to 2.0 V).

FIGURE 4-30   Logic-level voltage ranges for TTL and CMOS digital ICs.

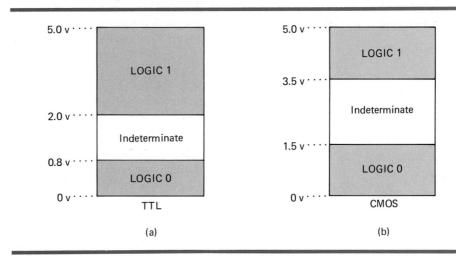

The logic level ranges for CMOS ICs operating with $V_{DD} = +5$ V are shown in Figure 4-30(b). Voltages between 0 and 1.5 V are defined as a logic 0, and voltages from 3.5 to 5 V as a logic 1. The indeterminate range includes voltages between 1.5 and 3.5 V.

Unconnected (Floating) Inputs

What happens when the input to a digital IC is left unconnected? An unconnected input is often called a "floating" input. The answer to this question will be different for TTL and CMOS.

A floating TTL input acts just like a logic 1. In other words, the IC will respond as if the input had a logic HIGH level applied to it. This characteristic is often used when testing a TTL circuit. A lazy technician might leave certain inputs unconnected instead of connecting them to a logic HIGH. Although this is logically correct, it is not a recommended practice, especially in final circuit designs, since the floating TTL input is extremely susceptible to picking up noise signals that can adversely affect the device's operation.

A floating TTL input will measure a dc level of between 1.4 and 1.8 V when checked with a VOM or oscilloscope. Even though this is in the indeterminate range for TTL, it will produce the same response as a logic 1. Being aware of this characteristic of a floating TTL input can be valuable when troubleshooting TTL circuits.

If a CMOS input is left floating, it may have disastrous results. The IC may become overheated and eventually destroy itself. For this reason all inputs to a CMOS IC must be connected to a LOW or a HIGH level or to the output of another IC. A floating CMOS input will not measure as a specific dc voltage, but will fluctuate randomly as it picks up noise. Thus, it does not act as logic 1 or logic 0, so its effect on the output is unpredictable. Sometimes the output will oscillate due to the noise picked up by the floating input.

Logic Circuit Connection Diagrams

A connection diagram shows *all* electrical connections, pin numbers, IC numbers, component values, signal names, and supply voltages. Figure 4-31 shows a typical connection diagram for a simple logic circuit. Examine it carefully and note the following important points:

1. The circuit uses logic gates from two different ICs. The two INVERTERs are part of a 7404 chip which has been given the designation Z1. The 7404 contains six INVERTERs; two of them are used in this circuit and each is labeled as being part of IC Z1. Likewise, the two NAND gates are part of a 7400 chip that contains four NAND gates. All of the gates on this chip are designated with the label Z2. By numbering each gate as Z1, Z2, Z3, and so on, we can keep track of which gate is part of which chip. This is especially valuable in more complex circuits containing many ICs with several gates per chip.

2. Each gate input and output pin number is indicated on the diagram. These pin numbers and the IC labels are used to easily reference any point in the circuit. For example, Z1-pin 2 prefers to the output pin of the top INVERTER. Likewise, we can say that Z1-pin 4 is connected to Z2-pin 9.

3. The power and ground connections to each IC are shown on the diagram. For

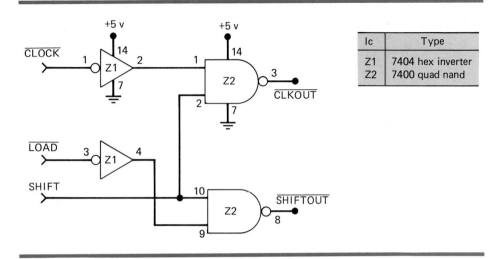

Ic	Type
Z1	7404 hex inverter
Z2	7400 quad nand

FIGURE 4-31 Typical logic circuit connection diagram.

example, Z1-pin 14 is connected to +5 V and Z1-pin 7 is connected to ground. These connections provide power to *all* of the six INVERTERs that are part of Z1.

Manufacturers of electronic equipment generally supply detailed schematics that use a format similar to that in Figure 4-31. These connection diagrams are a virtual necessity when troubleshooting a faulty circuit.

REVIEW QUESTIONS

1. What is the most common type of digital IC package?
2. Name the four common categories of digital ICs according to complexity.
3. *True or false:* A 7474 IC will contain the same logic and pin layout as the 74LS74.
4. *True or false:* A 74HC74 IC will contain the same logic and pin layout as the 7474 IC.
5. Which CMOS series is not pin-compatible with TTL?
6. What is the acceptable voltage range of a logic 0 for TTL? What is it for a logic 1?
7. Repeat question 6 for CMOS operating at $V_{DD} = 5$ V.
8. How does a TTL IC respond to a floating input?
9. How does a CMOS IC respond to a floating input?

4-10 TROUBLESHOOTING DIGITAL SYSTEMS

There are three basic steps in fixing a digital circuit or system that has a fault (failure):

1. *Fault detection.* Observe the circuit/system operation and compare it to the expected correct operation.
2. *Fault isolation.* Perform tests and make measurements to isolate the fault.
3. *Fault correction.* Replace the faulty component, repair faulty connection, remove short, and so on.

Although these steps may seem relatively apparent and straightforward, the actual troubleshooting procedure that is followed is highly dependent on the type and complexity of the circuitry, and on the kinds of troubleshooting tools and documentation that are available.

Good troubleshooting techniques can only be learned in a laboratory environment through experimentation and actual troubleshooting of faulty circuits and systems. There is absolutely no better way to become an effective troubleshooter than to do as much troubleshooting as possible, and no amount of textbook reading can provide that kind of experience. We can, however, help you to develop the analytical skills that are the most essential part of effective troubleshooting. We will describe the types of faults that are common to systems that are made primarily from digital ICs and tell you how to recognize them. We will then present typical case studies to illustrate the analytical processes involved in troubleshooting. In addition, there will be end-of-chapter problems to provide you with the opportunity to go through these analytical processes to reach conclusions about faulty digital circuits.

For all of the troubleshooting discussions throughout the text, we will assume that a technician always has a *logic probe* and *oscilloscope* available since these are standard tools in any digital lab. We will further assume that the logic probe has one or more indicator lights that can indicate the following different conditions at the probe tip:

1. Constant LOW level
2. Constant HIGH level
3. Repetitive pulse train (pulsing)
4. A LOW level that is momentarily pulsed HIGH
5. A HIGH level that is momentarily pulsed LOW
6. An open circuit or an indeterminate logic level

In the troubleshooting examples we will often state the logic probe indication at various points in the circuit.

4-11 INTERNAL DIGITAL IC FAULTS

The most common internal failures of digital ICs are:

1. Malfunction in the internal circuitry
2. Inputs or outputs shorted to ground or V_{CC}
3. Inputs or outputs open-circuited
4. Short between two pins (other than ground or V_{CC})

We will now describe each of these types of failure.

Malfunction in Internal Circuitry   This is usually caused by one of the internal components failing completely or operating outside its specifications. When this happens the IC outputs do not respond properly to its inputs. There is no way to predict what the outputs will do, because it depends on what internal component has failed. Examples of this type of failure would be a base-emitter short in transistor Q_4 or an extremely large resistance value for R_2 in the TTL INVERTER of Figure 4-29(a). This type of internal IC failure is not as common as the other three.

Input Internally Shorted to Ground or Supply   This type of internal failure will cause the input to be stuck in the LOW or HIGH state. Figure 4-32(a) shows input pin 2 of a NAND gate shorted to ground within the IC. This will cause pin 2 always to be in the LOW state. If this input pin is being driven by a logic signal B, it will effectively short B to ground. Thus, this type of fault will affect the output of the device that is generating the B signal.

Similarly, an IC input pin could be internally shorted to +5 V as in Figure 4-32(b). This would keep that pin stuck in the HIGH state. If this input pin is being driven by a logic signal A, it would effectively short A to +5 V.

Output Internally Shorted to Ground or Supply   This type of internal failure will cause the output pin to be stuck in the LOW or HIGH state. Figure 4-33(a) shows pin 3 of the NAND gate shorted to ground within the IC. This output is stuck LOW, and it will not respond to the conditions applied to input pins 1 and 2; in other words, logic inputs A and B will have no effect on output X.

An IC pin can also be shorted to +5 V within the IC as shown in Figure

FIGURE 4-32   **(a) IC input internally shorted to ground; (b) IC input internally shorted to supply voltage. This type of failure forces input signal at shorted pin to stay in one state.**

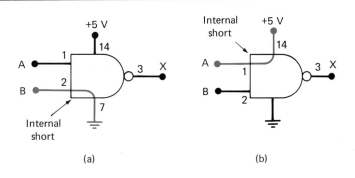

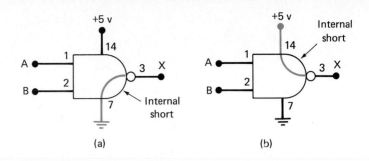

(a)   (b)

FIGURE 4-33   (a) IC output internally shorted to ground; (b) output internally shorted to supply voltage. This type of failure does not affect signals at the IC inputs.

4-33(b). This forces the output pin 3 to be stuck HIGH regardless of the state of the signals at the input pins. Note that this type of failure has no effect on the logic signals at the IC inputs.

EXAMPLE 4-24

Refer to the circuit of Figure 4-34. A technician uses a logic probe to determine the conditions at the various IC pins. The results are recorded in the figure. Examine these results and determine if the circuit is working properly. If not, suggest some of the possible faults.

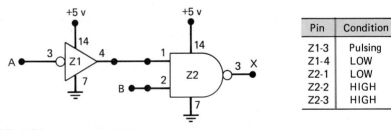

Pin	Condition
Z1-3	Pulsing
Z1-4	LOW
Z2-1	LOW
Z2-2	HIGH
Z2-3	HIGH

FIGURE 4-34   Example 4-24

SOLUTION
Output pin 4 of the INVERTER should be pulsing since its input is pulsing. The recorded results, however, show that pin 4 is stuck LOW. Since this is connected to Z2-pin 1, this keeps the NAND output HIGH. From our preceding discussion, we can list three possible faults that could produce this operation.

First, there could be an internal component failure in the INVERTER that prevents it from responding properly to its input. Second, pin 4 of the IN-VERTER could be internally shorted to ground, thereby keeping it stuck LOW. Third, pin 1 of Z2 could be shorted to ground internal to Z2. This would prevent the INVERTER output pin from changing.

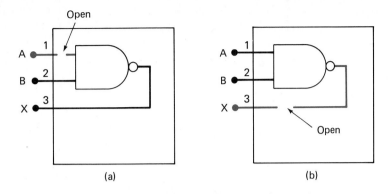

FIGURE 4-35 (a) IC with internally open input will not respond to signals applied to that input pin; (b) an internally open output pin will produce no output voltage.

Open-Circuited Input or Output Sometimes the very fine conducting wire that connects an IC pin to the IC's internal circuitry will break, producing an open circuit. Figure 4-35(a) shows this for an input pin, and Figure 4-35(b) for an output pin.

   If an input to an IC is internally open, the logic signal applied to that input will have no effect on the output. The open circuit leaves the input in the floating state. As we stated earlier, TTL and CMOS will respond differently to an open input. TTL ICs will respond as if the open input is a logic 1; CMOS ICs will respond erratically and may even become damaged due to overheating.

   When the output from an IC is internally open there will be no voltage present at the output pin regardless of the applied input conditions. If this output is connected to the input of one or more other ICs, this produces an open or floating input to those ICs.

EXAMPLE 4-25

Refer to the circuit of Figure 4-36 and the recorded logic probe indications. What are some of the possible faults that could produce the recorded results? Assume that the ICs are TTL.

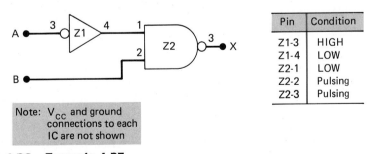

Pin	Condition
Z1-3	HIGH
Z1-4	LOW
Z2-1	LOW
Z2-2	Pulsing
Z2-3	Pulsing

Note: V_{CC} and ground connections to each IC are not shown

FIGURE 4-36 **Example 4-25**

SOLUTION

Examination of the recorded results indicates that the INVERTER appears to be working properly, but the NAND output is inconsistent with its inputs. The NAND output should be HIGH since its input pin 1 is LOW. This LOW should prevent the NAND gate from responding to the pulses at pin 2. It is probable that this LOW is not reaching the internal NAND gate circuitry because of an internal open. Because the IC is TTL, this open would produce the same effect as a logic HIGH at pin 1. If the IC had been CMOS, the internal open at pin 1 might have produced an indeterminate output and possible overheating and destruction of the chip.

From our earlier statement regarding open TTL inputs, you might have expected that the voltage at pin 1 of Z2 would be 1.4 to 1.8 V and should have registered as indeterminate by the logic probe. This would have been true if the open had been *external* to the NAND IC. There is no open between Z1-pin 4 and Z2-pin 1, so the voltage at Z1-pin 4 is reaching Z2-pin 1, but it gets disconnected *inside* the NAND chip.

Short Between Two Pins An internal short between two pins of an IC will force the logic signals at those pins always to be identical. Whenever two signals that are supposed to be different show the same logic-level variations, there is a good possibility that the signals are shorted together.

Consider the circuit in Figure 4-37 where pins 5 and 6 of the NOR gate are internally shorted together. The short causes the two INVERTER output pins to be connected together so that the signals at Z1-pin 2 and Z1-pin 4 have to be identical even when the two INVERTER input signals are trying to produce different outputs. To illustrate, consider the input waveforms shown in the diagram. Even though these input waveforms are different, the waveforms at outputs Z1-2 and Z1-4 are the same.

FIGURE 4-37 When two input pins are internally shorted, it forces the signals driving these pins to be identical, and usually produces a signal with three distinct levels.

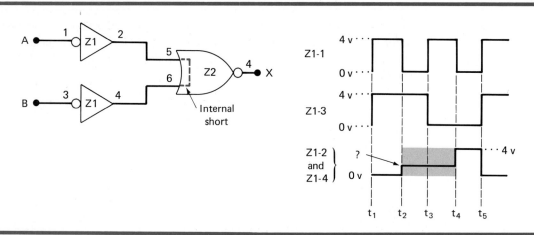

During the interval t_1–t_2, both INVERTERs have a HIGH input and both are trying to produce a LOW output, so their being shorted together makes no difference. During the interval t_4–t_5, both INVERTERs have a LOW input and are trying to produce a HIGH output, so again their being shorted has no effect. However, during the t_2–t_3 and t_3–t_4 intervals, one INVERTER is trying to produce a HIGH output while the other is trying to produce a LOW output. For this situation the actual voltage level that appears at the shorted outputs will depend on the internal IC circuitry. For TTL devices it will usually be a voltage in the high end of the logic 0 range (i.e., close to 0.8 V), although it may also be in the indeterminate range. For CMOS devices it will often be a voltage in the indeterminate range.

Whenever you see a waveform like the Z1-2, Z1-4 signal in Figure 4-37 with three different levels, you should suspect that two output signals may be shorted together.

REVIEW QUESTIONS

1. List the different internal digital IC faults.
2. Which internal IC fault can produce signals that show three different voltage levels?

4-12 EXTERNAL FAULTS

We have seen how to recognize the effects of various faults internal to digital ICs. There are many more things that can go wrong external to the ICs, and we will describe the most common ones in this section.

Open Signal Lines   This category includes any fault that produces a break or discontinuity in the conducting path such that a voltage level or signal is prevented from going from one point to another. Some of the causes of open signal lines are:

1. Broken wire
2. Poor solder connection; loose wire-wrap connection
3. Crack or cut trace on a printed circuit board (some of these are hairline cracks that are difficult to see without a magnifying glass)
4. Bent or broken pin on an IC
5. Faulty IC socket such that the IC pin does not make good contact with the socket

This type of circuit fault is easily verified by disconnecting power from the circuit and making an ohmmeter check between the two points in question.

EXAMPLE 4-26

Consider the CMOS circuit of Figure 4-38 and the accompanying logic probe indications. What is the most probable circuit fault?

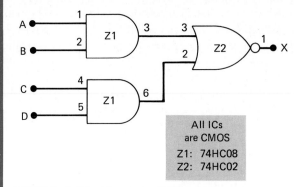

Pin	Condition
Z1-1	Pulsing
Z1-2	HIGH
Z1-3	Pulsing
Z1-4	LOW
Z1-5	Pulsing
Z1-6	LOW
Z2-3	Pulsing
Z2-2	Indeterminate
Z2-1	Indeterminate

All ICs
are CMOS
Z1: 74HC08
Z2: 74HC02

FIGURE 4-38   **Example 4-26**

SOLUTION

The indeterminate level at the NOR gate output is probably due to the indeterminate input at pin 2. Since there is a LOW at Z1-6, this LOW should also be at Z2-2. Clearly, the LOW from Z1-6 is not reaching Z2-2, and there must be an open in the signal path between these two points. The location of this open can be determined by starting at Z1-6 with the logic probe and tracing the LOW level toward Z2-2 until it changes into an indeterminate level.

Shorted Signal Lines   This type of fault has the same effect as an internal short between IC pins. It causes two signals to be exactly the same. The main causes are:

1.   *Sloppy wiring.* An example of this is stripping too much insulation from ends of wires that are in close proximity.

2.   *Solder bridges.* These are splashes of solder that short two or more points together. They commonly occur between points that are very close together, such as adjacent pins on a chip.

3.   *Incomplete etching.* The copper between adjacent conducting paths on a printed circuit board is not completely etched away.

An ohmmeter check can also be used to verify that two signal lines are shorted together.

Faulty Power Supply   All digital systems have one or more dc power supplies that supply the V_{CC} and V_{DD} voltages required by the chips. A faulty power supply or one that is overloaded (supplying more than its rated amount of current) will provide poorly regulated supply voltages to the ICs, and the ICs will either not operate, or will operate erratically.

A power supply may go out of regulation due to a fault in its internal circuitry, or because the circuits that it is powering are drawing more current than the supply is designed for. This can happen if a chip or a component has a fault that causes it to draw much more current than normal.

It is a good troubleshooting practice to check the voltage levels at each power supply in the system to see that they are within their specified ranges. It is also a good idea to check them on an oscilloscope to verify that there is no significant amount of ac ripple on the dc levels, and that the voltage levels stay regulated during the system operation.

One of the most common signs of a faulty power supply is one or more chips operating erratically or not at all. Some ICs are more tolerant of power supply variations and may operate properly, while others do not. You should always check the power and ground levels at each IC that appears to be operating incorrectly.

REVIEW QUESTIONS

1. What are the most common types of external faults?
2. List some of the causes of signal path opens.
3. What symptoms are caused by a faulty power supply?

4-13 TROUBLESHOOTING CASE STUDY

The following example will illustrate the analytical processes involved in troubleshooting digital circuits. Although the example is a fairly simple combinatorial logic circuit, the reasoning and troubleshooting procedures used can be applied to the more complex digital circuits that we encounter in subsequent chapters.

EXAMPLE 4-27

Consider the circuit of Figure 4-39. The output Y is supposed to go HIGH for either of the following conditions:

1. $A = 1, B = 0$ regardless of level on C.
2. $A = 0, B = 1, C = 1$.

You may wish to verify this for yourself.

When the circuit is tested, the technician observes that the output Y goes HIGH whenever A is HIGH or C is HIGH regardless of the level at B. She takes logic probe measurements for the condition where $A = B = 0$, $C = 1$ and comes up with the indications recorded in Figure 4-39.

Examine the recorded levels and list the possible causes for the malfunction. Then develop a step-by-step procedure to determine the exact fault.

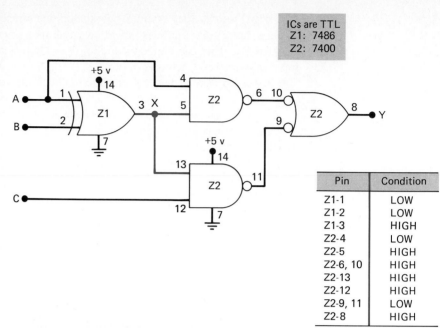

| ICs are TTL |
| Z1: 7486 |
| Z2: 7400 |

Pin	Condition
Z1-1	LOW
Z1-2	LOW
Z1-3	HIGH
Z2-4	LOW
Z2-5	HIGH
Z2-6, 10	HIGH
Z2-13	HIGH
Z2-12	HIGH
Z2-9, 11	LOW
Z2-8	HIGH

FIGURE 4-39 Example 4-27

SOLUTION

All of the NAND gate outputs are correct for the levels present at their inputs. The EX-OR gate, however, should be producing a LOW at output pin 3 since both of its inputs are at the same LOW level. It appears that Z1-3 is stuck HIGH even though its inputs should produce a LOW. There are several possible causes for this.

1. An internal component failure in Z1 that prevents its output from going LOW.

2. An external short to V_{CC} from any point along the conductors connected to node X (shaded in diagram?).

3. Pin 3 of Z1 internally shorted to V_{CC}.

4. Pin 5 of Z2 internally shorted to V_{CC}.

5. Pin 13 of Z2 internally shorted to V_{CC}.

All of these possibilities except for the first one will short node X (and every IC pin connected to it) directly to V_{CC}.

The following procedure can be used to isolate the fault. This procedure is not the only approach that can be used, and as we stated earlier, the actual troubleshooting procedure that a technician uses is very dependent on what test equipment is available.

1. Check the V_{CC} and ground levels at the appropriate pins of Z1. Although it is unlikely that the absence of either of these might cause Z1-3 to stay HIGH, it is a good idea to make this check on any IC that is producing an incorrect output.

2. Turn off power to the circuit and use an ohmmeter to check for a short (resistance less than 1 Ω) between node X and any point connected to V_{CC} (such as Z1-14 or Z2-14). If no short is indicated, the last four possibilities in our list can be eliminated. This means that it is very likely that Z1 has an internal failure and should be replaced.

3. If step 2 shows that there is a short from node X to V_{CC}, perform a thorough visual examination of the circuit board and look for solder bridges, unetched copper slivers, uninsulated wires touching each other, and any other possible cause of an external short to V_{CC}. A likely spot for a solder bridge would be between adjacent pins 13 and 14 of Z2. Pin 14 is connected to V_{CC} and pin 13 to node X. If an external short is found, remove it and perform an ohmmeter check to verify that node X is no longer shorted to V_{CC}.

4. If step 3 does not reveal an external short, the three possibilities that remain are internal shorts to V_{CC} at Z1-3, Z2-13, or Z2-5. One of these is shorting node X to V_{CC}.

To determine which of these IC pins is the culprit, we should disconnect each of them from node X *one at a time* and recheck for a short to V_{CC} after each disconnection. When the pin that is internally shorted to V_{CC} is disconnected, node X will no longer be shorted to V_{CC}.

The process of disconnecting each suspected pin from node X can be easy or difficult depending on how the circuit is constructed. If the ICs are in sockets, all you need to do is to pull the IC from its socket, bend out the suspected pin, and reinsert the IC into its socket. If the ICs are soldered into a printed circuit board, you will have to cut the trace that is connected to the pin (or cut the pin) and repair the cut trace (or pin) when you are finished.

There is a troubleshooting technique that makes it unnecessary to bend pins or cut traces when trying to isolate a short. It involves using a tool called a *current tracer* to trace the flow of current through the short circuit as the node is being pulsed. The current tracer senses the changing magnetic field around the conductor through which the current is being shorted.

Example 4-27, although fairly simple, shows you the kinds of thinking that a troubleshooter must employ in order to isolate a fault. You will have the opportunity to begin developing your own troubleshooting skills by working on Problems 4-33– 4-43.

Bipolar ICs Integrated digital circuits in which NPN and PNP transistors are the main circuit elements. [Sec. 4-9]

CMOS Complementary Metal-Oxide Semiconductor: A logic family that belongs to the category of unipolar digital ICs. [Sec. 4-9]

Combinatorial Logic Circuits Circuits made up of combinations of logic gates. [Sec. 4-4]

Current Tracer Digital troubleshooting tool which senses the changing magnetic field around a conductor. [Sec. 4-13]

DIP Dual-in-Line Package: The most common type of IC packages. [Sec. 4-9]

"Don't Care" Situation when a circuit's output level for a given set of input conditions can be assigned as either a 1 or a 0. [Sec. 4-5]

Exclusive-NOR Circuit Two-input logic circuit that produces a HIGH output only when the inputs are equal. [Sec. 4-6]

Exclusive-OR Circuit Two-input logic circuit that produces a HIGH output only when the inputs are different. [Sec. 4-6]

Floating Input Input signal that is left disconnected in a logic circuit. [Sec. 4-9]

Indeterminate Whenever a logic voltage level of a particular logic family falls out of the required range of voltages for either a logic 0 or a logic 1. [Sec. 4-9]

Inhibit Circuits Logic circuits that control the passage of an input signal through to the output. [Sec. 4-8]

Karnaugh Map Two-dimensional form of a truth table used to simplify a sum-of-products expression. [Sec. 4-5]

Logic Probe Digital troubleshooting tool which senses the logic level at a particular point in a circuit. [Sec. 4-10]

Looping When adjacent squares in a Karnaugh map containing 1s are combined for the purpose of simplification of a sum-of-products expression. [Sec. 4-5]

LSI Large-scale integration. [Sec. 4-9]

MSI Medium-scale integration. [Sec. 4-9]

Octets Group of eight 1s that are adjacent to each other within a Karnaugh map. [Sec. 4-5]

SSI Small-scale integration. [Sec. 4-9]

Substrate Piece of semiconductor material which is part of the building block of any digital IC. [Sec. 4-9]

Sum-of-Products Form Logic expression consisting of two or more AND terms (products) that are ORed together. [Sec. 4-1]

TTL Transistor-transistor logic. Logic family that belongs to the category of bipolar digital ICs. [Sec. 4-9]

ULSI Ultralarge-scale integration. [Sec. 4-9]

Unipolar ICs Integrated digital circuits where unipolar field-effect transistors (MOSFETs) are the main circuit elements. [Sec. 4-9]

VLSI Very large-scale integration. [Sec. 4-9]

PROBLEMS Sections 4-2 and 4-3

4-1. Simplify the following expressions using Boolean algebra.
(a) $x = \overline{A}\overline{B}\overline{C} + \overline{A}BC + ABC + A\overline{B}\overline{C} + A\overline{B}C$
(b) $z = (B + \overline{C})(\overline{B} + C) + \overline{\overline{A} + B + \overline{C}}$
(c) $y = \overline{(\overline{C} + D)} + \overline{A}C\overline{D} + A\overline{B}\overline{C} + \overline{A}\overline{B}CD + AC\overline{D}$

4-2. Simplify the circuit of Figure 4-40 using Boolean algebra.

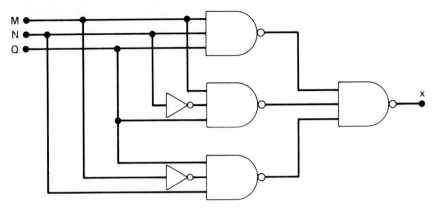

FIGURE 4-40

4-3. Change each gate in Figure 4-40 to a NOR gate and simplify the circuit using Boolean algebra.

4-4. Design the logic circuit corresponding to the following truth table.

A	B	C	x
0	0	0	1
0	0	1	0
0	1	0	1
0	1	1	1
1	0	0	1
1	0	1	0
1	1	0	0
1	1	1	1

4-5. Design a logic circuit whose output is HIGH *only* when a majority of inputs A, B, and C are LOW.

4-6. Three photocells are being illuminated by three different flashing lights. The lights are supposed to be flashing in sequence so that at no time should all three lights be on at the same time or off at the same time. Each photocell is used to monitor one of the lights, and each photocell is in a circuit that produces a LOW output voltage when the photocell is dark and a HIGH output voltage when the photocell is illuminated. Design a logic circuit that has as its inputs the photocell-circuit outputs and which produces a HIGH output whenever the three lights are *all* on or *all* off at the same time.

4-7. A 4-bit binary number is represented as $A_3A_2A_1A_0$, where A_3, A_2, A_1, and A_0 represent the individual bits with A_0 equal to the LSB. Design a logic circuit that will produce a HIGH output whenever the binary number is greater than 0010 and less than 1000.

4-8. Figure 4-41 shows a diagram for an automobile alarm circuit used to detect certain undesirable conditions. The three switches are used to indicate the

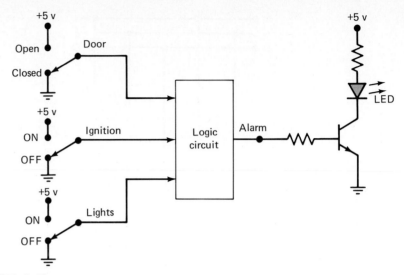

FIGURE 4-41

status of the door on the driver's seat, the ignition, and the headlights respectively. Design the logic circuit with these three switches as inputs so that the alarm will be activated whenever either of the following conditions exists:

- The headlights are ON while the ignition is OFF.
- The door is open while the ignition is ON.

4-9. Implement the circuit of Problem 4-4 using all NAND gates.

4-10. Implement the circuit of Problem 4-5 using all NAND gates.

4-11. Implement the expression $z = \overline{D} + ABC + \overline{AC}$ using ANDs, ORs, and INVERTERS; then convert to all NAND gates.

Section 4-5

4-12. Simplify the expression of Problem 4-1(a) using the K map.

4-13. Simplify the expression of Problem 4-1(c) using the K map.

4-14. Simplify the expression from Problem 4-7 using a K map.

4-15. Determine the minimum expressions for each K map in Figure 4-42.

4-16. Figure 4-43 shows a *BCD counter* that produces a 4-bit output representing the BCD code for the number of pulses that have been applied to the counter input. For example, after four pulses have occurred, the counter outputs are $DCBA = 0100_2 = 4_{10}$. The counter resets to 0000 on the tenth pulse and starts counting over again. In other words, the $DCBA$ outputs will never represent a number greater than $1001_2 = 9_{10}$. Design the logic circuit that produces a HIGH output whenever the count is 2, 3, or 9. Use K mapping and take advantage of the "don't care" conditions.

4-17. Figure 4-44 shows four switches that are part of the control circuitry in a copy machine. The switches are at various points along the path of the copy

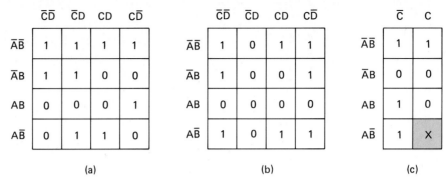

FIGURE 4-42

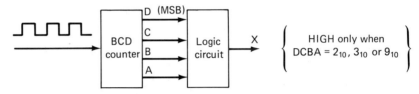

FIGURE 4-43

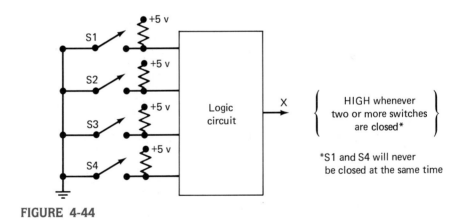

FIGURE 4-44

paper as the paper passes through the machine. Each switch is normally open, and as the paper passes over a switch, the switch closes. It is impossible for switches S1 and S4 to be closed at the same time. Design the logic circuit to produce a HIGH output whenever *two or more* switches are closed at the same time. Use K mapping and take advantage of the "don't care" conditions.

4-18. (a) Determine the output waveform for the circuit of Figure 4-45.

(b) Repeat with the *B* input held LOW.

(c) Repeat with *B* held HIGH.

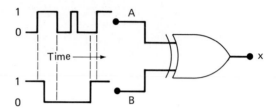

FIGURE 4-45

4-19. Determine the input conditions needed to produce $x = 1$ in Figure 4-46.

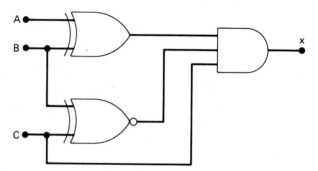

FIGURE 4-46

4-20. Redo Problem 4-6 using EX-OR gates and one other gate.

4-21. Figure 4-47 represents a multiplier circuit that takes 2-bit binary numbers x_1x_0 and y_1y_0 and produces an output binary number $Z_3Z_2Z_1Z_0$ that is equal to the arithmetic product of the two input numbers. Design the logic circuit for the multiplier. (*Hint:* The logic circuit will have four inputs and four outputs.)

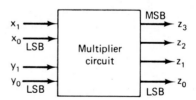

FIGURE 4-47

4-22. A BCD code is being transmitted to a remote receiver. The bits are A_3, A_2, A_1, A_0 with A_3 as the MSB. The receiver circuitry includes a *BCD error-detector* circuit that examines the received code to see if it is a legal BCD code (i.e., ≤ 1001). Design this circuit to produce a HIGH for any error condition.

4-23. Figure 4-48 represents a *relative-magnitude detector* that takes two 3-bit

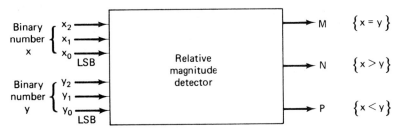

FIGURE 4-48

binary numbers $x_2x_1x_0$ and $y_2y_1y_0$ and determines whether they are equal and, if not, which one is larger. There are three outputs, defined as follows:

1. $M = 1$ only if the two input numbers are equal.

2. $N = 1$ only if $x_2x_1x_0$ is greater than $y_2y_1y_0$.

3. $P = 1$ only if $y_2y_1y_0$ is greater than $x_2x_1x_0$.

Design the logic circuitry for the comparator. The circuit has *six* inputs and *three* outputs and is therefore much too complex to handle using the truth-table approach. Refer to Example 4-16 as a hint to how you might start to solve this problem.

4-24. An EX-OR gate can be considered to be an *odd-parity detector* because it will produce a HIGH output only if an odd number of inputs are HIGH. Show how several EX-OR gates can be combined in a circuit which examines four bits A, B, C, D and produces a HIGH output only if an odd number of bits are 1. You are to use the 7486 quad EX-OR chip shown in Figure 4-49.

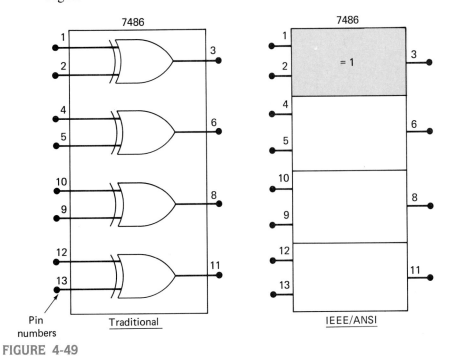

FIGURE 4-49

4-25. Design a logic circuit whose output is HIGH whenever *A* and *B* are both HIGH, while *C* and *D* are either both LOW or both HIGH. Try to do this without using a truth table. Then check your result by constructing a truth table for your circuit to see if it agrees with the problem statement.

4-26. Four large tanks at a chemical plant contain different liquids being heated. Liquid-level sensors are being used to detect whenever the level in tanks *A* and *B* rises above a predetermined level. Temperature sensors in tanks *C* and *D* detect when the temperature in these tanks drops below a prescribed temperature limit. Assume that the liquid-level sensor ouputs *A* and *B* are LOW when the level is satisfactory and HIGH when the level is too high. Also, the temperature-sensor outputs *C* and *D* are LOW when the temperature is satisfactory and HIGH when the temperature is to low. Design a logic circuit that will detect whenever the level in tank *A* or tank *B* is too high at the same time that the temperature in either tank *C* or tank *D* is too low.

4-27. Figure 4-50 shows the intersection of a main highway with a secondary access road. Vehicle-detection sensors are placed along lanes *C* and *D* (main road) and lanes *A* and *B* (access road). These sensor outputs are LOW (0) when no vehicle is present and HIGH (1) when a vehicle is present. The intersection traffic light is to be controlled according to the following logic.

1. The E-W traffic light will be green whenever *both* lanes *C* and *D* are occupied

2. The E-W light will be green whenever *either C* or *D* are occupied but lanes *A* and *B* are not *both* occupied.

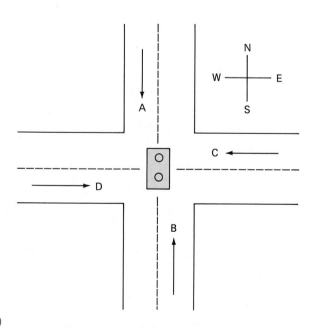

FIGURE 4-50

3. The N-S light will be green whenever *both* lanes A and B are occupied but C and D are not *both* occupied.

4. The N-S light will also be green when *either* A or B is occupied while C and D are *both* vacant.

5. The E-W will be green when *no* vehicles are present.

Using the sensor outputs A, B, C, and D as inputs, design a logic circuit to control the traffic light. There should be two outputs, N/S and E/W, which go HIGH when the corresponding light is to be *green*. Simplify the circuit as much as possible and show *all* steps.

Section 4.8

4-28. Design a logic circuit that will allow an input signal to pass through to the output only when control input B is LOW while control input C is HIGH; otherwise, the output is LOW.

4-29. Design a circuit that will *inhibit* the passage of an input signal only when control inputs B, C, and D are all HIGH; the output is to be HIGH in the inhibited condition.

4-30. Design a logic circuit that controls the passage of a signal A according to the following requirements.
1. Output X will equal A when control inputs B and C are the same.
2. X will remain HIGH when B and C are different.

4-31. Design a logic circuit that has two signal inputs A_1 and A_0 and a control input S so that it functions according to the requirements given in Figure 4-51. This type of circuit is called a *multiplexer* (covered in Chapter 9).

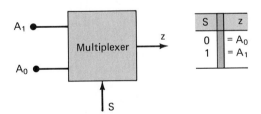

FIGURE 4-51

4-32. Use K mapping to design a circuit to meet the requirements of Example 4-16. Compare this circuit to the solution in Figure 4-22. This points out that the K-map method cannot take advantage of the EX-OR and EX-NOR gate logic. The designer has to be able to determine when these gates are applicable.

Sections 4-9–4-13

*****4-33.** The signals shown in Figure 4-52 are applied to the inputs of the circuit of Figure 4-31. Suppose there is an internal open at Z1-4.
(a) What will a logic probe indicate at Z1-4?

*Color indicates troubleshooting problems throughout the text.

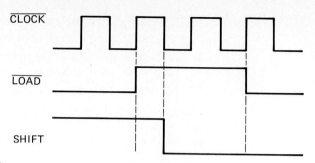

CLOCK

LOAD

SHIFT

FIGURE 4-52

      (b) What dc voltage reading would you expect a VOM to register at Z1-4? (Remember that the ICs are TTL.)

      (c) Sketch what you think the $\overline{CLKOUT}$ and $\overline{SHIFTOUT}$ signals will look like.

4.34. The signals of Figure 4.52 are applied to the inputs of the circuit of Figure 4.31. Suppose there is an internal short between pins 9 and 10 of Z2. Sketch the probable signal at Z2-10, $\overline{CLKOUT}$, and $\overline{SHIFTOUT}$.

4.35. Assume that the ICs in Figure 4-31 are CMOS. Describe how the circuit operation would be affected by an open in the conductor connecting Z2-2 and Z2-10.

4-36. In Example 4-24 we listed three possible faults for the situation of Figure 4-34. What procedure would you follow to determine which of the faults is the actual one?

4-37. Refer to the circuit of Figure 4-36. Assume that the devices are CMOS. Also assume that the logic probe indication at Z2-3 is "indeterminate" rather than "pulsing." List the possible faults and write a procedure to follow to determine the actual fault.

4.38. Refer to the logic circuit of Figure 4-39. Recall that output Y is supposed to be HIGH for either of the following conditioins.

    1. $A = 1, B = 0$, regardless of C

    2. $A = 0, B = 1, C = 1$

When testing the circuit, the technician observes that Y only goes HIGH for the first condition, but stays LOW for all other input conditions. Consider the following list of possible faults. For each one indicate "yes" or "no" as to whether or not it could be the actual fault. Explain your reasoning for each "no" response.

(a) An internal short to ground at Z2-13.

(b) An open in the connection to Z2-13.

(c) An internal short to V_{CC} at Z2-11.

(d) An open in the V_{CC} connection to Z2.

(e) An internal open at Z2-9.

(f) An open in the connection from Z2-11 to Z2-9.

(g) A solder bridge between pins 6 and 7 of Z2.

4-39. Develop a procedure for isolating the fault that is causing the malfunction described in Problem 4-38.

4-40. Assume that the gates in Figure 4-39 are all CMOS. When the technician tests the circuit he finds that it operates correctly except for the following conditions:

1. $A = 1, B = 0, C = 0$

2. $A = 0, B = 1, C = 1$

For these conditions, the logic probe indicates indeterminate levels at Z2-6, Z2-11, and Z2-8. What do you think is the probable fault in the circuit? Explain your reasoning.

4-41. Figure 4-53 is a combinatorial logic circuit that operates an alarm in a car whenever the driver and/or passenger seats are occupied and the seat belts are not fastened when the car is started. The active-HIGH signals *DRIV* and *PASS* indicate the presence of the driver and passenger, respectively, and are taken from pressure-actuated switches in the seats. The signal *IGN* is active-HIGH when the ignition switch is on. The signal $\overline{BELTD}$ is active-LOW and indicates that the driver's seat belt is *unfastened;* $\overline{BELTP}$ is the corresponding signal for the passenger seat belt. The alarm will be activated (LOW) whenever the car is started and either of the front seats is occupied and its seat belt is not fastened.

(a) Verify that the circuit will function as described.

(b) Describe how this alarm system would operate if Z1-2 were internally shorted to ground.

(c) Describe how it would operate if there were an open connection from Z2-6 to Z2-10.

4-42. Suppose the system of Figure 4-53 is functioning such that the alarm is activated as soon as the driver and/or passenger are seated and the car is started, regardless of the status of the seat belts. What are the possible faults? What procedure would you follow to find the actual fault?

4-43. Suppose the alarm system of Figure 4-53 is operating such that the alarm goes on continuously as soon as the car is started regardless of the state of the other inputs. List the possible faults and write a procedure to isolate the fault.

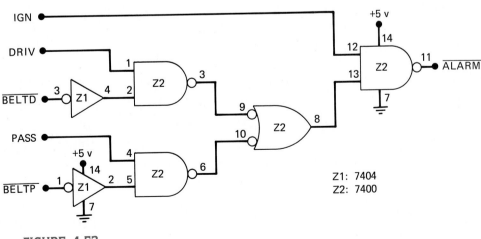

FIGURE 4-53

FLIP-FLOPS AND RELATED DEVICES

5

Upon completion of this chapter, you will be able to:

- Construct and analyze the operation of a latch flip-flop made from NAND or NOR gates.
- Debounce a mechanical switch by using a latch circuit.
- Describe the difference between synchronous and asynchronous systems.
- Understand several types of edge-triggered flip-flops, such as the J-K, D-type, and SET-CLEAR.
- Analyze and apply the various flip-flop timing parameters specified by the manufacturers.
- Describe a pulse-steering and an edge-detector circuit.
- Understand the major differences between parallel and serial data transfers.
- Draw the output timing waveforms of several types of flip-flops in response to a set of input signals.
- Analyze the various IEEE/ANSI flip-flop symbols.
- Explain the data lockout feature of certain flip-flops.
- Cite various flip-flop applications.
- Use flip-flops in synchronization circuits.
- Connect shift registers as data transfer circuits.
- Employ flip-flops as frequency-division and counting circuits.
- Understand the typical characteristics of Schmitt triggers.
- Apply two different types of one-shots in circuit design.
- Design a free-running oscillator using a 555 timer.
- Recognize and predict the effects of clock skew on synchronous circuits.
- Troubleshoot various types of flip-flop circuits.

INTRODUCTION

The logic circuits considered thus far have been combinatorial circuits whose output levels at any instant of time are dependent on the levels present at the inputs at that time. Any prior input-level conditions have no effect on the present outputs because combinatorial logic circuits have no memory. Most digital systems are made up of both combinatorial circuits and memory elements.

Figure 5-1 shows a block diagram of a general digital system that combines combinatorial logic gates with memory devices. The combinatorial portion accepts logic signals from external inputs and from the outputs of the memory elements.

FIGURE 5-1 General digital system diagram.

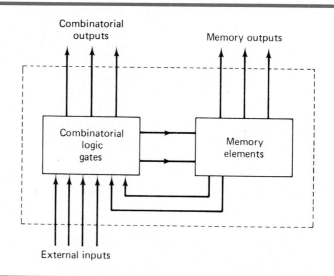

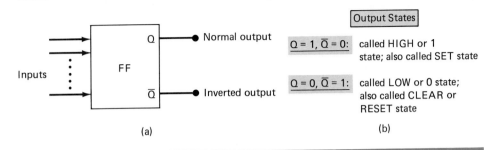

FIGURE 5-2 **General flip-flop symbol and definition of two possible output states.**

The combinatorial circuit operates on these inputs to produce various outputs, some of which are used to determine the binary values to be stored in the memory elements. The outputs of some of the memory elements, in turn, go to the inputs of logic gates in the combinatorial circuits. This process indicates that the external outputs of a digital system are a function of both its external inputs and the information stored in its memory elements.

The most important memory element is the *flip-flop*, which is made up of an assembly of logic gates. Even though a logic gate, by itself, has no storage capability, several can be connected together in ways that permit information to be stored. There are several different gate arrangements that are used to produce these flip-flops (abbreviated FF).

Figure 5-2(a) is the general type of symbol used for a flip-flop. It shows two outputs, labeled Q and $\overline{Q}$, that are the inverse of each other. Actually, any letter can be used, but Q is the one most often used. The Q output is called the *normal* FF output, and $\overline{Q}$ is the *inverted* FF output. Whenever we refer to the state of a FF, we are referring to the state of its normal (Q) output; it is understood that its inverted output ($\overline{Q}$) is in the opposite state. For example, if we say that a FF is in the HIGH(1) state, we mean that $Q = 1$; if we say that a FF is in the LOW (0) state, we mean that $Q = 0$. Of course, the $\overline{Q}$ state will always be the inverse of Q.

A FF, then, has the two allowed operating states indicated in Figure 5-2(b). Note the different ways that are used to refer to these two states. You should become familiar with each of these since they are all in common use.

As the symbol in Figure 5-2(a) implies, a FF can have one or more inputs. These inputs are used to cause the FF to switch back and forth ("flip-flop") between its possible output states. As we shall see, a FF input only has to be pulsed momentarily to cause a change in the FF output state, and the output will remain in that new state even after the input pulse is over. This is the FF's *memory* characteristic.

The flip-flop is known by other names, including *latch* and *bistable multivibrator*. The term "latch" is used for certain types of flip-flops that we will describe. The term "bistable multivibrator" is the more technical name for a flip-flop, but it is too much of a mouthful to be used regularly.

171

5-1 NAND GATE LATCH

The most basic FF circuit can be constructed from either two NAND gates or two NOR gates. The NAND gate version, called a *NAND gate latch* or simply a *latch*, is shown in Figure 5-3(a). The two NAND gates are cross-coupled so that the output of NAND-1 is connected to one of the inputs of NAND-2, and vice versa. The gate outputs, labeled Q and $\overline{Q}$, respectively, are the latch outputs. Under normal conditions, these outputs will always be the inverse of each other. The latch inputs are labeled SET and CLEAR, for reasons that will become clear.

The SET and CLEAR inputs are normally resting in the HIGH state and one of them will be pulsed LOW whenever we want to change the latch outputs. We begin our analysis by showing that there are two equally likely output states when SET = CLEAR = 1. One possibility is shown in Figure 5-3(a), where we have $Q = 0/\overline{Q} = 1$. With $Q = 0$, the inputs to NAND-2 are 0 and 1, which produce $\overline{Q} = 1$. The 1 from $\overline{Q}$ causes NAND-1 to have a 1 at both inputs to produce a 0 output at Q. In effect, what we have is the LOW at the NAND-1 output producing a HIGH at the NAND-2 output, which, in turn, keeps the NAND-1 output LOW.

The second possibility is shown in Figure 5-3(b), where $Q = 1/\overline{Q} = 0$. The HIGH from NAND-1 produces a LOW at the NAND-2 output, which, in turn, keeps the NAND-1 output HIGH. Thus, there are two possible output states when SET = CLEAR = 1; as we shall soon see, the one that actually exists will depend on what has occurred previously at the inputs.

Setting the Latch (FF) Now let's investigate what happens when the SET input is momentarily pulsed LOW while CLEAR is kept HIGH. Figure 5-4(a) shows what happens when $Q = 0$ prior to the occurrence of the pulse. As SET is pulsed LOW at time t_0, Q will go HIGH, and this HIGH will force $\overline{Q}$ to go LOW so that NAND-1 now has two LOW inputs. Thus, when SET returns to the 1 state at t_1, the NAND-1 output *remains* HIGH, which, in turn, keeps the NAND-2 output LOW.

FIGURE 5-3 **NAND latch has two possible resting states when SET = CLEAR = 1.**

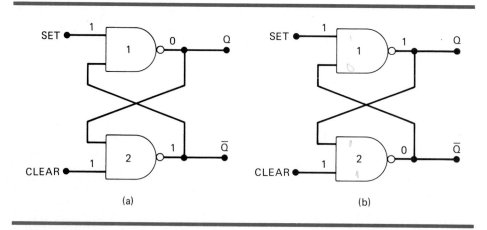

(a)   (b)

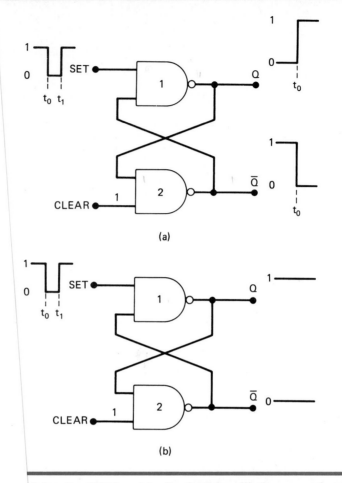

FIGURE 5-4 Pulsing the SET input to the 0 state will always produce the $Q = 1/\overline{Q} = 0$ output state: (a) $Q = 0$ prior to SET pulse; (b) $Q = 1$ prior to SET pulse.

Figure 5-4(b) shows what happens when $Q = 1/\overline{Q} = 0$ prior to the application of the SET pulse. Since $\overline{Q} = 0$ is already keeping the NAND-1 output HIGH, the LOW pulse at SET will not change anything. Thus, when SET returns HIGH, the latch outputs are still in the $Q = 1/\overline{Q} = 0$ state.

We can summarize Figure 5-4 by stating that a LOW pulse on the SET input will cause the latch to end up in the $Q = 1$ state. This operation is called *setting* the latch or FF. In fact, the $Q = 1$ state is also called the *set state*.

Clearing the Latch (FF)   Now let's consider what occurs when the CLEAR input is pulsed LOW while SET is kept HIGH. Figure 5-5(a) shows what happens when $Q = 0$ and $\overline{Q} = 1$ prior to the application of the pulse. Since $Q = 0$ is already keeping the NAND-2 output HIGH, the LOW pulse at CLEAR will not have any effect. When CLEAR returns HIGH, the latch outputs are still $Q = 0/\overline{Q} = 1$.

Figure 5-5(b) shows the situation where $Q = 1$ prior to the occurrence of the

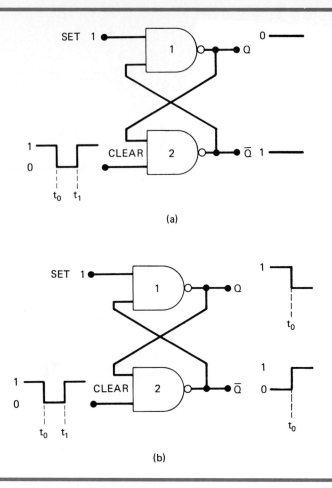

(a)

(b)

FIGURE 5-5   Pulsing the CLEAR input to the LOW state will always produce Q = 0/$\bar{Q}$ = 1: (a) Q = 0 prior to CLEAR pulse; (b) Q = 1 prior to CLEAR pulse.

CLEAR pulse. As CLEAR is pulsed LOW at t_0, $\bar{Q}$ will go HIGH, and this HIGH forces Q to go LOW so that NAND-2 now has two LOW inputs. Thus, when CLEAR returns HIGH at t_1, the NAND-2 output *remains* HIGH, which, in turn, keeps the NAND-1 output LOW.

Figure 5-5 can be summarized by stating that a LOW pulse on the CLEAR input will cause the latch to end up in the Q = 0 state. This operation is called *clearing* or *resetting* the latch, and the Q = 0 state is also called the *cleared* or *reset* state.

Simultaneous Setting and Clearing

The last case to consider is the case where the SET and CLEAR inputs are simultaneously pulsed LOW. This will produce HIGH levels at both NAND outputs so that Q = $\bar{Q}$ = 1. Clearly, this is an undesired condition, since the two outputs are supposed to be inverses of each other. Furthermore, when the SET and CLEAR inputs return HIGH, the resulting output state will depend on which input returns HIGH first. Simultaneous transitions back to the 1 state will produce unpredictable results. For these reasons the SET = CLEAR = 0 condition is normally not used for the NAND latch.

Summary of NAND Latch   The operation described above can be conveniently placed in a truth table (Figure 5-6) and is summarized as follows:

1.   **SET = CLEAR = 1:** This condition is the normal resting state and it has no effect on the output state. The Q and $\overline{Q}$ outputs will remain in whatever state they were prior to this input condition.
2.   **SET = 0, CLEAR = 1.** This will always cause the output to go to the $Q = 1$ state, where it will remain even after SET returns HIGH. This is called *setting* the latch.
3.   **SET = 1, CLEAR = 0.** This will always produce the $Q = 0$ state, where the output will remain even after CLEAR returns HIGH. This is called *clearing* or *resetting* the latch.
4.   **SET = CLEAR = 0.** This condition tries to set and clear the latch at the same time and can produce ambiguous results. It should not be used.

Alternate Representations   From the description of the NAND latch operation, it should be clear that the SET and CLEAR inputs are active-LOW. The SET input will set $Q = 1$ when SET goes LOW; the CLEAR input will clear $Q = 0$ when CLEAR goes LOW. For this reason, the NAND latch is often drawn using the alternate representation for each NAND gate, as shown in Figure 5-7(a). The bubbles on the SET and CLEAR inputs emphasize the fact that these inputs are active-LOW.

Figure 5-7(b) shows a simplified block representation that we will sometimes use. The S and C labels represent the SET and CLEAR inputs, and the bubbles indicate the active-LOW nature of these inputs. Whenever we use this block symbol, it represents a NAND latch.

FIGURE 5-6   (a) NAND latch; (b) truth table.

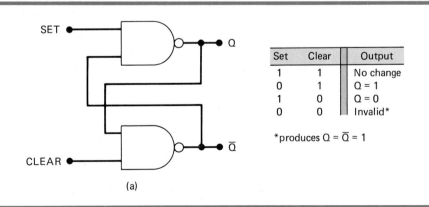

Set	Clear	Output
1	1	No change
0	1	Q = 1
1	0	Q = 0
0	0	Invalid*

*produces $Q = \overline{Q} = 1$

(a)

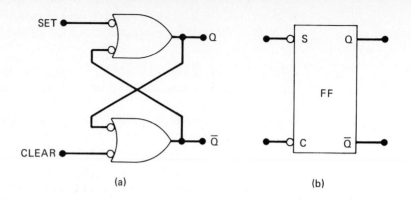

(a)   (b)

FIGURE 5-7   (a) NAND latch equivalent representation; (b) simplified block symbol.

Terminology   The action of *clearing* a FF or a latch is also called *resetting,* and both terms are used interchangeably in the digital field. In fact, a CLEAR input can also be called a RESET input, and a SET-CLEAR latch can be called a SET-RESET latch. We will use both terms throughout the text.

EXAMPLE 5-1

The waveforms of Figure 5-8 are applied to the inputs of a NAND gate latch. Assume that initially $Q = 0$, and determine the Q waveform.

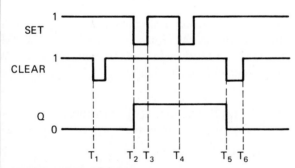

FIGURE 5-8   Example 5.1

SOLUTION
Initially, SET = CLEAR = 1 so that Q will remain in the 0 state. The LOW pulse that occurs on the CLEAR input at time T_1 will have no effect, since Q is already in the cleared (0) state.

   The only way that Q can go to the 1 state is by a LOW pulse on the SET input. This occurs at time T_2 when SET first goes LOW. When SET returns HIGH at T_3, Q will remain in its new HIGH state.

   At time T_4 when SET goes LOW again, there will be no effect on Q because Q is already set to the 1 state.

The only way to bring Q back to the 0 state is by a LOW pulse on the CLEAR input. This occurs at time T_5. When CLEAR returns to 1 at time T_6, Q remains in its now LOW state.

This example shows that the latch output "remembers" the last input that was activated and will not change states until the opposite input is activated.

EXAMPLE 5-2

It is virtually impossible to obtain a "clean" voltage transition from a mechanical switch because of the phenomenon of *contact bounce*. This is illustrated in Figure 5-9(a), where the action of moving the switch from contact position 1 to 2 produces several output voltage transitions as the switch bounces (makes and breaks contact with contact 2 several times) before coming to rest on contact 2.

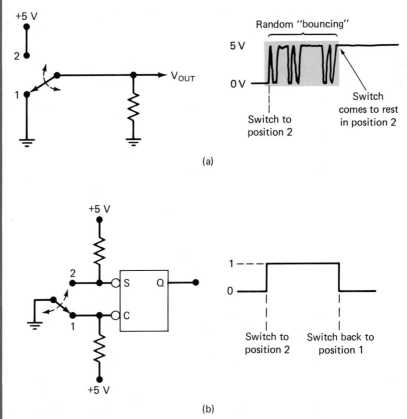

(a)

(b)

FIGURE 5-9 (a) Mechanical contact bounce will produce multiple transitions; (b) NAND latch used to debounce a mechanical switch.

The multiple transitions on the output signal generally last no longer than a couple of milliseconds, but they would be unacceptable in many applications. A NAND latch can be used to prevent the presence of switch bounce from affecting the output. Describe the operation of the "switch debouncing" circuit in Figure 5-9(b).

SOLUTION

Assume that the switch is resting in position 1 so that the CLEAR input is LOW and $Q = 0$. When the switch is moved to position 2, CLEAR will go HIGH, and a LOW will appear on the SET input as the switch first makes contact. This will set $Q = 1$ within a matter of a few nanoseconds (the response time of the NAND gate). Now if the switch bounces off of contact 2, SET and CLEAR will both be HIGH, and Q will not be affected; it will stay HIGH. Thus, nothing will happen at Q as the switch bounces on and off contact 2 before finally coming to rest in position 2.

Likewise, when the switch is moved from position 2 back to position 1, it will place a LOW on the CLEAR input as it first makes contact. This clears Q to the LOW state, where it will remain even if the switch bounces on and off contact 1 several times before coming to rest.

Thus, the output at Q will consist of a single transition each time the switch is moved from one position to the other.

REVIEW QUESTIONS

1. What is the normal resting state of the SET and CLEAR inputs? What is the active state of each input?
2. What will be the states of Q and $\overline{Q}$ after a FF has been cleared (reset)?
3. *True or false:* The SET input can never be used to make $Q = 0$.
4. When power is first applied to any FF circuit, it is impossible to predict the initial states of Q and $\overline{Q}$. What could be done to ensure that a NAND latch always started off in the $Q = 1$ state? (*Ans.* A momentary LOW pulse has to be applied to the SET input)

5-2 NOR GATE LATCH

Two cross-coupled NOR gates can be used as a NOR gate latch. The arrangement, shown in Figure 5-10(a), is similar to the NAND latch except the Q and $\overline{Q}$ outputs have reversed positions.

The analysis of the operation of the NOR latch can be performed in exactly the same manner as for the NAND latch. The results are given in the truth table in Figure 5-10(b) and are summarized as follows:

1. **SET = CLEAR = 0.** This is the normal resting state for the NOR latch and it has no effect on the output state. Q and $\overline{Q}$ will remain in whatever state they were prior to the occurrence of this input condition.
2. **SET = 1, CLEAR = 0.** This will always set $Q = 1$, where it will remain even after SET returns to 0.
3. **SET = 0, CLEAR = 1.** This will always clear $Q = 0$, where it will remain even after CLEAR returns to 0.

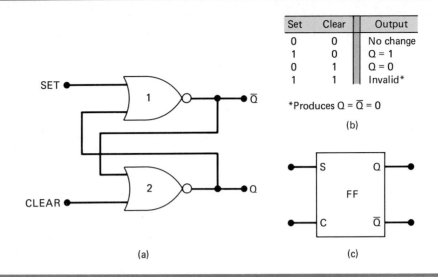

Set	Clear	Output
0	0	No change
1	0	Q = 1
0	1	Q = 0
1	1	Invalid*

*Produces $Q = \overline{Q} = 0$

(b)

(a)

(c)

FIGURE 5-10 **(a) NOR gate latch; (b) truth table; (c) simplified block symbol.**

4. **SET = 1, CLEAR = 1.** This condition tries to set and clear the latch at the same time, and it produces $Q = \overline{Q} = 0$. If the inputs are returned to 0 simultaneously, the resulting output state is unpredictable. This input condition should not be used.

The NOR gate latch operates exactly like the NAND latch except that the SET and CLEAR inputs are active HIGH rather than active LOW, and the normal resting state is SET = CLEAR = 0. Q will be set HIGH by a HIGH pulse on the SET input, and it will be cleared LOW by a HIGH pulse on the CLEAR input. The simplified block symbol for the NOR latch in Figure 5-10(c) is shown with no bubbles on the S and C inputs; this indicates that these inputs are active HIGH.

EXAMPLE 5-3

Assume that $Q = 0$ initially, and determine the Q waveform for the NOR latch inputs of Figure 5-11.

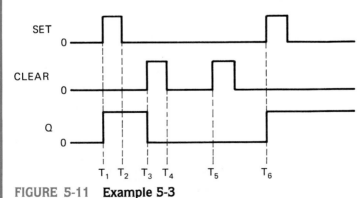

FIGURE 5-11 **Example 5-3**

SOLUTION

Initially, SET = CLEAR = 0, which has no effect on Q, and Q stays LOW. When SET goes HIGH at time T_1, Q will be set to 1 and will remain there even after SET returns to 0 at T_2.

At T_3 the CLEAR input goes HIGH and clears Q to the 0 state, where it remains even after CLEAR returns LOW at T_4.

The CLEAR pulse at T_5 has no effect on Q, since Q is already LOW. The SET pulse at T_6 again sets Q back to 1, where it will stay.

This example shows that the FF "remembers" the last input that was activated, and it will not change states until the opposite input is activated.

EXAMPLE 5-4

Figure 5-12 shows a simple circuit that can be used to detect the interruption of a light beam. The light is focused on a phototransistor that is connected in the common-emitter configuration to operate as a switch. Assume that the latch has previously been cleared to the 0 state by momentarily opening switch $S1$ and describe what happens if the light beam is momentarily interrupted.

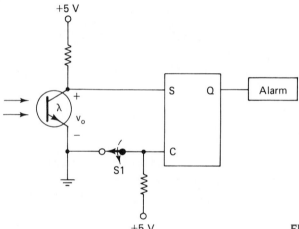

FIGURE 5-12   Example 5-4

SOLUTION

With light on the phototransistor, we can assume that it is fully conducting so that the resistance between the collector and emitter is very small. Thus, v_0 will be close to 0 V. This places a LOW on the SET input of the latch so that SET = CLEAR = 0.

When the light beam is interrupted, the phototransistor turns off and its collector-emitter resistance becomes very high (i.e., essentially an open circuit). This causes v_0 to rise to approximately 5 V; this activates the SET input and turns on the alarm.

Now, the alarm will remain on because Q will remain HIGH even if the light beam was only momentarily interrupted, and v_0 goes back to 0 V. The alarm can be turned off only by momentarily opening $S1$ to produce a HIGH on the CLEAR input.

In this application, the latch's memory characteristic is used to convert a momentary occurrence (beam interruption) into a constant output.

REVIEW QUESTIONS

1. What is the normal resting state of the NOR latch inputs? What is the active state?
2. When a FF is set, what are the states of Q and $\overline{Q}$?
3. What is the only way to cause the Q output of a NOR latch to change from 1 to 0?
4. If the NOR latch in Figure 5-12 were replaced by a NAND latch, why wouldn't the circuit work properly?

5-3 TROUBLESHOOTING CASE STUDY

The following two examples will present an illustration of the kinds of reasoning used in troubleshooting a circuit containing a latch.

EXAMPLE 5-5

Analyze and describe the operation of the circuit in Figure 5-13.

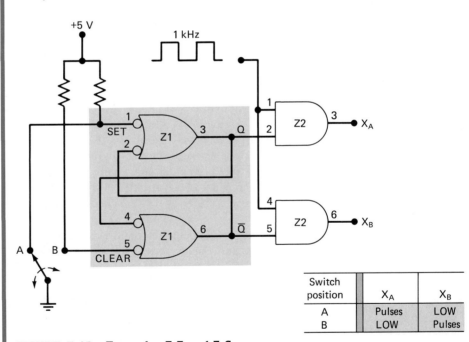

Switch position	X_A	X_B
A	Pulses	LOW
B	LOW	Pulses

FIGURE 5-13   Examples 5-5 and 5-6

SOLUTION
The switch is used to set or clear the NAND latch to produce clean bounce-free signals at Q and $\overline{Q}$. These latch outputs control the passage of the 1-kHz pulse signal through to the AND outputs at X_A and X_B.

When the switch moves to position A, the latch is set to $Q = 1$. This enables the 1-kHz pulses to pass through to X_A, while the LOW at $\overline{Q}$ keeps $X_B = 0$. When the switch moves to position B, the latch is cleared to $Q = 0$, which keeps $X_A = 0$, while the HIGH at $\overline{Q}$ enables the pulses to pass through to X_B.

EXAMPLE 5-6

A technician tests the circuit of Figure 5-13 and records the following observations:

SWITCH POSITION	SET (Z1-1)	CLEAR (Z1-5)	Q (Z1-3)	$\overline{Q}$ (Z1-6)	X_A (Z1-11)	X_B (Z1-8)
A	LOW	HIGH	LOW	HIGH	LOW	PULSES
B	HIGH	LOW	LOW	HIGH	LOW	PULSES

He notices that when the switch is in position B the circuit functions correctly, but in position A the latch does not set to the $Q = 1$ state. What are the possible faults that could produce this malfunction?

SOLUTION
There are several possibilities:

1.   An internal open at Z1-1. This would prevent Q from responding to the SET input.

2.   An internal component failure in NAND gate Z1-3 that prevents it from responding properly.

3.   The Q output is stuck LOW. This could be caused by:
    (a) Z1-3 internally shorted to ground.
    (b) Z1-4 internally shorted to ground.
    (c) Z2-2 internally shorted to ground.
    (d) The Q node externally shorted to ground.
    An ohmmeter check from Q to ground will determine if any of these conditions are present. A visual check should reveal any external short.

What about $\overline{Q}$ being internally or externally shorted to V_{CC}? A little thought will lead to the conclusion that this could not be the fault. If $\overline{Q}$ was shorted to V_{CC}, this would not prevent the Q output from going HIGH when SET goes LOW. Since Q *does not* go HIGH, this cannot be the fault. The reason that $\overline{Q}$ looks like it is stuck HIGH is because Q is stuck LOW and that keeps $\overline{Q}$ HIGH.

5-4   CLOCK SIGNALS AND CLOCKED FLIP-FLOPS

Digital systems can operate either *asynchronously* or *synchronously*. In asynchronous systems, the outputs of logic circuits can change state any time one or more of the inputs change. An asynchronous system is difficult to design and troubleshoot.
    In synchronous systems, the exact times at which any output can change states

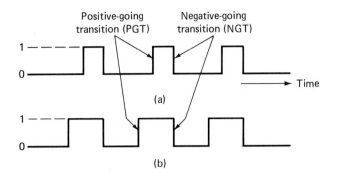

Positive-going transition (PGT)   Negative-going transition (NGT)

(a)

Time

(b)

FIGURE 5-14   Clock signals.

is determined by a signal commonly called the *clock*. This clock signal is generally a rectangular pulse train or squarewave as shown in Figure 5-14. The clock signal is distributed to all parts of the system, and most (if not all) of the system outputs can change state only when the clock makes a transition. The transitions (also called *edges*) are pointed out in Figure 5-14. When the clock changes from a 0 to a 1, this is called the *positive-going transition* (PGT); when the clock goes from 1 to 0, this is the *negative-going transition* (NGT). We will use the abbreviations PGT and NGT, since these terms appear so often throughout the text.

Most digital systems are principally synchronous (although there are always some asynchronous parts), since synchronous circuits are easier to design and troubleshoot. They are easier to troubleshoot because the circuit outputs can change only at specific instants of time. In other words, almost everything is synchronized to the clock-signal transitions.

The synchronizing action of the clock signals is accomplished through the use of *clocked flip-flops* that are designed to change states on one or the other of the clock's transitions.

Clocked Flip-Flops   There are several types of clocked FFs that are used in a wide range of applications. Before we begin our study of the different clocked FFs, we will describe the principal ideas that are common to all of them.

1.   Clocked FFs have a *clock* input that is typically labeled *CLK, CK,* or *CP*. We will use *CLK,* as shown in Figure 5-15. In most clocked FFs the *CLK* input is *edge-triggered,* which means that it is activated by a signal transition; this is indicated by the presence of a small triangle on the *CLK* input.

   In Figure 5-15(a) the *CLK* input is activated only when a positive-going transition (PGT) occurs; it is not affected at any other time. In Figure 5-15(b) the *CLK* input is activated only by a negative-going transition (NGT) as symbolized by the presence of the small bubble.

2.   Clocked FFs also have one or more *control* inputs that can have various names, depending on their operation. The control inputs will have no effect on *Q* until the active clock transition occurs. In other words, their effect is *synchronized* with the signal applied to *CLK*. For this reason they are called *synchronous* control inputs.

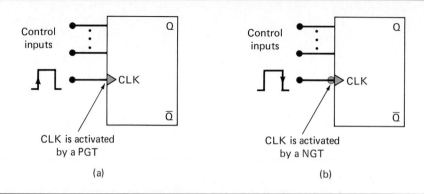

FIGURE 5-15 Clocked FFs have a clock input (CLK) that is active on either (a) PGT or (b) NGT. The control inputs determine the effect of the active clock transition.

For example, the control inputs of the FF in Figure 5-15(a) will have no effect on Q until the PGT of the clock signal occurs. Likewise, the control inputs in Figure 5-15(b) will have no effect until the NGT of the clock signal occurs.

3. In summary, we can say that the control inputs get the FF outputs ready to change, while the active transition at the *CLK* input actually *triggers* the change.

Setup and Hold Times   Two timing requirements must be met if a clocked FF is to respond reliably to its control inputs when the active *CLK* transition occurs. These requirements are illustrated in Figure 5-16 for a FF that triggers on a PGT.

The *setup time, t_s,* is the time interval immediately preceding the active transition of the *CLK* signal during which the synchronous input has to be maintained at the proper level. IC manufacturers usually specify the minimum allowable setup time. If this time requirement is not met, the FF may not respond reliably when the clock edge occurs.

The *hold time, t_H,* is the time interval immediately following the active transition of the *CLK* signal during which the synchronous input has to be maintained at the proper level. IC manufacturers usually specify the minimum acceptable value of hold time. If this requirement is not met, the FF will not trigger reliably.

Thus, to ensure that a clocked FF will respond properly when the active clock transition occurs, the synchronous inputs must be stable (unchanging) for a time interval equal to t_s *prior* to the clock transition, and for a time interval equal to t_H *after* the clock transition.

IC flip-flops will have t_S and t_H values in the nanosecond range. Setup times are usually in the range 5 to 50 ns while hold times are generally from 0 to 10 ns. Notice that these times are measured between the 50 percent points on the transitions.

These timing requirements are very important in synchronous systems, because, as we shall see, there will be many situations where the synchronous inputs to a FF are changing at approximately the same time as the *CLK* input.

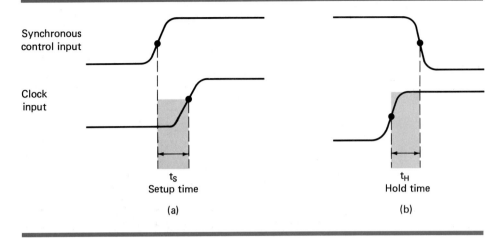

FIGURE 5-16 Control inputs have to be held stable for (a) a time, t_S, prior to active clock transition and for (b) a time, t_H, after the active clock transition.

REVIEW QUESTIONS

1. What two types of inputs does a clocked FF have?
2. What is meant by the term ''edge-triggered''?
3. *True or false:* The *CLK* input will affect the FF output only when the active transition of the control input occurs.
4. Define setup time and hold time for a clocked FF.

5-5 CLOCKED S-C FLIP-FLOP

Figure 5-17(a) shows the logic symbol for a *clocked S-C flip-flop* that is triggered by the positive-going edge of the clock signal. This means that the FF can change states *only* when a signal applied to its clock input makes a transition from 0 to 1. The *S* and *C* inputs control the state of the FF in the same manner as described earlier for the NOR-gate FF, but the FF does not respond to these inputs until the occurrence of the PGT of the clock signal.

The truth table in Figure 5-17(b) shows how the FF output will respond to the PGT at the *CLK* input for the various combinations of *S* and *C* inputs. This truth table uses some new nomenclature. The up arrow ($\uparrow$) indicates that a PGT is required at CLK; the label Q_0 indicates the level at Q prior to the PGT. This nomenclature is widely used by IC manufacturers on their IC data sheets.

The waveforms in Figure 5-17(c) illustrate the operation of the clocked S-C FF. If we assume that the setup and hold time requirements are being met in all cases, we can analyze these waveforms as follows:

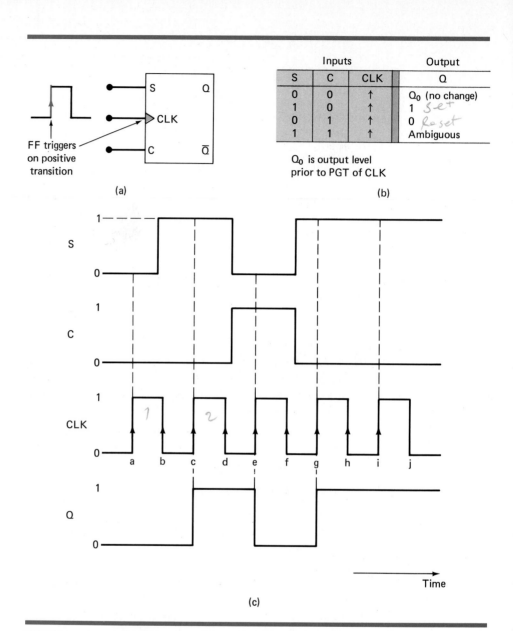

Inputs			Output
S	C	CLK	Q
0	0	↑	Q_0 (no change)
1	0	↑	1 Set
0	1	↑	0 Reset
1	1	↑	Ambiguous

FF triggers on positive transition

(a)

Q_0 is output level prior to PGT of CLK

(b)

(c)

FIGURE 5-17 (a) Clocked S-C FF that responds to the positive-going edge of clock pulse; (b) truth table; (c) typical waveforms.

1. Initially all inputs are 0 and the Q output is assumed to be 0.

2. When the PGT of the first clock pulse occurs (point a), the S and C inputs are both 0, so the FF is not affected and remains in the $Q = 0$ state (i.e., $Q = Q_0$).

3. At the occurrence of the PGT of the second clock pulse (point c), the S input is now high, with C still low. Thus, the FF sets to the 1 state at the rising edge of this clock pulse.

4. When the third clock pulse makes its positive transition (point *e*), it finds that $S = 0$ and $C = 1$, which causes the FF to clear to the 0 state.

5. The fourth pulse sets the FF once again to the $Q = 1$ state (point *g*) because $S = 1$ and $C = 0$ when the positive edge occurs.

6. The fifth pulse also finds that $S = 1$ and $C = 0$ when it makes its positive-going transition. However, Q is already high, so it remains in that state.

7. The $S = C = 1$ condition should not be used, because it results in an ambiguous condition.

It should be noted from these waveforms that the FF is not affected by the negative-going transitions of the clock pulses. Also, note that the *S* and *C* levels have no effect on the FF, except upon the occurrence of a positive-going transition of the clock signal. The *S* and *C* inputs are synchronous *control* inputs; they control which state the FF will go to when the clock pulse occurs; the *CLK* input is the *trigger* input that causes the FF to change states according to what the *S* and *C* inputs are when the active clock transition occurs.

Figure 5-18 shows the symbol for a clocked S-C flip-flop that triggers on the *negative*-going transition at its *CLK* input. The small circle and triangle on the *CLK* input indicates that this FF will trigger only when the *CLK* input goes from 1 to 0. This FF operates in the same manner as the positive-edge FF except that the output can change states only on the falling edge of the clock pulses (points *b*, *d*, *f*, *h*, and *j* in Figure 5-17). Both positive-edge and negative-edge triggering FFs are used in digital systems.

FIGURE 5-18 Clocked S-C FF that triggers on negative-going transitions.

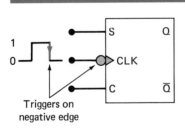

Triggers on
negative edge

Inputs			Output
S	C	CLK	Q
0	0	↓	Q_0 (no change)
1	0	↓	1
0	1	↓	0
1	1	↓	Ambiguous

Internal Circuitry of Edge-Triggered S-C Flip-Flop A detailed analysis of the internal circuitry of a clocked FF is not necessary since all types are readily available as ICs. Although our main interest is in the FF's external operation, our understanding of this external operation can be aided by taking a look at a simplified version of the FFs internal circuitry. Figure 5-19 shows this for an edge-triggered S-C FF.

The circuit contains three sections:

1. A basic NAND latch formed by NANDs 3 and 4

2. A *pulse-steering* circuit formed by NANDs 1 and 2

3. An *edge-detector* circuit

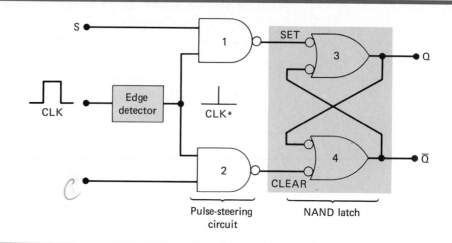

FIGURE 5-19 **Simplified version of the internal circuitry for an edge-triggered S-C FF.**

The edge-detector produces a narrow positive-going spike (*CLK*∗) that occurs coincident with the active transition of the *CLK* input pulse. The pulse-steering circuit "steers" the spike through to the SET or CLEAR input of the latch in accordance with the levels present at *S* and *C*. For example, with *S* = 1 and *C* = 0, the *CLK*∗ signal is inverted and passed through NAND-1 to produce a LOW pulse at the SET input of the latch that sets *Q* = 1. With *S* = 0, *C* = 1, the *CLK*∗ signal is inverted and passed through NAND-2 to produce a LOW pulse at the CLEAR input of the latch that resets *Q* = 0.

Figure 5-20(a) shows how the *CLK*∗ signal is generated for edge-triggered FFs that trigger on a PGT. The INVERTER produces a delay of a few nanoseconds

FIGURE 5-20 **Implementation of edge detector circuits used in edge-triggered FFs: (a) PGT; (b) NGT.**

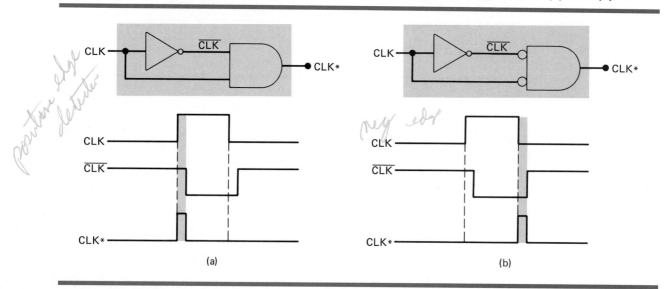

so that the transitions of $\overline{CLK}$ occur a little bit after those of CLK. The AND gate produces an output spike that is HIGH only for the few nanoseconds when CLK and $\overline{CLK}$ are both HIGH. The result is a narrow pulse at $CLK*$, which occurs on the PGT of CLK. The arrangement of Figure 5-20(b) likewise produces $CLK*$ on the NGT of CLK for FFs that are to trigger on a NGT.

Since the $CLK*$ signal is HIGH for only a few nanoseconds, Q is affected by the levels at S and C only for a short time during and after the occurrence of the active edge of CLK. This is what gives the FF its edge-triggered property.

REVIEW QUESTIONS

1. Suppose that the waveforms of Figure 5-17(c) are applied to the inputs of the FF of Figure 5-18. What will happen to Q at point b? (*Ans.* It will go HIGH)
2. Explain why the S and C inputs affect Q only during the active transition of CLK.

5-6 CLOCKED J-K FLIP-FLOP

Figure 5-21(a) shows a *clocked J-K flip-flop* that is triggered by the positive-going edge of the clock signal. The J and K inputs control the state of the FF in the same ways as the S and C inputs do for the clocked S-C FF except for one major difference: *the $J = K = 1$ condition does not result in an ambiguous output*. For this 1,1 condition, the FF will always go to its *opposite* state upon the positive transition of the clock signal. This is called the **toggle** mode of operation. In this mode, if both J and K are left HIGH, the FF will change states (toggle) for each clock pulse.

The truth table in Figure 5-21(a) summarizes how the *J-K* FF responds to the PGT for each combination of J and K. Notice that the truth table is the same as for the clocked S-C FF (Figure 5-17) except for the $J = K = 1$ condition. This condition results in $Q = \overline{Q}_0$, which means that the new value of Q will be the inverse of the value it had prior to the PGT; this is the toggle operation.

The operation of this FF is illustrated by the waveforms in Figure 5-21(b). Once again we assume that the set-up and hold time requirements are being met.

1. Initially all inputs are 0 and the Q output is assumed to be 1.
2. When the positive-going edge of the first clock pulse occurs (point a) the $J = 0$, $K = 1$ condition exists. Thus, the FF will be cleared to the $Q = 0$ state.
3. The second clock pulse finds $J = K = 1$ when it makes its positive transition (point c). This causes the FF to *toggle* to its opposite state, $Q = 1$.
4. At point e on the clock waveform, J and K are both 0, so the FF does not change states on this transition.

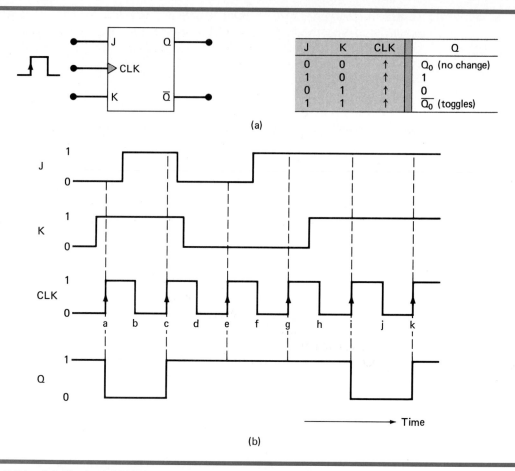

J	K	CLK	Q
0	0	↑	Q_0 (no change)
1	0	↑	1
0	1	↑	0
1	1	↑	$\overline{Q_0}$ (toggles)

(a)

(b)

FIGURE 5-21 (a) Clocked J-K FF that responds to positive edge of clock; (b) waveforms.

5. At point g, $J = 1$ and $K = 0$. This is the condition that sets Q to the 1 state. However, it is already 1, so it will remain there.

6. At point i, $J = K = 1$, so the FF toggles to its opposite state. The same thing occurs at point k.

It should also be noted from these waveforms that the FF is not affected by the negative-going edge of the clock pulses. Also, the J and K input levels have no effect except upon the occurrence of the PGT of the clock signal. The J and K inputs by themselves cannot cause the FF to change states.

Figure 5-22 shows the symbol for a clocked J-K flip-flop that triggers on the negative-going clock-signal transitions. The small circle on the *CLK* input indicates that this FF will trigger when the *CLK* input goes from 1 to 0. This FF operates in the same manner as the positive-edge FF of Figure 5-21 except that the output can change states only on negative-going clock-signal transitions (points b, d, f, h, and j). Both polarities of edge-triggered J-K FFs are in common usage.

The J-K FF is much more versatile than the S-C FF because it has no ambiguous states. The $J = K = 1$ condition, which produces the toggling operation,

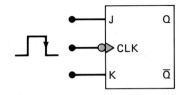

J	K	CLK	Q
0	0	↓	Q_0 (no change)
1	0	↓	1
0	1	↓	0
1	1	↓	$\overline{Q_0}$ (toggles)

FIGURE 5-22   J-K FF that triggers on negative-going transitions.

finds extensive use in all types of binary counters. In essence, the J-K FF can do anything the S-C FF can do *plus* operate in the toggle mode. Because of this, the J-K FF presently enjoys widespread use in almost all modern digital systems.

Internal Circuitry of Edge-Triggered J-K FF

A simplified version of the internal circuitry of an edge-triggered J-K FF is shown in Figure 5-23. It contains the same three sections as the edge-triggered S-C FF (Figure 5-19). In fact, the only difference between the two circuits is that the Q and $\overline{Q}$ outputs are fed back to the pulse-steering NAND gates. This feedback connection is what gives the J-K FF its toggle operation for the $J = K = 1$ condition.

Let's examine this toggle condition more closely by assuming that $J = K = 1$ and that Q is sitting in the LOW state when a *CLK* pulse occurs. With $Q = 0/\overline{Q} = 1$, NAND gate 1 will steer *CLK** (inverted) to the SET input of the NAND latch to produce $Q = 1$. If we assume that Q is HIGH when a *CLK* pulse occurs, NAND gate 2 will steer *CLK** (inverted) to the CLEAR input of the latch to produce $Q = 0$. Thus, Q always ends up in the opposite state.

In order for the toggle operation to work as described above, the *CLK** pulse

FIGURE 5-23   Internal circuit of edge-triggered J-K FF.

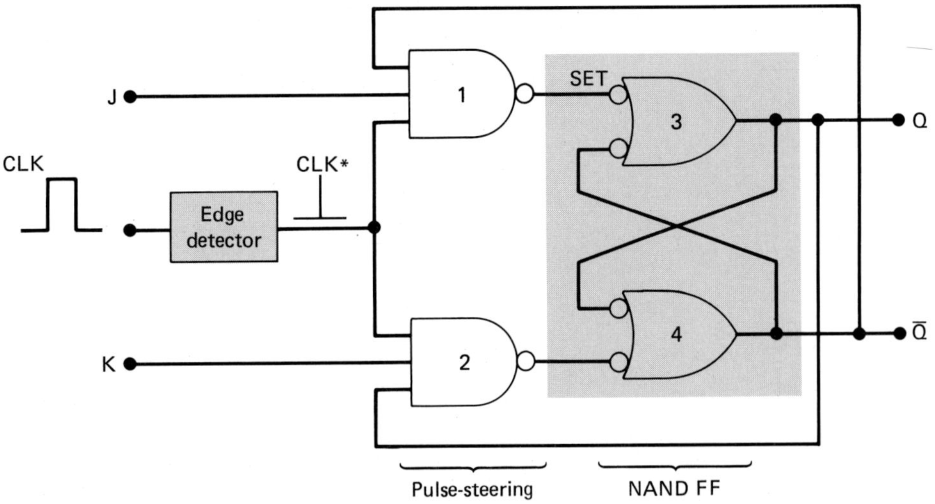

must be very narrow. It has to return to 0 before the Q and $\overline{Q}$ outputs toggle to their new values; otherwise the new values of Q and $\overline{Q}$ will cause the $CLK*$ pulse to toggle the latch outputs again.

REVIEW QUESTIONS

1. *True or false:* A J-K FF can be used as a S-C FF, but a S-C FF cannot be used as a J-K FF.
2. Does a J-K FF have any ambiguous input conditions?
3. What J-K input condition will always set Q upon the occurrence of the active *CLK* transition?

5-7   CLOCKED D FLIP-FLOP

Figure 5-24(a) shows the symbol and truth table for an edge-triggered D flip-flop that triggers on a PGT. Unlike the S-C and J-K FFs, this FF has only the one

FIGURE 5-24   (a) D FF that triggers on positive-going transitions; (b) waveforms.

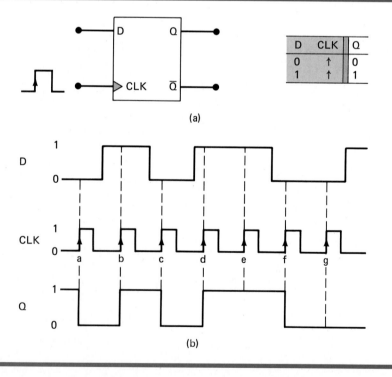

D	CLK	Q
0	↑	0
1	↑	1

(a)

(b)

synchronous control input, *D*, which stands for *data*. The operation of the D FF is very simple; *Q* will go to the same state that is present on the *D* input when a PGT occurs at CLK. In other words, the level present at *D* will be *stored* in the FF at the instant the PGT occurs. The waveforms in Figure 5-24(b) illustrate this operation.

Assume that *Q* is initially HIGH. When the first PGT occurs at point *a*, the *D* input is LOW; thus, *Q* will go to the 0 state. Even though the *D* input level changes between points *a* and *b*, it has no effect on *Q*; *Q* is storing the LOW that was on *D* at point *a*. When the PGT at *b* occurs, *Q* goes HIGH since *D* is HIGH at that time. *Q* stores this HIGH until the PGT at point *c* causes *Q* to go LOW since *D* is LOW at that time. In a similar manner, the *Q* output takes on the levels present at *D* when the PGTs occur at points *d*, *e*, *f*, and *g*. Note that *Q* stays HIGH at point *e* because *D* is still HIGH.

Again, it is important to remember that *Q* can change only when a PGT occurs. The *D* input has no effect between PGTs.

A negative-edge-triggered D FF operates in the same way just described except that *Q* will take on the value of *D* when a NGT occurs at *CLK*. The symbol for the D FF that triggers on NGTs will have a bubble on the *CLK* input.

Implementation of D FF   An edge-triggered D FF is easily implemented by adding a single INVERTER to the edge-triggered S-C FF as shown in Figure 5-25. If you try both values of *D*, you should see that *Q* takes on the level present at *D* when a PGT occurs.

FIGURE 5-25   Edge-triggered D FF implementation from S-C FF.

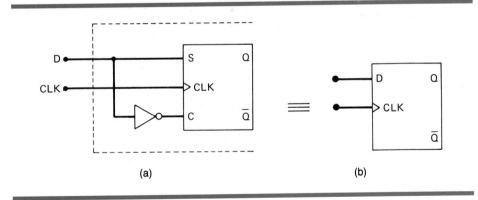

(a)   (b)

EXAMPLE 5-7

How can a J-K FF be modified to operate as a D FF?

SOLUTION
The modification, shown in Figure 5-26, is the same as was done for the S-C FF in Figure 5-25.

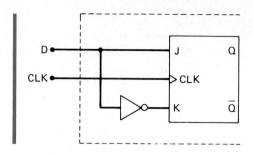

FIGURE 5-26 D FF implementation from a J-K FF.

Parallel Data Transfer   At this point you may well be wondering about the usefulness of the D FF, since it appears that the Q output is the same as the D input. Not quite; remember, Q takes on the value of D only at certain time instances, and so it is not identical to D (e.g., see the waveforms in Figure 5-24).

In most applications of the D FF, the Q output must take on the value at its D input only at precisely defined times. One example of this is illustrated in Figure 5-27. Logic-circuit outputs X, Y, Z are to be transferred to FFs Q_1, Q_2, and Q_3 for storage. Using the D FFs, the levels present at X, Y, and Z will be transferred to Q_1, Q_2, and Q_3 respectively, upon application of a TRANSFER pulse to the common CLK inputs. The FFs can store these values for subsequent processing. This is an example of *parallel* transfer of binary data; the three bits X, Y, and Z are all transferred *simultaneously*.

FIGURE 5-27   Parallel transfer of binary data using D FFs.

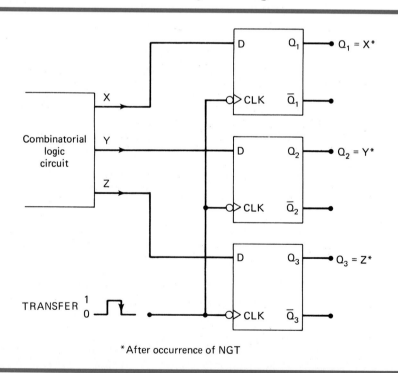

*After occurrence of NGT

ll happen to the Q waveform in Figure 5-24(b) if the D input is
anently LOW? (*Ans.* Q will go LOW and stay there)
false: The Q output will equal the level at the D input at all

red D FF uses an edge-detector circuit to ensure that the output will
D input *only* when the active transition of the clock occurs. If this
not used, the resultant circuit operates somewhat differently. It is
and has the arrangement shown in Figure 5-28(a).
it contains the NAND latch and the steering NAND gates 1 and 2.
put to the steering gates is called an *enable* input (abbreviated *EN*)
ock input because its effect on the Q and $\overline{Q}$ outputs is not restricted

b) truth table; (c) logic symbol.

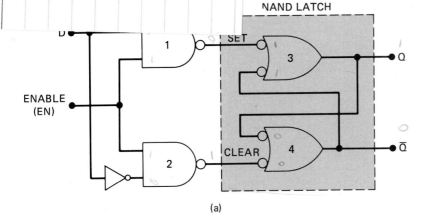

NAND LATCH

Inputs		Output
EN	D	Q
0	X	Q_0 (no change)
1	0	0
1	1	1

"X" indicates "don't care"
Q_0 is state Q just
prior to EN going LOW

(a)

(b)

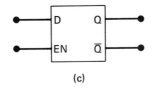

(c)

to occurring only on its transitions. The operation of the D latch is described as follows:

1. When *EN* is HIGH, the *D* input will produce a LOW at either the SET or CLEAR inputs of the NAND latch to cause *Q* to become the same level as *D*. If *D* changes while *EN* is HIGH, *Q* will follow the changes exactly. In other words, while *EN* = 1, the *Q* output will look exactly like *D*; in this mode, the *D* latch is said to be "transparent."

2. When *EN* goes LOW, the *D* input is inhibited from affecting the NAND latch since the outputs of both steering gates will be held HIGH. Thus, the *Q* and $\overline{Q}$ outputs will stay at whatever level they had just before *EN* went LOW. In other words, the outputs are "latched" to their current level and cannot change while *EN* is LOW even if *D* changes.

This operation is summarized in the truth table in Figure 5-28(b). The logic symbol for the *D* latch is given in Figure 5-28(c). Note that even though the *EN* input operates similar to the *CLK* input of an edge-triggered FF, there is no small triangle on the *EN* input. This is because the small triangle symbol is used strictly for inputs that can cause an output change only when a transition occurs.

EXAMPLE 5-8

Determine the *Q* waveform for a *D* latch with the *EN* and *D* inputs of Figure 5-29. Assume that *Q* = 0 initially.

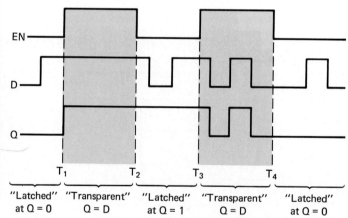

FIGURE 5-29 Waveforms for Example 5-8 showing the two modes for the *D* latch.

SOLUTION
Prior to time T_1, *EN* is LOW, so that *Q* is "latched" at the 0 level and cannot change even though *D* is changing. During the T_1–T_2 interval, *EN* is HIGH so that *Q* will take on the same level present at *D*. Thus, *Q* goes HIGH at T_1 and stays there since *D* is not changing. When *EN* returns LOW at T_2, *Q* will latch at the HIGH level that it has at T_2 and remain there while *EN* is LOW.

At T_3 when EN goes HIGH again, Q will follow the changes in the D input until T_4 when EN returns LOW. During the T_3–T_4 interval, the D latch is "transparent" since the variations in D go through to the output Q. At T_4 when EN goes LOW, Q will latch at the 0 level since that is its level at T_4. After T_4 the variations in D will have no effect on Q since it is latched (i.e., $EN = 0$).

REVIEW QUESTIONS

1. Describe how a D latch operates different from an edge-triggered D FF.
2. *True or false:* A D latch is in its transparent mode when $EN = 0$.
3. *True or false:* In a D latch, the D input can affect Q only when $EN = 1$.

5-9 ASYNCHRONOUS INPUTS

For the clocked flip-flops that we have been studying, the S, C, J, K, and D inputs have been referred to as *control* inputs. These inputs are also called *synchronous inputs,* because their effect on the FF output is synchronized with the *CLK* input. As we have seen, the synchronous control inputs must be used in conjunction with a clock signal to trigger the FF.

Most clocked FFs also have one or more *asynchronous inputs* which operate independently of the synchronous inputs and clock input. These asynchronous inputs can be used to set the FF to the 1 state or clear the FF to the 0 state at any time, regardless of the conditions at the other inputs. Stated in another way, the asynchronous inputs are *override* inputs, which can be used to override all the other inputs in order to place the FF in one state or the other.

Figure 5-30 shows a J-K FF with two asynchronous inputs designated as DC SET and DC CLEAR. These are active-LOW inputs, as indicated by the bubbles on the FF symbol. The accompanying truth table summarizes how they affect the FF output. Let's examine the various cases.

- **DC SET = DC CLEAR = 1.** The asynchronous inputs are inactive and the FF is free to respond to the J, K, and *CLK* inputs; in other words, the clocked operation can take place.
- **DC SET = 0; DC CLEAR = 1.** The DC SET is activated and Q is *immediately* set to 1 no matter what conditions are present at the J, K, and *CLK* inputs. The *CLK* input cannot affect the FF while DC SET = 0.
- **DC SET = 1; DC CLEAR = 0.** The DC CLEAR is activated and Q is *immediately* cleared to 0 independent of the conditions on the J, K, or *CLK* inputs. The *CLK* input has no effect while DC CLEAR = 0.
- **DC SET = DC CLEAR = 0.** This condition should not be used, since it can result in an ambiguous response.

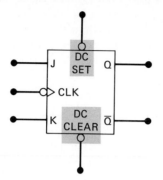

DC SET	DC CLEAR	FF response
1	1	Clocked operation*
0	1	$Q = 1$
1	0	$Q = 0$
0	0	Not used

*Q will respond to J, K, and CLK

FIGURE 5-30   Clocked J-K FF with asynchronous inputs.

It is important to realize that these asynchronous inputs respond to dc levels. This means that if a constant 0 is held on the DC SET input, the FF will remain in the $Q = 1$ state regardless of what is occurring at the other inputs. Similarly, a constant low on the DC CLEAR input holds the FF in the $Q = 0$ state. Thus, the asynchronous inputs can be used to hold the FF in a particular state for any desired interval. Most often, however, the asynchronous inputs are used to set or clear the FF to the desired state by application of a momentary pulse.

Many clocked FFs that are available as ICs will have both of these asynchronous inputs; some will have only the DC CLEAR input. Some FFs will have asynchronous inputs that are active-HIGH rather than active-LOW. For these FFs the FF symbol would not have a bubble on the asynchronous inputs.

Designations for Asynchronous Inputs   IC manufacturers have not agreed on what nomenclature to use for these asynchronous inputs. We have called them DC SET and DC CLEAR to emphasize their response to dc levels. The most commonly used designations are given below.

ASYNCHRONOUS **SET** INPUT	ASYNCHRONOUS **CLEAR** INPUT
DC SET	DC CLEAR
PRESET (*PRE*)	CLEAR (*CLR*)
SET	RESET
S_D (direct set)	C_D (direct clear)

From now on, we will use the labels *PRE* and *CLR* to represent the asynchronous inputs, since these seem to be the most commonly used labels. When these asynchronous inputs are active-LOW, as they generally are, we will use the overbar to indicate their active-LOW status, that is, $\overline{PRE}$ and $\overline{CLR}$.

Although most IC FFs have at least one or more asynchronous inputs, there are some circuit applications where they are not used. In such cases they are held permanently at their inactive level. Often, in our use of FFs throughout the remainder of the text, we will not show a FF's unused asynchronous inputs; it will be assumed that they are permanently connected to their inactive logic level.

EXAMPLE 5-9

Figure 5-31(a) shows an edge-triggered J-K FF with active-LOW asynchronous inputs. The J and K inputs are permanently held HIGH. Determine the Q output response to the input waveforms at CLK, $\overline{PRE}$, and $\overline{CLR}$. Assume that $Q = 1$ initially.

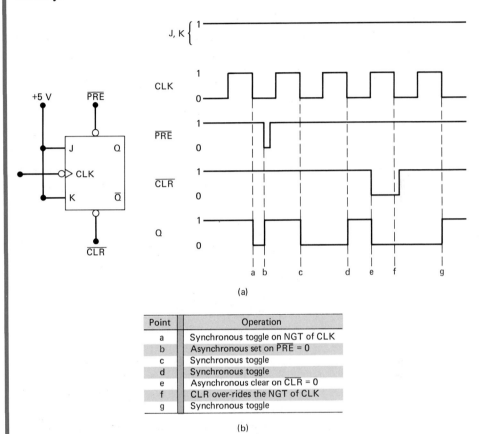

(a)

Point	Operation
a	Synchronous toggle on NGT of CLK
b	Asynchronous set on $\overline{PRE}$ = 0
c	Synchronous toggle
d	Synchronous toggle
e	Asynchronous clear on $\overline{CLR}$ = 0
f	CLR over-rides the NGT of CLK
g	Synchronous toggle

(b)

FIGURE 5-31   Waveforms for Example 5-9 showing how a clocked FF responds to asynchronous inputs.

SOLUTION

Initially, $\overline{PRE}$ and $\overline{CLR}$ are in their inactive HIGH state, so they will have no effect on Q. Thus, when the first NGT of the CLK signal occurs at point a, Q will toggle to its opposite state; remember, $J = K = 1$ produces the toggle operation.

At point b, the $\overline{PRE}$ input is pulsed to its active-LOW state. This will *immediately* set $Q = 1$. Note that $\overline{PRE}$ produces $Q = 1$ without waiting for a NGT at CLK. The asynchronous inputs operate independent of CLK.

At point c, the NGT of CLK will again cause Q to toggle to its opposite state. Note that $\overline{PRE}$ has returned to its inactive state prior to point c. Likewise, the NGT of CLK at point d will toggle Q back HIGH.

At point e, the $\overline{CLR}$ input is pulsed to its active-LOW state and will *immediately* clear $Q = 0$. Again, it does this independent of CLK.

The NGT of *CLK* at point *f will not toggle Q* because the $\overline{CLR}$ input is still active. The LOW at $\overline{CLR}$ overrides the *CLK* input and holds $Q = 0$.

When the NGT of *CLK* occurs at point *g*, it will toggle *Q* to the HIGH state since neither asynchronous input is active at that point.

These steps are summarized in Figure 5-31(b).

REVIEW QUESTIONS

1. How does the operation of an asynchronous input differ from that of a synchronous input?
2. Can a D FF respond to its *D* and *CLK* inputs while $\overline{PRE} = 1$? (*Ans.* Yes, since $\overline{PRE}$ is active-LOW)
3. List the conditions necessary for a positive-edge-triggered J-K FF with active-LOW asynchronous inputs to toggle to its opposite state. (*Ans. J = K = 1*, a PGT at *CLK*, and $\overline{PRE} = \overline{CLR} = 1$.

5-10   IEEE/ANSI SYMBOLS

We have been using the traditional symbols for each of the latches and FFs that have been introduced thus far, and we will continue to use these symbols in most of our circuit diagrams. In this section we will examine the new IEEE/ANSI symbols for these same devices so that you can become familiar with them.

Figure 5-32(a) shows the logic symbol for a single *D* latch. This is the new IEEE/ANSI symbol. It uses the letter "C" to denote the ENABLE input. As we shall see, the IEEE/ANSI symbology uses "C" for any input that *controls* when other inputs will have an effect on the output. As we know, the logic level applied to the ENABLE input controls when the *D* input is allowed to affect *Q* and $\overline{Q}$. Note that the outputs *Q* and $\overline{Q}$ are labeled outside the block, and note the right triangle on $\overline{Q}$ to indicate that it is the inverted output. This is standard for IEEE/ANSI. Recall that this right triangle is like the small bubble used in the older symbols.

Figure 5-32(b) shows the IEEE/ANSI symbol for a specific IC: the standard TTL 7475 quad latch. This IC contains four *D* latches that operate individually in the manner described earlier. This symbol also applies to the corresponding ICs in the other TTL and CMOS series: for example, the 74LS75, 74L75, 74C75, and 74HC75.

If we examine the logic symbol for the 7475 IC, several points should be noted. First, we can see that the overall symbol outline contains four smaller rectangles that represent the individual latches. Note how the inputs and outputs to each latch are labeled. For example, the *D* input to the top latch is labeled "1D," its enable input is labeled "C1," and its outputs are *1Q* and *1$\overline{Q}$*. Finally, note that the top two latches have a common enable input; that is, *C*1 and *C*2 are connected together internally and brought out to a single pin on the IC package.

Likewise, the bottom two latches share a common enable input. Figure 5-33(a) shows the IEEE/ANSI logic symbol for a negative-edge-triggered J-K FF with

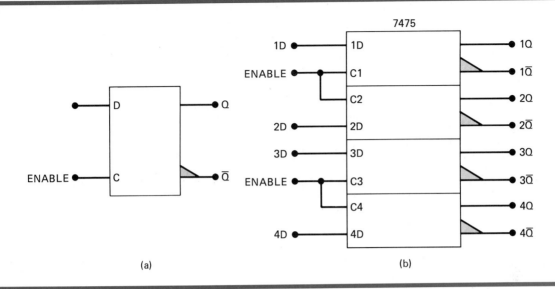

(a)

(b)

FIGURE 5-32 IEEE/ANSI symbols for (a) a single *D* latch and (b) an actual IC latch (7475 quad latch).

asynchronous inputs. The clock input is labeled "C" inside the symbol's rectangular outline. Note that there are two triangles on the clock input: the one inside the rectangle indicates that this input is edge-triggered; the one outside the rectangle indicates that it triggers on NGTs. The $\overline{PRE}$ and $\overline{CLR}$ inputs are active-LOW as

FIGURE 5-33 IEEE/ANSI symbols for (a) a single edge-triggered J-K FF and (b) an actual IC (74LS112 dual negative-edge-triggered J-K FF).

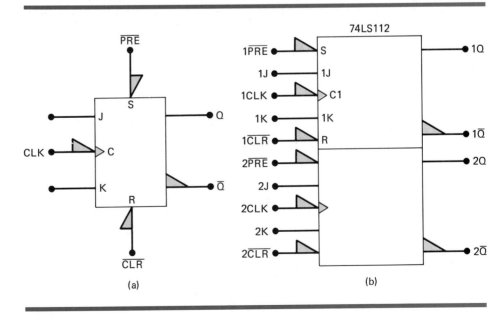

(a)

(b)

symbolized by the right triangles. Interestingly, the IEEE/ANSI standard uses the labels "S" and "R" inside the rectangle to indicate asynchronous SET and RESET, which are equivalent to PRESET and CLEAR, respectively.

Figure 5-33(b) shows the IEEE/ANSI logic symbol for an IC that is part of the 74LS series of TTL devices. The 74LS112 is a dual negative-edge-triggered J-K FF with preset and clear. It contains two J-K FFs, like the one symbolized in Figure 5-33(a). Note how the inputs and outputs are numbered. Also note that the input labels inside the rectangles are shown only for the top FF. It is understood that the inputs to the bottom FF are in the same arrangement as the top one. This same IC symbol applies to the 74S112 and 74HC112.

Figure 5-34(a) is the IEEE/ANSI symbol for a positive-edge-triggered D FF with asynchronous inputs. There is no right triangle on the clock input since this FF is clocked by PGTs.

Figure 5-34(b) is the IEEE/ANSI symbol for a 74175 IC that contains four D FFs that share a common *CLK* input and common $\overline{CLR}$ input. The FFs do not have a $\overline{PRE}$ input. This symbol contains a separate rectangle to represent each FF, and a special *common-control block* which is the notched rectangle on top. The common-control block is used whenever an IC has one or more inputs that are common to more than one of the circuits on the chip. For the 74175, the *CLK* and $\overline{CLR}$ inputs are common to all four of the D FFs on the IC. This means that a PGT on *CLK* will cause each of the *Q* outputs to take on the level present at its *D* input; it also means that a LOW on $\overline{CLR}$ will clear all *Q* outputs to the LOW state.

FIGURE 5-34 **IEEE/ANSI symbols for (a) a single edge-triggered D FF and (b) an actual IC (74175 quad D FF with common clock and clear).**

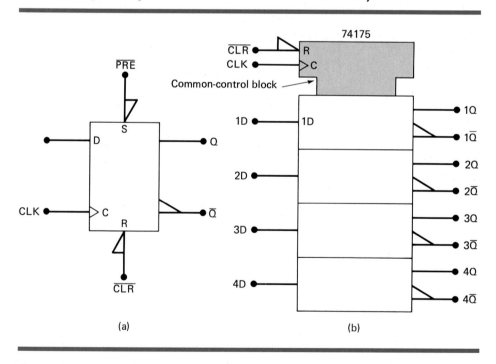

REVIEW QUESTIONS

1. Explain the meaning of the two different triangles that can be part of the IEEE/ANSI symbology at a clock input.
2. Describe the meaning of the common-control block.

5-11 FLIP-FLOP TIMING CONSIDERATIONS

Manufacturers of IC FFs will specify several important timing parameters and characteristics that must be considered before a FF is used in any circuit application. We will describe the most important of these and then give some actual examples of specific IC FFs from the TTL and CMOS logic families.

Setup and Hold Times   These have already been discussed, and you may recall from Section 5-4 that they represent requirements that must be met for reliable FF triggering. The manufacturer's IC data sheet will always specify the *minimum* values of t_S and t_H.

Propagation Delays   Whenever a signal is to change the state of a FF's output, there is a delay from the time the signal is applied to the time when the output makes its change. Figure 5-35 illustrates the propagation delays that occur in response to a positive transition on the *CLK* input. Note that these delays are measured between the 50 percent points on the input and output waveforms. The same types of delays occur in response to signals on a FF's asynchronous inputs (PRESET and CLEAR). The manufacturers' data sheets usually specify propagation delays in

FIGURE 5-35   **FF propagation delays.**

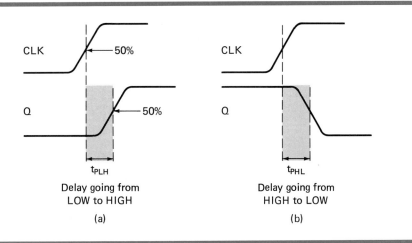

Delay going from
LOW to HIGH

(a)

Delay going from
HIGH to LOW

(b)

response to all inputs, and they usually specify the *maximum* values for t_{PLH} and t_{PHL}.

Modern IC flip-flops have propagation delays that range from a few nanoseconds to around 1 μs: The values of t_{PLH} and t_{PHL} are generally not the same and they increase in direct proportion to the number of loads being driven by the Q output. FF propagation delays play an important part in certain situations which we will encounter later.

Maximum Clocking Frequency, f_{MAX}

This is the highest frequency that may be applied to the *CLK* input of a FF and still have it trigger reliably. The f_{MAX} limit will vary from FF to FF even with FFs having the same device number. For example, the manufacturer of the 7470 J-K FF chip tests many of these FFs and may find that the f_{MAX} values fall in the range 20 to 35 MHz. He will then specify the *minimum* f_{MAX} as 20 MHz. This may seem confusing, but a little thought should make it clear that what the manufacturer is saying is that he cannot guarantee that the 7470 FF that you put in your circuit will work above 20 MHz; many of them will, but some of them will not. If you stay below 20 MHz, however, he guarantees that they will all work.

Clock Pulse HIGH and LOW Times

The manufacturer will also specify the *minimum* time duration that the *CLK* signal must remain LOW before it goes HIGH, sometimes called $t_w(L)$, and the *minimum* time that *CLK* must be kept HIGH before it returns LOW, sometimes called $t_w(H)$. These times are defined in Figure 5-36. Failure to meet these minimum time requirements can result in unreliable triggering.

Asynchronous Active Pulse Width

The manufacturer will also specify the *minimum* time duration that a PRESET or CLEAR input has to be kept in its active state in order to reliably set or clear the FF.

Clock Transition Times

For reliable triggering, the clock waveform transition times (rise and fall times) should be kept very short. If the clock signal takes too long to make the transitions from one level to the other, the FF may trigger erratically or not at all. Manufacturers usually do not list a maximum transition-time requirement for each FF IC. Instead, it is usually given as a general requirement for all ICs within a given logic family. For example, the transition times should generally be ≤ 50 ns for TTL devices and ≤ 200 ns for CMOS. These requirements will vary among the different manufacturers and among the various subfamilies within the broad TTL and CMOS logic families.

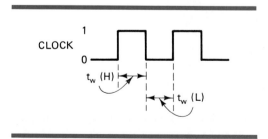

FIGURE 5-36 Definition of clock LOW and clock HIGH times.

Actual ICs As practical examples of these timing parameters, let's take a look at several actual IC FFs. In particular, we will look at the following ICs:*

- **7474**   Dual edge-triggered D FF (TTL family)
- **74LS112** Dual edge-triggered J-K FF (TTL)
- **4013B**   Dual edge-triggered D FF (CMOS family)
- **74HC112** Dual edge-triggered J-K FF (CMOS)

Table 5-1 lists the various timing values for each of these FFs as they appear in the manufacturers' data books. All the listed values are *minimum* values except for the propagation delays, which are *maximum* values.

TABLE 5-1   Flip-Flop Timing Values (in nanoseconds)

	TTL		CMOS	
	7474	**74LS112**	**4013B**	**74HC112**
t_S	20	20	60	25
t_H	5	0	0	0
t_{PHL}—from *CLK* to Q	40	24	200	31
t_{PLH}—from *CLK* to Q	25	16	200	31
t_{PHL}—from $\overline{CLR}$ to Q	40	24	225	41
t_{PLH}—from $\overline{PRE}$ to Q	25	16	225	41
$t_W(L)$—*CLK* LOW time	37	15	100	25
$t_W(H)$—*CLK* HIGH time	30	20	100	25
$t_W(L)$—at $\overline{PRE}$ or $\overline{CLR}$	30	15	60	25
f_{MAX}—in MHz	15	30	5	20

Examination of this table reveals two interesting points.

1. All the FFs have very low values of hold time; this is typical of most modern edge-triggered FFs.
2. The 74HC series of CMOS devices has timing values that are comparable to the TTL devices. The 4000 series is much slower than the 74HC series.

*We have not included an edge-triggered S-C FF IC in this list because there are none in the TTL or CMOS families. Edge-triggered S-C FFs are used in other ICs such as shift registers.

EXAMPLE 5-10

From Table 5-1 determine the following.

(a) Assume $Q = 0$. How long can it take for Q to go HIGH when a PGT occurs at the CLK input of a 7474?

(b) Assume $Q = 1$. How long can it take for Q to go LOW in response to the $\overline{CLR}$ input of a 74HC112?

(c) What is the narrowest pulse that should be applied to the DC CLEAR input of the 74LS112 FF to reliably clear Q?

(d) Which FF in Table 5-1 requires that the control inputs remain stable *after* the occurrence of the active clock transition?

SOLUTION

(a) The PGT will cause Q to go from LOW to HIGH. The delay from *CLK* to Q is listed as $t_{PLH} = 25$ ns for the 7474.

(b) For the 74HC112 the time required for Q to go from HIGH to LOW in response to the $\overline{CLR}$ input is listed at $t_{PHL} = 41$ ns.

(c) For the 74LS112 the narrowest pulse at the DC CLEAR input is listed as $t_W(L) = 15$ ns.

(d) The 7474 is the only FF in Table 5-1 that has a nonzero hold-time requirement.

REVIEW QUESTIONS

1. Which FF timing parameters indicate the time it takes the Q output to respond to an input?
2. *True or false:* A FF that has an f_{max} rating of 25 MHz can be reliably triggered by any *CLK* pulse waveform with a frequency below 25 MHz. (*Ans.* False; the waveform must also satisfy the $t_W(H)$ and $t_W(L)$ requirements)

5-12 POTENTIAL TIMING PROBLEM IN FF CIRCUITS

In many digital circuits, the output of one FF is connected either directly or through logic gates to the input of another FF, and both FFs are triggered by the same clock signal. This presents a potential timing problem. A typical situation is illustrated in Figure 5-37, where the output of Q_1 is connected to the J input of Q_2, and both FFs are clocked by the same signal at their *CLK* inputs.

The potential timing problem is this: since Q_1 will change on the NGT of the clock pulse, the J_2 input of Q_2 will be changing as it receives the same NGT. This could lead to an unpredictable response at Q_2.

Let's assume that initially $Q_1 = 1$ and $Q_2 = 0$. Thus, the Q_1 FF has $J_1 = K_1 = 1$, and Q_2 has $J_2 = Q_1 = 1$, $K_2 = 0$ prior to the NGT of the clock pulse. When the NGT occurs, Q_1 will toggle to the LOW state, but it will not actually go LOW until after its propagation delay, t_{PHL}. The same NGT will reliably clock Q_2 to the HIGH state provided that t_{PHL} is greater than Q_2's hold-time requirement, t_H. If this condition is not met, the response of Q_2 will be unpredictable.

Fortunately, all modern edge-triggered FFs have hold-time requirements that are 5 ns or less; most have $t_H = 0$, which means that they have no hold-time requirement. For these FFs, situations like that in Figure 5-37 will not be a problem.

Unless stated otherwise, in all of the FF circuits that we encounter throughout the text, we will assume that the FF's hold-time requirement is short enough to respond reliably according to the following rule:

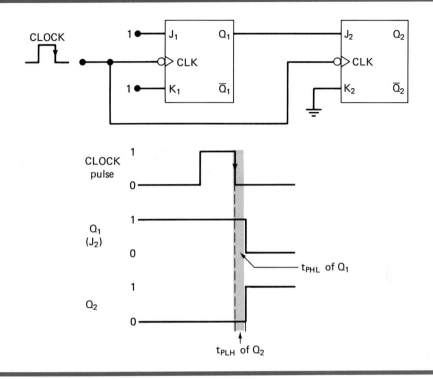

FIGURE 5-37 Q_2 will properly respond to the level present at Q_1 prior to the NGT of CLK, provided that Q_2's hold-time requirement, t_H, is less than Q_1's propagation delay.

> *The FF output will go to a state determined by the logic levels present at its synchronous control inputs just prior to the active clock transition.*

If we apply this rule to Figure 5-37, it says that Q_2 will go to a state determined by the $J_2 = 1$, $K_2 = 0$ condition that is present just prior to the NGT of the clock pulse. The fact that J_2 is changing in response to the same NGT has no effect.

EXAMPLE 5-11

Determine the Q output for a negative-edge-triggered J-K FF for the input waveforms shown in Figure 5-38. Assume that $t_H = 0$ and that $Q = 0$ initially.

SOLUTION
The FF will respond only at times T_2, T_4, T_6, and T_8. At T_2, Q will respond to the $J = K = 0$ condition present just prior to T_2. At T_4, Q will respond to the $J = 1$, $K = 0$ condition present just prior to T_4. At T_6, Q will respond to the $J = 0$, $K = 1$ condition present just prior to T_6. At T_8, Q responds to $J = K = 1$.

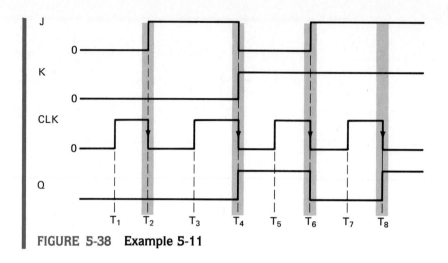

FIGURE 5-38　Example 5-11

5-13 MASTER/SLAVE FLIP-FLOPS

Before the development of edge-triggered FFs with small t_H values, timing problems such as that in Figure 5-37 were often handled by using a class of FFs called *master/slave* FFs (hereafter abbreviated M/S). They are so-called because of their internal structure, which actually contains two FFs—a master and a slave. All three types of edge-triggered FFs (S-C, J-K, and D) can also be implemented as M/S FFs. The importance of M/S FFs has seriously diminished because of the improved performance of the newer edge-triggered FFs. In fact, the newest IC technologies (74AS, 74ALS, 74HC, and 74HCT) do not include any M/S FF ICs in their series. Nevertheless, we will examine the M/S principle of operation since M/S FFs are still used in older equipment.

J-K Master/Slave Flip-Flop　Figure 5-39(a) shows a simplified circuit for a J-K master/slave (hereafter abbreviated M/S) flip-flop. It contains two almost identical stages. The first stage consists of a master that is a NAND latch with active-LOW inputs, and pulse-steering NAND gates 1 and 2. The second stage consists of a slave that is a NAND latch, and pulse-steering NAND gates 3 and 4. The Q_S and $\overline{Q}_S$ outputs of the slave FF serve as the overall FF outputs.

　　The *CLK* input is directly connected to NAND gates 1 and 2. When *CLK* = 0, there will be 1s at the S and C inputs of the master FF so that Q_M and $\overline{Q}_M$ cannot change states.

　　When *CLK* = 1, NAND gates 1 and 2 will be free to respond to *J*, *K*, and to the current slave FF outputs Q_S and $\overline{Q}_S$. For example, if $Q_S = 0/\overline{Q}_S = 1$ and $J = K = 1$, then when *CLK* goes HIGH, NAND 1 will go LOW and will set the master ($Q_M = 1$). Thus, *the master can change states only while CLK = 1.*

　　The *CLK* is inverted and fed to NAND gates 3 and 4. When *CLK* = 1, there will be 1s at the S and C inputs of the slave so that Q_S and $\overline{Q}_S$ cannot change.

　　When CLK = 0, NAND gates 3 and 4 will respond to the master outputs and will set or clear the slave accordingly. For example, if the master had been previously set to the $Q_M = 1$ state, then when $\overline{CLK}$ goes HIGH, the LOW at the NAND-

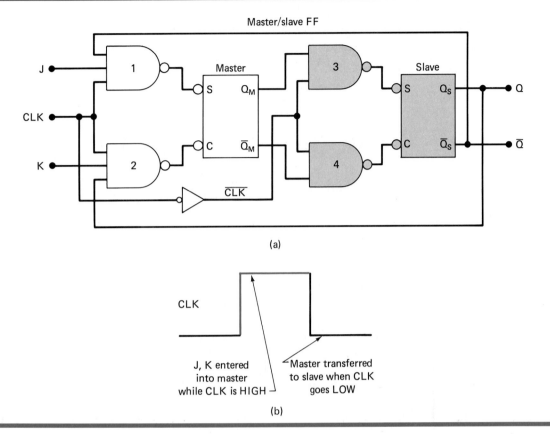

CLK

J, K entered
into master
while CLK is HIGH

Master transferred
to slave when CLK
goes LOW

(b)

FIGURE 5-39 **(a) Internal circuit for a M/S J-K FF; (b) two-step operation.**

3 output will set $Q_S = 1$. Conversely, if $Q_M = 0$ when $\overline{CLK}$ goes HIGH, a LOW will occur at the NAND-4 output to clear $Q_S = 0$. Thus, *the slave can change states only when CLK goes LOW, and Q_S takes on the current value of Q_M.*

This operation can be summarized in two basic steps:

1. **When *CLK* = 1:** The *J*, *K* levels are transferred into the master, which will go to a state determined by *J, K,* and the current Q_S. The slave cannot change states while *CLK* = 1.

2. **When *CLK* = 0:** The outputs of the master are transferred to the slave. The master cannot change states while *CLK* = 0.

LOGIC SYMBOLS FOR M/S FFs   It is important to understand that the slave can change states only when *CLK* goes LOW. As such, the external FF output *Q* can be considered to change states on the NGT of *CLK*. However, since *J* and *K* can affect the master while *CLK* is HIGH, the M/S FF is not really edge-triggered. In fact, it is often referred to as a *pulse-triggered FF* since it is affected by both states of the clock pulse. For this reason the logic symbol for the J-K M/S FF in Figure 5-40 does not use the edge-triggered indicator (▷) on the *CLK* input.

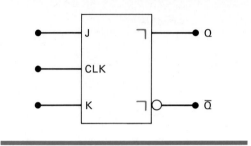

FIGURE 5-40 **Symbol for J-K master/slave FF. Note that there is no edge-triggered indicator (▷) inside the rectangle. Also note the postponed output indicator (⌐).**

The new IEEE/ANSI standard uses a special indicator as part of its logic symbol for a M/S FF. In Figure 5-40 the *postponed output* indicator (⌐) is used to represent the fact that the effects of the *J* and *K* inputs on the master do not reach the Q and $\overline{Q}$ outputs until *CLK* goes LOW. We will use this special indicator on a FF symbol whenever it is a M/S.

Advantage of Master/Slave FFs The master essentially stores the levels that are on J and K while *CLK* is HIGH, so that it is not necessary for J and K to remain stable once *CLK* goes LOW. In other words, *there is no hold-time requirement*. The master will transfer the correct data to the slave when *CLK* goes LOW even if J and K are changing at the same time. This allows us to use M/S FFs in situations like Figure 5-38, where the *J, K* inputs are changed by the same signal that is applied to the *CLK* input.

Limitation of M/S FF An *edge-triggered* FF is sensitive only to what the *J* and *K* inputs are at the time of the active clock transition. At all other times the *J* and *K* inputs are *disabled* (have no effect). In the M/S FF, the *J* and *K* inputs can affect the FF at any time while *CLK* = 1. This characteristic can cause problems if the *J, K* inputs are not stable or there is noise on either one of them, during the *CLK* HIGH interval. To illustrate, the waveforms of Figure 5-41 show what can

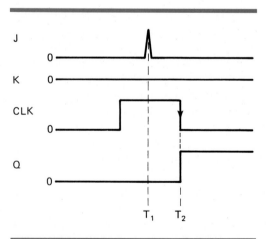

FIGURE 5-41 **M/S FF can respond to noise or changes in *J, K* inputs while *CLK* = 1.**

happen if a noise glitch appears on the J input at T_1 while $CLK = 1$. This glitch will set the master ($Q_M = 1$); then when CLK goes LOW at T_2, the 1 will be transferred from Q_M to output Q.

Master/Slave with Data Lockout Feature   This limitation of M/S FFs has been overcome by a modification called *data lockout*. The circuitry has been modified so that the J and K inputs affect the master for only a short time (20 to 30 ns) following the PGT of CLK; after that they are effectively disabled or "locked out" and have no effect on the state of the master.

The logic symbol for a J-K M/S FF with data lockout is shown in Figure 5-42. It is the same as the standard M/S with the addition of the edge-triggered indicator on CLK to indicate that the master is triggered *only* on the PGT of CLK. The postponed output indicators again indicate that the master is transferred to the slave on the NGT of CLK.

FIGURE 5-42   Symbol for M/S FF with data lockout. Note presence of edge-detector indicator on *CLK* input.

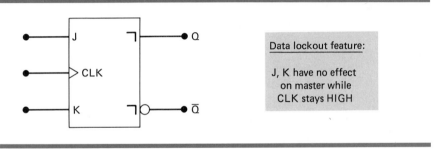

Data lockout feature:

J, K have no effect
on master while
CLK stays HIGH

Actual Devices   As we stated earlier, the newer IC technologies do not use M/S FFs. They have decreased in importance since the development of edge-triggered FFs with zero hold time. A few M/S FFs are still in use. The most common are the 7473, 7476, and 74107, all which are dual J-K FFs. The 74110 and 74111 are J-K M/S FFs with data lockout.

Comparing Operation of Various FF types   At this point it would not be surprising if you were somewhat fuzzy as to the differences in the operation of a latch, an edge-triggered FF, a master/slave (pulse-triggered) FF, and a M/S FF with data lockout. Perhaps the example in Figure 5-43 will help to make these differences more clear. In this example, the same signals are applied to the D and CLK (or EN) inputs of a D latch; an edge-triggered D FF; a M/S D FF; and a M/S D FF with data lockout. The Q output response is drawn for each device.

As we examine how the Q output is obtained for each FF, note the differences in the symbols. Each type has something distinctive about its symbol.

D Latch Q will change according to the variations in D only while EN is HIGH (see the shaded portions of the Q waveform). When EN goes LOW, Q latches at its current level and stays there until EN goes HIGH again. Note the absence of an edge-triggered indicator on the D latch symbol.

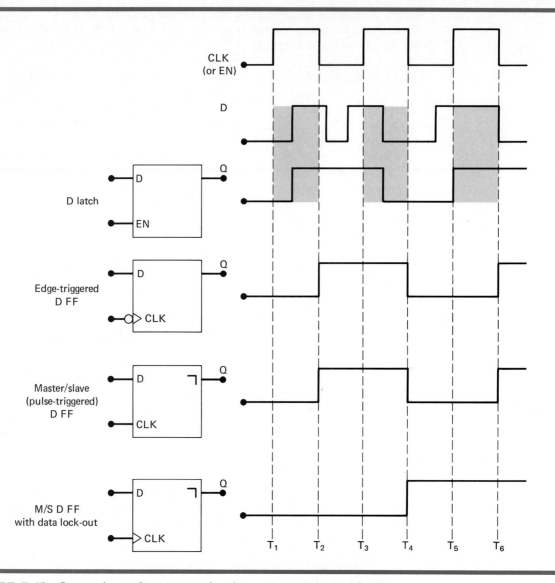

FIGURE 5-43 Comparison of response of various types of clocked D FFs.

Edge-triggered D FF Q can change only on the NGTs of *CLK*. At times T_2, T_4, and T_6, the level at D just prior to each of these NGTs will be transferred to Q.

Master/slave D FF The master will change while *CLK* is HIGH according to what is on the D input (much like the D latch). When a NGT occurs, the content of the master is transferred to the slave output Q. For example, between T_1 and T_2 the master goes from LOW to HIGH in accordance with D. At T_2, the HIGH is transferred to Q.

M/S D FF with Data Lockout The master can change only on the PGT of *CLK* according to the level at *D* at that time; it will not be affected by changes in *D* while *CLK* is HIGH. When *CLK* goes LOW, the content of the master is transferred to Q. For example, at T_3 the PGT will store a HIGH in the master. At T_4, this HIGH is transferred to Q even though *D* is LOW at that time.

REVIEW QUESTIONS

1. Describe the two steps in the M/S FF's response to a clock pulse.
2. *True or false:* The slave will respond to changes in *J* and *K* while *CLK* = 0.
3. Why should *J* and *K* be kept stable and noise-free while *CLK* = 1 for a M/S FF?
4. Explain the data lockout feature.

5-14 FLIP-FLOP APPLICATIONS

Earlier in this chapter we saw some examples of how the simple NAND FF and NOR FF were used to perform switch debouncing (Example 5-2) and event storage (Example 5-4). These simple unclocked FFs are somewhat limited in their applications. Clocked FFs offer the logic designer a group of versatile devices that have numerous applications. We will briefly introduce the more common applications in the following sections and will expand on them in subsequent chapters.

5-15 FLIP-FLOP SYNCHRONIZATION

Most digital systems are principally synchronous in their operation in that most of the signals will change states in synchronism with the clock transitions. In many cases, however, there will be an external signal that is not synchronized to the clock; in other words, it is asynchronous. Asynchronous signals often occur as a result of a human operator's actuating an input switch at some random time relative to the clock signal. This randomness can produce unpredictable and undesirable results. The following example illustrates how a FF can be used to synchronize the effect of an asynchronous input.

EXAMPLE 5-12

Figure 5-44(a) shows a situation where input signal *A* is generated from a de-bounced switch that is actuated by an operator (a debounced switch was first introduced in Example 5-2). *A* goes HIGH when the operator actuates the switch and goes LOW when the operator releases the switch. This *A* input is used to

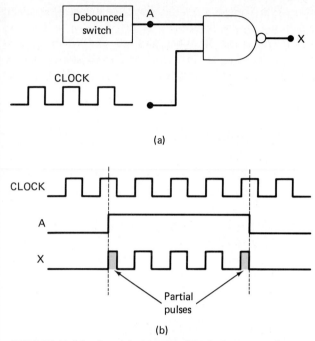

(a)

(b)

FIGURE 5-44 Asynchronous signal *A* can produce partial pulses at *X*.

FIGURE 5-45 An edge-triggered D FF is used to synchronize the enabling of the AND gate to the NGTs of the clock.

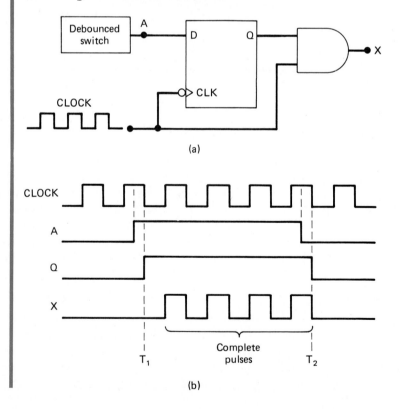

(a)

(b)

control the passage of the clock signal through the AND gate so that clock pulses appear at output X only as long as A is HIGH.

The problem with this circuit is that A is asynchronous; it can change states at any time relative to the clock signal because the exact times when the operator actuates or releases the switch are essentially random. This can produce *partial* clock pulses at output X if either transition of A occurs while the clock signal is HIGH, as shown in the waveforms of Figure 5-44(b).

This type of output is often not acceptable, so a method for preventing the appearance of partial pulses at X must be developed. One solution is shown in Figure 5-45(a). Describe how this circuit solves the problem, and draw the X waveform for the same situation as in Figure 5-44(b).

SOLUTION

The A signal is connected to the D input of FF Q which is clocked by the NGT of the clock signal. Thus, when A goes HIGH, Q will not go HIGH until the next NGT of the clock at time T_1. This HIGH at Q will enable the AND gate to pass subsequent *complete* clock pulses to X, as shown in Figure 5-45(b).

When A returns LOW, Q will not go LOW until the next NGT of the clock at T_2. Thus, the AND gate will not inhibit clock pulses until the clock pulse that ends at T_2 has been passed through to X.

Thus, output X contains only complete pulses.

5-16 DETECTING AN INPUT SEQUENCE

In many situations an output is to be activated only when the inputs are activated in a certain sequence. This cannot be accomplished using pure combinational logic but requires the storage characteristic of FFs.

For example, an AND gate can be used to determine when two inputs A and B are both HIGH, but its output will respond the same regardless of which input goes HIGH first. But suppose we want to generate a HIGH output *only* if A goes HIGH and then B goes HIGH some time later. One way to accomplish this is shown in Figure 5-46(a).

FIGURE 5-46 **Clocked J-K FF used to respond to a particular sequence of inputs.**

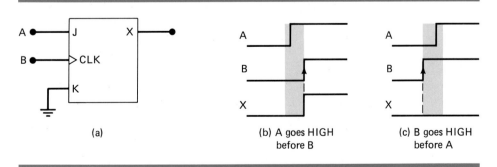

(a)

(b) A goes HIGH
before B

(c) B goes HIGH
before A

The waveforms in (b) and (c) show that X will go HIGH only if A goes HIGH before B goes HIGH. This is because A must be HIGH in order for X to go HIGH on the PGT of B.

In order for this circuit to work properly, A has to go HIGH prior to B by at least an amount of time equal to the setup time requirement of the FF.

5-17 DATA STORAGE AND TRANSFER

By far the most common use of flip-flops is for the storage of data or information. The data may represent numerical values (e.g., binary numbers, BCD-coded decimal numbers). These data are generally stored in groups of FFs called *registers*.

The operation most often performed on data that are stored in a FF or a register is the *transfer* operation. This involves the transfer of data from one FF or

FIGURE 5-47 Synchronous data transfer operation performed by various types of FFs.

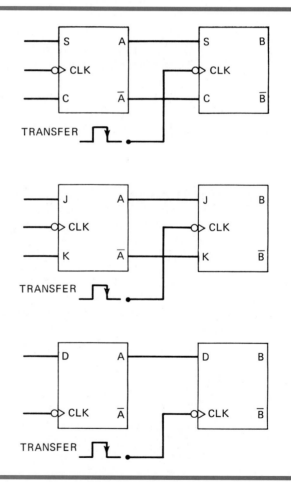

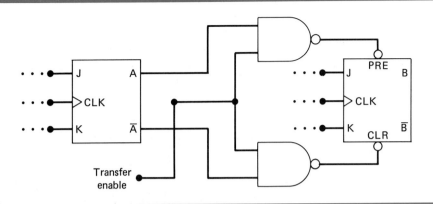

FIGURE 5-48   **Asynchronous data transfer operation.**

register to another. Figure 5-47 illustrates how data transfer can be accomplished between two FFs using clocked S-C, J-K, and D FFs. In each case, the logic value that is currently stored in FF A is transferred to FF B upon the NGT of the TRANSFER pulse. Thus, after this NGT, the B output will be the same as the A output.

The transfer operations in Figure 5-47 are examples of *synchronous transfer,* since the synchronous and *CLK* inputs are used to perform the transfer. A transfer operation can also be obtained using the asynchronous inputs of a FF. Figure 5-48 shows how an *asynchronous transfer* can be obtained using the PRESET and CLEAR inputs of any type of FF. Here, the asynchronous inputs respond to LOW levels. When the TRANSFER ENABLE line is held LOW, the two NAND outputs are kept HIGH, with no effect on the FF outputs. When the TRANSFER ENABLE line is made HIGH, one of the NAND outputs will go LOW, depending on the state of the A and $\overline{A}$ outputs. This LOW will either set or clear FF B to the same state as FF A. This asynchronous transfer is done independently of the synchronous and *CLK* inputs of the FF. Asynchronous transfer is also called *jam transfer* because the data can be "jammed" into FF B even if the synchronous inputs are active.

Parallel Data Transfer   Figure 5-49 illustrates data transfer from one register to another using D-type FFs. Register X consists of FFs X_1, X_2, and X_3; register Y consists of FFs Y_1, Y_2, and Y_3. Upon application of the TRANSFER pulse, the level stored in X_1 is transferred to Y_1, X_2 to Y_2, and X_3 to Y_3. The transfer of the contents of the X register into the Y register is a synchronous transfer. It is also referred to as a *parallel* transfer, since the contents of X_1, X_2, and X_3 are transferred *simultaneously* into Y_1, Y_2,, and Y_3. If a *serial* transfer were performed, the contents of the X register would be transferred to the Y register one bit at a time. This will be examined in the next section.

It is important to understand that parallel transfer does not change the contents of the register that is the source of data. For example, in Figure 5-49, the transfer of data from the X register to the Y register will leave both registers holding the same data.

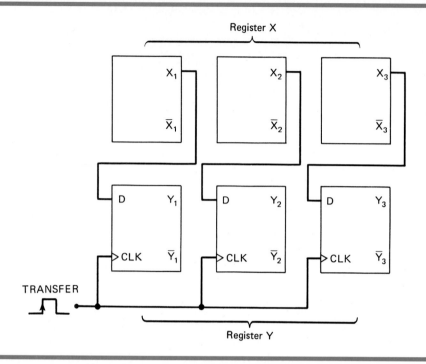

FIGURE 5-49 **Parallel transfer of contents of register *X* into register *Y*.**

REVIEW QUESTIONS

1. *True or false:* Asynchronous data transfer uses the *CLK* input.
2. Which type of FF is best suited for synchronous transfer because it requires the fewest number of interconnections from one FF to the other?
3. If J-K FFs were used in the registers of Figure 5-49, how many total interconnections would be required from register *X* to register *Y*? (*Ans.* Six)
4. *True or false:* Synchronous data transfer requires less circuitry than asynchronous transfer.

5-18 SHIFT REGISTERS

Shift registers are used to transfer the contents of one register into a second register one bit at a time. Before examining this operation of *serial* transfer, let us look at the operation of a basic shift register. Figure 5-50(a) shows four J-K FFs wired as a 4-bit shift register. Note that the FFs are connected so that the output of X_3 transfers into X_2, X_2 into X_1, and X_1 into X_0. What this means is that upon the

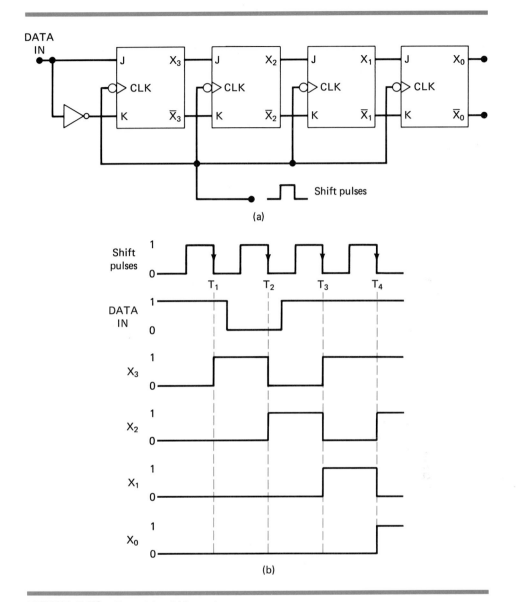

FIGURE 5-50   **Four-bit shift register.**

occurrence of the shift pulse, each FF takes on the value stored previously in the FF on its left. FF X_3 takes on a value determined by the conditions present on its J and K inputs when the shift pulse occurs. For now, we will assume that X_3's J and K inputs are fed by the DATA IN waveform shown in Figure 5-50(b). We will also assume that all FFs are in the 0 state before shift pulses are applied.

The waveforms in Figure 5-50(b) show how the input data are shifted from left to right from FF to FF as shift pulses are applied. When the first NGT occurs at T_1, each of the FFs X_2, X_1, and X_0 will have the $J = 0$, $K = 1$ condition present at its inputs because of the state of the FF on its left. FF X_3 will have $J = 1$, $K =$

0 because of DATA IN. Thus, at T_1 only X_3 will go HIGH while all the other FFs remain LOW. When the second NGT occurs at T_2, FF X_3 will have $J = 0$, $K = 1$ because of DATA IN. FF X_2 will have $J = 1$, $K = 0$ because of the current HIGH at X_3. FFs X_1 and X_0 will still have $J = 0$, $K = 1$. Thus, at T_2 only FF X_2 will go HIGH, FF X_3 will go LOW, and FFs X_1 and X_0 will remain LOW.

Similar reasoning can be used to determine how the waveforms change at T_3 and T_4. Note that on each NGT of the shift pulses, each FF output takes on the level that was present at the output of the FF on its left *prior* to the NGT. Of course, X_3 takes on the level that was present at DATA IN prior to the NGT.

Hold-Time Requirement

In this shift-register arrangement it is necessary that the FFs have a very small hold-time requirement, because there are times when the J, K inputs are changing at about the same time as the *CLK* transition. For example, the X_3 output switches from 1 to 0 in response to the NGT at T_2, causing the J, K inputs of X_2 to change while its *CLK* input is changing. Actually, because of the propagation delay of X_3, the J, K inputs of X_2 won't change for a short time after the NGT. For this reason, a shift register should be implemented either from M/S FFs or from edge-triggered FFs that have a t_H value less than one FF propagation delay (*CLK*-to-output). This latter requirement is easily satisfied by most modern edge-triggered FFs.

Serial Transfer between Registers

Figure 5-51 shows two 3-bit shift registers connected so that the contents of the X register will be serially transferred (shifted) into register Y. We are using D FFs for each shift register, since this requires fewer connections than J-K FFs. Notice how X_0, the last FF of register X, is connected to the input of Y_2, the first FF of register Y. Thus, as the shift pulses are applied, the information transfer takes place as follows: $X_2 \rightarrow X_1 \rightarrow X_0 \rightarrow Y_2 \rightarrow Y_1 \rightarrow Y_0$. The X_2 FF will go to a state determined by its D input. For now, D will be held LOW, so X_2 will go LOW on the first pulse and will remain there.

To illustrate, let us assume that before any shift pulses are applied, the contents of the X register is 1　0　1 (i.e., $X_2 = 1$, $X_1 = 0$, $X_0 = 1$) and the Y register is at 0　0　0. Refer to the table in Figure 5-51(b), which shows how the states of each FF change as shift pulses are applied. The following points should be noted:

1. On the falling edge of each pulse, each FF takes on the value that was stored in the FF on its left prior to the occurrence of the pulse.

2. After *three* pulses, the 1 that was initially in X_2 is in Y_2, the 0 initially in X_1 is in Y_1, and the 1 initially in X_0 is in Y_0. In other words, the 101 stored in the X register has now been shifted into the Y register. The X register is at 000; it has lost its original data.

3. The complete transfer of the *three* bits of data requires *three* shift pulses.

Parallel versus Serial Transfer

In parallel transfer, all the information is transferred simultaneously upon the occurrence of a *single* transfer command pulse (Figure 5-49), no matter how many bits are being transferred. In serial transfer, as

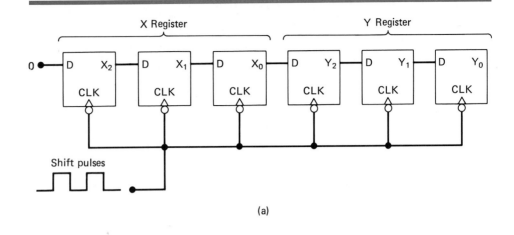

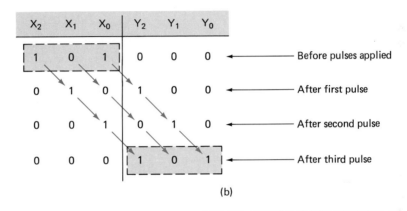

FIGURE 5-51   **Serial transfer of information from _X_ register into _Y_ register.**

exemplified by Figure 5-51, the complete transfer of N bits of information requires N clock pulses (3 bits requires three pulses, 4 bits requires four pulses, etc.). Parallel transfer, then, is obviously much faster than serial transfer using shift registers.

In parallel transfer, the output of each FF in register X is connected to a corresponding FF input in register Y. In serial transfer, only the last FF in register X is connected to register Y. In general, then, parallel transfer requires more interconnections between the sending register (X) and the receiving register (Y) than does serial transfer. This difference becomes more critical when a greater number of bits of information are being transferred. This is an important consideration when the sending and receiving registers are remote from each other, since it determines how many lines (wires) are needed for the transmission of the information.

The choice of either parallel or serial transmission depends on the particular system application and specifications. Often, a combination of the two types is used to take advantage of the *speed* of parallel transfer and the *economy and simplicity* of serial transfer. More will be said later about information transfer.

1. *True or false:* The fastest method for transferring data from one register to another is parallel transfer.
2. What is the major advantage of serial transfer over parallel transfer?
3. Refer to Figure 5-51. Assume that the initial contents of the registers is: $X_2 = 0$, $X_1 = 1$, $X_0 = 0$, $Y_2 = 1$, $Y_1 = 1$, $Y_0 = 0$. Also assume that the D input of X_2 is held HIGH. Determine the value of each FF output after the occurrence of the fourth shift pulse. (*Ans.* $X_2 = X_1 = X_0 = 1$, $Y_2 = 1$, $Y_1 = 0$, $Y_0 = 1$)

5-19 FREQUENCY DIVISION AND COUNTING

Refer to Figure 5-52(a). Each FF has its J and K inputs at the 1 level, so it will change states (toggle) whenever the signal on its *CLK* input goes from HIGH to LOW. The clock pulses are applied only to the *CLK* input of FF X_0. Output X_0 is

FIGURE 5-52 J-K FFs wired as a 3-bit binary counter (MOD-8).

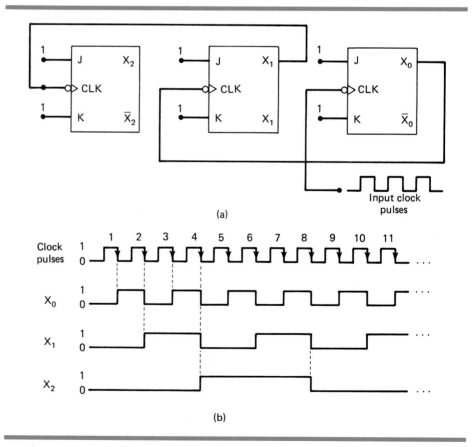

connected to the *CLK* input of FF X_1 and output X_1 is connected to the *CLK* input of FF X_2. The waveforms in Figure 5-52(b) show how the FFs change states as the pulses are applied. The following important points should be noted:

1. FF X_0 toggles on the negative-going transition of each input clock pulse. Thus, the X_0 output waveform has a frequency that is exactly 1/2 of the clock pulse frequency.

2. FF X_1 toggles each time the X_0 output goes from HIGH to LOW. The X_1 waveform has a frequency equal to exactly 1/2 the frequency of the X_0 output and therefore 1/4 of the clock frequency.

3. FF X_2 toggles each time the X_1 output goes from HIGH to LOW. Thus, the X_2 waveform has 1/2 the frequency of X_1 and therefore 1/8 of the clock frequency.

4. Each FF output is a squarewave (50 percent duty cycle).

As described above, each FF divides the frequency of its input by 2. Thus, if we were to add a fourth FF to the chain, it would have a frequency equal to 1/16 of the clock frequency, and so on. Using the appropriate number of FFs, this circuit could divide a frequency by any power of 2. Specifically, using N FFs would produce an output frequency from the last FF which is equal to $1/2^N$ of the input frequency.

Counting Operation   In addition to functioning as a frequency divider, the circuit of Figure 5-52 also operates as a *binary counter*. This can be demonstrated by examining the sequence of states of the FFs after the occurrence of each clock pulse. Figure 5-53 presents the results in tabular form. Let the $X_2X_1X_0$ values rep-

FIGURE 5-53   Sequence of FF states shows binary counting sequence.

2^2	2^1	2^0	
X_2	X_1	X_0	
0	0	0	Before applying clock pulses
0	0	1	After pulse #1
0	1	0	After pulse #2
0	1	1	After pulse #3
1	0	0	After pulse #4
1	0	1	After pulse #5
1	1	0	After pulse #6
1	1	1	After pulse #7
0	0	0	After pulse #8 recycles to 000
0	0	1	After pulse #9
0	1	0	After pulse #10
0	1	1	After pulse #11
.	.	.	. . .
.	.	.	. . .
.	.	.	. . .

resent a binary number where X_2 is the 2^2 position, X_1 is the 2^1 position, and X_0 is the 2^0 position. The first eight $X_2X_1X_0$ states in the table should be recognized as the binary counting sequence from 000 to 111. After the first pulse the FFs are in the 001 state ($X_2 = 0$, $X_1 = 0$, $X_0 = 1$), which represents 001_2 (equivalent to decimal 1); after the second pulse the FFs represent 010_2, which is equivalent to 2_{10}; after three pulses, $011_2 = 3_{10}$; after four pulses, $100_2 = 4_{10}$; and so on until after seven pulses, $111_2 = 7_{10}$. On the eighth pulse the FFs return to the 000 state, and the binary sequence repeats itself for succeeding pulses.

Thus, for the first seven input pulses, the circuit functions as a binary counter in which the states of the FFs represent a binary number equivalent to the number of pulses that have occurred. This counter can count as high as $111_2 = 7_{10}$, and then it returns to 000.

MOD Number   The counter of Figure 5-52 has $2^3 = 8$ different states (000 through 111). It would be referred to as a *MOD-8 counter,* where the MOD number indicates the number of states in the counting sequence. If a fourth FF were added, the sequence of states would count in binary from 0000 to 1111, a total of 16 states. This would be called a *MOD-16 counter.* In general, if N FFs are connected in the arrangement of Figure 5-52, the counter will have 2^N different states, so it is a MOD-2^N counter. It would be capable of counting up to $2^N - 1$ before returning to its zero state.

The MOD number of a counter also indicates the frequency division obtained from the last FF. For instance, a 4-bit counter has 4 FFs, each representing one binary digit (bit), so it is a MOD-2^4 = MOD-16 counter. It can therefore count up to 15 ($= 2^4 - 1$). It can also be used to divide the input pulse frequency by a factor of 16 (the MOD number).

We have looked only at the basic FF binary counter. We will examine counters in much more detail in Chapter 7.

EXAMPLE 5-13

Consider a counter circuit that contains six FFs wired in the arrangement of Figure 5-52 (i.e., X_5, X_4, X_3, X_2, X_1, X_0).

(a)   Determine the counter's MOD number.

(b)   Determine the frequency at the output of the last FF (X_5) when the input clock frequency is 1 MHz.

(c)   What is the range of counting states for this counter?

SOLUTION

(a)   MOD number $= 2^6 = 64$.

(b)   The frequency at the last FF will equal the input clock frequency divided by the MOD number. That is,

$$f(\text{at } X_5) = \frac{1 \text{ MHz}}{64} = 15.625 \text{ kHz}$$

(c)   The counter will count from 000000_2 to 111111_2 (0 to 63_{10}) for a total of 64 states. Note that the number of states is the same as the MOD number.

REVIEW QUESTIONS

1. A 20-kHz clock signal is applied to a J-K FF with $J = K = 1$. What is the frequency of the FF output waveform?
2. How many FFs are required for a counter that will count 0 to 255_{10}? (*Ans.* Eight)
3. What is the MOD number of this counter? (*Ans.* 256)
4. What is the frequency at the output of the eighth FF when the input clock frequency is 512 kHz? (*Ans.* 2 kHz)

5-20 SCHMITT-TRIGGER DEVICES

A *Schmitt-trigger* circuit is not classified as a flip-flop, but it does exhibit a type of memory characteristic that makes it useful in certain special situations. One of those situations is shown in Figure 5-54(a). Here a standard INVERTER is being driven by a logic input that has relatively slow transition times. When these transition times exceed the maximum allowed values (this depends on the particular logic family), the outputs of logic gates and INVERTERs may produce oscillations as the input signal passes through the indeterminate range. The same input conditions can also produce erratic triggering of FFs.

A device that has a Schmitt-trigger type input is designed to accept slow-changing signals and produce an output that has oscillation-free transitions. The output will generally have very rapid transition times (typically, 10 ns) that are independent of the input signal characteristics. Figure 5-54(b) shows a Schmitt-trigger INVERTER and its response to a slow-changing input.

If you examine the waveforms in Figure 5-54(b), you should note that the output does not change from HIGH to LOW until the input exceeds the *positive-going threshold* voltage, V_{T+}. Once the output goes LOW, it will remain there even when the input drops back below V_{T+} (this is its memory characteristic) until it drops all the way down below the *negative-going threshold* voltage, V_{T-}. The values of the two threshold voltages will vary from logic family to logic family, but V_{T-} will always be less than V_{T+}.

The Schmitt-trigger INVERTER, and all devices with Schmitt-trigger inputs, use the distinctive symbol shown in Figure 5-54(b) to indicate that they can reliably respond to slow-changing input signals. Logic designers use ICs with Schmitt-trigger inputs to convert slow-changing signals to clean, fast-changing signals that can drive standard IC inputs.

There are several ICs available with Schmitt-trigger inputs. The 7414, 74LS14, and 74HC14 are hex INVERTER ICs with Schmitt-trigger inputs. The 7413, 74LS13, and 74HC13 are dual four-input NANDs with Schmitt-trigger inputs.

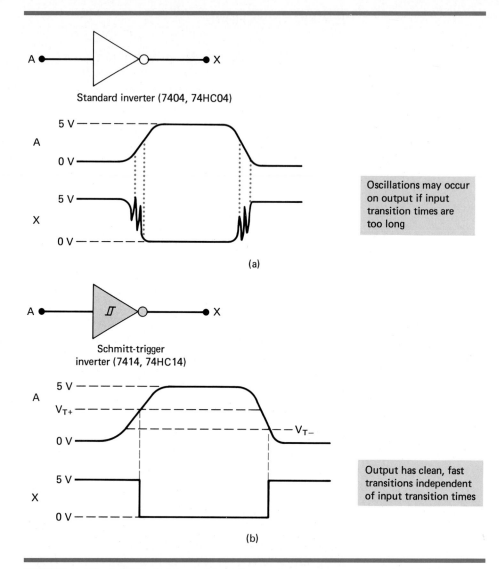

Standard inverter (7404, 74HC04)

Oscillations may occur on output if input transition times are too long

(a)

Schmitt-trigger inverter (7414, 74HC14)

Output has clean, fast transitions independent of input transition times

(b)

FIGURE 5-54 (a) If input transition times are too long, a standard logic device output might oscillate or change erratically; (b) a logic device with a Schmitt-trigger type of input will produce clean, fast output transitions.

REVIEW QUESTIONS

1. What could occur when a slow-changing signal is applied to a standard logic IC?
2. How does a Schmitt-trigger logic device operate different from a standard logic device?

ONE-SHOT (MONOSTABLE MULTIVIBRATOR)

A digital circuit that is somewhat related to the FF is the *one-shot* (abbreviated OS). Like the FF, the OS has two outputs, Q and $\overline{Q}$, which are inverses of each other. Unlike the FF, the OS has only one *stable* output state (normally, $Q = 0$, $\overline{Q} = 1$), where it remains until it is triggered by an input signal. Once triggered, the OS outputs switch to the opposite state ($Q = 1$, $\overline{Q} = 0$). It remains in this *quasi-stable* state for a fixed period of time, t_p, which is usually determined by an RC time constant which is connected to the OS. After a time t_p, the OS outputs return to their stable resting state until triggered again.

Figure 5-55(a) shows the logic symbol for a OS. The value of t_p is often indicated somewhere on the OS symbol. In practice, t_p can vary from several nanoseconds to several tens of seconds. The exact value of t_p is essentially determined by the values of external components R_T and C_T.

Two types of one-shots are available in IC form: the *nonretriggerable* and the *retriggerable*.

FIGURE 5-55 **OS symbol and typical waveforms for nonretriggerable operation.**

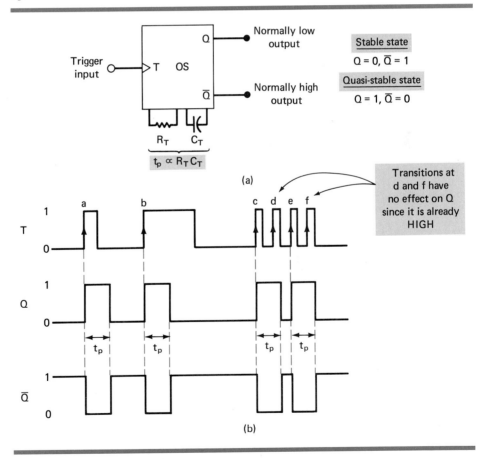

Nonretriggerable One-Shot The waveforms in Figure 5-55 illustrate the operation of a nonretriggerable OS that triggers on positive-going transitions at its trigger (T) input. The important points to note are:

1.   The PGTs at points a, b, c, and e will trigger the OS to its quasi-stable state for a time t_p, after which it automatically returns to the stable state.

2.   The PGTs at points d and f have no effect on the OS because it has already been triggered to the quasi-stable state. The OS must return to the stable state before it can be retriggered.

3.   The OS output-pulse duration is always the same regardless of the duration of the input pulses. As stated above, t_p depends only on R_T and C_T and the internal OS circuitry. A typical OS may have a t_p given by $t_p = 0.7R_TC_T$.

Retriggerable One-Shot The retriggerable OS operates much like the nonretriggerable OS except for one major difference: *it can be retriggered while it is in the quasi-stable state, and it will begin a new t_p interval.* This characteristic is illustrated in Figure 5-56(a) for a retriggerable OS with $t_p = 2$ ms.

   The first PGT at the T input occurs at $t = 1$ ms and triggers the Q output to its HIGH state for a time duration of $t_p = 2$ ms, after which it returns to 0. The second trigger pulse occurs at $t = 5$ ms and triggers Q to the HIGH state where it would normally stay until $t = 7$ ms; however, a third trigger pulse occurs at $t = 6$ ms and retriggers the OS to begin a new $t_p = 2$ ms interval. Thus, Q stays HIGH for 2 ms *after* this third trigger pulse.

   In effect, then, a retriggerable OS will begin a new t_p interval *each* time a

FIGURE 5-56 **Waveforms for a retriggerable OS with $t_p = 2$ms. Each trigger pulse begins a new 2-ms interval.**

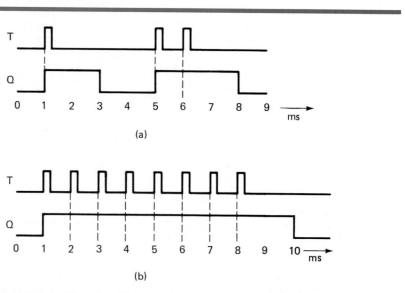

trigger pulse occurs, regardless of the state of Q. In fact, if a continuous train of trigger pulses having a frequency greater than $1/t_p$ is applied to the T input, the OS will trigger HIGH on the first pulse and remain HIGH as long as the pulses are applied. This is illustrated in Figure 5-56(b), where the trigger pulses occur every 1 ms ($f = 1$ kHz).

Actual Devices   There are several one-shot ICs available in both the retriggerable and nonretriggerable versions. The 74121 and 74L121 are single nonretriggerable one-shot ICs; the 74221, 74LS221, and 74HC221 are dual nonretriggerable one-shot ICs; the 74122, and 74LS122 are single retriggerable one-shot ICs; the 74123, 74LS123, and 74HC123 are dual retriggerable one-shot ICs.

Figure 5-57(a) shows the traditional symbol for the 74121 nonretriggerable one-shot IC. Note that it contains internal logic gates to allow inputs A_1, A_2, and B to trigger the OS in a variety of ways. The B input is a Schmitt-trigger type input that is allowed to have slow transition times and still reliably trigger the OS. The pins labeled R_{INT}, R_{EXT}/C_{EXT}, and C_{EXT} are used to connect an external resistor and capacitor to achieve the desired output pulse duration. Figure 5-57(b) is the IEEE/ANSI symbol for the 74121.

Monostable Multivibrator   One-shots are also called *monostable multivibrators* because they have only one stable state. The major applications of the OS are in timing circuits that utilize the predetermined t_p interval. Some of these applications will be introduced in the end-of-chapter problems and at various places throughout the text.

FIGURE 5-57   Logic symbols for the 74121 nonretriggerable one-shot: (a) traditional; (b) IEEE/ANSI.

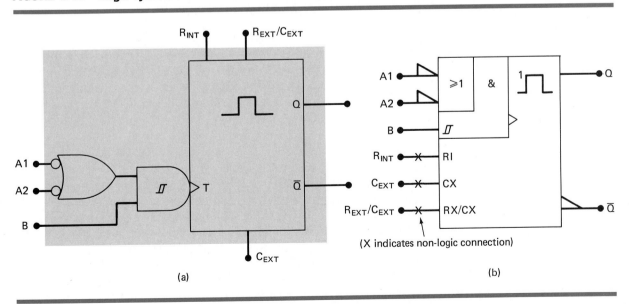

REVIEW QUESTIONS

1. In the absence of a trigger pulse, what will be the state of a OS output?
2. *True or false:* When a nonretriggerable OS is pulsed while it is in its quasi-stable state, the output is not affected.
3. What determines the t_p value for a OS?
4. Describe how a retriggerable OS operates differently from a nonretriggerable OS.

5-22 ANALYZING SEQUENTIAL CIRCUITS

Many logic circuits contain FFs, one-shots, and logic gates that are connected to perform a specific operation. Very often, a master clock signal is used to cause the logic levels in the circuit to go through a particular sequence of states. We can generally analyze these sequential circuits by following this step-by-step procedure:

1. Examine the circuit diagram and look for circuit structures such as counters or shift registers that you are familiar with. This can help to simplify the analysis.
2. Determine the logic levels that are present at the inputs of each FF *prior to* the occurrence of the first clock pulse.
3. Use these levels to determine how each FF output will change in response to the first clock pulse.
4. Repeat steps 2 and 3 for each successive clock pulse.

The following example will illustrate the procedure.

EXAMPLE 5-14

Consider the circuit of Figure 5-58. Initially, all the FFs and the OS are in the 0 state before the clock pulses are applied. These pulses have a repetition rate of 1 kHz. Determine the waveforms at X, Y, Z, W, $\overline{Q}$, A, and B for 16 cycles of the clock input.

SOLUTION
Initially, the FFs and OS are in the 0 state, so $X = Y = Z = W = Q = 0$. The inputs to the NAND gate are $X = 0$, $\overline{Y} = 1$, and $\overline{Z} = 1$, so its output $A = 1$. The inputs to the OR gate are $W = 0$ and $\overline{Q} = 1$, so its output is $B = 1$.

As long as B remains HIGH, FF Z will have its J and K inputs both HIGH, so it will operate in the toggle mode. FFs X and Y are kept in the toggle mode, since their J and K inputs are held at the 1 level permanently. It should be clear, then, that FFs X, Y, and Z will operate as a 3-bit counter as long as B stays HIGH. Thus, FF Z will toggle on the negative transition of each input clock pulse, FF Y will toggle on the negative transition of the Z output, and FF X will

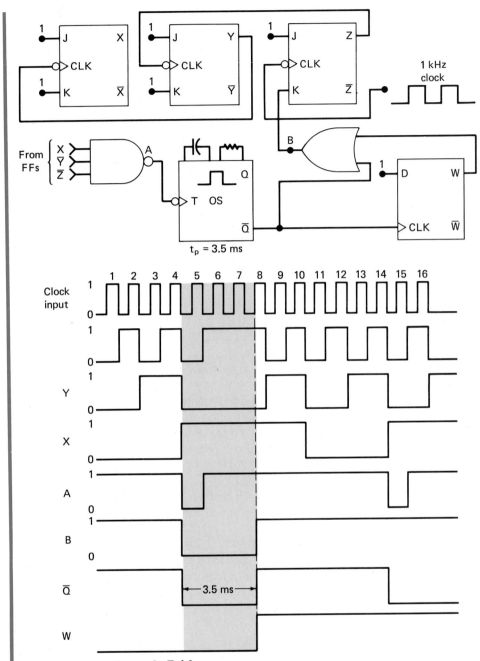

FIGURE 5-58 Example 5-14

toggle on the negative transition of FF Y. This operation continues for clock pulses 1–4.

When clock pulse 4 makes its negative transition, Z goes LOW, Y goes LOW, and X goes HIGH. Therefore, the inputs to the NAND gate become $X = 1$, $\overline{Y} = 1$, and $\overline{Z} = 1$, causing the NAND output A to go LOW. This negative transition at A will trigger the OS to its quasi-stable state, so $\overline{Q}$ goes LOW for

3.5 ms. This negative transition of $\overline{Q}$ will not affect FF W, since this D-type FF triggers on positive transitions at its CLK input. With $\overline{Q}$ and W both LOW, the OR gate output B goes LOW, so FF Z now has $J = 1$, $K = 0$. This means that Z will be *set* to the 1 state on the *next* clock pulse and will remain in this state for each successive clock pulse as long as $J = 1$, $K = 0$. Thus, Z goes HIGH on clock pulse 5 and stays HIGH for clock pulses 6 and 7. (*Note:* A goes back HIGH when Z goes HIGH on pulse 5 because $\overline{Z} = 0$.)

After 3.5 ms the OS returns to its stable state with $Q = 0$, $\overline{Q} = 1$. This positive transition of $\overline{Q}$ causes FF W to go HIGH since $D = 1$. The 1 at W produces a 1 at B so that FF Z is back in the toggle mode with $J = K = 1$. The counter will now operate properly in response to all succeeding clock pulses.

On the negative transition of clock pulse 14, the NAND output A gain goes LOW thereby triggering the OS. The OS, however, will not affect the K input of Z since W is keeping the OR output HIGH. Also, since its D input is HIGH, FF W will remain in the $W = 1$ state indefinitely, allowing the counter to count normally for all succeeding clock pulses.

5-23 ASTABLE MULTIBRATORS

Flip-flops have two stable states; therefore, we can say that they are *bistable multivibrators*. One-shots have one stable state; so we call them *monostable multivibrators*. A third type of multivibrator has no stable states; it is called an *astable* or *free-running multivibrator*. This type of logic circuit switches back and forth (oscillates) between two unstable output states. It is useful for providing clock signals for synchronous digital circuits.

There are several types of astable multivibrators that are in common use. We will present two of them without any attempt to analyze their operation. They are presented here so that you can construct a clock generator circuit if needed for a project or for testing digital circuits in the lab.

Schmitt-Trigger Oscillator   Figure 5-59 shows how a Schmitt-trigger INVERTER can be connected as an oscillator. The signal at v_{out} is an approximate squarewave with a frequency that depends on the R and C values. The relationship between the frequency and RC values is shown in Figure 5-59 for three different Schmitt-trigger INVERTERs. Note the maximum limits on the resistance value for each device. The circuit will fail to oscillate if R is not kept below these limits.

555 Timer Used as an Astable Multivibrator   The 555 timer IC is a TTL-compatible device that can operate in several different modes. Figure 5-60 shows how external components can be connected to a 555 so that it operates as a free-running oscillator. Its output is a repetitive rectangular waveform that switches between two logic levels with the time intervals at each logic level determined by the R and C values. The formulas for these time intervals, t_1 and t_2, and the overall period of the oscillations, T, are given in the figure. The frequency of the oscillations is, of course, the reciprocal of T. As the formulas in the diagram indicate, the

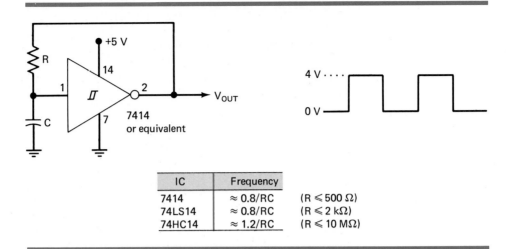

IC	Frequency	
7414	$\approx 0.8/RC$	($R \leqslant 500\ \Omega$)
74LS14	$\approx 0.8/RC$	($R \leqslant 2\ k\Omega$)
74HC14	$\approx 1.2/RC$	($R \leqslant 10\ M\Omega$)

FIGURE 5-59 **Schmitt-trigger oscillator. A 7413 Schmitt-trigger NAND may also be used.**

t_1 and t_2 intervals cannot be equal unless R_A is made zero. This cannot be done without producing excess current through the device. This means that it is impossible to produce a perfect 50% duty-cycle squarewave output. It is possible, however, to get very close to a 50% duty cycle by making $R_B \gg R_A$ (while keeping R_A greater than 500 Ω), so that $t_1 \approx t_2$.

FIGURE 5-60 **555 timer IC used as an astable multivibrator.**

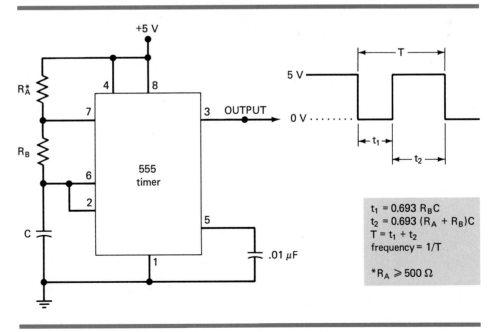

$t_1 = 0.693\ R_B C$
$t_2 = 0.693\ (R_A + R_B)C$
$T = t_1 + t_2$
frequency $= 1/T$

*$R_A \geqslant 500\ \Omega$

REVIEW QUESTIONS

1. Determine the approximate frequency of a Schmitt-trigger oscillator that uses a 74HC14, $R = 10$ kΩ, and $C = 0.005$ μF. (*Ans.* 24 kHz)
2. Determine the approximate frequency of the 555 output for $R_A = R_B = 2.2$ kΩ and $C = 1000$ pF. (*Ans.* 218.6 kHz)

5-24 TROUBLESHOOTING FLIP-FLOP CIRCUITS

Flip-flop ICs are susceptible to the same kinds of internal and external faults that occur in combinatorial logic circuits. All of the troubleshooting ideas that were discussed in Chapter 4 can readily be applied to circuits that contain FFs as well as logic gates.

Because of their memory characteristic, FF circuits with one or more faults will often exhibit symptoms that would not occur in combinatorial circuits. Some of these are described below.

Open Inputs   Unconnected or floating inputs of any logic circuit are particularly susceptible to picking up spurious voltage fluctuations called *noise*. If the noise is large enough in amplitude and long enough in duration, the logic circuit's output may change states in response to the noise. In a logic gate, the output will return to its original state when the noise signal subsides. In a FF, however, the output will remain in its new state because of its memory characteristic. Thus the effect of noise pickup at an open input is usually more critical for a FF or latch than it is for a logic gate.

The most susceptible FF inputs are those that can trigger the FF to a different state—such as the *CLK*, PRESET, and CLEAR. Whenever you see a FF output that is changing states erratically, you should consider the possibility of an open at one of these inputs.

EXAMPLE 5-15

Figure 5-61 shows a 3-bit shift register made up of TTL FFs. Initially, all of the FFs are in the LOW state before clock pulses are applied. As clock pulses are applied, each PGT will cause the information to shift from each FF to the one on its right. The diagram shows the "expected" sequence of FF states after each clock pulse. Since $J_2 = 1$ and $K_2 = 0$, FF X_2 will go HIGH on clock pulse 1 and will stay there for all subsequent pulses. This HIGH will shift into X_1, and then X_0 on clock pulses 2 and 3, respectively. Thus, after the third pulse, all FFs will be HIGH and should remain there as pulses are continually applied.

Now let's suppose that the "actual" response of the FF states is as shown in the diagram. Here the FFs change as expected for the first three clock pulses. From then on, FF X_0, instead of staying HIGH, alternates between HIGH and LOW. What possible circuit fault can produce this operation?

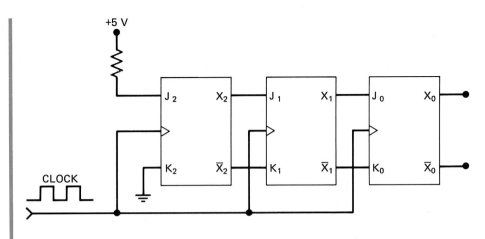

Clock pulse number	"Expected" X_2 X_1 X_0			"Actual" X_2 X_1 X_0		
0	0	0	0	0	0	0
1	1	0	0	1	0	0
2	1	1	0	1	1	0
3	1	1	1	1	1	1
4	1	1	1	1	1	0
5	1	1	1	1	1	1
6	1	1	1	1	1	0
7	1	1	1	1	1	1
8	1	1	1	1	1	0

FIGURE 5-61 **Example 5-15**

SOLUTION

On the second pulse, X_1 goes HIGH. This should make $J_0 = 1$, $K_0 = 0$ so that all subsequent clock pulses should set $X_0 = 1$. Instead, we see X_0 changing states (toggling) on all pulses after the second one. This toggle operation would occur if J_0 and K_0 were both HIGH. The most probable fault is an open in the connection between $\overline{X}$, and K_0. Recall that a TTL device responds to an open input as if it were a logic HIGH, so an open at K_0 is the same as a HIGH.

Shorted Outputs   The following example will illustrate how a fault in a FF circuit can cause a misleading symptom that may result in a longer time to isolate the fault.

EXAMPLE 5-16

Consider the circuit in Figure 5-62 and examine the logic probe indications shown in the accompanying table. There is a LOW at the D input of the FF when pulses are applied to its CLK input, but the Q output fails to go to the LOW state. The technician testing this circuit considers each of the following possible circuit faults:

1.   Z2-5 is internally shorted to V_{cc}.
2.   Z1-4 is internally shorted to V_{cc}.

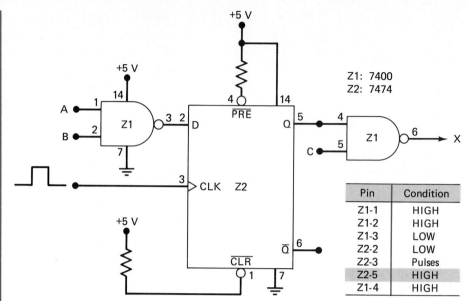

FIGURE 5-62   Example 5-16

3. Z2-5 or Z1-4 are externally shorted to V_{cc}.

4. Z2-4 is internally or externally shorted to GROUND. This would keep $\overline{PRE}$ activated and would override the *CLK* input.

5. There is an internal failure in Z2 that prevents Q from responding properly to its inputs.

The technician, after making the necessary ohmmeter checks, rules out the first four possibilities. He also checks Z2's V_{cc} and GROUND pins and finds that they are at the proper voltages. He is reluctant to unsolder Z2 from the circuit until he is certain it is faulty, so he decides to look at the clock signal. He uses an oscilloscope to check its amplitude, frequency, pulse width, and transition times. He finds that they are all within the specifications for the 7474 IC. Finally, he concludes that Z2 is faulty.

He removes the 7474 chip and replaces it with another 7474. To his dismay, the circuit with the new chip behaves in exactly the same way. After scratching his head, he decides to change the NAND gate chip, although he doesn't know why. As expected, he sees no change in the circuit operation.

Becoming more puzzled, he recalls his electronics lab instructor emphasizing the value of performing a thorough visual check on the circuit board, so he begins to examine it carefully. While he is doing that, he detects a solder bridge between pins 6 and 7 of Z2. He removes it, tests the circuit, and it functions correctly.

Explain how this fault produced the operation observed.

SOLUTION
The solder bridge was shorting the $\overline{Q}$ output to GROUND. This means that $\overline{Q}$ is permanently stuck LOW. Recall that in all latches and FFs, the $\overline{Q}$ and Q outputs

are internally cross-coupled so that the level on one will affect the other. For example, take another look at the internal circuitry for a J-K FF in Figure 5-23. Note how a constant LOW at $\overline{Q}$ would keep a LOW at one input of NAND gate 3 so that Q would have to stay HIGH regardless of the conditions at J, K and *CLK*.

The technician learned a valuable lesson about troubleshooting FF circuits. He learned that both outputs should be checked for faults, even those that are not connected to other devices.

Clock Skew One of the most common timing problems in synchronous circuits is *clock skew*. One type of clock skew occurs when a clock signal, because of propagation delays, arrives at the *CLK* inputs of different FFs at different times. In many situations the skew can cause a FF to go to a wrong state. This is best illustrated with an example.

Refer to Figure 5-63(a) where the signal CLOCK1 is connected directly to FF Q_1, and indirectly to Q_2 through a NAND gate and INVERTER. Both FFs are supposed to be clocked by the occurrence of a NGT of CLOCK1 provided that X is HIGH. If we assume that initially $Q_1 = Q_2 = 0$ and $X = 1$, the NGT of CLOCK1 should set $Q_1 = 1$ and have no effect on Q_2. The waveforms in Figure 5-63(b) show how clock skew can produce incorrect triggering of Q_2.

Because of the combined propogation delays of the NAND gate and INVERTER, the transitions of the CLOCK2 signal are delayed with respect to CLOCK1 by an amount of time t_1. The NGT of CLOCK2 arrives at Q_2's *CLK* input t_1 later than the NGT of CLOCK1 appears at Q_1's *CLK* input. This t_1 is the clock skew. The NGT of CLOCK1 will cause Q_1 to go HIGH after a time t_2 which is equal to Q_1's t_{PLH} propagation delay. If t_2 is less than the skew t_1, Q_1 will be HIGH when the NGT of CLOCK2 occurs, and this may incorrectly set $Q_2 = 1$ if its setup time requirement, t_S, is met.

For example, assume that the clock skew is 40 ns and the t_{PLH} of Q_1 is 25 ns. Thus, Q_1 will go HIGH 15 ns before the NGT of CLOCK2. IF Q_2's setup time requirement is smaller than 15 ns, Q_2 will respond to the HIGH at its D input when the NGT of CLOCK2 occurs and Q_2 will go HIGH. This, of course, is not the expected response of Q_2. It is supposed to remain LOW.

The effects of clock skew are not always easy to detect because the response of the affected FF may be intermittent (sometimes it works correctly, sometimes it doesn't). This is because the situation is dependent on circuit propagation delays and FF timing parameters, which vary with temperature, length of connections, power supply voltage, and loading. Sometimes just connecting an oscilloscope probe to a FF or gate output will add enough load capacitance to increase the device's propagation delay so that the circuit functions correctly; then when the probe is removed, the incorrect operation reappears. This is the kind of situation that explains why some technicians are prematurely gray.

Problems caused by clock skew can be eliminated by equalizing the delays in the various paths of the clock signal so that the active transition arrives at each FF at approximately the same time. Another soluton involves the use of M/S FFs with data lockout (see Problems 5-46 and 5-47).

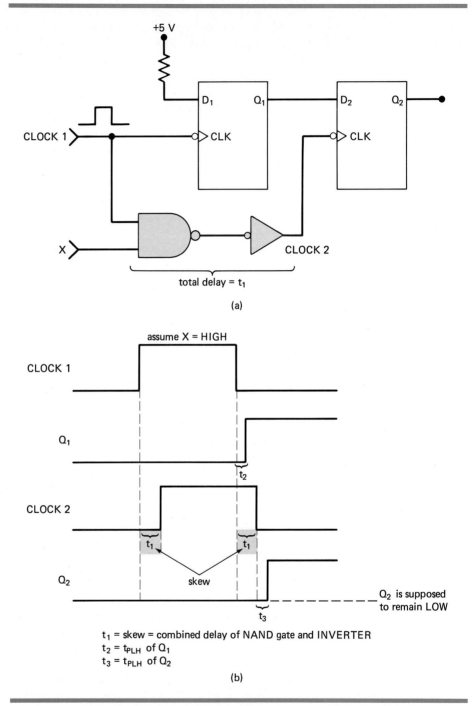

FIGURE 5-63 Clock skew occurs when two-flip-flops that are supposed to be clocked simultaneously are clocked at slightly different times due to a delay in the arrival of the clock signal at the second FF.

REVIEW QUESTIONS

1. What is clock skew? How can it cause a problem?

5-25 FLIP-FLOP SUMMARY

1. NOR gate LATCH (Figure 5-64)

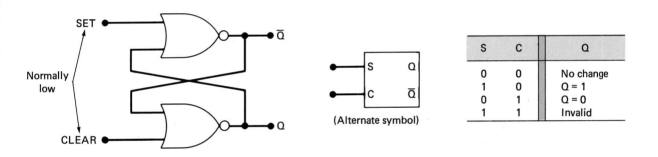

S	C	Q
0	0	No change
1	0	Q = 1
0	1	Q = 0
1	1	Invalid

FIGURE 5-64

2. NAND gate LATCH (Figure 5-65)

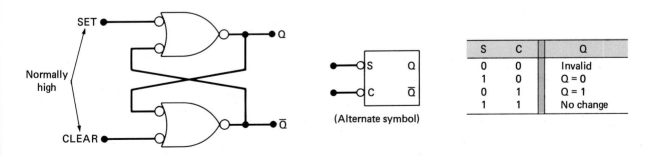

S	C	Q
0	0	Invalid
1	0	Q = 0
0	1	Q = 1
1	1	No change

FIGURE 5-65

3. Edge-triggered S-C FF (Figure 5-66)

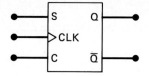

S	C		Q
0	0		Q_0 (no change)
1	0		1
0	1		0
1	1		Ambiguous

FIGURE 5-66

4. Edge-triggered J-K FF (Figure 5-67)

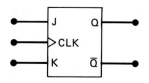

J	K		Q
0	0		Q_0 (no change)
1	0		1
0	1		0
1	1		$\overline{Q}_0$ (toggles)

FIGURE 5-67

5. Edge-triggered D FF (Figure 5-68)

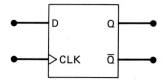

D		Q
0		0
1		1

FIGURE 5-68

6. D LATCH (Figure 5-69)

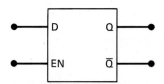

EN	D		Q*
0	X		No change
1	0		0
1	1		1

Q* follows D input while EN is HIGH

FIGURE 5-69

Asynchronous inputs (Figure 5-70)

PRESET	CLEAR	Q
1	1	No effect; FF can respond to J, K and CLK
1	0	Q = 0 independent of synchronous inputs
0	1	Q = 1 independent of synchronous inputs
0	0	Ambiguous (not used)

Master/slave FFs (Figure 5-71)

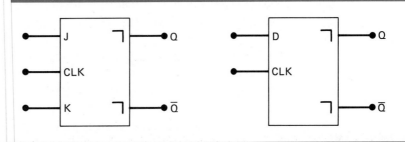

FIGURE 5-71 Data from control inputs (*J, K, D*) are entered into master when *CLK* is HIGH; contents of master is transferred to slave (Q) when *CLK* goes, LOW.

9. M/S FFs with data lockout (Figure 5-72)

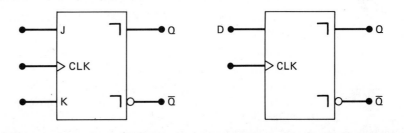

FIGURE 5-72 Data from control inputs (*J, K,* D) are entered into master on PGT of *CLK* input. Contents of master are transferred to slave (*Q*) when *CLK* goes LOW.

GLOSSARY

Astable Multivibrator Digital circuit that oscillates between two unstable output states. [Sec. 5.23]

Asynchronous Active Pulse Width Minimum time duration that a DC SET or DC CLEAR input has to be kept in its active state to reliably SET or CLEAR the flip-flop. [Sec. 5.11]

Asynchronous Inputs Flip-flop inputs that can affect the operation of the flip-flop independent of the synchronous and clock inputs. [Sec. 5.9]

Asynchronous Systems Systems in which outputs can change states any time that one or more of the inputs change. [Sec. 5.4]

Asynchronous Transfer Data transfer performed without the aid of the clock. [Sec. 5.17]

Binary Counter Group of flip-flops connected in a special arrangement in which the states of the flip-flops represent the binary number equivalent to the number of pulses that have occurred at the input of the counter. [Sec. 5.19]

Bistable Multivibrator Name that is sometimes used to describe a flip-flop. [Sec. 5.23]

Cleared State The $Q = 0$ state of a flip-flop. [Sec. 5.1]

Clock Digital signal in the form of a rectangular pulse train or a squarewave. [Sec. 5.4]

Clock Pulse High $[t_W(H)]$ Minimum time duration that a clock signal must remain HIGH before going LOW. [Sec. 5.11]

Clock Pulse Low $[t_W(L)]$ Minimum time duration that a clock signal must remain LOW before going HIGH. [Sec. 5.11]

Clock Skew When, because of propagation delays, a clock signal arrives at the clock inputs of different flip-flops at different times. [Sec. 5.24]

Clock Transition Times Specified by the manufacturer of a particular IC for the minimum rise and fall times of the clock signal transitions used by that IC. [Sec. 5.11]

Clocked D Flip-Flop Type of flip-flop where the D (data) input is the synchronous input. [Sec. 5.7]

Clocked Flip-Flops Flip-flops that have a clock input. [Sec. 5.4]

Clocked J-K Flip-Flop Type of flip-flop where the inputs J and K are the synchronous inputs. [Sec. 5.6]

Clocked S-C Flip-Flop Type of flip-flop where the inputs SET and CLEAR are the synchronous inputs. [Sec. 5.5]

Common-Control Block Symbol used by the IEEE/ANSI standard to describe when one or more inputs are common to more than one of the circuits in an IC. [Sec. 5.10]

Contact Bounce Random voltage transitions produced by operating a mechanical switch. [Sec. 5.1]

Control Inputs Control input signals synchronized with the active clock transition determine the output state of a flip-flop. [Sec. 5.4]

D Latch Circuit that contains a NAND gate latch and two steering NAND gates. [Sec. 5.8]

Data Lockout Feature on some master/slave flip-flops by which the master flip-flop is disabled a short time after the positive-going clock transition. [Sec. 5.13]

DC CLEAR Asynchronous flip-flop input used to clear Q immediately to 0. [Sec. 5.9]

DC SET Asynchronous flip-flop input used to set Q immediately to 1. [Sec. 5.9]

Edge Detector Circuit which produces a narrow positive spike that occurs coincident with the active transition of a clock input pulse. [Sec. 5.5]

Edge-Triggered Manner in which a flip-flop is activated by a signal transition. It may be either a positive or negative edge-triggered flip-flop. [Sec. 5.4]

555 Timer TTL-compatible IC which can be wired to operate in several different modes, such as a one-shot and an astable multivibrator. [Sec. 5.23]

Free-Running Multivibrator *See* Astable Multivibrator. [Sec. 5.23]

Hold Time (t_H) Time interval immediately following the active transition of the clock signal during which the control input has to be maintained at the proper level. [Secs. 5.4 and 5.11]

Jam Transfer *See* Asynchronous Transfer. [Sec. 5.17]

Latch A type of flip-flop. [Sec. 5.1]

Master/Slave Flip-Flops Flip-flops which have as their internal structure two flip-flops—a master and a slave. [Sec. 5.13]

Maximum Clocking Frequency (f_{MAX}) Highest frequency that may be applied to the clock input of a flip-flop and still have it trigger reliably. [Sec. 5.11]

MOD Number Number of different states that a counter can sequence through. [Sec. 5.19]

Monostable Multivibrator *See* One-Shot. [Sec. 5.21]

NAND-Gate Latch Flip-flop constructed from two NAND gates. [Sec. 5.1]

Negative-Going Threshold (V_{T-}) Voltage level inherent to a Schmitt-trigger circuit which if dropped below will cause the output to rapidly change. [Sec. 5.20]

Negative-Going Transition (NGT) When a clock signal changes from a logic 1 to a logic 0. [Sec. 5.4]

Noise Spurious voltage fluctuations that may be present in the environment and cause digital circuits to malfunction. [Sec. 5.24]

Nonretriggerable One-Shot Type of one-shot that will not respond to a trigger input signal while in its quasi-state. [Sec. 5.21]

NOR-Gate Latch Flip-flop constructed from two NOR gates. [Sec. 5.2]

One-Shot Circuit that belongs to the flip-flop family but which has only one stable state (normally $Q = 0$). [Sec. 5.21]

Override Inputs Synonymous with asynchronous inputs. [Sec. 5.9]

Parallel Data Transfer Operation by which the entire contents of a register are transferred simultaneously to another register. [Sec. 5.17]

Positive-Going Threshold (V_{T+}) Voltage level inherent to a Schmitt-trigger circuit which if exceeded will cause the output to rapidly change. [Sec. 5.20]

Positive-Going Transition (PGT) When a clock signal changes from a logic 0 to a logic 1. [Sec. 5.4]

Postponed Output Indicator Symbol used by the IEEE/ANSI standard to show the fact that the effects of the control inputs of a master/slave FF do not reach the output Q until the clock returns to the LOW state. [Sec. 5.13]

PRESET Term synonymous with DC SET. [Sec. 5.9]

Propagation Delays (t_{PLH}/t_{PHL}) Delay from the time a signal is applied to the time when the output makes its change. [Sec. 5.11]

Pulse-Triggered Flip-Flop Name sometimes used in reference to a master-slave flip-flop. [Sec. 5.13]

Quasi-Stable State to which a one-shot is temporarily triggered (normally, $Q = 1$) before returning to its normal state (normally, $Q = 0$). [Sec. 5.21]

Registers Group of flip-flops capable of storing data. [Sec. 5.17]

RESET Term synonymous with DC CLEAR. [Sec. 5.9]

Reset State The $Q = 0$ state of a flip-flop. [Sec. 5.1]

Retriggerable One-Shot Type of one-shot that will respond to a trigger input signal while in its quasi-stable state. [Sec. 5.21]

Schmitt Trigger Digital circuit that accepts a slow-changing input signal and produces a rapid oscillation-free transition at the output. [Sec. 5.20]

Serial Data Transfer When data are transferred from one place to another one bit at a time. [Secs. 5.17 and 5.18]

Set State The $Q = 1$ state of a flip-flop. [Sec. 5.1]

Setup Time (t_S) Time interval immediately preceding the active transition of the clock signal during which the control input has to be maintained at the proper level. [Secs. 5.4 and 5.11]

Shift Register Digital circuit that accepts binary data from some input source and then shifts these data through a chain of flip-flops one bit at a time. [Sec. 5.18]

Synchronous Inputs *See* Control Inputs. [Sec. 5.9]

Synchronous Systems Systems in which the circuit outputs can change states only on the transitions of a clock. [Sec. 5.4]

Synchronous Transfer Data transfer performed by using the synchronous and clock inputs of a flip-flop. [Sec. 5.17]

Toggle Mode When a flip-flop changes states for each clock pulse. [Sec. 5.6]

Transparent In a D latch when the Q output follows the D input, the device is said to be transparent. [Sec. 5.8]

Trigger Input signal to a flip-flop or one-shot which causes the output to change states depending on the conditions of the control signals. [Sec. 5.5]

PROBLEMS   Sections 5-1–5-3

5-1.   Assume that $Q = 0$ initially, apply the x and y waveforms of Figure 5-73 to the SET and CLEAR inputs of a NAND latch and determine the Q and $\overline{Q}$ waveforms.

5-2.   Invert the x and y waveforms of Figure 5-73, apply them to the SET and CLEAR inputs of a NOR latch and determine the Q and $\overline{Q}$ waveforms. Assume that $Q = 0$ initially.

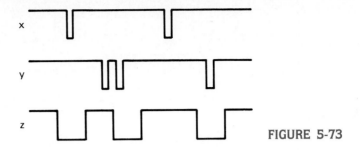

FIGURE 5-73

5-3. The waveforms of Figure 5-73 are connected to the circuit of Figure 5-74. Assume that $Q = 0$ initially, and determine the Q waveform.

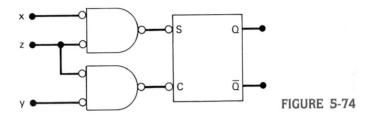

FIGURE 5-74

5-4. Modify the circuit of Figure 5-9 to use a NOR gate latch.

5-5. Modify the circuit of Figure 5-12 to use a NAND gate latch.

5-6. Refer to the circuit of Figure 5-13. A technician tests the circuit operation by observing the outputs with a storage oscilloscope while the switch is moved between A and B. When the switch is moved from A to B, the scope display of X_B appears as shown in Figure 5-75. What circuit fault could produce this result? (*Hint:* What is the function of the NAND latch?)

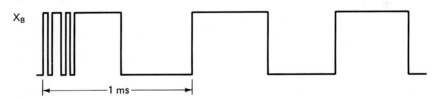

FIGURE 5-75

Sections 5-4 and 5-5

5-7. A certain clocked FF is specified to have $t_S = 20$ ns and $t_H = 5$ ns. How long must the control inputs be stable prior to the active clock transition?

5-8. Apply the S, C, and CLK waveforms of Figure 5-17 to the FF of Figure 5-18 and determine the Q waveform.

5-9. A *toggle* FF is one that has a single input and operates such that the FF output changes state for each pulse applied to its input. The clocked S-C FF can be wired to operate in the toggle mode, as shown in Figure 5-76. The waveform applied to the *CLK* input is a 1-kHz squarewave. Verify that this arrangement operates in the toggle mode, then determine the Q output waveform. Assume that $Q = 0$ initially.

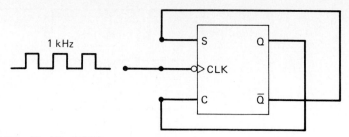

FIGURE 5-76   Problem 5-9

Section 5-6

5-10. Apply the *J*, *K*, and *CLK* waveforms of Figure 5-21 to the FF of Figure 5-22. Assume that $Q = 1$ initially and determine the Q waveform.

5-11. Show how the J-K FF can be operated as a toggle FF. Apply a 10-kHz squarewave to its input and determine its output waveform.

5-12. Connect the Q output of the FF from Problem 5-11 to the *CLK* input of a second J-K FF that also has $J = K = 1$. Determine the frequency of the waveform at the second FF output.

Section 5-7

5-13. A *D*-type FF is sometimes used to *delay* a binary waveform so that the binary information appears at the output a certain amount of time after it appears at the *D* input. Determine the Q waveform in Figure 5-77 and compare it to the input waveform. Note that it is delayed from the input by one clock period. How can a delay of two clock periods be obtained?

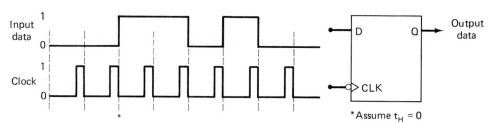

FIGURE 5-77

5-14. An edge-triggered D FF can be made to operate in the toggle mode by connecting it as shown in Figure 5-78. Assume that $Q = 0$ initially and determine the Q waveform.

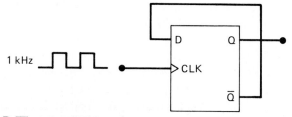

FIGURE 5-78   D FF connected toggle.

5-15. Change the circuit in Figure 5-78 so that Q is connected back to D. Then determine the Q waveform.

Section 5-8

5-16. Compare the operation of the D latch with a negative-edge triggered D FF by applying the waveforms of Figure 5-79 to each and determining the Q waveforms.

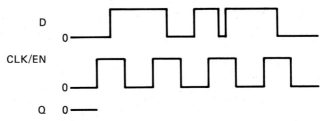

FIGURE 5-79

5-17. In Problem 5-14 we saw how an edge-triggered D FF can be operated in the toggle mode. Explain why this same idea will not work for a D latch.

Section 5-9

5-18. Determine the Q waveform for the FF in Figure 5-80. Assume that $Q = 0$ initially and remember that the asynchronous inputs override all other inputs.

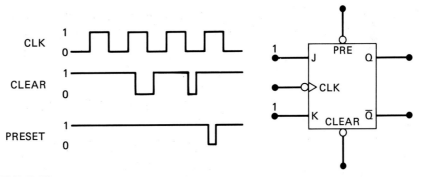

FIGURE 5-80

5-19. Apply the *CLK*, PRESET, and CLEAR waveforms of Figure 5-31 to a positive-edge-triggered D FF with active-LOW asynchronous inputs. Assume that D is kept HIGH, and Q is initially LOW. Determine the Q waveform.

Section 5-11

5-20. Use Table 5-1 in Section 5-11 to determine the following:
(a) How long can it take for the Q output of a 4013B to switch from 0 to 1 in response to an active *CLK* transition?

(b) Which FF in Table 5-1 requires its control inputs to remain stable for the longest time after the active *CLK* transition?

(c) What is the narrowest pulse that can be applied to the DC SET of a 7474 FF?

5-21. Refer to the circuit of Figure 5-81. It shows the two J-K FFs on a 74LS112 chip connected such that the Q_1 output serves as the *CLK* input for the Q_2 FF. Assume that $Q_1 = Q_2 = 1$ initially, and determine the *total* propagation delay between the NGT of the clock pulse and the NGT of Q_2.

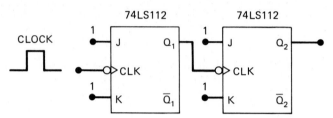

FIGURE 5-81

Section 5-13

5-22. The waveforms shown in Figure 5-82 are to be applied to four different FFs: (a) positive-edge-triggered J-K; (b) negative-edge-triggered J-K; (c) master/slave FF; and (d) M/S FF with data lockout. Draw the Q waveform response for each of these FFs, assuming that $Q = 0$ initially. Assume that each FF has $t_H = 0$.

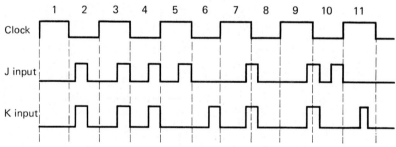

FIGURE 5-82

Sections 5-15 and 5-16

5-23. Modify the circuit of Figure 5-45 to use a J-K FF.

5-24. In the circuit of Figure 5-83 the inputs A, B, and C are all initially LOW. Output Y is supposed to go HIGH only when A, B, and C go HIGH in a certain sequence.

(a) Determine the sequence that will make Y go HIGH.

(b) Explain why the START pulse is needed.

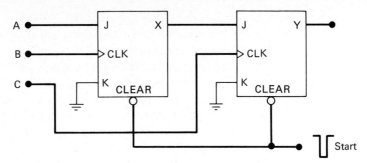

FIGURE 5-83

Sections 5-17 and 5-18

5-25. Draw the circuit diagram for the synchronous parallel transfer of data from one 3-bit register to another using J-K FFs.

5-26. Repeat Problem 5-25 for asynchronous parallel transfer.

5-27. A *recirculating* shift register is a shift register that keeps the binary information circulating through the register as clock pulses are applied. The shift register of Figure 5-50 can be made into a circulating register by connecting X_0 to the DATA IN line. No external inputs are used. Assume that this circulating register starts out with 1011 stored in it (that is, X_3, = 1, X_2 = 0, X_1 = 1, and X_0 = 1). List the sequence of states that the register FFs go through as eight shift pulses are applied.

5-28. Refer to Figure 5-51, where a 3-bit number stored in register X is serially shifted into register Y. How could the circuit be modified so that at the end of the transfer operation, the original number stored in X is present in both registers? (*Hint:* See Problem 5-27.)

Section 5-19

5-29. Refer to the binary counter of Figure 5-52. Change it by connecting $\overline{X}_0$ to the *CLK* of FF X_1, and $\overline{X}_1$ to the *CLK* of FF X_2. Start out with all FFs in the 1 state and draw the various FF output waveforms (X_0, X_1, X_2) for 16 input pulses. Then list the sequence of FF states as was done in Figure 5-53. This counter is called a *down counter*. Why?

5-30. Show how clocked D FFs can be used in a counter such as that in Figure 5-52. (*Hint:* See Problem 5-14.)

5-31. (a) How many FFs are required to build a binary counter circuit that counts from 0 to 1023?

(b) Determine the frequency at the output of the last FF of this counter for an input clock frequency of 2 MHz.

(c) What is the counter's MOD number?

5-32. A certain counter is being pulsed by a 256-kHz clock signal. The output frequency from the last FF is 2-kHz.

(a) Determine the MOD number.

(b) Determine the counting range.

5-33. A photodetecter circuit is being used to generate a pulse each time a customer walks into a certain establishment. The pulses are fed to an 8-bit counter. The counter is used to count these pulses as a means for determining how many customers have entered the store. After closing the store, the proprietor checks the counter and finds that it shows a count of 00001001_2 $= 9_{10}$. He knows that this is incorrect, because there were many more than nine people in his store. Assuming that the counter circuit is working properly, what could be the reason for the discrepancy?

Section 5-21

5-34. Determine the waveforms at Q_1, Q_2, and Q_3 in response to the single input pulse in Figure 5-84.

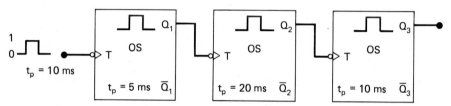

FIGURE 5-84

5-35. A retriggerable OS can be used as a pulse-frequency detector that detects when the frequency of a pulse input is below a predetermined value. A simple example of this application is shown in Figure 5-85. The operation begins by momentarily closing switch S_1.
(a) Describe how the circuit responds to input frequencies above 1 kHz.
(b) Describe how the circuit responds to input frequencies below 1 kHz.
(c) How would you modify the circuit to detect when the input frequency drops below 50 kHz?

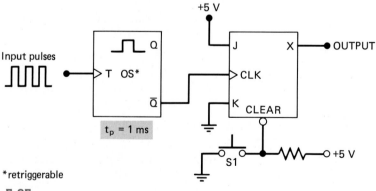

FIGURE 5-85

5-36. Refer to the logic symbol for a 74121 nonretriggerable one-shot in Figure 5-57(a).

(a) What input conditions are necessary for the OS to be triggered by a signal at the B input?

(b) What input conditions are necessary for the OS to be triggered by a signal at the A_1 input?

5-37. The output pulse width from a 74121 OS is given by the approximate formula

$$t_p \approx 0.7 R_T C_T$$

where R_T is the resistance connected between the R_{EXT}/C_{EXT} pin and V_{CC}, and C_T is the capacitance connected between the C_{EXT} pin and the R_{EXT}/C_{EXT} pin. The value for R_T can be varied between 2 and 40 kΩ, and C_T can be as large as 1000 μF.

(a) Show how a 74121 can be connected to produce a negative-going pulse with a 5-ms duration whenever either of two logic signals (E or F) makes a NGT. Both E and F are normally in the HIGH state.

(b) Modify the circuit so that a control input signal, G, will inhibit the OS output pulse regardless of what occurs at E or F.

Section 5-22

5-38. Consider the circuit of Figure 5-86. Initially all FFs are in the 0 state. The circuit operation begins with a momentary start pulse applied to the PRESET inputs of FFs X and Y. Determine the waveforms at A, B, C, X, Y, Z, and W for 20 cycles of the clock pulses after the start pulse. State all assumptions.

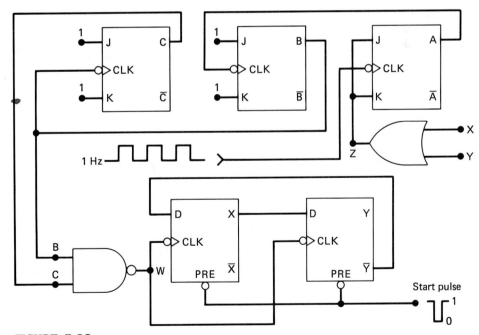

FIGURE 5-86

5-39. Show how to use a 74LS14 Schmitt-trigger INVERTER to produce an approximate squarewave with a frequency of 10 kHz.

5-40. Design a 555 free-running oscillator to produce an approximate squarewave at 40 kHz. *C* should be kept at 100 PF or greater.

5-41. A 555 oscillator can be combined with a J-K FF to produce a perfect (50% duty cycle) squarewave. Modify the circuit of Problem 5-40 to include a J-K FF. The final output is still to be a 40-kHz squarewave.

5-42. The circuit in Figure 5-87 can be used to generate two nonoverlapping clock signals at the same frequency. These clock signals are used in some microprocessor systems that require four different clock transitions to synchronize their operations. Draw the CP1 and CP2 waveforms in response to a 1-MHz input clock frequency.

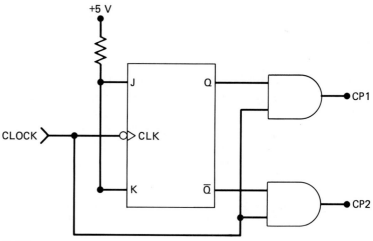

FIGURE 5-87

5-43. Refer to the counter circuit in Figure 5-52. Assume that all asynchronous inputs are connected to V_{CC}. When tested, the circuit waveforms appear as shown in Figure 5-88. Consider the following list of possible faults. For each one indicate "yes" or "no" as to whether it could cause the observed results. Explain each response.

(a) *CLR* input of X_2 is open.

(b) X_1 output's transition times are too long, possibly due to loading.

(c) X_2 output shorted to ground.

(d) X_2's hold-time requirement is not being met.

5-44. Refer to the circuit of Figure 5-51. All FFs are TTL ICs. Assume the following initial conditions: $X_2X_1X_0 = 100$ and $Y_2Y_1Y_0 = 011$. After four shift pulses, the conditions are $X_2X_1X_0 = 001$ and $Y_2Y_1Y_1 = 111$. Subsequent shift pulses produce no change in any of the FFs. What are some of the possible causes of this faulty operation?

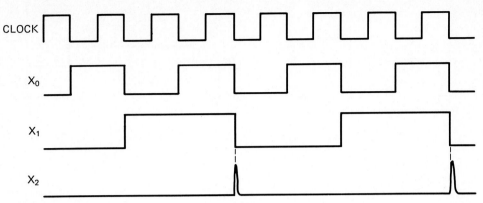

FIGURE 5-88

5-45. Consider the situation of Figure 5-63 for each of the following sets of timing values. For each indicate whether or not FF Q_2 will respond correctly.
    (a) *Each FF:* $t_{PLH} = 12$ ns; $t_{PHL} = 8$ ns; $t_S = 5$ ns; $t_H = 0$ ns
        NAND gate: $t_{PLH} = 8$ ns; $t_{PHL} = 6$ ns
        INVERTER: $t_{PLH} = 7$ ns; $t_{PHL} = 5$ ns
    (b) *Each FF:* $t_{PLH} = 10$ ns; $t_{PHL} = 8$ ns; $t_S = 5$ ns; $t_H = 0$ ns
        NAND gate: $t_{PLH} = 12$ ns; $t_{PHL} = 10$ ns
        INVERTER: $t_{PLH} = 8$ ns; $t_{PHL} = 6$ ns

5-46. Show and explain how the clock skew problem in Figure 5-63 can be eliminated by the appropriate insertion of two INVERTERS.

5-47. Explain how the use of M/S FFs with data lockout can eliminate the clock skew problem of Figure 5-63.

5-48. Refer to the circuit of Figure 5-58. Describe how the circuit operation will change for each of the following faults.
    (a) An internal short to ground at the NAND gate's top input.
    (b) An internal short between the OR gate inputs.
    (c) An open in the OS timing resistor.

5-49. Refer to the circuit of Figure 5-89. Assume that the ICs are of the TTL logic family. The Q waveform was obtained when the circuit was tested with the input signals shown and with the switch in the "up" position; it is not correct. Consider the following list of faults, and for each indicate "yes" or "no" as to whether it could be the actual fault. Explain each response.
    (a) Point X is always LOW due to a faulty switch.
    (b) Z1-pin 1 is internally shorted to V_{CC}.
    (c) The connection from Z1-3 to Z2-3 is broken.
    (d) There is a solder bridge between pins 6 and 7 of Z1.

5-50. The circuit of Figure 5-90 functions as a sequential combination lock. To operate the lock, proceed as follows:
    1. Momentarily activate the RESET switch.
    2. Set the switches SWA, SWB, and SWC to the first part of the combination. Then momentarily toggle the ENTER switch back and forth.

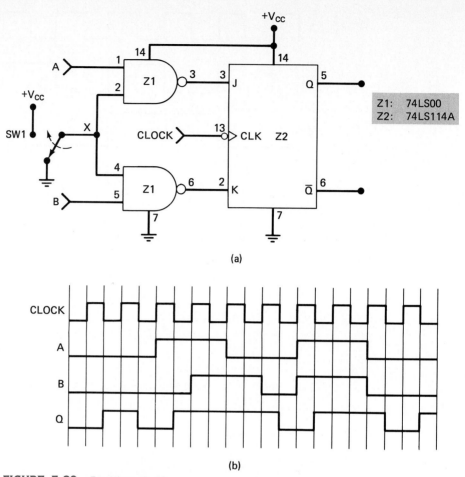

(a)

(b)

FIGURE 5-89   Problem 5-49

3. Set the switches to the second part of the combination, and toggle EN-TER again. This should produce a HIGH at Q_2 to open the lock.

   If the incorrect combination is entered in either step, the operator must start the sequence over. Analyze the circuit and determine the correct sequence of combinations that will open the lock.

5-51. When the combination lock of Figure 5-90 is tested, it is found that entering the correct combination does not open the lock. A logic probe check shows that entering the correct first combination sets Q_1 HIGH, but entering the correct second combination produces only a momentary pulse at Q_2. Consider each of the following faults and indicate which one(s) could produce the observed operation. Explain each choice.

(a) Switch bounce at SWA, SWB, or SWC.

(b) *CLR* input of Q_2 is open.

(c) Connection from NAND gate 4 output to NAND gate 3 input is open.

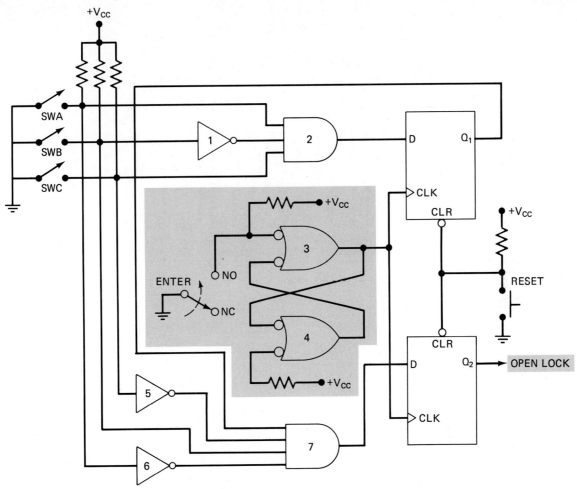

FIGURE 5-90   Problems 5-50 and 5-51

DIGITAL ARITHMETIC: OPERATIONS AND CIRCUITS

6

Digital computers and calculators perform the various arithmetic operations on numbers that are represented in binary form. The subject of digital arithmetic can be a very complex one if we want to understand all the various methods of computation and the theory behind them. Fortunately, this level of knowledge is not required by most technicians, at least not until they become experienced computer programmers. Our approach in this chapter will be to concentrate on those basic principles that are needed to understand how digital machines (i.e., computers) perform the basic arithmetic operations.

First we will see how the various arithmetic operations are performed on binary numbers using "pencil and paper," and then we will study the actual logic circuits that perform these operations in a digital system.

6-1 BINARY ADDITION

The addition of two binary numbers is performed in exactly the same manner as the addition of decimal numbers. In fact, binary addition is simpler, since there are fewer cases to learn. Let us first review decimal addition:

$$
\begin{array}{ccc}
3 & 7 & 6 \quad \text{LSD} \\
+4 & 6 & 1 \\
\hline
8 & 3 & 7
\end{array}
$$

The least-significant-digit (LSD) position is operated on first, producing a sum of 7. The digits in the second position are then added to produce a sum of 13, which produces a *carry* of 1 into the third position. This produces a sum of 8 in the third position.

The same general steps are followed in binary addition. However, only four cases can occur in adding the two binary digits (bits) in any position. They are:

$$0 + 0 = 0$$

$$1 + 0 = 1$$

$$1 + 1 = 10 = 0 + \text{carry of 1 into next position}$$

$$1 + 1 + 1 = 11 = 1 + \text{carry of 1 into next position}$$

The last case occurs when the two bits in a certain position are 1 and there is a carry from the previous position. Here are several examples of the addition of two binary numbers:

$$
\begin{array}{r}
011\ (3) \\
+\ 110\ (6) \\
\hline
1001\ (9)
\end{array}
\qquad
\begin{array}{r}
1001\ (9) \\
+\ 1111\ (15) \\
\hline
11000\ (24)
\end{array}
\qquad
\begin{array}{r}
11.011\ (3.375) \\
+\ 10.110\ (2.750) \\
\hline
110.001\ (6.125)
\end{array}
$$

It is not necessary to consider the addition of more than two binary numbers at a time, because in all digital systems the circuitry that actually performs the addition can handle only two numbers at a time. When more than two numbers are to be added, the first two are added together and then their sum is added to the third number; and so on. This is not a serious drawback, since modern digital computers can typically perform an addition operation in a few microseconds.

Addition is the most important arithmetic operation in digital systems. As we shall see, the operations of subtraction, multiplication, and division as they are performed in most modern digital computers and calculators actually use only addition as their basic operation.

Binary Addition versus Logical OR Addition   It is important to understand the difference between the OR-addition operation and binary addition. OR addition is the Boolean *logic* operation performed by an OR gate, which produces a 1 output whenever any one or more inputs are 1. Binary addition is an *arithmetic* operation that produces an arithmetic sum of two binary numbers. Although the $+$ sign is used for both OR addition and binary addition, the meaning of the $+$ sign will usually be apparent from the context in which it is being used. The major differences between OR addition and binary addition can be summarized as follows:

OR Addition	*Binary Addition*
$1 + 1 = 1$	$1 + 1 = 0 + \text{carry of 1}$
$1 + 1 + 1 = 1$	$1 + 1 + 1 = 1 + \text{carry of 1}$

REVIEW QUESTION

1. Add the following pairs of binary numbers: (a) 10110 + 00111, (b) 011.101 + 010.010, (c) 10001111 + 00000001. (*Ans.* 11101, 101.111, 10010000)

6-2 REPRESENTING SIGNED NUMBERS

In digital computers, the binary numbers are represented by a set of binary storage devices (usually flip-flops). Each device represents one bit. For example, a 6-bit FF register could store binary numbers ranging from 000000 to 111111 (0 to 63 in decimal). This represents the *magnitude* of the number. Since most digital computers and calculators handle negative as well as positive numbers, some means is required for representing the *sign* of the number ($+$ or $-$). This is usually done by adding another bit to the number called the *sign bit*. In general, the common convention which has been adopted is that a 0 in the sign bit represents a positive number and a 1 in the sign bit represents a negative number. This is illustrated in Figure 6-1. Register A contains the bits 0110100. The 0 in the leftmost bit (A_6) is the sign bit that represents $+$. The other six bits are the magnitude of the number 110100_2, which is equal to 52 in decimal. Thus, the number stored in the A register is $+52$. Similarly, the number stored in the B register is -31, since the sign bit is 1, representing $-$.

The sign bit is used to indicate whether a stored binary number is positive or negative. For *positive* numbers, the rest of the bits are always used to represent the magnitude of the number in binary form. For *negative* numbers, however, there are three possible ways to represent the magnitude: true-magnitude form, 1's-complement form, and 2's-complement form.

True-Magnitude Form The numbers in Figure 6-1 contain a sign bit and 6 magnitude bits. The magnitude bits are the true binary equivalent of the decimal values being represented. We call this the *true magnitude form* for representing signed binary numbers.

Although this true-magnitude system is straightforward and easy to understand, it is not as useful as two other systems for representing signed binary numbers. These other systems use the same true-magnitude form for *positive* numbers, but they use a different form for negative numbers.

FIGURE 6-1   Representation of signed numbers in true magnitude form.

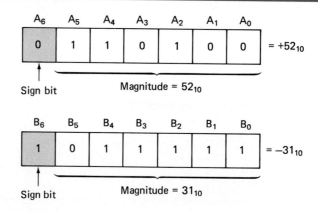

1's-Complement Form

The *1's-complement form* of any binary number is obtained simply by changing each 0 in the number to a 1 and each 1 in the number to a 0. In other words, change each bit to its complement. For example, the 1's complement of 101101 is 010010, and the 1's complement of 011010 is 100101.

When negative numbers are represented in 1's-complement form, the sign bit is made a 1 and the magnitude is converted from true binary form to its 1's complement. To illustrate, the number -57 would be represented as follows:

sign bits ─────────────────┐
                           ↓

$$-57 = \boxed{1}\ 111001 \quad \text{(true-magnitude form)}$$
$$ = \boxed{1}\ 000110 \quad \text{(1's-complement form)}$$

Note that the sign bit is not complemented but is kept as a 1 to indicate a negative number.

Some additional examples of representing negative numbers in 1's complement form are given below and should be verified by the reader. Again, the leftmost bit is the sign bit.

$$-14 = 10001$$
$$-326 = 1010111001$$
$$-7 = 1000$$

2's-Complement Form

The *2's-complement form* of a binary number is formed simply by taking the 1's complement of the number and adding 1 to the least-significant bit position. The procedure is illustrated below for converting 111001 (decimal 57) to its 2's-complement form.

```
  1   1   1   0   0   1      complement each bit to form 1's complement
  ↓   ↓   ↓   ↓   ↓   ↓
  0   0   0   1   1   0
                    +1        add 1 to LSB to form 2's complement
  ─────────────────────
  0   0   0   1   1   1
```

Thus, using the 2's-complement representation, -57_{10} would be represented as

$$-57_{10} = 1 \quad 0 \quad 0 \quad 0 \quad 1 \quad 1 \quad 1 \quad \text{(2's-complement form)}$$

sign bit ─────┘↑

where again the leftmost bit is the sign bit, and the other 6 bits are the 2's-complement form of the magnitude.

The three systems of representing signed numbers are summarized in Figure 6-2 for $+57$ and -57. In all three systems the representation of positive numbers is the same: a sign bit of 0 followed by the true binary equivalent of the magnitude of the number. The representation of negative number starts with a sign bit of 1 in all three systems, followed by either the true binary, 1's complement, or 2's complement of the magnitude. While all three systems have been used in the past, most modern computers use the 2's-complement system, so that is the one we will use from now on.

The 2's-complement system is used to represent negative numbers because, as we shall see, it allows us to perform the operation of subtraction by actually performing addition. This is significant because it means that a digital computer can use the same circuitry to both add and subtract, thereby realizing a savings of hardware.

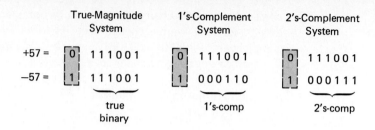

	True-Magnitude System	1's-Complement System	2's-Complement System
+57 =	0 111001	0 111001	0 111001
−57 =	1 111001	1 000110	1 000111
	true binary	1's-comp	2's-comp

FIGURE 6-2   Three systems for representing signed binary numbers.

EXAMPLE 6-1

Represent each of the following signed decimal numbers as a signed binary number in the 2's-complement system. Use a total of five bits including the sign bit:
(a) +13, (b) −9, (c) +3, (d) −2 (e) −8.

SOLUTION

(a) Since the number is positive, the magnitude (13) will be represented in its true-magnitude form—that is, $13 = 1101_2$. Attaching the sign bit of 0, we have

$$+13 = 01101$$

sign bit ⬏

(b) Since the number is negative, the magnitude (9) has to be represented in 2's-complement form:

$$9_{10} = 1001_2$$

0110	(1's complement)
+1	(add 1 to LSB)
0111	(2's complement)

When we attach the sign bit of 1, the complete signed number becomes

$$-9 = 10111$$

The procedure we have just followed required two steps. First we determined the 2's complement of the magnitude, then we attached the sign bit. This can be accomplished in one step if we include the sign bit in the 2's-complement process. For example, to find the representation for −9, we start with the representation for +9, *including the sign bit*, and we 2's-complement it in order to obtain the representation for −9.

$$+9 = \underline{01001}$$

10110	(1's comp. each bit incl. sign bit)
+1	(add 1 to LSB)
10111	(2's-comp. representation of −9)

The result is, of course, the same as before. In general, we can say that

the 2's-complement operation on a signed number will change a positive number to a negative number and vice versa.

(c) The decimal value 3 can be represented in binary using only 2 bits. However, the problem statement requires a 4-bit magnitude preceded by a sign bit. Thus, we have

$$+3_{10} = 00011$$

In many situations the number of bits is fixed by the size of the registers that will be holding the binary numbers, so that 0s may have to be added in order to fill the required number of bit positions.

(d) Start by writing $+2$ using 5 bits:

$$+2 = 00010$$

11101	(1's comp.)
+1	(add 1)
11110	(2's-comp. representation of -2)

(e) Start with $+8$:

$$+8 = 01000$$

10111	(complement each bit)
+1	(add 1)
11000	(2's-comp. representation of -8)

Converting from Complement Form to Binary

It is a relatively simple matter to take a number that is in its 1's- or 2's-complement form and convert it back to its true binary value. To go from 1's complement to true binary simply requires complementing each bit again. Similarly, to go from 2's complement to true binary requires complementing each bit and then adding 1 to the LSB. In both cases, the conversion back to binary is the same process that was used to produce the complement to begin with. The reader should verify this for a couple of binary numbers. Here's an example:

$$\text{Start with} \rightarrow 10110$$

$$\text{2's-comp.} \rightarrow 01010$$

$$\text{2's-comp. again} \rightarrow 10110$$

EXAMPLE 6-2

Each of the following numbers is a signed binary number in the 2's-complement system. Determine the decimal value in each case: (a) 01100, (b) 11010, (c) 10001.

SOLUTION

(a) The sign bit is 0 so the number is *positive,* and the other 4 bits represent the true magnitude of the number. That is, $1100_2 = 12_{10}$. Thus, the decimal number is **+12.**

(b) The sign bit is a 1 so the number is *negative,* and the other 4 bits represent the 2's complement of the magnitude. To find out what this magnitude is, we must 2's-complement it again. Remember, the 2's-complement operation will change the polarity of a signed number, so if we 2's-complement a negative binary number (including the sign bit), we will get the positive number.

```
11010      (original negative number)
00101      (1's comp.)
  +1       (add 1)
00110      (+6)
```

Since the result of 2's-complementing is a $+6$, the original number must have been -6. That is,

$$11010_2 = -6_{10}$$

(c) Follow the same procedure as in (b):

```
10001      (original neg. number)
01110      (1's comp.)
  +1       (add 1)
01111      (+15)
```

Thus, $10001 = -15$.

Special Case in 2's-Complement Representation

Whenever a signed number has a 1 in the sign bit and all 0s for the magnitude bits, then its decimal equivalent is -2^N, where N is the number of bits in the *magnitude*. For example,

$$1000 = -2^3 = -8$$

$$10000 = -2^4 = -16$$

$$100000 = -2^5 = -32$$

and so on.

Thus, we can state that the complete range of values that can be represented in the 2's-complement system having N magnitude bits is

$$-2^N \text{ to } +(2^N - 1)$$

There are a total of 2^{N+1} different values, *including* zero.

For example, here is the list of all signed numbers that can be represented in 4 bits using the 2's-complement system:

DECIMAL VALUE	SIGNED BINARY USING 2'S COMPLEMENT
$-8 = -2^3$	1000
-7	1001
-6	1010
-5	1011
-4	1100
-3	1101
-2	1110
-1	1111
$+0$	0000
$+1$	0001
$+2$	0010
$+3$	0011
$+4$	0100
$+5$	0101
$+6$	0110
$+7 = 2^3 - 1$	0111

Note that the sequence starts with $-8 = 1000$ and proceeds to $+7 = 0111$ by adding 0001 each time as in a 4-bit counter.

EXAMPLE 6-3

What is the range of *unsigned* decimal values that can be represented by 8 bits?

SOLUTION
With no sign bit, all 8 bits are used for the magnitude. Therefore, the values will range from

$$00000000_2 = 0_{10}$$

to

$$11111111_2 = 255_{10}$$

This is a total of 256 values.

EXAMPLE 6-4

What is the range of *signed* decimal values that can be represented by 8 bits?

SOLUTION
The largest negative value is

$$10000000_2 = -2^7 = -128_{10}$$

The largest positive value is

$$01111111_2 = +2^7 - 1 = +127_{10}$$

Thus, the range is -128 to $+127$; this is a total of 256 different values, including zero.

EXAMPLE 6-5

A certain computer is storing the following two signed numbers in its memory using the 2's-complement system. While executing a program, the computer is instructed to convert each number to its opposite sign: that is, change the $+31$ to -31 and change the -12 to $+12$. How will it do this?

$$00011111_2 = +31_{10}$$
$$11110100_2 = -12_{10}$$

SOLUTION
A signed number can have its polarity changed simply by performing the 2's-complement operation on the *complete* number, including the sign bit. The computer circuitry will take the signed number from memory, 2's-complement it, and put the result back in memory.

REVIEW QUESTIONS

1. Represent each of the following values as a 5-bit signed number in the 2's-complement system: (a) +13, (b) −7, (c) −16. (*Ans.* 01101, 11001, 10000)

2. Each of the following is a signed binary number in the 2's-complement system. Determine the decimal equivalent for each: (a) 100011, (b) 1000000, (c) 0111111. (*Ans.* −29, −64, +63)

3. What range of signed decimal values can be represented in 12 bits (including the sign bit)? (*Ans.* −2048 to +2047)

4. How many bits are required to represent decimal values ranging from −50 to +50? (*Ans.* Seven)

5. What is the largest negative decimal value that can be represented using a total of 16 bits? (*Ans.* −32,768)

6. Perform the 2's-complement operation on each of the following: (a) 10000, (b) 10000000, (c) 1000. (*Ans.* 10000, 10000000, 1000)

6-3 ADDITION IN THE 2'S-COMPLEMENT SYSTEM

The 1's-complement system and 2's-complement system are very similar. However, the 2's-complement system is generally in use because of several advantages it enjoys in circuit implementation. We will now investigate how the operations of addition and subtraction are performed in digital machines that use the 2's-complement representation for negative numbers. In the various cases to be considered, it is important to note that the sign bit of each number is operated on in the same manner as the magnitude bits.

Case I: Two Positive Numbers. The addition of two positive numbers is straight-forward. Consider the addition of +9 and +4:

$$
\begin{array}{rcl}
+9 \longrightarrow & 0 \;\; 1001 & \text{(augend)} \\
+4 \longrightarrow & 0 \;\; 0100 & \text{(addend)} \\
\hline
& 0 \;\; 1101 & \text{(sum } = +13) \\
& \uparrow & \\
& \text{sign bits} &
\end{array}
$$

Note that the sign bits of the *augend* and *addend* are both 0 and the sign bit of the sum is 0, indicating that the sum is positive. Also note that the augend and addend are made to have the same number of bits. This must *always* be done in the 2's complement system.

Case II: Positive Number and Smaller Negative Number. Consider the addition of +9 and −4. Remember that the −4 will be in its 2's-complement form. Thus, +4(00100) must be converted to −4(11100)

```
                    ┌─sign bits
                    │
  +9 ─────→    │0│ 1001      (augend)
  −4 ─────→    │1│ 1100      (addend)
              ───────────
       ⟋ │0│ 0101
        └────this carry is disregarded, so the result is 00101 (sum = +5)
```

In this case, the sign bit of the addend is 1. Note that the sign bits also participate in the addition process. In fact, a carry is generated in the last position of addition. *This carry is always disregarded,* so the final sum is 00101, which is equivalent to +5.

Case III: Positive Number and Larger Negative Number. Consider the addition of −9 and +4:

```
        −9 ─────→ 10111
        +4 ─────→ 00100
                 ──────
                  11011      -(sum = −5)
                  └────negative sign bit
```

The sum here has a sign bit of 1, indicating a negative number. Since the sum is negative, it is in 2's-complement form, so the last four bits, 1011, actually represent the 2's complement of the sum. To find the true magnitude of the sum, we must 2's complement 11011; the result is 00101 = +5. Thus, 11011 represents −5.

Case IV: Two Negative Numbers.

```
   −9 ─────→10111
   −4 ─────→11100
           ──────
    ⟋10011
    │   └────sign bit
    └────this carry is disregarded, so the result is 10011 (sum = −13)
```

This final result is again negative and in 2's-complement form with a sign bit of 1. Note that the 2's-complement of 10011 is 01101 = +13.

Case V: Equal and Opposite Numbers.

```
   −9 ─────→10111
   +9 ─────→01001
           ──────
    0    ⟋00000
         └────disregard, so the result is 00000 (sum = +0)
```

The result is obviously +0, as expected.

REVIEW QUESTIONS

1. *True or false:* Whenever the sum of two signed binary numbers has a sign bit of 1, the magnitude of the sum is in 2's-complement form.

2. Add the following pairs of signed numbers. Express the sum as a signed binary number and as a decimal number: (a) 100111 + 111011, (b) 100111 + 011001. [*Ans.* 100010 (−30), 000000 (+0)]

6-4 SUBTRACTION IN THE 2'S-COMPLEMENT SYSTEM

The subtraction operation using the 2's-complement system actually involves the operation of addition and is really no different than the various cases considered in Section 6-3. When subtracting one binary number (the *subtrahend*) from another binary number (the *minuend*), the procedure is as follows:

1. Take the 2's complement of the subtrahend, *including* the sign bit. If the subtrahend is a positive number, this will change it to a negative number in 2's-complement form. If the subtrahend is a negative number, this will change it to a positive number in true binary form. In other words, we are changing the sign of the subtrahend.

2. After taking the 2's complement of the subtrahend, it is *added* to the minuend. The minuend is kept in its original form. The result of this addition represents the required *difference*. The sign bit of this difference determines whether it is + or − and whether it is in true binary form or 2's-complement form.

3. Remember, both numbers have to have the same number of bits.

Let us consider the case where +4 is to be subtracted from +9.

$$\text{minuend } (+9) \longrightarrow 01001$$

$$\text{subtrahend } (+4) \longrightarrow 00100$$

Change the subtrahend to its 2's-complement form (11100), which represents −4. Now add this to the minuend.

$$
\begin{array}{ll}
01001 & (+9) \\
+\underline{11100} & (-4) \\
\cancel{1}\ 00101 & (+5)
\end{array}
$$

$\uparrow$⎿—disregard, so the result is 00101 = +5

When the subtrahend is changed to its 2's complement, it actually becomes −4, so we are *adding* +9 and −4, which is the same as subtracting +4 from +9. This is the same as case II of Section 6-3. Any subtraction operation, then, actually becomes one of addition when the 2's-complement system is used. This feature of the 2's-complement system has made it the most widely used of the methods available, since it allows addition and subtraction to be performed by the same circuitry.

The reader should verify the results of using the above procedure for the following subtractions: (a) +9 − (−4); (b) − 9 − (+4); (c) −9 − (−4); (d) +4 − (−4). Remember that when the result has a sign bit of 1, it is negative and in 2's-complement form.

Arithmetic Overflow   In each of the previous addition and subtraction examples, the numbers that were added consisted of a sign bit and 4 magnitude bits. The answers also consisted of a sign bit and 4 magnitude bits. Any carry into the sixth bit position was disregarded. In all of the cases considered, the magnitude of the answer was small enough to fit into 4 bits. Let's look at the addition of +9 and +8.

$$
\begin{aligned}
+9 &\longrightarrow \boxed{0} \quad 1001 \\
+8 &\longrightarrow \boxed{0} \quad 1000 \\
\hline
& \boxed{1} \quad 0001 \\
&\phantom{\longrightarrow \boxed{1} \quad} \uparrow
\end{aligned}
$$
sign bits

The answer has a negative sign bit, which is obviously incorrect. The answer should be $+17$, but the magnitude 17 requires more than 4 bits and therefore *overflows* into the sign-bit position. This overflow condition always produces an incorrect result, and its occurrence is detected by examining the sign bit of the result and comparing it to the sign bits of the numbers being added. In a computer, a special circuit is used to detect any overflow condition and to signal that the answer is erroneous. We will encounter such a circuit in one of the end-of-chapter problems.

REVIEW QUESTIONS

1. Perform the subtraction on the following pairs of signed numbers using the 2's-complement system. Express the results as signed binary numbers and as decimal values: (a) $01001 - 11010$, (b) $10010 - 10011$. (*Ans.* $01111 = +15$, $11111 = -1$)

2. How can arithmetic overflow be detected when signed numbers are being added?

6-5 MULTIPLICATION OF BINARY NUMBERS

The multiplication of binary numbers is done in the same manner as the multiplication of decimal numbers. The process is actually simpler, since the multiplier digits are either 0 or 1, so we are always multiplying by 0 or 1 and no other digits. The following example illustrates for unsigned binary numbers.

$$
\begin{array}{r}
1001 \quad \longleftarrow \text{ multiplicand} = 9_{10} \\
1011 \quad \longleftarrow \text{ multiplier} = 11_{10} \\
\hline
1001 \\
1001 \\
0000 \\
1001 \\
\hline
1100011\}
\end{array}
$$

partial products

final product $= 99_{10}$

In this example the multiplicand and multiplier are in true binary form and no sign bits are used. The steps followed in the process are exactly the same as in decimal multiplication. First, the LSB of the multiplier is examined; in our example it is a 1. This 1 multiplies the multiplicand to produce 1001, which is written down as the first partial product. Next, the second bit of the multiplier is examined. It is a 1, so 1001 is written for the second partial product. Note that this second partial product is *shifted* one place to the left relative to the first one. The third bit of the multiplier is 0, so 0000 is written as the third partial product; again, it is shifted one place to the left relative to the previous partial product. The fourth multiplier bit is 1, so the last partial product is 1001 shifted again one position to the left. The four partial products are then summed to produce the final product.

Most digital machines can add only two binary numbers at a time. For this reason, the partial products formed during multiplication cannot all be added together at the same time. Instead, they are added together two at a time; that is, the first is added to the second, their sum is added to the third, and so on. This process is now illustrated for the example above:

$$
\begin{array}{ll}
\text{Add} \left\{ \begin{array}{l} 1001 \\ 1001 \end{array} \right. & \begin{array}{l} \longleftarrow \text{ first partial product} \\ \longleftarrow \text{ second partial product shifted left} \end{array} \\[1em]
\text{Add} \left\{ \begin{array}{l} 11011 \\ 0000 \end{array} \right. & \begin{array}{l} \longleftarrow \text{ sum of first two partial products} \\ \longleftarrow \text{ third partial product shifted left} \end{array} \\[1em]
\text{Add} \left\{ \begin{array}{l} 011011 \\ 1001 \end{array} \right. & \begin{array}{l} \longleftarrow \text{ sum of first three partial products} \\ \longleftarrow \text{ fourth partial product shifted left} \end{array} \\[1em]
\quad\quad 1100011 & \begin{array}{l} \longleftarrow \text{ sum of four partial products which} \\ \quad\quad \text{equals final total product} \end{array}
\end{array}
$$

Multiplication in the 2's-Complement System   In machines that use the 2's-complement representation, multiplication is carried on in the manner described above provided that both the multiplicand and multiplier are put in true binary form. If the two numbers to be multiplied are positive, they are already in true binary form and are multiplied as they are. The resulting product is, of course, positive and is given a sign bit of 0. When the two numbers are negative, they will be in 2's-complement form. Each one is 2's-complemented to convert it to a positive number and then they are multiplied. The product is kept as a positive number and given a sign bit of 0.

When one of the numbers is positive and the other is negative, the negative number is first converted to a positive magnitude by taking its 2's complement. The product will be in true-magnitude form. However, the product has to be negative since the original numbers are of opposite sign. Thus, the product is then changed to 2's-complement form and given a sign bit of 1.

REVIEW QUESTION

1.   Multiply 0111 and 1110. (*Ans.* 1100010)

6-6   **BINARY DIVISION**

The process for dividing one binary number (the *dividend*) by another (the *divisor*) is the same as that which is followed for decimal numbers, that which we usually refer to as "long division." The actual process is simpler in binary because when we are checking to see how many times the divisor ''goes into'' the dividend, there are only two possibilities, 0 or 1. To illustrate, consider the following division examples

$$
\begin{array}{r}
0011 \\
11\overline{)1001} \\
011 \\
\hline
0011
\end{array}
\qquad (9 \div 3 = 3)
$$

$$
\begin{array}{r}
0010.1 \\
100\overline{)1010.0} \\
100 \\
\hline
100 \\
100 \\
\hline
0
\end{array}
\qquad (10 \div 4 = 2.5)
$$

In most modern digital machines the subtractions that are part of the division operation are usually carried out using 2's-complement subtraction—that is, 2's-complementing the subtrahend and then adding.

The division of signed numbers is handled in the same way as multiplication. Negative numbers are made positive by complementing and the division is then carried out. If the dividend and divisor are of opposite sign, the resulting quotient is changed to a negative number by 2's-complementing and is given a sign bit of 1. If the dividend and divisor are of the same sign, the quotient is left as a positive number and given a sign bit of 0.

6-7 BCD ADDITION

In Chapter 2 we stated that many computers and calculators use the BCD code to represent decimal numbers. Recall that this code takes *each* decimal digit and represents it by a 4-bit code ranging from 0000 to 1001. The addition of decimal numbers that are in BCD form can be best understood by considering the two cases that can occur when two decimal digits are added.

Sum Equals Nine or Less   Consider adding 5 and 4 using BCD to represent each digit:

$$
\begin{array}{rll}
5 & 0101 & \leftarrow \text{BCD for 5} \\
+4 & +\ 0100 & \leftarrow \text{BCD for 4} \\
\hline
9 & 1001 & \leftarrow \text{BCD for 9}
\end{array}
$$

The addition is carried out as in normal binary addition and the sum is 1001, which is the BCD code for 9. As another example, take 45 added to 33:

$$
\begin{array}{rlll}
45 & 0100 & 0101 & \leftarrow \text{BCD for 45} \\
+33 & +\ 0011 & 0011 & \leftarrow \text{BCD for 33} \\
\hline
78 & 0111 & 1000 & \leftarrow \text{BCD for 78}
\end{array}
$$

In this example the 4-bit codes for 5 and 3 are added in binary to produce 1000, which is BCD for 8. Similarly, adding the second-decimal-digit positions produces 0111, which is BCD for 7. The total is 0111 1000, which is the BCD code for 78.

In the examples above, none of the sums of the pairs of decimal digits exceeded 9; therefore, *no decimal carries were produced*. For these cases the BCD addition process is straightforward and is actually the same as binary addition.

Sum Greater Than Nine Consider the addition of 6 and 7 in BCD:

$$
\begin{array}{rl}
6 & \quad 0110 \leftarrow \text{BCD for 6} \\
+\ 7 & \quad +\ 0111 \leftarrow \text{BCD for 7} \\
\hline
+13 & \quad \overline{1101} \leftarrow \text{invalid code group for BCD}
\end{array}
$$

The sum 1101 does not exist in the BCD code; it is one of the six forbidden or invalid 4-bit code groups. This has occurred because the sum of the two digits exceeds 9. Whenever this occurs the sum has to be corrected by the addition of six (0110) to take into account the skipping of the six invalid code groups:

$$
\begin{array}{rl}
& 0110 \leftarrow \text{BCD for 6} \\
+ & 0111 \leftarrow \text{BCD for 7} \\
\hline
& 1101 \leftarrow \text{invalid sum} \\
& 0110 \leftarrow \text{add 6 for correction} \\
\hline
0001 & 0011 \leftarrow \text{BCD for 13} \\
\underbrace{}_{1} & \underbrace{}_{3}
\end{array}
$$

As shown above, 0110 is added to the invalid sum and produces the correct BCD result. Note that a carry is produced into the second decimal position. This addition of 0110 has to be performed whenever the sum of the two decimal digits is greater than 9.

As another example, take 47 plus 35 in BCD:

$$
\begin{array}{rl}
47 & \quad 0100 \quad 0111 \leftarrow \text{BCD for 47} \\
+\ 35 & \quad +\ 0011 \quad 0101 \leftarrow \text{BCD for 35} \\
\hline
82 & \quad 0111 \quad 1100 \leftarrow \text{invalid sum in first digit} \\
& \qquad\quad 1\leftarrow 0110 \leftarrow \text{add 6 to correct} \\
\hline
& \quad 1000 \quad 0010 \leftarrow \text{correct BCD sum} \\
& \underbrace{}_{8} \quad \underbrace{}_{2}
\end{array}
$$

The addition of the 4-bit codes for the 7 and 5 digits results in an invalid sum and is corrected by adding 0110. Note that this generates a carry of 1, which is carried over to be added to the BCD sum of the second-position digits.

Consider the addition of 59 and 38 in BCD:

$$
\begin{array}{rl}
& \qquad\qquad 1 \\
& \qquad\qquad \downarrow \\
59 & \quad 0101 \mid 1001 \leftarrow \text{BCD for 59} \\
+\ 38 & \quad +\ 0011 \mid 1000 \leftarrow \text{BCD for 38} \\
\hline
97 & \quad 1001 \quad 0001 \leftarrow \text{perform addition} \\
& \qquad\quad\ \ 0110 \leftarrow \text{add 6 to correct} \\
\hline
& \quad 1001 \quad 0111 \quad \text{BCD for 97} \\
& \underbrace{}_{9} \quad \underbrace{}_{7}
\end{array}
$$

Here, the addition of the least significant digits (LSDs) produces a sum of $17 = 10001$. This generates a carry into the next digit position to be added to the codes for 5 and 3. Since $17 > 9$, a correction factor of 6 has to be added to the LSD sum. Addition of this correction does not generate a carry; the carry was already generated in the original addition.

To summarize the BCD addition procedure:

1. Add, using ordinary binary addition, the BCD code groups for each digit position.
2. For those positions where the sum is 9 or less, no correction is needed. The sum is in proper BCD form.
3. When the sum of two digits is greater than 9, a correction of 0110 should be added to that sum to get the proper BCD result. This case always produces a carry into the next digit position, either from the original addition (step 1) or from the correction addition.

The procedure for BCD addition is clearly more complicated than straight binary addition. This is also true of the other BCD arithmetic operations. Readers should perform the addition of $275 + 641$. Then check the correct procedure below.

```
     275        0010   0111   0101  ← BCD for 275
  +  641     +  0110   0100   0001  ← BCD for 641
  -------     -------------------
     916        1000   1011   0110  ← perform addition
                +      0110         ← add 6 to correct 2nd digit
                -------------------
                1001   0001   0110  ← BCD for 916
```

REVIEW QUESTIONS

1. How can you tell when a correction is needed in BCD addition?
2. Represent 135_{10} and 265_{10} in BCD and then perform BCD addition. Check your work by converting result back to decimal.

6-8 HEXADECIMAL ARITHMETIC

Hex numbers are used extensively in machine-language computer programming and in conjunction with computer memories (i.e., addresses). When working in these areas, there will be situations where hex numbers have to be added or subtracted.

Hex Addition   This is done in much the same way as decimal addition as long as you remember that the largest hex digit is F instead of 9. The following procedure is suggested.

1. Add the two hex digits in decimal, mentally inserting the decimal equivalent for those digits larger than 9.

2. If the sum is 15 or less, it can be directly expressed as a hex digit.

3. If the sum is greater than or equal to 16, subtract 16 and carry a 1 to the next digit position.

The following examples will illustrate the procedure.

EXAMPLE 6-6

Add the hex numbers 58 and 24.

SOLUTION

$$
\begin{array}{r}
58 \\
+ \ 24 \\
\hline
7C
\end{array}
$$

Adding the LSDs (8 and 4) produces 12, which is C in hex. There is no carry into the next digit position. Adding the 5 and 2 produces 7.

EXAMPLE 6-7

Add the hex numbers 58 and 4B.

SOLUTION

$$
\begin{array}{r}
58 \\
+ \ 4B \\
\hline
A3
\end{array}
$$

Start by adding 8 and B, substituting decimal 11 for B. This produces a sum of 19. Since 19 is greater than 16, subtract 16 to get 3; write down the 3 and carry a 1 into next position. This carry is added to the 5 and 4 to produce a sum of 10_{10}, which is then converted to hexadecimal A.

EXAMPLE 6-8

Add 3AF to 23C.

SOLUTION

$$
\begin{array}{r}
3AF \\
+ \ 23C \\
\hline
5EB
\end{array}
$$

The sum of F and C is considered as $15 + 12 = 27_{10}$. Since this is greater than 16, subtract 16 to get 11_{10}, which is hexadecimal B, and carry a 1 into the second position. Add this carry to A and 3 to obtain E. There is no carry into the MSD position.

Hex Subtraction   Remember that hex numbers are just an efficient way to represent binary numbers. Thus we can subtract hex numbers using the same method we used for binary numbers. The hex subtrahend will be 2's-complemented and then *added* to the minuend, and any carry out of the MSD position will be disregarded.

How do we 2's-complement a hex number? One way is to convert it to binary, 2's-complement the binary equivalent, and then convert it back to hex. This process is illustrated below.

```
        ⟋73A⟍              ← hex number
 0111   0011   1010    ← convert to binary
 1000   1100   0110    ← 2's-complement it
        ⟍8C6⟋             ← convert back to hex
```

There is a quicker procedure. Subtract *each* hex digit from F, then add 1. Let's try this for the same hex number from the example above.

$$
\left. \begin{array}{ccc}
F & F & F \\
-\;7 & -\;3 & -\;A \\ \hline
8 & C & 5
\end{array} \right\} \leftarrow \text{subtract each digit from F}
$$

```
                   + 1      ← add 1
      ─────────────────
      8     C     6      ← hex. equivalent of 2's comp.
```

Try either of the above procedures on the hex number E63. The correct result for the 2's complement is 19D.

EXAMPLE 6-9

Subtract $3A5_{16}$ from 592_{16}.

SOLUTION
First, convert the subtrahend (3A5) to its 2's-complement form by using either method presented above. The result is C5B. Then add this to the minuend (592):

```
          592
    +     C5B
       ─────────
       ✗1ED
        ↑
        └──────── disregard carry
```

Ignoring the carry out of the MSD addition, the result is 1ED. We can prove that this is correct by adding 1ED to 3A5 and checking to see that it equals 592_{16}.

REVIEW QUESTIONS

1.   Add 67F + 2A4. (*Ans.* 923)
2.   Subtract 67F − 2A4. (*Ans.* 3DB)

6-9 ARITHMETIC CIRCUITS

One essential function of most computers and calculators is the performance of arithmetic operations. These operations are all performed in the arithmetic-logic unit of a computer, where logic gates and flip-flops are combined so that they can add, subtract, multiply, and divide binary numbers. These circuits perform arithmetic operations at speeds that are not humanly possible. Typically, an addition operation will take less than 1 μs.

We will now study some of the basic arithmetic circuits that are used to perform the arithmetic operations discussed earlier. In some cases we will go through the actual design process, even though the circuits may be commercially available in integrated-circuit form, to provide more practice in the use of the techniques learned in Chapter 4.

Arithmetic-Logic Unit   All arithmetic operations take place in the *arithmetic-logic unit* (ALU) of a computer. Figure 6-3 is a block diagram showing the major elements included in a typical ALU. The main purpose of the ALU is to accept binary data that are stored in the memory and to execute arithmetic operations on these data according to instructions from the control unit.

The arithmetic-logic unit contains at least two flip-flop registers: the *B register* and the *accumulator register*. It also contains combinatorial logic, which performs the arithmetic operations on the binary numbers that are stored in the *B* register and the accumulator. A typical sequence of operations may occur as follows:

1.   The control unit receives an instruction (from the memory unit) specifying that a number stored in a particular memory location (address) is to be added to the number presently stored in the accumulator register.

2.   The number to be added is transferred from memory to the *B* register.

FIGURE 6-3 **Functional parts of an ALU.**

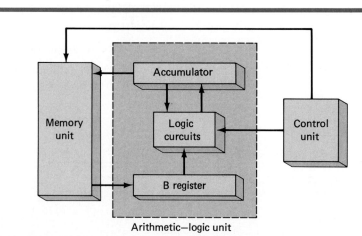

3. The number in the *B* register and the number in the accumulator register are added together in the logic circuits (upon command from control unit). The resulting sum is then sent to the accumulator to be stored.

4. The new number in the accumulator can remain there so that another number can be added to it or, if the particular arithmetic process is finished, it can be transferred to memory for storage.

These steps should make it apparent how the accumulator register derives its name. This register "accumulates" the sums that occur when performing successive additions between new numbers acquired from memory and the previously accumulated sum. In fact, for any arithmetic problem containing several steps, the accumulator always contains the results of the intermediate steps as they are completed as well as the final result when the problem is finished.

6-10 PARALLEL BINARY ADDER

Computers and calculators perform the addition operation on two binary numbers at a time, where each binary number can have several binary digits. Figure 6-4 illustrates the addition of two 5-bit numbers. The *augend* is stored in the accumulator register; that is, the accumulator contains five FFs, storing the values 10101 in successive FFs. Similarly, the *addend*, the number that is to be added to the augend, is stored in the *B* register (in this case, 00111).

The addition process starts by adding the least significant bits (LSBs) of the augend and addend. Thus, $1 + 1 = 10$, which means that the *sum* for that position is 0 with a *carry* of 1.

This carry has to be added to the next position along with the augend and addend bits in that position. Thus, in the second position, $1 + 0 + 1 = 10$, which is again a sum of 0 and a carry of 1. This carry is added to the next position together

FIGURE 6-4 **Typical binary addition process.**

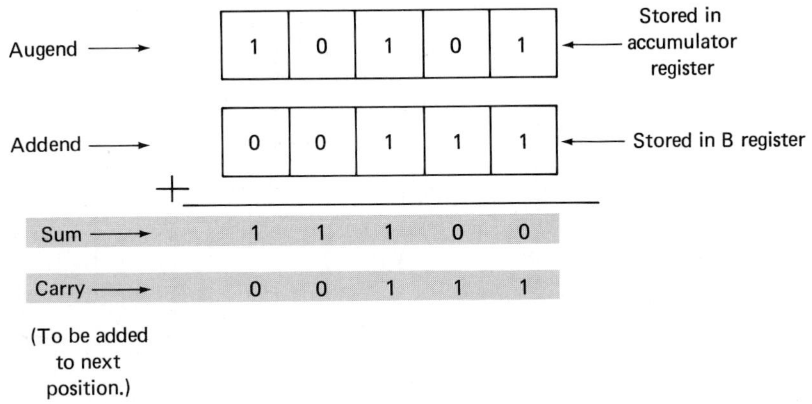

SEC. 6-9   ARITHMETIC CIRCUITS **277**

with the augend and addend bits in that position, and so on for the remaining positions.

At each step in this addition process we are performing the addition of 3 bits; the augend bit, the addend bit, and a carry bit from the previous position. The result of the addition of these 3 bits produces 2 bits: a *sum* bit, and a *carry* bit which is to be added to the next position. It should be clear that the same process is followed for each bit position. As such, if we can design a logic circuit that can duplicate this process, then we simply have to use identical circuits for each of the bit positions. This is illustrated in Figure 6-5.

In this diagram variables A_4, A_3, A_2, A_1, and A_0 represent the bits of the augend that are stored in the accumulator (which is also called the A register). Variables B_4, B_3, B_2, B_1, and B_0 represent the bits of the addend stored in the B register. Variables C_4, C_3, C_2, C_1, and C_0 represent the carry bits into the corresponding positions. Variables S_4, S_3, S_2, S_1, S_0 are the sum output bits for each position. Corresponding bits of the augend and addend are fed to a logic circuit called a *full adder,* along with a carry bit from the previous position. For example, bits A_1 and B_1 are fed into full adder 1 along with C_1, which is the carry bit produced by the addition of the A_0 and B_0 bits. Bits A_0 and B_0 are fed into full adder 0 along with C_0. Since A_0 and B_0 are the LSBs of the augend and addend, it appears that C_0 would always have to be 0, since there can be no carry into that position. However, we shall see that there will be situations when C_0 can also be 1.

The full-adder circuit used in each position has three inputs: an A bit, a B bit, and a C bit, and it produces two outputs: a sum bit and a carry bit. For example, full adder 0 has inputs A_0, B_0, and C_0, and it produces outputs S_0 and C_1. Full adder 1 has A_1, B_1, and C_1 as inputs and S_1 and C_2 as outputs; and so on. This arrangement is repeated for as many positions as there are in the augend and addend. Although

FIGURE 6-5   **Block diagram of 5-bit parallel adder circuit using full adders.**

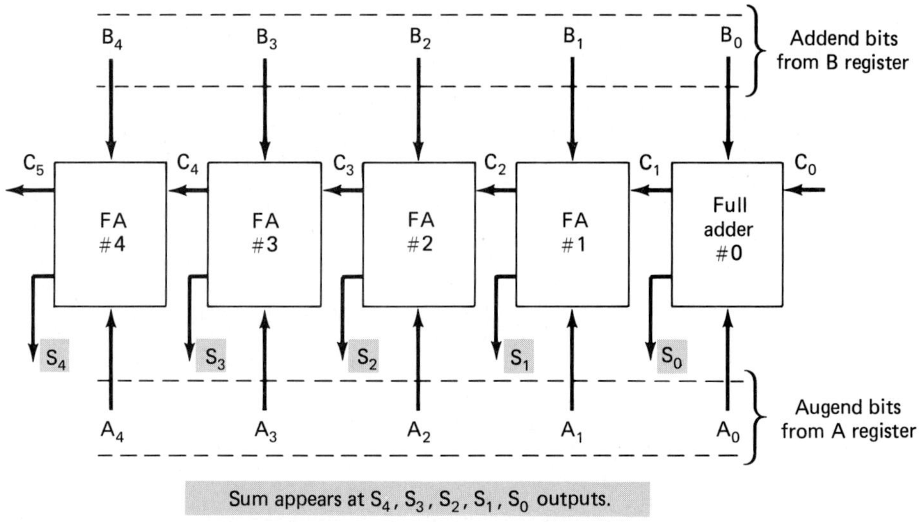

Sum appears at S_4, S_3, S_2, S_1, S_0 outputs.

this illustration is for 5-bit numbers, in modern computers the numbers usually range from 8 to 64 bits.

The arrangement in Figure 6-5 is called a *parallel adder* because all the bits of the augend and addend are present and are fed into the adder circuits *simultaneously*. This means that the additions in each position are taking place at the same time. This is different from how we add on paper, taking each position one at a time starting with the LSB. Clearly, parallel addition is extremely fast. More will be said about this later.

REVIEW QUESTIONS

1. How many inputs does a full adder have? How many outputs?
2. Assume the following input levels in Figure 6-5: $A_4A_3A_2A_1A_0 = 01001$; $B_4B_3B_2B_1B_0 = 00111$; $C_0 = 0$.
   (a) What are the logic levels at the outputs of FA #2? (*Ans.* $S_2 = 0$, $C_3 = 1$)
   (b) What is the logic level at the C_5 output? (*Ans.* $C_5 = 0$)

6-11 **DESIGN OF A FULL ADDER**

Now that we know the function of the full adder, we can proceed to design a logic circuit that will perform this function. First, we must construct a truth table showing the various input and output values for all possible cases. Figure 6-6 shows the truth table having three inputs A, B, and C_{IN}, and two outputs, S and C_{OUT}. There are

FIGURE 6-6 **Truth table for a full-adder circuit.**

Augend bit input	Addend bit input	Carry bit input	Sum bit output	Carry bit output
A	B	C_{IN}	S	C_{OUT}
0	0	0	0	0
0	0	1	1	0
0	1	0	1	0
0	1	1	0	1
1	0	0	1	0
1	0	1	0	1
1	1	0	0	1
1	1	1	1	1

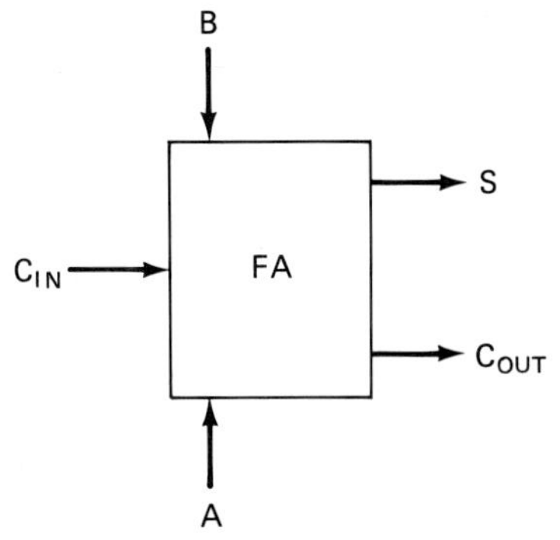

eight possible cases for the three inputs, and for each case the desired output values are listed. For example, consider the case $A = 1$, $B = 0$, and $C_{IN} = 1$. The full adder (hereafter abbreviated FA) must add these bits to produce a sum (S) of 0 and a carry (C_{OUT}) of 1. The reader should check the other cases to be sure they are understood.

Since there are two outputs, we will design the circuitry for each output individually, starting with the S output. The truth table shows that there are four cases where S is to be a 1. Using the sum-of-products method, we can write the expression for S as

$$S = \overline{A}\,\overline{B}C_{IN} + \overline{A}B\overline{C}_{IN} + A\overline{B}\,\overline{C}_{IN} + ABC_{IN} \qquad (6\text{-}1)$$

We can now try to simplify this expression by factoring. Unfortunately, none of the terms in the expression has two variables in common with any of the other terms. However, $\overline{A}$ can be factored from the first two terms and A can be factored from the last two terms:

$$S = \overline{A}(\overline{B}C_{IN} + B\overline{C}_{IN}) + A(\overline{B}\,\overline{C}_{IN} + BC_{IN})$$

The first term in parentheses should be recognized as the exclusive-OR combination of B and C_{IN}, which can be written as $B \oplus C_{IN}$. The second term in parentheses should be recognized as the exclusive-NOR of B and C_{IN}, which can be written as $\overline{B \oplus C_{IN}}$. Thus, the expression for S becomes

$$S = \overline{A}(B \oplus C_{IN}) + A(\overline{B \oplus C_{IN}})$$

If we let $X = B \oplus C_{IN}$, this can be written as

$$S = \overline{A} \cdot X + A \cdot \overline{X} = A \oplus X$$

which is simply the EX-OR of A and X. Replacing the expression for X, we have

$$S = A \oplus [B \oplus C_{IN}] \qquad (6\text{-}2)$$

Consider now the output C_{OUT} in the truth table of Figure 6-6. We can write the sum-of-products expression for C_{OUT} as follows:

$$C_{OUT} = \overline{A}BC_{IN} + A\overline{B}C_{IN} + AB\overline{C}_{IN} + ABC_{IN}$$

This expression can be simplified by factoring. We will employ the trick introduced in Chapter 3, whereby we will use the ABC_{IN} term *three* times since it has common factors with each of the other terms. Hence,

$$
\begin{aligned}
C_{OUT} &= BC_{IN}(\overline{A} + A) + AC_{IN}(\overline{B} + B) + AB(\overline{C}_{IN} + C_{IN}) \\
&= BC_{IN} + AC_{IN} + AB
\end{aligned}
\qquad (6\text{-}3)
$$

This expression cannot be simplified further.

Expressions (6-2) and (6-3) can be implemented as shown in Figure 6-7. There are several other implementations that can be used to produce the same expressions for S and C_{OUT}, none of which has any particular advantage over those shown. The complete circuit with inputs A, B, and C_{IN} and outputs S and C_{OUT} represents the full adder. Each of the FAs in Figure 6-5 contains the same circuitry (or its equivalent).

K-Map Simplification We simplified the expressions for S and C_{OUT} using algebraic methods. The K-map method can also be used. Figure 6-8(a) shows the K map for the S output. This map has no adjacent 1s, so there are no pairs or quads to loop. Thus, the expression for S cannot be simplified using the K map. This

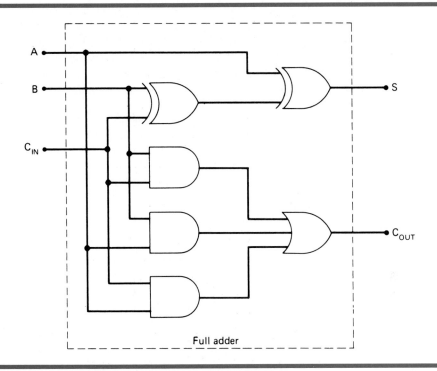

Full adder

FIGURE 6-7 **Complete circuitry for a full adder.**

points out a limitation of the K-map method as compared to the algebraic method. We were able to simplify the expression for S through factoring and the use of EX-OR and EX-NOR operations.

The K map for the C_{OUT} output is shown in Figure 6-8(b). The three pairs that are looped will produce the same expression obtained from the algebraic method.

FIGURE 6-8 **K mappings for the full-adder outputs.**

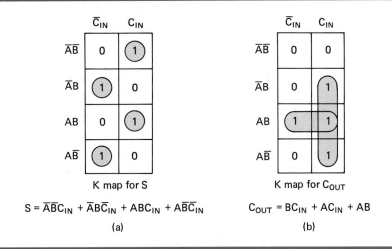

$$S = \overline{A}\,\overline{B}C_{IN} + \overline{A}B\overline{C}_{IN} + ABC_{IN} + A\overline{B}\,\overline{C}_{IN}$$

$$C_{OUT} = BC_{IN} + AC_{IN} + AB$$

(a) (b)

Half Adder   The FA operates on three inputs to produce a sum and carry output. In some cases a circuit is needed that will add only 2 input bits, to produce a sum and carry output. An example would be the addition of the LSB position of two binary numbers where there is no carry input to be added. A special logic circuit can be designed to take *two* input bits, A and B, and to produce sum (S) and carry (C_{OUT}) outputs. This circuit is called a *half adder* (HA). Its operation is similar to a FA except that it operates on only 2 bits. We shall leave the design of the HA as an exercise at the end of the chapter (see Problem 6-18). However, we will use the HA symbol whenever it is appropriate.

6-12 COMPLETE PARALLEL ADDER WITH REGISTERS

In a computer, the numbers that are to be added are stored in FF registers. Figure 6-9 shows the complete diagram of a 4-bit parallel adder including the storage registers. The augend bits A_3–A_0 are stored in the accumulator (A register); the addend bits B_3–B_0 are stored in the B register. Each of these registers is made up of D FFs for easy transfer of data.

The contents of the A register (i.e., the binary number stored in A_3–A_0) is added to the contents of the B register by the four FAs, and the sum is produced at outputs S_3–S_0. C_4 is the carry out of the fourth FA and it can be used as the carry input to a fifth FA, or as an *overflow* bit to indicate that the sum exceeds 1111.

Note that the sum outputs are connected to the D inputs of the A register. This will allow the sum to be parallel-transferred into the A register on the PGT of the ADD pulse. In this way, the sum can be stored in the A register.

Also note that D inputs of the B register are coming from the computer's memory, so that numbers from memory will be parallel-transferred into the B register on the PGT of the TRANSFER pulse. For now we will not be concerned with how the numbers come from memory.

Register Notation   Before we go through the complete process of how this circuit adds two binary numbers, it will be helpful to introduce some notation that makes it easy to describe the contents of a register and data transfer operations.

Whenever we want to give the levels that are present at each FF in a register or at each output of a group of outputs, we will use brackets as illustrated below:

$$[A] = 1011$$

This is the same as saying that $A_3 = 1$, $A_2 = 0$, $A_1 = 1$, $A_0 = 1$. In other words, think of [A] as representing "the contents of register A."

Whenever we want to indicate the transfer of data to or from a register, we will use an arrow as illustrated below:

$$[B] \longrightarrow [A]$$

This means that the contents of the B register has been transferred to the A register. The old contents of the A register will be lost as a result of this operation, and the B register will be unchanged.

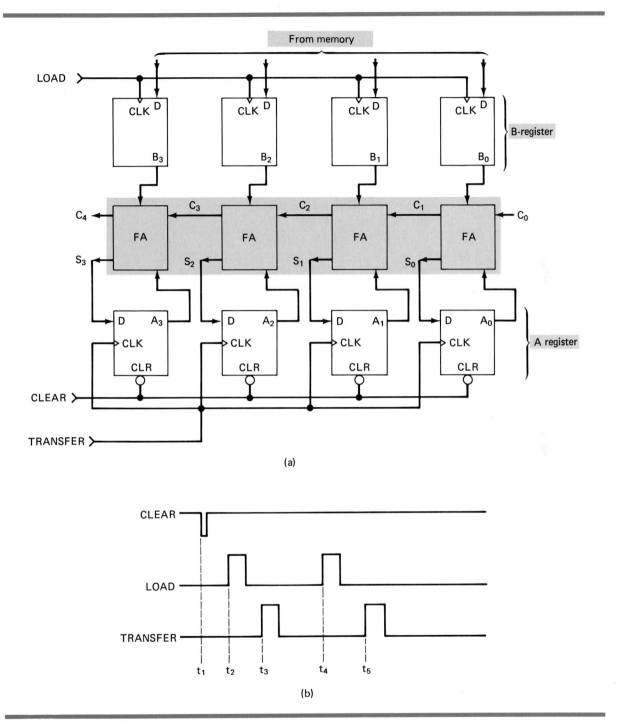

(a)

(b)

FIGURE 6-9   (a) Complete 4-bit parallel adder with registers; (b) signals used to add binary numbers from memory and store their sum in the accumulator.

Sequence of Operations We will now describe the process by which the circuit of Figure 6-9 will add the binary numbers 1001 and 0101. Assume $C_0 = 0$; that is, there is no carry into the LSB position.

1. **[A] = 0000.** A CLEAR pulse is applied to the asynchronous inputs (CLR) of each FF in register A. This occurs at time t_1.

2. **[M] → [B].** The first binary number is transferred from memory (M) to the B register. In this case, the binary number 1001 is loaded into register B on the PGT of the LOAD pulse at t_2.

3. **[S]* → [A].** With [B] = 1001 and [A] = 0000, the full adders produce a sum of 1001; that is, [S] = 1001. These sum outputs are transferred into the A register on the PGT of the TRANSFER pulse at t_3. This makes [A] = 1001.

4. **[M] → [B].** The second binary number, 0101, is transferred from memory into the B register on the PGT of the second LOAD pulse at t_4. This makes [B] = 0101.

5. **[S] → [A].** With [B] = 0101 and [A] = 1001, the FAs produce [S] = 1110. These sum outputs are transferred into the A register when the second TRANSFER pulse occurs at t_5. Thus, [A] = 1110.

6. At this point, the sum of the two binary numbers is present in the accumulator. In most computers the contents of the accumulator, [A], will usually be transferred to the computer's memory so that the adder circuit can be used for a new set of numbers. The circuitry that performs this [A]→[M] transfer is not shown in the diagram.

REVIEW QUESTIONS

1. Suppose that four different 4-bit numbers are to be taken from memory and added by the circuit of Figure 6-9. How many CLEAR pulses will be needed? How many TRANSFER pulses? How many LOAD pulses? (*Ans.* One, four, four)

2. Determine the contents of the A register after the following sequence of operations: [A] = 0000, [0110] → [B], [S] → [A], [1110] → [B], [S] → [A]. (*Ans.* 0100)

6-13 CARRY PROPAGATION

The parallel adder of Figure 6-9 performs additions at a relatively high speed, since it adds the bits from each position simultaneously. However, its speed is limited by

*Even though S is not a register, we will use [S] to represent the group of S outputs.

an effect called *carry propagation* or *carry ripple,* which can best be explained by considering the following addition:

$$0111$$

$$+ \ \underline{0001}$$

$$1000$$

Addition of the LSB position produces a carry into the second position. This carry, when added to the bits of the second position, produces a carry into the third position. This latter carry, when added to the bits of the third position, produces a carry into the last position. The key thing to notice in this example is that the sum bit generated in the *last* position (MSB) depended on the carry that was generated by the addition in the *first* position (LSB).

Looking at this from the viewpoint of the circuit of Figure 6-9, S_3 out of the last full adder depends on C_1 out of the first full adder. But the C_1 signal must pass through three FAs before it produces S_3. What this means is that the S_3 output will not reach its correct value until C_1 has propagated through the intermediate FAs. This represents a time delay that depends on the propagation delay produced in a FA. For example, if each FA is considered to have a propagation delay of 40 ns, then S_3 will not reach its correct level until 120 ns after C_1 is generated. This means that the add command pulse cannot be applied until 160 ns after the augend and addend numbers are present in the FF registers (the extra 40 ns is due to the delay of the LSB FA, which generates C_1).

Obviously, the situation becomes much worse if we extend the adder circuitry to add a greater number of bits. If the adder were handling 32-bit numbers, the carry propagation delay could be 1280 ns $= 1.28$ μs. The add pulse could not be applied until at least 1.28 μs after the numbers were present in the registers.

This magnitude of delay is prohibitive for high-speed computers. Fortunately, logic designers have come up with several ingenious schemes for reducing this delay. One of the schemes, called *look-ahead carry,* utilizes logic gates to look at the lower-order bits of the augend and addend to see if a higher-order carry is to be generated. For example, it is possible to build a logic circuit with B_2, B_1, B_0, A_2, A_1, and A_0 as inputs and C_3 as an output. This logic circuit would have a shorter delay than is obtained by the carry propagation through the FAs. This scheme requires a large amount of extra circuitry but is necessary to produce high-speed adders. The extra circuitry is not a significant consideration with the present use of integrated circuits. Many high-speed adders available in integrated-circuit form utilize the look-ahead carry or a similar technique for reducing overall propagation delays.

6-14 INTEGRATED-CIRCUIT PARALLEL ADDER

Several parallel adders are available as ICs. The most common is a 4-bit parallel adder IC that contains four interconnected FAs and the look-ahead carry circuitry needed for high-speed operation. The 7483A, 74LS83A, 74283, and 74LS283 are all TTL 4-bit parallel adder chips. The 283s are identical to the 83s except that they have V_{CC} and ground on pins 16 and 8, respectively; it has become standard on all new chips to have the power and ground pins at the corners of the chip. The 74HC283 is the high-speed CMOS version of the same 4-bit parallel adder.

Figure 6-10(a) shows the functional symbol for the 7483 4-bit parallel adder (and its equivalents). The inputs to this IC are two 4-bit numbers, $A_3A_2A_1A_0$ and $B_3B_2B_1B_0$, and the carry, C_0, into the LSB position. The outputs are the sum bits, $S_3S_2S_1S_0$, and the carry, C_4, out of the MSB position. The sum bits are often labeled $\Sigma_3\Sigma_2\Sigma_1\Sigma_0$, where Σ is the Greek capital letter *sigma*.

FIGURE 6-10 **(a) Block symbol for the 7483 4-bit parallel adder; (b) cascading two 2483s.**

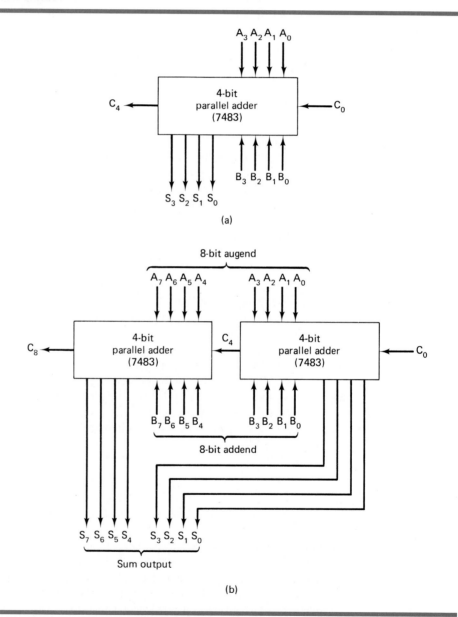

Cascading Parallel Adders   Two or more parallel adder blocks can be connected (cascaded) to accommodate the addition of larger binary numbers. To illustrate, Figure 6-10(b) shows how two 7483 adders can be connected to add two 8-bit numbers. The adder on the right adds the four least significant bits of the numbers. The C_4 output of this adder is connected as the input carry to the first position of the second adder, which adds the four most significant bits of the numbers. The eight sum outputs represent the resultant sum of the two 8-bit numbers. C_8 is the carry out of the last position (MSB) of the second adder. C_8 can be used as an overflow bit or as a carry into another adder state if larger binary numbers are to be handled.

EXAMPLE 6-10

Determine the logic levels at the inputs and outputs of the 8-bit adder in Figure 6-10(a) when 137_{10} is added to 72_{10}.

SOLUTION
First convert each number to an 8-bit binary number.

$$137 = 10001001$$
$$72 = 01001000$$

These two binary values will be applied to the A and B inputs; that is, the A inputs will be 10001001 from left to right, and the B inputs will be 01001000 from left to right. The adder will produce the binary sum of the two numbers.

$$[A] = 10001001$$
$$[B] = \underline{01001000}$$
$$[S] = 11010001$$

The sum outputs will read 11010001 from left to right. There is no overflow into the C_8 bit, so it will be a 0.

6-15   2'S-COMPLEMENT SYSTEM

Most modern computers use the 2's-complement system to represent negative numbers and to perform subtraction. The operations of addition and subtraction of signed numbers can be performed using only the addition operation if we use the 2's-complement form to represent negative numbers.

Addition   Positive and negative numbers, including the sign bits, can be added together in the basic parallel adder circuit when the negative numbers are in 2's-complement form. This is illustrated in Figure 6-11 for the addition of -3 and $+6$. The -3 is represented in its 2's-complement form as 1101, where the first 1 is the sign bit; the $+6$ is represented as 0110, with the first zero as the sign bit. These numbers are stored in their corresponding registers. The 4-bit parallel adder produces sum outputs of 0011, which represents $+3$. The C_4 output is 1 but is disregarded in the 2's-complement method.

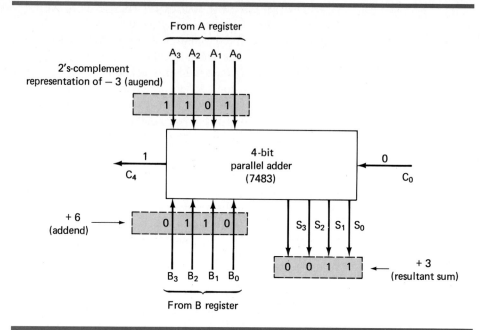

FIGURE 6-11 **Parallel adder used to add + and − numbers in 2's-complement system.**

Subtraction When the 2's-complement system is used, the number to be subtracted (the subtrahend) is 2's-complemented and then *added* to the minuend (the number the subtrahend is being subtracted from). For example, we can assume that the minuend is already stored in the accumulator (*A* register). The subtrahend is then placed in the *B* register (in a computer it would be transferred here from memory) and is changed to its 2's-complement form before it is added to the number in the *A* register. The sum outputs of the adder circuit now represent the *difference* between the minuend and subtrahend.

The parallel-adder circuit that we have been discussing can be adapted to perform the subtraction described above if we provide a means for taking the 2's complement of the *B* register number. The 2's complement of a binary number is obtained by complementing (inverting) each bit and then adding 1 to the LSB. Figure 6-12 shows how this can be accomplished. The *inverted* outputs of the *B* register are used rather than the normal outputs; that is, $\overline{B}_0$, $\overline{B}_1$, $\overline{B}_2$, and $\overline{B}_3$ are fed to the adder inputs (remember, B_3 is the sign bit). This takes care of complementing each bit of the *B* number. Also, C_0 is made a logical 1, so it adds an extra 1 into the LSB of the adder; this accomplishes the same effect as adding 1 to the LSB of the *B* register for forming the 2's complement.

The S_3–S_0 outputs represent the results of the subtraction operation. Of course, S_3 is the sign bit of the result and indicates whether the result is + or −. The carry output C_4 is again disregarded.

To help clarify this operation, study the following steps for subtracting +6 from +4:

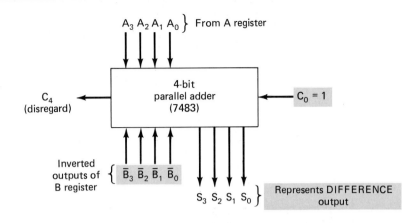

FIGURE 6-12 Parallel adder used to perform subtraction ($A - B$) using 2's-complement system. The bits of the subtrahend (B) are inverted, and $C_0 = 1$ to produce the 2's complement.

1. $+4$ is stored in the A register as 0100.

2. $+6$ is stored in the B register as 0110.

3. The inverted outputs of the B-register FFs are fed to the adder—that is, 1001.

4. The 1001 is added to 0100 by the parallel adder along with a 1 added to the LSB position by making $C_0 = 1$. This produces sum output bits 1110 and a $C_4 = 1$, which is disregarded. This 1110 represents the required difference. Since the sign bit $= 1$, it is a negative result and is in 2's-complement form. We can verify that 1110 represents -2_{10} by 2's-complementing it and obtaining $+2_{10}$:

$$\begin{array}{r} 1110 \\ 0001 \\ + \quad 1 \\ \hline 0010 \end{array} = \; +2_{10}$$

Combined Addition and Subtraction It should now be clear that the basic parallel adder circuit can be used to perform addition or subtraction depending on whether the B number is left unchanged or is 2's-complemented. A complete circuit that can perform *both* addition and subtraction in the 2's-complement system is shown in Figure 6-13.

   This adder/subtractor circuit is controlled by the two control signals ADD and SUB. When the ADD level is HIGH, the circuit performs addition of the numbers stored in the A and B registers. When the SUB level is HIGH, the circuit subtracts the B-register number from the A-register number. The operation is described as follows:

1. Assume that ADD $= 1$ and SUB $= 0$. The SUB $= 0$ *disables* (inhibits) AND gates 2, 4, 6, and 8, holding their outputs at 0. The ADD $= 1$ *enables* AND gates 1, 3, 5, and 7, allowing their outputs to pass the B_0, B_1, B_2, and B_3 levels, respectively.

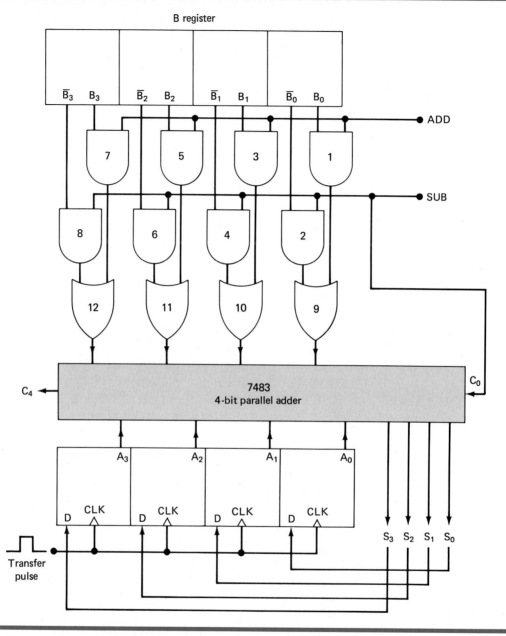

FIGURE 6-13   Parallel adder/subtractor using 2's-complement system.

2. The B_0–B_3 levels pass through the OR gates into the 4-bit parallel adder to be added to the A_0–A_3 bits. The *sum* appears at the S_0–S_3 outputs.

3. Note that SUB = 0 causes C_0 = 0 into the adder.

4. Now assume that ADD = 0 and SUB = 1. The ADD = 0 inhibits AND gates 1, 3, 5, and 7. The SUB = 1 enables AND gates 2, 4, 6, and 8, so their outputs pass the $\overline{B}_0$, $\overline{B}_1$, $\overline{B}_2$, and $\overline{B}_3$ levels, respectively.

5. The $\overline{B}_0$–$\overline{B}_3$ levels pass through the OR gates into the adder to be added to the A_0–A_3 bits. Note also that C_0 is now 1. Thus, the B-register number has essentially been 2's-complemented.

6. The *difference* appears at the S_0–S_3 outputs.

Circuits like the adder/subtractor of Figure 6-13 are used in computers because they provide a relatively simple means for adding and subtracting signed binary numbers. In most computers the outputs present at the S output lines are usually transferred into the A register (accumulator), so the results of the addition or subtraction always end up stored in the A register. This is accomplished by applying a TRANSFER pulse to the CLK inputs of register A.

REVIEW QUESTIONS

1. Why does C_0 have to be a 1 in order to use the adder circuit in Figure 6-12 as a subtractor?
2. Assume that $[A]$ = 0011 and $[B]$ = 0010 in Figure 6-13. If ADD = 1 and SUB = 0, determine the logic levels at the OR-gate outputs. (*Ans.* 0010)
3. Repeat question 2 for ADD = 0, SUB = 1. (*Ans.* 1101)
4. *True or false:* When the adder/subtractor circuit is used for subtraction, the 2's complement of the subtrahend appears at the input of the adder. (*Ans.* False)

6-16 BCD ADDER

The BCD addition process was discussed in Section 6-7 and is reviewed below:

1. Add the BCD code groups for each decimal digit position; use ordinary binary addition.
2. For those positions where the sum is 9 or less, the sum is in proper BCD form and no correction is needed.
3. When the sum of two digits is greater than 9, a correction of 0110 should be added to that sum to produce the proper BCD result. This will produce a carry to be added to the next decimal position.

A BCD adder circuit must be able to operate in accordance with the above steps. In other words, the circuit must be able to do the following:

1. Add two 4-bit BCD code groups, using straight binary addition.
2. Determine if the sum of this addition is greater than 1001 (decimal 9); if it is, add 0110 (6) to this sum and generate a carry to the next decimal position.

The first requirement is easily met by using a 4-bit binary parallel adder such as the 7483 IC. For example, if the two BCD code groups represented by $A_3A_2A_1A_0$ and $B_3B_2B_1B_0$, respectively, are applied to a 4-bit parallel adder, the adder will perform the following operation:

$$A_3A_2A_1A_0 \longleftarrow \text{BCD code group}$$
$$+\,B_3B_2B_1B_0 \longleftarrow \text{BCD code group}$$
$$S_4S_3S_2S_1S_0 \longleftarrow \text{straight binary sum}$$

S_4 is actually C_4, the carry out of the MSB.

The sum outputs $S_4S_3S_2S_1S_0$ can range anywhere from 00000 to 10010 (when both BCD code groups are $1001 = 9$). The circuitry for a BCD adder must include the logic needed to detect whenever the sum is greater than 01001, so that the correction can be added in. These cases where the sum is greater than 01001 are listed below:

S_4	S_3	S_2	S_1	S_0	
0	1	0	1	0	(10)
0	1	0	1	1	(11)
0	1	1	0	0	(12)
0	1	1	0	1	(13)
0	1	1	1	0	(14)
0	1	1	1	1	(15)
1	0	0	0	0	(16)
1	0	0	0	1	(17)
1	0	0	1	0	(18)

Let's define X as a logic output that will go HIGH only when the sum is greater than 01001 (i.e., for the cases listed above). If we examine these cases, it can be reasoned that X will be HIGH for either of the following conditions:

1. Whenever $S_4 = 1$ (sums greater than 15).

2. Whenever $S_3 = 1$ and either S_2 or S_1 or both are 1 (sums 10–15).

This can be expressed as

$$X = S_4 + S_3(S_2 + S_1)$$

Whenever $X = 1$, it is necessary to add the correction 0110 to the sum bits and to generate a carry. Figure 6-14 shows the complete circuitry for a BCD adder, including the logic-circuit implementation for X.

The circuit consists of three basic parts. The two BCD code groups $A_3A_2A_1A_0$ and $B_3B_2B_1B_0$ are added together in the upper 4-bit adder to produce the sum $S_4S_3S_2S_1S_0$. The logic gates implement the expression for X. The lower 4-bit adder will add the correction 0110 to the sum bits *only* when $X = 1$, producing the final BCD sum output represented by $\Sigma_3\Sigma_2\Sigma_1\Sigma_0$. X is also the carry output that is produced when the sum is greater than 01001. Of course, when $X = 0$, there is no carry and no addition of 0110. In such cases, $\Sigma_3\Sigma_2\Sigma_1\Sigma_0 = S_3S_2S_0S_0$.

To help in the understanding of the BCD adder, the reader should try several

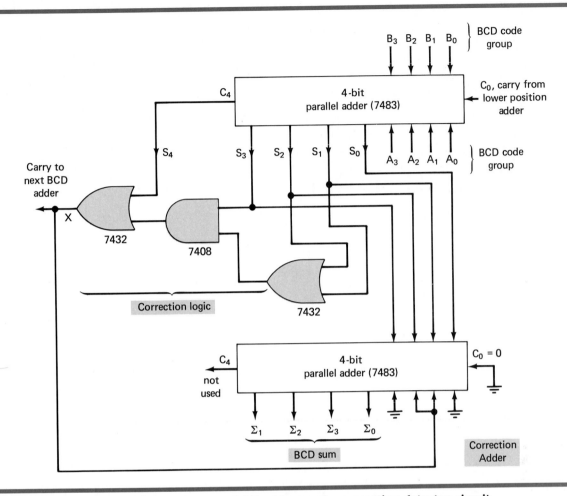

FIGURE 6-14 **A BCD adder contains two 4-bit adders and a correction-detector circuit.**

cases by following them through the circuit. The following cases would be particularly instructive:

Inputs

(a) $[A] = 0101$, $[B] = 0011$, $C_0 = 0$

(b) $[A] = 0111$, $[B] = 0110$, $C_0 = 0$

Outputs

(a) $[S] = 01000$, $X = 0$, $[\Sigma] = 1000$, CARRY $= 0$

(b) $[S] = 01101$, $X = 1$, $[\Sigma] = 0011$, CARRY $= 1$

Cascading BCD Adders   The circuit of Figure 6-14 is used for adding two decimal digits that have been encoded in BCD code. When decimal numbers with several digits are to be added together, it is necessary to use a separate BCD adder

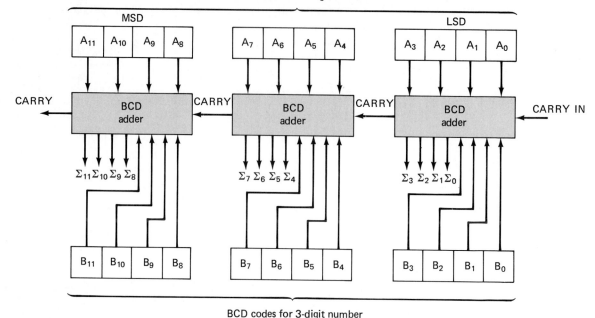

BCD codes for 3-digit number

FIGURE 6-15 **Cascading BCD adders to add two three-digit decimal numbers.**

for each digit position. Figure 6-15 illustrates for the addition of two three-digit decimal numbers. The A register contains 12 bits, which are the three BCD code groups for one of the three-digit decimal numbers, similarly, the B register contains the BCD representation of the other three-digit decimal number. The A_3–A_0 and B_3–B_0 code groups representing the least significant digits are fed to the first BCD adder. Each BCD adder block is assumed to contain the circuitry of Figure 6-14. This first BCD adder produces sum outputs $\Sigma_3\Sigma_2\Sigma_1\Sigma_0$, which is the BCD code for the least significant digit of the sum. It also produces a carry output that is sent to the second BCD adder, which is adding A_7–A_4 and B_7–B_4, the BCD code groups for the second-decimal-digit position. The second BCD adder produces $\Sigma_7\Sigma_6\Sigma_5\Sigma_4$, the BCD code for the second digit of the sum, and so on. This arrangement can, of course, be extended to decimal numbers of any size by simply adding more FFs to the registers and including a BCD adder for each digit position.

EXAMPLE 6-11

Determine the inputs and outputs when the circuit of Figure 6-15 is used to add 247_{10} to 638_{10}.

SOLUTION
First, the decimal numbers are represented in BCD.

$$247 = 0010 \quad 0100 \quad 0111 \quad \text{(BCD)}$$
$$538 = 0101 \quad 0011 \quad 1000 \quad \text{(BCD)}$$

These BCD numbers will be placed in the A and B registers, respectively, so that

$$[A] = 0010 \quad 0100 \quad 0111$$
$$[B] = 0101 \quad 0011 \quad 1000$$

The CARRY IN to the LSD adder will be a 0.

Once the data is in the registers, the BCD adders will begin to produce the correct BCD sums at their outputs. The LSD adder will add the 0111(7) and 1000(8) to produce a sum of 0101(5) and a CARRY of 1 into the middle adder. The middle adder will add the 0100(4) and 0011(3) and the CARRY of 1 to produce a sum of 1000(8) and a CARRY of 0 into the MSD adder. The MSD adder will add 0010(2) and 0101(5) for a sum of 0111(7) and no CARRY out. Thus, at the sum outputs we have

$$[\Sigma] = 0111 \quad 1000 \quad 0101$$

and there is a CARRY output of 0 from the MSD adder.

REVIEW QUESTIONS

1. What are the three basic parts of a BCD adder circuit?
2. Describe how the BCD adder circuit detects the need for a correction and executes it.

6-17 BINARY MULTIPLIERS

The multiplication of two binary numbers is done with paper and pencil by performing successive additions and shifting. To illustrate:

```
      1011      multiplicand (11)
    × 1101      multiplier (13)
      1011
     0000
    1011
   1011
   10001111     product (143)
```

This process consists of examining the successive bits of the multiplier, beginning with the LSB. If the multiplier bit is a 1, the multiplicand is copied down; if it is a 0, zeros are written down. The numbers written down in successive lines are shifted one position to the left relative to the previous line. When all the multiplier bits have been examined, the various lines are *added* to produce the final product.

In the digital machines this process is modified somewhat because the binary adder is designed to add only two binary numbers at a time. Instead of adding all the lines at the end, they are added two at a time and their sum is accumulated in a register (the accumulator register). In addition, when the multiplier bit is 0, there is no need to write down and add zeros since it does not affect the final result. The previous example is redone here showing the modified process.

multiplicand: 1011
multiplier: 1101

1011	LSB of multiplier = 1; write down multiplicand; shift multiplicand one position to the left (10110)
1011	2nd multiplier bit = 0; write down previous result; shift multiplicand to the left again (101100)
+ 101100	3rd multiplier bit = 1; write down multiplicand (101100)
110111	add
	shift new multiplicand to the left (1011000)
+ 1011000	4th multiplier bit = 1; write down new multiplicand (1011000)
10001111	add to obtain final product

This multiplication process can be performed by the *serial multiplier* circuit shown in Figure 6-16, which will multiply two 4-bit numbers to produce an 8-bit product. The circuit consists of the following elements:

- *X register.* A 4-bit shift register that stores the multiplier. It will shift right on the NGT of the clock. Note that 0s are shifted in from the left.
- *B register.* An 8-bit shift register that stores the multiplicand. It will shift left on the NGT of the clock. Note that 0s are shifted in from the right.
- *A register.* An 8-bit register that is the accumulator register that accumulates the partial products.
- *Adder.* An 8-bit parallel adder that produces the sum of the A and B registers. The adder outputs S_7–S_0 are connected to the D inputs of the accumulator so that the sum can be transferred to the accumulator *only when a clock pulse gets through the AND gate.*

The circuit operation can best be described by going through each step in the multiplication of 1011 and 1101. The complete process requires four clock cycles. Refer to Figure 6-17 for the contents of each register and the adder outputs as we describe the sequence of steps.

1. *Before first clock pulse.* Prior to the occurrence of the first clock pulse, the A register is loaded with 00000000, the B register with the multiplicand 00001011, and the X register with the multiplier 1101. We can assume that each of these registers was loaded using its asynchronous inputs (i.e., PRE-SET and CLEAR). The adder outputs will be the sum of A and B, that is, 00001011.

2. *First clock pulse.* Since the LSB of the multiplier (X_0) is a 1, the first clock pulse gets through the AND gate, and its PGT transfers the sum outputs into the accumulator. The subsequent NGT causes the X and B registers to shift right and left, respectively. This, of course, produces a new sum of A and B.

3. *Second clock pulse.* The second bit of the original multiplier is now in X_0. Since this bit is a 0, the second clock pulse is inhibited from reaching the

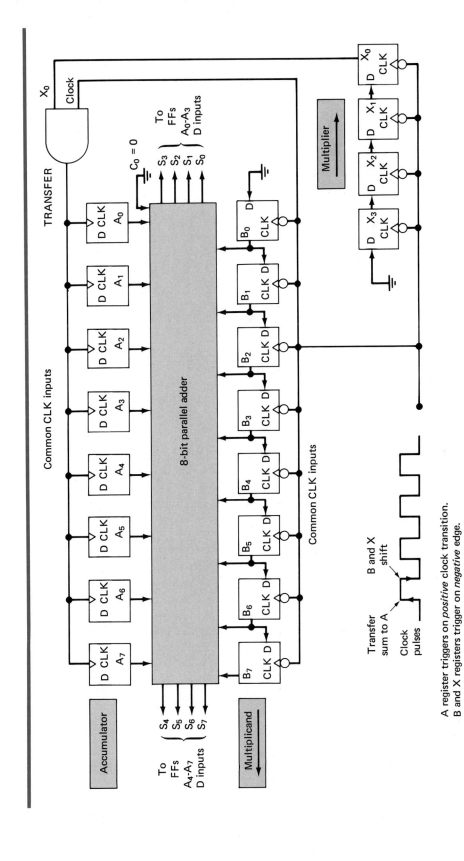

FIGURE 6-16 Circuit for a 4-bit serial multiplier that uses shift registers for the multiplier and multiplicand.

297

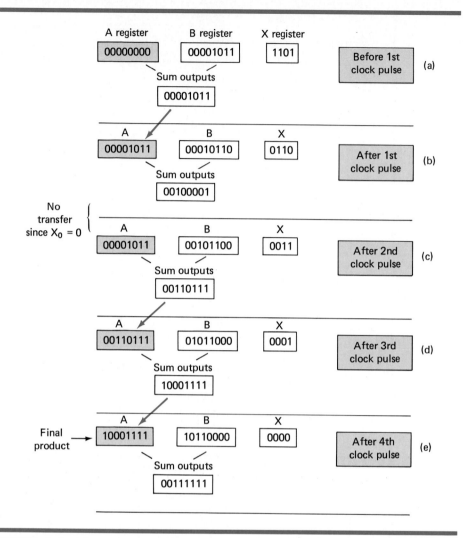

FIGURE 6-17 **Contents of various registers during multiplication of 1011 × 1101.**

accumulator. Thus, the sum outputs are not transferred into the accumulator and the number in the accumulator does not change. The NGT of the clock pulse will again shift the X and B registers.

4. *Third clock pulse.* The third bit of the original multiplier is now in X_0. Since it is a 1, the PGT of the third clock pulse transfers the sum into the A register. The NGT shifts X and B again.

5. *Fourth clock pulse.* The last bit of the original multiplier is now in X_0, and since it is a 1, the PGT of the fourth clock pulse transfers the sum into the accumulator. The accumulator now holds the final product. The NGT shifts X and B again. Note that X is now 0000 since all of the multiplier bits have been shifted out.

REVIEW QUESTIONS

1. Describe the sequence of events that occurs in the multiplier circuit during one clock cycle.
2. *True or false:* The outputs of the adder are transferred to the accumulator on *every* PGT of the clock.

6-18 COMPLEX ARITHMETIC INTEGRATED CIRCUITS

There are several complex ICs available in the TTL and CMOS families that can perform one or more types of arithmetic operation. We will briefly describe a few of them here to illustrate the range of available devices.

The 74LS381 and 74HC381 ICs are called *arithmetic-logic units/function generators*. They can perform eight different binary arithmetic and logic operations on two 4-bit inputs. These operations, which are selected by a set of three SELECT inputs, include addition, subtraction, ORing, EX-ORing, and ANDing.

The 74284 and 74285 are *parallel binary multipliers*. Like the serial multiplier circuit of Figure 6-16, these ICs contain the circuitry for generating the product of two 4-bit binary numbers. Unlike the serial multiplier circuit, these ICs perform parallel multiplication using only combinatorial logic to look at the two 4-bit inputs and generate the correct 8-bit output product. The total multiplication process requires only about 40 ns. The serial multiplier would take at least four times that long because of the need to shift the multiplier and multiplicands four times.

The AM9511 is an LSI device called an *arithmetic processing unit* or APU. It is a 24-pin chip that performs many complex arithmetic operations, including *addition, subtraction, multiplication, division, square roots, trigonometric functions (e.g., sin, cos, tan, etc.),* and *logarithms.* The APU can perform these and other operations on 16-bit or 32-bit input numbers. It uses a clock input signal (like the serial multiplier in Figure 6-16) to sequence its operations. The more complex operations require more clock cycles.

The APU is designed to receive its data and commands from a *microprocessor,* which is an LSI chip that is the central processing unit (CPU) of a microcomputer. Once it receives the data and commands, the APU performs the requested operation in the required number of clock cycles. When it is done, it sends a signal to the microprocessor signifying the completion of the operation. The microprocessor will then typically take the output data (i.e., the answer) from the APU's output pins and put it into one of its internal registers for subsequent processing.

6-19 IEEE/ANSI SYMBOLS

Figure 6-18(a) shows the IEEE/ANSI symbol for a 1-bit adder (full adder). Note that the symbol Σ is used to indicate the addition operation. Figure 6-18(b) is the IEEE/ANSI symbol for a 4-bit parallel adder like the 7483 and 74283. Note how

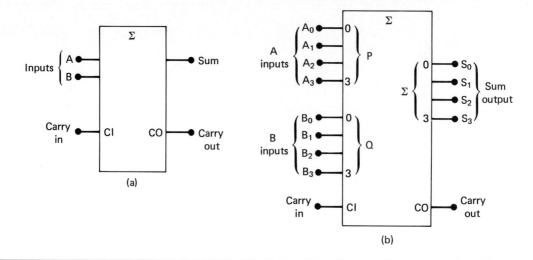

FIGURE 6-18   IEEE/ANSI symbols for (a) a full adder and (b) a 4-bit parallel adder IC (7483/74283).

the letters P and Q are used to represent the two 4-bit inputs, and Σ is used for the 4-bit output sum. The P, Q, and Σ are specified by the IEEE/ANSI standard and must be used inside the symbol outline. The letters used for the inputs and outputs external to the symbol outline are not specified by the standard, so we are free to use A, B, and S, as we did in our earlier discussions.

6-20   TROUBLESHOOTING CASE STUDY

A technician is testing the adder/subtractor circuit of Figure 6-13 and records the following test results for the various operating modes:

- **Mode 1: ADD = 0, SUB = 0.** The sum outputs are always equal to the number in the A register *plus one*. For example, when $[A] = 0110$, the sum is $[S] = 0111$. This is incorrect since the OR outputs and C_0 should all be 0 in this mode to produce $[S] = [A]$.
- **Mode 2: ADD = 1, SUB = 0.** The sum is always one more than it should be. For example, with $[A] = 0010$ and $[B] = 0100$, the sum output is 0111 instead of 0110.
- **Mode 3: ADD = 0, SUB = 1.** The S outputs are always equal to $[A] - [B]$, as expected.

When he examines these test results, the technician sees that the sum outputs exceed the expected results by 1 for the first two modes of operation. At first he suspects a possible fault in one of the LSB inputs to the adder, but he dismisses this because such a fault would also affect the subtraction operation which is working correctly. Eventually, he realizes that there is another fault that could add an extra 1 to the results for the first two modes without causing an error in the subtraction mode.

Recall that C_0 is made a 1 in the subtraction mode as part of the 2's-complement operation on $[B]$. For the other modes, C_0 is to be a 0. The technician checks the connection between the SUB signal and the C_0 input to the adder and finds that it is open due to a bad solder connection. This open explains the observed results since the TTL adder responds as if C_0 were a constant logic 1, causing an extra 1 to be added to the result in modes 1 and 2. The open would have no effect on mode 3, because C_0 is supposed to be a 1 anyway.

EXAMPLE 6-12

Consider again the adder/subtractor circuit. Suppose that there is a break in the connection path between the SUB input and the AND gates as shown in Figure 6-19. Describe the effects of this open on the circuit operation for each mode.

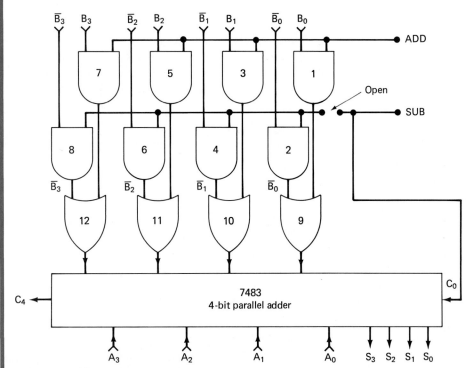

FIGURE 6-19   Example 6-12

SOLUTION
First, realize that this fault will produce a logic 1 at the affected input of AND gates 2, 4, 6 and 8, which will permanently enable each one to pass its $\overline{B}$ input to the following OR gate as shown.

Mode 1: ADD = 0, SUB = 0: The fault will cause the circuit to perform subtraction—almost. The 1's complement of [B] will reach the OR gate outputs and be applied to the adder along with [A]. With $C_0 = 0$, the 2's complement of [B] will not be complete; it will short by 1. Thus, the adder will produce [A] − [B] − 1. To illustrate, let's try [A] = +6 = 0110 and [B] = +3 = 0011. The adder will add as follows:

$$\begin{array}{rr} \text{1's comp. of } [B] = & 1100 \\ [A] = & \underline{0110} \\ \text{result} = & \cancel{1}\,0010 \end{array}$$

$$\uparrow\text{———disregard carry}$$

The result is $0010 = +2$ instead of $0011 = +3$ as it would be for normal subtraction.

Mode 2: ADD = 1, SUB = 0: With ADD = 1, AND gates 1, 3, 5, and 7 will pass the B inputs to the following OR gate. Thus, each OR gate will have a $\overline{B}$ and a B at its inputs, thereby producing a 1 output. For example, the inputs to OR gate 9 will be $\overline{B}_0$ coming from AND gate 2 (because of the fault), and B_0 coming from AND gate 1 (because ADD = 1). Thus, OR gate 9 will produce an output of $\overline{B}_0 + B_0 = 1$.

The adder will add the 1111 from the OR gates to the [A] to produce a sum that is 1 less than [A]. Why? Because $1111_2 = -1_{10}$.

Mode 3: ADD = 0, SUB = 1: This mode will work correctly since SUB = 1 is supposed to enable AND gates 2, 4, 6, and 8 anyway.

GLOSSARY **Accumulator Register** Principal register of an arithmetic-logic unit. [Sec. 6-9]

Addend Number to be added to another. [Sec. 6-3]

Arithmetic-Logic Unit (ALU) Digital circuit used in computers to perform various arithmetic and logic operations. [Secs. 6-9 and 6-18]

Arithmetic Processing Unit (APU) Large-scale-integration IC capable of performing many complex arithmetic operations on 16-bit or 32-bit binary numbers. [Sec. 6-18]

Augend Number to which an addend is added. [Sec. 6-3]

BCD Adder Special adder containing two 4-bit parallel adders and a correction detector circuit. Whenever the addition of two BCD code groups is greater than 1001_2 (9_{10}), the correction detector circuit senses it, adds to the result the correction factor 0110_2 (6_{10}), and generates a carry to the next decimal position. [Sec. 6-16]

Binary Multiplier Special digital circuit capable of performing the arithmetic operation of multiplication on two binary numbers. Binary multipliers can be either serial or parallel. [Sec. 6-17]

Carry Digit or bit that is generated when two numbers are added and the result is greater than that of the base for that number system. [Sec. 6-1]

Carry Propagation It is the intrinsic circuit delay of some parallel adders that prevents the carry bit (C_{OUT}) and the result of the addition from appearing at the output simultaneously. [Sec. 6-13]

Carry Ripple *See* Carry Propagation. [Sec. 6-13]

Dividend Number to be divided. [Sec. 6-6]

Divisor Number by which a dividend is divided. [Sec. 6-6]

Full Adder Logic circuit with three inputs and two outputs. The inputs are a carry bit (C_{IN}) from a previous stage, a bit from the augend, and a bit from the addend, respectively. The outputs are the sum bit produced by the addition of the bit from the addend with the bit from the augend and the resulted carry (C_{OUT}) bit which will be added to the next stage. [Sec. 6-10]

Function Generator *See* Arithmetic-Logic Unit (ALU). [Sec. 6-18]

Half Adder Logic circuit with two inputs and two outputs. The inputs are a bit from the augend and a bit from the addend, respectively. The outputs are the sum bit produced by the addition of the bit from the addend with the bit from the augend and the resultant carry (C_{OUT}) bit, which will be added to the next stage. [Sec. 6-11]

Look-Ahead Carry Ability of some parallel adders to predict, without having to wait for the carry to propagate through the full adders, whether or not a carry bit (C_{OUT}) will be generated as a result of the addition, thus reducing the overall propagation delays. [Sec. 6-13]

Minuend Number from which the subtrahend is to be subtracted. [Sec. 6-4]

1's-Complement Form: Result obtained when each bit of a binary number is complemented. [Sec. 6-2]

OR Addition Boolean logic operation performed by an OR gate. [Sec. 6-1]

Overflow When in the process of adding signed binary numbers a 1 is generated from the MSB position of the number into the sign bit position. [Secs. 6-4 and 6-12]

Parallel Adder Digital circuit made from full adders and used to add all the bits from the addend and the augend together simultaneously. [Sec. 6-10]

Sigma (Σ) Greek letter that represents addition and is often used to label the sum output bits of a parallel adder. [Secs. 6-14 and 6-19]

Sign Bit Binary bit that is added to the leftmost position of a binary number to indicate whether that number represents a positive or a negative quantity. [Sec. 6-2]

Subtrahend Number that is to be subtracted from a minuend. [Sec. 6-4]

True-Magnitude Form: Representation of a binary number where the magnitude bits are the true binary equivalent of the decimal numbers. [Sec. 6-2]

2's-Complement Form Result obtained when a 1 is added to the least significant bit position of a binary number in the 1's-complement form. [Sec. 6-2]

6-1.   Add the following groups of binary numbers using binary addition.
(a) 1010 + 1011   (b) 1111 + 0011   (c) 1011.1101 + 11.1
(d) 0.1011 + 0.1111   (e) 10011011 + 10011101

Section 6-2

6-2.   Represent each of the following signed decimal numbers in the 2's-complement system. Use a total of eight bits including sign bit.
(a) +32   (b) −14   (c) +63   (d) −104
(e) −1   (f) −128   (g) +169   (h) 0

6-3.   Each of the following numbers represents a signed decimal number in the 2's-complement system. Determine the decimal value in each case.
(a) 01101   (b) 11101   (c) 01111011   (d) 10011001
(e) 01111111   (f) 100000   (g) 11111111   (h) 10000001

6-4.   (a) What range of signed decimal values can be represented using 12 bits including sign bit?
(b) How many bits would be required to represent decimal numbers from −32,768 to +32,767?

6-5.   List, in order, all of the signed numbers that can be represented in 5 bits using the 2's-complement system.

6-6.   A certain computer has the following binary number stored in its memory: 10100100. Only the programmer who put this data in memory knows what it represents. Here are some of the possibilities.
(a) An unsigned decimal number.
(b) A signed decimal number using the 2's-complement system.
(c) A BCD coded decimal number.
(d) An ASCII code with a parity bit.
(e) A signed decimal number using true-magnitude system.
For each of these possibilities, determine what the data represents.

6-7.   What is the range of unsigned decimal values that can be represented in 10 bits? What is the range of signed decimal values using the same number of bits?

Sections 6-3 and 6-4

6-8.   The reason why the true-magnitude method for representing signed numbers is not used in most computers can readily be illustrated by performing the following.
(a) Represent +12 in 5 bits using true magnitude and a sign bit.
(b) Represent −12 in 5 bits using true magnitude and a sign bit.
(c) Add the two binary numbers and note that the sum does not look anything like zero.

6-9. Perform the following operations in the 2's-complement system. Use 8 bits (including the sign bit) for each number. Check your results by converting the binary result back to decimal.

(a) Add $+9$ to $+6$. (b) Add $+14$ to -17.

(c) Add $+19$ to -24. (d) Add -48 to -80.

(e) Subtract $+16$ from $+17$. (f) Subtract $+21$ from -13.

(g) Subtract $+47$ from $+47$. (h) Subtract -36 from -15.

(i) Add $+17$ to -17. (j) Subtract $+17$ from $+17$.

6-10. Repeat Problem 6-9 for the following cases, and show that overflow occurs in each case.

(a) Add $+37$ to $+95$. (b) Subtract $+37$ from -95.

Sections 6-5 and 6-6

6-11. Multiply the following pairs of binary numbers.

(a) 111×101 (b) 1011×1011 (c) 101.101×110.010

(d) $.1101 \times .1011$

6-12. Perform the following divisions.

(a) $1100 \div 100$ (b) $111111 \div 1001$ (c) $10111 \div 100$

(d) $10110.1101 \div 1.1$

Sections 6-7 and 6-8

6-13. Add the following decimal numbers after converting each to its BCD code.

(a) $74 + 23$ (b) $58 + 37$ (c) $147 + 380$ (d) $385 + 118$

6-14. Find the sum of each of the following pairs of hex numbers.

(a) $3E91 + 2F93$ (b) $91B + 6F2$ (c) $ABC + DEF$

6-15. Perform the following subtractions on the pairs of hex numbers.

(a) $3E91 - 2F93$ (b) $91B - 6F2$ (c) $0300 - 005A$

6-16. The owner's manual for a certain personal computer states that the computer has usable memory locations at the following hex addresses: 0200 through 03FF, and 4000 through 7FD0. What is the total number of available memory locations?

Section 6-11

6-17. Convert the FA circuit of Figure 6-7 to all NAND gates.

6-18. Write the truth table for a half adder (inputs A and B; outputs SUM and CARRY). From the truth table design a logic circuit that will act as a half adder.

6-19. A full adder can be implemented in many different ways. Figure 6-20 shows how one may be constructed from two half adders. Construct a truth table for this arrangement and verify that it operates as a FA.

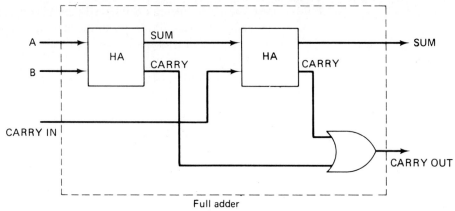

FIGURE 6-20

Section 6-12

6-20. Refer to Figure 6-9. Determine the contents of the *A* register after the following sequence of operations: [*A*] = 0000, [0101] ⟶ [*B*], [*S*] ⟶ [*A*], [1011] ⟶ [*B*], [*S*] ⟶ [*A*].

6-21. Refer to Figure 6-9. Assume that each FF has $t_{PLH} = t_{PHL} = 30$ ns, a setup time of 10 ns, and each FA has a propagation delay of 40 ns. What is the minimum time allowed between the PGT of the LOAD pulse and the PGT of the TRANSFER pulse for proper operation?

6-22. In the adder and subtractor circuits discussed in this chapter we gave no consideration to the possibility of *overflow*. Overflow occurs when the two numbers being added or subtracted produce a result that contains more bits than the capacity of the accumulator. For example, using 4-bit registers, including a sign bit, numbers ranging from $+7$ to -8 (in 2's complement) can be stored. Therefore, if the result of an addition or subtraction exceeds $+7$ or -8, we would say that an overflow has occurred. When an overflow occurs, the results are useless since they cannot be stored correctly in the accumulator register. To illustrate, add $+5$ (0101) and $+4$ (0100), which results in 1001. This 1001 would be interpreted incorrectly as a negative number since there is a 1 in the sign-bit position.

In computers and calculators there are usually circuits that are used to detect an overflow condition. There are several ways to do this. One method that can be used for the adder that operates in the 2's-complement system works as follows:

1. Examine the sign bits of the two numbers being added.

2. Examine the sign bit of the result.

3. Overflow occurs whenever the numbers being added are *both positive* and the sign bit of the result is 1 *or* when the numbers are *both negative* and the sign bit of the result is 0.

This method can be verified by trying several examples. The reader should try the following cases for his own clarification: (1) 5 + 4; (2) −4 + (−6); (3) 3 + 2. Cases 1 and 2 will produce an overflow and case 3 will not. Thus, by examining the sign bits a logic circuit can be designed that will produce a 1 output whenever the overflow condition occurs. Design this overflow circuit for the adder of Figure 6-9.

Section 6-13

6-23. Design a look-ahead carry circuit for the adder of Figure 6-9 which generates the carry C_3 to be fed to the FA of the MSB position based on the values of A_0, B_0, C_0, A_1, B_1, A_2, and B_2. In other words, derive an expression for C_3 in terms of A_0, B_0, C_0, A_1, B_1, A_2, and B_2. (*Hint:* Begin by writing the expression for C_1 in terms of A_0, B_0, and C_0. Then write the expression for C_2 in terms of A_1, B_1, and C_1. Substitute the expression for C_1 into the expression for C_2. Then write the expression for C_3 in terms of A_2, B_2, and C_2. Substitute the expression for C_2 into the expression for C_3. Simplify the final expression for C_3 and put it in sum-of-products form. Implement the circuit.)

Section 6-14

6-24. Show the logic levels at each input and output of Figure 6-10(a) when 354_8 is added to 103_8.

Section 6-15

6-25. For the circuit of Figure 6-13, determine the sum outputs for the following cases.
(a) *A* register = 0101 (+5), *B* register = 1110 (−2); SUBTRACT = 1, ADD = 0.
(b) *A* register = 1100 (−4), *B* register = 1110 (−2); SUBTRACT = 0, ADD = 1.

6-26. Modify the circuit of Figure 6-13 so that a single control input, *X*, is used in place of ADD and SUB. The circuit is to function as an adder when *X* = 0, and as a subtractor when *X* = 1. Then simplify each set of gates. (*Hint:* Note that now each set of gates is functioning as a controlled inverter.)

Section 6-16

6-27. Assume the following inputs in Figure 6-14: [*A*] = 0101, [*B*] = 1001, $C_0 = 0$. Determine the logic levels at [*S*], *X*, [Σ], and CARRY.

6-28. Would it make any difference in the BCD adder of Figure 6-14 if the C_0 of the upper adder was held LOW, while the C_0 of the lower adder was used as the carry input? Explain.

6-29. Assume that the *A* register in Figure 6-15 holds the BCD code for 376, and the *B* register holds the BCD code for 469. Determine the outputs.

6-30. Refer to the circuit of Figure 6-16. Assume that the B register initially holds the multiplicand 0111, and the X register holds the multiplier 1001. Also assume $[A] = 00000000$ prior to the first clock pulse. Show the contents of registers A, B, and X after each clock pulse.

Section 6-20

6-31. The serial multiplier circuit of Figure 6-16 is tested using various values for the multiplicand and multiplier. The results are recorded below.

MULTIPLICAND	MULTIPLIER	RESULT
4	5	60
3	4	45
15	2	225
2	15	30
1	7	15

Examination of this test data shows that the circuit is operating incorrectly. Consider each of the following possible faults, and for each indicate whether or not it could be the actual fault. Explain each answer. (*Hint:* See if there is a pattern in the recorded data.)

(a) The AND gate inputs are shorted together.

(b) There is a break in the X_0 connection to the AND gate.

(c) The D input of FF X_3 is open.

6-32. Assume that there is a broken connection at C_0 in the serial multiplier circuit of Figure 6-16. Show how the operation will be affected by this fault.

6-33. The serial multiplier circuit of Figure 6-16 is tested by a technician who records the following results.

MULTIPLICAND	MULTIPLIER	RESULT
2	3	6
3	4	12
4	6	8
6	4	8
9	3	11
12	8	0
7	7	1

Examine the recorded data and try to determine one or more possible causes of the faulty operation.

6-34. The BCD adder of Figure 6-14 is tested and the results are recorded below.

	$B_3B_2B_1B_0$	$A_3A_2A_1A_0$	$\Sigma_3\Sigma_2\Sigma_1\Sigma_0$	CARRY (X)
(1)	0 0 1 1	0 1 1 0	1 0 0 1	0
(2)	0 1 1 1	1 0 0 0	1 1 1 1	0
(3)	1 0 0 1	1 0 0 1	0 0 1 0	0

Consider each of the following possible faults, and indicate whether or not it could be the actual fault. Explain each answer.

(a) The A_1 and A_0 inputs of the correction adder are internally shorted together.

(b) There is an open from X to the correction adder.

(c) The upper OR gate inputs are internally shorted together.

(d) The AND gate output is stuck LOW.

COUNTERS AND REGISTERS

7

Upon completion of this chapter, you will be able to:

- Understand the operation and characteristics of synchronous and asynchronous counters.
- Construct counters with MOD numbers less than 2^N.
- Identify IEEE/ANSI symbols used in IC counters and registers.
- Construct both up and down counters.
- Analyze and evaluate various types of presettable counters.
- Understand several types of schemes used to decode different types of counters.
- Eliminate decoder spikes by employing a technique called strobing.
- Compare the major differences between the ring and Johnson counters.
- Analyze the theory of operation of a frequency counter and of a digital clock.
- Recognize and understand the operation of various types of registers.
- Apply existing troubleshooting techniques used for combinatorial logic systems to troubleshoot sequential logic systems.

In Chapter 5 we saw how flip-flops could be connected to function as counters and registers. At that time we studied only the basic counter and register circuits. Digital systems employ many variations of these basic circuits, mostly in integrated-circuit form. In this chapter we will look at how FFs and logic gates can be combined to produce different types of counters and registers.

We will present several practical applications to illustrate the many ways in which these logic circuits can be used in digital systems. We will also look at several of the numerous counters and registers that are available as ICs.

7-1 ASYNCHRONOUS (RIPPLE) COUNTERS

Figure 7-1 shows a 4-bit binary counter circuit such as the one discussed in Chapter 5. Recall the following points concerning its operation:

1. The clock pulses are applied only to the *CLK* input of FF *A*. Thus, FF *A* will toggle (change to its opposite state) each time the clock pulses make a negative (HIGH to LOW) transition. Note that $J = K = 1$ for all FFs.

2. The normal output of FF *A* acts as the *CLK* input for FF *B*, so FF *B* will toggle each time the *A* output goes from 1 to 0. Similarly, FF *C* will toggle when *B* goes from 1 to 0 and FF *D* will toggle when *C* goes from 1 to 0.

3. The table in Figure 7-1 shows the sequence of binary states that the FFs will follow as clock pulses are continuously applied. If we let the FF outputs *D, C, B,* and *A* represent a binary number, with *D* being the MSB and *A* the LSB, then a binary counting sequence from 0000 to 1111 is produced.

4. After the 15th clock pulse has occurred, the counter FFs are in the 1111 condition. On the 16th clock pulse FF *A* goes from 1 to 0, which causes FF *B* to go from 1 to 0, and so on until the counter is in the 0000 state. In other words, the counter has gone through one complete cycle (0000 through 1111)

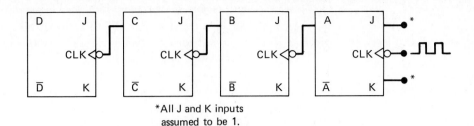

*All J and K inputs
assumed to be 1.

D	C	B	A	Number of clock pulses
0	0	0	0	0
0	0	0	1	1
0	0	1	0	2
0	0	1	1	3
0	1	0	0	4
0	1	0	1	5
0	1	1	0	6
0	1	1	1	7
1	0	0	0	8
1	0	0	1	9
1	0	1	0	10
1	0	1	1	11
1	1	0	0	12
1	1	0	1	13
1	1	1	0	14
1	1	1	1	15
0	0	0	0	16 (recycles)
0	0	0	1	17
0	0	1	0	18
0	0	1	1	19
0	1	0	0	20
0	1	0	1	21
.	.	.	.	.
.	.	.	.	.
.	.	.	.	.

FIGURE 7-1   Four-bit asynchronous (ripple) counter.

and has *recycled* back to 0000, from where it will begin a new counting cycle as subsequent clock pulses are applied.

This type of counter, where each FF output serves as the *CLK* input signal for the next FF, is referred to as an *asynchronous counter*. This is because all the FFs do *not* change states in exact synchronism with the clock pulses; only FF *A* responds to the clock pulses. FF *B* has to wait for FF *A* to change states before it is toggled; FF *C* has to wait for FF *B;* and so on. Thus, there is a delay between the responses of each FF. In modern FFs this delay may be very small (typically 10 to 40 ns), but in some cases, as we shall see, it can be troublesome. Because of the manner in which this type of counter operates, it is also commonly referred to as a *ripple counter*. In the following discussions we will use the terms ''asynchronous counter'' and ''ripple counter'' interchangeably.

EXAMPLE 7-1

The counter in Figure 7-1 starts off in the 0000 state and then clock pulses are applied. Some time later the clock pulses are removed and the counter FFs read 0011. How many clock pulses occurred?

SOLUTION

The apparent answer seems to be 3, since 0011 is the binary equivalent of 3. However, with the information given there is no way to tell whether the counter has recycled or not. This means that there could have been 19 clock pulses; the first 16 pulses bring the counter back to 0000 and the last 3 bring it to 0011. There could have been 35 pulses (two complete cycles and then 3 more), or 51 pulses, and so on.

MOD Number   The counter in Figure 7-1 has 16 distinctly different states (0000 through 1111). Thus, it is a *MOD-16 ripple counter*. Recall that the MOD number is always equal to the number of states which the counter goes through in each complete cycle before it recycles back to its starting state. The MOD number can be increased simply by adding more FFs to the counter. That is,

$$\text{MOD number} = 2^N \tag{7-1}$$

where N is the number of FFs connected in the arrangement of Figure 7-1.

EXAMPLE 7-2

A counter is needed that will count the number of items passing on a conveyor belt. A photocell and light source combination is used to generate a single pulse each time an item crosses its path. The counter has to be able to count as many as one thousand items. How many FFs are required?

SOLUTION

It is a simple matter to determine what value of N is needed so that $2^N \geq 1000$. Since $2^9 = 512$, 9 FFs will not be enough. $2^{10} = 1024$, so 10 FFs would produce a counter that could count as high as $1111111111_2 = 1023_{10}$. Therefore, we should use 10 FFs. We could use more than 10, but it would be a waste of FFs, since any FF past the tenth one will never be toggled.

Frequency Division   In Chapter 5 we saw that in the basic counter each FF provides an output waveform that is exactly *half* the frequency of the waveform at its *CLK* input. To illustrate, suppose that the clock signal in Figure 7-1 is 16 kHz. Figure 7-2 shows the FF output waveforms. The waveform at output A is an 8-kHz *squarewave,* at output B it is 4 kHz, at output C it is 2 kHz, and at output D it is 1 kHz. Notice that the output of FF D has a frequency equal to the original clock frequency divided by 16. In general, *for any counter the output from the last FF*

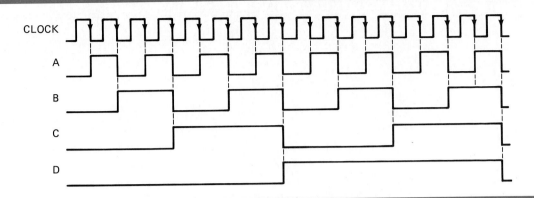

FIGURE 7-2 Counter waveforms showing frequency-division-by-2 for each FF.

(i.e., the MSB) divides the input clock frequency by the MOD number of the counter. For example, a MOD-16 counter could also be called a *divide-by-16 counter.*

EXAMPLE 7-3

The first step involved in building a digital clock* is to take the 60-Hz power-line waveform and feed it into a shaping circuit to produce a squarewave as illustrated in Figure 7-3. The 60-Hz squarewave is then put into a MOD-60 counter, which is used to divide the 60-Hz frequency by exactly 60 to produce a 1-Hz waveform. This 1-Hz waveform is fed to a series of counters, which then count seconds, minutes, hours, etc. How many FFs are required for the MOD-60 counter?

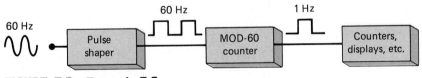

FIGURE 7-3 **Example 7-3**

SOLUTION
There is no integer power of 2 that will equal 60. The closest is $2^6 = 64$. Thus, a counter using six FFs would act as a MOD-64. Obviously, this will not satisfy the requirement. It seems that there is no solution using a counter of the type shown in Figure 7-1. This is partly true; in the next section we will see how to modify this basic binary counter so that virtually *any* MOD number can be obtained and we will not be limited to values of 2^N.

*Here we are talking about a clock that indicates time in hours, minutes, etc.

REVIEW QUESTIONS

7-2 COUNTERS WITH MOD NUMBERS $< 2^N$

The basic ripple counter of Figure 7-1 is limited to MOD numbers that are equal to 2^N, where N is the number of FFs. This value is actually the maximum MOD number that can be obtained using N FFs. The basic counter can be modified to produce MOD numbers less than 2^N by allowing the counter to *skip states* that are normally part of the counting sequence. One of the most common methods for doing this is illustrated in Figure 7-4 where a 3-bit ripple counter is shown. Disregarding the NAND gate for a moment we can see that the counter is a MOD-8 binary counter which will count in sequence from 000 to 111. However, the presence of the NAND gate will alter this sequence as follows:

1. The NAND output is connected to the asynchronous CLEAR inputs of each FF. As long as the NAND output is HIGH, it will have no affect on the counter. When it goes LOW, however, it will clear all the FFs so that the counter immediately goes to the 000 state.

2. The inputs to the NAND gate are the outputs of the B and C FFs, so the NAND output will go LOW whenever $B = C = 1$. This condition will occur when the counter goes from the 101 state to the 110 state (input pulse 6 on waveforms). The LOW at the NAND output will immediately (generally within a few nanoseconds) clear the counter to the 000 state. Once the FFs have been cleared, the NAND output goes back HIGH, since the $B = C = 1$ condition no longer exists.

3. The counting sequence is, therefore,

```
C B A
0 0 0 ←
0 0 1  |
0 1 0  |
0 1 1  |
1 0 0  |
1 0 1  |
1 1 0 ─┘    (temporary state needed to clear counter)
```

Although the counter does go to the 110 state, it remains there for only a few nanoseconds before it recycles to 000. Thus, we can essentially say that this

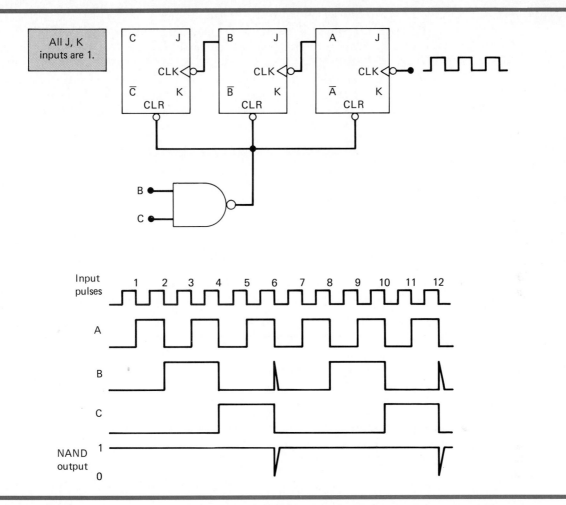

FIGURE 7-4 **MOD-6 counter produced by clearing a MOD-8 counter when count of six (110) occurs.**

counter counts from 000 (zero) to 101 (five) and then recycles to 000. It essentially skips 110 and 111 so that it only goes through six different states; thus, it is a MOD-6 counter.

Notice that the waveform at the *B* output contains a *spike* or *glitch* caused by the momentary occurrence of the 110 state before clearing. This glitch is very narrow and so would not produce any visible indication on indicator lights or numerical displays. It could, however, cause a problem if the *B* output is being used to drive other circuitry outside the counter. It should also be noted that the *C* output has a frequency equal to ⅙ of the input frequency; in other words, this MOD-6 counter has divided the input frequency by *six*. The waveform at *C* is *not* a symmetrical squarewave (50% duty cycle) because it is only HIGH for two clock cycles while it is LOW for four cycles.

Changing the MOD Number   The counter of Figure 7-4 is a MOD-6 because of the choice of inputs to the NAND gate. Any desired MOD number can be obtained by changing these inputs. For example, using a three-input NAND gate with inputs *A*, *B*, and *C*, the counter would function normally until the 111 condition was reached, at which point it would immediately reset to the 000 state. Ignoring the temporary excursion into the 111 state, the counter would go from 000 through 110 and then recycle back to 000, resulting in a MOD-7 counter (seven states).

EXAMPLE 7-4

Determine the MOD number of the counter in Figure 7-5(a). Also determine the frequency at the *D* output.

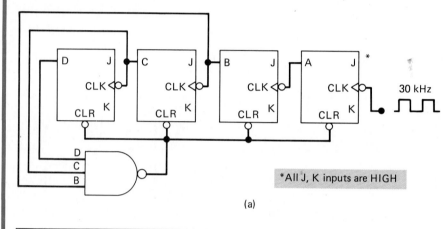

(a)

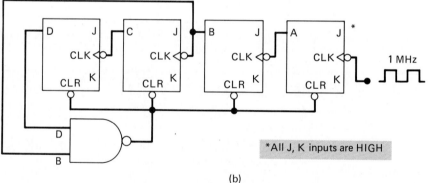

(b)

FIGURE 7-5   (a) MOD-14 ripple counter; (b) MOD-10 (decade) ripple counter.

SOLUTION
This is a 4-bit counter, which would normally count from 0000 through 1111. The NAND inputs are *D*, *C*, and *B* which means that the counter will immediately recycle to 0000 when the 1110 (decimal 14) count is reached. Thus, the counter actually has 14 stable states 0000 through 1101 and is therefore a **MOD-14** counter. Since the input frequency is 30 kHz, the frequency at output *D* will be

$$\frac{30 \text{ kHz}}{14} = 2.14 \text{ kHz}$$

General Procedure   To construct a counter that starts counting from all 0s and has a MOD number of X:

1. Find the smallest number of FFs such that $2^N \geq X$ and connect them as a counter.
2. Connect a NAND gate to the asynchronous CLEAR inputs of all the FFs.
3. Determine which FFs will be in the HIGH state at a count $= X$; then connect the normal outputs of these FFs to the NAND gate inputs.

EXAMPLE 7-5

Construct a MOD-10 counter that will count from 0000 (zero) through 1001 (decimal 9).

SOLUTION
$2^3 = 8$ and $2^4 = 16$; thus, four FFs are required. Since the counter is to have stable operation up to the count of 1001, it must be reset to zero when the count of 1010 is reached. Therefore, FF outputs D and B must be connected as the NAND gate inputs. Figure 7-5(b) shows the arrangement.

Decade Counters/BCD Counters   The MOD-10 counter of Example 7-5 is also referred to as a *decade counter*. In fact, a decade counter is any counter that has 10 distinct states, no matter what the sequence. A decade counter such as the one in Figure 7-5(b), which counts in sequence from 0000 (zero) through 1001 (decimal 9), is also commonly called a *BCD counter* because it uses only the 10 BCD code groups 0000, 0001, . . . , 1000, and 1001. To reiterate, any MOD-10 counter is a decade counter; and any decade counter that counts in binary from 0000 to 1001 is a BCD counter.

Decade counters, especially the BCD type, find widespread use in applications where pulses or events are to be counted and the results displayed on some type of decimal numerical readout. We shall examine this later in more detail. A decade counter is also often used for dividing a pulse frequency *exactly* by 10. The input pulses are applied to FF A and the output pulses are taken from the output of FF D, which has $\frac{1}{10}$ the frequency of the input.

EXAMPLE 7-6

In Example 7-3 a MOD-60 counter was needed to divide the 60-Hz line frequency down to 1 Hz. Construct an appropriate MOD-60 counter.

SOLUTION
$2^5 = 32$ and $2^6 = 64$, so we need six FFs, as shown in Figure 7-6. The counter is to be cleared when it reaches the count of sixty (111100). Thus, the outputs of FFs C, D, E, and F must be connected to the NAND gate. The output of FF F will have a frequency of 1 Hz.

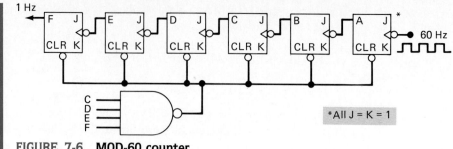

FIGURE 7-6 MOD-60 counter.

REVIEW QUESTIONS

1.   What FF outputs should be connected to the clearing NAND gate to form a MOD-13 counter? (*Ans.* D, C, and A)
2.   *True or false:* All BCD counters are decade counters.
3.   What is the output frequency of a decade counter that is clocked from a 50-kHz signal? (*Ans.* 5 kHz)

7-3 IC ASYNCHRONOUS COUNTERS

There are several TTL and CMOS asynchronous counter ICs. One of them is the TTL 7493 and its logical equivalent, the 74293. The 74293 is preferred in new designs because it has its power and ground at the corner pins on the chip. Figure 7-7(a) shows the logic diagram for the 74293 as it would appear in the manufacturer's TTL data book. Some of the nomenclature is different from what we have been using, but it should be easy to figure out. Note the following points:

1.   The 74293 has four J-K FFs with outputs Q_0, Q_1, Q_2, Q_3, where Q_0 is the LSB and Q_3 is the MSB. The FFs are shown arranged with the LSB on the left. This is done to satisfy the convention that the circuit input signals appear on the left. We have been drawing our counters with LSB on the right so that the order of the FFs is the same as the order of the bits in the binary count. We will continue doing it this way.

2.   Each FF has a *CP* (clock pulse) input, which is just another name for the *CLK* input. The clock inputs to Q_0 and Q_1, *labeled* $\overline{CP}_0$ *and* $\overline{CP}_1$, respectively, are externally accessible. The inversion bars over these inputs indicates that they are activated by a NGT.

3.   Each FF has an asynchronous CLEAR input, C_D. These are connected together to the output of a two-input NAND gate with inputs *MR*, and MR_2, where *MR* stands for *master reset*.

4.   FFs Q_1, Q_2, and Q_3 are already connected as a 3-bit ripple counter. FF Q_0 is not connected to anything internally. This allows the user the option of either connecting Q_0 to Q_1 to form a 4-bit counter, or using Q_0 separately if desired.

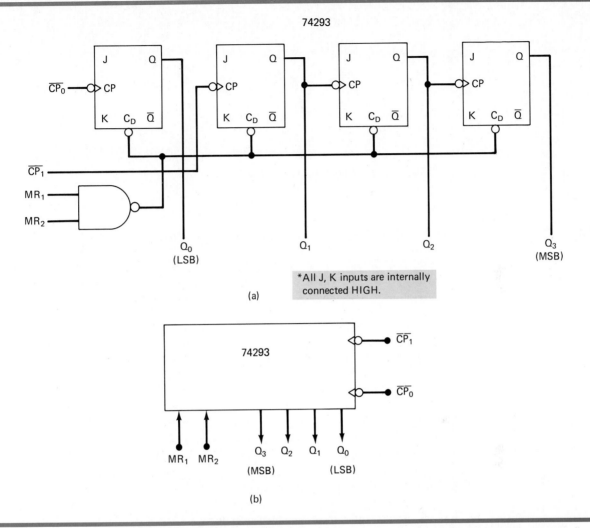

74293

*All J, K inputs are internally connected HIGH.

(a)

74293

$\overline{CP_1}$

$\overline{CP_0}$

MR₁ MR₂   Q₃ Q₂ Q₁ Q₀

(MSB)   (LSB)

(b)

FIGURE 7-7   (a) Logic diagram for 7493 and 74293 asynchronous counter IC; (b) simplified symbol.
(Courtesy of Fairchild, a Schlumberger company)

The following examples will illustrate some of the ways the 74293 can be wired to produce different counters. In these examples we will use the simplified logic symbol shown in Figure 7-7(b).

EXAMPLE 7-7

Show how the 74293 should be connected to operate as a MOD-16 counter wth a 10-kHz clock input.

SOLUTION
A MOD-16 requires four FFs, so we have to connect the Q_0 output to $\overline{CP_1}$, the clock input of FF Q_1 (see Figure 7-8). The 10-kHz pulses are applied to $\overline{CP_0}$, the clock input of Q_0.

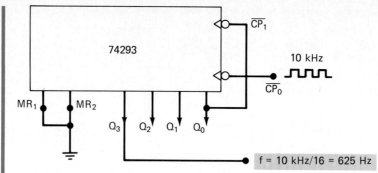

FIGURE 7-8   **74293 wired as a MOD-16.**

In order for the counter to function as a MOD-16 counter, it must be allowed to count through its complete sequence 0000 to 1111. This means that the reset NAND gate has to be disabled; that is, its output should be kept permanently HIGH so that it has no effect on the counter operation. This can be accomplished by connecting MR_1, MR_2, or both to a constant LOW.

EXAMPLE 7-8

Show how to wire the 74293 as a MOD-10 counter.

SOLUTION
A MOD-10 requires four FFs, so again we need to connect Q_0 to $\overline{CP}_1$. This time, however, we want the counter to recycle back to 0000 when it tries to go to the count of 1010 (ten). Thus, the Q_3 and Q_1 outputs have to be connected to the master reset inputs; when they both go HIGH at the count of 1010, the NAND output will immediately reset the counter to 0000.

The circuit wiring is shown in Figure 7-9.

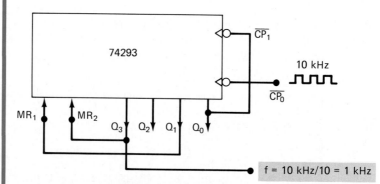

FIGURE 7-9   **74293 wired as a MOD-10.**

EXAMPLE 7-9

Show how to wire a 74293 as a MOD-14.

SOLUTION

When the counter reaches the count of 1110 (14), the Q_3, Q_2, and Q_1 outputs are all HIGH. Unfortunately, the 74293's built-in reset NAND gate has only two inputs. Thus, we have to add some extra logic to ensure that the counter will reset back to 0000 when $Q_3 = Q_2 = Q_1 = 1$. In fact, all we need is a two-input AND gate as shown in Figure 7-10.

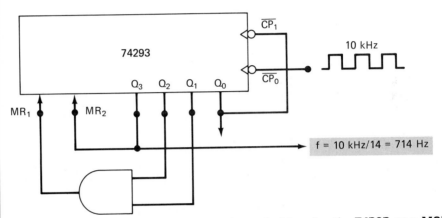

FIGURE 7-10 **An external AND gate is needed to wire the 74293 as a MOD-14 counter.**

EXAMPLE 7-10

In Example 7-6 we divided the input frequency by 60 with a MOD-60 counter using six J-K FFs and a NAND gate. Another way to get a MOD-60 is shown in Figure 7-11. Explain how this circuit works.

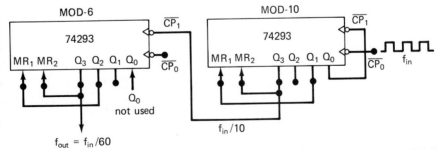

FIGURE 7-11 **Two 74293s can be combined to provide a frequency division of 60.**

SOLUTION
This circuit divides the input frequency by 60 in two steps. The 74293 counter on the right is wired as a MOD-10 so that its output Q_3 has a frequency $= f_{in}/10$. This signal is connected to the $\overline{CP}_1$ input of the second 74293 counter which is wired as a MOD-6 (note that Q_0 is not being used). Thus, the Q_3 output of the second counter will have a frequency

$$f_{out} = \frac{f_{in}/10}{6} = \frac{f_{in}}{60}$$

This last example shows that two (or more) counters can be cascaded to produce an overall MOD number equal to the *product* of their individual MOD numbers. This can be very useful in applications where a large amount of frequency division is required.

IEEE/ANSI Symbol for 74293 Counter

Figure 7-12 shows the new IEEE/ANSI symbol for the 74293. This symbol contains several new aspects of the IEEE/ANSI standard. As we describe these, you should continue to appreciate how the new IEEE/ANSI symbology is designed to tell us a lot about the IC's operation.

The symbol contains three distinct blocks. The top block (with the notches) is the common-control block. The notation "CTR" identifies this IC as a counter. Recall from our discussion in Chapter 5 (Figure 5-34) that the common-control block is used whenever an IC has one or more inputs that are common to more than one of the circuits on the chip. For the 74293, the MR_1 and MR_2 inputs are common to all the FFs in the counter.

MR_1 and MR_2 are shown as active-HIGH inputs that are internally combined using the AND operation as indicated by the "&" notation. This indicates that *both* MR_1 and MR_2 must be in their active states in order to clear the counter. The notation "$CT = 0$" tells us that the action of the MR inputs is to make the count equal zero.

The middle block is labeled "DIV2" to indicate that it is a MOD-2 counter, which of course is a single FF. DIV2 means that the counter will divide its clock

FIGURE 7-12   IEEE/ANSI symbol for the 74293 IC.

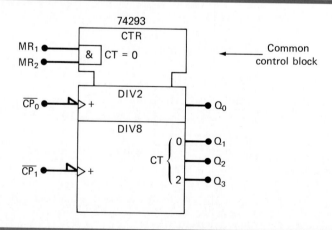

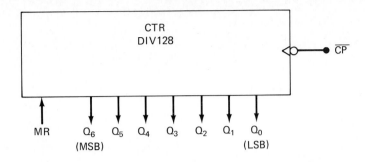

input frequency by 2. The bottom block is labeled "DIV8" to indicate that it is a MOD-8 counter. The clock inputs to each of these blocks are shown as activated by negative-going transitions. The " + " notation on each clock input indicates that the NGT of the clock will cause the count to be *incremented by 1*. In other words, the counter *counts up* on each NGT.

While we will continue to use the traditional symbol for the 74293 and other counters, we will use some of the notation from the IEEE/ANSI symbology. For example, we will sometimes indicate a counter's MOD number using the DIVn notation, where n is the MOD number.

CMOS Asynchronous Counters   There are several asynchronous counters in the CMOS family. Most of them are equivalents to the TTL versions. There are, however, some CMOS asynchronous counter ICs that do not have a TTL counterpart. One of these is the 74HC4024; its logic symbol is shown in Figure 7-13. It is a 7-bit counter with one asynchronous master reset input. The seven FFs are internally connected as a ripple counter. The *MR* input is active-HIGH and can be used to reset all the FFs to the 0 state. Note that we have used the notation "CTR DIV128" to signify that this is a MOD-128 counter.

Another CMOS ripple counter that has no TTL counterpart is the 74HC4040, which is a 12-bit counter with a single active-HIGH master reset input. The clock input to this counter is a Schmitt-trigger type of input that permits the use of slow-changing signals without producing erratic counting.

REVIEW QUESTIONS

1. A 2-kHz clock signal is applied to $\overline{CP}_1$ of a 74293. What is the frequency at Q_3? (*Ans.* 250 Hz)
2. What would be the final output frequency if the order of the counters were reversed in Figure 7-11? (*Ans.* $f_{in}/60$)
3. What is the MOD number of a 74HC4040 counter? (*Ans.* 4096)
4. What would the notation "DIV64" mean on a counter symbol?

7-4 ASYNCHRONOUS DOWN COUNTER

All the counters we have looked at thus far have counted *upward* from zero; that is, they were *up counters*. It is a relatively simple matter to construct asynchronous (ripple) *down counters*, which will count downward from a maximum count to zero. Before looking at a ripple down counter, let us examine the count-down sequence for a 3-bit down counter:

	CBA	*CBA*	*CBA*
(7)	1 1 1	1 1 1	1 1 1
(6)	1 1 0	1 1 0	1 1 0
(5)	1 0 1	1 0 1	. . .
(4)	1 0 0	1 0 0	. . .
(3)	0 1 1	0 1 1	. . .
(2)	0 1 0	0 1 0	etc.
(1)	0 0 1	0 0 1	
(0)	0 0 0	0 0 0	etc.

A, B, and *C* represent the FF output states as the counter goes through its sequence. It can be seen that the A FF (LSB) changes states (toggles) at each step in the sequence just as it does in the up counter. The B FF changes states each time *A* goes from LOW to HIGH; *C* changes states each time *B* goes from LOW to HIGH. Thus, in a down counter each FF, except the first, must toggle when the preceding FF goes from LOW to HIGH. If the FFs have *CLK* inputs that respond to negative transitions (HIGH to LOW), then an inverter can be placed in front of each *CLK*

FIGURE 7-14   MOD-8 down counter.

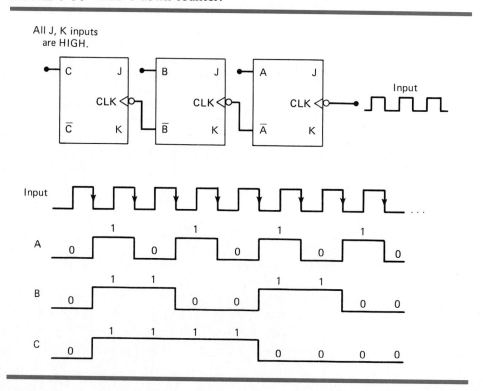

input; however, the same effect can be accomplished by driving each FF *CLK* input from the *inverted* output of the preceding FF. This is illustrated in Figure 7-14 for a MOD-8 down counter.

The input pulses are applied to the A FF; the $\overline{A}$ output serves as the *CLK* input for the B FF; the $\overline{B}$ output serves as the *CLK* input for the *C* FF. The waveforms at *A*, *B*, and *C* show that *B* toggles whenever *A* goes LOW to HIGH (so $\overline{A}$ goes HIGH to LOW) and *C* toggles whenever *B* goes LOW to HIGH. This results in the desired down-counting sequence at the *C*, *B*, and *A* outputs.

Down counters are not as widely used as up counters. Their major application is in situations where it must be known when a desired number of input pulses has occurred. In these situations the down counter is *preset* to the desired number and then allowed to count down as the pulses are applied. When the counter reaches the *zero* state it is detected by a logic gate whose output then indicates that the preset number of pulses has occurred. We shall discuss presettable counters in a later section.

REVIEW QUESTIONS

1. What is the difference between the counting sequence of an up counter and a down counter?
2. Describe how an asynchronous down-counter circuit differs from an up-counter circuit.

7-5 PROPAGATION DELAY IN RIPPLE COUNTERS

Ripple counters are the simplest type of binary counters, since they require the fewest components to produce a given counting operation. They do, however, have one major drawback, which is caused by their basic principle of operation. Each FF is triggered by the transition at the output of the preceding FF. Because of the inherent propagation delay time (t_{pd}) of each FF, this means that the second FF will not respond until a time t_{pd} after the first FF receives an active clock transition; the third FF will not respond until a time equal to $2 \times t_{pd}$ after that clock transition; and so on. In other words, the propagation delays of the FFs accumulate so that the *N*th FF cannot change states until a time equal to $N \times t_{pd}$ after the clock transition occurs. This is illustrated in Figure 7-15, where the waveforms for a 3-bit ripple counter are shown.

The first set of waveforms in Figure 7-15(a) shows a situation where an input pulse occurs every 1000 ns (the clock period T = 1000 ns) and it is assumed that each FF has a propagation delay of 50 ns (t_{pd} = 50 ns). Notice that the A FF output toggles 50 ns after the NGT of each input pulse. Similarly, B toggles 50 ns after A goes from 1 to 0 and C toggles 50 ns after *B* goes from 1 to 0. As a result, when the fourth input pulse occurs, the *C* output goes HIGH after a delay of 150 ns. In this situation the counter does operate properly in the sense that the FFs do even-

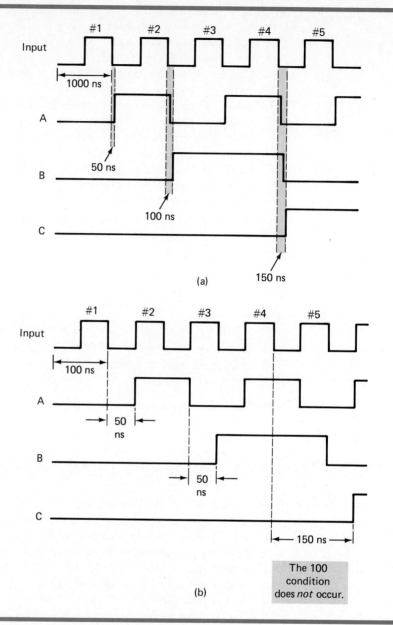

The 100 condition does *not* occur.

FIGURE 7-15 **Waveforms of 3-bit ripple counter illustrating effects of FF propagation delays for different input pulse frequencies.**

tually get to their correct states, representing the binary count. However, the situation worsens if the input pulses are applied at a much higher frequency.

The waveforms in Figure 7-15(b) show what happens if the input pulses occur once every 100 ns. Again, each FF output responds 50 ns after the 1-to-0 transition at its *CLK* input (note the change in the relative time scale). Of particular interest is the situation after the falling edge of the fourth input pulse where the *C* output does not go HIGH until 150 ns later, which is the same time that the A output goes HIGH in response to the fifth input pulse. In other words, the condition $C = 1$, $B = A = 0$ (count of 100), never appears because the input frequency is too high. This could cause a serious problem if this condition were supposed to be used to control some other operation in a digital system. Problems such as this can be avoided if the period between input pulses is made longer than the total propagation delay of the counter. That is,

$$T_{\text{clock}} \geq N \times t_{\text{pd}} \qquad (7\text{-}2)$$

where N = number of FFs. Stated in terms of input-clock frequency, the maximum frequency that can be used is given by

$$f_{\text{max}} = \frac{1}{N \times t_{\text{pd}}} \qquad (7\text{-}3)$$

For example, suppose a 4-bit ripple counter is constructed using the 74LS112 J-K FF. Table 5-1 shows that the 74LS112 has $t_{\text{PLH}} = 16$ ns and $t_{\text{PHL}} = 24$ ns as the propagation delays from CLK to Q. To calculate f_{max}, we will assume the "worst case;" that is, we will use $t_{\text{pd}} = t_{\text{PHL}} = 24$ ns, so that

$$f_{\text{max}} = \frac{1}{4 \times 24 \text{ ns}} = 10.4 \text{ MHz}$$

Clearly, as the number of bits in the counter increases, the total propagation delay increases and f_{max} decreases. For example, a ripple counter that uses six 74LS112 FFs will have

$$f_{\text{max}} = \frac{1}{6 \times 24 \text{ ns}} = 6.9 \text{ MHz}$$

Thus, asynchronous counters are not useful at very high frequencies, especially for large numbers of bits. Another problem caused by propagation delays in asynchronous counters occurs when the counter outputs are *decoded*. This problem is discussed in a later section. Despite these problems, the simplicity of asynchronous counters makes them useful for applications where their frequency limitation is not critical.

REVIEW QUESTIONS

1. Explain why a ripple counter's maximum frequency limitation decreases as more FFs are added to the counter.
2. A certain J-K FF has $t_{\text{pd}} = 12$ ns. What is the largest MOD counter that can be constructed from these FFs and still operate up to 10 MHz? (*Ans.* MOD-256)

7-6 SYNCHRONOUS (PARALLEL) COUNTERS

The problems encountered with ripple counters are caused by the accumulated FF propagation delays; stated another way, the FFs do not all change states simultaneously in synchronism with the input pulses. These limitations can be overcome with the use of *synchronous* or *parallel* counters in which all the FFs are triggered simultaneously (in parallel) by the clock input pulses. Since the input pulses are applied to all the FFs, some means must be used to control when a FF is to toggle and when it is to remain unaffected by a clock pulse. This is accomplished by using the J and K inputs and is illustrated in Figure 7-16 for a 4-bit, MOD-16 synchronous counter.

If we compare the circuit arrangement for this synchronous counter with its asynchronous counterpart in Figure 7-1, we can see the following notable differences:

- The *CLK* inputs of all the FFs are connected together so that the input clock signal is applied to each FF simultaneously.
- Only FF A, the LSB, has its J and K inputs permanently at the HIGH level. The J, K inputs of the other FFs are driven by some combination of FF outputs.
- The synchronous counter requires more circuitry than does the asynchronous counter.

Circuit Operation   The basic principle of operation of the synchronous counter is this:

> *The J and K inputs of the FFs are connected so that only those FFs that are supposed to toggle on a given NGT will have J = K = 1 when that NGT occurs*

Let's examine this principle for each of the FFs with the help of the counting sequence shown in Figure 7-16(b)

The counting sequence shows that the A FF has to change states at each NGT. For this reason its J and K inputs are permanently HIGH so that it will toggle on each NGT of the clock input.

The counting sequence shows that FF B has to change states on each NGT that occurs while $A = 1$. For example, when the count is 0001, the next NGT has to toggle B to the 1 state; when the count in 0011, the next NGT has to toggle B to the 0 state; and so on. This operation is accomplished by connecting output A to the J and K inputs of FF B.

The counting sequence shows that FF C has to change states on each NGT that occurs while $A = B = 1$. For example, when the count is 0011, the next NGT has to toggle C to the 1 state; when the count is 0111, the next NGT has to toggle C to the 0 state; and so on. This operation is ensured by connecting the signal AB to the J and K inputs of FF C.

In a like manner, we can see that FF D has to toggle on each NGT that occurs while $A = B = C = 1$. When the count is 0111, the next NGT has to toggle D to the 1 state; when the count is 1111, the next NGT has to toggle D to the 0 state. This is accomplished by connecting ABC to the J and K inputs of FF D.

Advantage of Synchronous Counters over Asynchronous   In a parallel counter all the FFs will change states simultaneously; that is, they are all synched to the NGTs of the input clock pulses. Thus, unlike the asynchronous counters, the

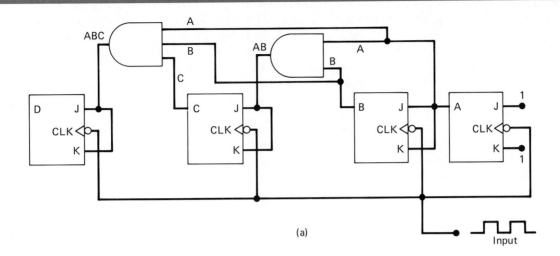

(a)

Count	D	C	B	A
0	0	0	0	0
1	0	0	0	1
2	0	0	1	0
3	0	0	1	1
4	0	1	0	0
5	0	1	0	1
6	0	1	1	0
7	0	1	1	1
8	1	0	0	0
9	1	0	0	1
10	1	0	1	0
11	1	0	1	1
12	1	1	0	0
13	1	1	0	1
14	1	1	1	0
15	1	1	1	1
0	0	0	0	0
.	.	.	.	.
.	.	.	.	.
.	.	etc.	.	.

(b)

FIGURE 7-16 Synchronous MOD-16 counter. Each FF is clocked by the NGT of the input signal so that all FFs will be toggling at the same time.

propagation delays of the FFs do not add together to produce the overall delay. Instead, the total response time of a synchronous counter like the one in Figure 7-16 is the time it takes *one* FF to toggle plus the time for the new logic levels to propagate through a *single* AND gate to reach the *J, K* inputs. That is

$$\text{total delay} = \text{FF } t_{pd} + \text{AND gate } t_{pd}$$

This means that a synchronous counter can operate at a much higher input frequency than an asynchronous counter with the same number of FFs. Of course, the synchronous counter requires more circuitry than the asynchronous counter.

This ability to operate at higher frequencies is the major advantage of synchronous counters. They have another advantage related to the *decoding* process that we will be discussing shortly.

EXAMPLE 7-11

(a) Determine f_{max} for the counter of Figure 7-16(a) if t_{pd} for each FF is 50 ns and t_{pd} for each AND gate is 20 ns. Compare this to f_{max} for a MOD-16 ripple counter.

(b) What has to be done to convert this counter to MOD-32?

(c) Determine f_{max} for the MOD-32 parallel counter.

SOLUTION

(a) The total delay that must be allowed between input clock pulses is equal to FF t_{pd} + AND gate t_{pd}. Thus, $T_{clock} \geq 50 + 20 = 70$ ns, so the parallel counter has

$$f_{max} = \frac{1}{70 \text{ ns}} = 14.3 \text{ MHz (parallel counter)}$$

A MOD-16 ripple counter uses four FFs with $t_{pd} = 50$ ns. Thus, f_{max} for the ripple counter is

$$f_{max} = \frac{1}{4 \times 50 \text{ ns}} = 5 \text{ MHz (ripple counter)}$$

(b A fifth FF must be added, since $2^5 = 32$. The *CLK* input of this FF is also fed by the input pulses. Its *J* and *K* inputs are fed by the output of a four-input AND gate those inputs are *A*, *B*, *C*, and *D*.

(c) f_{max} is still determined as in (a) regardless of the number of FFs in the parallel counter. Thus, f_{max} is still 14.3 MHz.

REVIEW QUESTIONS

1. What is the advantage of a synchronous counter over an asynchronous counter? What is the disadvantage?

2. How many logic devices are required for a MOD-64 parallel counter? (*Ans.* Six FFs and four AND gates)

7-7 PARALLEL DOWN AND UP/DOWN COUNTERS

In Section 7-4 we saw that a ripple counter could be made to count down by using the inverted outputs of each FF to drive the next FF in the counter. A parallel down counter can be constructed in a similar manner—that is, by using the inverted FF outputs to drive the following J, K inputs. For example, the parallel up counter of Figure 7-15 can be converted to a down counter by connecting the $\bar{A}$, $\bar{B}$, and $\bar{C}$ outputs in place of A, B, and C, respectively. The counter will then proceed through the following sequence as input pulses are applied.

$$
\begin{array}{lcccc}
(15) & 1 & 1 & 1 & 1 \leftarrow \\
(14) & 1 & 1 & 1 & 0 \\
(13) & 1 & 1 & 0 & 1 \\
(12) & 1 & 1 & 0 & 0 \\
 & \cdot & \cdot & \cdot & \cdot \\
 & \cdot & \cdot & \cdot & \cdot \quad \text{recycle} \\
 & \cdot & \cdot & \cdot & \cdot \\
(3) & 0 & 0 & 1 & 1 \\
(2) & 0 & 0 & 1 & 0 \\
(1) & 0 & 0 & 0 & 1 \\
(0) & 0 & 0 & 0 & 0 \rightarrow
\end{array}
$$

To form a parallel up/down counter (see Figure 7-17) the control inputs (COUNT-UP and COUNT-DOWN) are used to control whether the normal FF outputs or the inverted FF outputs are fed to the J and K inputs of the following FFs. The counter in Figure 7-17 is a MOD-8 up/down counter that will count from 000 up to 111 when the COUNT-UP control input is 1 and from 111 down to 000 when the COUNT-DOWN control input is 1.

FIGURE 7-17   Parallel up/down counter (MOD-8).

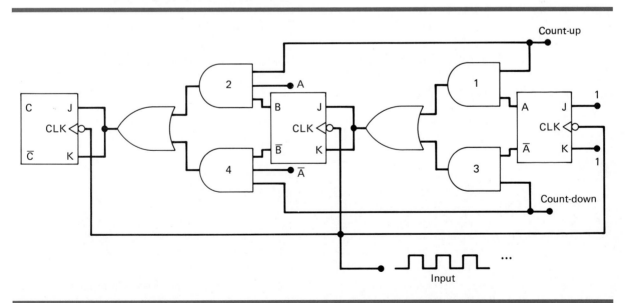

A logical 1 on the COUNT-UP line while COUNT-DOWN = 0 enables AND gates 1 and 2 and disables AND gates 3 and 4. This allows the A and B outputs through to the J and K inputs of the following FFs so that the counter will count up as pulses are applied. The opposite action takes place when COUNT-UP = 0 and COUNT-DOWN = 1.

EXAMPLE 7-12

Describe how the circuit of Figure 7-17 will operate when COUNT-UP and COUNT-DOWN are both LOW, and when they are both HIGH.

SOLUTION

When both these inputs are LOW, each AND-gate and each OR-gate output will be LOW. Thus, the J-K inputs of FFs B and C will be LOW, and neither of these FFs can change states. The A FF will still toggle on each input pulse.

When both inputs are HIGH, each OR-gate output will be HIGH because one of the AND outputs feeding each OR gate will be HIGH, depending on the state of the corresponding FF. Thus, all three FFs will have $J = K = 1$ and will toggle on each input pulse.

7-8 PRESETTABLE COUNTERS

Many synchronous (parallel) counters that are available as ICs are designed to be *presettable;* in other words, they can be preset to any desired starting count either asynchronously (independent of the clock signal) or synchronously (on the active transition of the clock signal). This presetting operation is also referred to as *loading* the counter.

Figure 7-18 shows the logic circuit for a 3-bit presettable parallel up counter. The J, K, and CLK inputs are wired for operation as a parallel up counter. The asynchronous PRESET and CLEAR inputs are wired to perform asynchronous presetting. The counter is loaded with any desired count at any time by doing the following:

1. Apply the desired count to the parallel data inputs, P_2, P_1, and P_0.
2. Apply a LOW pulse to the PARALLEL LOAD input, $\overline{PL}$.

This procedure will perform an asynchronous transfer of the P_2, P_1, and P_0 levels into FFs Q_2, Q_1, and Q_0, respectively (Section 5-17). This *jam transfer* occurs independent of the J, K, and CLK inputs. The effect of the CLK input will be disabled as long as $\overline{PL}$ is in its active-LOW state since each FF will have one of its asynchronous inputs activated while $\overline{PL} = 0$. Once $\overline{PL}$ returns HIGH, the FFs can respond to their CLK inputs and can resume the counting-up operation starting from the count that was loaded into the counter.

For example, let's say that $P_2 = 1$, $P_1 = 0$, and $P_0 = 1$. While $\overline{PL}$ is HIGH, these parallel data inputs have no effect. If clock pulses are present, the counter will perform the normal count-up operation. Now let's say that $\overline{PL}$ is pulsed LOW

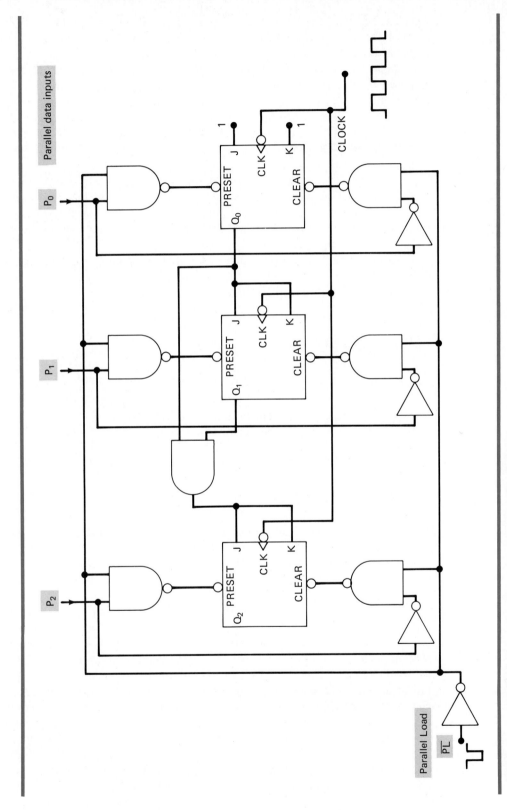

FIGURE 7-18 Presettable parallel counter with asynchronous preset.

when the counter is at the 010 count (i.e., $Q_2 = 0$, $Q_1 = 1$, and $Q_0 = 0$). This LOW at $\overline{PL}$ will produce LOWs at the CLEAR input of Q_1 and at the PRESET inputs of Q_2 and Q_0 so that the counter will go to the 101 count *regardless of what is occurring at the CLK input*. The count will hold at 101 until $\overline{PL}$ is deactivated (returned HIGH); at that time the counter will resume counting up clock pulses from the count of 101.

This asynchronous presetting is used by several IC counters, such as the TTL 74190, 74191, 74192, and 74193 and the CMOS equivalents, 74HC190, 74HC191, 74HC192, and 74HC193.

Synchronous Presetting   Many IC parallel counters use *synchronous presetting* whereby the counter is preset on the active transition of the same clock signal that is used for counting. The logic level applied to the $\overline{PL}$ input determines whether the active clock transition will preset the counter or whether it will be counted as in the normal counting operation.

Examples of IC counters that use synchronous presetting include the TTL 74160, 74161, 74162, and 74163 and their CMOS equivalents, 74HC160, 74HC161, 74HC162, and 74HC163.

REVIEW QUESTIONS

1. What is meant when we say that a counter is presettable?
2. Describe the difference between asynchronous and synchronous presetting.

7-9   THE 74193 COUNTER

Figure 7-19 shows the logic symbol and the input/output description for the 74193 counter. This counter can be described as a MOD-16, presettable up/down counter with synchronous counting, asynchronous preset, and asynchronous master reset. Let us look at the function of each input and output.

Clock Inputs CP_U and CP_D   The counter will respond to the positive-going transitions at one of two clock inputs. CP_U is the *count-up clock* input. When pulses are applied to this input, the counter will increment (count up) on each PGT to a maximum count of 1111; then it recycles to 0000 and starts over. CP_D is the *count-down clock* input. When pulses are applied to this input, the counter will decrement (count down) on each PGT to a minimum count of 0000; then it recycles to 1111 and starts over. Thus, one clock input or the other will be used for counting while the other clock input is inactive (kept HIGH).

Master Reset *(MR)*   This is an active-HIGH asynchronous input that resets the counter to the 0000 state. *MR* is a dc reset, so it will hold the counter at 0000 as long as *MR* = 1. It also overrides *all* other inputs.

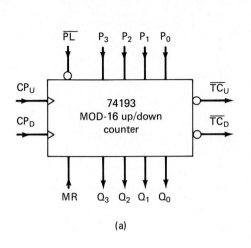

Pin Names	Description
CP_U	Count-up clock input (active rising edge)
CP_D	Count-down clock input (active rising edge)
MR	Asynchronous master reset input (active HIGH)
$\overline{PL}$	Asynchronous parallel load input (active LOW)
P_0–P_3	Parallel data inputs
Q_0–Q_3	Flip-flop outputs
$\overline{TC}_D$	Terminal count-down (borrow) output (active LOW)
$\overline{TC}_U$	Terminal count-up (carry) output (active LOW)

(b)

Mode Select

MR	$\overline{PL}$	CP_U	CP_D	Mode
H	X	X	X	Asynch. reset
L	L	X	X	Asynch. preset
L	H	H	H	No change
L	H	↑	H	Count up
L	H	H	↑	Count down

H = HIGH; L = LOW
X = Don't care; ↑ = PGT

(c)

FIGURE 7-19 **74193 presettable up/down counter: (a) logic symbol; (b) input/output description; (c) mode select table.** (Courtesy of Fairchild, a Schlumberger company)

Preset Inputs The counter FFs can be preset to the logic levels present on the parallel data inputs P_3–P_0 by momentarily pulsing the parallel load input $\overline{PL}$ from HIGH to LOW. This is an asynchronous preset that overrides the counting operation. $\overline{PL}$ will have no effect, however, if the *MR* input is in its active HIGH state.

Count Outputs The current count is always present at the FF outputs Q_3–Q_0, where Q_0 is the LSB, and Q_3 is the MSB.

Terminal Count Outputs These outputs are used when two or more 74193s are connected as a multistage counter to produce a larger MOD number. In the count-up mode, the $\overline{TC}_U$ output of the lower-order counter is connected to the CP_U input of the next higher-order counter. In the count-down mode, the $\overline{TC}_D$ output of the lower-order counter is connected to the CP_D input of the next higher-order counter.

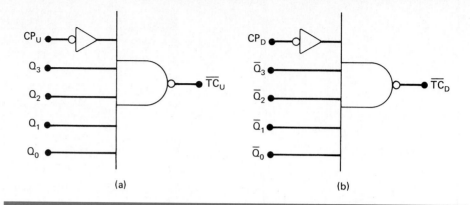

(a)   (b)

FIGURE 7-20   (a) Logic on 74193 for generating $\overline{TC}_U$; (b) logic for generating $\overline{TC}_D$.

$\overline{TC}_U$ is the *terminal count-up* (also called the *carry*) output. It is generated on the 74193 chip using the logic shown in Figure 7-20(a). Clearly, $\overline{TC}_U$ will be LOW only when the counter is in the 1111 state and CP_U is LOW. Thus, $\overline{TC}_U$ will remain HIGH as the counter counts up from 0000 to 1110. On the next PGT of CP_U, the count goes to 1111, but $\overline{TC}_U$ does not go LOW until CP_U returns LOW. The next PGT at CP_U recycles the count to 0000 and also causes $\overline{TC}_U$ to return HIGH. This PGT at TC_U occurs when the counter recycles from 1111 to 0000, and can be used to clock a second 74193 up counter to its next higher count.

$\overline{TC}_D$ is the *terminal count-down* (also called the *borrow*) output. It is generated as shown in Figure 7-20(b). It is normally HIGH and does not go LOW until the counter has counted down to the 0000 state and CP_D is LOW. When the next PGT at CP_D recycles the counter to 1111, it causes $\overline{TC}_D$ to return HIGH. This PGT at $\overline{TC}_D$ can be used to clock a second 74193 down counter to its next lower count.

EXAMPLE 7-13

Refer to Figure 7-21(a), where a 74193 is wired as an up counter. The parallel data inputs are permanently connected as 1011, and the CP_U, $\overline{PL}$, and MR input waveforms are shown in Figure 7-21(b). Assume that the counter is initially in the 0000 state, and determine the counter output waveforms.

SOLUTION
Initially (at t_0) the counter FFs are all LOW. This causes $\overline{TC}_U$ to be HIGH. Just prior to time t_1 the $\overline{PL}$ input is pulsed LOW. This immediately loads the counter with 1011 to produce $Q_3 = 1$, $Q_2 = 0$, $Q_1 = 1$, and $Q_0 = 1$. At t_1 the CP_U input makes a PGT, but the counter cannot respond to this because $\overline{PL}$ is still active at that time. At times t_2, t_3, t_4, and t_5 the counter counts up on each PGT at CP_U. After the PGT at t_5 the counter is in the 1111 state, but $\overline{TC}_U$ does not go LOW until CP_u goes LOW at t_6. When the next PGT occurs at t_7, the counter recycles to 0000, and $\overline{TC}_U$ returns HIGH.

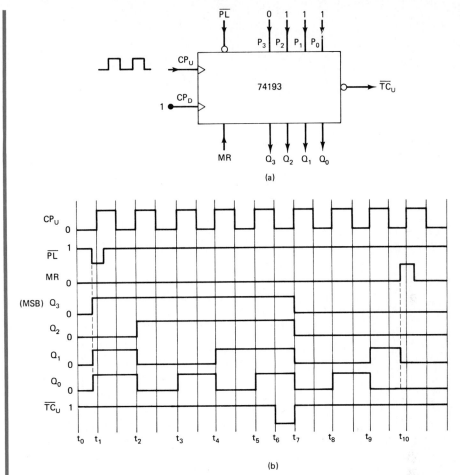

FIGURE 7-21 **Example 7-13**

The counter will count up in response to the PGTs at t_8 and t_9. The PGT at t_{10} will have no effect because the MR goes HIGH prior to t_{10} and remains active at t_{10}. This will reset all FFs to 0 and overrides the CP_U signal.

EXAMPLE 7-14

Figure 7-22(a) shows the 74193 wired as a down counter. The parallel data inputs are permanently wired as 0111, and the CP_D and $\overline{PL}$ waveforms are shown in Figure 7-22(b). Assume that the counter is initially in the 0000 state, and determine the output waveforms.

SOLUTION

At t_0 all the FF outputs are LOW and CP_D is LOW. These are the conditions that produce $\overline{TC}_D = 0$. Prior to t_1 the $\overline{PL}$ input is pulsed LOW. This immediately presets the counter to 0111 and therefore causes $\overline{TC}_D$ to go HIGH. The PGT of CP_D at t_1 will have no effect, since $\overline{PL}$ is still active. The counter will respond

THE 74193 COUNTER **339**

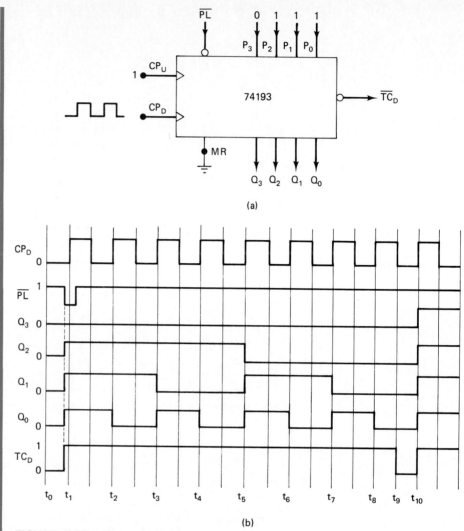

(a)

(b)

FIGURE 7-22 Example 7-14

to the PGTs at t_2–t_8 and counts down to 0000 at t_8. $\overline{TC}_D$ does not go LOW until t_9, when CP_D goes LOW. At t_{10} the PGT of CP_D causes the counter to recycle to 1111 and also drives $\overline{TC}_D$ back HIGH.

Variable MOD Number Using the 74193   Presettable counters can be easily wired for different MOD-numbers without the need for additional logic circuitry. We will demonstrate this for the 74193 using the circuit of Figure 7-23(a). Here, the 74193 is used as a down counter with its parallel load inputs permanently connected at 0101 (5_{10}). Note that the $\overline{TC}_D$ output is connected back to the $\overline{PL}$ input.

We will begin our analysis by assuming that the counter has been counting down and is in the 0101 state at time t_0. Refer to Figure 7-23(b) for the counter waveforms.

The counter will decrement (count down) on the PGTs of CP_D at times t_1–t_5. At

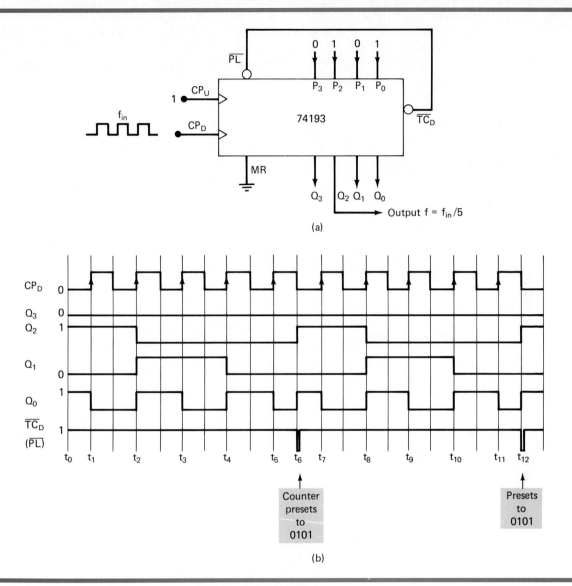

FIGURE 7-23   (a) 74193 wired as a MOD-5; (b) waveforms.

t_5 the counter is in the 0000 state. When CP_D goes LOW at t_6, it drives $\overline{TC}_D$ LOW. This immediately activates the $\overline{PL}$ input and presets the counter back to the 0101 state. Note that $\overline{TC}_D$ stays LOW for only a short interval because once the counter outputs go to 0101 in response to $\overline{PL} = 0$, the condition needed to keep $\overline{TC}_D = 0$ is removed. Thus, there is only a narrow glitch at $\overline{TC}_D$.

This same sequence is repeated at times t_7–t_{12} and at equal intervals thereafter. If we examine the Q_2 waveform, we can see that it goes through one complete cycle for every *five* cycles of CP_D. For example, there are *five* clock cycles between the PGT of Q_2 at t_6 and the PGT of Q_2 at t_{11}. Thus, the frequency of the Q_2 waveform is ⅕ of the clock frequency.

It is no coincidence that the frequency-division ratio (5) is the same as the number applied to the parallel data inputs (0101 = 5). In fact, we can vary the frequency division by changing the logic levels applied to the parallel data inputs.

A *variable frequency-divider* circuit can be easily implemented by connecting switches to the parallel data inputs of the circuit in Figure 7-23. The switches can be set to a value equal to the desired frequency-division ratio.

REVIEW QUESTIONS

1. Describe the function of the $\overline{PL}$ and P_0–P_3 inputs.
2. Describe the function of the MR input.
3. *True or false:* The 74193 cannot be preset while MR is active.
4. What logic levels must be present at CP_D, $\overline{PL}$, and MR in order for the 74193 to count pulses that appear at CP_U? (*Ans.* 1, 1, 0, respectively)

7-10 MORE ON THE IEEE/ANSI DEPENDENCY NOTATION*

We can learn more about the dependency notation that is such an important part of the new IEEE/ANSI symbology by examining the IEEE/ANSI symbol for the 74193 IC shown in Figure 7-24. Each type of IC that we examine in this way will add to your understanding of the new symbology and will help to prepare you for the more extensive use of these symbols in the future.

Once again, it should be stated that only the labels inside the rectangular outlines are specified by the IEEE/ANSI standard. The names or labels shown out-

FIGURE 7-24 IEEE/ANSI symbol for the 74193 IC.

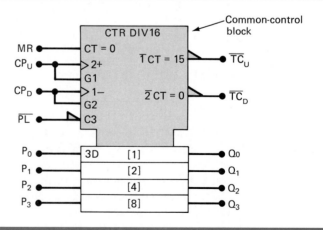

*This section may be omitted without affecting the rest of the book.

side the outlines are not standard, and in fact will vary from one IC manufacturer to another.

Some of the notation used in Figure 7-24 should be familiar. The symbol outline is divided into the common-control block that affects all of the counter FFs, and the four narrow rectangles representing the individual FFs. The bracketed numbers inside each FF rectangle denote their relative weights in the counter. The label CTR DIV16 signifies that this device, when operated normally, is a counter (CTR) with sixteen states (i.e., a divide-by-16 counter). The MR input to the common-control block has the notation $CT = 0$ to indicate that the counter will reset to zero when MR is HIGH.

Control Dependency (C)

The letter C in the label for an input denotes that that input *controls* the entry of data into a storage element (i.e., a FF). Usually, C is used for clock inputs that clock data into a FF on its active transition. We saw this when we looked at the IEEE/ANSI symbols for FFs in Chapter 5. In Figure 7-24, C is used for the parallel/load input $\overline{PL}$ since this input controls the entry of data into the four counter FFs. Specifically, the label C3 indicates that this input will control any other input that has the digit 3 as a prefix in its label. In this case that includes inputs P_0, P_1, P_2, and P_3 since they all have the label 3D (it is shown only on the top FF block, but it is assumed to be the same for the other FF blocks). The "D" part of the label denotes "data."

What this all means is that when $\overline{PL}$ is in its active-LOW state, data from P_0–P_3 will be entered into FFs Q_0–Q_3. Since there is no edge-triggered symbol at $\overline{PL}$, it is understood that $\overline{PL}$ is in effect as long as it stays at its active-LOW level.

Counting Direction (+ or −)

The CP_U and CP_D inputs are shown as having two separate labels because they have several distinct internal effects. Let's first consider the upper label. This label for the CP_U input is 2+. The plus sign (+) indicates that a PGT at this input will increment the count by 1, in other words, cause the counter to count up. Likewise, the upper label for the CP_D input has a minus sign (−) to show that this input will decrement the count by 1, in other words, cause it to count down. The significance of the digits in front of the + and − minus signs will be explained in the following paragraphs.

AND Dependency (G)

The letter G in the label for an input denotes AND dependency. This means that an input designated by a G followed by a digit is internally ANDed with any other input or output that has the same digit as a prefix in its label. In Figure 7-24 we see that the lower label for the CP_U input is G1. This means that CP_U is internally ANDed with any input or output that has a 1 in its label. The upper label for CP_D is 1, so there must be an AND dependency between CP_D and CP_U. Specifically, this AND dependency tells us that CP_U must be HIGH in order for CP_D to perform its count-down function.

The lower label for CP_D is G2, which indicates that there is an AND dependency between CP_D and any input or output that has a 2 in its label. For example, the upper label for CP_U is 2+, which tells us that CP_D must be HIGH in order for CP_U to perform its count-up function.

Now let's look at the $\overline{TC}_D$ output label. It is $\overline{2}CT = 0$. It includes a 2 in its label, indicating that it has an AND dependency with CP_D. Actually, since it is a $\overline{2}$, the AND dependency is with $\overline{CP}_D$. Thus, the label for $\overline{TC}_D$ tells us that $\overline{TC}_D$ will go to its active-LOW state when CP_D is LOW *and* the count is zero ($CT = 0$). In a like manner, the label for $\overline{TC_U}$ tells us that $\overline{TC_U}$ will go to its active-LOW state when CP_U is LOW *and* the count is fifteen ($CT = 15$).

REVIEW QUESTIONS

1. Explain the meaning of control dependency and AND dependency.
2. Give the meaning of the following input labels: (a) +, (b) G4, (c) C5, (d) 5D.

7-11 DECODING A COUNTER

Digital counters are often used in applications where the count represented by the states of the FFs must somehow be determined or displayed. One of the simplest means for displaying the contents of a counter involves just connecting the output of each FF to a small indicator LED. In this way the states of the FFs are visibly represented by the LEDs (bright = 1, dark = 0) and the count can be mentally determined by decoding the binary states of the LEDs. For instance, suppose that this method is used for a BCD counter and the states of the LEDs are respectively dark-bright-bright-dark. This would represent 0110, which we would mentally decode as decimal 6. Other combinations of LED states would represent the other possible counts.

The indicator LED method becomes inconvenient as the size (number of bits) of the counter increases, because it is much harder to mentally decode the displayed results. For this reason it would be preferable to develop a means for *electronically* decoding the contents of a counter and displaying the results in a form that would be immediately recognizable and would require no mental operations.

An even more important reason for electronic decoding of a counter occurs because of the many applications in which counters are used to control the timing or sequencing of operations *automatically* without human intervention. For example, a certain system operation might have to be initiated when a counter reaches the 101100 state (count of 44_{10}). A logic circuit can be used to decode for or detect when this particular count is present and then initiate the operation. Many operations may have to be controlled in this manner in a digital system. Clearly, human intervention in this process would be undesirable except in extremely slow systems.

Active-HIGH Decoding   A MOD-X counter has X different states; each state is a particular pattern of 0s and 1s stored in the counter FFs. A decoding network is a logic circuit that generates X different outputs, each of which detects (decodes) the presence of one particular state of the counter. The decoder outputs can be designed to produce either a HIGH or a LOW level when the detection occurs. An active-HIGH decoder produces HIGH outputs to indicate detection. Figure 7-25 shows the complete active-HIGH decoding logic for a MOD-8 counter. The decoder consists of eight three-input AND gates. Each AND gate produces a HIGH output for one particular state of the counter.

For example, AND gate 0 has as its inputs the FF outputs $\overline{C}$, $\overline{B}$, and $\overline{A}$. Thus its output will be LOW at all times *except* when $A = B = C = 0$—that is, on the count of 000 (zero). Similarly, AND gate 5 has as its inputs the FF outputs C, $\overline{B}$, and A, so its output will go HIGH only when $C = 1$, $B = 0$, and $A = 1$—that is, on the count of 101 (decimal 5). The rest of the AND gates perform in the same manner for the other possible counts. At any one time only one AND-gate output is HIGH, the one which is decoding for the particular count that is present in the counter. The waveforms in Figure 7-25 show this clearly.

FIGURE 7-25   Using AND gates to decode a MOD-8 counter.

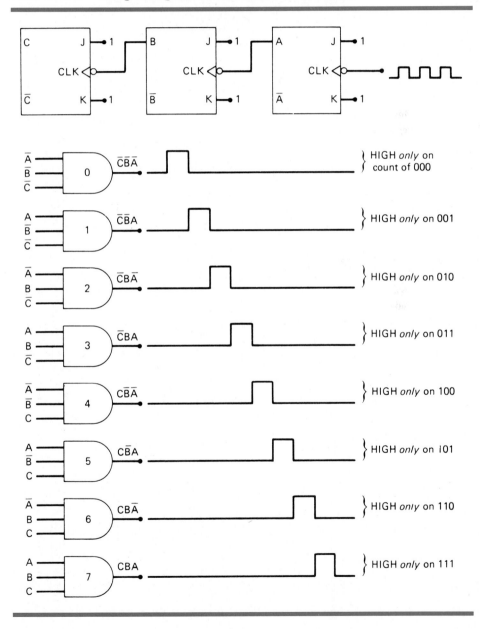

The eight AND outputs can be used to control eight separate indicator LEDs, which represent the decimal numbers 0–7. Only one LED will be on at a given time, indicating the proper count.

The AND gate decoder can be extended to counters with any number of states. The following example illustrates.

EXAMPLE 7-15

How many AND gates are required to completely decode all the states of a MOD-32 binary counter? What are the inputs to the gate that decodes for the count of 21?

SOLUTION

A MOD-32 counter has 32 possible states. One AND gate is needed to decode for each state; therefore, the decoder requires 32 AND gates. Since $32 = 2^5$, the counter contains five FFs. Thus, each gate will have five inputs, one from each FF. To decode for the count of 21 that is 10101_2 requires AND-gate inputs of E, $\overline{D}$, C, $\overline{B}$, and A, where E is the MSB flip-flop.

Active-LOW Decoding   If NAND gates are used in place of AND gates, the decoder outputs will produce a normally HIGH signal, which goes LOW only when the number being decoded occurs. Both types of decoders are used, depending on the type of circuits being driven by the decoder outputs.

EXAMPLE 7-16

Figure 7-26 shows a common situation in which a counter is used to help generate a control waveform which could be applied to devices such as a motor,

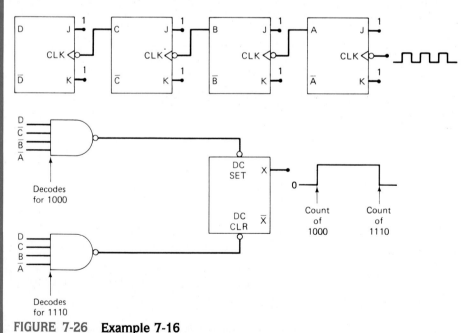

FIGURE 7-26   Example 7-16

solenoid valve, or heater. The MOD-16 counter cycles and recycles through its counting sequence. Each time it goes to the count of 8 (1000), the upper NAND gate will produce a LOW output, which sets FF X to the 1 state. FF X stays HIGH until the counter reaches the count of 14 (1110), at which time the lower NAND gate decodes it and produces a LOW output to clear X to the 0 state. Thus, the X output is HIGH between the counts of 8 and 14 for each cycle of the counter.

BCD Counter Decoding A BCD counter has 10 states, which can be decoded using the techniques previously described. BCD decoders provide 10 outputs corresponding to the decimal digits 0 through 9 represented by the states of the counter FFs. These 10 outputs can be used to control 10 indicator LEDs for a visual display. More often, instead of using 10 separate LEDs, a single display device is used to display the decimal numbers 0 through 9. One such device, called a *nixie tube,* contains 10 very thin numerically shaped filaments stacked on top of each other. The BCD decoder outputs control which filament is illuminated. A newer class of decimal displays contains seven small segments made of a material (usually LEDs or liquid crystal displays) which either emits light or reflects ambient light. The BCD decoder outputs control which segments are illuminated in order to produce a pattern representing one of the decimal digits.

We will go into more detail concerning these types of decoders and displays in Chapter 9. However, since BCD counters and their associated decoders and displays are very commonplace, we will use the decoder/display unit (see Figure 7-27) to represent the complete circuitry used to visually display the contents of a BCD counter as a decimal digit.

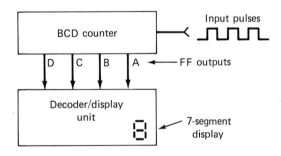

FIGURE 7-27 **BCD counters usually have their count displayed on a single display device.**

REVIEW QUESTION

1. Describe the decoding gate needed to produce a LOW output when a MOD-64 counter is at the count of 23. (*Ans.* A six-input NAND gate with inputs A, B, C, $\overline{D}$, E, and $\overline{F}$)

7-12 DECODING GLITCHES

Section 7-5 discussed the effects of FF propagation delays in ripple counters. As we saw then, the accumulated propagation delays serve to essentially limit the frequency response of ripple counters. The delays between FF transitions can also cause problems when decoding a ripple counter. The problem occurs in the form of glitches or spikes at the outputs of some of the decoding gates. This is illustrated in Figure 7-28 for a MOD-4 ripple counter.

The waveforms at the outputs of each FF and decoding gate are shown in the figure. Notice the propagation delay between the clock waveform and the A output waveform and between the A waveform and the B waveform. The glitches in the X_0 and X_2 decoding waveforms are caused by the delay between the A and B waveforms. X_0 is the output of the AND gate decoding for the normal 00 count. The 00 condition also occurs momentarily as the counter goes from the 01 to the 10 count, as shown by the waveforms. This is because B cannot change states until A goes *LOW*. This momentary 00 condition only lasts for several nanoseconds (depending on t_{pd} of FF B) but can be detected by the decoding gate if the gate's response is fast enough. Hence, the spike at the X_0 output.

A similar situation produces a glitch at the X_2 output. X_2 is decoding for the 10 condition, and this condition occurs momentarily as the counter goes from 11 to 00 in response to the fourth clock pulse, as shown in the waveforms. Again, this is due to the delay of FF B's response after A has gone LOW.

Although the situation is illustrated for a MOD-4 counter, the same type of situation can occur for *any* ripple counter. This is because ripple counters work on the "chain-reaction" principle, whereby each FF triggers the next one and so on. The spikes at the decoder outputs may or may not present a problem, depending on how the counter is being used. When the counter is being used only to count pulses and display the results, the decoding spikes are of no consequence because they are very short in duration and will not even show up on the display. However, when the counter is used to control other logic circuits, such as was done in Figure 7-26, the spikes can cause improper operation. For example, in Figure 7-26, a spike at the output of either decoding NAND gate would cause FF X to be set or cleared at the wrong time.

In situations where the decoding spikes cannot be tolerated, there are two basic solutions to the problem. The first possibility is to use a parallel counter instead of a ripple counter. Recall that in a parallel counter the FFs are all triggered at the same time by the clock pulses so that it appears that the conditions which produced the decoder spikes cannot occur. However, even in a parallel counter the spikes may occur because the FFs will not all necessarily have the same t_{pd}, especially when some FFs may be loaded more heavily than others.

Strobing   A more reliable method for eliminating the decoder spikes is to use a technique called *strobing*. This technique uses a signal called a *strobe signal* to keep the decoding AND gates disabled (outputs at 0) until all the FFs have reached a stable state in response to the negative clock transition. This is illustrated in Figure 7-29(a), where the strobe signal is connected as an input to each decoding gate. The accompanying waveforms show that the strobe signal goes LOW when the

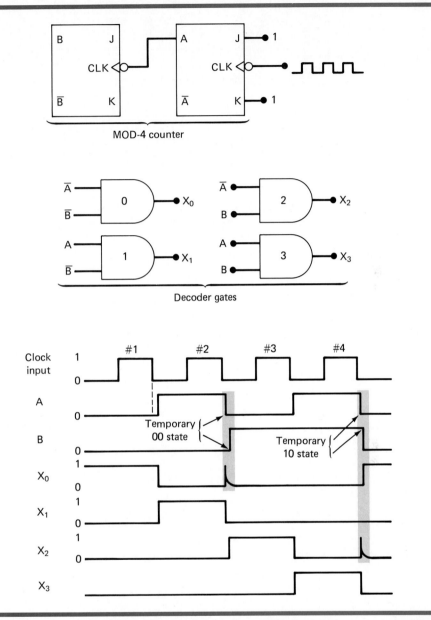

FIGURE 7-28 FF and decoding waveforms for a MOD-4 ripple counter showing glitches at X_0 and X_2 outputs.

clock pulse goes HIGH. During the time that the strobe is LOW, the decoding gates are kept LOW. The strobe signal goes HIGH some time t_D *after* the clock pulse goes LOW to enable the decoding gates. t_D is chosen to be greater than the total time it takes the counter to reach a stable count and depends, of course, on the FF delays and the number of FFs in the counter. In this way the decoding gate outputs

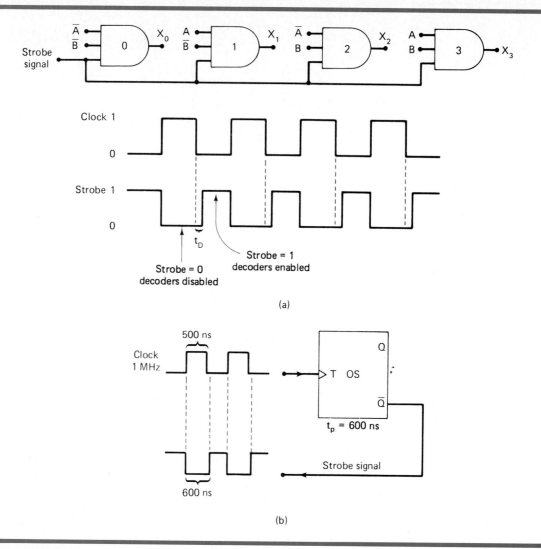

FIGURE 7-29 (a) Use of strobe signal to eliminate decoding spikes; (b) one possible way to generate strobe waveform.

will not contain any spikes because they are disabled during the time the FFs are in transition.

Figure 7-29(b) shows a simple circuit used to generate the strobe signal for a typical situation. The OS is triggered by the *positive* transition of the 1-MHz clock and its $\overline{Q}$ output goes LOW for $t_p = 600$ ns. Thus, the strobe output will stay LOW until 100 ns after the clock goes LOW. While the strobe signal is HIGH, the decoder outputs will be activated.

The strobe method is not used if a counter is only used for display purposes, since the decoding spikes are too narrow to affect the display. The strobe signal is used when the counter is used in control applications like that of Figure 7-26, where the spikes could cause erroneous operation.

1. Explain why the decoding gates for an asynchronous counter may have glitches on their outputs.
2. How does strobing eliminate decoding glitches?

7-13 CASCADING BCD COUNTERS

BCD counters are used whenever pulses are to be counted and the results displayed in decimal. A single BCD counter can count from 0 through 9 and then recycles to 0. To count to larger decimal values, we can cascade BCD counters as illustrated in Figure 7-30. This arrangement operates as follows:

1. Initially all counters are cleared to the zero state. Thus, the decimal display is 000.

2. As input pulses arrive, the units BCD counter advances one count per pulse. After nine pulses have occurred, the hundreds and tens BCD counters are still at zero and the units counter is at 9 (binary 1001). Thus, the decimal display reads 009.

3. On the tenth input pulse the units counter recycles to zero, causing its D FF output to go from 1 to 0. This 1-to-0 transition acts as the clock input for the tens counter and causes it to advance one count. Thus, after 10 input pulses, the decimal readout is 010.

4. As additional pulses occur, the units counter advances one count per pulse, and each time the units counter recycles to zero, it advances the tens counter one count. Thus, after 99 input pulses have occurred, the tens counter is at 9, as is the units counter. The decimal readout is thus 099.

FIGURE 7-30 Cascading BCD counters to count and display numbers from 000 to 999.

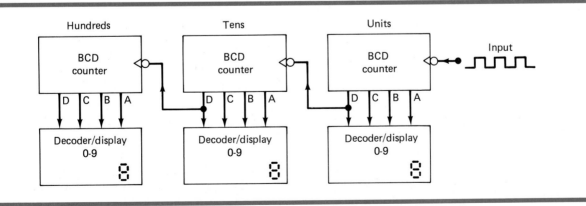

5. On the hundredth input pulse, the units counter recycles to zero, which in turn causes the tens counter to recycle to zero. The D FF output of the tens counter thus makes a 1-to-0 transition which acts as the clock input for the hundreds counter and causes it to advance one count. Thus, after 100 pulses the decimal readout is 100.

6. This process continues up until 999 pulses. On the 1000th pulse, all the counters recycle back to zero.

It should be obvious that this arrangement can be expanded to any desired number of decimal digits simply by adding on more stages. For example, to count up to 999,999 will require six BCD counters and associated decoders and displays. In general, then, we need one BCD counter per decimal digit.

Incidentally, the BCD counters used in Figure 7-30 could be 74293s wired as MOD-10 counters, or they could be IC counters such as the 7490 or 74192 that are internally wired as BCD counters.

7-14 SHIFT-REGISTER COUNTERS

Shift registers can be arranged to form several types of counters. All *shift-register counters* use *feedback,* whereby the output of the last FF in the shift register is in some way connected to the first FF. The most widely used shift-register counters are the ring counter and the Johnson counter.

Ring Counter   The simplest shift-register counter is essentially a *circulating* shift register connected so that the last FF shifts its value into the first FF. This arrangement is shown in Figure 7-31 using D-type FFs (J-K FFs can also be used). The FFs are connected so that information shifts from left to right and back around from Q_0 to Q_3. In most instances only a single 1 is in the register and it is made to circulate around the register as long as clock pulses are applied. For this reason it is called a *ring counter.*

The waveforms and sequence table in Figure 7-31 show the various states of the FFs as pulses are applied, assuming a starting state of $Q_3 = 1$ and $Q_2 = Q_1 = Q_0 = 0$. After the first pulse, the 1 has shifted from Q_3 to Q_2 so that the counter is in the 0100 state. The second pulse produces the 0010 state, and the third pulse produces the 0001 state. On the *fourth* clock pulse the 1 from Q_0 is transferred to Q_3, resulting in the 1000 state, which is, of course, the initial state. Subsequent pulses cause the sequence to repeat.

This counter functions as a MOD-4 counter, since it has *four* distinct states before the sequence repeats. Although this circuit does not progress through the normal binary counting sequence, it is still a counter because each count corresponds to a particular state of the FFs. Note that each FF output waveform has a frequency equal to ¼ of the clock frequency, since this is a MOD-4 ring counter.

Ring counters can be constructed for any desired MOD number; a MOD-N ring counter uses N FFs connected in the arrangement of Figure 7-31. In general, a ring counter will require more FFs than a binary counter for the same MOD number; for example, a MOD-8 ring counter requires eight FFs while a MOD-8 binary counter requires only three.

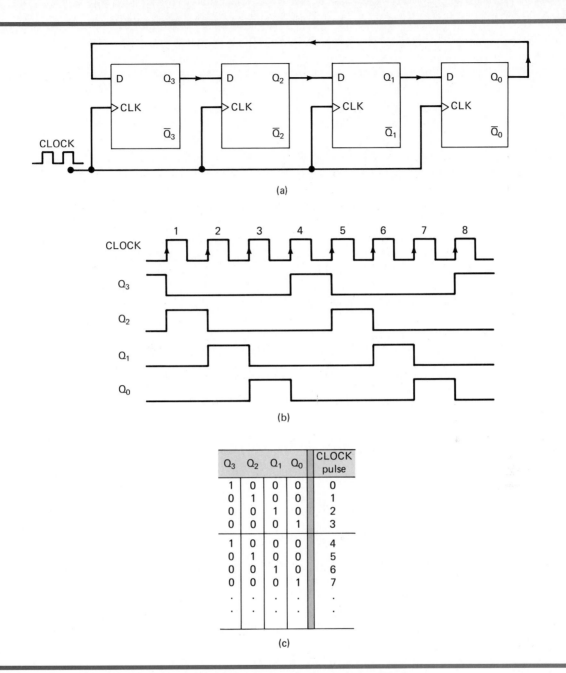

FIGURE 7-31 (a) Four-bit ring counter; (b) waveforms; (c) sequence table.

Despite the fact that it is less efficient in the use of FFs, a ring counter is still useful because it can be decoded without the use of decoding gates. The decoding signal for each state is obtained at the output of its corresponding FF. Compare the FF waveforms of the ring counter with the decoding waveforms in Figure 7-25. In some cases a ring counter might be a better choice than a binary counter with its associated decoding gates. This is especially true in applications where the counter is being used to control the sequencing of operations in a system.

Starting a Ring Counter

To operate properly a ring counter must start off with only one FF in the 1 state and all the other FFs at 0. When power is first applied to the circuit, there is only a remote possibility that the FFs will come up in such a state. Thus, it is necessary to preset the counter to the required starting state before clock pulses are applied. A simple way to accomplish this is to apply a momentary pulse to the PRESET input of one of the FFs and to the CLEAR inputs of all the others. This will place a single 1 into the ring counter.

Johnson Counter

The basic ring counter can be modified slightly to produce another type of shift register counter, which will have somewhat different properties. The *Johnson* or *twisted-ring counter* is constructed exactly like a normal ring counter except that the *inverted* output of the last FF is connected to the input of the first FF. A 3-bit Johnson counter is shown in Figure 7-32. Note that the $\overline{Q_0}$ output is connected back to the D input of Q_2. This means that the *inverse* of the level stored in Q_0 will be transferred to Q_3 on the clock pulse.

The Johnson-counter operation is easy to analyze if we realize that on each positive clock-pulse transition the level at Q_2 shifts into Q_1, the level at Q_1 shifts into Q_0, and the *inverse* of the level at Q_0 shifts into Q_2. Using these ideas and assuming that all FFs are initially 0, the waveforms and sequence table of Figure 7-32 can be generated.

Examination of the waveforms and sequence table reveals the following important points:

1. This counter has six distinct states: 000, 100, 110, 111, 011, and 001 before it repeats the sequence. Thus, it is a MOD-6 Johnson counter. Note that it does not count in a normal binary sequence.

2. The waveform of each FF is a squarewave (50% duty cycle) at ⅙ the frequency of the clock. In addition, the FF waveforms are shifted by one clock period with respect to each other.

The MOD number of a Johnson counter will always be equal to *twice* the number of FFs. For example, if we connect five FFs in the arrangement of Figure 7-32, the result is a MOD-10 Johnson counter, where each FF output waveform is a squarewave at ⅒ the clock frequency. Thus it is possible to construct a MOD-N counter (where N is an even number) by connecting $N/2$ FFs in a Johnson-counter arrangement.

Decoding a Johnson Counter

For a given MOD number, a Johnson counter requires only half the number of FFs that a ring counter requires. However, a Johnson counter requires decoding gates whereas a ring counter does not. As in the binary counter, the Johnson counter uses one logic gate to decode for each count, but each gate requires only two inputs, regardless of the number of FFs in the counter. Figure 7-33 shows the decoding gates for the six states of the Johnson counter of Figure 7-32.

Notice that each decoding gate has only two inputs, even though there are

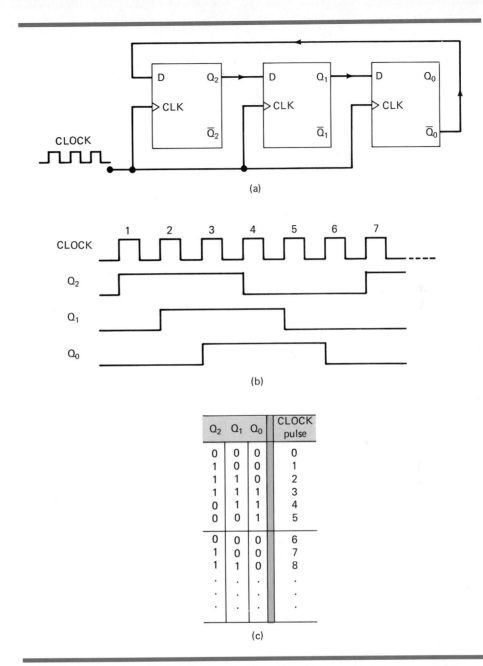

FIGURE 7-32 (a) MOD-6 Johnson counter; (b) waveform; (c) sequence table.

three FFs in the counter. This is because for each count, two of the three FFs are in a unique combination of states. For example, the combination $Q_2 = Q_0 = 0$ occurs only once in the counting sequence, at the count of 0. Thus, AND gate 0 with inputs $\overline{Q}_2$ and $\overline{Q}_0$ can be used to decode for this count. This same characteristic is shared by all the other states in the sequence, as the reader can verify. In fact, for *any size* Johnson counter, only two-input decoding gates are required.

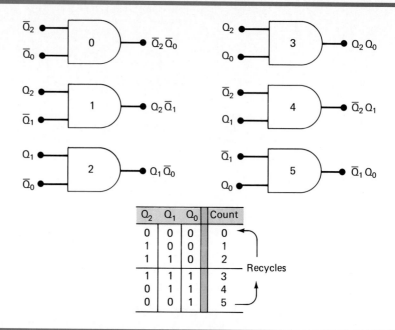

Q_2	Q_1	Q_0	Count
0	0	0	0
1	0	0	1
1	1	0	2
1	1	1	3
0	1	1	4
0	0	1	5

Recycles

FIGURE 7-33 **Decoding logic for MOD-6 Johnson counter.**

Johnson counters represent a middle ground between ring counters and binary counters. A Johnson counter requires fewer FFs than a ring counter but generally more than a binary counter; it has more decoding circuitry than a ring counter but less than a binary counter. Thus, they sometimes represent a logical choice for certain applications.

IC Shift-Register Counters   There are very few ring counters or Johnson counters available as ICs. The reason is that it is relatively simple to take a shift-register IC and to wire it as either a ring or Johnson counter. A couple of CMOS Johnson-counter ICs (74HC4017, 74HC4022) include the complete decoding circuitry on the same chip as the counter.

REVIEW QUESTIONS

1. Which shift-register counter requires the most FFs for a given MOD number?
2. Which shift-register counter requires the most decoding circuitry?
3. How can a ring counter be converted to a Johnson counter?
4. *True or false:*
   (a) The outputs of a ring counter are always squarewaves.
   (b) The decoding circuitry for a Johnson counter is simpler than for a binary counter.
   (c) Ring and Johnson counters are synchronous counters.

7-15 COUNTER APPLICATIONS: FREQUENCY COUNTER

There are numerous applications for the many types of counters we have been discussing. In this section and the next we will look at two representative applications that illustrate the uses of counters in digital systems.

A *frequency counter* is a circuit that can measure and display the frequency of a pulse signal. One of the most straightforward methods for constructing a frequency counter is shown in Figure 7-34(a) in simplified form. It contains a counter with its associated decoder/display circuitry and an AND gate. The AND gate inputs include the pulses with unknown frequency, f_x, and a SAMPLE pulse that controls how long these pulses are allowed to pass through the AND gate into the counter. The counter is usually made up of cascaded BCD counters (Figure 7-30), and the decoder/display unit converts the BCD outputs into a decimal display for easy monitoring.

The waveforms in Figure 7-34(b) show that a CLEAR pulse is applied to the counter at t_0 to start the counter at zero. Prior to t_1 the SAMPLE pulse waveform is LOW, so the AND output, Z, will be LOW and the counter will not be counting. The SAMPLE pulse goes HIGH from t_1 to t_2; this is called the *sampling interval*. During this sampling interval the unknown frequency pulses will pass through the AND gate and will be counted by the counter. After t_2 the AND output returns LOW and the counter stops counting. Thus, the counter will have counted the number of pulses that occurred during the sampling interval, and the resulting contents of the counter is a direct measure of the frequency of the pulse waveform.

FIGURE 7-34   **Basic frequency-counter method.**

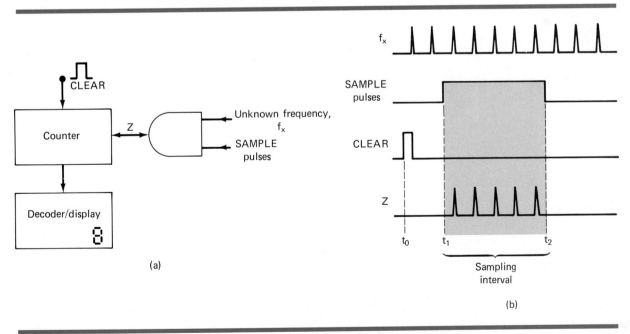

(a)

(b)

EXAMPLE 7-17

The unknown frequency is 3792 pps. The counter is cleared to the zero state prior to t_1. Determine the counter reading after a sampling interval of (a) 1 s, (b) 0.1 s, and (c) 10 ms.

SOLUTION

(a) Within a sampling interval of 1 s there will be 3792 pulses entering the counter, so after t_2 the contents of the counter will read 3792.

(b) With a 0.1-s sampling interval the number of pulses passing through the AND gate into the counter will be 3792 pulses/second $\times$ 0.1 s = 379.2. This means that either 379 or 380 pulses will be counted, depending on what part of a pulse cycle that t_1 occurs.

(c) With a 10-ms = 0.01-s sampling interval, the counter will read either 37 or 38.

The accuracy of this method depends almost entirely on the duration of the sampling interval, which must be very accurately controlled. A commonly used method for obtaining very accurate sample pulses is shown in Figure 7-35. A crystal-controlled oscillator is used to generate a very accurate 100-kHz waveform, which is shaped into square pulses and fed to a series of decade counters that are being used to successively divide this 100-kHz frequency by 10. The frequencies at the outputs of each decade counter are as accurate (percentagewise) as the crystal frequency.

The switch is used to select one of the decade-counter output frequencies to be fed to a single FF to be divided by 2. For example, in switch position 1 the 1-Hz pulses are fed to FF Q, which is acting as a toggle FF so that its output will be a squarewave with a period of $T = 2$ s and a pulse duration of $t_p = T/2 = 1$ s. This pulse duration is the desired 1-s sampling interval. In position 2 the sampling interval would be 0.1 s, and so on for the other positions.

EXAMPLE 7-18

Assume that the counter in Figure 7-34 is made up of three cascaded BCD counters and their associated displays. If the unknown input frequency is between 1 and 9.99 kpps, what is the best setting for the switch position in Figure 7-35?

SOLUTION

With 3 BCD counters the total capacity of the counter is 999. A 9.99 kpps frequency would produce a count of 999 if a 0.1-s sample interval were used. Thus, in order to use the full capacity of the counter, the switch should be set to position 2. If a 1-s sampling interval were used, the counter capacity would always be exceeded for frequencies in the specified range. If a shorter sample interval were used, the counter would count only between 0 and 99; this would give a reading to only two significant figures and would be a waste of the counter's capacity.

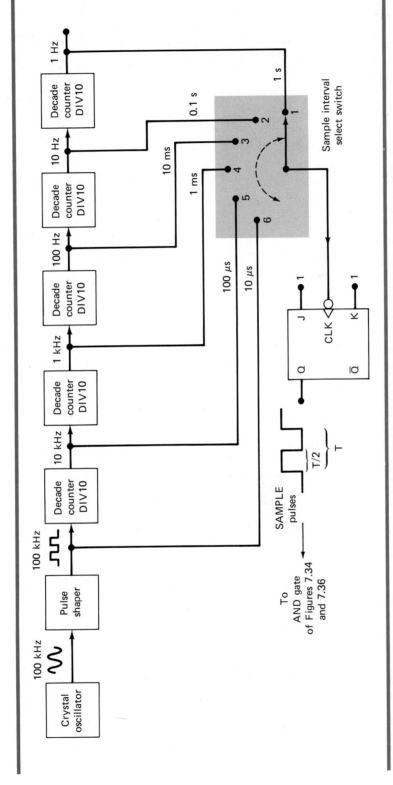

FIGURE 7-35 Example 7-18: method for obtaining accurate sampling intervals for frequency counter.

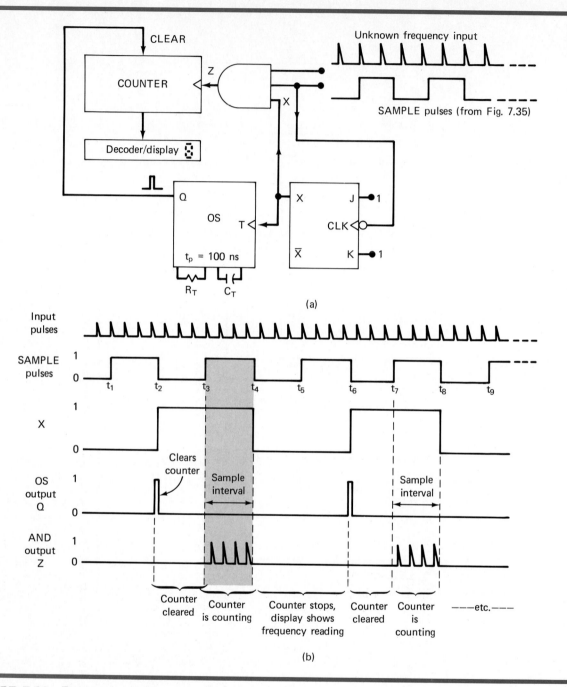

FIGURE 7-36 Frequency counter.

Complete Frequency Counter We will now look at a more complete frequency-counter circuit in Figure 7-36(a). The circuit now contains a one-shot and a J-K FF operating in the toggle mode, and the AND gate has three inputs, one of which is the FF output X. The SAMPLE pulses are connected to the AND gate and also to the *CLK* input of the FF. These SAMPLE pulses would be generated from a circuit such as that in Figure 7-35. The following step-by-step description refers to the waveforms in Figure 7-36(b).

1. Assume that FF X is in the 0 state (it has toggled to 0 on the falling edge of the previous sample pulse).

2. This LOW from X is fed to the AND gate, disabling its output, so no pulses are fed to the counter even when the first sample pulse occurs between t_1 and t_2.

3. At t_2 the NGT of the first sample pulse toggles FF X to the 1 state (note that $J = K = 1$). This positive transition at X triggers the OS, which generates a 100-ns pulse to *clear* the counter. The counter now displays *zero*.

4. At t_3 the second sample pulse enables the AND gate (since X is now 1) and allows the unknown frequency into the counter to be counted until t_4.

5. At t_4 the sample pulse returns LOW and toggles X LOW, disabling the AND gate. The counter stops counting.

6. Between t_4 and t_6 the counter holds and displays the count that it had reached at t_4. Note that the third sample pulse does not enable the AND gate because FF X is LOW.

7. At t_6 the NGT of the sample pulse toggles X HIGH and the operation follows the same sequence that began at t_2.

This frequency counter, then, goes through a repetitive sequence of clearing to zero, counting, holding for display, clearing to zero, counting, an so on. For example, let's assume that the counter has three BCD stages and a three-digit display. If we use a sample interval of 1 s and the unknown frequency is 237 pps, the counter and display will go through the following sequence over and over:

- Clear to zero and display zero for 1 s [t_2–t_3 in Figure 7-36(b)].
- Starting at zero, count unknown frequency pulses during the 1-s sample interval (t_3–t_4); the count will stop at 237.
- Hold and display the count of 237 for 2 s (t_4–t_6).

Since the display is connected directly to the counter outputs, the display will show the clearing and counting action of the counter. This makes it very difficult to read the display to determine the unknown frequency except at very slow sample intervals. This problem can be solved by inserting a *buffer register* between the counter and the decoder/display unit. We will consider this feature in Problem 7-33.

REVIEW QUESTIONS

1. What is the best sample interval setting to use if the pulse counter has four BCD stages and the input frequency is between 2 and 8 Mpps? (*Ans.* 1 ms)

2. Describe the sequence of operations of the complete frequency counter of Figure 7-36.

One of the most common applications of counters is the digital clock—a time clock which displays the time of day in hours, minutes, and sometimes seconds. In order to construct an accurate digital clock, a very closely controlled basic clock frequency is required. For battery-operated digital clocks (or watches) the basic frequency is normally obtained from a quartz-crystal oscillator. Digital clocks operated from the ac power line can use the 60-Hz power frequency as the basic clock frequency. In either case, the basic frequency has to be divided down to a frequency of 1 Hz or 1 pulse per second (pps). Figure 7-37 shows the basic block diagram for a digital clock operating from 60 Hz.

The 60-Hz signal is sent through a shaping circuit to produce square pulses at the rate of 60 pps. This 60-pps waveform is fed into a MOD-60 counter which is used to divide the 60 pps down to 1 pps. The 1-pps signal is fed into the SECONDS section, which is used to count and display seconds from 0 through 59. The BCD counter advances one count per second. After 9 seconds the BCD counter recycles to 0, which triggers the MOD-6 counter and causes it to advance one count. This continues for 59 s when the MOD-6 counter is at the 101 (5) count and the BCD counter is at 1001 (9), so the display reads 59 s. The next pulse recycles the BCD counter to 0, which in turn recycles the MOD-6 counter to 0 (remember: the MOD-6 counts from 0 through 5).

The output of the MOD-6 counter in the SECONDS section has a frequency of 1 pulse per minute (the MOD-6 recycles every 60 s). This signal is fed to the MINUTES section, which counts and displays minutes from 0 through 59. The MINUTES section is identical to the SECONDS section and operates in exactly the same manner.

The output of the MOD-6 counter in the MINUTES section has a frequency of 1 pulse per hour (the MOD-6 recycles every 60 min). This signal is fed to the HOURS section, which counts and displays hours from 1 through 12. This HOURS section is different from the SECONDS and MINUTES sections in that it never goes to the zero state. The circuitry in this section is sufficiently unusual to warrant a closer investigation.

Figure 7-38 shows the detailed circuitry contained in the HOURS section. It includes a BCD counter to count units of hours, and a single FF (MOD-2) to count tens of hours. The BCD counter is a 74192 which operates exactly like the 74193 that we studied earlier except that it counts only between 0000 and 1001. In other words, the 74192 can either count up in BCD fashion (i.e., 0 to 9 and recycles to 0) or count down in BCD fashion (i.e., 9 to 0 and recycles to 9). Here it is used to count up in response to the 1-pulse/hour signal coming from the MINUTES section. The INVERTER on the CP_U input is needed because the 74192 responds to PGTs, and we want it to respond to the NGT that occurs when the MINUTES section recycles back to zero.

The incoming pulses advance the BCD counter once per hour. For example, at 7 o'clock this counter will be at 0111, and its decoder/display circuitry will display the numeral 7. At the same time, X will be LOW and its display will show a zero. Thus, the two displays will show "07". When the BCD counter is in the 1001 (9) state and the next input pulse occurs, it will recycle back to 0000. The

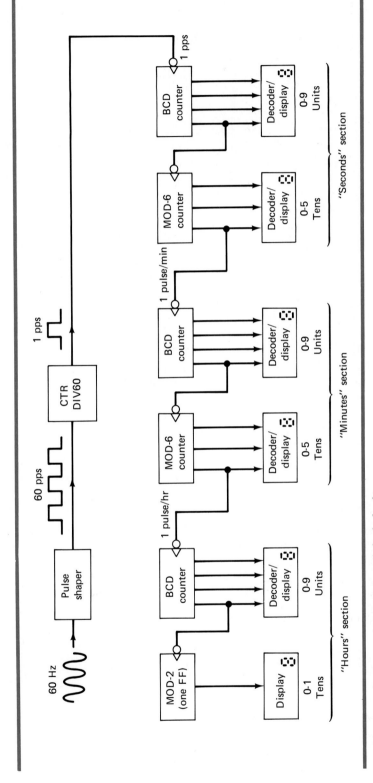

FIGURE 7-37 Block diagram for digital clock.

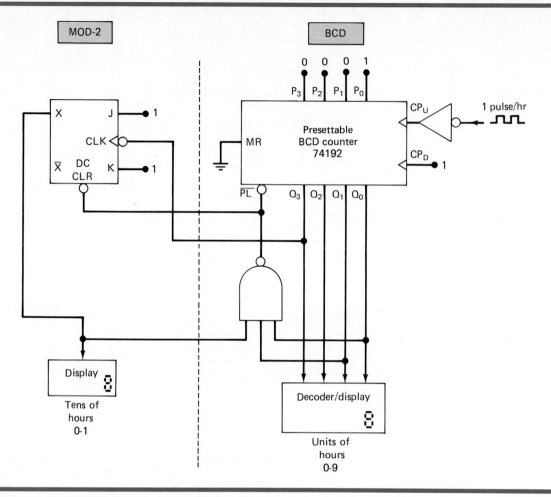

FIGURE 7-38   **Detailed circuitry for the hours section.**

NGT at Q_3 will toggle FF X from 0 to 1. This produces a numeral 1 on the X display and a numeral 0 on the BCD display so that the combined displays show "10" for 10 o'clock.

The next two pulses advance the BCD counter so that "11" and "12" are displayed at 11 o'clock and 12 o'clock, respectively. The next pulse advances the BCD counter to 0011 (3). In this state, the counter's Q_1 and Q_0 outputs are both HIGH, and X is still HIGH. Thus, the NAND gate output goes LOW and activates the DC CLEAR of FF X and the $\overline{PL}$ input of the 74192. This clears X to 0 and presets the BCD counter to 0001. The result is a display of "01" for 1 o'clock. Several of the end-of-chapter problems will provide more details on the digital clock circuit.

7-17 INTEGRATED-CIRCUIT REGISTERS

The various types of registers can be classified according to the manner in which data can be entered into the register for storage and the manner in which data are output from the register. The various classifications are listed below.

1. Parallel in/parallel out.
2. Serial in/serial out.
3. Parallel in/serial out.
4. Serial in/parallel out.

Each of these types is available in IC form so that a logic designer can usually find exactly what is required for a given application. In the following sections we will examine a representative IC from each of the above categories.

7-18 PARALLEL IN/PARALLEL OUT—THE 74174 AND 74178

There are actually two types of registers in this category; one is strictly parallel, the other is actually a shift register that can be loaded with parallel data and has parallel outputs available.

The 74174   Figure 7-39(a) shows the logic diagram for the 74174, a 6-bit register that has parallel inputs D_5–D_0 and parallel outputs Q_5–Q_0. Parallel data are loaded into the register on the PGT of the clock input CP. A master reset input $\overline{MR}$ can be used to asynchronously reset all the register FFs to 0.

The logic symbol for the 74174 is shown in Figure 7-39(b). This symbol is used in circuit diagrams to represent the circuitry of Figure 7-39(a).

The 74178   Figure 7.40(a) shows the logic diagram for the 74178, a 4-bit shift register that has parallel data entry (P_0–P_3) and parallel outputs (Q_0–Q_3). It has a serial data input, D_S, and two enable inputs: PE (parallel enable) and SE (serial enable).

Figure 7-40(b) is the mode-select table that describes the various modes of operation for this IC. The first entry gives the input conditions necessary for the

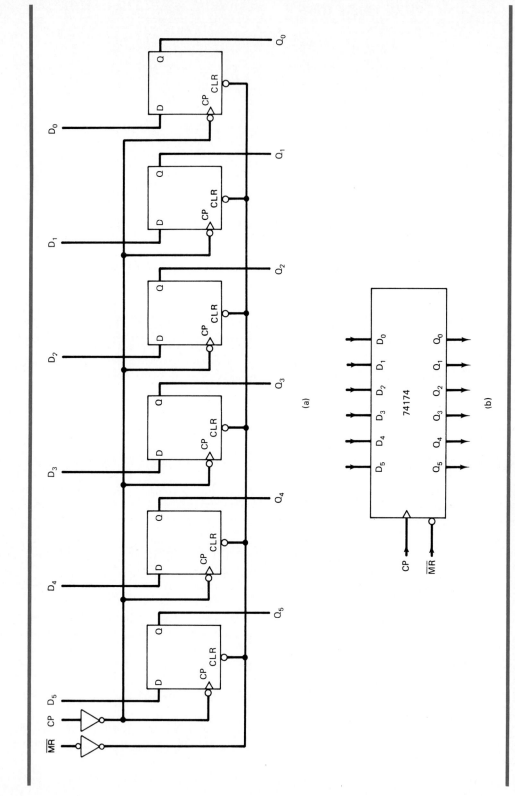

FIGURE 7-39 **(a) Circuit diagram of 74174; (b) logic symbol.** (Courtesy of Fairchild, a Schlumberger company)

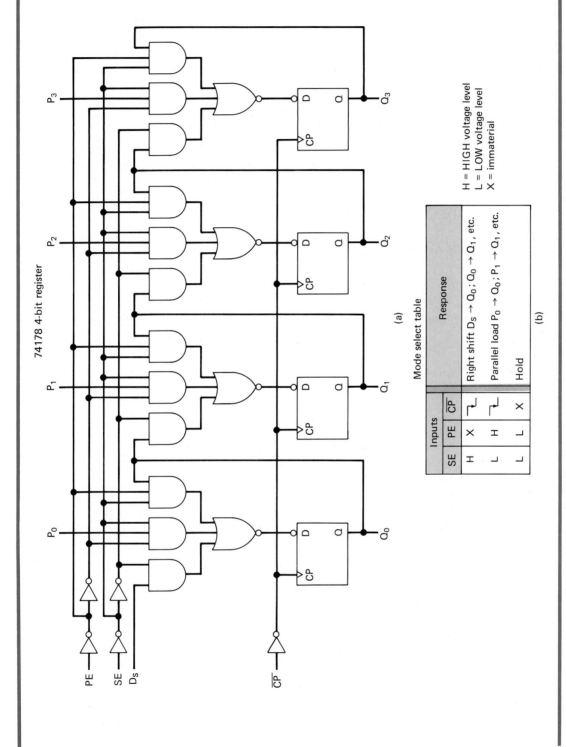

Mode select table

Inputs			Response
SE	PE	$\overline{CP}$	
H	X	⎍	Right shift $D_s \to Q_0$; $Q_0 \to Q_1$, etc.
L	H	⎍	Parallel load $P_0 \to Q_0$; $P_1 \to Q_1$, etc.
L	L	X	Hold

(b)

H = HIGH voltage level
L = LOW voltage level
X = immaterial

FIGURE 7-40 (a) Logic diagram for the 74178; (b) mode-select table. (Courtesy of Fairchild, a Schlumberger Company)

shift-right operation. With $SE = 1$, the data will shift from left to right on the NGT of clock input $\overline{CP}$ regardless of the logic level at the PE input (recall that X represents the "don't care" condition). You can varify this by tracing through the logic diagram and noting that when $SE = 1$, the D_S input passes through the logic gates and appears at the D input of the Q_0 FF. Likewise, Q_0 will appear at the D input of Q_1; Q_1 will appear at the D input of Q_2; and Q_2 will appear at the D input of Q_3.

The second entry in the table gives the conditions needed to produce parallel transfer from the parallel data inputs (P_0–P_3) to the outputs (Q_0–Q_3). When $SE = 0$ and $PE = 1$, this parallel transfer occcurs on the NGT of $\overline{CP}$. Note that this is a *synchronous* transfer. Again, by tracing through the logic diagram, you can see that when $SE = 0$ and $PE = 1$ the parallel data inputs will get through the logic gates and will appear at the D inputs of their respective FFs.

The final entry in the table indicates that the $SE = 0$, $PE = 0$ condition will cause the register FFs to hold their current levels regardless of what happens at the clock input. For this input condition, each FF output is allowed to pass through its respective logic gates and appear at the D input of the same FF. Thus, a NGT on $\overline{CP}$ will not change the state of the FF.

Figure 7-41 shows the logic symbol for the 74178.

FIGURE 7-41   Logic symbol for the 74178.

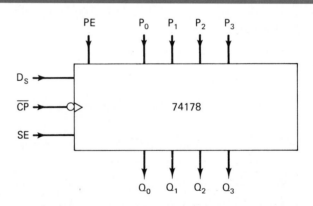

EXAMPLE 7-19

Describe how the 74178 can be connected as a ring counter.

SOLUTION
The correct connections are shown in Figure 7-42. The Q_3 output is connected back around to the D_S input. The parallel data inputs are permanently set at 0001 and the PE input is connected permanently HIGH. The SE input is initially connected LOW. This allows each NGT at $\overline{CP}$ to parallel-load the shift register with 0001. Then, SE is switched to the HIGH state (shift mode), where it remains as the NGTs at $\overline{CP}$ cause the single 1 to shift and recirculate through the register.

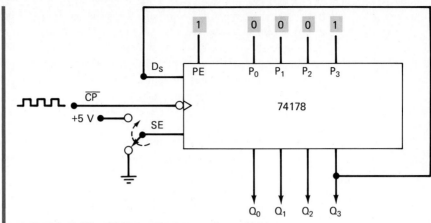

FIGURE 7-42 **74178 wired as a ring counter.**

7-19 SERIAL IN/SERIAL OUT—THE 4731B

The 4731B is a CMOS *quad* 64-bit shift register. It contains four identical 64-bit shift registers on one chip. Figure 7-43 shows the logic diagram for one of the 64-

FIGURE 7-43 **Logic diagram for one of four 64-bit shift registers on a 4731B.** (Courtesy of Fairchild, a Schlumberger company)

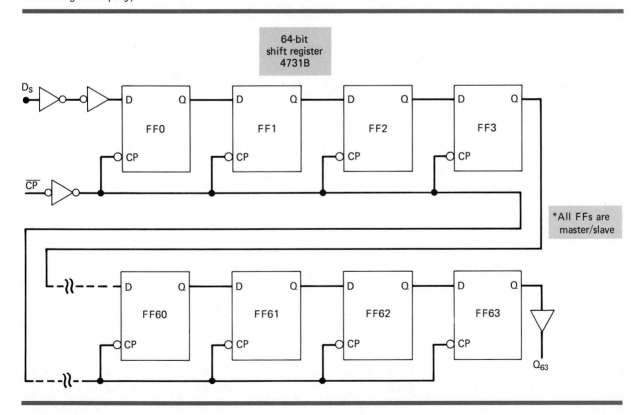

bit registers. It has a serial input, D_S, a clock input $\overline{CP}$ that responds to NGTs, and a serial output from the last FF, Q_{63}. This is the only output that is externally accessible. Note that this output goes through a *buffer* circuit (triangle symbol with no inversion bubble). A buffer does not change the signal's logic level; it is used to provide a greater output-current capability than normal. Also note that there is no means for parallel data entry into the register FFs.

EXAMPLE 7-20

How can the 4731B chip be wired as a single 256-bit shift register?

SOLUTION
The four shift registers can be strung together as one long register by connecting the Q_{63} output of one to the D_S input of the next, and by connecting all $\overline{CP}$ inputs together as a common clock input. This is shown in Figure 7-44 where a simplified logic symbol for each shift register is used for convenience.

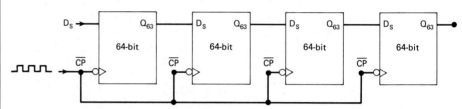

FIGURE 7-44   **The four 64-bit shift registers on the 4731B chip can be connected together as a 256-bit register.**

7-20   PARALLEL IN/SERIAL OUT—THE 74165/74HC165

The 74165, an 8-bit register, is shown in Figure 7-45(a). It actually has both serial data entry via D_S and parallel data entry via P_0–P_7. The only accessible FF outputs are Q_7 and $\overline{Q}_7$.

Note that the FFs are the clocked *SR* type (same as clocked *SC*) that respond to NGTs at their clock inputs. Each FF has asynchronous inputs, PRESET and *CL* (same as CLEAR), that are used for parallel data entry.

Also note that there are two clock inputs, CP_1 and CP_2, either of which can be used to produce the shift operation on PGTs.

EXAMPLE 7-21

Determine the necessary input conditions if pulses at CP_1 are to produce the shift right operation.

SOLUTION
In order for the CP_1 pulses to get through the AND gate, the $\overline{PL}$ input has to be HIGH. The CP_1 pulses out of the AND gate will be inverted by the NOR gate

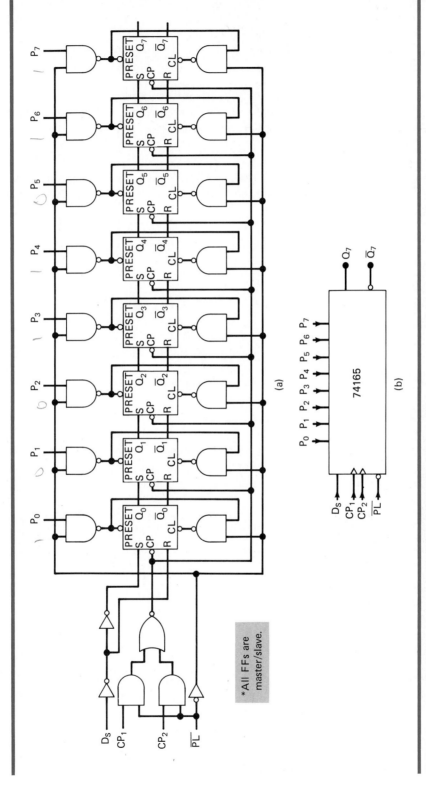

FIGURE 7-45 (a) Logic diagram of the 74165; (b) logic symbol. (Courtesy of Fairchild, a Schlumberger company)

only if the other NOR input is LOW. Thus, CP_2 must be held LOW. The inverted CP_1 pulses will trigger the FFs. The FFs trigger on the NGT of the inverted CP_1 pulses, which means that they trigger on the PGT of the pulses applied to the CP_1 input.

EXAMPLE 7-22

How can the 74165 be placed in the hold mode so that the register contents does not change as pulses are applied to CP_1?

SOLUTION

If CP_2 is kept HIGH while $\overline{PL}$ is HIGH, the bottom AND gate output will be HIGH and will keep the NOR output at a constant LOW regardless of the CP_1 input.

7-21   SERIAL IN/PARALLEL OUT—THE 74164/74HC164

The logic diagram for the 74164 is shown in Figure 7-46(a). It is an 8-bit shift register with each FF output externally accessible. Instead of a single serial input, an AND gate combines inputs A and B to produce the serial input to FF Q_0.

The shift operation occurs on the PGTs of the clock input CP. The $\overline{MR}$ input provides asynchronous resetting of all FFs on a LOW level.

EXAMPLE 7-23

Assume that the initial contents of the 74164 register in Figure 7-47(a) is 00000000. Determine the sequence of states as clock pulses are applied.

SOLUTION

The correct sequence is given in Figure 7-47(b). With $A = B = 1$, the serial input is 1, so that 1s will shift into the register on each PGT of CP. Since Q_7 is initially at 0, the $\overline{MR}$ input is inactive.

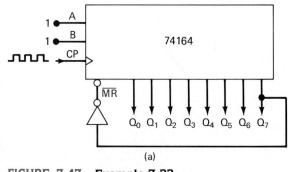

(a)

FIGURE 7-47   Example 7-23

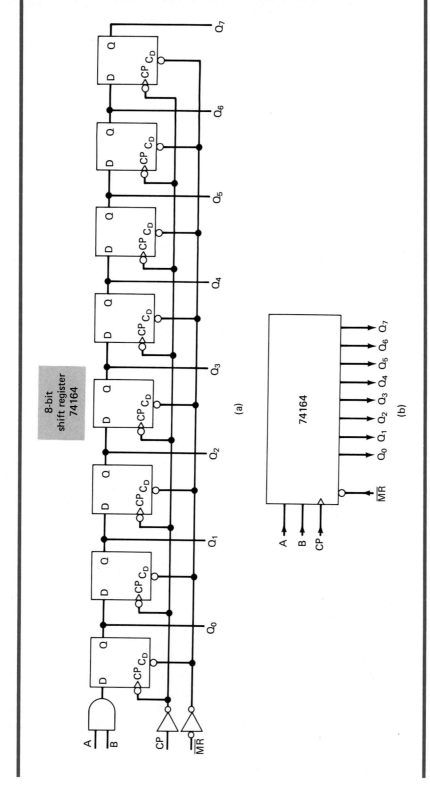

FIGURE 7-46 (a) Logic diagram for the 74164; (b) logic symbol. (Courtesy of Fairchild, a Schlumberger company)

Input pulse number	Q_0	Q_1	Q_2	Q_3	Q_4	Q_5	Q_6	Q_7	
0	0	0	0	0	0	0	0	0	
1	1	0	0	0	0	0	0	0	
2	1	1	0	0	0	0	0	0	
3	1	1	1	0	0	0	0	0	
4	1	1	1	1	0	0	0	0	Recycles
5	1	1	1	1	1	0	0	0	
6	1	1	1	1	1	1	0	0	
7	1	1	1	1	1	1	1	0	
8	1	1	1	1	1	1	1	1	

Temporary state

(b)

FIGURE 7-47   Example 7-23 (*cont.*)

On the eighth pulse, the register tries to go to the 11111111 state as the 1 from Q_6 shifts into Q_7. This state occurs only momentarily because $Q_7 = 1$ produces a LOW at $\overline{MR}$ that immediately resets the register back to 00000000. The sequence is then repeated on the next eight clock pulses.

The IC registers that have been presented here are representative of the various types that are commercially available. Although there are many variations on these basic registers, most of them should now be relatively easy to understand from the manufacturers' data sheets.

We will present several register applications in the end-of-chapter problems and in the material covered in subsequent chapters.

7-22   IEEE/ANSI REGISTER SYMBOLS*

We will present two examples of the IEEE/ANSI symbols for register ICs. First, let's consider the 74174 parallel-in/parallel-out IC whose internal logic and traditional logic symbol was shown in Figure 7-39. The IEEE/ANSI symbol for the 74174 is given in Figure 7-48(a). Its outline consists of the notched common-control block and the six narrow rectangles representing the six FFs.

The common-control block has the inputs that are common to all of the elements in the IC; in this case, the $\overline{MR}$ and CP inputs are common to the six FFs Q_0–Q_5 that make up the register. The internal label for the $\overline{MR}$ input is shown as an R to indicate that its function is to *reset* each FF. The internal label for the CP input is C1, which tells us that this input controls the entry of data into any storage element that has a prefix of 1 in its input label. Each FF's D input has an internal label of 1D (shown only for Q_0, but assumed the same for each FF). The "1" in

*This section may be omitted without affecting the remainder of the book.

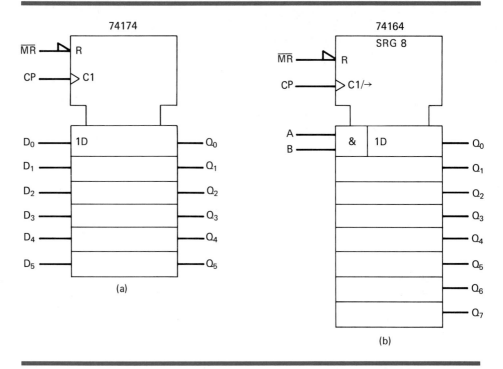

FIGURE 7-48 IEEE/ANSI symbols for the (a) 74174 parallel-in/parallel-out register and (b) 74164 serial-in/parallel-out shift register.

C1 and 1D establishes the dependency of the FF D inputs on the common clock input CP.

The IEEE/ANSI symbol for the 74164 serial-in/parallel-out shift register is presented in Figure 7-48(b). Its outline consists of the common-control block and the eight FFs that make up the register. The notation SRG 8 identifies this IC as being an 8-bit shift register.

The CP input has the internal label C1/→. The slash (/) is used to separate the two functions C1 and → performed by this input. The C1 indicates that CP controls the data entry into FF Q_0 since Q_0 has the input label 1D. Note that the data entered into Q_0 is indicated as the AND combination of inputs A and B. Also note that since there are no external data inputs to $Q_1–Q_7$, CP does not control data entry into these FFs. The → denotes that the active transition of CP will produce the shift right operation (from Q_0 toward Q_7).

REVIEW QUESTIONS

1. What does the slash (/) mean when it appears in an input label?
2. What notation would be used to describe the function performed by one of the registers on the 4731B IC of Figure 7-43? (*Ans.* SRG 64)

7-23 TROUBLESHOOTING

Flip-flops, counters, and registers are the major components in *sequential logic systems*. A sequential logic system, because of it storage devices, has the characteristic that its outputs and sequence of operations depend on both the present inputs and the inputs that occurred earlier. Even though sequential logic systems are generally more complex than combinatorial logic systems, the essential procedures for troubleshooting apply equally well to both types of systems. Sequential systems suffer from the same types of failures (opens, shorts, internal IC faults, etc.) as do combinatorial systems.

Many of the same steps used to isolate faults in a combinatorial system can be applied to sequential systems. One of the most effective troubleshooting techniques begins with the troubleshooter observing the system operation, and by analytical reasoning, determining the possible causes of the system malfunction. Then he/she uses available test instruments to isolate the exact fault. The following examples will show the kinds of analytical reasoning that are used in troubleshooting sequential systems. After studying these examples, you should be ready to tackle the troubleshooting problems at the end of the chapter, starting with Problem 7-44.

EXAMPLE 7-24

Figure 7-49(a) shows a 74293 wired as a MOD-10 counter. A technician tests the counter operation by applying a 1-kHz clock signal and observing the Q outputs with an oscilloscope. The displayed waveforms are shown in Figure 7-49(b). Determine the possible causes for the incorrect circuit behavior.

SOLUTION

The waveforms show that Q_0 is toggling in response to the NGTs of the clock, but all other FFs are stuck in the LOW state. There are several possible faults that could produce this operation.

1. The Q_1 output is internally or externally shorted to ground. Referring to Figure 7-7(a), we can see that this would prevent Q_2 and Q_3 from toggling since Q_1 is the clock signal for Q_2, and Q_2 is the clock signal for Q_3.

2. The connection from Q_0 to $\overline{CP}_1$ is open so that Q_1 receives no clock signal.

3. There is an internal fault in the IC that prevents Q_1 from toggling.

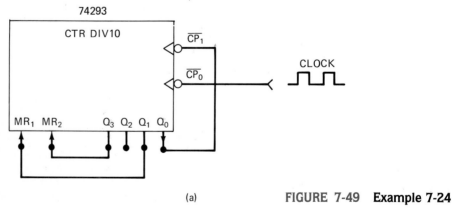

(a)

FIGURE 7-49 Example 7-24

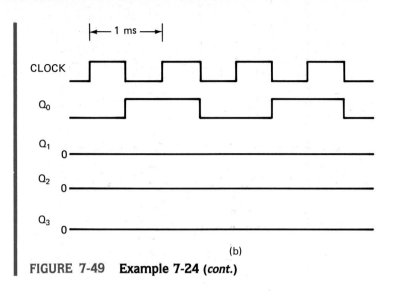

(b)

FIGURE 7-49   **Example 7-24 (*cont.*)**

EXAMPLE 7-25

After analyzing the situation described in Example 7-24, the technician proceeds to isolate the fault. He performs ohmmeter checks and verifies that Q_1 is not shorted to ground and that Q_0 is connected to $\overline{CP}_1$. This eliminated the first two possible faults. Concluding that the IC is bad, he replaces it. To his surprise, the circuit operation exhibits the same symptoms. Scratching his head, he decides to take a closer look at the FF waveforms by displaying them using a 10-ns/cm time scale. On this scale he can see a very narrow glitch occurring on the Q_1 signal at the time when Q_0 makes a NGT (see Figure 7-50). What is the probable fault?

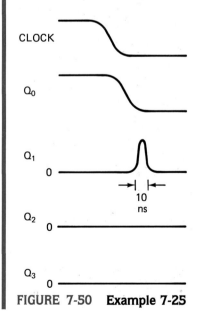

FIGURE 7-50   **Example 7-25**

SOLUTION

When the counter is operating correctly there should be a glitch at Q_1 when the counter goes to the 1010 (ten) state, at which point the HIGHs at Q_3 and Q_1 cause the MR inputs to clear the count back to 0000. The waveforms in Figure 7-50, however, do not show Q_3 being HIGH when the glitch at Q_1 occurs. The most probable fault is an open at MR_2 since this would be interpreted as a constant logic HIGH by the TTL IC. Thus, as soon as Q_1 goes HIGH, the MR inputs are both HIGH and the counter resets to 0000.

EXAMPLE 7-26

A technician tests the frequency counter of Figure 7-36 for various settings of the sample interval and for different unknown input frequencies. In all cases she finds that the displayed frequency is exactly *twice* what it should be. What is the probable cause of the malfunction?

SOLUTION

Referring to Figure 7-36, we see that the unknown frequency is allowed through the AND gate into the counter during the $t_3–t_4$ interval while SAMPLE and X are both HIGH. If the middle input to the AND gate were open, it would act like a permanent HIGH (assuming TTL devices). This would allow the unknown frequency pulses through the gate while X is HIGH during $t_2–t_4$. This is twice the normal interval, so the counter will count to twice the normal value.

EXAMPLE 7-27

A technician wires up the digital clock of Figures 7-37 and 7-38. He observes that the SECONDS sections is counting properly. In order to quickly test the operation of the MINUTES and HOURS sections, he bypasses the MOD-60 counter so that the counters will be pulsed at a rate that is 60 times faster than normal. He observes that the MINUTES section is working correctly, but the HOURS section counts and displays in the following manner:

TENS OF HOURS	UNITS OF HOURS
0	1
0	2
0	3
0	4
0	5
0	6
0	7
1	8
1	9
1	0
1	1
1	2

recycle and repeat

What is the probable cause of this incorrect sequence?

SOLUTION

Since the problem is in the HOURS section, we need to refer to Figure 7-38. The sequence above is correct except that the TENS digit is incremented from 0 to 1 when the UNITS digit goes from 7 to 8 instead of when it goes from 9 to 0. This operation would occur if the *CLK* input of FF *X* were mistakenly connected to Q_2 rather than Q_3 of the BCD counter. If this were the case, then when the BCD counter increments from 7 to 8, its Q_2 FF makes a NGT that will toggle FF *X* prematurely.

GLOSSARY

Active-HIGH (LOW) Decoder Decoder that produces a logic HIGH (LOW) at the output when detection occurs. [Sec. 7-11]

& When used inside an IEEE/ANSI symbol, it indicates an AND gate or AND function. [Sec. 7-3]

Asynchronous Counter Type of counter where each flip-flop output serves as the clock input signal for the next flip-flop in the chain. [Sec. 7-1]

BCD Counter Binary counter that counts from 0000_2 to 1001_2 before it recycles. [Sec. 7-2]

Buffer Register Register that holds digital data temporarily. [Sec. 7-15]

C When used inside an IEEE/ANSI symbol, the letter C in the label for an input indicates that the input controls the entry of data into a storage element. [Sec. 7-10]

Circulating Shift Register Shift register where one of the outputs of the last flip-flop in the shift register is connected to the input of the first flip-flop in the shift register. [Sec. 7-14]

C1 When used inside an IEEE/ANSI symbol, the designation C1 in the label indicates that this input controls the entry of data into any storage element that has the prefix 1 in its label. [Sec. 7-22]

$CT = 0$ When used inside an IEEE/ANSI symbol, $CT = 0$ in the label for an input indicates that the counter will clear when that input goes active. [Sec. 7-3]

CTR When used inside an IEEE/ANSI symbol, CTR indicates that the IC is a counter. [Sec. 7-3]

D When used inside an IEEE/ANSI symbol, the letter D in the label indicates data. [Sec. 7-10]

Decade Counter Any counter that is capable of going through 10 different logic states. [Sec. 7-2]

DIV*n* When used inside an IEEE/ANSI symbol, DIV*n* indicates that it is a MOD*n* counter. [Secs. 7-3 and 7-10]

Down Counter Counter that counts downward from a maximum count to zero. [Sec. 7-4]

Frequency Counter Circuit that can measure and display the frequency of a signal. [Sec. 7-15]

G When used inside an IEEE/ANSI symbol, the letter G in the label for an input indicates AND dependency. [Sec. 7-10]

Glitch Momentary, narrow, spurious, and sharply defined change in voltage. [Sec. 7-2]

Johnson Counter Shift register where the inverted output of the last flip-flop in the shift register is connected to the input of the first flip-flop in the shift register. [Sec. 7-14]

LED Light-emitting diode. [Sec. 7-11]

− When used inside an IEEE/ANSI symbol and on a clock input, a minus indicates that the counter will be decremented by 1 when clocked. [Secs. 7-3 and 7-10]

Parallel Counter *See* Synchronous Counter. [Sec. 7-6]

Parallel In/Parallel Out Register Type of register that can be loaded with parallel data and has parallel outputs available. [Sec. 7-18]

Parallel In/Serial Out Type of register that can be loaded with parallel data and has only one serial output. [Sec. 7-20]

+ When used inside an IEEE/ANSI symbol and on a clock input, a plus indicates that the counter will be incremented by 1 when clocked. [Secs. 7-3 and 7-10]

Presettable Counter Counter that can be preset to any starting count either synchronously or asynchronously. [Sec. 7-8]

R When used inside an IEEE/ANSI symbol, the letter R in the label for an input indicates a reset function. [Sec. 7-22]

Ring Counter Shift register in which the output of the last flip-flop is connected to the input of the first flip-flop in the shift register. [Sec. 7-14]

Ripple Counter *See* Asynchronous Counter. [Sec. 7-1]

Sampling Interval Time window during which a frequency counter samples and thereby determines the unknown frequency of a signal. [Sec. 7-15]

Sequential Logic System Logic system in which the logic output states and sequence of operations depend on both the present and past input conditions. [Sec. 7-23]

Serial In/Parallel Out Type of register that can be loaded with data serially and has parallel outputs available. [Sec. 7-21]

Serial In/Serial Out Type of register that can be loaded with data serially and has only one serial output. [Sec. 7-19]

/ When used inside an IEEE/ANSI symbol, a slash (/) in the label for an input indicates the separation of two functions. [Sec. 7-22]

Spike *See* Glitch [Sec. 7-2]

SRGn When used inside an IEEE/ANSI symbol, SRG 8 in the common-control block indicates that this IC is an n-bit shift register. [Sec. 7-22]

Strobing Technique often used to eliminate decoding spikes. [Sec. 7-12]

Synchronous Counter Counter in which all the flip-flops are clocked simultaneously. [Sec. 7-6]

Twisted Ring Counter *See* Johnson Counter. [Sec. 7-14]

Up Counter Counter that counts upward from zero to a maximum count. [Sec. 7-4]

→ When used inside an IEEE/ANSI symbol, → in the label for an input indicates that when an active transition exists at that input, a shift-right operation will result. [Sec. 7-22]

7-1. An 8-MHz squarewave clocks a 5-bit ripple counter. What is the frequency of the last FF? What is the duty cycle of this output waveform?

7-2. Repeat Problem 7-1 if the input has a 20% duty cycle.

7-3. Assume that the 5-bit binary counter starts in the 00000 state. What will be the _____ pulses?

_____ her necessary logic to construct a MOD-24 asyn-

_____ all the FFs in the decade counter of Figure 7-5(b) _____ ock frequency. Show any glitches that might ap- _____ puts. Determine the frequency at the *D* output.

_____ ed to clear a counter back to zero as in the count- _____ ere may be a possible problem because the prop- _____ AR input to FF output may vary from FF to FF. _____ one FF clears in 10 ns and another clears in 50 _____ ears to 0 it will cause the NAND gate output to _____ rminating the CLEAR pulse and the slower FF _____ roblem is especially prevalent when the FF out- _____ since, in general, the propagation delay of a FF

_____ eliminating this problem is shown in Figure 7- _____ he SET-CLEAR FF is the cross-coupled NAND _____ Analyze the operation of this circuit and explain

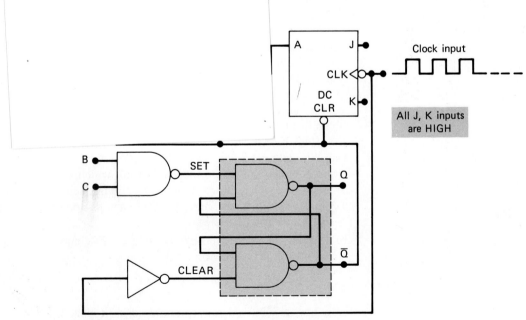

FIGURE 7-51

how the clearing function is performed without the problem described above.

7-7. Refer to the counter shown in Figure 7-52. How can you tell that it is a down counter? It has been modified so that it does *not* count through the entire binary sequence 111 to 000. Determine the actual sequence it counts through.

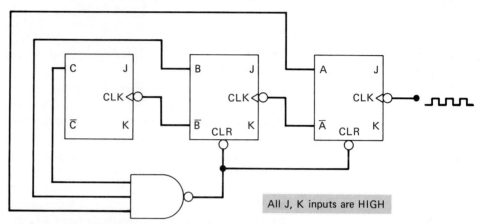

FIGURE 7-52

7-8. A counter or group of counters is often used to divide a high-frequency clock signal down to a lower-frequency output. When these counters are binary counters (i.e., they count in the binary sequence), the output will not be a symmetrical squarewave if the binary sequence has been shortened in order to produce the desired MOD number. For example, refer to the *C* waveform of the MOD-6 counter in Figure 7-4.

When a counter is being used only for frequency division, it is not necessary that it count in a binary sequence as long as it has the desired MOD number. A symmetrical squarewave output can be obtained for any *even* MOD number by breaking the MOD number into the product of two MOD numbers, one of which is a power of 2. For example, a MOD-6 counter can be formed from a MOD-3 counter and a MOD-2 counter as shown in Figure 7-53.

Here FFs *A*, *B*, and the NAND gate make up the MOD-3 counter whose *B* output has one-third the frequency of the input pulses. This *B* output is connected to the input of FF *C*, which is acting as a MOD-2 to divide the frequency down to one-sixth the frequency of the input pulses.
(a) Assume all FFs are initially LOW, and sketch the waveforms at each FF output for 12 cycles of the input.
(b) List the sequence of FFs states in a table and show that it is not a normal binary sequence.

Section 7-3

7-9. Show how a 74293 counter can be used to produce a 1.2-kpps output from an 18-kpps input.

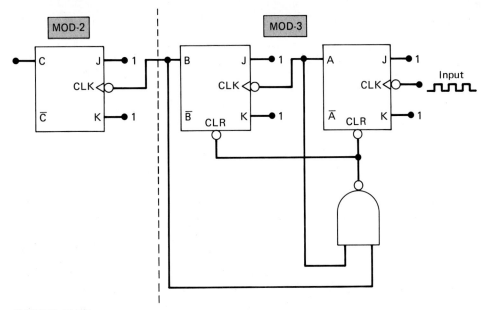

FIGURE 7-53

7-10. Show how two 74293s can be connected to divide an input frequency by 60 while producing a symmetrical squarewave output.

7-11. Determine the frequency at output X in Figure 7-54.

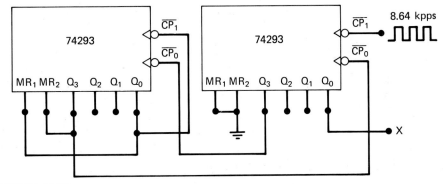

FIGURE 7-54

Section 7-5

7-12. A 4-bit ripple counter is driven by a 20-MHz clock signal. Draw the waveforms at the output of each FF if each FF has $t_{pd} = 20$ ns. Determine which counter states, if any, will not occur because of the propagation delays.

7-13. What is the maximum clock frequency that can be used with the counter of Problem 7-12? What would be f_{max} if the counter were expanded to 6 bits?

Section 7-6

7-14. (a) Draw the circuit diagram for a MOD-64 parallel counter.
(b) Determine f_{max} for this counter if each FF has $t_{pd} = 20$ ns and each gate has $t_{pd} = 10$ ns.

384

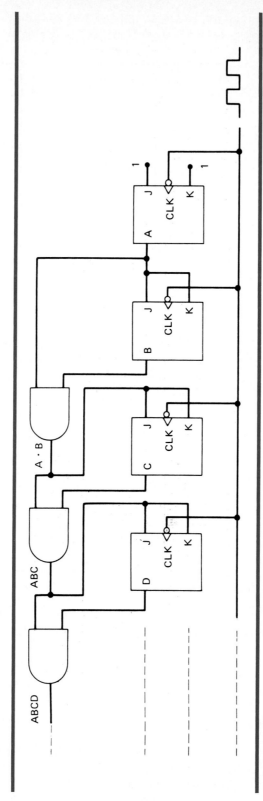

FIGURE 7-55

(c) A more efficient gating arrangement can be used for this counter by deriving the inputs for each AND gate from the output of the preceding gate (see Figure 7-55). For example, the output of the first AND gate (equal to $A \cdot B$) can be fed to the input of the second AND gate, together with C, to produce $A \cdot B \cdot C$. Similarly, the output of the second AND gate (equal to $A \cdot B \cdot C$) is fed to the third AND gate along with D to produce $A \cdot B \cdot C \cdot D$, and so on. This arrangement requires only two-input AND gates. However, it increases the total propagation delay, since the FF output signals must now propagate through more than one gate. Calculate the total delay for this arrangement and then calculate $f_{\max}$ and compare it to the value calculated in part (b). Assume the same delays as given in part (b).

7-15. Figure 7-56 shows a 4-bit parallel counter which is designed so that it does not sequence through the entire 16 binary states. Analyze its operation by drawing the waveforms at each FF output. Then determine the sequence which the counter goes through. Assume that all FFs are initially 0.

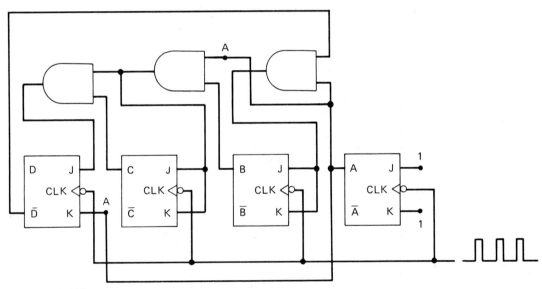

FIGURE 7-56

7-16. Refer to Appendix II for the 7490 data sheets.
(a) Show how to connect the 7490 as a BCD counter.
(b) Show how to connect the 7490 so that it divides the input frequency by 10 and produces a symmetrical squarewave output.

Sections 7-8 and 7-9

7-17. Figure 7-57 shows how a presettable down counter can be used in a *programmable timer* circuit. The input clock frequency is an accurate 1 Hz derived from the 60-Hz line frequency after division by 60. Switches S1–S4 are used to preset the counter to a desired starting count when a momentary pulse is applied to $\overline{PL}$. The timer operation is initiated by depress-

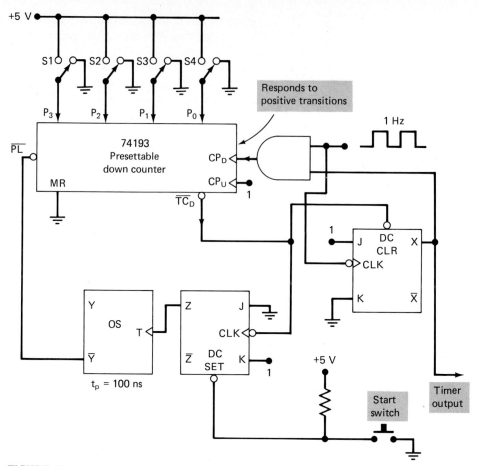

FIGURE 7-57

ing the START pushbutton switch. FF Z is used to eliminate effects of bounce in the START switch. The OS is used to provide a very narrow pulse to the $\overline{PL}$ input. The output of FF X will be a waveform that goes HIGH for a number of seconds equal to the number set on the switches.

(a) Assume that all FFs and the counter are in the 0 state and analyze and explain the circuit operation, showing waveforms when necessary, for the case where S1 and S4 are LOW and S2 and S3 are HIGH. Be sure to explain the function of FF X.

(b) Why can't the timer output be taken at the $\overline{TC}_D$ output?

(c) Why can't the START switch be used to trigger the OS directly?

(d) What will happen if the START switch is held down too long? Add the necessary logic needed to ensure that holding the START switch down will not affect the timer operation.

7-18. Modify the circuit of Figure 7-23 so that it functions as a MOD-10 counter. The frequency at the Q_3 output should be one-tenth the frequency of the CP_D input. Draw the waveforms at Q_3, Q_2, Q_1, Q_0, and $\overline{TC}_D$.

7-19. Change the parallel data inputs in Figure 7-23 to 1001. Draw the waveforms at Q_3, Q_2, Q_1, Q_0, and $\overline{TC}_D$. What is the MOD number?

Section 7-10

7-20. Figure 7-58 shows the IEEE/ANSI symbol for a 7490 or 74290 counter IC. Examine the symbol and determine the following.
(a) The overall MOD number.
(b) The function performed by the MR inputs.
(c) The function performed by the MS inputs.

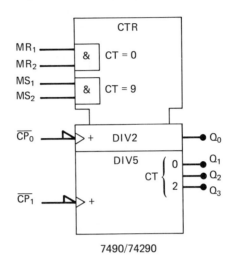

7490/74290

FIGURE 7-58

7-21. The 74192 counter IC operates exactly like the 74193 except for the following differences.
- The 74192 is a BCD counter that either counts up from 0 to 9, or counts down from 9 to 0.
- The TC_U output is activated when the count is 9 and the CP_U input is LOW.

Modify the IEEE/ANSI symbol of Figure 7-24 so that it represents the 74192.

Sections 7-11 and 7-12

7-22. Draw the gates necessary to decode all the states of a MOD-16 counter using active-LOW outputs.

7-23. Draw the AND gates necessary to decode the 10 states of the BCD counter of Figure 7-5(b).

7-24. Figure 7-59 shows a counter being used to help generate control waveforms. Control waveforms 1 and 2 could be used for many purposes, including control of motors, solenoids, valves, and heaters. Determine the control waveforms, assuming that all FFs are initially LOW. Ignore decoding glitches. Assume that clock frequency = 1 kpps.

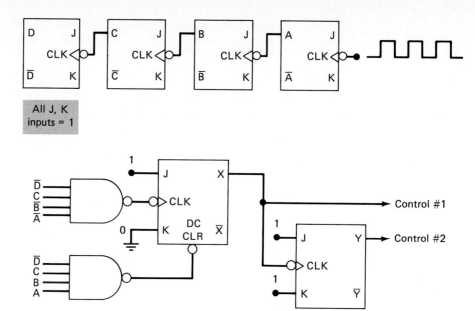

FIGURE 7-59

7-25   Draw the complete waveforms at the output of the decoding gates of a MOD-16 *ripple* counter, including any glitches or spikes that can occur due to the FF delays. Why are the gates which are decoding for *even* numbers the only ones that have glitches?

7-26.   The circuit of Figure 7-59 might malfunction because of glitches at the outputs of the decoding NAND gates.
(a) Determine at what point(s) the glitches can cause erroneous operation.
(b) What are two ways that can be used to eliminate the possibility of erroneous operation?

Section 7-13

7-27.   How many FFs are used in Figure 7-30? Indicate the states of each of these FFs after 795 pulses have occurred.

7-28.   How many cascaded BCD counters are needed to be able to count up to 8000? How many FFs does this require? Compare this to the number of FFs required for a normal binary counter to count up to 8000. Since it uses more FFs, why is the cascaded BCD method used?

Section 7-14

7-29.   Draw the diagram for a 5-bit ring counter using J-K flip flops.

7-30.   Combine the ring counter of Problem 7-29 with a *single* J-K FF to produce a MOD-10 counter. Determine the sequence of states for this counter. This is an example of a decade counter that is not a BCD counter.

7-31.   Draw the diagram for a MOD-10 Johnson counter using J-K flip-flops and determine its counting sequence. Draw the decoding circuit needed to decode each of the 10 states. This is another example of a decade counter that is not a BCD counter.

7-32. Determine the frequency of the pulses at points w, x, y, and z in the circuit of Figure 7-60.

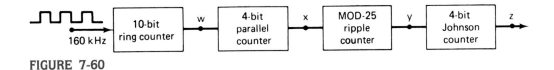

FIGURE 7-60

Section 7-15

7-33. As pointed out in the text, the frequency counter of Figure 7-36 has the disadvantage that the display shows all of the counter operations (resetting, counting, holding) and would therefore be confusing, if not unreadable. This can be overcome by the addition of buffer registers to store the contents of the counter at the end of the counting interval (t_3–t_4 in Figure 7-36) and hold it for display until the end of the next counting interval (t_7–t_8). Figure 7-61 shows this modification. A register consisting of four D FFs has been inserted between each BCD counter and its decoder/display unit.
   (a) Analyze this circuit and describe its operation, particularly the transferring of data from the counters to the display.
   (b) What would you see on a three-digit display if the unknown frequency were constant at 2570 pps and the sample interval were 0.1 s?
   (c) What would you see on this display if the unknown frequency were suddenly changed to 3230 pps?

7-34. The frequency counter of Figure 7-61 uses three BCD counters and a sampling interval of 100 μs. Determine the readings on the three frequency-counter displays for each of the following input frequencies.
   (a) 220 kkps
   (b) 4.5 Mpps
   (c) 750 pps

Section 7-16

7-35. Design the complete circuit for the SECONDS section of the digital clock circuit of Figure 7-37. Use a 74293 for the MOD-6 and a 7490 for the BCD.

7-36. The digital clock of Figure 7-37 has to have some means for manually setting the HOURS and MINUTES section to the correct starting time. For example, this can be done by switching the 1-pps signal into the input of the MINUTES section when a SET MINUTES pushbutton is activated. A similar operation can be done with a SET HOURS pushbutton. Design the necessary logic to provide this capability using two pushbutton switches.

7-37. Modify the HOURS section of the digital clock (Figure 7-36) so that it counts and displays military time (i.e., 00 to 23 hours).

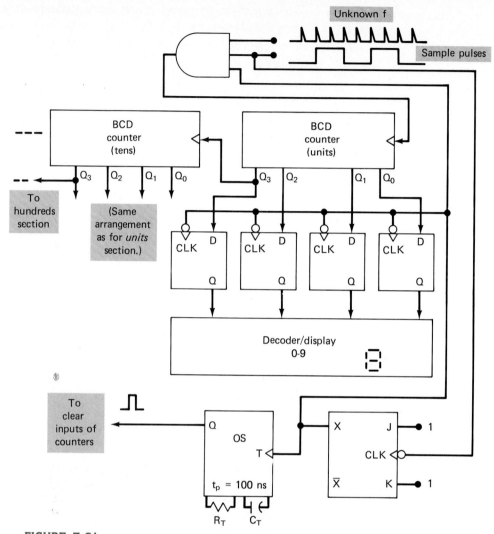

FIGURE 7-61

Sections 7-18 and 7-19

7-38. Modify the frequency counter of Figure 7-61 so that it uses 74174 ICs. Assume that the counter contains three BCD counters and a three-digit display.

7-39. Show how to connect two 74178s as an 8-bit ring counter.

7-40. Figure 7-62 shows how a 74178 can be used as a *parallel-to-serial converter*. The parallel data that are entered at P_0–P_3 are shifted out serially so that a serial waveform appears at Q_3. The shifting out of the parallel data is controlled by the occurrence of the START pulse. This START pulse occurs asynchronously to the clock pulses, so the D FFs are used to synchronize the parallel loading and shifting of the 74178.

Assume that the START pulse has been LOW and that clock pulses have been continually applied to the circuit for a long time before time t_0

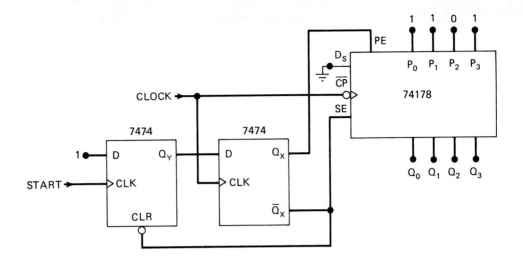

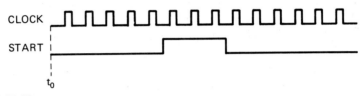

FIGURE 7-62

(see waveforms). Draw the waveforms that appear at Q_X, $\overline{Q}_X$, Q_Y, and Q_3 in response to the START pulse shown in the figure.

7-41. A shift register is often used to delay a logic signal by an integral number of clock cycles. If the waveforms in Figure 7-63 are applied to the inputs of one of the 4731B shift registers, how long will it take the D_S waveform to appear at the Q_{63} output? (Assume that the t_H requirement of the register FFs is met by the D waveform.)

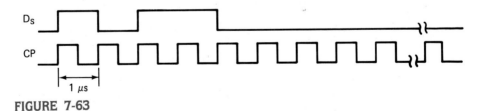

FIGURE 7-63

Sections 7-20 and 7-21

7-42. Refer to the circuit of Figure 7-64.
(a) Draw the waveforms at each FF output in response to the input waveforms shown.
(b) Add the necessary logic to produce a timing-signal output that goes HIGH only during time intervals t_1-t_2 and t_8-t_9.

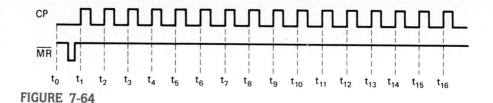

FIGURE 7-64

(c) Add the necessary logic to produce a timing signal that goes LOW only during the t_4–t_7 interval.

7-43. Assume that pulses have been applied continuously to CP_1 prior to t_0 in Figure 7-65. Draw the waveform at Q_7 in response to the CP_1 and $\overline{PL}$ waveforms shown.

Section 7-23

7-44. A technician tests the counter of Figure 7-49(a) by applying a low-frequency clock signal and monitoring the FF outputs on indicator LEDs. He observes the following repetitive sequence indicated by the LEDs.

Q_3	Q_2	Q_1	Q_0	
0	0	0	0	
0	0	0	1	
0	0	1	0	
0	0	1	1	recycle and
0	1	0	0	repeat
0	1	0	1	
0	1	1	0	
0	1	1	1	

What are the possible reasons why the counter is not counting properly?

7-45. Refer to the digital clock circuit of Figures 7-37 and 7-38. A technician testing the circuit observes that the SECONDS and MINUTES sections count properly, but the HOURS section counts as follows: 01, 02, 03, 04, 05, 06, 07, 08, 09, 10, 11, 12, 11, 12, 11, 12, ... What is the probable cause of the malfunction?

7-46. A technician tests the digital clock circuit (Figures 7-37 and 7-38) and observes that the HOURS section does not count and the MINUTES section counts from 00 to 39, then recycles to 00 and repeats. What are the possible causes of this incorrect behavior?

7-47. Refer to the modified frequency counter of Figure 7-61. Assume that there are three BCD counter stages with associated buffer registers. The sample interval is set at 1s and the unknown frequency is 125 pps. Describe what will appear on the display for each of the following circuit faults.
(a) An open at the top input of the AND gate.
(b) A burned-out resistor, R_T.

7-48. A technician tests the frequency counter of Figure 7-61 using a sample interval of 1s and an unknown frequency of 125 pps. She expects to see a display of 125, but instead sees the display changing every few seconds as

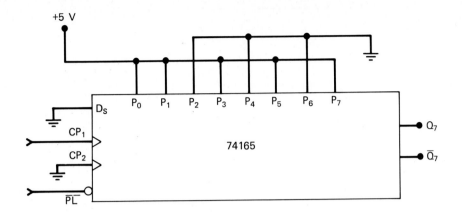

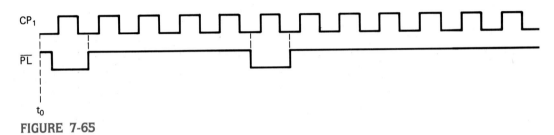

FIGURE 7-65

follows: 125, 250, 375, 500, 625, 750, 875, 000, 125, 250, etc. What can be the cause of this malfunction?

7-49. Refer to the up/down counter of Figure 7-17. Describe how each of the following circuit faults will affect the count-up and count-down operations.
(a) AND gate 4 output is internally shorted to V_{CC}.

(b) A solder bridge is shorting AND gate 1 output to ground.

7-50. A technician runs a test on the timer circuit of Figure 7-57 and records the following results:

S1	S2	S3	S4	TIMER OUTPUT(S)
5 V	0 V	0 V	0 V	10
0 V	0 V	5 V	5V	3
5 V	0 V	5 V	5V	11
5 V	5 V	5 V	0V	14
0 V	5 V	0 V	5V	7
5 V	5 V	0 V	0V	14

Examine the recorded data and determine the possible causes of the faulty operation.

7-51 A technician wires up the counter circuit of Figure 7-54. He applies an accurate 8.64 kpps signal to the input and measures a frequency of 54 pps at X instead of the expected 60 pps. What is the most probable wiring error that he made?

INTEGRATED-CIRCUIT LOGIC FAMILIES

8

Upon completion of this chapter, you will be able to:

- Read and understand digital IC terminology as specified in manufacturer's data sheets.
- Compare the characteristics of standard TTL and the various TTL series.
- Determine the fan-out for a particular logic device.
- Use logic devices with open-collector outputs in a wired-AND arrangement.
- Analyze circuits containing tristate devices.
- Describe the major characteristics and differences among TTL, ECL, MOS, and CMOS logic families.
- Cite and implement the various considerations that are required when interfacing digital circuits from different logic families.
- Use a logic pulser and current tracer as digital circuit troubleshooting tools.

As we described in Chapter 4, digital IC technology has advanced rapidly from small-scale integration (SSI), with fewer than 12 gates per chip, through medium-scale integration (MSI), with 12 to 99 equivalent gates per chip, to large-scale and very-large-scale integration (LSI and VLSI), which can have tens of thousands of gates per chip.

Most of the reasons that modern digital systems use integrated circuits are obvious. ICs pack a lot more circuitry in a small package, so that the overall size of almost any digital system is reduced. The cost is dramatically reduced because of the economies of mass-producing large volumes of similar devices. Some of the other advantages are not so apparent.

ICs have made digital systems more reliable by reducing the number of external interconnections from one device to another. Before we had ICs, every circuit connection was from one discrete component (transistor, diode, resistor, etc.) to another. Now most of the connections are internal to the ICs, where they are protected from poor soldering, breaks or shorts in connecting paths on a circuit board, and other physical problems. ICs have also drastically reduced the amount of electrical power needed to perform a given function since its miniature circuitry typically requires less power than its discrete counterpart. In addition to the savings in power-supply costs, this reduction in power has also meant that a system does not require as much cooling.

There are some things that ICs cannot do. They cannot handle very large currents or voltages because the heat generated in such small spaces would cause temperatures to rise beyond acceptable limits. In addition, ICs cannot easily implement certain electrical devices such as inductors, transformers, and large capacitors. For these reasons, ICs are principally used to perform low-power circuit operations that are commonly called *information processing*. The operations that require high power levels or devices that cannot be integrated are still handled by discrete components.

With the widespread use of ICs comes the necessity to know and understand the electrical characteristics of the most common IC logic families. Remember that the various logic families differ in the major components that they use in their circuitry. TTL and ECL use bipolar transistors as their major circuit element; PMOS, NMOS, and CMOS use unipolar MOSFET transistors as their principal component.

In this chapter we will present the important characteristics of each of these IC families and their subfamilies. Once these are understood, you will be much better prepared to do analysis, troubleshooting, and some design of digital circuits that contain any combination of IC families.

8-1 DIGITAL IC TERMINOLOGY

Although there are many digital IC manufacturers, much of the nomenclature and terminology is fairly standardized. The most useful terms are defined and discussed below.

Current and Voltage Parameters (see Figure 8-1)

V_{IH} (min)—**High-Level Input Voltage** The voltage level required for a logical 1 at an *input*. Any voltage below this level will not be accepted as a HIGH by the logic circuit.

V_{IL} (max)—**Low-Level Input Voltage** The voltage level required for a logic 0 at an *input*. Any voltage above this level will not be accepted as a LOW by the logic circuit.

V_{OH} (min)—**High-Level Output Voltage** The voltage level at a logic circuit *output* in the logical 1 state. The minimum value of V_{OH} is usually specified.

V_{OL} (max)—**Low-Level Output Voltage** The voltage level at a logic circuit *output* in the logical 0 state. The maximum value of V_{OL} is usually specified.

I_{IH}—**High-Level Input Current** The current that follows into an input when a specified high-level voltage is applied to that input.

I_{IL}—**Low-Level Input Current** The current that flows into an input when a specified low-level voltage is applied to that input.

FIGURE 8-1 **Currents and voltages in the two logic states.**

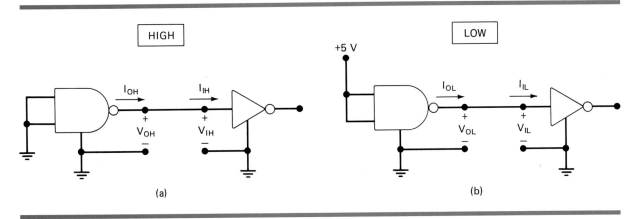

(a) (b)

I_{OH}—**High-Level Output Current** The current that flows from an output in the logical 1 state under specified load conditions.

I_{OL}—**Low-Level Output Current** The current that flows from an output in the logical 0 state under specified load conditions.

Fan-Out In general, a logic-circuit output is required to drive several logic inputs. The *fan-out* (also called *loading factor*) is defined as the *maximum* number of standard logic inputs that an output can drive reliably. For example, a logic gate that is specified to have a fan-out of 10 can drive 10 standard logic inputs. If this number is exceeded, the output logic-level voltages cannot be guaranteed.

Propagation Delays A logic signal always experiences a delay in going through a circuit. The two propagation delay times are defined as

t_{PLH}: delay time in going from logical 0 to logical 1 state (LOW to HIGH).

t_{PHL}: delay time in going from logical 1 to logical 0 state (HIGH to LOW).

Figure 8-2 illustrates these propagation delays for an INVERTER. Note that t_{PHL} is the delay in the output's response as it goes from HIGH to LOW. It is measured between the 50 percent points on the input and output transitions. The t_{PLH} value is the delay in the output's response as it goes from LOW to HIGH.

In general, t_{PHL} and t_{PLH} are not the same value, and both will vary depending on loading conditions. The values of propagation times are used as a measure of the relative speed of logic circuits. For example, a logic circuit with values of 10 ns is a faster logic circuit than one with values of 20 ns.

Power Requirements Every IC requires a certain amount of electrical power to operate. This power is supplied by one or more power-supply voltages connected to the power pin(s) on the chip. Usually there is only one power supply terminal on the chip, and it is labeled V_{CC} (for TTL) or V_{DD} (for MOS devices).

FIGURE 8-2 Propagation delays.

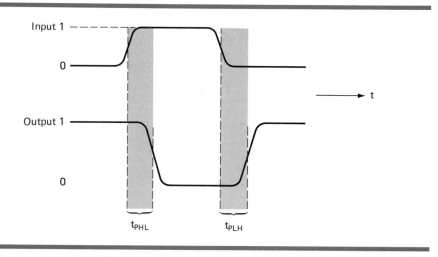

The amount of power that an IC requires is usually specified in terms of the current, I_{CC}, that it draws from the V_{CC} supply, and the actual power is the product $I_{CC} \times V_{CC}$. For many ICs the current drain on the supply will vary depending on the logic states of the circuits on the chip. For example, Figure 8-3(a) shows a NAND chip where *all* the gate *outputs* are HIGH. The current drain on the V_{CC} supply for this case is called I_{CCH}. Likewise, Figure 8-3(b) shows the current drain when *all* the gate *outputs* are LOW. This current is called I_{CCL}.

In general, I_{CCH} and I_{CCL} will be different values. The average current is

$$I_{CC}(\text{avg}) = \frac{I_{CCH} + I_{CCL}}{2}$$

and it can be used to calculate average power drain as

$$P_D(\text{avg}) = I_{CC}(\text{avg}) \times V_{CC}$$

Speed–Power Product   Digital IC families have historically been characterized for both speed and power. It is generally more desirable to have shorter gate propagation delays (higher speed) and lower values of power dissipation. As we shall soon see, the various logic families and subfamilies provide a wide spectrum of speed and power ratings. A common means for measuring and comparing the overall performance of an IC family is the *speed–power product* which is obtained by multiplying the gate propagation delay by the gate power dissipation. For ex-

FIGURE 8-3 I_{CCH} and I_{CCL}.

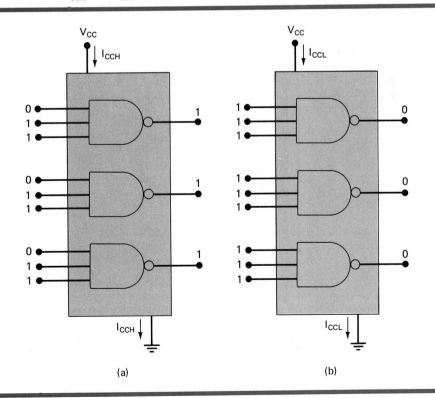

(a)   (b)

ample, suppose an IC family has an average propagation delay of 10 ns and an average power dissipation of 5 mW. The speed–power product is

$$10 \text{ ns} \times 5 \text{ mW} = 50 \times 10^{-12} \text{ watt-second}$$

$$= 50 \text{ picojoules}$$

$$= 50 \text{ pJ}$$

Note that when propagation delay is in nanoseconds and power is in milliwatts, the speed–power product is in picojoules.

Clearly, a low value of speed–power product is desirable. IC designers are continually striving to reduce the speed–power product by increasing the speed of an IC (i.e., reducing propagation delay) or by decreasing its power dissipation. Because of the nature of transitor switching circuits, it is difficult to do both.

Noise Immunity Stray electrical and magnetic fields can induce voltages on the connecting wires between logic circuits. These unwanted, spurious signals are called *noise* and can sometimes cause the voltage at the input to a logic circuit to drop below $V_{IH}(\text{min})$ or rise above $V_{IL}(\text{max})$, which could produce unpredictable operation. The *noise immunity* of a logic circuit refers to the circuit's ability to tolerate noise voltages on its inputs. A quantitative measure of noise immunity is called *noise margin* and is illustrated in Figure 8-4.

Figure 8-4(a) is a diagram showing the range of voltages that can occur at a logic circuit output. Any voltages greater than $V_{OH}(\text{min})$ are considered a logic 1, and any voltages lower than $V_{OL}(\text{max})$ are considered a logic 0. Voltages in the indeterminate range should not appear at a logic circuit output under normal conditions. Figure 8-4(b) shows the voltage requirements at a logic circuit input. The logic circuit will respond to any input greater than $V_{IH}(\text{min})$ as a logic 1, and will

FIGURE 8-4 DC noise margins.

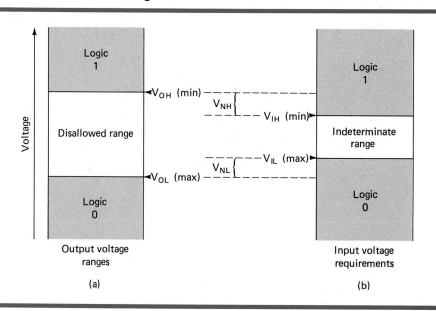

respond to voltages lower than V_{IL}(max) as a logic 0. Voltages in the indeterminate range will produce an unpredictable response and should not be used.

The *high-state noise margin* V_{NH} is defined as

$$V_{NH} = V_{OH}(min) - V_{IH}(min) \qquad (8\text{-}1)$$

as illustrated in Figure 8-4. V_{NH} is the difference between the lowest possible HIGH output and the minimum input voltage required for a HIGH. When a HIGH logic output is driving a logic circuit input, any negative noise spikes greater than V_{NH} appearing on the signal line can cause the voltage to drop into the indeterminate range, where unpredictable operation can occur.

The *low-state noise margin* V_{NL} is defined as

$$V_{NL} = V_{IL}(max) - V_{OL}(max) \qquad (8\text{-}2)$$

and it is the difference between the largest possible LOW output and the maximum input voltage required for a LOW. When a LOW logic output is driving a logic input, any positive noise spikes greater than V_{NL} can cause the voltage to rise into the indeterminate range.

EXAMPLE 8-1

The input/output voltage specifications for the standard TTL family are listed below. Use these values to determine

(a) The maximum-amplitude noise spike that can be tolerated when a HIGH output is driving an input.

(b) The maximum-amplitude noise spike that can be tolerated when a LOW output is driving an input.

PARAMETER	MIN (V)	TYPICAL (V)	MAX (V)
V_{OH}	2.4	3.4	
V_{OL}		0.1	0.4
V_{IH}	2.0*		
V_{IL}			0.8*

*Normally only the minimum V_{IH} and maximum V_{IL} values are given.

SOLUTION

(a) When an output is HIGH, it may be as low as V_{OH} (min) = 2.4 V. The minimum voltage that an input will respond to as a HIGH is V_{IH} (min) = 2.0 V. A negative noise spike can drive the actual voltage below 2.0 V if its amplitude is greater than

$$V_{NH} = V_{OH}(min) - V_{IH}(min)$$
$$= 2.4 \text{ V} - 2.0 \text{ V} = 0.4 \text{ V}$$

(b) When an output is LOW, it may be as high as V_{OL}(max) = 0.4 V. The maximum voltage that an input will respond to as a LOW is V_{IL}(max) = 0.8 V. A positive noise spike can drive the actual voltage above the 0.8-V level if its amplitude is greater than

$$V_{NL} = V_{IL}(max) - V_{OL}(max)$$
$$= 0.8 \text{ V} - 0.4 \text{ V} = 0.4 \text{ V}$$

AC Noise Margin Strictly speaking, the noise margins predicted by expressions (8-1) and (8-2) are termed dc noise margins. The term "dc noise margin" might seem somewhat inappropriate when dealing with noise, which is generally thought of as an ac signal of the transient variety. However, in today's high-speed integrated circuits, a pulse width of 1 μs is extremely long and may be treated as dc as far as the response of a logic circuit is concerned. As pulse widths decrease to the low-nanosecond region, a limit is reached where the *pulse duration* is too short for the circuit to respond. At this point, the pulse amplitude would have to be increased appreciably to produce a change in the circuit output. What this means is that a logic circuit can tolerate a large noise amplitude if the noise pulse is of a very short duration compared to the circuit's response time (i.e., propagation delays). In other words, a logic circuit's *ac noise margins* are generally substantially greater than its dc noise margins given by (8-1) and (8-2).

Current-Sourcing and Current-Sinking Logic Logic families can be described according to how current flows between the output of one logic circuit and the input of another. Figure 8-5(a) illustrates *current-sourcing* action. When the output of gate 1 is in the HIGH state, it supplies a current I_{IH} to the input of gate 2, which acts essentially as a resistance to ground. Thus, the output of gate 1 is acting as a *source* of current for the gate 2 input.

Current-sinking action is illustrated in Figure 8-5(b). Here the input circuitry of gate 2 is represented as a resistance tied to $+V_{CC}$, the positive terminal of a power supply. When the gate 1 output goes to its LOW state, current will flow in the direction shown from the input circuit of gate 2 back through the output resis-

FIGURE 8-5 **Comparison of current-sourcing and current-sinking actions.**

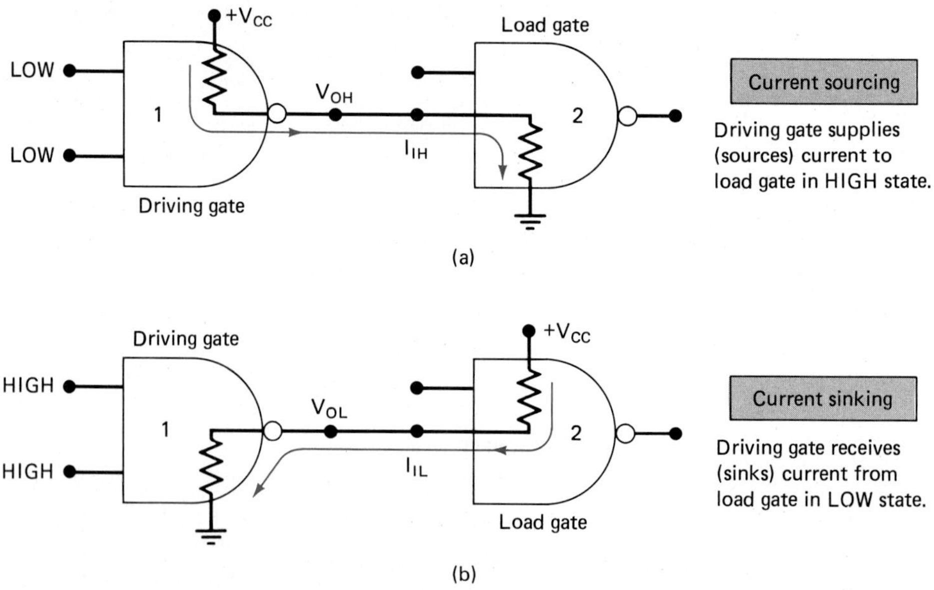
+VCC
Load gate
LOW
LOW
1
VOH
IIH
2
Driving gate
Current sourcing
Driving gate supplies (sources) current to load gate in HIGH state.
(a)
Driving gate
+VCC
HIGH
HIGH
1
VOL
IIL
2
Load gate
Current sinking
Driving gate receives (sinks) current from load gate in LOW state.
(b)

tance of gate 1 to ground. In other words, in the LOW state the circuit driving an input of gate 2 must be able to *sink* a current, I_{IL}, coming from that input.

The distinction between current sourcing and current sinking is an important one which will become more apparent as we examine the various logic families.

REVIEW QUESTIONS

1. Define each of the following: V_{OH}, V_{IL}, I_{OL}, I_{IH}, t_{PLH}, t_{PHL}, I_{CCL}, I_{CCH}.
2. *True or false:* If a logic circuit has a fan-out of 5, the circuit has five outputs.
3. Define V_{NL} and V_{NH}.
4. *True or false:* A logic family with $t_{pd}(avg) = 12$ ns and P_D (avg) = 15 mW has a greater speed–power product than one with 8 ns and 30 mW.
5. Describe the difference between current sinking and current sourcing.

8-2   THE TTL LOGIC FAMILY

At this writing, the transistor-transistor logic (TTL) family still enjoys the most widespread use in applications that require SSI and MSI devices. The basic TTL logic circuit is the NAND gate. Its detailed circuit diagram, shown in Figure 8-6(a), has several distinctive characteristics. First, note that transistor Q_1 has two emitters; thus, it has two emitter–base (E-B) junctions that can be used to turn Q_1 ON. This *multiple-emitter* input transistor can have up to eight emitters for an eight-input NAND gate.

Also note that on the output side of the circuit, transistors Q_3 and Q_4 are in a *totem-pole* arrangement. As we will see shortly, in normal operation either Q_3 or Q_4 will be conducting, depending on the logic state of the output.

Circuit Operation—LOW State
Although this circuit looks extremely complex, we can simplify its analysis somewhat by using the diode equivalent of the multi-emitter transistor Q_1 as shown in Figure 8-6(b). Diodes D_2 and D_3 represent the two E-B junctions of Q_1, and D_4 is the collector-base (C-B) junction. In the following analysis we will use this representation for Q_1.

First, let's consider the case where the output is LOW. Figure 8-7(a) shows this situation with inputs A and B both at +5 V. The +5 V at the cathodes of D_2 and D_3 will turn these diodes OFF and they will conduct almost no current. The +5-V supply will push current through R_1 and D_4 into the base of Q_2, which turns ON. Current from Q_2's emitter will flow into the base of Q_4 and turn Q_4 ON. At the same time, the flow of Q_2 collector current produces a voltage drop across R_2 that reduces Q_2's collector voltage to a low value that is insufficient to turn ON Q_3.

The voltage at Q_2's collector is shown as approximately 0.8 V. This is because Q_2's emitter is at 0.7 V relative to ground due to Q_4's E-B forward voltage, and

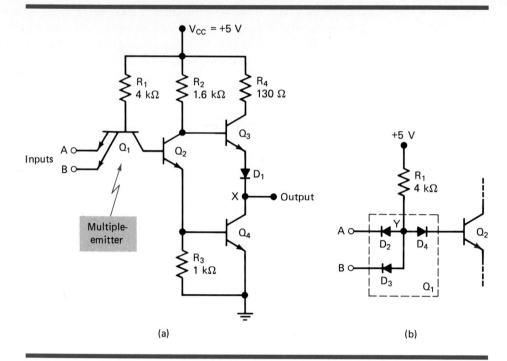

FIGURE 8-6   (a) Basic TTL NAND gate; (b) diode equivalent for Q_1.

Q_2's collector is at 0.1 V relative to its emitter due to V_{CE}(sat). This 0.8 V at Q_3's base is not enough to forward-bias both Q_3's E-B junction and diode D_1. In fact, D_1 is needed to keep Q_3 OFF in this situation.

With Q_4 ON, the output terminal, X, will be at a very low voltage, since Q_4's ON-state resistance will be low (1–25 Ω). Actually, the output voltage, V_{OL}, will depend on how much collector current Q_4 conducts. With Q_3 OFF, there is no current coming from the +5-V terminal through R_4. As we shall see, Q_4's collector current will come from the TTL inputs that terminal X is connected to.

It is important to note that the HIGH inputs at A and B will have to supply only a very small diode leakage current. Typically, this current I_{IH} is only around 10 µA at room temperature.

Circuit Operation—HIGH State
Figure 8-7(b) shows the situation where the circuit output is HIGH. This situation can be produced by connecting either or both inputs LOW. Here, input B is connected to ground. This will forward-bias D_3 so that current will flow from the +5-V source terminal, through R_1 and D_3, and through terminal B to ground. The forward voltage across D_3 will hold point Y at approximately 0.7 V. This voltage is not enough to forward-bias D_4 and the E-B junction of Q_2 sufficiently for conduction.

With Q_2 OFF, there is no base current for Q_4, and it turns OFF. Since there is no Q_2 collector current, the voltage at Q_3's base will be large enough to forward-bias Q_3 and D_1, so Q_3 will conduct. Actually, Q_3 acts as an emitter-follower, because output terminal X is essentially at its emitter. With no load connected from point X to ground, V_{OH} will be around 3.4–3.8 V, because two 0.7-V diodes drops

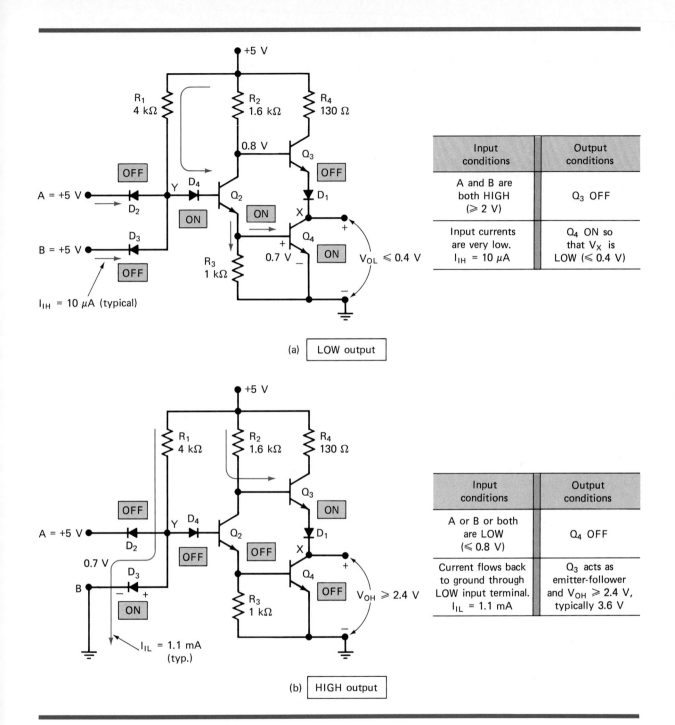

FIGURE 8-7 **TTL NAND gate in its two output states.**

(E-B of Q_3, and D_1) subtract from the 5 V applied to Q_3's base. This voltage will decrease under load because the load will draw emitter current from Q_3 which draws base current through R_2, thereby increasing the voltage drop across R_2.

It's important to note that there is a substantial current flowing back through input terminal B to ground. This current, I_{IL}, is typically around 1.1 mA. The LOW B input acts as a *sink* to ground.

Current-Sinking Action

A TTL output acts as a current sink in the LOW state in that it *receives* current from the input of the gate it is driving. Figure 8-8 shows one TTL gate driving the input of another gate (the load) for both output voltage states. In the output LOW state situation depicted in Figure 8-8(a), transistor Q_4 of the driving gate is ON and essentially "shorts" point X to ground. This LOW voltage at X forward-biases the emitter–base of Q_1 and current flows, as shown, back through Q_4. Thus, Q_4 is performing a current-sinking action that derives its current from the input current (I_{IL}) of the load gate. We will often refer to Q_4 as the *current-sinking transistor*. We will sometimes refer to it as the *pull-down transistor* because it brings the output voltage down to its LOW state.

Current-Sourcing Action

A TTL output acts as a current source in the HIGH state. This is shown in Figure 8-8(b), where transistor Q_3 is supplying the input current, I_{IH}, required by the Q_1 transistor of the load gate. As stated above, this current is a small reverse-bias leakage current. We will often refer to Q_3 as the *current-sourcing transistor* or *pull-up transistor*.

FIGURE 8-8 **(a) When TTL output is in LOW state, Q_4 acts as a current sink deriving its current from the load; (b) in the output HIGH state, Q_3 acts as a current source providing current to the load gate.**

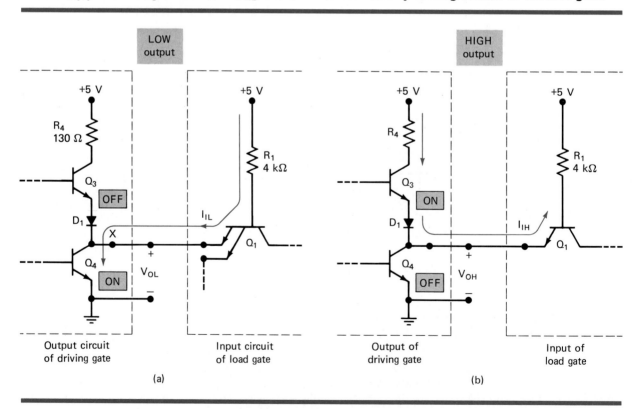

Totem-Pole Output Circuit   Several points should be mentioned concerning the totem-pole arrangement of the TTL output circuit, since it is not readily apparent why it is used. The same logic could be accomplished by eliminating Q_3 and D_1 and connecting the bottom of R_4 to the collector of Q_4. But this would mean that Q_4 would conduct a fairly heavy current in its saturation state (5 V/130 $\Omega \approx$ 40 mA). With Q_3 in the circuit, there will be no current through R_4 in the output LOW state. This is important because it keeps the circuit power dissipation down.

Another advantage of this arrangement occurs in the output HIGH state. Here Q_3 is acting as an emitter-follower with its associated low output impedance (typically 10 Ω). This low output impedance provides a short time constant for charging up any capacitive load on the output. This action (commonly called *active pull-up*) provides very fast rise-time waveforms at TTL outputs.

A disadvantage of the totem-pole output arrangement occurs during the transition from LOW to HIGH. Unfortunately, Q_4 turns OFF more slowly than Q_3 turns ON, so there is a period of a few nanoseconds during which both transistors are conducting and a relatively large current (30–40 mA) will be drawn from the 5-V supply. This can present a problem that will be examined later.

SUMMARY

All TTL circuits have a structure similar to this basic NAND gate. The input of any TTL circuit will be the cathode (N-region) of a P-N junction, so that a HIGH input voltage will turn OFF the device and only a small leakage current (I_{IH}) will flow. Conversely, a LOW input voltage turns ON the device and a relatively large current (I_{IL}) will flow. Most, but not all, TTL circuits will have some type of totem-pole output configuration. There are some exceptions that will be discussed later.

REVIEW QUESTIONS

1. *True or false:* A TTL output acts as a current sink in the LOW state.
2. In which TTL input state does the largest amount of input current flow?
3. State the advantages and disadvantages of a totem-pole output.
4. Which TTL transistor is the pull-up transistor?

8-3   STANDARD TTL SERIES CHARACTERISTICS

In 1964 Texas Instruments Corporation introduced the first line of standard TTL ICs. The 54/74 series, as it is called, has been one of the most widely used IC logic families. We will simply refer to it as the 74 series since the only difference between the 54 and 74 versions is that devices in the 54 series can operate over a wider range of temperatures and power supply voltages. Many semiconductor manufacturers now produce TTL ICs. Fortunately, they all use the same numbering system, so that the basic IC number is the same from one manufacturer to another. Each manufacturer, however, usually attaches its own special prefix to the IC number. For example, Texas Instruments uses the prefix SN, National Semiconductor uses DM, and Signetics uses S. Thus, depending on the manufacturer, you may see a quad NOR-gate chip labeled as a DM7402, SN7402, S7402 or some other similar designation. The important part is the number 7402, which is the same for all manufacturers.

As we learned in Chapter 4, there are several series in the TTL family of logic devices (74LS, 74S, 74L, etc.). We will first examine the electrical characteristics of the standard 74 series. Later we will introduce the other TTL series and compare their characteristics to the standard series.

Manufacturers' Data Sheets

To illustrate the characteristics of the standard TTL series, we will use the 7400 quad NAND gate IC. We can find all of the information we need on any IC by consulting the manufacturer's data manual for that particular IC family. Figure 8-9 is the manufacturer's data sheet for the 5400/7400 NAND gate IC showing the recommended operating conditions, electrical characteristics, and switching characteristics. Most of the quantities discussed in the following paragraphs in this section can be found on this data sheet. As we discuss each quantity, you should refer to this data sheet to see where the information came from.

Supply Voltage and Temperature Range

Both the 74 series and 54 series use a nominal supply voltage (V_{CC}) of 5 V. The 74 series will operate reliably over the range 4.75 to 5.25 V, while the 54 series can tolerate a supply variation of 4.5 to 5.5 V. The 74 series is designed to operate properly in ambient temperatures ranging from 0 to 70°C, while the 54 series can handle -55 to $+125$°C. Because of its greater tolerance of voltage and temperature variations, the 54 series is more expensive. It is employed only in applications where reliable operation must be maintained over an extreme range of conditions. Examples are military and space applications.

Voltage Levels

Table 8-1 lists the required input and output voltage levels for the standard 74 series. The minimum and maximum values shown are for worst-case conditions of power supply, temperature, and loading conditions. Inspection of the table reveals a guaranteed maximum logical 0 output $V_{OL} = 0.4$ V, which is 400 mV less than the logical 0 voltage needed at the input $V_{IL} = 0.8$ V. This means that the guaranteed LOW-state dc noise margin is 400 mV. That is,

TABLE 8-1 Standard 74 Series Voltage Levels

	MINIMUM	TYPICAL	MAXIMUM
V_{OL}	—	0.1	0.4
V_{OH}	2.4	3.4	—
V_{IL}	—	—	0.8
V_{IH}	2.0	—	—

recommended operating conditions

		SN5400			SN7400			UNIT
		MIN	NOM	MAX	MIN	NOM	MAX	
V_{CC}	Supply voltage	4.5	5	5.5	4.75	5	5.25	V
V_{IH}	High-level input voltage	2			2			V
V_{IL}	Low-level input voltage			0.8			0.8	V
I_{OH}	High-level output current			−0.4			−0.4	mA
I_{OL}	Low-level output current			16			16	mA
T_A	Operating free-air temperature	−55		125	0		70	°C

electrical characteristics over recommended operating free-air temperature range (unless otherwise noted)

PARAMETER	TEST CONDITIONS†		SN5400			SN7400			UNIT
			MIN	TYP‡	MAX	MIN	TYP‡	MAX	
V_{IK}	V_{CC} = MIN,	I_I = −12 mA			−1.5			−1.5	V
V_{OH}	V_{CC} = MIN,	V_{IL} = 0.8 V, I_{OH} = −0.4 mA	2.4	3.4		2.4	3.4		V
V_{OL}	V_{CC} = MIN,	V_{IH} = 2 V, I_{OL} = 16 mA		0.2	0.4		0.2	0.4	V
I_I	V_{CC} = MAX,	V_I = 5.5 V			1			1	mA
I_{IH}	V_{CC} = MAX,	V_I = 2.4 V			40			40	μA
I_{IL}	V_{CC} = MAX,	V_I = 0.4 V			−1.6			−1.6	mA
I_{OS}§	V_{CC} = MAX		−20		−55	−18		−55	mA
I_{CCH}	V_{CC} = MAX,	V_I = 0 V		4	8		4	8	mA
I_{CCL}	V_{CC} = MAX,	V_I = 4.5 V		12	22		12	22	mA

† For conditions shown as MIN or MAX, use the appropriate value specified under recommended operating conditions.
‡ All typical values are at V_{CC} = 5 V, T_A = 25°C.
§ Not more than one output should be shorted at a time.

switching characteristics, V_{CC} = 5 V, T_A = 25°C (see note 2)

PARAMETER	FROM (INPUT)	TO (OUTPUT)	TEST CONDITIONS		MIN	TYP	MAX	UNIT
t_{PLH}	A or B	Y	R_L = 400 Ω,	C_L = 15 pF		11	22	ns
t_{PHL}						7	15	ns

NOTE 2: See General Information Section for load circuits and voltage waveforms.

FIGURE 8-9   **Data sheet for 7400 NAND-gate IC.** (Courtesy of Texas Instruments)

$$V_{NL} = V_{IL}(\text{max}) - V_{OL}(\text{max}) = 0.8 \text{ V} - 0.4 \text{ V} = 0.4 \text{ V} = 400 \text{ mV}$$

Similarly, the logical 1 output V_{OH} is a guaranteed minimum of 2.4 V, which is 400 mV greater than the logical 1 voltage needed at the input V_{IH} = 2.0 V. Thus, the HIGH-state dc noise margin is 400 mV.

$$V_{NH} = V_{OH}(\text{min}) - V_{IH}(\text{min}) = 2.4 \text{ V} - 2.0 \text{ V} = 0.4 \text{ V} = 400 \text{ mV}$$

Thus, the *guaranteed worst-case* dc noise margins for the 74 series are both 400 mV. In actual operation the *typical* dc noise margins are somewhat higher (V_{NL} = 1 V and V_{NH} = 1.6 V).

Maximum Voltage Ratings

The voltage values in Table 8-1 *do not include* the absolute maximum ratings beyond which the useful life of the IC may be impaired. The voltages applied to any input of a standard 74 series IC must never exceed $+5.5$ V. A voltage greater than $+5.5$ V applied to an input emitter can cause reverse breakdown of the E-B junction of Q_1.

There is also a limit on the maximum *negative* voltage that can be applied to a TTL input. This limit, -0.5 V, is caused by the fact that most TTL circuits employ protective shunt diodes on each input, as illustrated in Figure 8-10. These diodes were purposely left out of our earlier analysis, since they do not enter into the normal circuit operation. They are connected from each input to ground to limit the negative input voltage excursions that often occur when logic signals have excessive ringing. With these diodes, we should not apply more than -0.5 V to an input, because the protective diodes would begin to conduct and draw substantial current.

Power Dissipation

A standard TTL NAND gate draws an average power of 10 mW. This is a result of $I_{CCH} = 4$ mA and $I_{CCL} = 12$ mA, which produces $I_{CC}(\text{avg}) = 8$ mA and $P_D(\text{avg}) = 8$ mA $\times$ 5 V $= 40$ mW. This 40 mW is the total power required by all four gates on the chip. Thus, one NAND gate requires an average power of 10 mW.

Propagation Delays

The standard TTL NAND gate has typical propagation delays of $t_{PLH} = 11$ ns and $t_{PHL} = 7$ ns, which is an *average* propagation delay $t_{pd}(\text{avg})$ of 9 ns.

Fan-Out

A standard TTL output can typically drive 10 standard TTL inputs. Some IC data sheets give the device's fan-out explicitly, but this is not the case for the data sheet in Figure 8-9. Later we will see how the fan-out can be determined from the input and output current values given in Figure 8-9.

Table 8-2 summarizes the characteristics of the standard 74 series NAND gate.

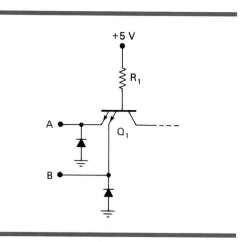

FIGURE 8-10 **Protective diodes on TTL inputs.**

TABLE 8-2   Standard 74 Series Characteristics

Noise margins (worst-case)	$V_{NL} = V_{NH} = 400$ mV
Average power dissipation (basic gate)	$P_D = 10$ mW
Average propagation delay (basic gate)	9 ns
Typical fan-out	10

EXAMPLE 8-2

Refer to the data sheet for the 7400 quad two-input NAND IC in Figure 8-9. Determine the maximum average power dissipation and maximum average propagation delay of a *single* gate.

SOLUTION

Look under the electrical characteristics for the maximum I_{CCH} and I_{CCL} values. The values are 8 mA and 22 mA, respectively. The average I_{CC} is therefore $(8 + 22)/2 = 15$ mA. The average power is obtained by multiplying by V_{CC}. The data sheet indicates that these I_{CC} values were obtained when V_{CC} was at its maximum value (5.25 V for the 74 series). Thus we have

$$P_D(\text{avg}) = 15 \text{ mA} \times 5.25 \text{ V} = 78.75 \text{ mW}$$

as the power drawn by the *complete* IC. We can determine the power drain of one NAND gate by dividing this by 4:

$$P_D(\text{avg}) = 19.7 \text{ mW per gate}$$

Since this average power drain was calculated using the maximum current and voltage values, it is the maximum average power that a 7400 NAND gate will draw under "worst-case" conditions. Designers often use worst-case values to ensure that their circuits will work under all conditions.

The maximum propagation delays for a 7400 NAND gate are listed as

$$t_{PLH} = 22 \text{ ns}, \qquad t_{PHL} = 15 \text{ ns}$$

so that the average propagation delay is

$$t_{pd}(\text{avg}) = \frac{22 + 15}{2} = 18.5 \text{ ns}$$

Again, this is a worst-case maximum possible average propagation delay.

8-4   OTHER TTL SERIES

The standard 74 series ICs offer a combination of speed and power dissipation suited for many applications. ICs offered in this series include a wide variety of gates, flip-flops, and one-shots in the small-scale integration (SSI) line and shift registers, counters, decoders, memories, and arithmetic circuits in the medium-scale integration (MSI) line.

Several other TTL series have been developed since the introduction of the standard 74 series. These other series provide a wide choice of speed and power characteristics. We will describe these series in the following paragraphs. Notice that whenever we use the term "TTL," we are usually referring to the standard 74 series.

Low-Power TTL, 74L Series

Low-Power TTL circuits designated as the 74L series have essentially the same basic circuit as the standard 74 series except that all the resistor values are *increased*. The larger resistors reduce the power requirements but at the expense of longer propagation delays. A typical NAND gate in this series has an average power dissipation of 1 mW and an average propagation delay of 33 ns.

The 74L series is well suited for applications in which power dissipation is more critical than speed. Low-frequency, battery-operated circuits such as calculators are well suited for this TTL series. This series has the lowest power dissipation of all the TTL series.

For the most part, the 74L series has been rendered obsolete by the development of the 74LS, 74ALS, and CMOS series, each of which offers low power dissipation at a higher speed of operation than that of the 74L devices. For this reason, the 74L series is not recommended for use in any new circuit designs.

High-Speed TTL, 74H Series

The 74H series is a high-speed TTL series. The basic circuitry for this series is essentially the same as the standard 74 series except that *smaller* resistor values are used and the emitter-follower transistor Q_3 is replaced by a Darlington pair. These differences result in a much faster switching speed with an average propagation delay of 6 ns. However, the increased speed is accomplished at the expense of increased power dissipation. The basic NAND gate in this series has an average P_D of 23 mW. The 74H series has essentially become obsolete since the development of the Schottky TTL series described in the following paragraphs, and is not recommended for new designs.

Schottky TTL, 74S Series

The 74, 74H, and 74L series all operate using saturated switching in which many of the transistors, when conducting, will be in the saturated condition. This operation causes a storage-time delay, t_S, when the transistors switch from ON to OFF, and limits the circuit's switching speed.

The 74S series reduces this storage-time delay by not allowing the transistor to go as deeply into saturation. It accomplishes this by using a Schottky barrier diode (SBD) connected between the base and collector of each transistor as shown in Figure 8-11(a). The SBD has a forward voltage of only 0.25 V. Thus, when the C-B junction becomes forward-biased at the onset of saturation, the SBD will conduct and divert some of the input current away from the base. This reduces the excess base current and decreases the storage-time delay at turn-OFF.

As shown in Figure 8-11(a), the transistor/SBD combination is given a special symbol. This symbol is used for all the transistors in the circuit diagram for the 74S00 NAND gate shown in Figure 8-11(b). This 74S00 NAND gate has an average propagation delay of only 3 ns, which is twice as fast as the 74H00. Note the presence of shunt diodes D_1 and D_2 to limit negative input voltages.

Circuits in the 74S series also use smaller resistor values to help improve switching times. This increases the circuit average power dissipation to about 20 mW, about the same as for 74H. The 74S circuits also use a Darlington pair (Q_3 and Q_4) to provide a more rapid output rise time when switching from ON to OFF.

Thus, 74S has twice the speed of 74H at about the same power requirement. This is why 74H has become obsolete.

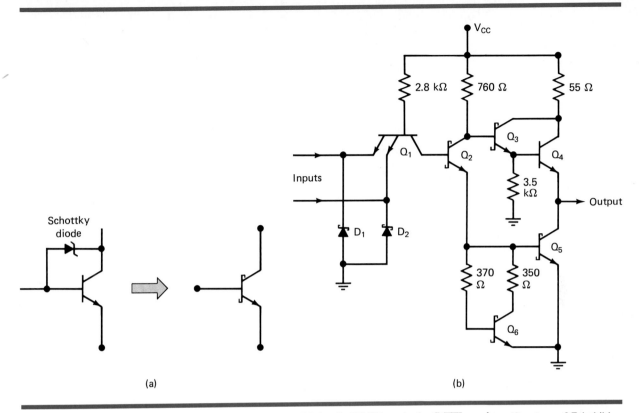

FIGURE 8-11 **(a) Schottky-clamped transistor; (b) basic NAND gate in S-TTL series.** (Courtesy of Fairchild, a Schlumberger company)

Low-Power Schottky TTL, 74LS Series (LS-TTL)   This series is a lower-powered, slower-speed version of the 74S series. It uses the Schottky-clamped transistor, but with larger resistor values than the 74S series (see Figure 8-12). The larger resistor values reduce the circuit power requirement, but at the expense of an increase in switching times.

A NAND gate in the 74LS series will typically have an average propagation delay of 9.5 ns and an average power dissipation of 2 mW. Since it has about the same switching speed as the standard TTL series at a much lower power requirement, the 74LS series has gradually replaced the 74 series in those applications where relatively high-speed operation is required at minimum power consumption. It has become the "mainstay" of the TTL family, and it can be found in almost all new designs that do not require maximum speed. Its leading position, however, will gradually be taken over by the new, improved 74ALS series.

Note that the 74LS NAND gate in Figure 8-12 does not use the multiple-emitter input transistor. Instead, it uses diodes D_3 and D_4, but the circuit operation remains essentially the same.

Advanced Schottky TTL, 74AS Series (AS-TTL)   Recent innovations in integrated-circuit design have led to the development of two new improved TTL series: advanced Schottky (74AS) and advanced low-power Schottky (74ALS). The

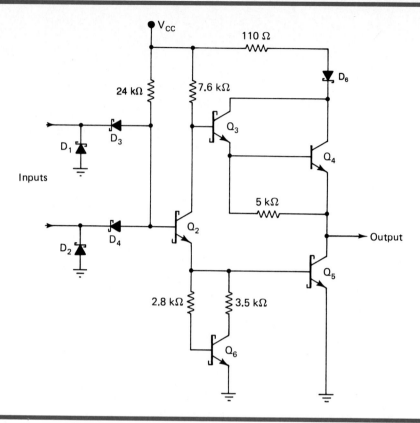

FIGURE 8-12 **74LS NAND gate.** (Courtesy of Fairchild, a Schlumberger company)

74AS series provides a considerable improvement in speed over the 74S series at a much lower power requirement. The comparison is shown below for a NAND gate in each series.

	74S	74AS
Propagation delay	**3 ns**	**1.7 ns**
Power dissipation	**20 mW**	**8 mW**
Speed–power product	**60 pJ**	**13.6 pJ**

This comparison clearly shows the advantage of the 74AS series. It is the fastest TTL series and its speed–power product is significantly lower than the 74S series. The 74AS has other improvements, including lower input current requirements (I_{IL}, I_{IH}), that result in a greater fan-out than the 74S series.

Because of these advantages, the 74AS has begun to replace the 74S series in all high-speed applications. As the cost of 74AS devices continues to come down, and as more functions become available in this series, the 74AS will eventually make the 74S series obsolete.

Advanced Low-Power Schottky TTL, 74ALS Series

This series offers an improvement over the 74LS series in both speed and power dissipation, as the following numbers illustrate.

	74LS	74ALS
Propagation delay	9.5 ns	4 ns
Power dissipation	2 mW	1.2 mW
Speed–power product	19 pJ	4.8 pJ

The 74ALS series has the lowest speed–power product of all the TTL series and is very close to having the lowest gate power dissipation (74L has 1 mW). It also has a much higher fan-out than that of the 74LS series. For these reasons we will see the 74ALS eventually replacing the 74LS as the most widely used TTL series.

Comparison of TTL Series Characteristics

Table 8-3 gives the typical values for some of the more important characteristics of each of the TTL series. All of the performance ratings, except for maximum clock rate, are for a NAND gate in each series. The maximum clock rate is specified as the maximum frequency that can be used to toggle a J-K FF. This gives a useful measure of the frequency range over which each IC series can be operated.

TABLE 8-3 Typical TTL Series Characteristics

	74	74L	74H	74S	74LS	74AS	74ALS
Performance ratings							
Propagation delay (ns)	9	33	6	3	9.5	1.7	4
Power dissipation (mW)	10	1	23	20	2	8	1.2
Speed-power product (pJ)	90	33	138	60	19	13.6	4.8
Max. clock rate (MHz)	35	3	50	125	45	200	70
Fan-out (same series)	10	20	10	20	20	40	20
Voltage parameters							
V_{OH}(min)	2.4	2.4	2.4	2.7	2.7	2.5	2.5
V_{OL}(max)	0.4	0.4	0.4	0.5	0.5	0.5	0.4
V_{IH}(min)	2.0	2.0	2.0	2.0	2.0	2.0	2.0
V_{IL}(max)	0.8	0.7	0.8	0.8	0.8	0.8	0.8

EXAMPLE 8-3

Use Table 8-3 to calculate the dc noise margins for a typical 74LS IC. How does this compare to the standard TTL noise margins obtained in Section 8-3?

SOLUTION

$$V_{NH} = V_{OH}(\text{min}) - V_{IH}(\text{min})$$
$$= 2.7 \text{ V} - 2.0 \text{ V}$$
$$= 0.7 \text{ V}$$

as compared to $V_{NH} = 0.4$ V for standard TTL.

$$V_{NL} = V_{OL}(\text{max}) - V_{IL}(\text{max})$$
$$= 0.8\ V - 0.5\ V$$
$$= 0.3\ V$$

as compared to $V_{NL} = 0.4$ V for TTL.

EXAMPLE 8-4

Which TTL series can drive the most number of device inputs of the same series?

SOLUTION

The 74AS series has the highest fan-out (40). This means, for example, that a 74AS00 NAND gate can drive 40 inputs of other 74AS devices. If we want to determine the number of inputs of a *different* TTL series that an output can drive, we will need to know the input and output currents of the two series. This will be dealt with in the next section.

REVIEW QUESTIONS

1. (a) Which TTL series is the best at high frequencies?
   (b) Which TTL series has the largest HIGH-state noise margin?
   (c) Which series have essentially become obsolete?
   (d) Which series use a special diode to reduce switching time?
   (e) Which series would be best for a battery-powered circuit operating at 10 MHz?
   (*Ans.* 74AS; 74S and 74LS; 74H and 74L; 74S, 74LS, 74AS, 74ALS; 74ALS)

2. Assuming the same cost for each, why should you choose to use a 74ALS193 counter over a 74LS193 or 74AS193 in a circuit operating from a 40-MHz clock?

8-5 TTL LOADING AND FAN-OUT

It is important to understand what determines the fan-out or load drive capability of an IC output. Figure 8-13(a) shows a standard TTL output in the LOW state connected to drive several standard TTL inputs. Transistor Q_4 is ON and is acting as a current sink for an amount of current I_{OL} that is the sum of the I_{IL} currents from each input. In its ON state, Q_4's collector–emitter resistance is very small, but it is not zero, so the current I_{OL} will produce a voltage V_{OL}. This voltage must not exceed the $V_{OL}(\text{max})$ limit of the IC. This limits the maximum value of I_{OL} and thus the number of loads that can be driven.

To illustrate, suppose the ICs are in the 74 series, and each I_{IL} is 1.6 mA. From Table 8-3 we see that the 74 series has $V_{OL}(\text{max}) = 0.4$ V and $V_{IL}(\text{max}) =$

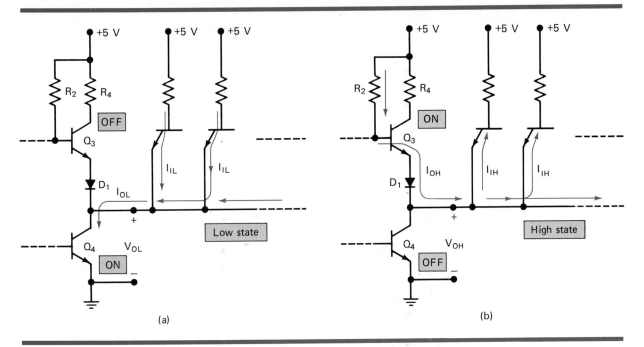

(a)   (b)

FIGURE 8-13   Currents when a TTL output is driving several inputs.

0.8 V. Let's suppose further that Q_4 can sink up to 16 mA before its output voltage reaches $V_{OL}(\text{max}) = 0.4$ V. This means that it can sink the current from up to 16 mA/1.6 mA $= 10$ loads. If it is connected to more than 10 loads, its I_{OL} will increase and cause V_{OL} to increase above 0.4 V. This is usually undesirable because it reduces the noise margin at the IC inputs [remember, $V_{NL} = V_{IL}(\text{max}) - V_{OL}(\text{max})$]. In fact, if V_{OL} rises above $V_{IL}(\text{max}) = 0.8$ V, it will be in the indeterminate range.

A similar situation occurs in the HIGH state depicted in Figure 8-13(b). Here Q_3 is acting as an emitter-follower that is sourcing (supplying) a total current I_{OH} that is the sum of the I_{IH} currents of each TTL input. If too many loads are being driven, this current I_{OH} will become large enough to cause the voltage drops across R_2, Q_3's emitter–base junction, and D_1 to bring V_{OH} below $V_{OH}(\text{min})$. This too is undesirable since it reduces the HIGH-state noise margin and could even cause V_{OH} to go into the indeterminate range.

What this all means is that a TTL output has a limit, $I_{OL}(\text{max})$, on how much current it can sink in the LOW state. It also has a limit, $I_{OH}(\text{max})$, on how much current it can source in the HIGH state. These output current limits must not be exceeded if the output voltage levels are to be maintained within their specified ranges.

Determining the Fan-Out   To determine how many different inputs an IC output can drive, you need to know the current drive capability of the output [that is, $I_{OL}(\text{max})$ and $I_{OH}(\text{max})$] and the current requirements of each input (that is, I_{IL} and I_{IH}). This information is always presented in some form on the manufacturer's IC data sheet. The following example will illustrate one type of situation.

EXAMPLE 8-5

How many 7400 NAND gate inputs can be driven by a 7400 NAND gate output?

SOLUTION

We will consider the LOW state first as depicted in Figure 8-14. Refer to the 7400 data sheet in Figure 8-9 and find

$$I_{OL}(\text{max}) = 16 \text{ mA}$$
$$I_{IL}(\text{max}) = 1.6 \text{ mA}$$

This says that a 7400 output can sink a maximum of 16 mA and that each 7400 input will source a maximum of 1.6 mA back through the driving gate's output. Thus, the number of inputs that can be driven in the LOW state is obtained as

$$\text{fan-out(LOW)} = \frac{I_{OL}(\text{max})}{I_{IL}(\text{max})}$$
$$= \frac{16 \text{ mA}}{1.6 \text{ mA}}$$
$$= 10$$

The HIGH state is analyzed in the same manner. Refer to the data sheet to find

$$I_{OH}(\text{max}) = 0.4 \text{ mA} = 400 \text{ }\mu\text{A}$$
$$I_{IH}(\text{max}) = 40 \text{ }\mu\text{A}$$

Thus, the number of inputs that can be driven in the HIGH state is

$$\text{fan-out (HIGH)} = \frac{I_{OH}(\text{max})}{I_{IH}(\text{max})}$$
$$= \frac{400 \text{ }\mu\text{A}}{40 \mu\text{A}}$$
$$= 10$$

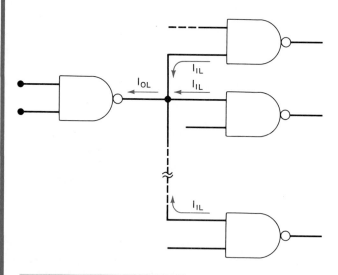

*All gates are 7400 NAND gates.

FIGURE 8-14   Example 8-5

These results indicate that the fan-out is 10 in both voltage states. Thus, the 7400 NAND gate can drive up to 10 other 7400 NAND gates. If fan-out (LOW) and fan-out (HIGH) are not the same, as will sometimes occur, the fan-out is chosen as the smaller of the two.

EXAMPLE 8-6

Refer to the data sheet in Appendix II and determine how many 74ALS20 NAND gates can be driven by the output of another 74ALS20.

SOLUTION
The 74ALS20 data sheet gives the following values:

$$I_{OH}(max) = 0.4 \text{ mA} = 400 \text{ } \mu A$$
$$I_{OL}(max) = 8 \text{ mA}$$
$$I_{IH}(max) = 20 \text{ } \mu A$$
$$I_{IL}(max) = 0.1 \text{ mA}$$

Considering the HIGH state first, we have

$$\text{fan-out (HIGH)} = \frac{400 \text{ } \mu A}{20 \text{ } \mu A} = 20$$

For the LOW state we have

$$\text{fan-out (LOW)} = \frac{8 \text{ mA}}{0.1 \text{ mA}} = 80$$

In this case, the overall fan-out is chosen to be 20 since it is the lower of the two values. Thus, one 74ALS20 can drive 20 other 74ALS20 inputs.

Unit Loads   Some manufacturers specify the device input and output currents in terms of a unit load (UL) where a unit load is defined as follows:

$$1 \text{ unit load (UL)} = \begin{cases} 40 \text{ } \mu A \text{ in the HIGH state} \\ 1.6 \text{ mA in the LOW state} \end{cases}$$

This simply means that when we talk about the HIGH state, UL is the same as 40 μA; when we talk about the LOW state UL is the same as 1.6 mA. For example, if an IC is specified as having a fan-out of 10 UL in both states, this is the same as saying that

$$I_{OH}(max) = 10 \times 40 \text{ } \mu A = 400 \text{ } \mu A$$

$$I_{OL}(max) = 10 \times 1.6 \text{ mA} = 16 \text{ mA}$$

Similarly, if an IC input is rated as 1 UL in both states, this is the same as saying that

$$I_{IH}(max) = 1 \times 40 \text{ } \mu A = 40 \text{ } \mu A$$

$$I_{IL}(max) = 1 \times 1.6 \text{ mA} = 1.6 \text{ mA}$$

Table 8-4 gives the typical input and output factors for the various TTL series. These values are typical, so there may be some variations, depending on the particular device or manufacturer. The device data sheet should be consulted to determine exact values.

TABLE 8-4

TTL SERIES	INPUT LOADING		FAN-OUT	
	HIGH	LOW	HIGH	LOW
7400	1 UL	1 UL	10 UL	10 UL
74H00	1.25 UL	1.25 UL	12.5 UL	12.5 UL
74L00	0.5 UL	≈0.1 UL	10 UL	2.5 UL
74S00	1.25 UL	1.24 UL	25 UL	12.5 UL
74LS00	0.5 UL	≈0.25 UL	10 UL	5 UL
74AS	0.5 UL	≈0.3 UL	50 UL	12.5 UL
74ALS	0.5 UL	≈0.06 UL	10 UL	5 UL

$$1 \text{ UL} = \begin{cases} 40 \ \mu\text{A (HIGH)} \\ 1.6 \text{ mA (LOW)} \end{cases}$$

EXAMPLE 8-7

Determine the input and output loading factors for the 7404 INVERTER chip (Appendix II).

SOLUTION

This information can be found on the 7404 data sheet under the heading "INPUT LOADING/FAN-OUT." Look under "54/74(UL)" and you will find the first entry, 1.0/1.0. This means that the input requirements are 1 UL in each state. In other words, the 7404 input draws a maximum of 40 μA of current from the input signal source in the HIGH state, and sources a maximum of 1.6 mA back through a LOW input. These are the maximum worst-case currents that flow under extreme conditions. In practice, you might measure $I_{IH} = 10 \ \mu$A and $I_{IL} = 1.1$ mA at the input. Most designers assume the worst-case values that the manufacturer provides on the data sheets.

The second entry under "54/74(UL)" is 20/10. This means that the 7404 output is rated at 20 UL in the HIGH state and 10 UL in the LOW state. In other words, the 7404 output can *supply* up to

$$20 \times 40 \ \mu\text{A} = 800 \ \mu\text{A}$$

to load devices in the HIGH state without its V_{OH} dropping below $V_{OH}(\text{min}) = 2.4$ V, and it can *sink* up to

$$10 \times 1.6 \text{ mA} = 16 \text{ mA}$$

in the LOW state without its V_{OL} rising above $V_{OL}(\text{max}) = 0.4$ V.

EXAMPLE 8-8

Repeat for the 74LS04 IC.

SOLUTION

Again, look under the heading "INPUT LOADING/FAN-OUT" for the 54/74LS entry, and we see that the inputs are rated at 0.5/0.25. Thus, the 74LS04 has an input loading factor of 0.5 UL in the HIGH state, and 0.25 UL in the LOW state.

The outputs are shown rated at 10/5.0, which means that a 74LS04 output can drive 10 UL in the HIGH state, and 5 UL in the LOW state.

EXAMPLE 8-9

The output of a 7404 INVERTER is providing the clock signal to a parallel register made up of 74107 J-K FFs. What is the maximum number of FFs that this clock signal can drive?

SOLUTION

This type of problem is solved by first determining the unit load capabilities of a 7404 output and then determining the unit load input requirements of a 74107's *CLK* input. We already know the 7404's fan-out from Example 8.4 as 20 UL (HIGH) and 10 UL (LOW). The 74107 data sheet shows that its $\overline{CP}$ input requirements are 2 UL in both states. Thus, the number of 74107 $\overline{CP}$ loads that the 7404 output can drive is

$$\text{number of loads} = \frac{\text{output rating}}{\text{input rating}} = \frac{10\ \text{UL}}{2\ \text{UL}} = \mathbf{5}$$

Note that we used the 7404's LOW-state output rating because it is more restrictive than its HIGH-state output rating.

EXAMPLE 8-10

Repeat Example 8-9 using a 74LS04 and a 74LS107.

SOLUTION

The 74LS04 has output ratings of 10 UL (HIGH) and 5 UL (LOW), while the 74LS107 clock inputs require 2 UL (HIGH) and 0.5 UL (LOW). Here it will be necessary to calculate the number of 74LS107 loads that a 74LS04 can drive in each state, and then take the smaller of the two numbers.

$$\text{number of loads (HIGH)} = \frac{10\ \text{UL}}{2\ \text{UL}} = 5$$

$$\text{number of loads (LOW)} = \frac{5\ \text{UL}}{0.5\ \text{UL}} = 10$$

Thus, a 74LS04 can drive **5** 74LS107 $\overline{CP}$ inputs.

EXAMPLE 8-11

A certain IC output is rated at $I_{OH}(\text{max}) = 800\ \mu\text{A}$ and $I_{OL} = 48\ \text{mA}$. Express the IC's fan-out in terms of unit loads.

SOLUTION
HIGH state: fan-out $= 800\ \mu\text{A}/40\ \mu\text{A} = 20\ \text{UL}$
LOW state: fan-out $= 48\ \text{mA}/1.6\ \text{mA} = 30\ \text{UL}$

EXAMPLE 8-12

A certain TTL FF's CLK input is rated at $I_{IL} = 0.8$ mA and $I_{IH} = 10$ μA. Express its input requirements in ULs.

SOLUTION

HIGH state: input requirement = 10 μA/40 μA = 0.25 UL

LOW state: input requirement = 0.8 mA/1.6 mA = 0.5 UL

These examples should make it clear that the UL rating is just a convenient way of specifying an IC's input and output current ratings. Some manufacturers prefer to list these current ratings explicitly without using ULs; others use ULs; still others provide the information in both forms. Since you will typically use ICs from several manufacturers, you should be able to convert between ULs and currents.

REVIEW QUESTIONS

1. What factors determine a device's I_{OL}(max) rating?
2. A certain TTL output has a fan-out of 5 UL in both states. How much current can it supply to loads in the HIGH state? (*Ans.* 200 μA)
3. What can happen if a TTL output is connected to more unit loads than its output rating specification?
4. How many 74107 $\overline{CP}$ inputs can be driven by a 74LS04 output? (*Ans.* 2)

8-6 OTHER TTL CHARACTERISTICS

Several other characteristics of TTL logic must be understood if one is to intelligently use TTL in a digital-system application.

Unconnected Inputs (Floating) Any input to a TTL circuit that is left disconnected (open) acts exactly like a logical 1 applied to that input, because in either case the emitter-base junction or diode at the input will not be forward-biased. This means that on *any* TTL IC, *all* the inputs are 1s if they are not connected to some logic signal or to ground. When an input is left unconnected, it is said to be "floating."

Unused Inputs Frequently, all the inputs on a TTL IC are not being used in a particular application. A common example is when all the inputs to a logic gate are not needed for the required logic function. For example, suppose that we needed the logic operation $\overline{AB}$ and we were using a chip that had a three-input NAND gate. The possible ways of accomplishing this are shown in Figure 8-15.

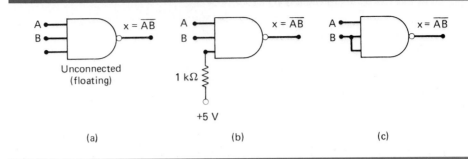

FIGURE 8-15   Three ways to handle unused logic inputs.

In Figure 8-15(a) the unused input is left disconnected, which means that it acts as a logical 1. The NAND-gate output is therefore $x = \overline{A \cdot B \cdot 1} = \overline{A \cdot B}$, which is the desired result. Although the logic is correct, it is usually undesirable to leave an input disconnected because it will act like an antenna, which is liable to pick up stray radiated signals that could cause the gate to operate improperly. A better technique is shown in Figure 8-15(b). Here the unused input is connected to $+5$ V through a 1-kΩ resistor, so the logic level is a 1. The 1-kΩ resistor is simply for current protection of the emitter-base junctions of the gate inputs in case of spikes on the power-supply line. This same technique can be used for AND gates, since a 1 on an unused input will not affect the output. As many as 30 unused inputs can share the same 1-kΩ resistor tied to V_{CC}.

A third possibility is shown in Figure 8-15(c), where the unused input is tied to a used input. This is satisfactory provided that the circuit driving input B is not going to have its fan-out exceeded. This technique can be used for *any* type of gate.

For OR gates and NOR gates, the unused inputs cannot be left disconnected or tied to $+5$ V since this would produce a constant-output logic level (1 for OR, 0 for NOR) regardless of the other inputs. Instead, for these gates the unused inputs must either be connected to ground (0 V) for a logic 0 or they can be tied to a used input as in Figure 8-15(c).

Tied-Together Inputs   When two or more TTL-gate inputs are connected together to form a common input as in Figure 8-15(c), the common input will generally have an input loading factor that is the sum of the input loading factors for each input. The only exception is for NAND and AND gates. For these gates, the LOW-state input loading factor will be the *same as a single input* no matter how many inputs are tied together.

To illustrate, assume that each input of the three-input NAND gate in Figure 8-15(c) is rated at 1 UL in each state. The common B input will therefore present an input loading factor of 2 UL in the HIGH state but only 1 UL in the LOW state. The same would be true if this were an AND gate. If it were an OR or NOR gate, the common B input would present an input loading factor of 2 UL in both states.

The reason for this characteristic can be found by looking back at the circuit diagram of the TTL NAND gate in Figure 8-7(b). The current I_{IL} is limited by the resistance R_1. Even if inputs A and B were tied together and grounded, this current would not change; it would merely divide up and flow through the parallel paths provided by diodes D_2 and D_3. The situation is different for OR and NOR gates, since they do not use multiple-emitter transistors, but rather have a separate input transistor for each input.

EXAMPLE 8-13

Determine the number of ULs that the X output is driving in Figure 8-16. Assume each gate input is rated at 1 UL in each state.

SOLUTION

The loading on X will be different for its two different logic states as shown in the figure. Note that the tied-together inputs of NAND gate 2 are counted as only 1 UL in the LOW state. The NOR gate inputs are treated as separate loads for either state.

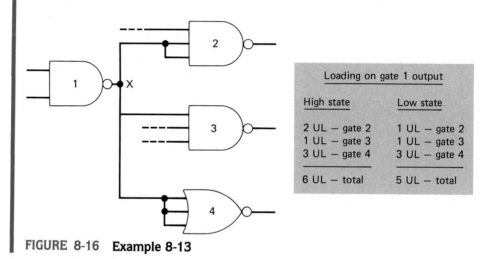

Loading on gate 1 output	
High state	Low state
2 UL — gate 2	1 UL — gate 2
1 UL — gate 3	1 UL — gate 3
3 UL — gate 4	3 UL — gate 4
6 UL — total	5 UL — total

FIGURE 8-16   Example 8-13

Biasing TTL Inputs Low   Occasionally, the situation arises where a TTL input must be held normally LOW and then caused to go HIGH by the actuation of a mechanical switch. This is illustrated in Figure 8-17 for the input to a one-shot. This OS triggers on a positive transition that occurs when the switch is momentarily closed. The resistor R serves to keep the T input LOW while the switch is open. Care must be taken to keep the value of R low enough so that the voltage developed across it by the current I_{IL} that flows out of the OS input to ground will not exceed $V_{IL}(\text{max})$. Thus, the largest value of R is given by

$$I_{IL} \times R_{\max} = V_{IL}(\text{max})$$

$$R_{\max} = \frac{V_{IL}(\text{max})}{I_{IL}}$$

(8.3)

R must be kept below this value to ensure that the OS input will be at an acceptable LOW level while the switch is open. The minimum value of R is determined by the current drain on the 5-V supply when the switch is closed. In practice, this current drain should be minimized by keeping R just slightly below $R_{\max}$.

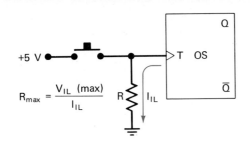

$$R_{max} = \frac{V_{IL}\ (max)}{I_{IL}}$$

FIGURE 8-17

EXAMPLE 8-14

Determine an acceptable value for R if the OS is a standard TTL IC with an input rating of 1 UL.

SOLUTION
At 1 UL, the value of I_{IL} will be a maximum of 1.6 mA. This maximum value should be used to calculate R_{max}. For standard TTL, $V_{IL}(max) = 0.8$ V. Thus, we have

$$R_{max} = \frac{0.8\ \text{V}}{1.6\ \text{mA}} = 500\ \Omega$$

A good choice here would be $R = 470\ \Omega$, a standard resistor value.

EXAMPLE 8-15

Repeat Example 8-14 if the OS belongs to the 74LS series with an input rating of 0.25 UL.

SOLUTION
At 0.25 UL, we have $I_{IL}(max) = 0.4$ mA. For 74LS, $V_{IL}(max) = 0.8$ V. Thus,

$$R_{max} = \frac{0.8\ \text{V}}{0.4\ \text{ma}} = 2\ \text{k}\Omega$$

A good choice here would be 1.8 kΩ, a standard value.

Current Transients   TTL logic circuits suffer from internally generated current transients or spikes because of the totem-pole output structure. When the output is switching from the LOW state to the HIGH state (see Figure 8-18), the two output transistors are changing states; Q_3 OFF to ON and Q_4 ON to OFF. Since Q_4 is changing from the saturated condition, it will take longer than Q_3 to switch states. Thus, there is a short interval of time (about 2 ns) during the switching transition where both transistors are conducting and a relatively large surge of current (30–50 mA) is drawn from the +5-V supply. The duration of this current transient

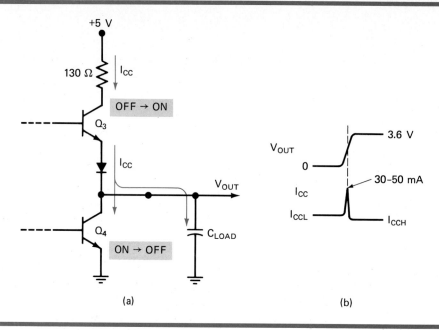

FIGURE 8-18 A large current spike is drawn from V_{CC} when a totem-pole output switches from LOW to HIGH.

is extended by the effects of any load capacitance on the circuit output. This capacitance consists of stray wiring capacitance and the input capacitance of any load circuits and must be charged up to the HIGH-state output voltage. This overall effect can be summarized as follows: *Whenever a totem-pole TTL output goes from LOW to HIGH, a high-amplitude current spike is drawn from the V_{CC} supply.*

In a complex digital circuit or system there may be many TTL outputs switching states at the same time, each one drawing a narrow spike of current from the power supply. The accumulative effect of all these current spikes will be to produce a voltage spike on the common V_{CC} line, mostly due to the distributed inductance on the supply line [remember: $V = L(di/dt)$ for inductance and di/dt is very large for a 2-ns current spike]. This voltage spike can cause serious malfunctions during switching transitions unless some type of filtering is used. The most common technique uses small *RF* capacitors connected from V_{CC} to GROUND to essentially "short out" these high-frequency spikes. This is called *power-supply decoupling*.

It is standard practice to connect a 0.01-μF or 0.1-μF low-inductance, ceramic disk capacitor between V_{CC} and ground near each TTL IC on a circuit board. The capacitor leads are kept very short to minimize inductance.

In addition, it is standard practice to connect a single large capacitor (2 to 20 μF) between V_{CC} and ground on each board to filter out possible variations in V_{CC} caused by the large changes in I_{CC} levels as outputs switch states.

1. What will be the logic output of a TTL NAND gate that has all its inputs unconnected?
2. What are two acceptable ways to handle unused inputs to an AND gate?
3. Repeat for a NOR gate.
4. *True or false:* When NAND gate inputs are tied together, they are always treated as a single load on the signal source.
5. What is power-supply decoupling? Why is it used?

8-7 TTL OPEN-COLLECTOR OUTPUTS

Consider the logic circuit of Figure 8-19(a). NAND gates 4 and 5 provide the AND function, which is ANDing the outputs of NAND gates 1, 2, and 3, so the final output X has the expression

$$X = \overline{AB} \cdot \overline{CD} \cdot \overline{EF}$$

The circuit of Figure 8-19(b) shows the same logic operation obtained by simply tying together the outputs of NAND gates 1, 2, and 3. In other words, the AND operation is performed by tying the outputs together. This can be reasoned as follows: With all outputs tied together, when any one of the gate outputs goes to the LOW state, the common output point must go LOW as a result of the "shorting-to-ground" action of the Q_4 transistor in that gate. The common output point will be HIGH only when all gate outputs are in the HIGH state. Clearly, this is the AND operation.

FIGURE 8-19 Wired-AND operation.

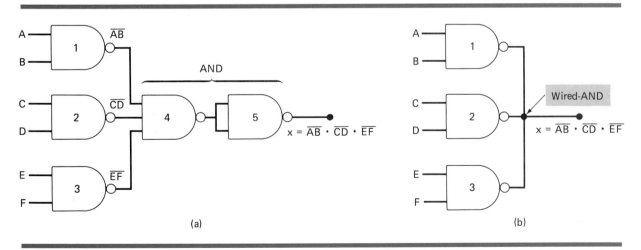

(a)   (b)

The arrangement in Figure 8-19(b) has an advantage over the conventional arrangement of Figure 8-19(a). It requires fewer gates to produce the desired output. This configuration is called the *wired-AND* operation because it produces the AND operation by connecting output wires together. It is sometimes misleadingly called the *wired-OR* operation.

Totem-Pole Outputs Cannot Be Wired-ANDed
In order to take advantage of the wired-AND configuration the outputs of two or more gates must be tied together without harmful effects. Unfortunately, the totem-pole output circuitry of conventional TTL circuits prohibits tying outputs together. This is illustrated in Figure 8-20, where the totem-pole outputs of two separate gates are connected together at point X. Suppose that the gate-A output is in the HIGH state (Q_{3A} ON, Q_{4A} OFF) and the gate-B output is in the LOW state (Q_{3B} OFF, Q_{4B} ON). In this situation Q_{4B} is a very low resistance load on Q_{3A} and will draw a current which can go as high as 55 mA. This current can easily damage Q_{3A} or Q_{4B}. The situation is even worse when more than two TTL outputs are tied together.

Sometimes totem-pole outputs are unintentionally tied together because of wiring errors or accidental shorting on a printed-circuit board. When this happens, the signal at the common point will usually be the AND combination of the two output signals that have been shorted together. This is true only of TTL circuits; MOS and CMOS behave less predictably when outputs are tied together.

FIGURE 8-20 Totem-pole outputs tied together can produce harmful current through Q_4.

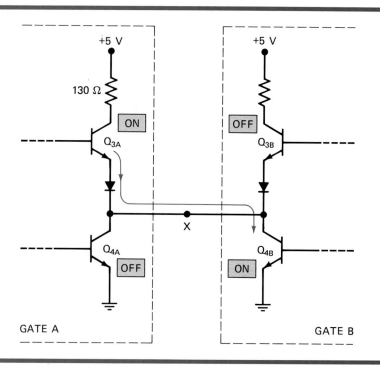

Open-Collector Outputs   To permit wire-ANDing, some TLL circuits are designed with *open-collector* outputs. As shown in Figure 8-21(a) the open-collector-type circuit eliminates Q_3, D_1, and R_4. The output is taken at Q_4's collector, which is unconnected. In the output LOW state, Q_4 is ON (has base current), and in the HIGH state it is OFF (essentially an open circuit). For proper operation an external *pull-up* collector resistor R_P should be connected as shown in Figure 8-21(b), so a high voltage level will appear at the output in the HIGH state. Without R_P, there would be no output voltage when Q_4 is OFF.

With open-collector outputs, the wired-AND operation can be accomplished safely. Figure 8-22 shows three two-input *open-collector* NAND gates (7401s) that are wired-ANDed together. Notice that the open-collector NAND gates have no special symbol. Also notice the presence of the external pull-up resistor (R_P) and the sometimes-used wired-AND symbol.

Value of R_P   The value of R_P must be chosen so that when one gate output goes LOW while the others are HIGH, the sink current through the LOW output does not exceed its I_{OL}(max) limit. A value of $R_P = 1\ k\Omega$ will produce a sink current of about 5 mA through the LOW gate's output transistor. Of course, the output is usually driving other TTL loads that will add to this sink current, whose total must not exceed I_{OL}(max). It might appear that a large value of R_P would therefore be advisable. However, it must be realized that any load capacitance will be charged up through R_P, so that R_P should be made as small as possible to enhance switching

FIGURE 8-21   (a) Open-collector TTL circuit; (b) with external pull-up resistor.

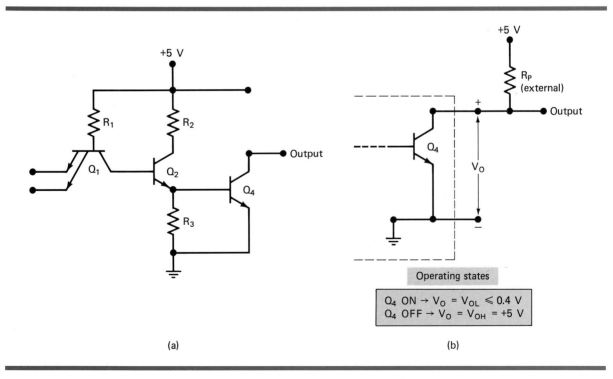

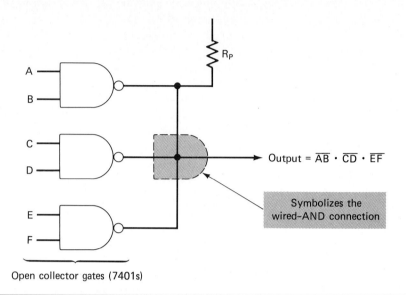

Output = $\overline{AB} \cdot \overline{CD} \cdot \overline{EF}$

Symbolizes the wired-AND connection

Open collector gates (7401s)

FIGURE 8-22   Wired-AND operation using open-collector gates.

speed. Even with R_P minimized, this open-collector arrangement is much slower than totem-pole TTL outputs, where Q_3 acts as a low-impedance emitter-follower to rapidly charge up load capacitance. For this reason, open-collector circuits should not be used in applications where switching speed is a principal consideration.

EXAMPLE 8-16

The 7405 IC (see Appendix II) contains six inverters with open-collector outputs. These six inverters are connected in a wired-AND arrangement in Figure 8-23(a).

(a)   Determine the logic expression for output X.

(b)   Determine a value for R_P assuming that output X is to drive other circuits with a total loading factor of 4 UL.

SOLUTION

(a)   Each inverter output is the inverse of its input. The wired-AND connection simply ANDs each inverter output. Thus,

$$X = \overline{A} \cdot \overline{B} \cdot \overline{C} \cdot \overline{D} \cdot \overline{E} \cdot \overline{F}$$

Using DeMorgan's theorem, this is equivalent to

$$A = \overline{A + B + C + D + E + F}$$

which is the NOR operation.

(b)   If we assume that only one of the INVERTERs has a HIGH input, then the output transistor of that INVERTER has to be able to sink the currents I_{RP} and I_{IL} as shown in Figure 8-23(b). Referring to the data sheet for the 7405, we find that it has a fan-out of 10 in the LOW state. Thus, it can sink a total current $I_{OL}(\text{max}) = 16$ mA. Summing the currents at X, we have

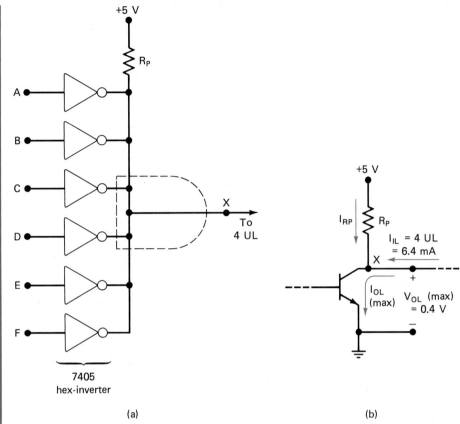

FIGURE 8-23   **Example 8-16**

$$I_{OL}(\text{max}) = I_{RP} + I_{IL}$$
$$16 \text{ mA} = I_{RP} + 6.4 \text{ mA}$$
$$\therefore I_{RC} = 9.6 \text{ mA}$$

The voltage across R_P will be $V_{CC} - V_{OL}(\text{max})$; we use $V_{OL}(\text{max})$ since the transistor is sinking its maximum current. For standard TTL, we know that $V_{OL}(\text{max}) = 0.4$ V. Thus, we have

$$R_P(\text{min}) = \frac{V_{CC} - V_{OL}(\text{max})}{I_{RC}} = \frac{5 \text{ V} - 0.4 \text{ V}}{9.6 \text{ mA}} = 480 \ \Omega$$

In practice, we would use a value slightly larger than this to be on the safe side. Remember, we want to keep R_P as small as possible to avoid increasing the switching time. A good choice would be the standard value of 560 Ω.

EXAMPLE 8-17

A technician breadboards the circuit of Figure 8-23 with $R_P = 10 \text{ k}\Omega$. He figures that since the circuit is operating at a very low frequency, there is no need to worry about the decrease in switching speed produced by increasing R_P. Besides, he wants to minimize the current drawn from V_{CC}.

He tests the circuit and finds that the voltage at point X is slightly less than 2 V in the HIGH state. He replaces the 7405 chip and observes the same result. What is causing V_{CC} to be so low?

SOLUTION

In order for X to go HIGH, each INVERTER input has to be LOW. This will turn OFF the output transistor of each open-collector output. Unfortunately, these transistors are not ideal, and they do have some leakage current in the OFF state. This leakage current, I_{OH}, is specified as a *maximum* of 250 μA for the standard TTL series, and 100 μA for LS-TTL. Under *normal* conditions it will usually be about 50 μA for TTL and 20 μA for LS-TTL. These leakage currents, however, are additive, so the total leakage current of the six output transistors can be substantial. This current has to be supplied by V_{CC} through R_P. In addition V_{CC} has to supply the I_{IH} for each of the four ULs. As a result, there may typically be anywhere from 300 to 400 μA flowing from V_{CC} through R_P. With $R_P = 10$ kΩ, it is not surprising that around 3 V is dropped across R_P, leaving only about 2 V at point X.

Because the open-collector leakage currents and the I_{IH} of the loads have to flow through R_P, it is best to keep R_P as close as possible to the minimum value that is calculated based on the open-collector output's maximum sink current, $I_{OL}(\text{max})$.

Open-Collector Buffers/Drivers Any logic circuit that is called a *buffer, a driver,* or a *buffer/driver* is designed to have a greater output current and/or voltage capability than an ordinary logic circuit. Buffer/driver ICs are available with totem-pole outputs and with open-collector outputs.

The 7406 is a popular open-collector buffer/driver IC that contains six IN-VERTERs with open-collector outputs that can sink up to 40 mA in the LOW state. In addition, the 7406 can handle output voltages up to 30 V. This means that the output transistor can be connected to a voltage greater than 5 V.

This is illustrated in Figure 8-24, where a 7406 is used as a buffer between a 74107 FF and an incandescent indicator lamp that is rated at 24 V, 25 mA. The 7406 controls the lamp's ON/OFF status to indicate the state of FF output Q. Note that the lamp is powered from +24 V, and it acts as the pull-up resistor for the open-collector output.

When $Q = 1$, the 7406 output goes LOW and its output transistor sinks the 25 mA of lamp current supplied by the 24-V source, and the lamp is ON. When $Q = 0$, the 7406 output transistor turns OFF; there is no path for current and the lamp turns OFF. In this state, the full 24 V will appear across the OFF output transistor so that $V_{OH} = 24$ V, which is lower than the 7406 maximum V_{OH} rating.

Open-collector outputs are often used to drive indicator LEDs as shown in Figure 8-25. Here either a 7405 or a 7406 can be used, depending on the LED's required current. The resistor is used to limit the current to a safe value. When the INVERTER output is LOW, its output transistor will provide a low-resistance path to ground for the LED current, so the LED will be ON. When the INVERTER output is HIGH, its output transistor will be OFF and there will be no path for LED current; in this state, the LED will be OFF.

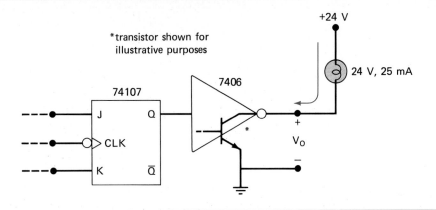

*transistor shown for
illustrative purposes

FIGURE 8-24 An open-collector buffer-driver drives a high-current, high-voltage load.

IEEE/ANSI Symbol for Open-Collector Outputs

The traditional symbols for logic circuits with open-collector outputs are the same as for totem-pole outputs. The new IEEE/ANSI symbology, however, does use a distinctive notation to identify open-collector outputs. Figure 8-26 shows the standard IEEE/ANSI designation for an open-collector output. It is an underlined diamond. Although we will not normally use the complete IEEE/ANSI symbology in this book, we will use this underlined diamond to indicate open-collector outputs.

FIGURE 8-25 An open-collector output can be used to drive an LED indicator.

FIGURE 8-26 IEEE/ANSI notation for open-collector outputs.

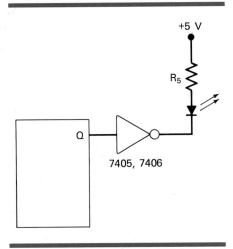

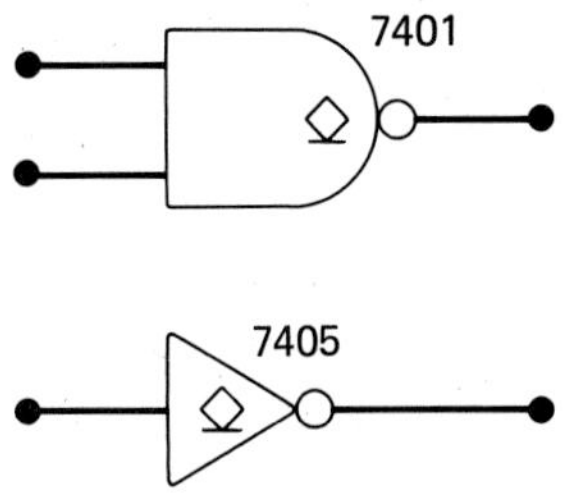

REVIEW QUESTIONS

1. What is wired-ANDing?
2. Why shouldn't logic devices with totem-pole outputs be wired-ANDed?
3. How do open-collector outputs differ from totem-pole outputs?
4. Why do open-collector outputs need a pull-up resistor?
5. What factors are involved in determining a value for the pull-up resistor?
6. Why are open-collector outputs generally slower than totem-pole?
7. What is the IEEE/ANSI symbol for open-collector outputs?

8-8 TRISTATE (3-STATE) TTL

This is a third type of TTL output configuration. It utilizes the high-speed operation of the totem-pole arrangement while permitting outputs to be wire-ANDed (connected together). It is called tristate TTL because it allows *three* possible output states: HIGH, LOW, and high-impedance (Hi-Z). The Hi-Z state is a condition where both transistors in the totem-pole arrangement are turned OFF so that the output terminal is a high impedance to ground and to V_{CC}. In other words, the output is an open or floating terminal that is neither a LOW nor a HIGH. In practice, the output terminal is not an exact open circuit, but has a resistance of several megohms or more relative to ground and V_{CC}.

The tristate operation is obtained by modifying the basic totem-pole circuit. Figure 8-27(a) shows the circuit for a tristate *INVERTER* where the portion enclosed in dotted lines has been added to the basic circuit. The circuit has two inputs: A is the normal logic input, E is an ENABLE input that can produce the Hi-Z state. We will examine the operation for both states of E.

The Enabled State   With $E = 1$ the circuit operates as a normal INVERTER because the HIGH voltage at E has no effect on Q_1 or D_2. In this enabled condition, the output is simply the inverse of logic input A.

The Disabled State (Hi-Z)   When $E = 0$ the circuit goes into its Hi-Z state regardless of the state of logic input A. The LOW at E forward-biases the emitter–base junction of Q_1 and shunts the R_1 current away from Q_2 so that Q_2 turns OFF, which turns Q_4 OFF. The LOW at E also forward-biases diode D_2 to shunt current away from the base of Q_3, so that Q_3 also turns OFF.

With both totem-pole transistors in the nonconducting state, the output terminal is essentially an open circuit. This is shown symbolically in the table of Figure 8-27(c).

The logic symbol for the tristate INVERTER is shown in Figure 8-27(b). Note where the ENABLE input is placed on the INVERTER symbol. Also note that E is active-HIGH; that is, the INVERTER is enabled when $E = 1$.

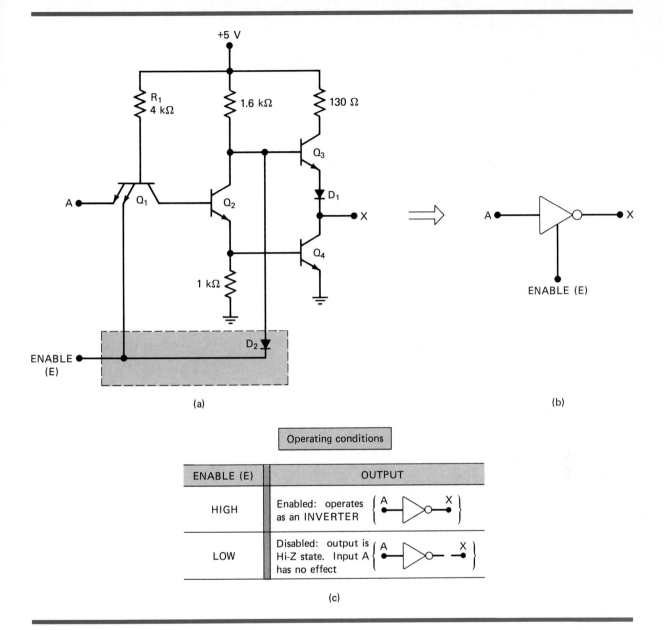

+5 V

R₁ 4 kΩ   1.6 kΩ   130 Ω

Q₃

A   Q₁   Q₂   D₁

X

Q₄

1 kΩ

D₂

ENABLE (E)

(a)

A

X

ENABLE (E)

(b)

Operating conditions

ENABLE (E)	OUTPUT
HIGH	Enabled: operates as an INVERTER A ▷○ X
LOW	Disabled: output is Hi-Z state. Input A has no effect A ▷○ X

(c)

FIGURE 8-27   Tristate TTL INVERTER.

Advantage of Tristate   The outputs of tristate ICs can be connected together (paralleled) without sacrificing switching speed. This is because a tristate output, when enabled, operates as a totem-pole output with its associated low-impedance, high-speed characteristic. It is important to realize, however, that when tristate outputs are paralleled, only one of them should be enabled at one time. Otherwise, two active totem-pole outputs would be connected together and damaging currents could flow (Figure 8-20). We will elaborate on this in our discussion of tristate buffers.

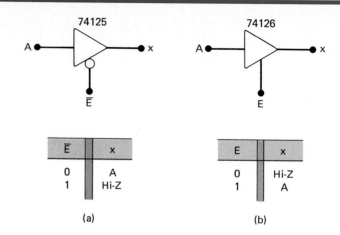

FIGURE 8-28 Tristate buffers.

Tristate Buffers

Tristate Buffers A *tristate buffer* is a circuit that is used to control the passage of a logic signal from input to output. Some tristate buffers also invert the signal as it goes through. The circuit in Figure 8-27 can be called an *inverting tristate buffer*.

Two of the most commonly used tristate buffer ICs are the 74125 and 74126. Both contain four *noninverting* tristate buffers like those shown in Figure 8-28. The

FIGURE 8-29 (a) Tristate buffers used to connect several signals to a common bus; (b) conditions for transmitting B to the bus.

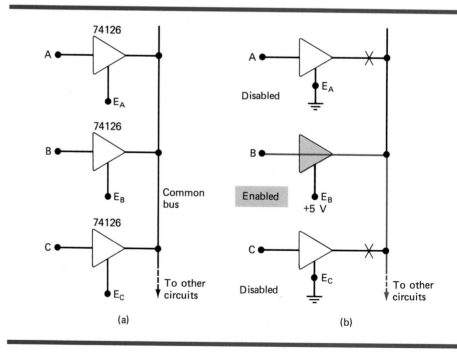

74125 and 74126 differ only in the active state of their ENABLE inputs. The 74125 allows the input signal A to reach the output when $E = 0$, while the 74126 passes the input when $E = 1$.

Tristate buffers have many applications in circuits where several signals are connected to common lines (buses). We will examine some of these applications in Chapter 9, but we can get the basic idea from Figure 8-29(a). Here we have three logic signals A, B, and C connected to a common bus line through 74126 tristate buffers. This arrangement permits us to transmit any one of these signals over the bus line to other circuits by enabling the appropriate buffer.

For example, consider the situation in Figure 8-29(b) where $E_B = 1$ and $E_A = E_C = 0$. This disables the upper and lower buffers so that their outputs are in the Hi-Z state and are essentially disconnected from the bus. This is symbolized by the X's on the diagram. The middle buffer is enabled so that its input, B, is passed through to its output and onto the bus, from where it is routed to other circuits connected to the bus. When tristate outputs are connected together as in Figure 8-29, it is important to remember that no more than one output should be enabled at one time. Otherwise, two or more active totem-pole outputs would be connected, which could produce damaging currents. Even if damage did not occur, this situation would produce a signal on the bus that is a combination of more than one signal. This is commonly referred to as *bus contention*. In tristate bus systems the designer has to make sure that the enable signals do not allow bus contention to occur.

Tristate ICs   In addition to tristate buffers, there are many ICs that are designed with tristate outputs. For example, the 74LS374 is an octal D-type FF register IC with tristate outputs. This means that it is an 8-bit register made up of D-type FFs whose outputs are connected to tristate buffers. This type of register can be connected to common bus lines along with the outputs from other similar devices to allow efficient transfer of data over the bus. We will examine this *tristate data bus* arrangement in Chapter 9. Other types of logic devices that are available with tristate outputs include decoders, multiplexers, analog-to-digital converters, memory chips, and microprocessors.

IEEE/ANSI Symbol for Tristate Outputs   The traditional logic symbology has no special notation for tristate ouputs. Figure 8-30 shows the notation used in the IEEE/ANSI symbology to indicate a tristate output. It is a triangle that points downward. Although it is not part of the traditional symbology, we will use this triangle to designate tristate outputs throughout the remainder of the book.

FIGURE 8-30   **IEEE/ANSI notation for tristate outputs.**

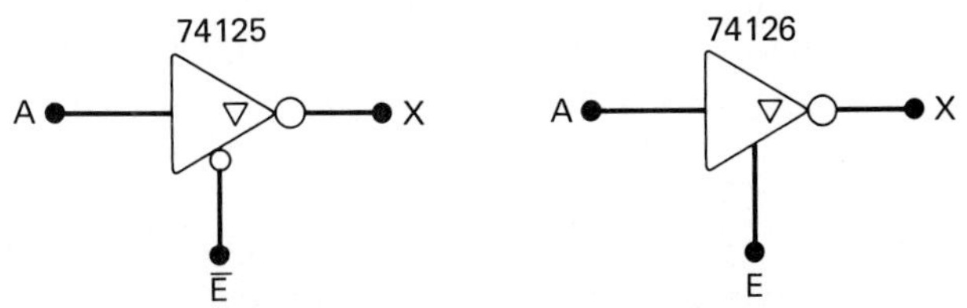

REVIEW QUESTIONS

1. What are the three possible output states of a tristate IC?
2. What is the state of a tristate output when it is disabled?
3. What is bus contention?
4. What conditions are necessary to transmit signal C onto the bus in Figure 8-29? (*Ans*, $E_A = E_B = 0$, $E_C = 1$)
5. What is the IEEE/ANSI designation for tristate outputs?

8-9 THE ECL DIGITAL IC FAMILY

The TTL family uses transistors operating in the saturated mode. As a result, their switching speed is limited by the storage delay time associated with a transistor that is driven into saturation. Another *bipolar* logic family has been developed that prevents transistor saturation, thereby increasing overall switching speed. This logic family is called *emitter-coupled logic* (ECL), and it operates on the principle of current switching whereby a fixed bias current less than $I_{C(sat)}$ is switched from one transistor's collector to another. Because of this current-mode operation, this logic form is also referred to as *current-mode logic* (CML).

Basic ECL Circuit   The basic circuit for emitter-coupled logic is essentially the differential amplifier configuration of Figure 8-31(a). The V_{EE} supply produces an essentially fixed current I_E which remains around 3 mA during normal operation. This current is allowed to flow through either Q_1 or Q_2, depending on the voltage level at V_{IN}. In other words, this current will switch between Q_1's collector and Q_2's collector as V_{IN} switches between its two logic levels of -1.7 V (logical 0 for ECL) and -0.8 V (logical 1 for ECL). The table in Figure 8-31(a) shows the resulting output voltages for these two conditions at V_{IN}. Two important points should be noted: (1) V_{C1} and V_{C2} are the *complements* of each other, and (2) the output voltage levels are not the same as the input logic levels.

The second point noted above is easily taken care of by connecting V_{C1} and V_{C2} to emitter-follower stages (Q_3 and Q_4), as shown in Figure 8-31(b). The emitter-followers perform two functions: (1) they subtract approximately 0.8 V from V_{C1} and V_{C2} to shift the output levels to the correct ECL logic levels, and (2) they provide a very low output impedance (typically 7 Ω), which provides for large fanout and fast charging of load capacitance. This circuit produces two complementary outputs: V_{OUT1}, which equals $\overline{V_{IN}}$, and V_{OUT2}, which is equal to V_{IN}.

ECL OR/NOR Gate   The basic ECL circuit of Figure 8-31(b) can be used as an inverter if the output is taken at V_{OUT1}. This basic circuit can be expanded to more than one input by paralleling transistor Q_1 with other transistors for the other inputs, as in Figure 8-32(a). Here either Q_1 or Q_3 can cause the current to be switched out of Q_2, resulting in the two outputs V_{OUT1} and V_{OUT2} being the logical NOR and OR operations, respectively. This OR/NOR gate is symbolized in Figure 8-32(b) and is the fundamental ECL gate.

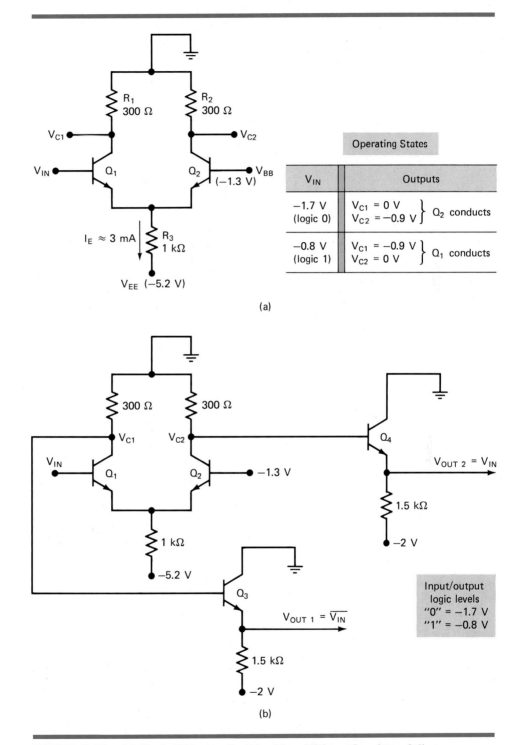

The table within figure (a):

	Operating States	
V_{IN}		Outputs
−1.7 V (logic 0)	$V_{C1} = 0$ V $V_{C2} = −0.9$ V	Q_2 conducts
−0.8 V (logic 1)	$V_{C1} = −0.9$ V $V_{C2} = 0$ V	Q_1 conducts

Input/output logic levels
"0" = −1.7 V
"1" = −0.8 V

FIGURE 8-31 (a) Basic ECL circuit; (b) with addition of emitter-followers.

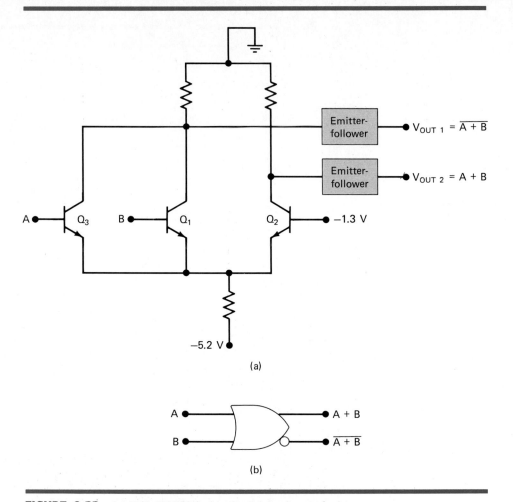

FIGURE 8-32   (a) ECL NOR/OR circuit; (b) logic symbol.

ECL Characteristics   The following are the most important characteristics of the ECL family of logic circuits:

1.   The transistors never saturate, so switching speed is very high. Typical propagation delay time is 1 ns, which makes ECL a little faster than advanced Schottky TTL (74AS series).

2.   The logic levels are nominally -0.8 V and -1.70 V for the logical 1 and 0, respectively.

3.   Worst-case ECL noise margins are approximately 250 mV. These low noise margins make ECL somewhat unreliable for use in heavy industrial environments.

4.   An ECL logic block usually produces an output and its complement. This eliminates the need for inverters.

5.   Fan-outs are typically around 25, owing to the low-impedance emitter-follower outputs.

6. Typical power dissipation for a basic ECL gate is 40 mW, somewhat higher than the 74AS series.

7. The total current flow in an ECL circuit remains relatively constant regardless of its logic state. This helps to maintain an unvarying current drain on the circuit power supply even during switching transitions. Thus, no noise spikes will be internally generated like those produced by TTL totem-pole circuits.

Table 8-5 shows how ECL compares to the important TTL logic families.

TABLE 8-5

LOGIC FAMILY	t_{pd} (ns)	P_D (mW)	WORST-CASE NOISE MARGIN (mV)	MAXIMUM CLOCK RATE (MHz)
74	9	10	400	35
74AS	1.7	8	300	200
74ALS	4	1.2	400	70
74S	3	20	300	125
74LS	9.5	2	300	45
ECL	1	40	250	300

The ECL family is not as widely used as the TTL and MOS families except in very high frequency applications where its speed is superior. Its relatively low noise margins and high power drain are disadvantages compared to other logic families. Another drawback is its negative supply voltage and logic levels, which are not compatible with the other logic families; this makes it difficult to use ECL in conjunction with TTL and MOS circuits.

REVIEW QUESTION

1. *True or false:*
   (a) ECL obtains high-speed operation by preventing transistor saturation.
   (b) ECL circuits have two outputs.
   (c) The noise margins for ECL circuits are larger than TTL noise margins.
   (d) ECL circuits do not generate noise spikes during state transitions.
   (e) ECL devices require less power than standard TTL.

8-10   MOS DIGITAL INTEGRATED CIRCUITS

MOS (metal-oxide-semiconductor) technology derives its name from the basic MOS structure of a metal electrode over an oxide insulator over a semiconductor substrate. The transistors of MOS technology are field-effect transistors called MOSFETs. Most of the MOS digital ICs are constructed entirely of MOSFETs and no other components.

The chief advantages of the MOSFET are that it is relatively simple and inexpensive to fabricate, it is small, and it consumes very little power. The fabrication of MOS ICs is approximately one-third as complex as the fabrication of bipolar ICs (TTL, ECL, etc.). In addition, MOS devices occupy much less space on a chip than bipolar transistors; typically, a MOSFET requires 1 square mil of chip area while a bipolar transistor requires about 50 square mils. More importantly, MOS digital ICs normally do not use the IC resistor elements, which take up so much of the chip area of bipolar ICs.

All of this means that MOS ICs can accommodate a much larger number of circuit elements on a single chip than bipolar ICs. This advantage is evidenced by the fact that MOS ICs have dominated bipolar ICs in the area of large-scale integration (LSI, VLSI). The high packing density of MOS ICs results in a greater system reliability because of the reduction in the number of necessary external connections.

The principal disadvantage of MOS ICs is their relatively slow operating speed when compared to the bipolar IC families. In many applications this is not a prime consideration, so MOS logic offers an often superior alternative to bipolar logic. We will examine the MOS logic families after a brief discussion of MOSFETs.

8-11 THE MOSFET

There are presently two general types of MOSFETs: *depletion* and *enhancement*. MOS digital ICs use enhancement MOSFETs exclusively, so only this type will be considered in the following discussion. Furthermore, we will concern ourselves only with the operation of these MOSFETs as ON/OFF switches.

Figure 8-33 shows the schematic symbols for the *N*-channel and *P*-channel enhancement MOSFETs, where the direction of the arrow indicates either *P*- or *N*-channel. The symbols show a broken line between the *source* and *drain* to indicate that there is *normally* no conducting channel between these electrodes. The symbol also shows a separation between the *gate* and the other terminals to indicate the very high resistance (typically greater than 10,000 MΩ) between the gate and channel.

FIGURE 8-33 **Schematic symbols for enhancement MOSFETs.**

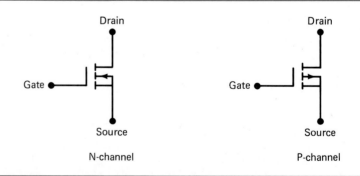

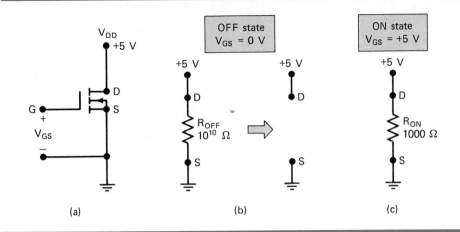

(a)   (b)   (c)

FIGURE 8-34   *N*-channel MOSFET switching states.

Basic MOSFET Switch   Figure 8-34 shows the switching operation of an *N*-channel *MOSFET*. For the *N*-channel device the drain is always biased positive relative to the source. The gate-to-source voltage V_{GS} is the input voltage, which is used to control the resistance between drain and source (i.e., the channel resistance) and therefore determines whether the device is ON or OFF.

When $V_{GS} = 0$ V, there is no conductive channel between source and drain, and the device is OFF. Typically the channel resistance in this OFF state is 10^{10} Ω, which for most purposes is an *open circuit*. The MOSFET will remain OFF as long as V_{GS} is zero or negative. As V_{GS} is made positive (gate positive relative to source), a threshold voltage (V_T) is reached, at which point a conductive channel begins to form between source and drain. Typically $V_T = +1.5$ V for N-MOSFET, so any $V_{GS} \geq 1.5$ V will cause the MOSFET to conduct. Generally, a value of V_{GS} much larger than V_T is used to turn ON the MOSFET more completely. As shown in Figure 8-34(b), when $V_{GS} = +5$ V, the channel resistance between source and drain has dropped to a value of $R_{ON} = 1000$ Ω.

Table 8-6 summarizes the *P*- and *N*-channel switching characteristics.

In essence, then, the N-MOSFET will switch from a very high resistance to a low resistance as the gate voltage switches from a LOW voltage to a HIGH voltage. It is helpful simply to think of the MOSFET as a switch that is either opened or closed between source and drain.

TABLE 8-6

	DRAIN-TO-SOURCE BIAS	GATE-TO-SOURCE VOLTAGE (V_{GS}) NEEDED FOR CONDUCTION	R_{ON} (Ω)	R_{OFF} (Ω)
P-channel	Negative	Typically more negative than -1.5 V	1000 (typical)	10^{10}
N-channel	Positive	Typically more positive than $+1.5$ V	1000 (typical)	10^{10}

The *P*-channel MOSFET operates in exactly the same manner as the *N*-channel except that it uses voltages of the opposite polarity. For P-MOSFETs the drain is connected to $-V_{DD}$ so that it is biased negative relative to the source. To turn the P-MOSFET ON, a negative voltage that exceeds V_T must be applied to the gate.

8-12 DIGITAL MOSFET CIRCUITS

Digital circuits employing MOSFETs are broken down into three categories: (1) P-MOS, which uses *only* P-channel enhancement MOSFETs; (2) N-MOS, which uses *only* N-channel enhancement MOSFETs; and (3) CMOS (complementary MOS), which uses both *P*- and *N*-channel devices.

P-MOS and N-MOS digital ICs have a greater packing density (more transistors per chip) and are therefore more economical than CMOS. N-MOS has about twice the packing density of P-MOS. In addition, N-MOS is also about twice as fast as P-MOS, owing to the fact that free electrons are the current carrier in N-MOS while holes (slower-moving positive charges) are the current carriers for P-MOS. CMOS has the greatest complexity and lowest packing density of the MOS families, but it possesses the important advantages of higher speed and much lower power dissipation.

In this section we will look at some of the basic N-MOS logic circuits, keeping in mind that the P-MOS circuits would be the same except for the voltage polarities. Since P-MOS and N-MOS find their widest applications in LSI and VLSI (microprocessors, memories, ROMs, etc.), we will defer any applications of these families until later. CMOS, which, like TTL, is widely used in MSI applications, will be covered in more detail beginning with Section 8-14.

FIGURE 8-35   N-MOS INVERTER.

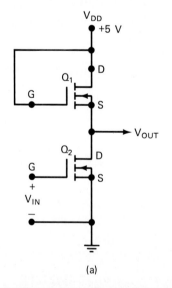

V_{IN}	Q_1	Q_2	$V_{OUT} = \overline{V_{IN}}$
0 V (logic 0)	$R_{ON} =$ 100 kΩ	$R_{OFF} =$ 10^{10} Ω	+5 V (logic 1)
+5 V (logic 1)	$R_{ON} =$ 100 kΩ	$R_{ON} =$ 1 kΩ	+0.05 V (logic 0)

(a)   (b)

N-MOS INVERTER   Figure 8-35 shows the basic N-MOS INVERTER circuit. It contains two N-channel MOSFETs: Q_1 is called the *load* MOSFET and Q_2 the *switching* MOSFET. Q_1 has its gate *permanently* connected to $+5$ V, so it is *always* in the ON state and essentially acts as a load resistor of value R_{ON}. Q_2 will switch from ON to OFF in response to V_{IN}. The Q_1 MOSFET is designed to have a narrower channel than Q_2, so Q_1's R_{ON} is much greater than Q_2's. Typically, R_{ON} for Q_1 is 100 kΩ and R_{ON} for Q_2 is 1 kΩ. R_{OFF} for Q_2 is usually around 10^{10} Ω.

The two states of the INVERTER are summarized in Figure 8-35(b). The best way to analyze this circuit is to consider each MOSFET channel as a resistance so that the output voltage is taken from a voltage divider formed by the two resistances. With $V_{IN} = 0$ V, transistor Q_2 is OFF, with a very large resistance of 10^{10} Ω. Since Q_1 has $R_{ON} = 100$ kΩ, the voltage-divider output will be essentially $+5$ V.

FIGURE 8-36   **(a) N-MOS NAND gate; (b) NOR gate.**

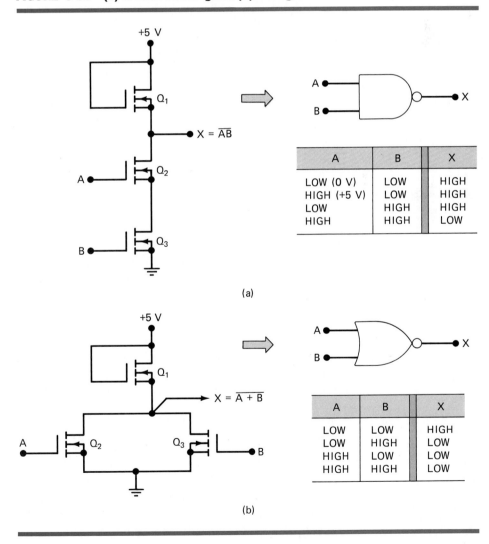

(a)

A	B	X
LOW (0 V)	LOW	HIGH
HIGH (+5 V)	LOW	HIGH
LOW	HIGH	HIGH
HIGH	HIGH	LOW

(b)

A	B	X
LOW	LOW	HIGH
LOW	HIGH	LOW
HIGH	LOW	LOW
HIGH	HIGH	LOW

With $V_{IN} = +5$ V, Q_2 is ON, with $R_{ON} = 1$ kΩ. The voltage divider is now 100 kΩ and 1 kΩ, so $V_{OUT} = 1/101 \times (+5$ V$) \approx 0.05$ V.

The circuit functions as an INVERTER since a LOW input produces a HIGH output and vice versa. This basic INVERTER can be modified to form NAND and NOR logic gates.

N-MOS NAND Gate

The NAND operation is performed by the circuit of Figure 8-36(a) (see page 445), where Q_1 is again acting as a load resistance while Q_2 and Q_3 are switches controlled by input levels A and B. If either A or B is at 0 V (logical 0), the corresponding FET is OFF, thereby presenting a high resistance from the output terminal to ground so that output X is HIGH ($+5$ V). When both A and B are $+5$ V (logical 1), both Q_2 and Q_3 are ON, so output X is LOW. Clearly, the output equals the NAND of the inputs ($x = \overline{AB}$).

N-MOS NOR Gate

The NOR gate of Figure 8-36(b) uses Q_2 and Q_3 as parallel switches with Q_1 again acting as a load resistance. When either input A or B is at $+5$ V, the corresponding MOSFET is ON, forcing the output to be LOW. When both inputs are at 0 V, both Q_2 and Q_3 are OFF, so the output goes HIGH. Clearly, this is the NOR operation with $X = \overline{A + B}$.

N-MOS OR gates and AND gates are easily formed by combining the NOR or NAND with inverters.

N-MOS Flip-Flops

Flip-flops are formed by using two cross-coupled NOR gates or NAND gates. The MOS FF is very important in MOS memories, which will be discussed later.

8-13 CHARACTERISTICS OF MOS LOGIC

Compared to the bipolar logic families the MOS logic families are slower in operating speed, require much less power, have a better noise margin, a greater supply voltage range, and a higher fan-out, and, as was mentioned earlier, require much less "real estate" (chip area).

Operating Speed

A typical N-MOS NAND gate has a propagation delay time of 50 ns. This is due to *two* factors: the relatively high output resistance (100 kΩ) in the HIGH state and the capacitive loading presented by the inputs of the logic circuits being driven. MOS logic inputs have very high input resistance ($> 10^{12}$ Ω), and they have a reasonably high gate capacitance (MOS capacitor), typically 2 to 5 picofarads. This combination of large R_{OUT} and large C_{LOAD} serves to increase switching time.

Noise Margin

Typically N-MOS noise margins are around 1.5 V when operated from $V_{DD} = 5$ V and will be proportionally higher for larger values of V_{DD}.

Fan-Out

Because of the extremely high input resistance at each MOSFET input, one would expect that the fan-out capabilities of MOS logic would be virtually unlimited. This is essentially true for dc or low-frequency operation. However, for frequencies greater than around 100 kHz, the gate input capacitances cause a deterioration in switching time which increases in proportion to the number of loads being driven. Even so, MOS logic can easily operate at a fan-out of 50, which is somewhat better than most of the bipolar families.

Power Drain   MOS logic circuits draw small amounts of power because of the relatively large resistances being used. To illustrate, we can calculate the power dissipation of the INVERTER of Figure 8-35 for its two operating states.

1. $V_{IN} = 0$ V: $R_{ON(Q_1)} = 100$ kΩ; $R_{OFF(Q_2)} = 10^{10}$ Ω. Therefore, I_D, current from V_{DD} supply, ≈ 0.05 nA, and $P_D = 5$ V $\times$ 0.05 nA $= 0.25$ nW.

2. $V_{IN} = +5$ V: $R_{ON(Q_1)} = 100$ kΩ; $R_{ON(Q_2)} = 1$ kΩ. Therefore, $I_D = 5$ V/101 k$\Omega \approx 50$ μA and $P_D = 5$ V $\times$ 50 $\mu A = 0.25$ mW.

This gives an *average* P_D of a little over 0.1 mW for the INVERTER. The low power drain of MOS logic makes it suitable for LSI and VLSI where many gates, FFs, etc., can be on one chip without causing overheating that can damage the chip.

Process Complexity   MOS logic is the simplest logic family to fabricate since it uses only one basic element, an N-MOS (or P-MOS) transistor. It requires no other elements, such as resistors, diodes, etc. This characteristic, together with its lower P_D, makes it ideally suited for LSI (large memories, calculator chips, microprocessors) and this is where MOS logic has made its greatest impact in the digital field. The operating speed of P-MOS and N-MOS is not comparable with TTL, so very little has been done with them in SSI and MSI applications. In fact, there are very few MOS logic circuits in the SSI or MSI categories (gates, FFs, counters, etc.) CMOS, however, is competitive in the MSI area, which was until recently dominated by TTL.

REVIEW QUESTIONS

1.   Describe the relative advantages and disadvantages of N-MOS circuits as compared to TTL.
2.   What determines the fan-out limitations of MOS logic?
3.   What factors limit the switching speed of N-MOS circuits?
4.   What makes MOS circuitry so well suited for LSI?

8-14   COMPLEMENTARY MOS LOGIC

The *complementary MOS* (CMOS) logic family uses *both P-* and *N*-channel MOS-FETs in the same circuit to realize several advantages over the P-MOS and N-MOS

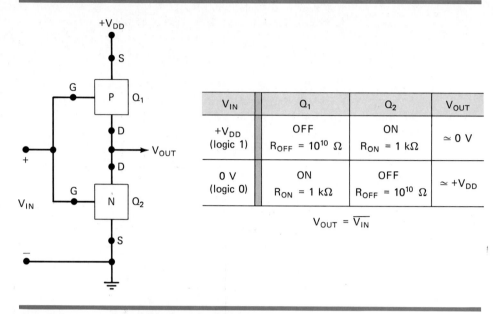

V_{IN}	Q_1	Q_2	V_{OUT}
$+V_{DD}$ (logic 1)	OFF $R_{OFF} = 10^{10}$ Ω	ON $R_{ON} = 1$ kΩ	$\simeq 0$ V
0 V (logic 0)	ON $R_{ON} = 1$ kΩ	OFF $R_{OFF} = 10^{10}$ Ω	$\simeq +V_{DD}$

$$V_{OUT} = \overline{V_{IN}}$$

FIGURE 8-37   Basic CMOS INVERTER.

families. Generally speaking, CMOS is faster and consumes even less power than the other MOS families. These advantages are offset somewhat by the increased complexity of the IC fabrication process and a lower packing density. Thus, CMOS cannot yet hope to compete with MOS in applications requiring the utmost in LSI.

However, CMOS logic has undergone a constant growth in the MSI area, mostly at the expense of TTL, with which it is directly competitive. The CMOS fabrication process is simpler than TTL and has a greater packing density, therefore permitting more circuitry in a given area and reducing the cost per function. CMOS uses only a fraction of the power needed for even the low-power TTL series (74L) and is thus ideally suited for applications using battery power or battery back-up power. CMOS is generally slower than TTL, although the new high-speed CMOS series can compete with the 74 and 74LS series.

CMOS Inverter   The circuitry for the basic CMOS INVERTER is shown in Figure 8-37. For this diagram and those that follow, the standard symbols for the MOSFETs have been replaced by blocks labeled P and N to denote a P-MOSFET and N-MOSFET, respectively. This is done simply for convenience in analyzing the circuits. The CMOS INVERTER has two MOSFETs in series such that the P-channel device has its source connected to $+V_{DD}$ (a positive voltage) and the N-channel device has its source connected to ground.* The gates of the two devices are connected together as a common input. The drains of the two devices are connected together as the common output.

*Most manufacturers label this terminal as V_{SS}.

The logic levels for CMOS are essentially $+V_{DD}$ for logical 1 and 0 V for logical 0. Consider, first, the case where $V_{IN} = +V_{DD}$. In this situation the gate of Q_1 (P-channel) is at 0 V relative to the source of Q_1. Thus, Q_1 will be in the OFF state with $R_{OFF} \approx 10^{10}$ Ω. The gate of Q_2 (N-channel) will be at $+V_{DD}$ relative to its source. Thus, Q_2 will be ON with typically $R_{ON} = 1$ kΩ. The voltage divider between Q_1's R_{OFF} and Q_2's R_{ON} will produce $V_{OUT} \approx 0$ V.

Next, consider the case where $V_{IN} = 0$ V. Q_1 now has its gate at a negative potential relative to its source while Q_2 has $V_{GS} = 0$ V. Thus, Q_1 will be ON with $R_{ON} = 1$ kΩ and Q_2 OFF with $R_{OFF} = 10^{10}$ Ω, producing a V_{OUT} of approximately $+V_{DD}$. These two operating states are summarized in Figure 8-37(b), showing that the circuit does act as a logic INVERTER.

CMOS NAND Gate Other logic functions can be constructed by modifying the basic INVERTER. Figure 8-38 shows a NAND gate formed by adding a parallel P-channel MOSFET and a series N-channel MOSFET to the basic INVERTER. To analyze this circuit it helps to realize that a 0-V input turns ON its corresponding P-MOSFET and turns OFF its corresponding N-MOSFET, and vice versa for a $+V_{DD}$ input. Thus, it can be seen that the only time a LOW output will occur is

FIGURE 8-38 **CMOS NAND gate.**

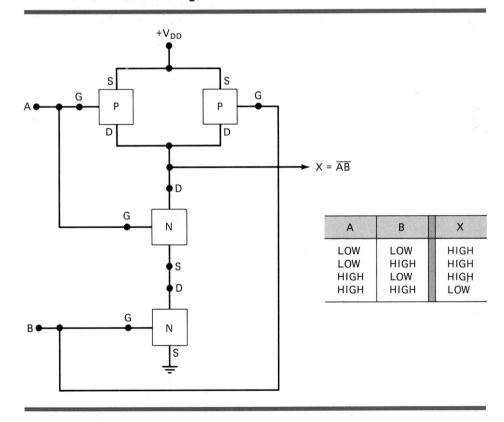

A	B	X
LOW	LOW	HIGH
LOW	HIGH	HIGH
HIGH	LOW	HIGH
HIGH	HIGH	LOW

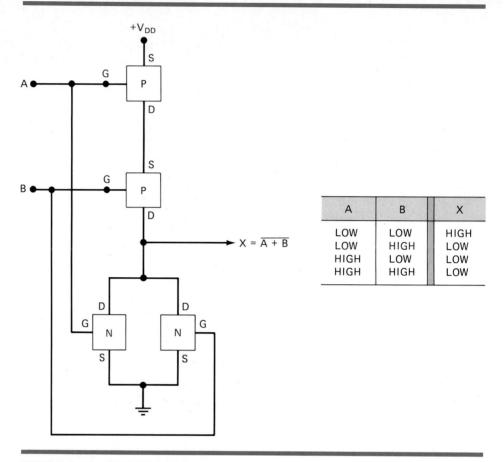

A	B	X
LOW	LOW	HIGH
LOW	HIGH	LOW
HIGH	LOW	LOW
HIGH	HIGH	LOW

FIGURE 8-39   **CMOS NOR gate.**

when inputs A and B are both HIGH ($+V_{DD}$) to turn ON both N-MOSFETs, thereby providing a low resistance from the output terminal to ground. For all other input conditions, at least one P-MOSFET will be ON while at least one N-MOSFET will be OFF. This produces a HIGH output.

CMOS NOR Gate   A CMOS NOR gate is formed by adding a series P-MOSFET and a parallel N-MOSFET to the basic inverter as shown in Figure 8-39. Once again this circuit can be analyzed by realizing that a LOW at any input turns ON its corresponding P-MOSFET and turns OFF its corresponding N-MOSFET, and vice versa for a HIGH input. It is left to the reader to verify that this circuit operates as a NOR gate.

CMOS AND and OR gates can be formed by combining NANDs and NORs with inverters.

CMOS SET-CLEAR FF   Two CMOS NOR gates or NAND gates can be cross-coupled to form a simple SET-CLEAR FF. Additional gating circuitry is used to convert the basic SET-CLEAR FF to clocked D and J-K FFs.

8-15 CMOS SERIES CHARACTERISTICS

4000 Series   There are several different series in the CMOS family of digital ICs. The 4000 series, which was introduced by RCA, was the first CMOS series. The original 4000 series is the 4000A series; an improved version is the 4000B series, which has higher output current capabilities. Both are still widely used despite the emergence of the new CMOS series. The 4000 series has been around much longer and has many functions not yet available in the newer series.

74C Series   This CMOS series is pin-for-pin and function-for-function compatible with TTL devices having the same number. For example, a 74C74 is a dual edge-triggered D FF that has the same pin configuration as the TTL 7474 dual edge-triggered D FF IC. Many, but not all functions that are available in TTL are also available in this CMOS series. This makes it possible to replace some TTL circuits by an equivalent CMOS design. The performance characteristics of the 74C series are about the same as the 4000 series.

74HC Series (High-Speed CMOS)   This is an improved version of the 74C series. The main improvement is about a tenfold increase in switching speed. The speed of devices in this series is comparable to that of the 74LS TTL series. Another improvement is its higher output current capability.

74HCT Series   This is also a high-speed CMOS series. The major difference between this series and the 74HC is that it is designed to be voltage-compatible with TTL devices. In other words, it can be directly driven by a TTL output. As we shall see, this is not true for the other CMOS series.
   The following paragraphs will discuss some of the operating and performance characteristics that are common to all CMOS devices.

Power-Supply Voltage   The 4000 series and 74C series will operate with V_{DD} values ranging from 3 to 15 V, so that power supply regulation is not critical. The 74HC and 74HCT series can operate with a supply voltage range of 2 to 6 V. When CMOS and TTL are being used together, the supply voltage is usually made 5 V so that a single 5-V supply can serve as both the V_{DD} for CMOS and the V_{CC} for TTL. In those situations where the CMOS devices are operated at a supply voltage that is greater than 5 V, special steps have to be taken to permit the CMOS and TTL devices to work together. We will examine these steps later.

Voltage Levels   When CMOS outputs drive only CMOS inputs, the output voltage levels will be very close to 0 V for the LOW state, and $+V_{DD}$ for the HIGH state. This is because the very high CMOS input resistance draws very little current from the CMOS output that is driving it. The input voltage requirements for both logic states are expressed as a percentage of the supply voltage. $V_{IL}(\text{max})$ is specified as 30 percent of V_{DD}, and $V_{IH}(\text{min})$ is 70% of V_{DD}.*

---

*These percentages apply to the 4000B series. The V_{IH} and V_{IL} values for the 74HC and 74HCT series will be presented in Section 8-18.

TABLE 8-7 CMOS Voltage Levels (4000B)

$$V_{OL}(\text{max}) = 0 \text{ V}$$
$$V_{OH}(\text{min}) = V_{DD}$$
$$V_{IL}(\text{max}) = 30\% \ V_{DD}$$
$$V_{IH}(\text{min}) = 70\% \ V_{DD}$$

These voltage levels are summarized in Table 8-7.

To illustrate, when a CMOS IC is operated from $V_{DD} = 5$ V, it will accept any input voltage less than $V_{IL}(\text{max}) = 1.5$ V as a LOW, and any input voltage greater than $V_{IH}(\text{min}) = 3.5$ V as a HIGH.

Noise Margins The CMOS dc noise margins can be determined from Table 8-7 as follows:

$$V_{NH} = V_{OH}(\text{min}) - V_{IH}(\text{min})$$

$$= V_{DD} - 70\% \ V_{DD}$$

$$= 30\% \ V_{DD}$$

$$V_{NL} = V_{IL}(\text{max}) - V_{OL}(\text{max})$$

$$= 30\% \ V_{DD} - 0$$

$$= 30\% \ V_{DD}$$

The noise margins are the same in both states and depend on V_{DD}. At $V_{DD} = 5$ V, the noise margins are both 1.5 V. This is substantially better than TTL and ECL. This makes CMOS attractive for applications that are exposed to a high-noise environment. Of course, the noise margins can be made even better by using a larger V_{DD}. This improvement in noise immunity, however, would be obtained at the expense of a higher power drain because of the larger supply voltage.

Power Dissipation When a CMOS logic circuit is in a static state (not changing), its power dissipation is extremely low. We can see the reason by examining each of the circuits shown in Figures 8-37–8-39. Note that regardless of the state of the output, there is always a very high resistance between the V_{DD} terminal and ground because there is always an OFF MOSFET in the current path. This results in a typical CMOS dc power dissipation of only 2.5 nW per gate when $V_{DD} = 5$ V; even at $V_{DD} = 10$ V this power increases to only 10 nW. With these values for P_D, it is easy to see why CMOS is widely used in applications where power drain is a prime concern.

P_D Increases with Frequency The power dissipation of a CMOS IC will be very low as long as it is in a dc condition. Unfortunately, P_D will increase in proportion to the frequency at which the circuits are switching states. For example, a CMOS NAND gate that has $P_D = 10$ nW under dc conditions will have $P_D = 0.1$ mW at a frequency of 100 kpps, and 1 mW at 1 MHz. The reason for this dependence on frequency is illustrated in Figure 8-40.

Each time a CMOS output switches from LOW to HIGH, a transient charging current has to be supplied to the load capacitance. This capacitance consists of the

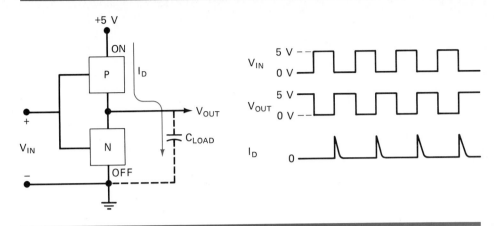

FIGURE 8-40 Current spikes are drawn from the V_{DD} supply each time the output switches from LOW to HIGH. This is due mainly to charging current of load capacitance.

combined input capacitances of any loads being driven and the device's own output capacitance. These narrow spikes of current are supplied by V_{DD} and can have a typical amplitude of 5 mA and a duration of 20 to 30 ns. Clearly, as the switching frequency increases, there will be more of these current spikes occurring per second, and the average current drawn from V_{DD} will increase.

Thus, at higher frequencies, CMOS begins to lose some of its advantage over other logic families. As a general rule, a CMOS gate will have the same average P_D as a 74LS gate at frequencies near 2-3 MHz. For MSI chips, the situation is somewhat more complex than stated here, and a logic designer must do a detailed analysis to determine whether or not CMOS has a power-dissipation advantage at a particular frequency of operation.

Fan-Out   Like N-MOS and P-MOS, CMOS inputs have an extremely large resistance (10^{12} ohms) that draws essentially no current from the signal source. Each CMOS input, however, typically presents a 5-pF load to ground. This input capacitance limits the number of CMOS inputs that one CMOS output can drive (see Figure 8-41). The CMOS output has to charge and discharge the parallel combination of each input capacitance, so that the output switching time will be increased in proportion to the number of loads being driven. Typically, each CMOS load increases the driving circuit's propagation delay by 3 ns. For example, NAND gate 1 in Figure 8-41 might have a t_{PLH} of 25 ns if it were driving no loads; this would increase to 25 ns + 20(3 ns) = 85 ns if it were driving *twenty* loads.

Thus, CMOS fan-out depends on the permissible maximum propagation delay. Typically, CMOS outputs are limited to a fan-out of 50 for low-frequency operation ($\leq$ 1 MHz). Of course, for higher-frequency operation the fan-out would have to be less.

Switching Speed   Although CMOS, like N-MOS and P-MOS, has to drive relatively large load capacitances, its switching speed is somewhat faster because of its low output resistance in each state. Recall that an N-MOS output has to charge

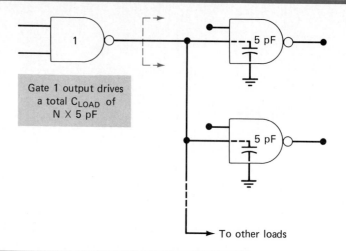

FIGURE 8-41 Each CMOS input adds to the total load capacitance seen by the driving gate's output.

the load capacitance through a relatively large (100 kΩ) resistance. In the CMOS circuit, the output resistance in the HIGH state is the R_{ON} of the P-MOSFET, which is typically 1 kΩ or less. This allows more rapid charging of load capacitance.

A 4000 series NAND gate will typically have an average t_{pd} of 50 ns at V_{DD} = 5 V, and 25 ns at V_{DD} = 10 V. The reason for the improvement in t_{pd} as V_{DD} is increased is that the R_{ON} on the MOSFETs decreases significantly at higher supply voltages. Thus, it appears that V_{DD} should be made as large as possible for operation at higher frequencies. Of course, the larger V_{DD} will result in increased power dissipation.

A typical NAND gate in the 74HC or 74HCT series has an average t_{pd} of around 8 ns when operated at V_{DD} = 5 V. This is comparable to the speed of the 74LS series.

Unused Inputs
CMOS inputs should never be left disconnected. All CMOS inputs have to be tied either to a fixed voltage level (0 V or V_{DD}) or to another input. This rule applies even to the inputs of extra unused logic gates on a chip. An unconnected CMOS input is susceptible to noise and static charges that could easily bias both the *P*- and *N*-Channel MOSFETs in the conductive state, resulting in increased power dissipation and possible overheating.

Static-Charge Susceptibility
The high input resistance of CMOS inputs makes them especially prone to static-charge buildup that can produce voltages large enough to break down the dielectric insulation between the MOSFET gate and channel. This static charge can result from improper handling such as pushing a chip into a styrofoam carrier. CMOS and MOS chips require careful handling precautions such as (1) storing them in conductive foam or metal containers and (2) using a grounded-tip soldering iron when soldering them into a circuit.

Most of the CMOS-series devices are now protected against static-charge damage by the inclusion of protective diodes on each input. Even so, it is best to observe the rules stated above to prevent static-charge damage.

TABLE 8-8 Digital IC Series Comparison*

	74HC	4000B	74	74S	74LS	74AS	74ALS	ECL
Power dissipation per gate (mW)								
Static	2.5×10^{-6}	0.001	10	20	2	8	1.2	40
At 100 kHz	0.17	0.1	10	20	2	8	1.2	40
Propagation delay (ns)	8	50	9	3	9.5	1.7	4	1
Speed–power (at 100 kHz) (pJ)	1.4	5	90	60	19	13.6	4.8	40
Maximum clock rate (MHz)	40	12	35	12.5	45	200	70	300
Worst-case noise margin (V)	0.9	1.5	0.4	0.3	0.3	0.3	0.4	0.25

*All CMOS values are for $V_{DD} = 5$ V.

Comparison of CMOS and TTL Series Table 8-8 compares the typical characteristics of the principal series of digital ICs.

REVIEW QUESTIONS

1. Which CMOS series are pin-for-pin compatible with TTL?
2. What are the improvements of the 74HC over the 74C series?
3. How does the 74HCT series differ from the 74HC?
4. Describe what happens to each of the following CMOS characteristics as V_{DD} is increased: (a) noise margin, (b) power dissipation, (c) switching speed.
5. What factors limit CMOS fan-out?
6. What should be done with unused CMOS inputs?
7. What precautions should be taken when handling CMOS devices?
8. Which IC series is the best for battery-operated circuitry?
9. Which series is the best for operation in high-noise environments?

8-16 CMOS OPEN-DRAIN AND TRISTATE OUTPUTS

Conventional CMOS outputs should never be connected together. Figure 8.42 illustrates what can happen when two CMOS INVERTER outputs are shorted together. The output on the left is trying to go to the HIGH state and its *P*-channel MOSFET is conducting with $R_{ON} = 1$ kΩ. The other output is trying to go to the LOW state and its *N*-channel MOSFET is conducting with $R_{ON} = 1$ kΩ. The common output terminal X will be at a voltage of approximately $V_{DD}/2$ because of the voltage-divider action between the two 1-kΩ resistances.

This voltage is in the indeterminate range (recall that $V_{IL}(\text{max}) = 30\% \ V_{DD}$ and $V_{IH}(\text{min}) = 70\% \ V_{DD}$) and is therefore unacceptable for driving other devices. Furthermore, the current through the two conducting MOSFETs will be much greater than normal, especially at higher values of V_{DD}, and can damage the ICs.

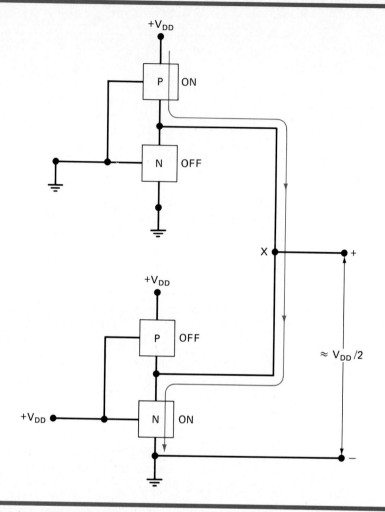

FIGURE 8-42 When CMOS outputs are shorted together, the common output terminal will be approximately $V_{DD}/2$ if the outputs are trying to be at different levels.

It is relatively easy to recognize when two CMOS outputs have been unintentionally wired or shorted together because the signals at the outputs will have three different levels: (1) HIGH when both outputs are trying to be HIGH; (2) LOW when both are trying to be LOW; and (3) appoximately $V_{DD}/2$ when the outputs are trying to be at different levels. This is illustrated in Figure 8-43.

Open-Drain Outputs Some CMOS devices are available with open-drain outputs that are the counterpart to TTL open-collector outputs. In these devices, the output stage consists only of an *N*-channel MOSFET whose drain is unconnected since the upper *P*-channel MOSFET has been eliminated. An external pull-up resistor is needed to produce a HIGH-state voltage level. Like open-collector outputs, open-drain outputs can be wired-ANDed. Figure 8-44(a) shows three 74HC05 open-drain INVERTERs connected in a wired-AND arrangement.

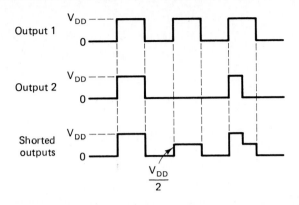

FIGURE 8-43 **When two ordinary CMOS outputs are shorted together, the shorted outputs will be at $\approx V_{DD}/2$ when the outputs are trying to be different.**

Tristate Outputs Several CMOS ICs have tristate outputs whose operation is similar to TTL tristate outputs. Everything we said about tristate TTL can be applied to CMOS tristate. CMOS tristate outputs can be connected together in a bus arrangement provided that only one output is enabled at one time. Figure 8-44(b) shows three 74HC125 tristate buffers connected in a bus arrangement.

FIGURE 8-44 **(a) CMOS open-drain devices in wired-AND connection; (b) CMOS tristate outputs in bus arrangement.**

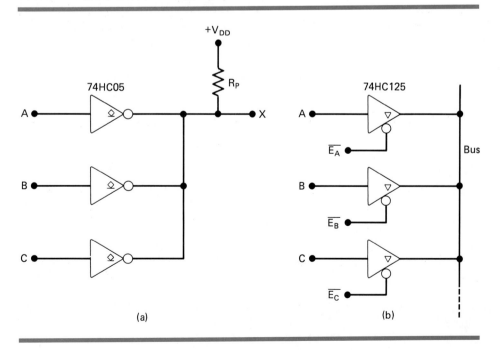

1. Why shouldn't conventional CMOS outputs be tied together?
2. What is an open-drain output?

8-17 CMOS TRANSMISSION GATE

A special CMOS circuit that has no TTL or ECL counterpart is the *transmission gate* or *bilateral switch,* which acts essentially as a single-pole, single-throw switch controlled by an input logic level. This transmission gate will pass signals in both directions and is useful for digital and analog applications.

Figure 8-45(a) is the basic arrangement for the bilateral switch. It consists of a P-MOSFET and N-MOSFET in parallel so that both polarities of input voltage can be switched. The CONTROL input and its inverse are used to turn the switch ON (closed) and OFF (open). When the CONTROL is HIGH, both MOSFETS are turned ON and the switch is closed. When CONTROL is LOW, both MOSFETs are turned OFF and the switch is open. Ideally, this circuit operates like an electromechanical relay. In practice, however, it is not a perfect short circuit when the switch is closed; the switch resistance R_{ON} is typically 200 Ω. In the open state, the switch resistance is very large, typically 10^{12} Ω, which for most purposes is an open circuit. The symbol in Figure 8-45(b) is used to represent the bilateral switch.

This circuit is called a *bilateral* switch because the input and output terminals can be interchanged. The signals applied to the switch input can be either digital or analog signals, provided that they stay within the limits of 0 to V_{DD} volts.

FIGURE 8-45 **CMOS bilateral switch (transmission gate).**

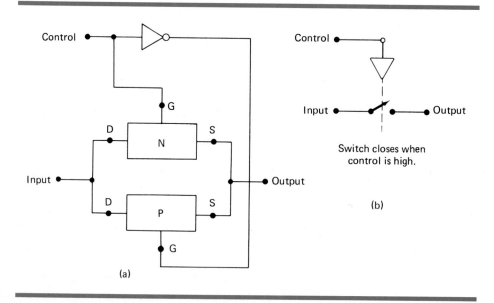

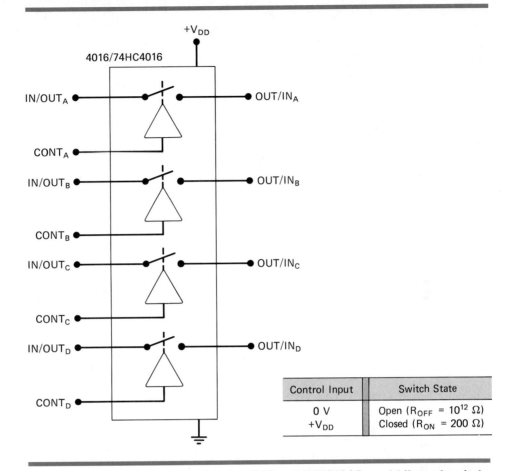

Control Input	Switch State
0 V	Open (R_{OFF} = 10^{12} Ω)
+V_{DD}	Closed (R_{ON} = 200 Ω)

FIGURE 8-46 Logic diagram for the 4016 and 74HC4016 quad bilateral switch.

Figure 8-46 shows the logic diagram for a 4016 quad bilateral switch IC. It is also available in the 74HC series as a 74HC4016. The IC contains four bilateral switches that operate as described above. Each switch is independently controlled by its own control input. A typical application is shown in Figure 8-47.

EXAMPLE 8-18

Describe the operation of the circuit of Figure 8-47. (See following page.)

SOLUTION
Here two of the bilateral switches are connected so that a common analog input signal can be switched to either output X or output Y, depending on the logic state of the OUTPUT SELECT input. When OUTPUT SELECT is LOW, the upper switch is closed and the lower one is open so that V_{IN} is connected to output X. When OUTPUT SELECT is HIGH, the upper switch is open and the lower one is closed so that V_{IN} is connected to output Y. Figure 8-47(b) shows some typical waveforms. Note that for proper operation, V_{IN} must be within the range 0 V to +V_{DD}.

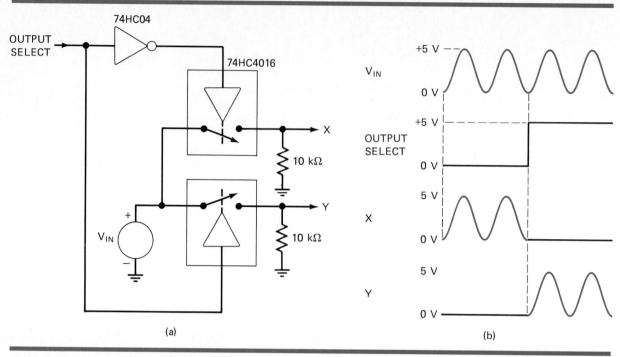

FIGURE 8-47 Example 8-18; 74HC4016 bilateral switches used to switch an analog signal to two different outputs.

The 4016/74HC4016 bilateral switch can only switch input voltages that lie between 0 V and V_{DD}, so it could not be used for signals that were both positive and negative relative to ground. The 4316 and 74HC4316 ICs are quad bilateral switches that can switch bipolar analog signals. These devices have a second power-supply terminal called V_{EE}, which can be made negative with respect to ground. This permits input signals that can range from V_{EE} to V_{DD}. For example, with V_{EE} = -5 V and V_{DD} = $+5$ V, the analog input signal can be anywhere from -5 V to $+5$ V.

REVIEW QUESTIONS

1. Describe the operation of a CMOS transmission gate.
2. *True or false:* There is no TTL transmission gate.

8-18 IC INTERFACING

Interfacing means connecting the output(s) of one circuit or system to the input(s) of another circuit or system that has different electrical characteristics. Often a direct connection cannot be made because of the difference in the electrical characteristics of the *driver* circuit that is providing the output signal and the *load* circuit that is receiving the signal.

An *interface* circuit is a circuit connected between the driver and the load; its function is to take the driver output signal and condition it so that it is compatible with requirements of the load.

In the following sections we will address the problems involved in interfacing devices from one logic family to those of a different logic family. This type of interfacing occurs quite often in the more complex digital systems, where designers utilize different logic families for different parts of the system in order to take advantage of the strong points of each family. For example, high-speed TTL (74AS, 74S) might be used in those parts of the system that are operating at the highest frequency, LS-TTL in the slower parts of the system, and N-MOS for the LSI and VLSI parts of the system.

ICs from the same logic series are designed to be connected together without any special considerations, provided that the fan-out limitation of each output is not exceeded. When you connect the output of an IC to the input of an IC from a different logic family or a different series within the same logic family, you generally have to be concerned with the voltage and current parameters of the two devices. This usually involves checking the device data sheets for values of input and output current/voltage parameters. We will use Table 8-9 to help us examine the various interfacing cases in the following sections. This table contains the worst-case values of the input and output parameters for standard devices in the CMOS and TTL series. It should be pointed out that the values in this table apply to standard devices with no special input or output circuitry. The output values may be different for devices such as buffers that use special output circuitry for increased current and voltage capabilities. The input current values may also be different for devices where an external input is internally connected to more than one gate. While the values in this table are valid for a large majority of devices, it would be advisable to check the specific IC data sheets.

8-19 TTL DRIVING CMOS

When interfacing different types of ICs, we must check that the driving device can meet the current and voltage requirements of the load device. Examination of Table 8-9 (see page 462) indicates that the input current values for CMOS are extremely low compared to the output current capabilities of any TTL series. Thus, TTL has no problem meeting the CMOS input current requirements.

There is a problem, however, when we compare the TTL output voltages with the CMOS input voltage requirements. We see that the $V_{OH}(min)$ of every TTL series is too low when compared to the $V_{IH}(min)$ requirement of the 4000B and 74HC series. For these situations, something must be done to raise the TTL output voltage to an acceptable level for CMOS.

The most common solution to this interface problem is shown in Figure 8-48, where the TTL output is connected to $+5$ V with a pull-up resistor. The presence of the pull-up resistor will cause the TTL output to rise to approximately 5 V in the HIGH state, thereby providing an adequate CMOS input. We determine the value of the pull-up resistor in the same way as we did for the open-collector pull-up

TABLE 8-9   Worst-Case Values for CMOS/TTL Interfacing*

PARAMETER	CMOS			TTL			
	4000B	74HC	74HCT	74	74LS	74AS	74ALS
V_{IH}(min)	3.5 V	3.5 V	2.0 V	2.0 V	2.0 V	2.0 V	2.0 V
V_{IL}(max)	1.5 V	1.0 V	0.8 V	0.8 V	0.8 V	0.8 V	0.8 V
V_{OH}(min)	4.95 V†	4.9 V†	4.9 V†	2.4 V	2.7 V	2.7 V	2.7 V
V_{OL}(max)	0.05 V†	0.1 V†	0.1 V†	0.4 V	0.5 V	0.5 V	0.4 V
I_{IH}(max)	1 μA	1 μA	1 μA	40 μA	20 μA	200 μA	20 μA
I_{IL}(max)	1 μA	1 μA	1 μA	1.6 mA	0.4 mA	2 mA	100 μA
I_{OH}(max)	0.4 mA	4 mA	4 mA	0.4 mA	0.4 mA	2 mA	400 μA
I_{OL}(max)	0.4 mA	4 mA	4 mA	16 mA	8 mA	20 mA	8 mA

*Supply voltage = 5 V.
†CMOS driving only CMOS inputs.

resistor. If the TTL output is driving only CMOS, a value from 1 to 10 kΩ is normally used.

TTL Driving 74HCT

As we stated earlier, the 74HCT series is designed so that it can be driven directly by TTL outputs. In other words, its input voltage requirements are the same as standard TTL devices. This eliminates the need for the pull-up resistor of Figure 8-48. This ease of interfacing TTL to a 74HCT device comes at the cost of increased power dissipation in the CMOS device.

TTL Driving High-Voltage CMOS

If the CMOS IC is operating with V_{DD} greater than 5 V, the situation becomes somewhat more difficult. For example, with $V_{DD} = 10$ V, the CMOS input requires V_{IH}(min) = 7 V. The outputs of many TTL devices cannot be operated at more than 5 V, so a pull-up resistor connected to +10 V is prohibited. The LS-TTL series from certain manufacturers (e.g., Fairchild) can operate with an output pull-up to 10 V. In general, the device's data sheet should be checked before using a pull-up to more than 5 V.

When the TTL output cannot be pulled up to V_{DD}, there are some alternatives. One common solution is shown in Figure 8-49, where a 7407 open-collector buffer

FIGURE 8-48   **External pull-up resistor is used when TTL drives CMOS.**

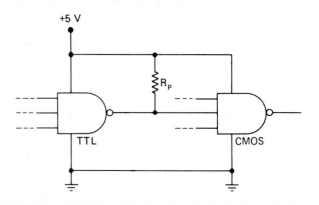

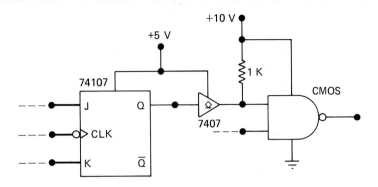

FIGURE 8-49 **A 7407 open-collector buffer can be used to interface TTL to high-voltage CMOS.**

is used as the interface between a TTL totem-pole output and CMOS operating at $V_{DD} > 5$ V. The 7407 is the noninverting counterpart of the 7406 and has an output voltage rating of 30 V.

Another solution is to utilize a *level-translator* circuit such as the 40104. This is a CMOS chip that is designed to take a low-voltage input (e.g., from TTL) and translate it to a high-voltage output for CMOS.

REVIEW QUESTIONS

1. What has to be done to interface a conventional TTL output to a 4000B or 74HC input?
2. What are the various ways to interface TTL to high-voltage CMOS?

8-20 CMOS DRIVING TTL

Before we consider the problem of interfacing CMOS outputs to TTL inputs, it will be helpful to review the CMOS output characteristics for the two logic states.

Figure 8-50(a) shows the equivalent output circuit in the HIGH state. The R_{ON} of the P-MOSFET connects the output terminal to V_{DD} (remember, the N-MOSFET is OFF). Thus, the CMOS output circuit acts like a V_{DD} source with a source resistance of R_{ON}. The value of R_{ON} typically ranges from 100 to 1000 ohms.

Figure 8-50(b) shows the equivalent output circuit in the LOW state. The R_{ON} of the N-MOSFET connects the output terminal to ground (remember, the P-MOS-FET is OFF). Thus, the CMOS output acts as a low resistance to ground; that is, it acts as a current sink.

CMOS Driving TTL in the HIGH State   Table 8-9 shows that CMOS outputs can easily supply enough voltage (V_{OH}) to satisfy the TTL input requirement in the HIGH state (V_{IH}). It also shows that CMOS outputs can supply more than enough

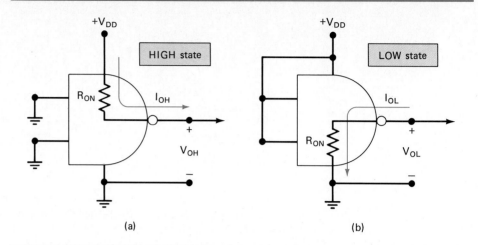

(a) (b)

FIGURE 8-50   Equivalent CMOS output circuits for both logic states.

current (I_{OH}) to meet the TTL input current requirements (I_{IH}). Thus, no special consideration is needed for the HIGH state.

CMOS Driving TTL in the LOW State   The table shows that TTL inputs have a relatively high input current in the LOW state, ranging from 100 μA to 2 mA. The 74HC and 74HCT families can sink up to 4 mA, so they would have no trouble driving a *single* TTL load of any series. The 4000B series, however, is much more limited. Its low I_{OL} capability is not sufficient to drive even one input of the 74 or 74AS series.

EXAMPLE 8-19

How many 74LS inputs can be driven by a 74HC output? Repeat for a 4000B output.

SOLUTION
The 74LS series has I_{IL}(max) = 0.4 mA. The 74HC can sink up to I_{OL}(max) = 4 mA. Thus, the 74HC can drive *ten* 74LS loads (4 mA/0.4 mA = 10).
    The 4000B can sink only 0.4 mA, so it can drive only *one* 74LS input.

EXAMPLE 8-20

How many 74ALS inputs can be driven by a 74HC output? Repeat for a 4000B output.

SOLUTION
The 74ALS series has I_{IL}(max) = 100 μA. Thus, the 74HC can drive *forty* 74ALS inputs (4mA/100 μA = 40). A 4000B can drive *four* 74ALS inputs (0.4 mA/100 μA = 4).

EXAMPLE 8-21

What's wrong with the circuit in Figure 8-51(a)?

SOLUTION
The 74HC00 can sink 4 mA, but the three 74AS inputs require 3 × 2 mA = 6 mA.

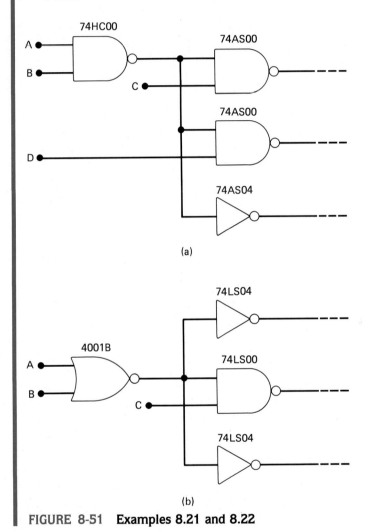

(a)

(b)

FIGURE 8-51   Examples 8.21 and 8.22

EXAMPLE 8-22

What's wrong with the circuit in Figure 8-51(b)?

SOLUTION
The 4001B NOR gate can sink 0.4 mA, but the three 74LS inputs require 3 × 0.4 mA = 1.2 mA.

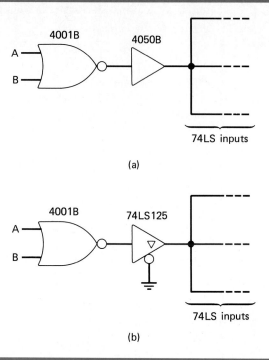

(a)

(b)

FIGURE 8-52 Buffers used to interface low-current CMOS to 74LS inputs.

For situations like those in Figure 8-51, some type of interface circuit is needed between the CMOS and TTL devices. The interface circuit should have a low input current requirement and a sufficiently high output current rating to drive the loads. Figure 8.52 shows two possible interfaces for the case of Figure 8-51(b). In Figure 8-52(a) the CMOS 4050B is a noninverting buffer that has an output current rating of $I_{OL}(\text{max}) = 3$ mA so that it can easily drive the three 74LS loads. In Figure 8-52(b) the 74LS125 is a permanently enabled noninverting tristate buffer which can be driven by the 4000B. Its output can easily drive the 74LS loads. In both of these circuits, the interface buffer simply passes the 4000B output signal to the 74LS loads.

FIGURE 8-53 A 4050B buffer can also serve as a level translator between high-voltage CMOS and TTL.

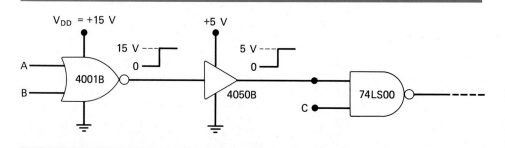

High-Voltage CMOS Driving TTL   Some IC manufacturers have produced several 74LS TTL devices that can withstand input voltages as high as 15 V. These devices can be driven directly from CMOS outputs operating at $V_{DD} = 15$ V. Most TTL inputs cannot handle more than 7 V before becoming damaged, so an interface is necessary if they are to be driven from high-voltage CMOS. The interface functions as a *voltage-level translator* that converts the high-voltage input to a 5-V output that can be connected to TTL. Figure 8-53 shows how the 4050B performs this level translation between 15 V and 5 V.

REVIEW QUESTIONS

1. What is the function of an *interface* circuit?
2. *True or false:* All CMOS outputs can drive TTL in the HIGH state.
3. *True or false:* Any CMOS output can drive any single TTL input.
4. Which TTL series can drive CMOS inputs without a pull-up resistor?
5. How many 7400 inputs can be driven from a 74HCT00 output? (*Ans.* two)
6. How is high-voltage CMOS interfaced to TTL?

8-21   TROUBLESHOOTING

A *logic pulser* is a testing and troubleshooting tool that generates a short-duration pulse when manually actuated, usually by pressing a pushbutton. The logic pulser shown in Figure 8-54 has a needle-shaped tip that is touched to the circuit node that

FIGURE 8-54   A logic pulser can inject a pulse at any node that is not shorted directly to ground or V_{CC}.

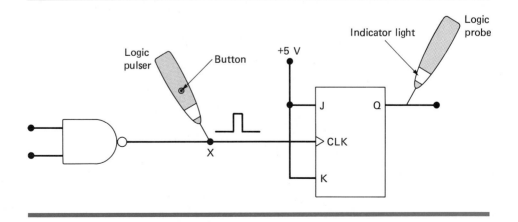

is to be pulsed. The logic pulser is designed so that it senses the existing voltage level at the node and produces a voltage pulse in the opposite direction. In other words, if the node is LOW, the logic pulser produces a narrow positive-going pulse; if the node is HIGH, it produces a narrow negative-going pulse.

The logic pulser is used to momentarily change the logic level at a circuit node even though the output of another device may be connected to that same node. In Figure 8-54 the logic pulser is contacting node X, which is also connected to the output of the NAND gate. The logic pulser has a very low output impedance (typically 2 Ω or less), so that it can overcome the NAND gate's output and can change the voltage at the node. The logic pulser, however, cannot produce a voltage pulse at a node that is shorted directly to ground or V_{CC} (e.g., as through a solder bridge).

Using Logic Pulser and Probe to Test a Circuit

A logic pulser can be used to manually inject a pulse or series of pulses into a circuit in order to test the circuit's response. A logic probe is almost always used to monitor the circuit's response to the logic pulser. In Figure 8-54, the J-K FF's toggle operation is being tested by applying pulses from the logic pulser and monitoring Q with the logic probe. This logic pulser/logic probe combination is very useful for checking the operation of a logic device while it is wired into a circuit. Note that the logic pulser is applied to the circuit node *without* disconnecting the output of the NAND gate that is driving that node.

Finding Shorted Nodes

The logic pulser and logic probe can be used to check for nodes that are shorted directly to ground or V_{CC}. When you touch a logic pulser and logic probe to the same node and press the logic pulser button, the logic probe should indicate the occurrence of a pulse at the node. If the probe indicates a constant LOW, the node is shorted to ground; if the probe indicates a constant HIGH, the node is shorted to V_{CC}.

The Current Tracer

This is a troubleshooting tool that can detect a *changing* current in a wire or printed-circuit-board trace without breaking the circuit. The current tracer has an insulated tip that contains a magnetic pickup coil. When the tip is placed at a point in the circuit, it senses a changing magnetic field produced by a changing current and causes a small indicator LED to flash. The current tracer does not respond to static current levels, no matter how great the current may be.

A current tracer is used with a logic pulser to trace the exact location of shorts to ground or V_{CC}. This is illustrated in Figure 8-55, where node X is shorted to ground through the internal short at gate 2's input. If the logic pulser is touched to X and its button is pressed, no voltage pulse will be detected, because of the short to ground. However, there will be a pulse of current flowing from the pulser's output to ground through the short. This current pulse can be detected by a current tracer.

In Figure 8-55(a), the current tracer is placed to the left of X and the logic pulser is pulsed. The tracer will indicate no current pulse through the path it is monitoring. In Figure 8-55(b), the tracer is moved to the other side of X. This time when the logic pulser is pulsed, the current tracer will indicate the occurrence of a current pulse through the path it is monitoring. This proves that the short to ground is inside gate 2's input rather than gate 1's output.

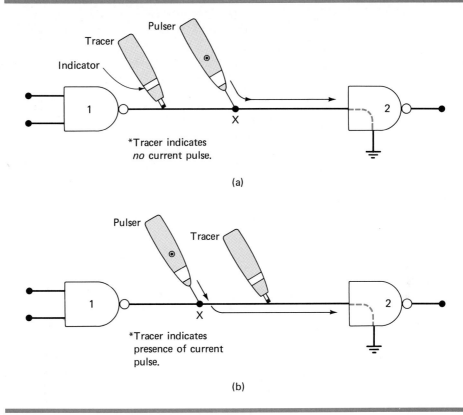

(a)

(b)

FIGURE 8-55 Logic pulser and current tracer can be used to trace shorted nodes.

The current tracer/logic pulser combination is very useful in troubleshooting circuits where many open-collector, open-drain, or tristate outputs are connected to a common point that is stuck LOW or HIGH. Any one of the outputs can be producing the fault, and the current tracer/logic pulser technique used above can be employed to determine the faulty output.

REVIEW QUESTIONS

1. What is the function of a logic pulser?
2. *True or false:* A logic pulser will produce a voltage pulse at any node.
3. What is the function of a current tracer?
4. *True or false:* A current tracer only detects a changing current.

AC Noise Margin The ac noise margin of a digital circuit is substantially greater than its dc noise margin, and is caused by input pulse durations which are too short for the circuit to respond. [Sec. 8-1]

Bilateral Switch CMOS circuit that acts like a single-pole, single-throw switch (SPST) controlled by an input logic level. [Sec. 8-17]

Buffer/Driver Circuit designed to have a greater output current and/or voltage capability than an ordinary logic circuit. [Sec. 8-7]

Bus Contention Situation in which the outputs of two or more active devices are placed on the same bus line at the same time. [Sec. 8-8]

CML (Current-Mode Logic) Also referred to as emitter-coupled logic. [Sec. 8-9]

CMOS Complementary metal-oxide semiconductor. [Sec. 8-14]

Current Tracer Testing tool that detects a changing current in a wire or PC-board trace. [Sec. 8-21]

Current Transients Current spikes generated by the totem-pole output structure of a TTL circuit, and caused when both transistors are simultaneously turned ON. [Sec. 8-6]

Current-Sinking Logic Logic family in which the output of a logic circuit sinks current from the input of a different logic circuit. [Sec. 8-1]

Current-Sinking Transistor Name given to the output transistor (Q_4) of a TTL circuit. This transistor is turned on when the output logic level is LOW. [Sec. 8-2]

Current-Sourcing Logic Logic family in which the output of a logic circuit sources, or supplies, current to the input of a different logic circuit. [Sec. 8-1]

Current-Sourcing Transistor Name given to the output transistor (Q_3) of most TTL circuits. This transistor is conducting when the output logic level is HIGH. [Sec. 8-2]

ECL (Emitter-Coupled Logic) Also referred to as current-mode logic. [Sec. 8-9]

Fan-Out Maximum number of standard logic inputs that the output of a digital circuit can drive reliably. [Sec. 8-1]

Floating Inputs Any digital circuit input that is left disconnected. [Sec. 8-6]

High-Speed TTL TTL subfamily which uses the basic TTL standard circuit except that all the resistor values are decreased and the emitter-follower transistor (Q_3) is replaced by a Darlington pair. [Sec. 8-4]

Interfacing When the output of a system is connected to the input of a different system with different electrical characteristics. [Sec. 8-18]

Loading Factor *See* Fan-Out. [Sec. 8-1]

Logic Pulser Testing tool that generates a short-duration pulse when manually actuated. [Sec. 8-21]

Low-Power Schottky TTL (LS-TTL) TTL subfamily that uses the identical Schottky TTL circuit but with larger resistor values. [Sec. 8-4]

Low-Power TTL TTL subfamily that uses the basic TTL standard circuit except that all the resistor values are increased. [Sec. 8-4]

MOS Metal-oxide semiconductor. [Sec. 8-10]

Noise Immunity A circuit's ability to tolerate noise voltages on its inputs. [Sec. 8-1]

Noise Margin Quantitative measure of noise immunity. [Sec. 8-1]

Open-Collector Output Type of output structure of some TTL circuits in which only one transistor with a floating collector is used. [Sec. 8-7]

Power-Supply Decoupling When a small RF capacitor is connected between ground and V_{CC} near each TTL IC on a circuit board. [Sec. 8-6]

Pull-Down Transistor *See* Current-Sinking Transistor. [Sec. 8-2]

Pull-Up Transistor *See* Current-Sourcing Transistor. [Sec 8-2]

SBD Schottky barrier diode. [Sec. 8-4]

Schottky TTL TTL subfamily that uses the basic TTL standard circuit except that it uses a Schottky barrier diode (SBD) connected between the base and collector of each transistor. [Sec. 8-4]

Speed–Power Product Numerical value (in joules) often used to compare different logic families. It is obtained by multiplying the propagation delay by the power dissipation of a logic circuit. [Sec. 8-1]

Totem-Pole Output Term used to describe the way in which two bipolar transistors are arranged at the output of most TTL circuits. [Sec. 8-2]

Transmission Gate *See* Bilateral Switch. [Sec. 8-17]

Tristate TTL Type of TTL output structure that allows three types of output states: HIGH, LOW, and high-impedance (Hi-Z). [Sec. 8-8]

Unit Load (UL) Way that some manufacturers specify a device's input and output currents. In a standard TTL circuit, 1UL in the high state is equal to 40 μA, and in the low state is equal to 1.6 mA. [Sec. 8-5]

Voltage-Level Translator Circuit that takes one set of input voltage levels and translates it to different set of output levels. [Secs. 8-19 and 8-20]

Wired-AND Term used to describe the logic function created when open-collector outputs are tied together. [Sec. 8-7]

PROBLEMS   Section 8-1–8-3

(Data sheets for the ICs referred to in these problems may be found in Appendix II)

8-1. Two different logic circuits have the following characteristics:

	CIRCUIT A	CIRCUIT B
V_{supply}	6 V	5 V
$V_{IH}(\text{min})$	1.6 V	1.8 V
$V_{IL}(\text{max})$	0.9 V	0.7 V
$V_{OH}(\text{min})$	2.2 V	2.5 V
$V_{OL}(\text{max})$	0.4 V	0.3 V
t_{PLH}	10 ns	18 ns
t_{PHL}	8 ns	14 ns
P_D	16 mW	10 mW

(a) Which circuit has the best LOW-state dc noise immunity? The best HIGH-state dc noise immunity?

(b) Which circuit can operate at higher frequencies?

(c) Which circuit draws the most supply current?

8-2. Refer to Appendix II for the IC data sheets and use maximum values to determine P_D(avg), t_{pd}(avg) and speed–power product for one gate on each of the following TTL ICs.
(a) 7432 (b) 74S32 (c) 74LS20 (d) 74ALS20 (e) 74AS20

8-3. A certain logic family has the following voltage parameters:

$$V_{IH}(\text{min}) = 3.5 \text{ V} \qquad V_{IL}(\text{max}) = 1.0 \text{ V}$$
$$V_{OH}(\text{min}) = 4.9 \text{ V} \qquad V_{OH}(\text{max}) = 0.1 \text{ V}$$

(a) What is the largest positive-going noise spike that can be tolerated?
(b) What is the largest negative-going noise spike that can be tolerated?

Sections 8-5 and 8-6

8-4. Refer to the data sheet for the 74LS112 J-K FF.
(a) Determine the input loading factor at the J and K inputs.
(b) Determine the input loading factor at the clock and clear inputs.
(c) How many other 74LS112s can the output of one 74LS112 drive at the clock input?

8-5. Figure 8-56(a) shows a 74107 J-K FF whose output is required to drive a total of 14 UL. Since this exceeds the fan-out of the 74107, a buffer of some type is needed. Figure 8-56(b) shows one possibility using one of the NAND gates from the 7437 quad NAND buffer, which has a much higher fan-out than the 74107. Note that $\overline{Q}$ is used since the NAND is acting as an INVERTER. Refer to the data sheet for the 7437.
(a) Determine its fan-out.
(b) Determine its maximum sink current in the LOW state.

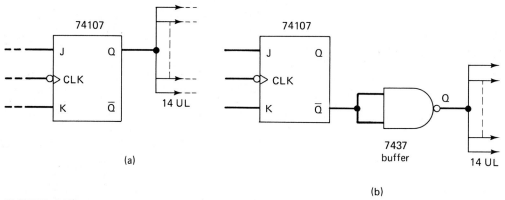

(a)

(b)

FIGURE 8-56

8-6. Buffer gates are generally more expensive than ordinary gates, and sometimes there are several unused gates available that can be used to solve a loading problem such as that in Figure 8-56(a). Show how 7400 NAND gates can be used to solve this problem.

8-7. Refer to the logic diagram of Figure 8-57, where the 7486 exclusive-OR output is driving several 7420 inputs. Determine whether the fan-out of the 7486 is being exceeded, and explain.

8-8. For the circuit of Figure 8-57 determine the longest time it will take for a change in the *A* input to be felt at output *W*. Use all worst-case conditions and maximum values of gate propagation delays. (*Hint:* Remember that NAND gates are inverting gates.)

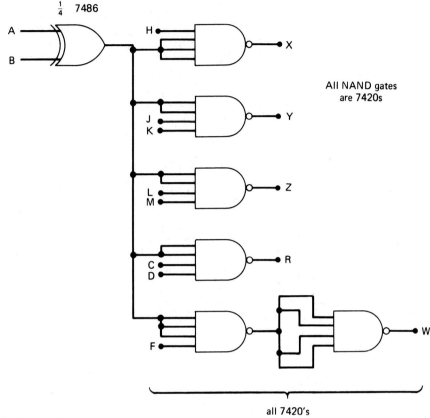

FIGURE 8-57

8-9. Figure 8-58 shows a 74121 one-shot being triggered by the closing of the switch. What maximum value of *R* should be used to ensure that the *B* input is biased LOW while the switch is open?

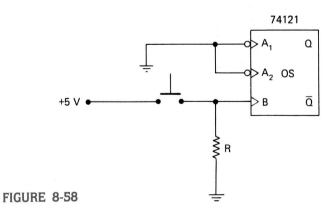

FIGURE 8-58

8-10. Figure 8-59(a) shows a circuit that is used to convert a 60-Hz sinewave to a 60-pps signal that can reliably trigger FFs and counters, etc. This type of circuit might be used in a digital clock.

(a) Explain the circuit operation.

(b) A technician is testing this circuit and observes that the 74LS14 output stays LOW. He checks the waveform at the INVERTER input and it appears as shown in Figure 8-59(b). Thinking that the INVERTER is faulty, he replaces the chip and observes the same results. What do you think is causing the problem, and how can it be fixed? (*Hint:* Examine the v_x waveform carefully.)

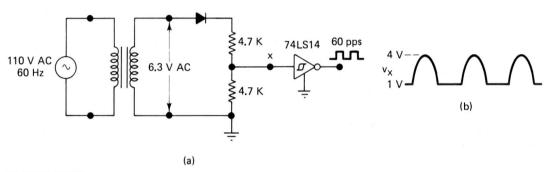

FIGURE 8-59

8-11. For each waveform in Figure 8-60, determine *why* it will *not* reliably trigger a 74LS107 FF at its *CLK* input.

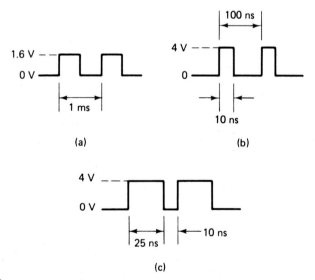

FIGURE 8-60

8-12. A technician breadboards a logic circuit for testing. As she tests the circuit's operation, she finds that many of the FFs and counters are triggering erratically. Like any good technician, she checks the V_{CC} line with a dc meter and reads 4.97 V, which is acceptable for TTL. She then checks all circuit wiring and replaces each IC one by one, but the problem persists. Finally she decides to observe V_{CC} on the scope and sees the waveform shown in Figure 8-61. What is the probable cause of the noise on V_{CC}? What did the technician forget to include when she breadboarded the circuit?

V_{CC} 4.97 V

1.3 V

FIGURE 8-61

Sections 8-7 and 8-8

8-13. The 7409 TTL IC is a quad two-input AND with open collector outputs. Show how 7409s can be used to implement the operation $x = A \cdot B \cdot C \cdot D \cdot E \cdot F \cdot G \cdot H \cdot I \cdot J \cdot K \cdot M$.

8-14. Determine the logic expression for output X in Figure 8-62.

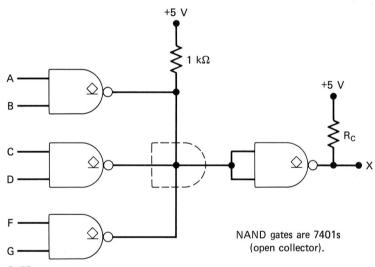

+5 V

1 kΩ

A

B

+5 V

C

R_C

D

X

F

NAND gates are 7401s
(open collector).

G

FIGURE 8-62

8-15. Determine a value for R_P in Figure 8-62 if output X is driving the clear inputs of four 74LS112 FFs.

8-16. Figure 8-63 shows a 7406 open-collector inverting buffer used to control the ON/OFF status of an LED to indicate the state of FF output Q. The LED's nominal specification is $V_F = 2.4$ V @ $I_F = 20$ mA, and I_F(max) = 30 mA.

(a) What voltage will appear at the 7406 output when $Q = 0$?

(b) Choose an appropriate value for the series resistor for proper operation.

8-17. Figure 8-64 shows how two tristate buffers can be used to construct a *bidirectional transceiver* that allows digital data to be transmitted in either direction (*A* to *B,* or *B* to *A*). Describe the circuit operation for the two states of the DIRECTION input.

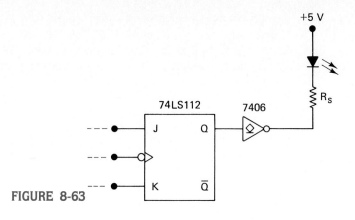

FIGURE 8-63

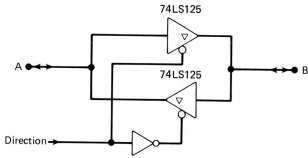

FIGURE 8-64

8-18. The circuit of Figure 8-65 is used to provide the enable inputs for the circuit of Figure 8-29.
    (a) Determine which of the data inputs (A, B, or C) will appear on the bus for each combination of inputs X and Y.
    (b) Explain why the circuit will not work if the NOR is changed to an EX-NOR.

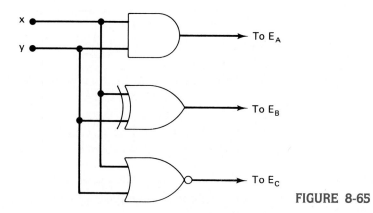

FIGURE 8-65

8-19. A 7406 output is driving 5 ULs of TTL. What is the largest value of pull-up resistor that can be used and still maintain the TTL noise margin? (*Hint:* See Example 8-17 and 7406 data sheet.)

8-20. The circuit of Figure 8-66 is an N-MOS logic gate. Determine what type of gate it is. Use $+5$ V $=$ logic 1 and 0 V $=$ logic 0.

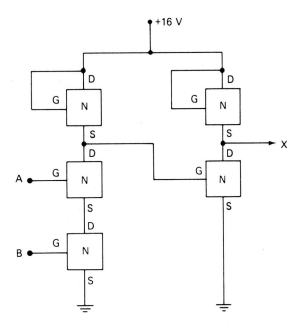

FIGURE 8-66

8-21. Draw the circuitry for an NMOS OR gate.

8-22. Which of the following are advantages that P-MOS and N-MOS logic have over TTL?
(a) Greater packing density.
(b) Greater operating speed.
(c) Greater fan-out.
(d) More suitable for LSI.
(e) Lower P_D.
(f) Complementary outputs.
(g) Greater noise immunity.
(h) Simpler fabrication process.
(i) More SSI and MSI functions.
(j) Uses lower supply voltage.

8-23. Which of the following are advantages that CMOS has over TTL?
(a) Greater packing density.
(b) Higher speed.
(c) Greater fan-out.
(d) Lower output impedance.

(e) Simpler fabrication process.

(f) More suited for LSI.

(g) Lower p_D (below 1MHz).

(h) Uses transistors as only circuit element.

(i) Lower input capacitance.

8-24. Which of the following operating conditions will probably result in the lowest average P_D for a CMOS logic system? Explain.
(a) $V_{DD} = 10$ V, switching frequency $f_{max} = 1$ MHz
(b) $V_{DD} = 5$ V, $f_{max} = 10$ kHz
(c) $V_{DD} = 10$ V, $f_{max} = 10$ kHz

8-25. What are the dc noise margins of the 4000B series that is operating at $V_{DD} = 12$ V?

8-26. Refer to the data sheet for the 74HC20 NAND gate IC in Appendix II. Use maximum values to calculate $P_D(avg)$, $t_{pd}(avg)$, and speed–power product. Compare to the values calculated in Problem 8-2 for TTL.

Sections 8-16 and 8-17

8-27. Determine the waveform at output X in Figure 8-67 for the given input waveforms. Assume $R_{ON} \approx 200$ Ω for the bilateral switch.

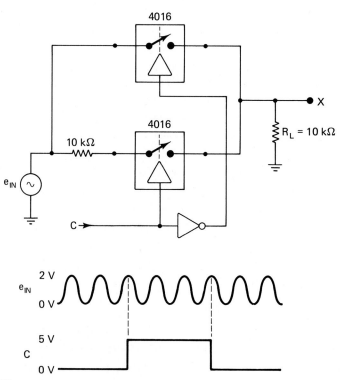

FIGURE 8-67

8-28. Determine the gain of the op-amp circuit of Figure 8-68 for the two states of the GAIN SELECT input. This circuit shows the basic principle of digitally controlled signal amplification.

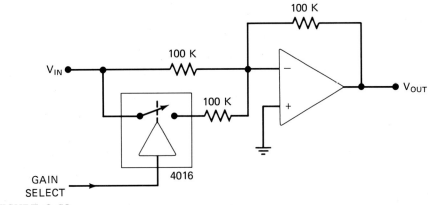

FIGURE 8-68

Sections 8-18–8-20

8-29. Refer to Figure 8-69(a) where a standard TTL output, Q, is driving a CMOS INVERTER operating at $V_{DD} = 10$ V. The waveforms at Q and X appear as shown in Figure 8-69(b). Which of the following is a possible reason why X stays HIGH?
(a) The 10-V supply is faulty.
(b) The pull-up resistor is too large.

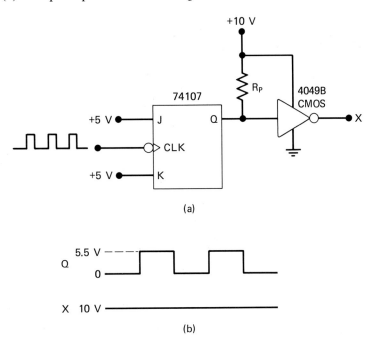

FIGURE 8-69

(c) The 74107 output breaks down at well below 10 V and maintains a 5.5-V level in the HIGH state. This is in the indeterminate range for the CMOS input.

(d) The CMOS input is loading down the TTL output.

8-30. How many 74AS inputs can a 4000B output drive?

8-31. How many 74AS inputs can a 74HC output drive?

8-32. Figure 8-70 is a logic circuit that was poorly designed. It contains at least eight instances where the characteristics of the ICs have not been properly taken into account. Find as many of these as you can.

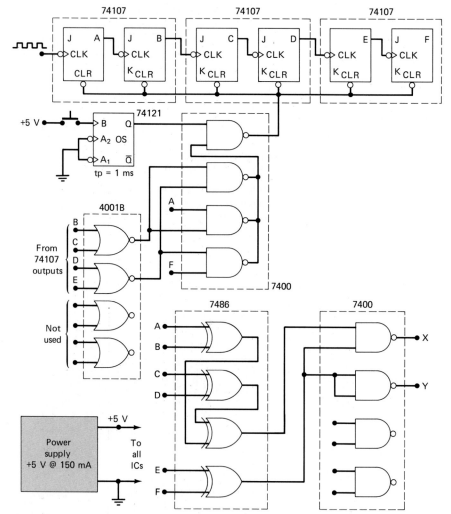

FIGURE 8-70

8-33. Repeat the last problem with the following changes in the circuit:

*Each TTL IC is replaced with its 74LS equivalent.

*The 4001B is replaced with a 74HCT02.

8-34. The circuit in Figure 8-71 uses a 74HC05 IC which contains six open-drain INVERTERs. The INVERTERs are connected in a wired-AND arrangement. Using a pulser and logic probe, it is determined that node X is struck in the HIGH state. Describe a procedure for using a current tracer to isolate the fault.

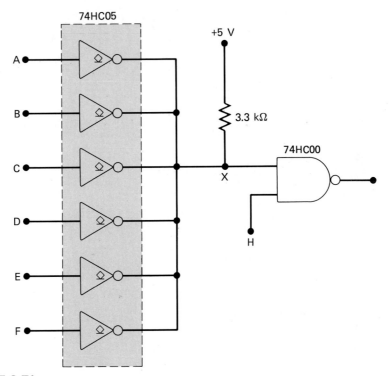

FIGURE 8-71

8-35. The circuit of Figure 8-54 has a solder bridge to ground somewhere between the output of the NAND gate and the input of the FF. Describe a procedure for tracking down the location of the solder bridge.

8-36. In Figure 8-48 a logic probe indicates that the lower end of the pull-up resistor is stuck in the LOW state. Using a logic pulser and current tracer a technician detects a current in the wire connecting the TTL output to the resistor. Which of the following is the possible fault?
(a) R_P is open.
(b) The TTL gate's current sinking transistor has a collector–emitter short.
(c) There is a break in the connection from R_P to the CMOS gate.

8-37. In Figure 8-48 a logic probe indicates that the lower end of R_P is stuck HIGH. A logic pulser and current tracer is used to detect a current in the path connecting R_P to the CMOS gate. What is the probable fault?

MSI LOGIC CIRCUITS

9

nd use both decoders and encoders in various types of circuit
ns.

the advantages and disadvantages between LEDs and LCDs.

- Utilize the observation/analysis technique for troubleshooting digital circuits.

- Understand the operation of multiplexers and demultiplexers by analyzing several circuit applications.

- Compare two binary numbers by using the magnitude comparator circuit.

- Cite the precautions that must be considered when connecting digital circuits using the data-bus concept.

- Interpret the notation used on the IEEE/ANSI symbols for various MSI devices.

INTRODUCTION

Digital systems contain binary-coded data and information that is continuously being operated on in some manner. Some of the operations include: (1) *decoding and encoding*—changing the data from one type of code to another, (2) *multiplexing*—selecting one out of several groups of data, (3) *demultiplexing*—distributing data to one of several destinations, and (4) *data busing*—transmitting data among several devices on a common bus. All these operations and others have been facilitated by the availability of numerous ICs in the MSI (medium-scale-integration) category.

   In this chapter we will study many of the common types of MSI devices. For each type, we will start with a brief discussion of its basic operating principle, then introduce specific ICs. We will then show how they can be used alone or in combination with other ICs in various applications.

9-1   DECODERS

A *decoder* is a logic circuit that converts an N-bit binary input code into M^* output lines such that each output line will be activated for only one of the possible combinations of inputs. Figure 9-1 shows the general decoder diagram with N inputs and M outputs. Since each of the N inputs can be 0 or 1, there are 2^N possible input combinations or codes. For each of these input combinations only one of the M outputs will be active HIGH; all the other outputs are LOW. Many decoders are designed to produce active LOW outputs, where only the selected output is LOW while all others are HIGH. This would be indicated by the presence of small circles on the output lines in the decoder diagram.

   Some decoders do not utilize all of the 2^N possible input codes but only certain ones. For example, a BCD-to-decimal decoder has a 4-bit input code and *ten* output lines that correspond to the *ten* BCD code groups 0000 through 1001. Decoders of

*N can be any integer and M is an integer that is less than or equal to 2^N.

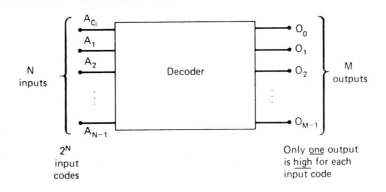

FIGURE 9-1　**General decoder diagram.**

this type are often designed such that if any of the unused codes are applied to the input, *none* of the outputs will be activated.

In Chapter 7 we saw how decoders are used in conjunction with counters to detect the various states of the counter. In that application it was the FFs in the counter that provided the binary code inputs for the decoder. The same basic decoder circuitry is used no matter where the inputs come from. Figure 9-2 shows the circuitry for a decoder with three inputs and $2^3 = 8$ outputs. It uses all AND gates, so the outputs are active HIGH. For active LOW outputs, NAND gates would be used. Note that for a given input code, the only output which is active (HIGH) is the one corresponding to the decimal equivalent of the binary input code (e.g., output O_6 goes HIGH only when CBA $= 110_2 = 6_{10}$).

This decoder can be referred to in several ways. It can be called a *3-line-to-8-line decoder,* because it has three input lines and eight output lines. It could also be called a *binary-to-octal decoder or convertor* because it takes a 3-bit binary input code and activates the one of the eight (octal) outputs corresponding to that code. It is also referred to as a 1-*of*-8 *decoder,* because only 1 of the 8 outputs is activated at one time.

ENABLE Inputs　Some decoders have one or more ENABLE inputs that are used to control the operation of the decoder. For example, refer to the decoder in Figure 9-2 and visualize having a common ENABLE line connected to a fourth input of each gate. With this ENABLE line held HIGH the decoder will function normally and the *A, B, C* input code will determine which output is HIGH. With ENABLE held LOW, however, *all* the outputs will be forced to the LOW state regardless of the levels at the *A, B, C* inputs. Thus, the decoder is ENABLED only if ENABLE is HIGH.

Figure 9-3(a) shows the logic diagram for the 74LS138 decoder as it appears in the Fairchild *TTL Data Book.* By examining this diagram carefully, we can determine exactly how this decoder functions. First, notice that it has NAND-gate outputs, so that its outputs are active LOW. Another indication is the labeling of the outputs as $\overline{O}_7$, $\overline{O}_6$, $\overline{O}_5$, etc.; the inverting overbar indicates active LOW outputs.

The input code is applied at A_2, A_1, and A_0, where A_2 is the MSB. With three inputs and eight outputs, this is a 3-to-8 decoder or, equivalently, a 1-of-8 decoder.

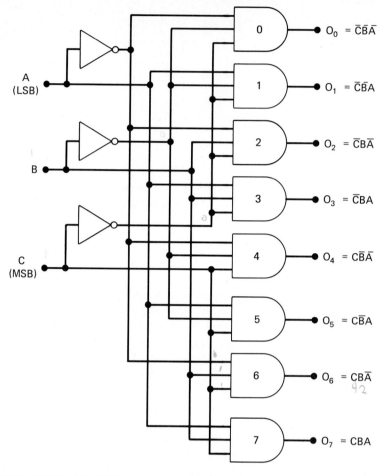

C	B	A	O_0	O_1	O_2	O_3	O_4	O_5	O_6	O_7
0	0	0	1	0	0	0	0	0	0	0
0	0	1	0	1	0	0	0	0	0	0
0	1	0	0	0	1	0	0	0	0	0
0	1	1	0	0	0	1	0	0	0	0
1	0	0	0	0	0	0	1	0	0	0
1	0	1	0	0	0	0	0	1	0	0
1	1	0	0	0	0	0	0	0	1	0
1	1	1	0	0	0	0	0	0	0	1

FIGURE 9-2   Three-line-to-8-line (or 1-of-8) decoder.

Inputs $\overline{E}_1$, $\overline{E}_2$, and E_3 are separate enable inputs that are combined in the AND gate. In order to enable the output NAND gates to respond to the input code at $A_2A_1A_0$, this AND-gate output has to be HIGH. This will occur only when $\overline{E}_1 = \overline{E}_2 = 0$ and $E_3 = 1$. In other words, $\overline{E}_1$ and $\overline{E}_2$ are active LOW, E_3 is active HIGH, and all three have to be in their active states to activate the decoder outputs. If one or more of the enable inputs is in its inactive state, the AND output will be LOW,

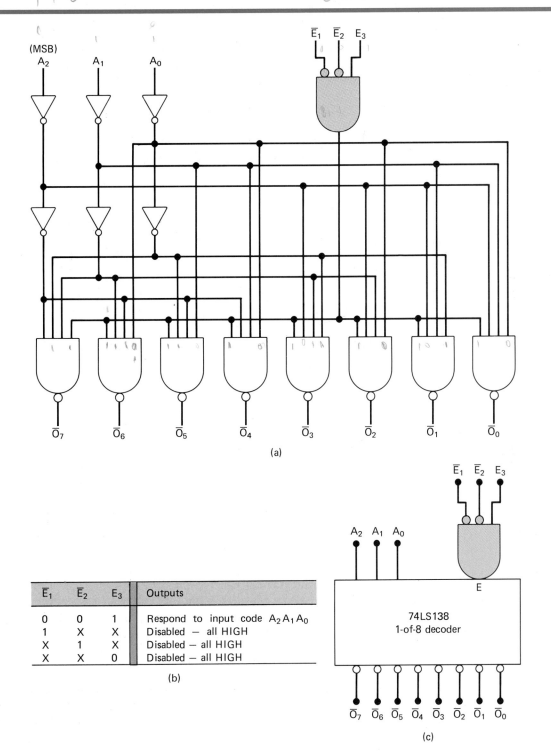

(a)

$\bar{E}_1$	$\bar{E}_2$	E_3	Outputs
0	0	1	Respond to input code $A_2 A_1 A_0$
1	X	X	Disabled — all HIGH
X	1	X	Disabled — all HIGH
X	X	0	Disabled — all HIGH

(b)

(c)

FIGURE 9-3   (a) Logic diagram for 74LS138 decoder; (b) truth table; (c) logic symbol. (Courtesy of Fairchild, a Schlumberger company)

which will force all NAND outputs to their inactive HIGH state regardless of the input code. This operation is summarized in the truth table in Figure 9-3(b). Recall that "x" represents the "don't care" condition.

The logic symbol for the 74LS138 is shown in Figure 9-3(c). Note how the active LOW outputs are represented, and how the enable inputs are represented. Even though the enable AND gate is shown as external to the decoder block, it is part of the IC's internal circuitry.

EXAMPLE 9-1

Indicate the states of the 74LS138 outputs for each of the following sets of inputs.

(a) $E_3 = \overline{E}_2 = 1, \overline{E}_1 = 0, A_2 = A_1 = 1, A_0 = 0$.
(b) $E_3 = 1, \overline{E}_2 = \overline{E}_1 = 0, A_2 = 0, A_1 = A_0 = 1$.

SOLUTION

(a) With $\overline{E}_2 = 1$, the decoder is disabled and all of its outputs will be in their inactive-HIGH state. This can be determined from the truth table or by following the input levels through the circuit logic.

(b) All of the enable inputs are activated, so that the decoding portion is enabled. It will decode the input code $011_2 = 3_{10}$ to activate output $\overline{O}_3$. Thus, $\overline{O}_3$ will be LOW and all other outputs will be HIGH.

EXAMPLE 9-2

Figure 9-4 shows how four 74LS138s and an INVERTER can be arranged to function as a 1-of-32 decoder. The decoders are labeled Z1–Z4 for easy reference, and the eight outputs from each one are combined into 32 outputs. Z1's outputs are $\overline{O}_0$–$\overline{O}_7$, Z2's $\overline{O}_0$–$\overline{O}_7$ outputs are renamed $\overline{O}_8$–$\overline{O}_{15}$, respectively, Z3's outputs are renamed $\overline{O}_{16}$–$\overline{O}_{23}$, and Z4's are renamed $\overline{O}_{24}$–$\overline{O}_{31}$. A 5-bit input code $A_4A_3A_2A_1A_0$ will activate only one of these 32 outputs for each of the 32 possible input codes.

(a) Which output will be activated for $A_4A_3A_2A_1A_0 = 01101$?
(b) What range of input codes will activate the Z4 chip?

SOLUTION

(a) The 5-bit input code has two distinct portions. The A_4 and A_3 bits determine which one of the decoder chips Z1–Z4 will be enabled, while $A_2A_1A_0$ determine which output of the enabled chip will be activated. With $A_4A_3 = 01$, only Z2 has all its enable inputs activated. Thus Z2 responds to the $A_2A_1A_0 = 101$ code and activates its $\overline{O}_5$ output, which has been renamed $\overline{O}_{13}$. Thus the input code 01101, which is the binary equivalent of decimal 13, will cause output $\overline{O}_{13}$ to go LOW, while all others stay HIGH.

(b) To enable Z4, both A_4 and A_3 have to be HIGH. Thus all input codes ranging from 11000(24_{10}) to 11111(31_{10}) will activate Z4. This corresponds to outputs $\overline{O}_{24}$ to $\overline{O}_{31}$.

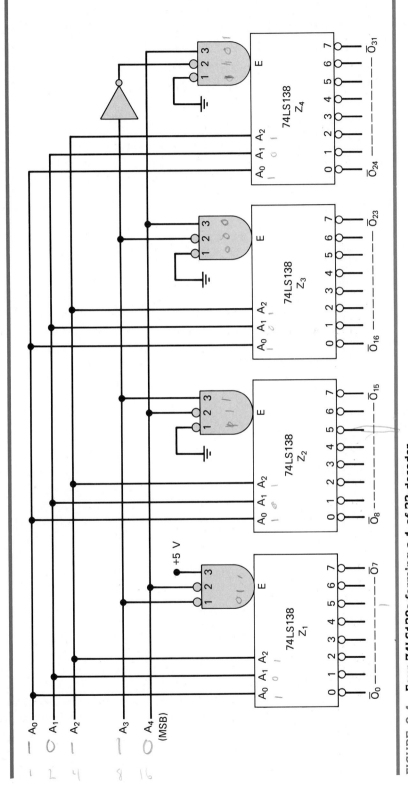

FIGURE 9-4 Four 74LS138s forming a 1-of-32 decoder.

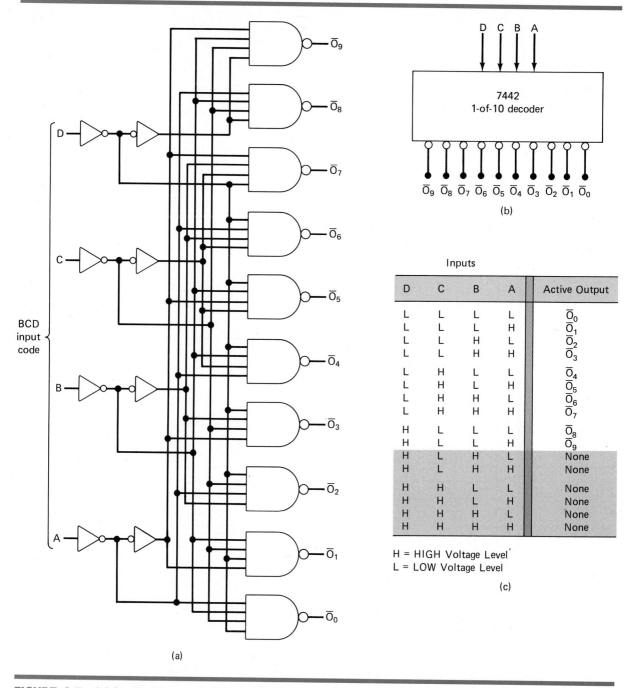

The following detail labels appear within the figure:

$\overline{O}_9$
$\overline{O}_8$
$\overline{O}_7$
$\overline{O}_6$
$\overline{O}_5$
$\overline{O}_4$
$\overline{O}_3$
$\overline{O}_2$
$\overline{O}_1$
$\overline{O}_0$

D
C
B
A

BCD input code

D C B A

7442
1-of-10 decoder

$\overline{O}_9$ $\overline{O}_8$ $\overline{O}_7$ $\overline{O}_6$ $\overline{O}_5$ $\overline{O}_4$ $\overline{O}_3$ $\overline{O}_2$ $\overline{O}_1$ $\overline{O}_0$

(b)

Inputs				
D	C	B	A	Active Output
L	L	L	L	$\overline{O}_0$
L	L	L	H	$\overline{O}_1$
L	L	H	L	$\overline{O}_2$
L	L	H	H	$\overline{O}_3$
L	H	L	L	$\overline{O}_4$
L	H	L	H	$\overline{O}_5$
L	H	H	L	$\overline{O}_6$
L	H	H	H	$\overline{O}_7$
H	L	L	L	$\overline{O}_8$
H	L	L	H	$\overline{O}_9$
H	L	H	L	None
H	L	H	H	None
H	H	L	L	None
H	H	L	H	None
H	H	H	L	None
H	H	H	H	None

H = HIGH Voltage Level
L = LOW Voltage Level

(c)

(a)

FIGURE 9-5 **(a) Logic diagram for the 7442 BCD-to-decimal decoder; (b) logic symbol; (c) truth table.**
(Courtesy of Fairchild, a Schlumberger company)

BCD-to-Decimal Decoders   Figure 9-5(a) shows the logic diagram for a 7442 *BCD-to-decimal decoder*. It is also available as a 74LS42 and 74HC42. Each output goes LOW only when its corresponding BCD input is applied. For example, $\overline{O}_5$ will go LOW only when inputs DCBA = 0101; $\overline{O}_8$ will go LOW only when DCBA = 1000. For input combinations that are invalid for BCD, none of the outputs will be activated. This decoder can also be referred to as a *4-to-10 decoder* or a *1-of-10 decoder*. The logic symbol and truth table for the 7442 are also shown in the figure. Note that this decoder does not have an enable input. In Problem 9-6 we will see how the 7442 can be used as a 3-to-8 decoder with the *D* input used as an enable input.

BCD-to-Decimal Decoder/Driver   The TTL 7445 is a BCD-to-decimal decoder/*driver*. The term ''driver'' is added to its description because this IC has open-collector outputs that can operate at higher current and voltage limits than a normal TTL output. The 7445's outputs can sink up to 80 mA in the LOW state, and can be pulled up to 30 V in the HIGH state. This makes them suitable for directly driving loads such as indicator LEDs or lights, relays, or dc motors.

Decoder Applications   Decoders are used whenever an output or group of outputs is to be activated only on the occurrence of a specific combination of input levels. These input levels are often provided by the outputs of a counter or register. When the decoder inputs come from a counter that is being continually pulsed, the decoder outputs will be activated sequentially, and they can be used as timing or sequencing signals to turn devices on or off at specific times. An example of this operation is shown in Figure 9-6 using the 74293 counter and the 7445 decoder described above. See page 492 for Figure 9-6.

EXAMPLE 9-3

Describe the operation of the circuit in Figure 9-6(a). (See page 492.)

SOLUTION

The counter is being pulsed by a l-pps signal so that it will sequence through the binary counts at the rate of 1 count/s. The counter FF outputs are connected as the inputs to the decoder. The 7445 open-collector outputs $\overline{O}_3$ and $\overline{O}_6$ are used to switch relays K1 and K2 on and off. For instance, when $\overline{O}_3$ is in its inactive HIGH state, its output transistor will be OFF (nonconducting) so that no current can flow through relay K1 and it will be deenergized. When $\overline{O}_3$ is in its active-LOW state, its output transistor is ON and acts as a current sink for current through K1 so that K1 is energized. Note that the relays operate from +24 V. Also note the presence of the diodes across the relay coils; these protect the decoder's output transistors from the large ''inductive kick'' voltage that would be produced when coil current is abruptly interrupted.

The timing diagram in Figure 9-6(b) shows the sequence of events. If we assume that the counter is in the 0000 state at time 0, then both outputs $\overline{O}_3$ and $\overline{O}_6$ are initially in the inactive-HIGH state, where their output transistors are OFF and both relays are deenergized. As clock pulses are applied, the counter will be incremented once per second. On the NGT of the third pulse (time 3), the counter will go to the 0011(3) state. This will activate decoder output $\overline{O}_3$ and thereby energize K1. On the NGT of the fourth pulse, the counter goes to the 0100(4) state. This will deactivate $\overline{O}_3$ and deenergize relay K1.

Similarly, at time 6 the counter will go to the 0110(6) state; this will make $\overline{O}_6 = 0$ and energize K2. At time 7 the counter goes to 0111(7) and deactivates $\overline{O}_6$ to deenergize K2.

The counter will continue counting as pulses are applied. After 16 pulses, the sequence just described will start over.

Decoders are widely used in the memory system of a computer, where they respond to the address code input from the central processor to activate the memory storage location specified by the address code. We will examine this application in detail when we study memories in Chapter 11.

FIGURE 9-6   Example 9-3; counter/decoder combination used to provide timing and sequencing operations.

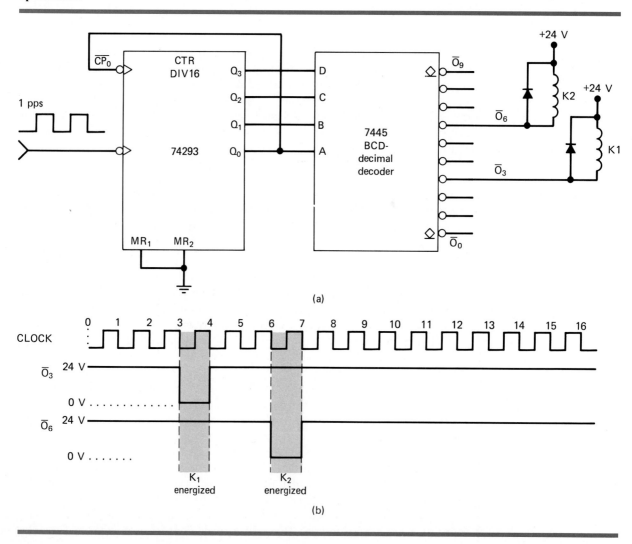

Another major application area for decoders is in the conversion of binary data to a form suitable for displaying on decimal readouts. We will look closely at this application in the following sections.

REVIEW QUESTIONS

1. Can more than one decoder output be activated at one time?
2. What is the function of a decoder's enable input(s)?
3. How does the 7445 differ from the 7442?
4. The 74154 is a 4-to-16 decoder with two active-LOW enable inputs. How many pins (including power and ground) does this IC have? (*Ans.* 24 pins, including 2 enables, 4 for the input code, 16 outputs, ground, and V_{CC}).

9-2 BCD-TO-7-SEGMENT DECODER/DRIVERS

Many numerical displays use a 7-segment configuration [Figure 9-7(a)] to produce the decimal characters 0–9 and sometimes the hex characters A–F. Each segment is made up of a material that emits light when current is passed through it. Most commonly used materials include light-emitting diodes (LEDs) and incandescent filaments. Figure 9-7(b) shows the patterns of segments which are used to display the various digits. For example, to display "6," segments c, d, e, f, and g are bright while segments a and b are dark.

A *BCD-to-7-segment decoder/driver* is used to take a 4-bit BCD input and provide the outputs that will pass current through the appropriate segments to display the decimal digit. The logic for this decoder is more complicated than those

FIGURE 9-7 **(a) Seven-segment arrangement; (b) active segments for each digit.**

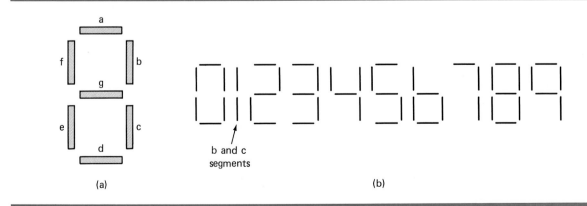

(a)

b and c
segments

(b)

we looked at previously because each output is activated for more than one combination of inputs. For example, the e segment must be activated for any of the digits 0, 2, 6, and 8, which means whenever any of the codes 0000, 0010, 0110, or 1000 occurs.

Figure 9-8(a) shows a BCD-to-7-segment decoder/driver (TTL 7446 or 7447) being used to drive a 7-segment LED readout. Each segment consists of one or two LEDs. The anodes of the LEDs are all tied to V_{CC} (+5 V). The cathodes of the LEDs are connected through current-limiting resistors to the appropriate outputs of the decoder/driver. The decoder/driver has active LOW outputs which are open-collector driver transistors that can sink a fairly large current. This is because LED readouts may require 10 mA to 40 mA per segment, depending on their type and size.

To illustrate the operation of this circuit, let us suppose that the BCD input is $D = 0$, $C = 1$, $B = 0$, $A = 1$, which is BCD for 5. With these inputs the decoder/driver outputs $\bar{a}$, $\bar{f}$, $\bar{g}$, $\bar{c}$, and $\bar{d}$ will be driven LOW (connected to ground),

FIGURE 9-8 **(a) BCD-to-7 segment decoder/driver driving a common-anode 7-segment LED display; (b) segment patterns for all possible input codes.**

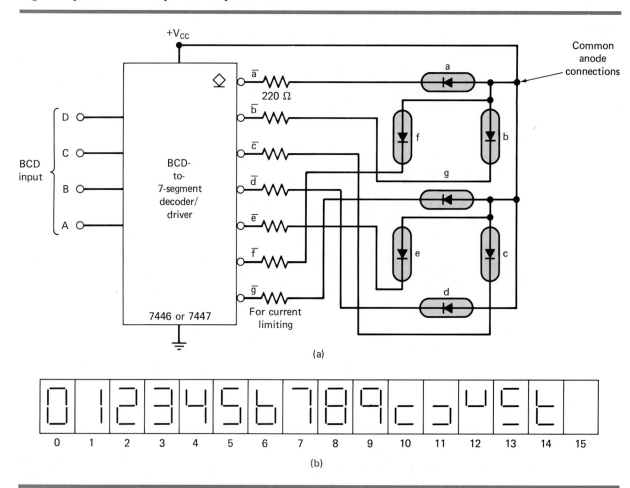

allowing current to flow through the a, f, g, c, and d LED segments and thereby displaying the numeral 5. The $\overline{b}$ and $\overline{e}$ outputs will be HIGH (open), so LED segments b and e cannot conduct.

The 7446 and 7447 decoder/drivers are designed to activate specific segments even for input codes greater than 1001 (9). Figure 9.8(b) shows which segments are activated for each of the input codes from 0000 to 1111 (15). Note that an input code of 1111 will blank out all the segments.

The LED display used in Figure 9-8 is a *common-anode* type because the anodes of each segment are tied together to V_{CC}. Another type of 7-segment LED display uses a *common-cathode* arrangement where the cathodes of each segment are tied together and connected to ground. This type of display has to be driven by a BCD-to-7-segment decoder/driver with active-HIGH outputs that apply a HIGH voltage to the anodes of those segments that are to be activated. The 7448 is a decoder/driver that can be used for this purpose.

EXAMPLE 9-4

Each segment of a typical 7-segment LED display is rated to operate at 10 mA at 2.7 V for normal brightness. Calculate the value of the current-limiting resistor needed to produce approximately 10 mA per segment.

SOLUTION
Referring to Figure 9-8(a), we can see that the series resistor will have to have a voltage drop equal to the difference between $V_{CC} = 5$ V and the segment voltage of 2.7 V. This 2.3 V across the resistor must produce a current of about 10 mA. Thus we have

$$R_S = \frac{2.3 \text{ V}}{10 \text{ mA}} = 230 \ \Omega$$

A standard resistor value close to this can be used. A 220-Ω resistor would be a good choice.

REVIEW QUESTIONS

1. Which LED segments will be ON for a decoder/driver input of 1001? (*Ans.* Segments a, b, c, f, and g)
2. *True or false:* More than one output of a BCD-to-7-segment decoder/driver can be active at one time?

9-3   LIQUID CRYSTAL DISPLAYS

Basically, LCDs operate from a low-voltage (typically 3 to 15 V rms), low-frequency (25 to 60 Hz) ac signal and draw very little current. They are often arranged

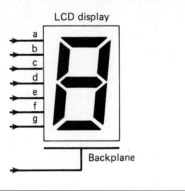

LCD display

Backplane

FIGURE 9-9 **Liquid crystal display.**

as 7-segment displays for numerical readouts as shown in Figure 9-9. The ac voltage needed to turn ON a segment is applied between the segment and the *backplane,* which is common to all segments. The segment and backplane form a capacitor that draws very little current as long as the ac frequency is kept low. It is generally not lower than 25 Hz, because this would produce visible flicker.

LCDs draw much less current than LED displays and are widely used in battery-powered devices such as calculators and watches. An LCD does not emit light energy like an LED, and so it requires an external source of light.

Driving an LCD
An LCD segment will turn ON when an ac voltage is applied between the segment and the backplane, and will turn OFF when there is no voltage between the two. Rather than generating an ac signal, it is common practice to produce the required ac voltage by applying out-of-phase squarewaves to the segment and backplane. This is illustrated in Figure 9-10 for one segment. A 40-Hz squarewave is applied to the backplane and also to the input of a CMOS 4070 EXCLUSIVE-OR. The other input to the EX-OR is a CONTROL input that will control whether the segment is ON or OFF.

When the CONTROL input is LOW, the EX-OR output will be exactly the same as the 40-Hz squarewave, so that the signals applied to the segment and back-

FIGURE 9-10 **Method for driving an LCD segment. When CONTROL is LOW, segment is OFF. When CONTROL is HIGH, segment is ON.**

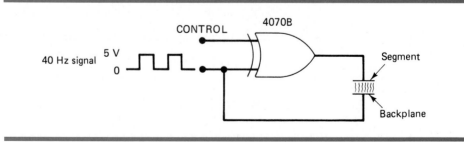

plane are equal. Since there is no difference in voltage, the segment will be OFF. When the CONTROL input is HIGH, the EX-OR output will be the INVERSE of the 40-Hz squarewave, so that the signal applied to the segment is out of phase with the signal applied to the backplane. As a result, the segment voltage will alternately be at $+5$ V and at -5 V relative to the backplane. This ac voltage will turn ON the segment.

This same idea can be extended to a complete 7-segment LCD display as shown in Figure 9-11. Here the CMOS 4511 BCD-to-7-segment decoder/driver supplies the CONTROL signals to each of seven EX-ORs for the seven segments. The 4511 has active HIGH outputs, since a HIGH is required to turn ON a segment.

In general, CMOS devices are used to drive LCDs for two reasons: (1) they require much less power than TTL and are more suited to the battery-operated applications where LCDs are used; (2) the TTL LOW-state voltage is not exactly 0 V and can be as much as 0.4 V. This will produce a dc component of voltage between the segment and backplane that considerably shortens the life of an LCD.

FIGURE 9-11   Method for driving 7-segment LCD.

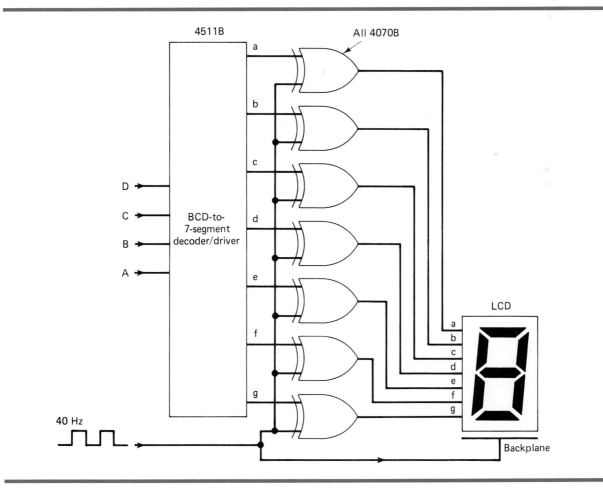

LIQUID CRYSTAL DISPLAYS

REVIEW QUESTION

9-4 ENCODERS

A decoder accepts an N-bit input code and produces a HIGH (or LOW) at *one and only one* output line. In other words, we can say that a decoder identifies, recognizes, or detects a particular code. The opposite of this decoding process is called *encoding* and is performed by a logic circuit called an *encoder*. An encoder has a number of input lines, only one of which is activated at a given time, and produces an N-bit output code, depending on which input is activated. Figure 9-12 is the general diagram for an encoder with M inputs and N outputs. Here the inputs are active-HIGH, which means they are normally LOW.

We saw that a *binary-to-octal decoder (3-line-to-8-line decoder)* accepts a 3-bit input code and activates one of eight output lines corresponding to that code. An *octal-to-binary encoder (8-line-to-3-line encoder)* performs the opposite function; it accepts eight input lines and produces a 3-bit output code corresponding to the activated input. Figure 9-13 shows the logic circuit and truth table for an octal-to-binary encoder with active-LOW inputs.

By following through the logic, you can verify that a LOW at any single input will produce the output binary code corresponding to that input. For instance, a LOW at $\overline{A}_3$ (while all other inputs are HIGH) will produce $\overline{O}_2 = 0$, $\overline{O}_1 = 1$, and $\overline{O}_0 = 1$, which is the binary code for 3. Notice that $\overline{A}_0$ is not connected to the logic gates because the encoder outputs will normally be at 000 when none of the $\overline{A}_1$–$\overline{A}_9$ inputs is LOW.

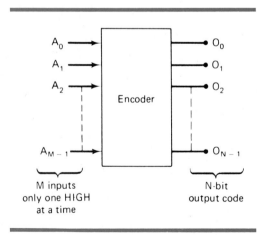

M inputs
only one HIGH
at a time

N-bit
output code

FIGURE 9-12 **General encoder diagram.**

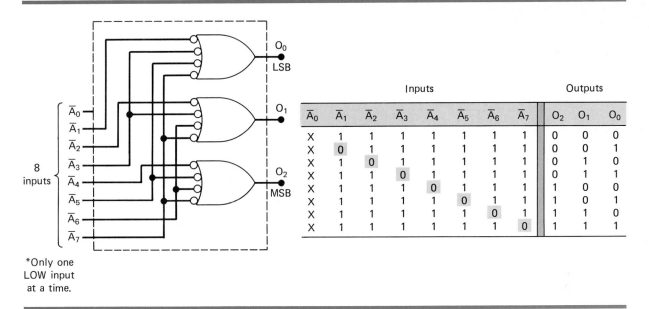

Inputs								Outputs		
$\overline{A}_0$	$\overline{A}_1$	$\overline{A}_2$	$\overline{A}_3$	$\overline{A}_4$	$\overline{A}_5$	$\overline{A}_6$	$\overline{A}_7$	O_2	O_1	O_0
X	1	1	1	1	1	1	1	0	0	0
X	0	1	1	1	1	1	1	0	0	1
X	1	0	1	1	1	1	1	0	1	0
X	1	1	0	1	1	1	1	0	1	1
X	1	1	1	0	1	1	1	1	0	0
X	1	1	1	1	0	1	1	1	0	1
X	1	1	1	1	1	0	1	1	1	0
X	1	1	1	1	1	1	0	1	1	1

8 inputs: $\overline{A}_0$, $\overline{A}_1$, $\overline{A}_2$, $\overline{A}_3$, $\overline{A}_4$, $\overline{A}_5$, $\overline{A}_6$, $\overline{A}_7$

*Only one LOW input at a time.

FIGURE 9-13 **Logic circuit for an octal-to-binary (8-line-to-3-line) encoder. For proper operation, only one input should be active at one time.**

EXAMPLE 9-5

Determine the outputs of the encoder in Figure 9-13 when $\overline{A}_3$ and $\overline{A}_5$ are simultaneously LOW.

SOLUTION
Following through the logic gates, we see that the LOWs at these two inputs will produce HIGHs at each output, in other words, the binary code 111. Clearly, this is not the code for either activated input.

Priority Encoders   This last example points up a drawback of the simple encoder circuit of Figure 9-13 when more than one input is activated at one time. A modified version of this circuit, called a *priority encoder* includes the necessary logic to ensure that when two or more inputs are activated, the output code will correspond to the highest-numbered input. For example, when both $\overline{A}_3$ and $\overline{A}_5$ are LOW, the output code will be 101(5). Similarly, when $\overline{A}_6$, $\overline{A}_2$, and $\overline{A}_0$ are all LOW, the output code is 110(6). The 74148, 74LS148, and 74HC148 are all octal-to-binary priority encoders.

74147 Decimal-to-BCD Priority Encoder   Figure 9-14 shows the logic symbol and truth table for the 74147 (74LS147, 74HC147), which functions as a decimal-to-BCD priority encoder. It has nine active-LOW inputs representing the decimal digits 1–9, and produces the *inverted* BCD code corresponding to the highest-numbered activated input.

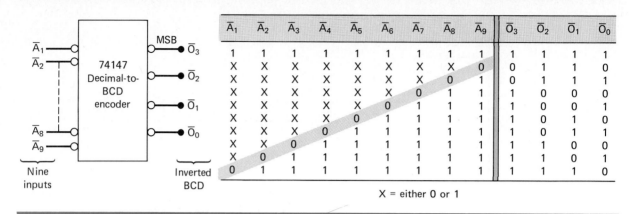

$\overline{A}_1$	$\overline{A}_2$	$\overline{A}_3$	$\overline{A}_4$	$\overline{A}_5$	$\overline{A}_6$	$\overline{A}_7$	$\overline{A}_8$	$\overline{A}_9$	$\overline{O}_3$	$\overline{O}_2$	$\overline{O}_1$	$\overline{O}_0$
1	1	1	1	1	1	1	1	1	1	1	1	1
X	X	X	X	X	X	X	X	0	0	1	1	0
X	X	X	X	X	X	X	0	1	0	1	1	1
X	X	X	X	X	X	0	1	1	1	0	0	0
X	X	X	X	X	0	1	1	1	1	0	0	1
X	X	X	X	0	1	1	1	1	1	0	1	0
X	X	X	0	1	1	1	1	1	1	0	1	1
X	X	0	1	1	1	1	1	1	1	1	0	0
X	0	1	1	1	1	1	1	1	1	1	0	1
0	1	1	1	1	1	1	1	1	1	1	1	0

X = either 0 or 1

FIGURE 9-14   **74147 decimal-to-BCD priority encoder.**

Let's examine the truth table to see how this IC works. The first line in the table shows all inputs in their inactive-HIGH state. For this condition the outputs are 1111, which is the inverse of 0000, the BCD code for 0. The second line in the table indicates that a LOW at $\overline{A}_9$, regardless of the states of the other inputs, will produce an output code of 0110 which is the inverse of 1001, the BCD code for 9. The third line shows that a LOW at $\overline{A}_8$, provided that $\overline{A}_9$ is HIGH, will produce an output code of 0111, the inverse of 1000, the BCD code for 8. In a similar manner, the remaining lines in the table show that a LOW at any input, provided that all higher-numbered inputs are HIGH, will produce the inverse of the BCD code for that input.

The 74147 outputs will be normally HIGH when none of the inputs are activated. This corresponds to the decimal 0 input condition. There is no $\overline{A}_0$ input, since the encoder assumes the decimal 0 input state when all other inputs are HIGH. The 74147 inverted BCD outputs can be converted to normal BCD by putting each one through an INVERTER.

Switch Encoder   Figure 9-15 shows how a 74147 can be used as a *switch encoder*. The 10 switches might be the keyboard switches on a calculator representing digits 0–9. The switches are normally open types, so that the encoder inputs are all normally HIGH, and the BCD output is 0000 (note the INVERTERs). When a digit key is depressed, the circuit will produce the BCD code for that digit. Since the 74147 is a priority encoder, simultaneous key depressions will produce the BCD code for the higher-numbered key.

The switch encoder of Figure 9-15 can be used whenever BCD data have to be manually entered into a digital system. A prime example would be in an electronic calculator, where the operator depresses several keyboard switches in succession to enter a decimal number. In a simple, basic calculator the BCD code for each decimal digit is entered into a 4-bit storage register. In other words, when the first key is depressed, the BCD code for that digit is sent to a 4-bit FF register; when the second switch is depressed, the BCD code for that digit is sent to *another* 4-bit FF register, and so on. Thus, a calculator that can handle eight digits will have eight 4-bit registers to store the BCD codes for these digits. Each 4-bit register

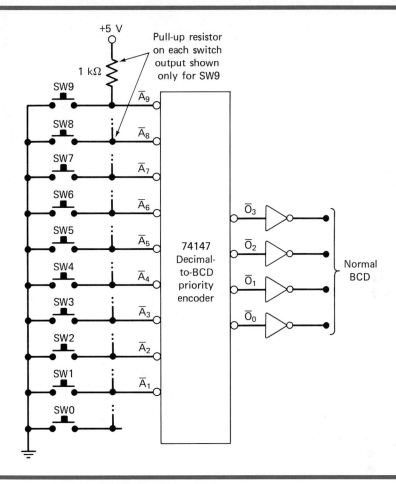

FIGURE 9-15 Decimal-to-BCD switch encoder.

drives a decoder/driver and numerical display so that the eight-digit number can be displayed.

The operation described above can be accomplished with the circuit in Figure 9-16. This circuit will take three decimal digits entered from the keyboard in sequence, encode them in BCD, and store the BCD in three FF output registers. The 12 D-type FFs Q_0–Q_{11} are used to receive and store the BCD codes for the digits. Q_8–Q_{11} stores the BCD code for the most significant digit (MSD), which is the first one entered on the keyboard. Q_4–Q_7 stores the second entered digit and Q_0–Q_3 stores the third entered digit. The X, Y, and Z FFs form a ring counter (Chapter 7) which controls the transfer of data from the encoder outputs to the appropriate output register. The OR gate produces a HIGH output any time one of the keys is depressed. This output may be affected by switch contact bounce, which would produce several pulses before settling down to the HIGH state. The OS is used to neutralize the switch bounce by triggering on the first positive transition from the OR gate and remaining HIGH for 20 ms, well past the time duration of the switch bounce. The OS output clocks the ring counter.

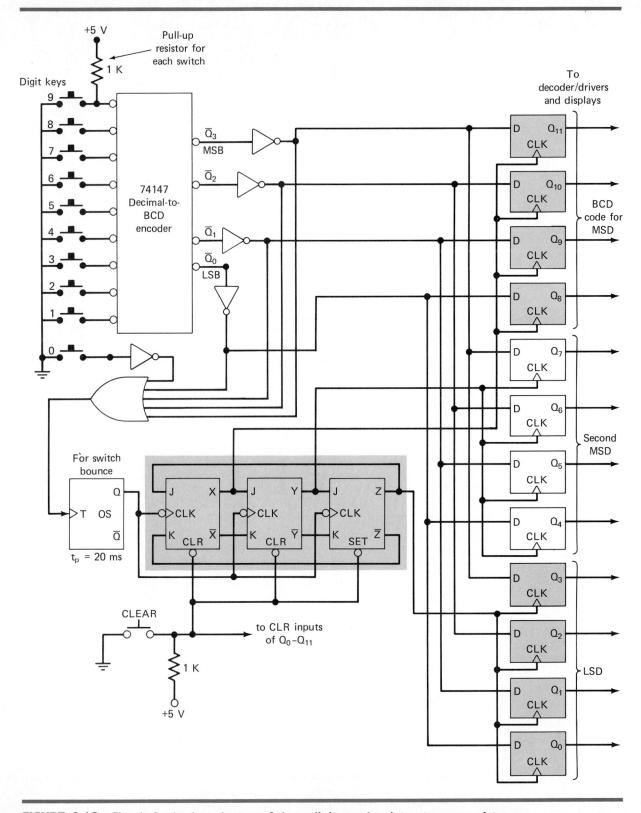

FIGURE 9-16 Circuit for keyboard entry of three-digit number into storage registers.

The circuit operation is described as follows for the case where the decimal number 309 is being entered:

1. The CLEAR key is depressed. This clears all the Q_0–Q_{11} storage FFs to 0. It also clears FFs X and Y and presets FF Z to 1, so the ring counter begins in the 001 state.

2. The CLEAR key is released and the "3" key is depressed. The encoder outputs 1100 are inverted to produce 0011, the BCD code for 3. These binary values are sent to the D inputs of the three 4-bit output registers.

3. The OR output goes HIGH (since two inputs have gone HIGH) and triggers the OS output $Q = 1$ for 20 ms. After 20 ms Q returns LOW and clocks the ring counter to the 100 state (X goes HIGH). The positive transition at X is fed to the CLK inputs of FFs Q_8–Q_{11}, so the encoder outputs are transferred to these FFs. That is, $Q_{11} = 0$, $Q_{10} = 0$, $Q_9 = 1$, and $Q_8 = 1$. Note that FFs Q_0–Q_7 are not affected because their *CLK* inputs have not received a positive transition.

4. The "3" key is released and the OR gate output returns LOW. The "0" key is then depressed. This produces the BCD code of 0000, which is fed to the inputs of the storage FFs.

5. The OR output goes HIGH in response to the "0" key (note the inverter) and triggers the OS for 20 ms. After 20 ms the ring counter shifts to the 010 state (Y goes HIGH). The positive transition at Y is fed to the *CLK* inputs of Q_4–Q_7 and transfers 0000 to these FFs. Note that FFs Q_0–Q_3 and Q_8–Q_{11} are not affected by the Y transition.

6. The "0" key is released and the OR output returns LOW. The "9" key is depressed, producing BCD outputs 1001, which are fed to the storage FFs.

7. The OR output goes HIGH again, triggering the OS, which in turn clocks the ring counter to the 001 state (Z goes HIGH). The positive transition at Z is fed to the *CLK* inputs of Q_0–Q_3 and transfers the 1001 into these FFs. The other storage FFs are unaffected.

8. At this point the storage register contains 0011 0000 1001, beginning with Q_{11}. This is the BCD code of 309. These register outputs feed decoder/drivers which drive appropriate displays for indicating the decimal digits 309.

9. The storage FF outputs are also fed to other circuits in the system. In a calculator, for example, these outputs would be sent to the arithmetic section to be processed.

Several problems at the end of the chapter will deal with some other aspects of this circuit including troubleshooting exercises.

REVIEW QUESTIONS

1. How does an encoder differ from a decoder?
2. How does a priority encoder differ from an ordinary encoder?

3. What will be the outputs of the 74147 encoder if all its inputs are LOW? (*Ans.* 0110)
4. Describe the functions of each of the following parts of the keyboard entry circuit of Figure 9-16: (a) OR gate; (b) 74147 encoder; (c) one-shot; (d) FFs, *X, Y, Z;* (e) FFs $Q_0–Q_{11}$.

FIGURE 9-17   IEEE/ANSI symbols for several decoders.

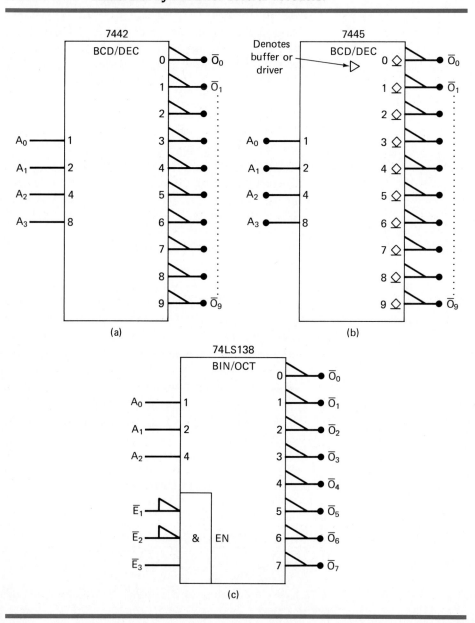

(a)

(b)

(c)

9-5 IEEE/ANSI SYMBOLS

We will now look at the IEEE/ANSI symbols for several of the decoders and encoders. Figure 9-17(a) shows the IEEE/ANSI symbol for the 7442 decoder. It is fairly straightforward. Note the label BCD/DEC, which denotes that it is a BCD-to-decimal decoder. Also note the manner in which the inputs and outputs are numbered inside the symbol block.

The IEEE/ANSI symbol for the 7445 in Figure 9-17(b) is similar to the 7442 with two differences. First, each output has the underlined diamond symbol to indicate its open-collector structure. Second, the triangle in the middle of the symbol denotes that this device is a buffer or a driver with greater-than-normal current and/or voltage capabilities.

Figures 9-17(c) is the symbol for the 74LS138 decoder. Note the BIN/OCT label to denote its binary-to-octal decoding function. Also note how the three enable inputs are combined in an AND block to produce an overall internal enable signal, EN.

Figure 9-18 shows the IEEE/ANSI symbol for the 74147 encoder IC. The label HPRI/BCD denotes that the function of this IC is to convert the active input with the highest priority to its BCD code. Again, note how the inputs and outputs are numbered inside the symbol block.

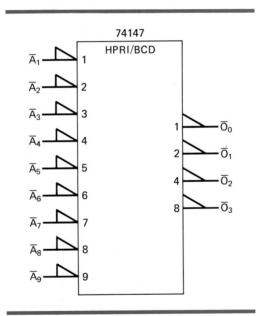

FIGURE 9-18 IEEE/ANSI symbol for the 74147 encoder.

REVIEW QUESTIONS

1. What does the label BCD/DEC mean?
2. What does the label HPRI mean?
3. What does the symbol $\triangleright$ mean inside a symbol block?

9-6 TROUBLESHOOTING

As circuits and systems become more complex, the number of possible causes of failure obviously increases. While the procedure for fault isolation and correction remains essentially the same, the application of the *observation/analysis* process is more important for complex circuits because it helps the troubleshooter narrow the location of the fault to a small area of the circuit. This reduces to a reasonable amount the testing steps and resulting data that has to be analyzed. By understanding the circuit operation, observing the symptoms of the failure, and reasoning through the operation, the troubleshooter can often predict the possible faults before ever picking up a logic probe or oscilloscope. This observation/analysis process is one that inexperienced troubleshooters are hesitant to apply, probably because of the great variety and capabilities of modern test equipment available to them. It is easy to become overly reliant on these tools while not adequately utilizing the human brain's reasoning and analytical skills.

The following examples illustrate how the observation/analysis process can be applied. Many of the end-of-chapter troubleshooting problems will provide you with the opportunity to develop your skill at applying this process.

EXAMPLE 9-6

A technician tests the circuit of Figure 9-4 by using a set of switches to apply the input code at A_4–A_0. She runs through each possible input code and checks the corresponding decoder output to see if it is activated. She observes that all of the odd-numbered outputs respond correctly, but all of the even-numbered outputs fail to respond when their code is applied. What are the most probable faults?

SOLUTION

In a situation where so many outputs are failing, it is unreasonable to expect that each output has a fault. It is much more likely that some faulty input condition is causing the output failures. What do all of the even-numbered outputs have in common? Let's look at the input codes for several of them.

OUTPUT	INPUT CODE
$\overline{O}_0$	00000
$\overline{O}_4$	00100
$\overline{O}_{14}$	01110
$\overline{O}_{18}$	10010

Clearly, each even-numbered output requires an input code with an $A_0 = 0$ in order to be activated. Thus, the most probable faults would be those that prevent A_0 from going LOW. This includes:

- A faulty switch connected to the A_0 output.
- An open in the path between the switch and the A_0 line.

- An external short from the A_0 line to V_{CC}.
- An internal short to V_{CC} at the A_0 inputs of any one of the decoder chips.

Thus, the fault is narrowed to a specific area of the circuit. The exact fault can be traced with the testing and measurement techniques that we are already familiar with.

EXAMPLE 9-7

A technician wires the outputs from a BCD counter to the inputs of the decoder/driver of Figure 9-8. He applies pulses to the counter at a very slow rate and observes the LED display as the counter counts up from 0000 to 1001. The observed sequence is recorded below. Examine it carefully and try to predict the most probable fault.

COUNT	0	1	2	3	4	5	6	7	8	9
Observed display	0	1	2	3	4	5	6	7	8	9
Expected display	0	1	2	3	4	5	6	7	8	9

SOLUTION

Comparing the observed display with the expected display for each count, we see several important points:

- For those counts where the observed display is incorrect, the observed display is not one of the segment patterns that correspond to counts greater than 1001. This rules out a faulty counter or faulty wiring from the counter to the decoder/driver.
- The correct segment patterns (0, 1, 3, 6, 7, and 8) have the common property that segments e and f are either both on or both off.
- The incorrect segment patterns have the common property that segments e and f are in opposite states, and if we interchange the states of these two segments, the correct pattern is obtained.

Giving some thought to those points should lead us to conclude that the technician has probably "crossed" the connections to the e and f segments.

9-7 MULTIPLEXERS (DATA SELECTORS)

A *multiplexer* or *data selector* is a logic circuit that accepts several data inputs and allows only *one* of them at a time to get through to the output. The routing of the desired data input to the output is controlled by SELECT inputs (sometimes referred to as ADDRESS inputs). Figure 9-19 shows the functional diagram of a general

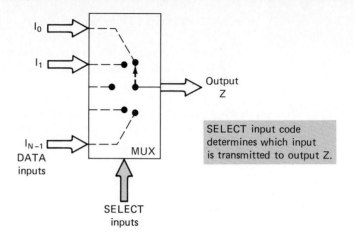

FIGURE 9-19   **Functional diagram of a digital multiplexer (MUX).**

multiplexer (MUX). In this diagram the inputs and outputs are drawn as large arrows to indicate that they may be one or more signal lines.

The multipilexer acts like a digitally controlled multiposition switch where the digital code applied to the SELECT inputs controls which data inputs will be switched to the output. For example, output Z will equal data input I_0 for some particular SELECT input code; Z will equal I_1 for another particular SELECT input code; and so on. Stated another way, a multiplexer selects 1 out of N input data sources and transmits the selected data to a single output channel. This is called *multiplexing*.

Basic Two-Input Multiplexer
Figure 9-20 shows the logic circuitry for a two-input multiplexer with data inputs I_0 and I_1 and SELECT input S. The logic level applied to the S input determines which AND gate is enabled so that its data input passes through the OR gate to output Z. Looking at it another way, the Boolean expression for the output is

$$Z = I_0\bar{S} + I_1S$$

With $S = 0$, this expression becomes

$$Z = I_0 \cdot 1 + I_1 \cdot 0$$

$$= I_0$$

which indicates that Z will be identical to input signal I_0, which can be a fixed logic level or a time-varying logic signal. With $S = 1$, the expression becomes

$$Z = I_0 \cdot 0 + I_1 \cdot 1 = I_1$$

showing that output Z will be identical to input signal I_1.

Four-Input Multiplexer
The same basic idea can be used to form the four-input multiplexer shown in Figure 9-21. Here there are four inputs, which are selectively transmitted to the output based on the four possible combinations of the S_1S_0 select inputs. Each data input is gated with a different combination of select input levels. I_0 is gated with $\bar{S}_1\bar{S}_0$ so that I_0 will pass through its AND gate to output

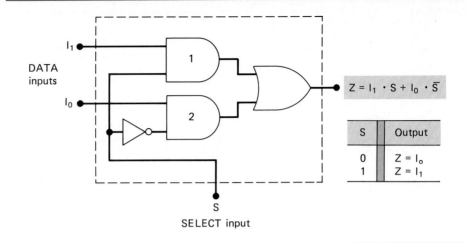

$$Z = I_1 \cdot S + I_0 \cdot \bar{S}$$

S	Output
0	$Z = I_0$
1	$Z = I_1$

DATA inputs

I_1

I_0

S

SELECT input

FIGURE 9-20 Two-input multiplexer.

Z only when $S_1 = 0$ and $S_0 = 0$. The table in the figure gives the outputs for the other three input select codes.

Two-, four-, eight-, and 16-input multiplexers are readily available in the TTL and CMOS logic families. These basic ICs can be combined for multiplexing a larger number of inputs.

FIGURE 9-21 Four-input multiplexer.

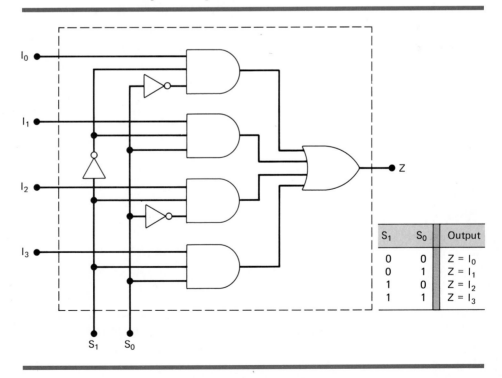

S_1	S_0	Output
0	0	$Z = I_0$
0	1	$Z = I_1$
1	0	$Z = I_2$
1	1	$Z = I_3$

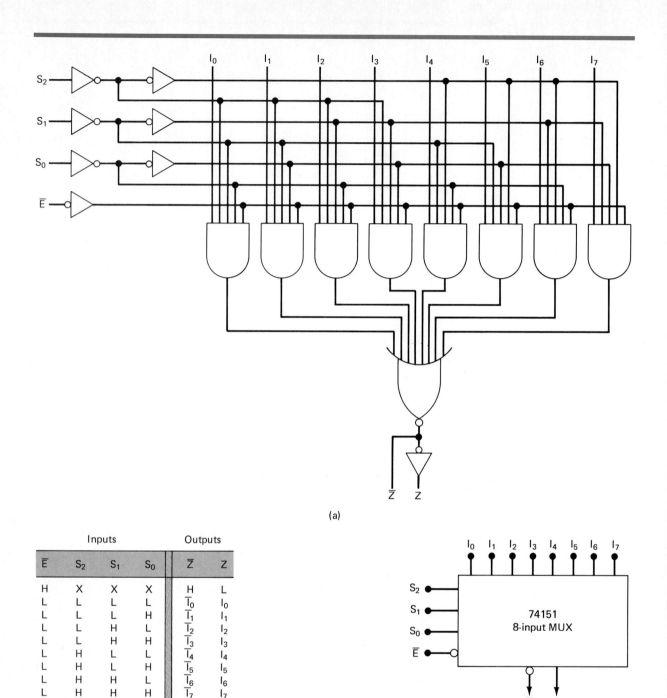

(a)

Inputs				Outputs	
$\bar{E}$	S_2	S_1	S_0	$\bar{Z}$	Z
H	X	X	X	H	L
L	L	L	L	$\bar{I_0}$	I_0
L	L	L	H	$\bar{I_1}$	I_1
L	L	H	L	$\bar{I_2}$	I_2
L	L	H	H	$\bar{I_3}$	I_3
L	H	L	L	$\bar{I_4}$	I_4
L	H	L	H	$\bar{I_5}$	I_5
L	H	H	L	$\bar{I_6}$	I_6
L	H	H	H	$\bar{I_7}$	I_7

(b)

(c)

FIGURE 9-22 **(a) Logic diagram for 74151 multiplexer; (b) truth table; (c) logic symbol.** (Courtesy of Fairchild, a Schlumberger company)

Eight-Input Multiplexer Figure 9-22(a) shows the logic diagram for the 74151 (74HC151) eight-input multiplexer. This multiplexer has an enable input, $\overline{E}$, and provides both the normal and inverted outputs. When $\overline{E} = 0$, the select inputs $S_2 S_1 S_0$ will select one data input (I_0–I_7) for passage to output Z. When $\overline{E} = 1$, the multiplexer is disabled so that $Z = 0$ regardless of the select input code. This operation is summarized in Figure 9-22(b), and the 74151 logic symbol is shown in Figure 9-22(c).

EXAMPLE 9-8

The circuit in Figure 9-23 uses two 74151s, an INVERTER, and an OR gate. Describe this circuit's operation.

SOLUTION
This circuit has a total of 16 data inputs, eight applied to each multiplexer. The two multiplexer outputs are combined in the OR gate to produce a single output

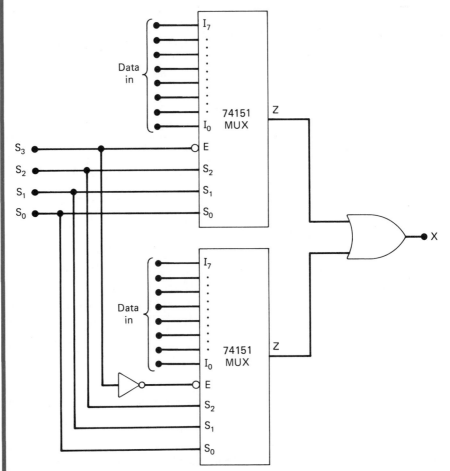

FIGURE 9-23 **Example 9.8 showing two 74151s combined to form a 16-input multiplexer.**

X. The circuit functions as a 16-input multiplexer. The four select inputs $S_3 S_2 S_1 S_0$ will select one of the 16 inputs to pass through to X.

The S_3 input determines which multiplexer is enabled. When $S_3 = 0$, the top multiplexer is enabled, and the $S_2 S_1 S_0$ inputs determine which of its data inputs will appear at its output and pass through the OR gate to X. When $S_3 = 1$, the bottom multiplexer is enabled and the $S_2 S_1 S_0$ inputs select one of its data inputs for passage to output X.

Quad Two-Input Multiplexer (74157)

This is a very useful multiplexer IC that contains four two-input multiplexers like the one in Figure 9-20. The logic diagram for the 74157 is shown in Figure 9-24(a). Note the manner in which the data inputs and outputs are labeled.

EXAMPLE 9-9

Determine the input conditions required for each Z output to take on the logic level of its corresponding I_0 input. Repeat for I_1.

SOLUTION
First of all, the enable input has to be active; that is, $\overline{E} = 0$. In order for Z_a to equal I_{0a}, the select input has to be LOW. These same conditions will produce $Z_b = I_{0b}$, $Z_c = I_{0c}$, and $Z_d = I_{0d}$.

With $\overline{E} = 0$ and $S = 1$, the Z outputs will follow the set of I_1 inputs; that is, $Z_a = I_{1a}$, $Z_b = I_{1b}$, $Z_c = I_{1c}$, and $Z_d = I_{1d}$.

All the outputs will be disabled (LOW) when $\overline{E} = 1$.

It is helpful to think of this multiplexer as being a simple two-input multiplexer, but where each input is four lines and the output is four lines. The four output lines switch back and forth between the two sets of four input lines under the control of the select input. This operation is represented by the 74157's logic symbol in Figure 9-24(b).

REVIEW QUESTIONS

1. What is the function of a multiplexer's select inputs?
2. A certain multiplexer can switch one of 32 data inputs to its output. How many different inputs does this MUX have? (*Ans.* 32 data inputs and five select inputs)

9-8 MULTIPLEXER APPLICATIONS

Multiplexer circuits find numerous and varied applications in digital systems of all types. These applications include data selection, data routing, operation sequencing, parallel-to-serial conversion, waveform generation, and logic-function generation. We shall look at some of these applications here and several more in the problems at the end of the chapter.

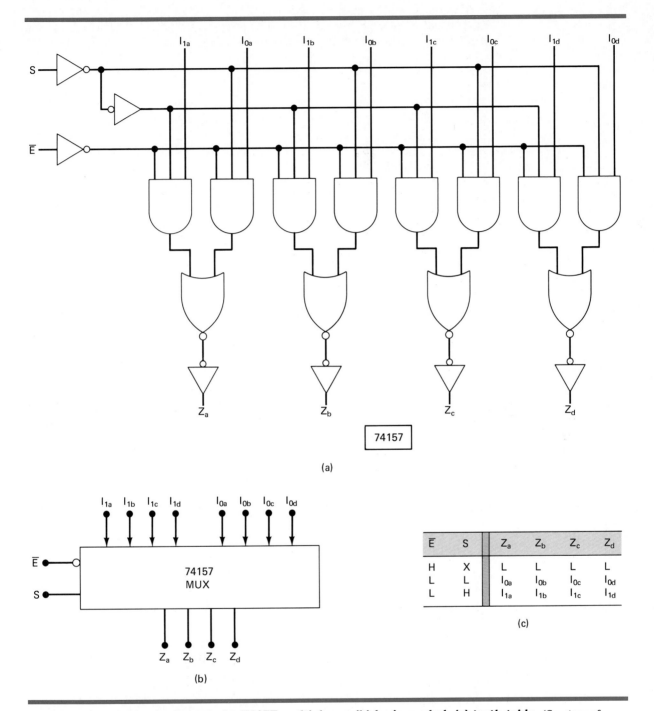

FIGURE 9-24 **(a) Logic diagram for 74157 multiplexer; (b) logic symbol; (c) truth table.** (Courtesy of Fairchild, a Schlumberger company)

MULTIPLEXER APPLICATIONS 513

Data Routing   Multiplexers can route data from one of several sources to one destination. One typical application uses 74157 multiplexers to select and display the contents of either of two BCD counters using a *single* set of decoder/drivers and LED displays. The circuit arrangement is shown in Figure 9-25.

FIGURE 9-25   System for displaying two multidigit BCD counters one at a time.

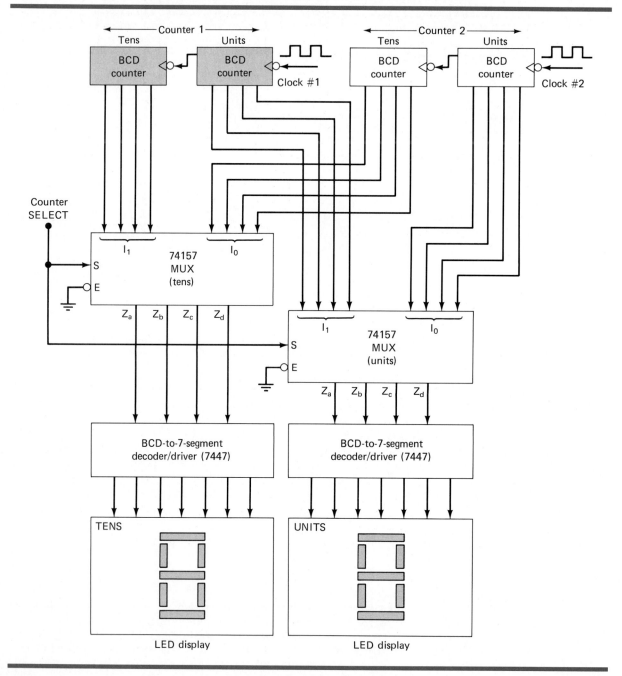

Each counter consists of two cascaded BCD stages, and each one is driven by its own clock signal. When the COUNTER SELECT line is HIGH, the outputs of counter 1 will be allowed to pass through the multiplexers to the decoder/drivers to be displayed on the LED readouts. When COUNTER SELECT = 0, the outputs of counter 2 will pass through the multiplexers to the displays. In this way the decimal contents of one counter or the other will be displayed under the control of the COUNTER SELECT input. A common situation where this might be used is in a digital watch. The digital watch circuitry contains many counters and registers that keep track of seconds, minutes, hours, days, months, alarm settings, and so on. A multiplexing scheme such as this one allows different data to be displayed on the limited number of decimal readouts.

The purpose of the multiplexing technique, as it is used here, is to *time-share* the decoder/drivers and display circuits between the two counters rather than have a separate set of decoder/drivers and displays for each counter. This results in a significant saving in the number of wiring connections, especially when more BCD stages are added to each counter. Even more importantly, it represents a significant decrease in power consumption, because decoder/drivers and LED readouts typically draw relatively large amounts of current from the V_{CC} supply. Of course, this technique has the limitation that only one counter's contents can be displayed at a time. However, in many applications this is not a drawback. A mechanical switching arrangement could have been used to perform the function of switching first one counter and then the other to the decoder/drivers and displays, but the number of required switch contacts, the complexity of wiring, and the physical size could all be disadvantages over the completely logic method of Figure 9-25.

Parallel-to-Serial Conversion   Many digital systems process binary data in parallel form (all bits simultaneously) because it is faster. When these data are to be transmitted over relatively long distances, however, the parallel arrangement is undesirable because it requires a large number of transmission lines. For this reason, binary data or information that are in parallel form are often converted to serial form before being transmitted to a remote destination. One method for performing this *parallel-to-serial conversion* uses a multiplexer, as illustrated in Figure 9-26.

The data are present in parallel form at the outputs of the X register and are fed to the eight-input multiplexer. A 3-bit (MOD-8) counter is used to provide the select code bits $S_2 S_1 S_0$ so that they cycle through from 000 to 111 as clock pulses are applied. In this way, the output of the multiplexer will be X_0 during the first clock period, X_1 during the second clock period, and so on. The output Z is a waveform which is a serial representation of the parallel input data. The waveforms in the figure are for the case where $X_7 X_6 X_5 X_4 X_3 X_2 X_1 X_0 = 10110101$. This conversion process takes a total of eight clock cycles. Note that X_0 (the LSB) is transmitted first and the X_7 (MSB) is transmitted last.

Operation Sequencing   The circuit of Figure 9-27 uses an eight-input multiplexer as part of a control sequencer that steps through seven steps, each of which actuates some portion of the physical process being controlled. This process could, for example, be a large high-temperature oven which is energized by seven different heaters, which must be activated one at a time. The circuit also uses a

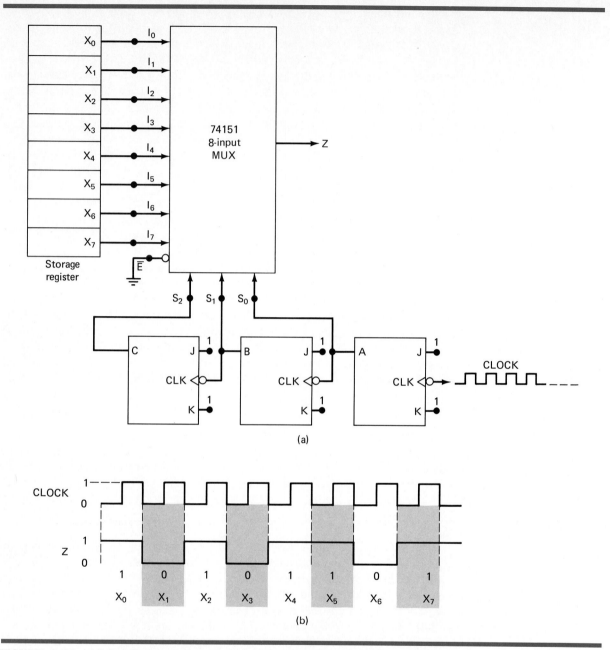

FIGURE 9-26 (a) Parallel-to-serial converter; (b) waveforms for $X_7X_6X_5X_4X_3X_2X_1X_0$ = 10110101.

3-line-to-8-line decoder and a MOD-8 binary counter. The operation is described as follows:

1. Initially the counter is reset to the 000 state. The counter outputs are fed to the select inputs of the multiplexer and to the inputs of the decoder. Thus, the decoder output $\overline{O}_0$ = 0 and the others are all 1, so all the ACTUATOR inputs

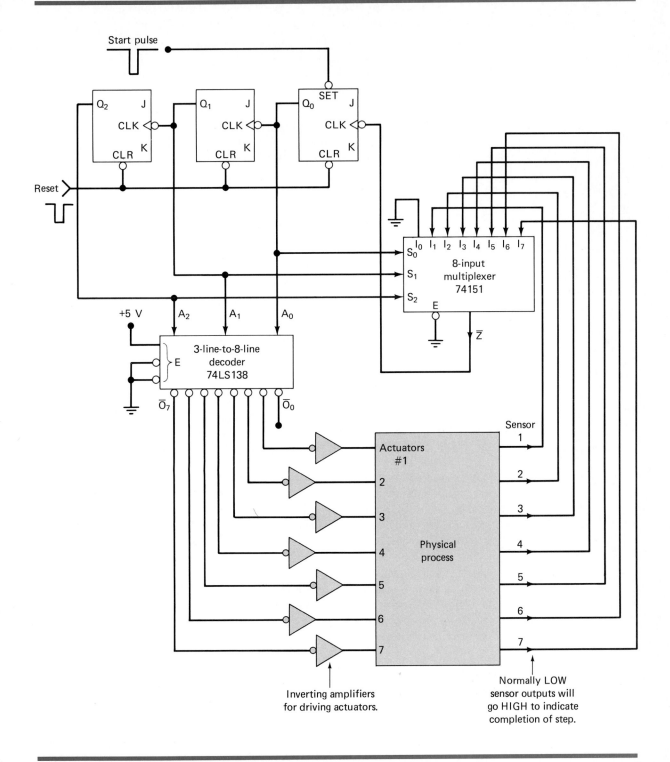

FIGURE 9-27 **Seven-step control sequencer.**

of the process are LOW. The SENSOR outputs of the process all start out LOW. The multiplexer output $\overline{Z} = \overline{I}_0 = 1$, since the S inputs are 000.

2. The START pulse initiates the sequencing operation by setting FF Q_0 HIGH, bringing the counter to the 001 state. This causes decoder output $\overline{O}_1$ to go LOW, thereby activating actuator 1, which is the first step in the process (it could be the turning on of a heater).

3. Some time later SENSOR output 1 goes HIGH, indicting the completion of the first step (it could be the reaching of a certain temperature level). This HIGH is now present at the I_1 input of the multiplexer. It is inverted and reaches the $\overline{Z}$ output since the select code from the counter is 001.

4. The LOW at $\overline{Z}$ is fed to the CLK of FF Q_0. This negative transition advances the counter to the 010 state.

5. Decoder output $\overline{O}_2$ now goes LOW, activating actuator 2, which is the second step in the process. $\overline{Z}$ now equals $\overline{I}_2$ (select code is 010). Since SENSOR output 2 is still LOW, $\overline{Z}$ will go HIGH.

6. When the second process step is complete, SENSOR output 2 goes HIGH, producing a LOW at $\overline{Z}$ and advancing the counter to 011.

7. This same action is repeated for each of the other steps. When the seventh step is completed, SENSOR output 7 goes HIGH, causing the counter to go from 111 to 000, where it will remain until another START pulse reinitiates the sequence.

Logic Function Generation   Multiplexers can be used to implement logic functions directly from a truth table without the need for simplification. When used for this purpose the select inputs are used as the logic variables and each data input is connected permanently HIGH or LOW as necessary to satisfy the truth table.

Figure 9-28 illustrates how an eight-input multiplexer can be used to implement the logic circuit that satisfies the given truth table. The input variables A, B, C are connected to S_0, S_1, S_2, respectively, so that the levels on these inputs determine which data input appears at output Z. According to the truth table, Z is supposed to be LOW when $CBA = 000$. Thus, multiplexer input I_0 should be connected LOW. Likewise, Z is supposed to be LOW for $CBA = 011$, 100, 101, and 110, so that inputs I_3, I_4, I_5, and I_6 should also be connected LOW. The other sets of CBA conditions must produce $Z = 1$, so multiplexer inputs I_1, I_2, and I_7 are connected permanently HIGH.

It is easy to see that any three-variable truth table can be implemented with this eight-input multiplexer. This method of implementation is often more efficient than using separate logic gates. For example, if we write the sum-of-products expression for the truth table in Figure 9-28, we have

$$Z = A\overline{B}\overline{C} + \overline{A}B\overline{C} + ABC$$

This *cannot* be simplified either algebraically or by K-mapping, so its gate implementation would require three INVERTERs and four NAND gates, for a total of two ICs.

There is an even more efficient method for using multiplexers to implement logic functons. This method will allow the logic designer to use a multiplexer with three select inputs (e.g., the 74151) to implement a *four-variable* logic function. We will present this method in Problem 9-33.

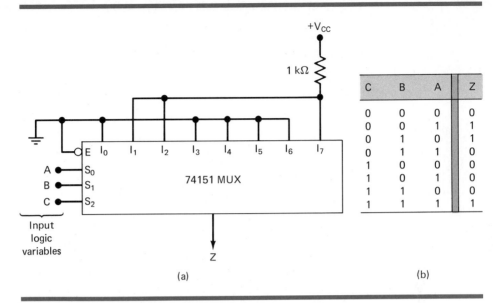

C	B	A	Z
0	0	0	0
0	0	1	1
0	1	0	1
0	1	1	0
1	0	0	0
1	0	1	0
1	1	0	0
1	1	1	1

(a)   (b)

FIGURE 9-28 **Multiplexer used to implement a logic function described by truth table.**

REVIEW QUESTIONS

1. What are some of the major applications of multiplexers?
2. *True or false:* When a multiplexer is used to implement a logic function, the logic variables are applied to the multiplexer's data inputs.
3. What type of circuit provides the select inputs when a MUX is used as a parallel-to-serial converter?

9-9 DEMULTIPLEXERS (DATA DISTRIBUTORS)

A multiplexer takes several inputs and transmits *one* of them to the output. A *de-multiplexer* performs the reverse operation; it takes a single input and distributes it over several outputs. Figure 9-29 shows the functional diagram for a demultiplexer (DEMUX). The large arrows for inputs and outputs can represent one or more lines. The select input code determines to which output the DATA input will be transmitted. In other words, the demultiplexer takes one input data source and selectively distributes it to 1 of N output channels just like a multiposition switch.

1-Line-to-8-Line Demultiplexer Figure 9-30 shows the logic diagram for a demultiplexer that distributes one input line to eight output lines. The single data input line I is connected to all eight AND gates, but only one of these gates will be

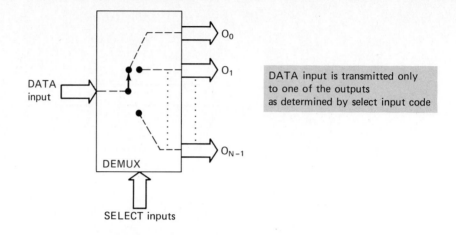

DATA input is transmitted only to one of the outputs as determined by select input code

FIGURE 9-29   **General demultiplexer.**

enabled by the SELECT input lines. For example, with $S_2S_1S_0$ = 000, only AND gate 0 will be enabled, and data input I will appear at output O_0. Other SELECT codes cause input I to reach the other outputs. The truth table summarizes the operation.

The demultiplexer circuit of Figure 9-30 is very similar to the 3-line-to-8-line decoder circuit in Figure 9-2 except that a fourth input (I) has been added to each gate. It was pointed out earlier that many IC decoders have an ENABLE input, which is an extra input added to the decoder gates. This type of decoder chip can therefore be used as a demultiplexer, with the binary code inputs (e.g., A, B, C in Figure 9-2) serving as the SELECT inputs and the ENABLE input serving as the data input I. For this reason, IC manufacturers often call this type of device a *decoder/demultiplexer,* and it can be used for either function.

We saw earlier how the 74LS138 is used as a 1-of-8 decoder. Figure 9-31 shows how it can be used as a demultiplexer. The enable input $\overline{E}_1$ is used as the data input I, while the other two enable inputs are held in their active states. The $A_2A_1A_0$ inputs are used as the select code. To illustrate the operation, let's assume that the select inputs are 000. With this input code, the only output that can be activated is $\overline{O}_0$, while all other outputs are HIGH. $\overline{O}_0$ will go LOW only if $\overline{E}_1$ goes LOW and will be HIGH if $\overline{E}_1$ goes HIGH. In other words, $\overline{O}_0$ *will follow the signal on $\overline{E}_1$* (i.e., the data input, I) while all other outputs stay HIGH. In a similar manner, a different select code applied to $A_2A_1A_0$ will cause the corresponding output to follow the data input, I.

Figure 9-31(b) shows typical waveforms for the case where $A_2A_1A_0$ = 000 selects output $\overline{O}_0$. For this case, the data signal applied to $\overline{E}_1$ will be transmitted to $\overline{O}_0$ and all other outputs will remain in their inactive-HIGH state.

Clock Demultiplexer   Many applications of the demultiplexing principle are possible. Figure 9-32 shows the 74LS138 demultiplexer being used as a *clock demultiplexer.* Under control of the SELECT lines, the clock signal is routed to the desired destination. For example, with $S_2S_1S_0$ = 000, the clock signal applied to I will appear at output $\overline{O}_0$. With $S_2S_1S_0$ = 101, the clock will appear at $\overline{O}_5$.

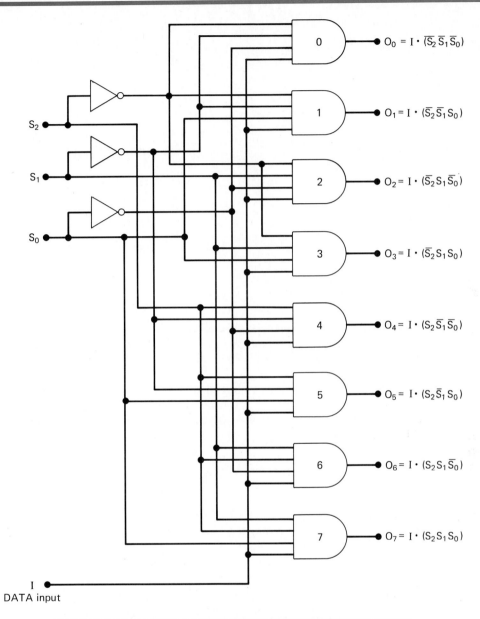

SELECT code			Outputs							
S_2	S_1	S_0	O_7	O_6	O_5	O_4	O_3	O_2	O_1	O_0
0	0	0	0	0	0	0	0	0	0	I
0	0	1	0	0	0	0	0	0	I	0
0	1	0	0	0	0	0	0	I	0	0
0	1	1	0	0	0	0	I	0	0	0
1	0	0	0	0	0	I	0	0	0	0
1	0	1	0	0	I	0	0	0	0	0
1	1	0	0	I	0	0	0	0	0	0
1	1	1	I	0	0	0	0	0	0	0

FIGURE 9-30   One-line-to-8-line demultiplexer.

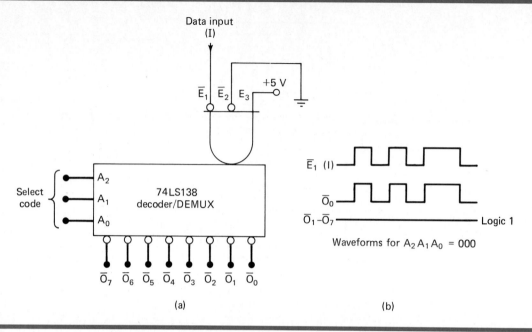

(a)

(b)

FIGURE 9-31   (a) The 74LS138 decoder can function as a demultiplexer with $\bar{E}_1$ used as the data input; (b) typical waveforms for a select code of $A_2A_1A_0 = 000$ shows $\bar{O}_0$ identical to data input.

FIGURE 9-32   Clock demultiplexer transmits clock signal to a destination determined by select code inputs.

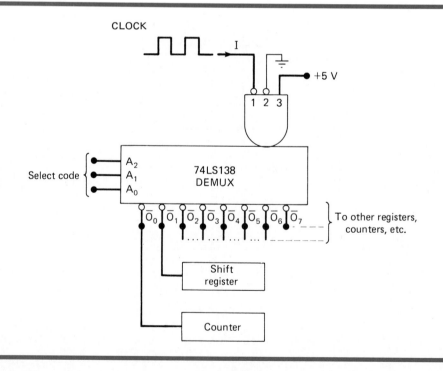

Security Monitoring System   Consider the case of a security monitoring system in an industrial plant where the open/closed status of many access doors is to be monitored. Each door controls the state of a switch, and it is necessary to display the state of each switch on LEDs that are mounted on a remote monitoring panel at the security guard's station. One way to do this would be to run a separate signal from each door switch to an LED on the monitoring panel. This would require running many wires over a long distance. A better approach that would reduce the amount of wiring to the monitoring panel uses a multiplexer/demultiplexer combination. Figure 9-33 shows a system that can handle eight doors, but the basic idea can be expanded to any number.

FIGURE 9-33   **Security monitoring system.**

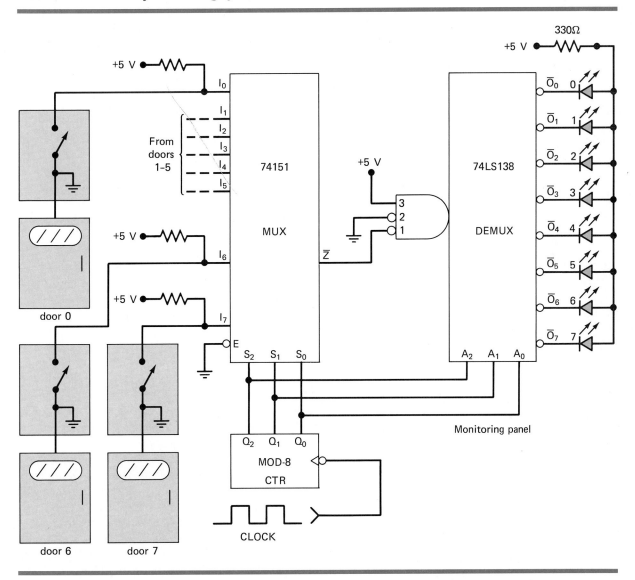

DEMULTIPLEXERS (DATA DISTRIBUTORS)   **523**

EXAMPLE 9-10

Examine Figure 9-33 carefully and describe the complete operation.

SOLUTION

The eight door switches are the data inputs to the MUX; they produce a HIGH when a door is open and a LOW when it is closed. The MOD-8 counter provides the select inputs to the MUX and also to the DEMUX on the remote monitoring panel. Each DEMUX output is connected to an indicator LED that will be ON when the output is LOW. Clock pulses applied to the counter will cause the select inputs to sequence through all possible states 000–111. At each number of the counter, the switch status for the door of the same number will be inverted by the MUX and passed to output $\overline{Z}$. From there it is transmitted to the DEMUX input, which passes it through to the output corresponding to the same number.

For example, let's say that the counter is at the count of 110(6). While the counter is in this state, let's say that door 6 is closed. The LOW at I_6 will pass through the MUX and be inverted to produce a HIGH at $\overline{Z}$. This HIGH will be passed through the DEMUX to output $\overline{O}_6$ so that LED 6 will be OFF, indicating that door 6 is closed. Now let's say that door 6 is open. A LOW will appear at $\overline{Z}$ and $\overline{O}_6$ so that LED 6 will be ON to signal that door 6 is open. Of course, all other LEDs will be OFF during this time since $\overline{O}_6$ is the only active output.

As the counter is clocked through its eight states 000–111, the LEDs will sequentially indicate the status of the eight doors. If all the doors are closed, none of the LEDs will be ON even when the corresponding DEMUX output is selected. If a door is open, its LED will turn ON only during the time interval that the counter is at the appropriate count; it will be OFF at all other counts. Thus, the LED will be flashing ON and OFF if its door is open. The flashing rate can be adjusted by changing the frequency of the clock.

Note that there are only four signal lines going from the "door sensing" circuitry to the remote monitoring panel: the Z output and the three select lines. This is a savings of four lines when compared to the alternative of having one line per door. The MUX/DEMUX combination is used to transmit the status of each door to its LED one at a time (serially) instead of all at once (parallel).

Synchronous Data Transmission System

Figure 9-34 shows the logic diagram for a synchronous data transmission system that is used to serially transmit four 4-bit data words from a transmitter to a remote receiver. Let's look at the transmitter circuitry first. The data words are stored in registers A, B, C, and D that are connected as recirculating shift registers with a common SHIFT (clock) input. Each register will shift right on the PGT of the SHIFT pulses from AND gate 2. The LSB of each register is connected as a data input to the 4-input multiplexer.

The two MOD-4 counters control the transmission of the data register contents to the multiplexer output Z. The *word counter* selects the register data that will appear at Z. As this counter cycles from 00 to 11, the data from each register will sequentially appear at Z. The *bit counter* makes sure that 4 data bits from each register are transmitted through the multiplexer before advancing to the next register. The bit counter advances one count for each SHIFT pulse, so that after four SHIFT pulses, it recycles to 00. The NGT at the Q_1 output of the bit counter will

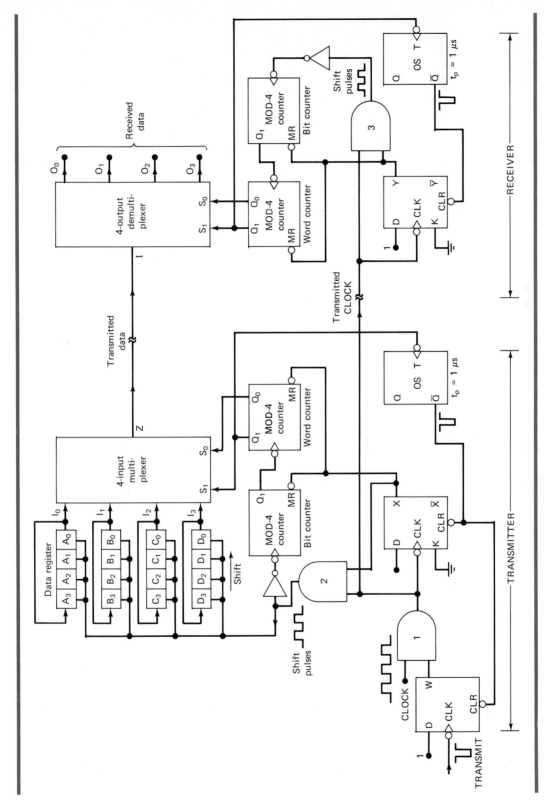

FIGURE 9-34 Synchronous data transmission system.

cause the word counter to be incremented to the next count to select the next data register for transmission. In this way the contents of each of the data registers will be transmitted to Z, one bit at a time, starting with register A (for $S_1S_0 = 00$) and proceeding through each register as the word counter advances one count for every four SHIFT pulses. The Z signal will thus contain 16 bits of serial data, 4 bits from each register. These data are said to be *time-multiplexed* because four different sets of data are appearing on the same output line at different times.

The transmission process is controlled by the two D-FFs, AND gates 1 and 2, and the one-shot. The operation of this control logic will be described in a later paragraph.

The Receiver   The receiver circuitry contains a 1-to-4 demultiplexer that receives the Z signal from the transmitter's multiplexer and *demultiplexes* it; that is, it separates the four different sets of data and distributes them to four different outputs so that the data that came from register A will appear serially at O_0, the data from register B at O_1, and so on. The end result is almost the same as having each transmitter data register connected to its corresponding output in the receiver, except that the data are sent from one register at a time over the serial data transmission path.

The MOD-4 counters in the receiver have the same function as their counterparts in the transmitter. The word counter selects which demultiplexer output will be receiving data, and the bit counter allows 4 bits of data to reach each output before advancing the word counter to its next state. The functions of the FF, OS, and AND gate are described below.

Complete Operation   It should be clear that in order for this data transmission to work properly, there has to be some means for synchronizing the selection of the multiplexer inputs in the transmitter with the selection of the demultiplexer outputs in the receiver. We will go through a complete operation cycle to see how this synchronization is accomplished. For this illustration we will assume the following register data:

$$[A] = 0110, \quad [B] = 1001, \quad [C] = 1011, \quad [D] = 0100$$

As we go through the following steps, we will refer to the waveforms in Figure 9-35.

1. FFs W and X in the transmitter and FF Y in the receiver are normally LOW. The LOWs from X and Y will keep both sets of counters in the zero state. The LOW at W prevents CLOCK pulses from getting through AND gate 1.

2. With both word counters at 00, the LOW at A_0 passes through the multiplexer to Z into the demultiplexer to output O_0. All other demultiplexer outputs are LOW, since they are not selected.

3. This is the situation prior to time t_0. At t_0, the TRANSMIT pulse sets W = 1 to enable AND gate 1 to pass CLOCK pulses. These CLOCK pulses also become the TRANSMITTED CLOCK that is sent to the receiver along with the TRANSMITTED DATA.

4. The NGT of the first CLOCK pulse out of AND gate 1 will set X and Y both HIGH at time t_2. This removes the resets from the counters and also enables

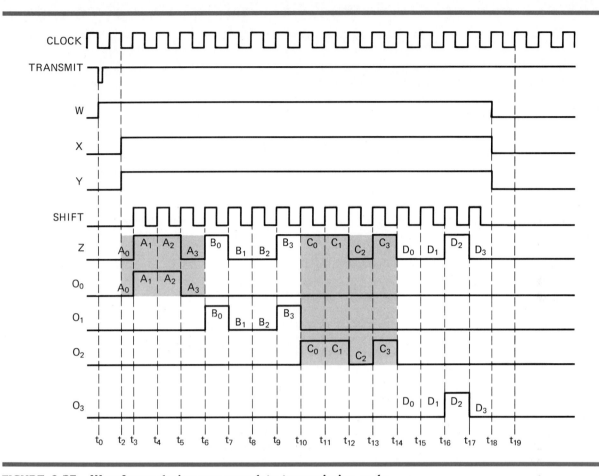

FIGURE 9-35 Waveforms during one complete transmission cycle.

AND gates 2 and 3 to pass CLOCK pulses starting at t_3. The pulses out of AND gate 2 are the SHIFT pulses. The pulses out of AND gate 3 are exactly the same as the SHIFT pulses out of AND gate 2.

5. The PGTs of the SHIFT pulses at t_3, t_4, and t_5 will shift A_1, A_2, and A_3 into the multiplexer, out of Z, into the demultiplexer, and out of output O_0. These three PGTs are also counted by both bit counters.

6. The PGT at t_6 will shift all the registers back to their original data and will recycle the bit counters to 00. The NGT at Q_1 of the bit counters will increment the transmitter and receiver word counters to 01 to select I_1 and O_1. Thus, the HIGH at B_0 will pass through the multiplexer into the demultiplexer and out of O_1.

7. The SHIFT pulses at t_7, t_8, and t_9 will shift B_1, B_2, and B_3 into the multiplexer and out of O_1. At t_{10} the bit counters will recycle and increment the word counters to 10 to select I_2 and O_2. This places the HIGH from C_0 at Z and at O_2.

8. The SHIFT pulses at t_{11}, t_{12}, and t_{13} will shift C_1, C_2, and C_3 into the multi-

plexer and out of O_2. At t_{14} the bit counters recycle and increment the word counters to 11 to select I_3 and O_3. This places the LOW from D_0 at Z and at O_3.

9. The SHIFT pulses at t_{15}, t_{16}, and t_{17} will shift D_1, D_2, and D_3 into the multiplexer and out of O_3. At t_{18} the bit counters recycle and increment the word counters to 00. The NGT at Q_1 of the word counters will trigger their respective one-shots to produce narrow clearing pulses for FFs W, X, and Y. With these FFs all LOW, all CLOCK pulses and SHIFT pulses are inhibited, and all the counters remain in the zero state.

10. The circuit conditions are back to their original state. No more data will be transmitted until the next TRANSMIT pulse occurs.

The waveforms at Z and O_0–O_3 show how the register data are multiplexed onto the Z signal and then demultiplexed so that each output receives the correct data.

REVIEW QUESTIONS

1. Explain the difference between a DEMUX and a MUX.
2. *True or false:* The circuit for a DEMUX is basically the same as for a decoder.
3. For the system of Figure 9-33, what will the security guard see on the monitoring panel when all the doors are open? (*Ans.* The LEDs will go ON and OFF in sequence)

9-10 MORE IEEE/ANSI SYMBOLOGY*

We will now look at the IEEE/ANSI symbols for the MUX and DEMUX ICs that we have been using. First consider the symbol for the 74151 shown in Figure 9-36(a). The G_7^0 label within the block denotes the AND dependency between the select inputs and each of the data inputs 0 through 7. In other words, each data input is ANDed with one specific combination of select inputs, and when that combination occurs, the data input is routed to the output.

Figure 9-36(b) is the IEEE/ANSI symbol for the 74157 multiplexer. Recall from Figure 9-24 that this IC contains four two-input MUXes that operate identically. The common-control block shows that the enable and select inputs are common to all the MUXes. The G1 notation on the select input and the $\bar{1}$ and 1 labels on the data inputs indicate the AND dependency between the select input and the data inputs. The $\bar{1}$ label for I_{0a} means that this input will be routed to output Z_a only

*This section may be skipped without affecting the rest of the book.

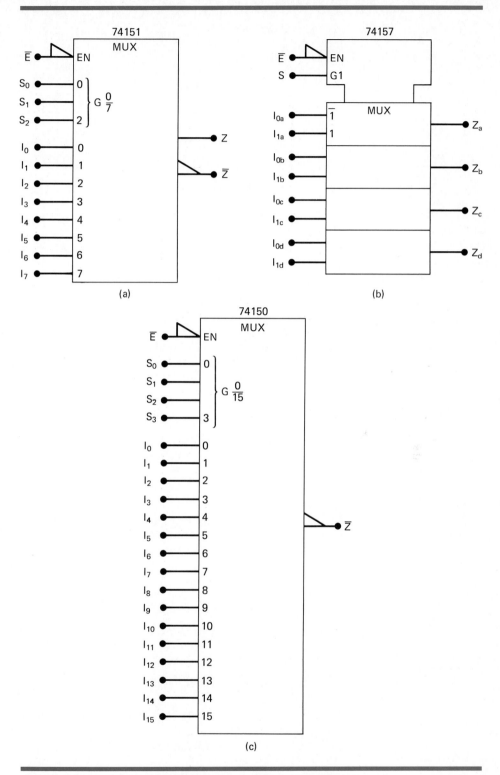

FIGURE 9-36   IEEE/ANSI symbols for various multiplexers.

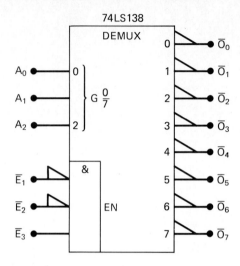

FIGURE 9-37 IEEE/ANSI symbol for the 74LS138 DEMUX.

when the select input is a 0. Similarly, the 1 on the I_{1a} input indicates that this input will be routed to Z_a only when the select input is a 1. The other three MUXes operate the same way.

EXAMPLE 9-11

The IEEE/ANSI for the 74150 IC is shown in Figure 9-36(c). What is the function of this device?

SOLUTION
The 74150 is a 16-input multiplexer. Note the four select inputs required to select one of the sixteen data inputs for transmission to the output. Also note that the output is inverted, so it will be the inverse of the selected data input.

Figure 9-37 is the IEEE/ANSI symbol for the 74LS138 demultiplexer. Compare this symbol to that in Figure 9-17(c), where the 74LS138 is used as a decoder.

9-11 MORE TROUBLESHOOTING

Here are two more examples to illustrate the observation/reasoning process that is such an important initial step when troubleshooting. For each case, try to determine the circuit fault before looking at the solutions.

EXAMPLE 9-12

Consider the circuit of Figure 9-25. A test performed on this circuit yields the following results:

		ACTUAL COUNT	DISPLAYED COUNT
Case 1	Counter 1	25	25
	Counter 2	37	35
Case 2	Counter 1	49	49
	Counter 2	72	79
Case 3	Counter 1	96	96
	Counter 2	14	16

What is the probable circuit fault?

SOLUTION

In each of the test cases, the display of counter 1 matches the counter's actual count. This indicates that the I_1 inputs, all MUX outputs, and both displays are probably working correctly. On the other hand, each test case shows that counter 2's *tens* digit is displayed correctly but its *units* digit is displayed incorrectly. This could mean that there is a fault somewhere between the output of the units section of counter 2 and the I_0 inputs of the units MUX. We should compare the bit patterns of the actual and displayed values of the units for counter 2.

	ACTUAL UNITS	DISPLAYED UNITS
Case 1	0111(7)	0101(5)
Case 2	0010(2)	1001(9)
Case 3	0100(4)	0110(6)

The idea is to look for things like a bit that does not change (stuck LOW or HIGH) or two bits that are reversed (crossed connections). The data above reveals no obvious pattern.

If we take another look at the recorded test results, we see that the displayed units digit of counter 2 is always the same as the units digit of counter 1. This symptom is probably the result of a constant logic HIGH at the select input of the units MUX since that would continually pass the units digit of counter 1 to the units MUX output. This constant HIGH at the select input is most likely caused by an open somewhere between the select input of the tens MUX and the select input of the units MUX. It could not be caused by a short to V_{CC} since that would also keep the select input of the tens MUX at a constant HIGH, and we know that the tens MUX is working.

EXAMPLE 9-13

The security monitoring system of Figure 9-33 is tested and the results are recorded below.

CONDITION	LEDS
All doors closed	All LEDs off
Door 0 open	LED 4 flashing
Door 1 open	LED 5 flashing
Door 2 open	LED 6 flashing
Door 3 open	LED 7 flashing
Door 4 open	LED 4 flashing
Door 5 open	LED 5 flashing
Door 6 open	LED 6 flashing
Door 7 open	LED 7 flashing

What are the possible faults that could produce these results?

SOLUTION

Again, the data should be reviewed to see if there is some pattern that could help to narrow down the search for the fault to a small area of the circuit. The data above reveals that the correct LEDs flash for the open doors 4 through 7. It also shows that for open doors 0 through 3 the number of the flashing LED is *four* more than the number of the door, and LEDs 0 through 3 are always OFF. This is most probably caused by a constant logic HIGH at A_2, the MSB of the select input of the DEMUX, since this would always make the select code 4 or greater, and it would add 4 to the select codes 0 through 3.

Thus we have two possibilities: A_2 is somehow shorted to V_{CC}, or there is an open at A_2. A little thought will eliminate the first choice as a possibility since this would also mean that S_2 of the MUX would also be stuck HIGH. If that were so, then the status of doors 0 through 3 would not get through the MUX and into the DEMUX. We know that this is not true because the data shows that when any of these doors is open, it affects one of the DEMUX outputs.

EXAMPLE 9-14

A technician is troubleshooting a malfunctioning synchronous data transmission system (Figure 9-34). An oscilloscope is used to monitor the MUX and DEMUX outputs during a transmission cycle with the results shown in Figure 9-38. What are the possible causes of the malfunction?

SOLUTION

Examination of the waveforms produces the following observations:

- All of the signals appear to be correct between t_0 and t_9, with the O_0 signal containing the serial data from register A, and O_1 containing the serial data from register B.
- The O_2 and O_3 outputs are never activated.

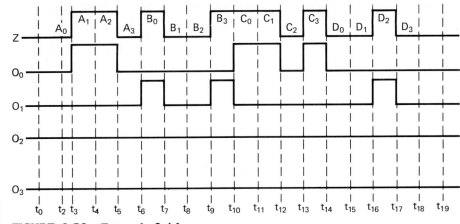

FIGURE 9-38   Example 9-14

- Between t_{10} and t_{14}, the O_0 output, rather than O_2, contains the serial data from register C.
- Between t_{14} and t_{18}, the O_1 output, rather than O_3, contains the serial data from register D.

It appears that the select inputs of the receiver's DEMUX are selecting only O_0 and O_1. This would happen if S_1 were stuck in the LOW state. This stuck node could be caused by an internal short to ground at S_1, Q_1 or the T input of the one-shot, or an external short to ground. The actual location of the short can be determined using the techniques that have been presented in previous chapters.

9-12   MAGNITUDE COMPARATOR

Another useful member of the MSI category of ICs is the *magnitude comparator*. It is a combinatorial logic circuit that compares two input binary quantities and generates outputs to indicate which word has the greater magnitude. Figure 9-39 shows the logic symbol and truth table for the 74LS85 4-bit magnitude comparator which is also available as a 7485 and 74HC85.

Data Inputs   The 74LS85 compares two *unsigned* 4-bit binary numbers. One of them is $A_3A_2A_1A_0$ which is called word A; the other is $B_3B_2B_1B_0$, which is called word B. The term "word" is used in the digital computer field to designate a group of bits that represents some specific type of information. Here word A and word B represent numerical quantities.

Outputs   The 74LS85 has three active-HIGH outputs. Output $O_{A>B}$ will be HIGH when the magnitude of word A is greater than the magnitude of word B. Output $O_{A<B}$ will be HIGH when the magnitude of word A is less than the magnitude of word B. Output $O_{A=B}$ will be HIGH when word A and word B are identical.

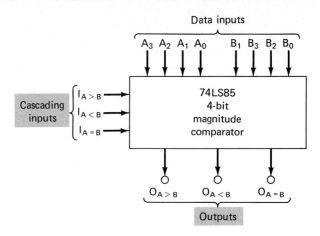

TRUTH TABLE

COMPARING INPUTS				CASCADING INPUTS			OUTPUTS		
A_3, B_3	A_2, B_2	A_1, B_1	A_0, B_0	$I_{A>B}$	$I_{A<B}$	$I_{A=B}$	$O_{A>B}$	$O_{A<B}$	$O_{A=B}$
$A_3 > B_3$	X	X	X	X	X	X	H	L	L
$A_3 < B_3$	X	X	X	X	X	X	L	H	L
$A_3 = B_3$	$A_2 > B_2$	X	X	X	X	X	H	L	L
$A_3 = B_3$	$A_2 < B_2$	X	X	X	X	X	L	H	L
$A_3 = B_3$	$A_2 = B_2$	$A_1 > B_1$	X	X	X	X	H	L	L
$A_3 = B_3$	$A_2 = B_2$	$A_1 < B_1$	X	X	X	X	L	H	L
$A_3 = B_3$	$A_2 = B_2$	$A_1 = B_1$	$A_0 > B_0$	X	X	X	H	L	L
$A_3 = B_3$	$A_2 = B_2$	$A_1 = B_1$	$A_0 < B_0$	X	X	X	L	H	L
$A_3 = B_3$	$A_2 = B_2$	$A_1 = B_1$	$A_0 = B_0$	H	L	L	H	L	L
$A_3 = B_3$	$A_2 = B_2$	$A_1 = B_1$	$A_0 = B_0$	L	H	L	L	H	L
$A_3 = B_3$	$A_2 = B_2$	$A_1 = B_1$	$A_0 = B_0$	X	X	H	L	L	H
$A_3 = B_3$	$A_2 = B_2$	$A_1 = B_1$	$A_0 = B_0$	L	L	L	H	H	L
$A_3 = B_3$	$A_2 = B_2$	$A_1 = B_1$	$A_0 = B_0$	H	H	L	L	L	L

H = HIGH Voltage Level
L = LOW Voltage Level
X = Immaterial

FIGURE 9-39   Logic symbol and truth table for a 74LS85 (7485, 74HC85) 4-bit magnitude comparator.

Cascading Inputs   These inputs provide a means for expanding the comparison operation to more than 4 bits by cascading two or more 4-bit comparators. Note that the cascading inputs are labeled the same as the outputs. When a 4-bit comparison is being made as in Figure 9-40(a), the cascading inputs should be connected as shown in order for the comparator to produce the correct outputs.

When two comparators are to be cascaded, the outputs of the lower-order comparator are connected to the corresponding inputs of the higher-order compara-

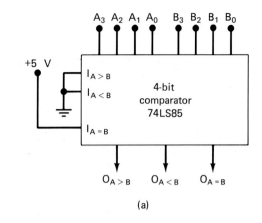

(a)

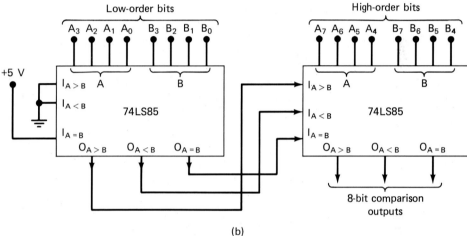

(b)

FIGURE 9-40   (a) 74LS85 wired as a 4-bit comparator; (b) two 74LS85s cascaded to perform an 8-bit comparison.

tor. This is shown in Figure 9-40(b), where the comparator on the left is comparing the lower-order 4 bits of the two 8-bit words: $A_7A_6A_5A_4A_3A_2A_1,A_0$ and $B_7B_6B_5B_4B_3B_2B_1B_0$. Its outputs are fed to the cascade inputs of the comparator on the right, which is comparing the high-order bits. The outputs of the high-order comparator are the final outputs that indicate the result of the 8-bit comparison.

EXAMPLE 9-15

Describe the operation of the 8-bit comparison arrangement in Figure 9-40(b) for the following cases.

(a) $A_7A_6A_5A_4A_3A_2A_1A_0$ = 10101111; $B_7B_6B_5B_4B_3B_2B_1B_0$ = 10110001

(b) $A_7A_6A_5A_4A_3A_2A_1A_0$ = 10101111; $B_7B_6B_5B_4B_3B_2B_1B_0$ = 10101001

SOLUTION

(a) The high-order comparator compares its inputs $A_7A_6A_5A_4 = 1010$ and $B_7B_6B_5B_4 = 1011$ and produces $O_{A<B} = 1$ regardless of what levels are applied to its cascade inputs from the low-order comparator. In other words, once the high-order comparator senses a difference in the high-order bits of the two 8-bit words, it knows which 8-bit word is greater without having to look at the results of the low-order comparison.

(b) The high-order comparator sees $A_7A_6A_5A_4 = B_7B_6B_5B_4 = 1010$, and so it has to look at its cascade inputs to see the result of the low-order comparison. The low-order comparator has $A_3A_2A_1A_0 = 1111$ and $B_3B_2B_1B_0 = 1001$, which produces a 1 at its $O_{A>B}$ output and the $I_{A>B}$ input of the high-order comparator. The high-order comparator senses this 1, and since its data inputs are equal, produces a HIGH at its $O_{A\rangle B}$ to indicate the result of the 8-bit comparison.

Applications   Magnitude comparators are often used as part of the address decoding circuitry used in a computer to select a specific input/output device or area of memory for the storage or retrieval of data. The comparators compare an address code generated by the central processer (CPU) with a hard-wired address code; if they are equal, the $O_{A=B}$ output of the comparator will activate the device corresponding to that address. Magnitude comparators are also useful in control applications where a binary number representing the physical variable being controlled (e.g., position, speed) is compared to a reference value. The comparator outputs are used to actuate circuitry to drive the physical variable toward the reference value. We will examine a typical comparator application in Problem 9-46.

IEEE/ANSI Symbol   Figure 9-41 shows the IEEE/ANSI symbol for the 74LS85 comparator. Note that the symbol uses P and Q to represent the input variables. This is the preferred designation used by the IEEE/ANSI standard.

FIGURE 9-41   IEEE/ANSI symbol for comparator.

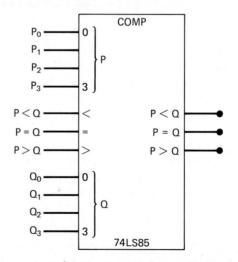

1. What is the purpose of the cascading inputs of the 74LS85?
2. What are the outputs of a 74LS85 with the following inputs: $A_3A_2A_1A_0 = B_3B_2B_1B_0 = 1001$, $I_{A>B} = I_{A<B} = 0$, and $I_{A=B} = 1$? (Ans, $O_{A<B} = O_{A>B} = 0$, $O_{A=B} = 1$)

9-13 TRISTATE REGISTERS

In most modern computers the transfer of data takes place over a common set of connecting lines called a *data bus*. In these bus-organized computers, many different devices can have their outputs and inputs tied to the common data bus lines. Because of this, the devices that are tied to the data bus will often have tristate outputs, or they will be tied to the data bus through tristate buffers.

Some of the devices that are commonly connected to a data bus are (1) microprocessors, which will be discussed in Chapter 12; (2) semiconductor memory chips, which will be covered in Chapter 11; (3) digital-to-analog (D/A) converters and analog-to-digital (A/D) converters, to be covered in Chapter 10.

Almost always, the devices connected to a data bus will contain registers that hold their data. The outputs of these registers will have tristate buffers to allow data busing. There are many IC registers available that include the tristate buffers on the same chip. One of these is the TTL 74173 (and its CMOS counterparts, the 4076B and 74HC173), whose logic diagram and truth tables are shown in Figure 9-42.

The 74173 is a 4-bit register with parallel in/parallel out capability. Note that the FF outputs are connected to tristate buffers that provide outputs O_0–O_3. Also note that the data inputs D_0–D_3 are connected to the D inputs of the register FFs through logic circuitry. This logic allows two modes of operation: (1) *load*, where the data at the D_0–D_3 inputs are transferred into the FFs on the PGT of the clock pulse at CP; (2) *hold*, where the data in the register does not change when the PGT of CP occurs.

EXAMPLE 9-16

(a) What input conditions will produce the load operation?

(b) What input conditions will produce the hold operation?

(c) What input conditions will allow the internal register outputs to appear at O_0–O_3?

SOLUTION

(a) The last two entries in the truth table show that each Q output takes on the value present at its D input when a PGT occurs at CP provided that MR is LOW, and both Input Enable inputs, $\overline{IE}_1$ and $\overline{IE}_2$, are LOW.

(b) The third and fourth lines of the truth-table state that when either $\overline{IE}$ input is HIGH, the D inputs have no effect, and the Q outputs will retain their current values when the PGT occurs.

Inputs					Output
MR	CP	$\overline{IE}_1$	$\overline{IE}_2$	D_n	Q
H	X	X	X	X	L
L	L	X	X	X	Q_0
L	⌐	H	X	X	Q_0
L	⌐	X	H	X	Q_0
L	⌐	L	L	L	L
L	⌐	L	L	H	H

When either $\overline{OE}_1$ or $\overline{OE}_2$ is HIGH the output is in the OFF
state (high impedance); however this does not affect the
contents or sequential operating of the register

H = HIGH voltage level Q_0 = output prior to PGT
L = LOW voltage level
X = immaterial

Logic Diagram

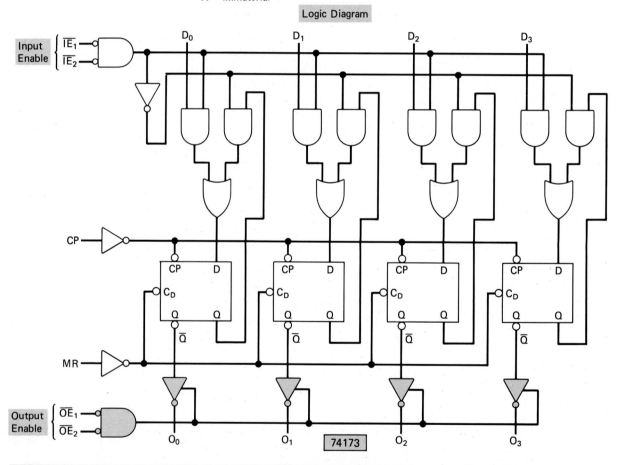

FIGURE 9-42 **Truth table and logic diagram for 74173 tristate register.** (Courtesy of Fairchild, a Schlumberger company)

(c) The output buffers are enabled when both Output Enable inputs, $\overline{OE}_1$ and $\overline{OE}_2$, are LOW. This will pass the register outputs through to the external outputs O_0–O_3. If either Output Enable input is HIGH, the buffers will be disabled, and the outputs will be in the Hi-Z state.

Note that the $\overline{OE}$ inputs have no effect on the data load operation. They are only used to control whether or not the register outputs are passed to the external outputs.

Figure 9-43 is the traditional logic symbol for the 74173. We will use this symbol in the following circuit diagrams.

9-14 DATA BUSING

The data bus is very important in computer systems, and its significance will not be appreciated until our later studies of memories and microprocessors. For now, we will illustrate the data-bus concept for register-to-register data transfer. Figure 9-44 shows a bus-organized system for three 74173 tristate registers. Note that each register has its pair of $\overline{OE}$ inputs tied together as one $\overline{OE}$ input, and likewise for the $\overline{IE}$ inputs. Also note that the registers will be referred to as registers A, B, and C from top to bottom. This is indicated by the subscripts on each input and output.

In this arrangement, the data bus consists of four lines labeled DB_0–DB_3. Corresponding outputs of each register are connected to the same data-bus line (e.g., O_{3A}, O_{3B}, and O_{3C} are connected to DB_3). Since the three registers have their outputs connected together, it is imperative that only one register has its outputs enabled while the other two register outputs are in the Hi-Z state. Otherwise, there

FIGURE 9-43 Logic symbol for 74173.

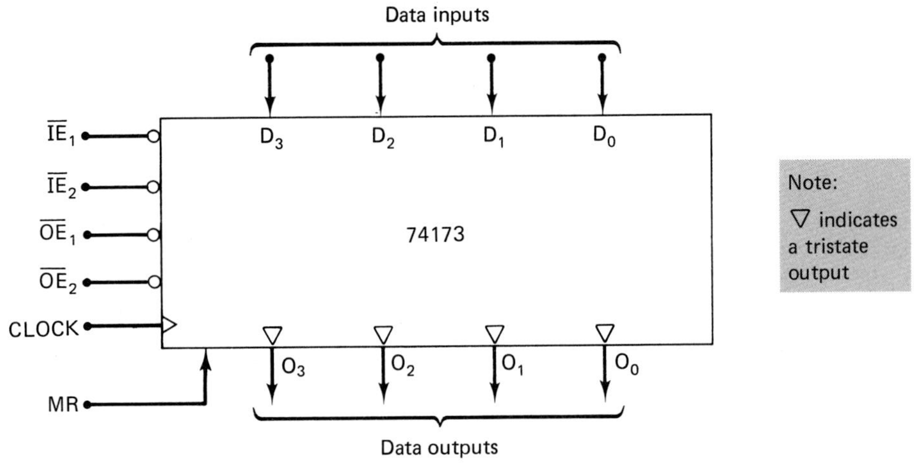

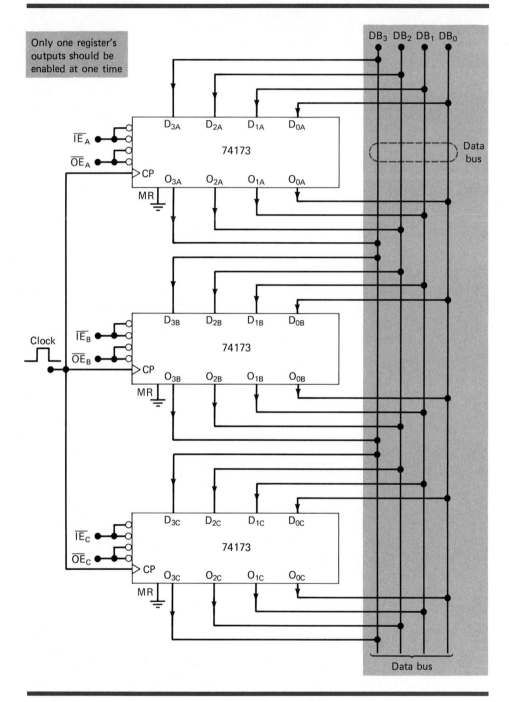

FIGURE 9-44   Tristate registers connected to data bus.

will be "bus contention" (two or more sets of outputs fighting each other) producing uncertain levels on the bus and possible damage to the register outputs.

Corresponding register inputs are also tied to the same bus line (e.g., D_{3A}, D_{3B}, and D_{3C} are tied to DB_3). Thus, the levels on the bus will always be ready to be transferred to one or more of the registers depending on the $\overline{IE}$ inputs.

Data Transfer Operation The contents of any one of the three registers can be parallel-transferred over the data bus to one of the other registers through the proper application of logic levels to the register enable inputs. In a typical system, the control unit of a computer (i.e., the CPU) will generate the signals that select which register will put its data on the data bus, and which one will take the data from the data bus. The following example will illustrate this.

EXAMPLE 9-17

Describe the input signal requirements for transferring [A] → [C].

SOLUTION
First of all, only register A should have its outputs enabled. That is, we need
$$\overline{OE}_A = 0, \qquad \overline{OE}_B = \overline{OE}_C = 1$$
This will place the contents of register A onto the data-bus lines.

Next, only register C should have its inputs enabled. For this, we want
$$\overline{IE}_C = 0, \qquad \overline{IE}_A = \overline{IE}_B = 1$$
This will allow only register C to accept data from the data bus when the PGT of the clock signal occurs.

Finally, a clock pulse is required to actually transfer the data from the bus into the register-C FFs.

Bus Signals The timing diagram in Figure 9-45 shows the various signals involved in the transfer of the data 1011 from register A to register C. The $\overline{IE}$ and $\overline{OE}$ lines that are not shown are assumed to be in their inactive-HIGH state. Prior to time t_1, the $\overline{IE}_C$ and $\overline{OE}_A$ lines are also HIGH, so that all of the register outputs are disabled, and none of the registers will be placing their data on the bus lines. In other words, the data bus lines are in the Hi-Z or "floating" state as represented by the hashed lines on the timing diagram. The Hi-Z state does not correspond to any particular voltage level.

At t_1 the $\overline{IE}_C$ and $\overline{OE}_A$ inputs are activated. The outputs of register A are enabled, and they start changing the data bus lines DB_3–DB_0 from the Hi-Z state to the logic levels 1011. After allowing time for the logic levels to stabilize on the bus, the PGT of the clock is applied at t_2. This PGT will transfer these logic levels into register C since $\overline{IE}_C$ is active. If the PGT occurs before the data bus has valid logic levels, unpredictable data will be transferred into C.

At t_3, the $\overline{IE}_C$ and $\overline{OE}_A$ lines return to the inactive state. As a result, register A's outputs go to the Hi-Z state. This removes the register A output data from the bus lines and the bus lines return to the Hi-Z state.

Note that the data bus lines show valid logic levels only during the time interval when register A's outputs are enabled. At all other times, the data bus lines are

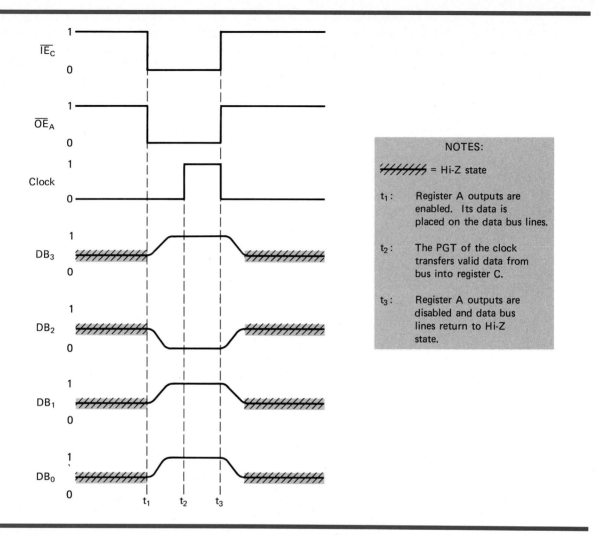

FIGURE 9-45 Signal activity during the transfer of the data 1011 from register A to register C.

indeterminate, and there is no way to easily predict what they would look like if displayed on an oscilloscope. A logic probe would give an "indeterminate" indication if it were monitoring a Hi-Z bus line. Also note the relatively slow rate at which the signals on the data bus lines are changing. Although this effect has been somewhat exaggerated in the diagram, it is a characteristic common to bus systems and is caused by the capacitive load on each line. This load consists of a combination of parasitic capacitance and the capacitances contributed by each input and output connected to the line.

Expanding the Bus It is easy to see how we can add more registers to the data bus so that a greater number of register-to-register data transfers are possible. Of course, each register adds two more enable inputs that have to be controlled for each data transfer. Each register also adds data inputs that have to be driven by the outputs of the register that is putting data on the bus. In systems where a large number of registers (or other devices) are connected to the bus, the register outputs may have to be buffered by circuits that are called *bus drivers*. Even when the devices are all MOS or CMOS, the total capacitive load connected to each bus line will deteriorate the transition times of the level changes on each bus line. A bus-

FIGURE 9-46   **Simplified representation of bus arrangement.**

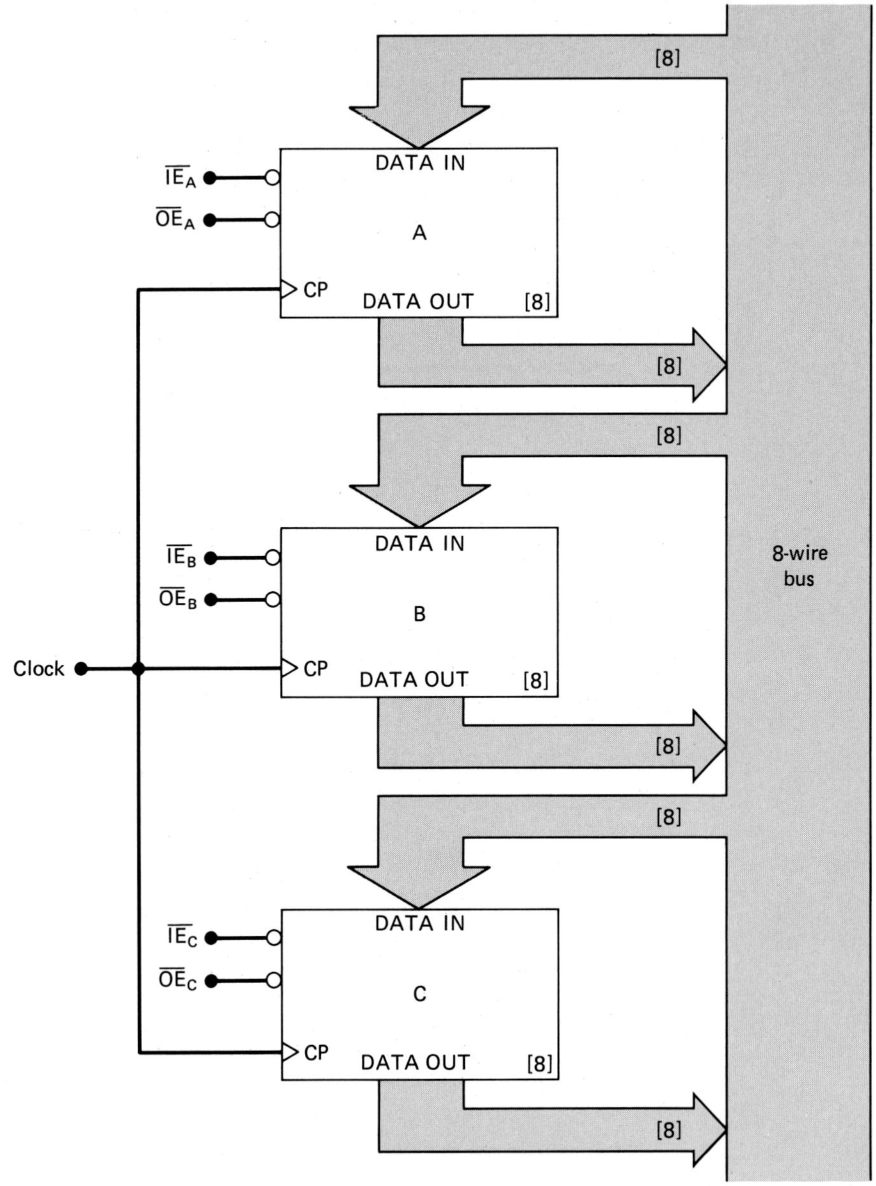

driver IC will have very low output resistance in either logic state and will be able to charge and discharge the bus capacitances more rapidly than a normal IC.

For an illustration of bus expansion and the circuitry for selecting which devices will send and receive data, refer to the discussions in Problems 9-51–9-53.

Simplified Bus Representation   Usually many devices are connected to the same data bus. On a circuit schematic this can produce a confusing array of lines and connections. For this reason, a more simplified representation of data-bus connections is often used on block diagrams and in all but the most detailed circuit schematics. One type of simplified representation is shown in Figure 9-46 for an eight-line data bus. (Refer to page 543 for Figure 9-46.)

The connections to and from the data bus are represented by wide arrows. The numbers in brackets indicate the number of bits that each register contains, as well as the number of lines connecting the register inputs and outputs to the bus.

An even simpler bus representation that is becoming standard on many circuit diagrams is shown in Figure 9-47. It uses a single line to represent the bus and the parallel connections to the bus. The slash (/) through the line indicates that it actually represents a bus containing the designated number of lines.

FIGURE 9-47   **Simplified bus representation that uses a single line. The "/8" denotes an 8-line bus.**

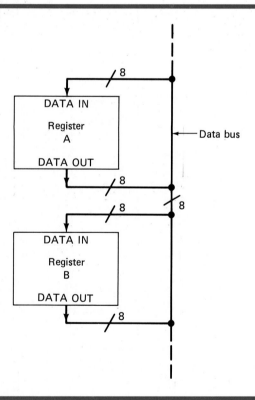

Bidirectional Busing   Each register in Figure 9-44 has both its inputs and its outputs connected to the data bus, so that corresponding inputs and outputs are shorted together. For example, each register has output O_2 connected to input D_2 because of their common connection to DB_2. This, of course, would not be true if bus drivers were connected between the register outputs and the data bus.

Because inputs and outputs are often connected together in bus systems, IC manufacturers have developed ICs that connect inputs and outputs together *internal* to the chip in order to reduce the number of IC pins. Figure 9–48 illustrates this for a 4-bit register. The separate data input lines (D_0–D_3) and output lines (O_0–O_3) have been replaced by input/output lines (I/O_0–I/O_3).

Each I/O line will function as either an input or an output depending on the states of the enable inputs. Thus, they are called *bidirectional data lines*. The 74LS299 is an 8-bit register with common I/O lines. Many memory ICs and microprocessors have bidirectional transfer of data.

We will return to the important topic of data busing in our comprehensive coverage of memory systems in Chapter 11 and in our introduction to microprocessors and microcomputers in Chapter 12.

FIGURE 9-48   Bidirectional register connected to data bus.

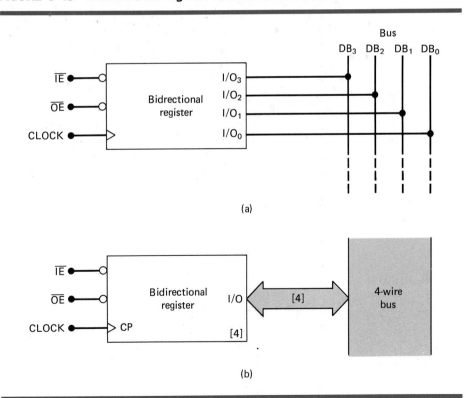

(a)

(b)

REVIEW QUESTIONS

1. What is meant by the term "data bus"?
2. What is "bus contention," and what must be done to prevent it from occurring?
3. What is the function of a bus driver?
4. What is the principal reason for having registers with common I/O lines?
5. Redraw Figure 9-48(a) using the single-line bus representation.

GLOSSARY

Backplane Electrical connection common to all segments of an LCD. [Sec. 9-3]

BCD/DEC When used inside an IEEE/ANSI symbol, it indicates a BCD-to-decimal decoding function. [Sec. 9-5]

BCD-to-Decimal Decoder Decoder that converts a BCD input into a single decimal output equivalence. [Sec. 9-1]

BCD-to-7-Segment Decoder/Driver Digital circuit that takes a 4-bit BCD input and activates the required outputs to display the equivalent decimal digit on an LED. [Sec. 9-2]

Bidirectional Data Line Term used when a data line functions either as an input or an output line depending on the states of the enable inputs. [Sec. 9-14]

BIN/OCT When used inside an IEEE/ANSI symbol, it indicates a binary-to-octal decoding function. [Sec. 9-5]

Bus Drivers When a large number of devices have to be connected to a common bus, the devices' outputs have to be buffered by circuits called bus drivers. [Sec. 9-14]

Data Distributors *See* Demultiplexer. [Sec. 9-9]

Data Selectors *See* Multiplexer. [Sec. 9-7]

Decoder Digital circuit that converts an input binary code into a corresponding single numeric output. [Sec. 9-1]

Demultiplexer (DEMUX) Logic circuit that depending on the status of its select inputs will channel its data input to one of several data outputs. [Sec. 9-9]

Driver Technical term sometimes added to IC descriptions to indicate that the IC's outputs can operate at higher current and/or voltage limits than a normal standard IC. [Sec. 9-1]

8-Line-to-3-Line Encoder Digital circuit that generates a different 3-bit code depending on which one of the eight inputs is activated. [Sec. 9-4]

Encoder Digital circuit that produces an output code depending on which of its inputs is activated. [Sec. 9-4]

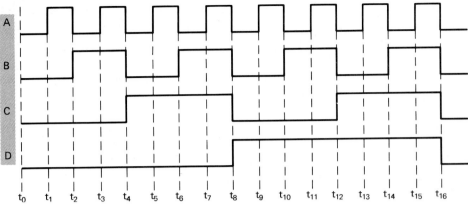

FIGURE 9-51

9-8. Modify the circuit of Figure 9-6 so that relay K1 stays energized from t_3 to t_5, and K2 stays energized from t_6 to t_9. (*Hint:* This modification requires no additional circuitry.)

Sections 9-2 and 9-3

9-9. Show how to connect BCD-to-7 segment decoder/drivers and LED 7-segment displays to the clock circuit of Figure 7-37 to display minutes and hours. Assume that each segment is to operate at approximately 10 mA at 2.5 V.

9-10. (a) Refer to Figure 9-10 and draw the segment and backplane waveforms relative to ground for CONTROL = 0. Then draw the waveform of segment voltage relative to backplane voltage.

(b) Repeat for CONTROL = 1.

9-11. The BCD-to-7-segment decoder/driver of Figure 9-8 contains the logic for activating each segment for the appropriate BCD inputs. Design the logic for activating the *g* output.

Section 9-4

9-12. Determine the output levels for the 74147 encoder when $\overline{A}_8 = \overline{A}_4 = 0$ and all other inputs are HIGH.

9-13. Apply the signals of Figure 9-51 to the inputs of a 74147 as follows:
$$A \rightarrow \overline{A}_7, \qquad B \rightarrow \overline{A}_4, \qquad C \rightarrow \overline{A}_2, \qquad D \rightarrow \overline{A}_1$$
Draw the waveforms for the encoder's outputs.

9-14. Figure 9-52 shows the block diagram of a logic circuit used to control the number of copies made by a copy machine. The machine operator selects the number of desired copies by closing one of the selector switches $S_1 - S_9$. This number is encoded in BCD by the encoder and sent to a comparator circuit. The operator then hits a momentary-contact START switch, which clears the counter and initiates a HIGH OPERATE output that is sent to the machine to signal it to make copies. As the machine makes each copy, a copy pulse is generated and fed to the BCD counter. The counter

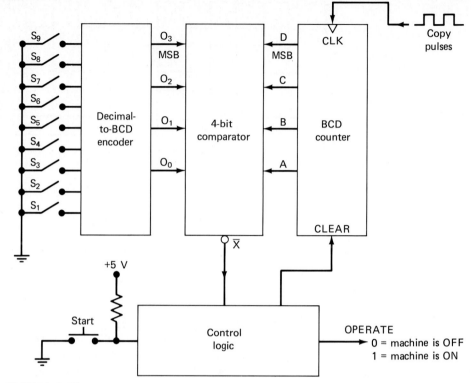

FIGURE 9-52

outputs are continually compared with the switch encoder outputs by the comparator. When the two BCD numbers match, indicating that the desired number of copies have been made, the comparator output $\overline{X}$ goes LOW; this causes the OPERATE level to return LOW and stop the machine so that no more copies are made. Activating the START switch will cause this process to be repeated. Design the complete logic circuitry for the comparator and control sections of this system.

9-15. The keyboard circuit of Figure 9-16 is designed to accept a three-digit decimal number. What would happen if *four* digit keys were activated (e.g., 3095)? Design the necessary logic to be added to this circuit so that after three digits have been entered, any further digits will be ignored until the CLEAR key is depressed. In other words, if 3095 is entered on the keyboard, the output registers will display 309 and ignore the 5 and any subsequent digits until the circuit is CLEARED.

Section 9-6

9-16. A technician breadboards the keyboard entry circuit of Figure 9-16 and tests its operation by trying to enter a series of three-digit numbers. He finds that sometimes the digit "0" is entered instead of the digit he pressed. He also observes that it happens with all the keys more or less randomly, although it is worse for some keys than others. He replaces all the ICs, and the malfunction persists. Which of the following circuit faults would explain his observations? Explain each choice.

(a) The technician forgot to ground the unused inputs of the OR gate.

(b) He has mistakenly used $\overline{Q}$ instead of Q from the one-shot.

(c) The switch bounce from the digit keys lasts longer than 20 ms.

(d) The Y and Z outputs are shorted together.

9-17. Repeat Problem 9-16 with the following symptom: the registers and displays stay at 0 no matter how many times a key is pressed.

9-18. While testing the circuit of Figure 9-16, a technician finds that every odd-numbered key results in the correct digit being entered, but every even-numbered key results in the wrong digit being entered as follows: the "0" key causes a "1" to be entered, the "2" key causes a "3" to be entered, the "4" key causes a "5" to be entered, and so on. Consider each of the following faults as possible causes of the malfunction. For each one explain why it can or cannot be the actual cause.

(a) An open in the connection from the output of the LSB inverter to the D inputs of the FFs.

(b) The D input of FF Q_8 is internally shorted to V_{CC}.

(c) A solder bridge is shorting $\overline{O}_0$ to ground.

9-19. A technician tests the circuit of Figure 9-4 as described in Example 9-6, and obtains the following results: all of the outputs work except $\overline{O}_{16}$–$\overline{O}_{19}$ and O_{24}–$\overline{O}_{27}$, which are permanently HIGH. What is the most probable circuit fault?

9-20. A technician tests the circuit of Figure 9-4 as described in Example 9-6, and finds that the correct output is activated for each possible input code except those listed in the table below. Examine this table and determine the probable cause of the malfunction.

INPUT CODE $A_4A_3A_2A_1A_0$	ACTIVATED OUTPUTS
1 0 0 0 0	$\overline{O}_{16}$ and $\overline{O}_{24}$
1 0 0 0 1	$\overline{O}_{17}$ and $\overline{O}_{25}$
1 0 0 1 0	$\overline{O}_{18}$ and $\overline{O}_{26}$
1 0 0 1 1	$\overline{O}_{19}$ and $\overline{O}_{27}$
1 0 1 0 0	$\overline{O}_{20}$ and $\overline{O}_{28}$
1 0 1 0 1	$\overline{O}_{21}$ and $\overline{O}_{29}$
1 0 1 1 0	$\overline{O}_{22}$ and $\overline{O}_{30}$
1 0 1 1 1	$\overline{O}_{23}$ and $\overline{O}_{31}$

9-21. Suppose that a 22-Ω resistor was mistakenly used for the g segment in Figure 9-8. How would this affect the display? What possible problems could occur?

9-22. Repeat Example 9-7 with the observed sequence shown below.

Count	0	1	2	3	4	5	6	7	8	9
Observed display	*0*	*1*	*2*	*3*	*8*	*9*	*c*	*⊐*	*4*	*5*

9-23. Repeat Example 9-7 with the observed sequence shown below.

Count	0	1	2	3	4	5	6	7	8	9
Observed display	*0*	*7*	*2*	*3*	*9*	*9*	*8*	*7*	*8*	*9*

9-24. To test the circuit of Figure 9-11, a technician connects a BCD counter to the 4511B inputs and pulses the counter at a very slow rate. She notices that the f segment works erratically, and no particular pattern is evident. What are some of the possible causes of the malfunction? (*Hint:* Remember, the ICs are CMOS.)

Sections 9-7 and 9-8

9-25. The circuit in Figure 9-53 uses 3 two-input multiplexers (Figure 9-20). Determine the function performed by this circuit.

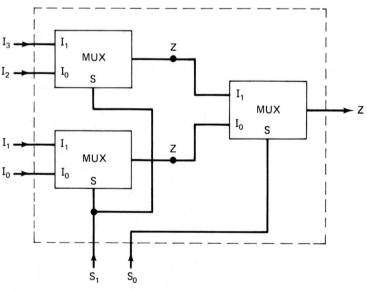

FIGURE 9-53

9-26. Use the idea from the last problem to arrange several 74151 1-of-8 multiplexers to form a 1-of-64 multiplexer.

9-27. Show how two 74157s and a 74151 can be arranged to form a 1-of-16 multiplexer with no other required logic. Label the inputs I_0–I_{15} to show how they correspond to the select code.

9-28. (a) Expand the circuit of Figure 9-25 to display the contents of 2 three-stage BCD counters.

    (b) Count the number of connections in this circuit and compare it to the number required if a separate decoder/driver and display were used for each stage of each counter.

9-29. Figure 9-54 shows how a multiplexer can be used to generate logic waveforms with any desirable pattern. The pattern is programmed using eight SPDT switches, and the waveform is repetitively produced by pulsing the MOD-8 counter. Draw the waveform at Z for the given switch positions.

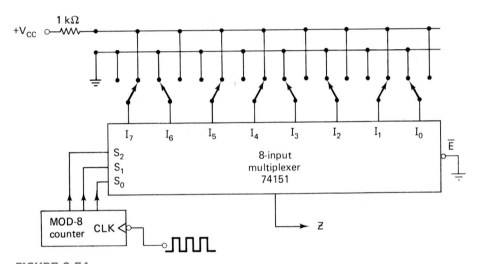

FIGURE 9-54

9-30. Change the MOD-8 counter in Figure 9-54 to a MOD-16 counter and connect the MSB to the multiplexer $\overline{E}$ input. Draw the Z waveform.

9-31. Show how a 74151 can be used to generate the logic function: $Z = AB + BC + AC$.

9-32. Show how a 16-input multiplexer such as the 74150 (Figure 9-36) is used to generate the function $Z = \overline{AB}\overline{C}D + BCD + A\overline{B}\overline{D} + AB\overline{C}D$.

9-33. The circuit of Figure 9-55 shows how an 8-input MUX can be used to generate a four-variable logic function even though the MUX has only three SELECT inputs. Three of the logic variables A, B, and C are connected to the SELECT inputs. The fourth variable D and its inverse $\overline{D}$ are connected to selected data inputs of the MUX as required by the desired logic function. The other MUX data inputs are tied to a LOW or a HIGH as required by the function.

    (a) Set up a truth table showing the output Z for the sixteen possible combinations of input variables.

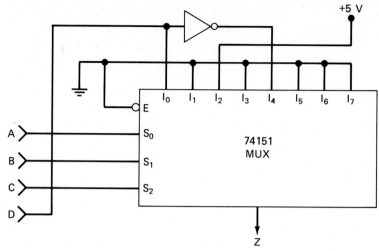

FIGURE 9-55

(b) Write the sum-of-products expression for Z and simplify it to verify that

$$Z = \overline{C}B\overline{A} + D\overline{C}\overline{B}\overline{A} + \overline{D}C\overline{B}\overline{A}$$

9-34. The method used in Figure 9-55 can be used to generate any four-variable logic function by following these steps:

1. Set up the truth table for the desired function with Z as the output.

2. Write the sum-of-products expression for Z; do not simplify it. For example, $Z = DC\overline{B}A + \overline{D}CB\overline{A} + DCB\overline{A} + D\overline{C}BA + \overline{D}\overline{C}BA + \overline{D}\overline{C}B\overline{A}$.

3. Look for terms that have the same combination of C, B, and A, and factor.

$$Z = DC\overline{B}A + CB\overline{A}(\overline{D} + D) + \overline{C}BA(\overline{D} + D) + \overline{D}\overline{C}B\overline{A}$$

$$= DC\overline{B}A + CB\overline{A} + \overline{C}BA + \overline{D}\overline{C}B\overline{A}$$

4. Consider those terms that contain only C, B, and A. For each of these, connect the corresponding MUX data input to a HIGH.

$$CB\overline{A} \rightarrow \text{connect HIGH to input } I_6$$

$$\overline{C}BA \rightarrow \text{connect HIGH to input } I_3$$

5. Consider the terms that contain the D variable. Connect the D or $\overline{D}$ variable to the MUX input that corresponds to the CBA variables.

$$DC\overline{B}A \rightarrow \text{connect } D \text{ to input } I_5$$

$$\overline{D}\overline{C}B\overline{A} \rightarrow \text{connect } \overline{D} \text{ to input } I_5$$

6. Connect the remaining MUX inputs to a LOW.

(a) Verify the design of Figure 9-55 using this method.

(b) Use the method above to implement a function that will produce a HIGH only when the four input variables are at the same level or when the B and C variables are at different levels.

Section 9-9

9-35. Show how the 7442 decoder can be used as 1-to-8 demultiplexer. (*Hint:* See Problem 9-6.)

9-36. Apply the waveforms of Figure 9-51 to the inputs of the 74LS138 DEMUX of Figure 9-31(a) as follows:

$$D \rightarrow A_2, \qquad C \rightarrow A_1, \qquad B \rightarrow A_0, \qquad A \rightarrow \overline{E}_1$$

Draw the waveforms at the DEMUX outputs.

9-37. Consider the system of Figure 9-33. Assume that the clock frequency is 10 pps. Describe what the monitoring panel indications will be for each of the following cases.
(a) All doors closed.
(b) All doors open.
(c) Doors 2 and 6 are open.

9-38. Modify the system of Figure 9-33 to handle 16 doors. Use a 74150 16-input MUX and two 74LS138 DEMUXes. How many lines are going to the remote monitoring panel?

9-39. Draw the waveforms at Z, O_0, O_1, O_2, and O_3 in Figure 9-34 for the following register data: $[A] = 0011$, $[B] = 0110$, $[C] = 1001$, $[D] = 0111$.

Section 9-11

9-40. Consider the control sequencer of Figure 9-27. Describe how each of the following faults will affect the operation.
(a) The I_3 input of the MUX is shorted to ground.
(b) The connections from sensors 3 and 4 to the MUX are reversed.

9-41. Consider the circuit of Figure 9-25. A test of the circuit yields the following results:

		ACTUAL COUNT	DISPLAYED COUNT
Case 1	Counter 1	33	33
	Counter 2	47	47
Case 2	Counter 1	82	02
	Counter 2	64	64
Case 3	Counter 1	63	63
	Counter 2	95	15

What are the possible causes of the malfunction?

9-42. A test of the security monitoring system of Figure 9-33 produces the results recorded below.

CONDITION	LEDs
All doors closed	All LEDs off
Door 0 open	LED 0 flashing
Door 1 open	LED 2 flashing
Door 2 open	LED 1 flashing
Door 3 open	LED 3 flashing
Door 4 open	LED 4 flashing
Door 5 open	LED 6 flashing
Door 6 open	LED 5 flashing
Door 7 open	LED 7 flashing

What are the possible faults that could cause this operation?

9-43. A test of the security monitoring system of Figure 9-33 produces the results recorded below.

CONDITION	LEDs
All doors closed	All LEDs off
Door 0 open	LED 0 flashing
Door 1 open	LED 1 flashing
Door 2 open	LED 2 flashing
Door 3 open	LED 3 flashing
Door 4 open	LED 4 flashing
Door 5 open	LED 5 flashing
Door 6 open	No LED flashing
Door 7 open	No LED flashing
Doors 6 and 7 open	LEDs 6 and 7 flashing

What are the possible faults that could cause this operation?

9-44. Suppose that the synchronous data transmission system of Figure 9-34 is malfunctioning as follows: the Z waveform is correct, but the O_0 waveform is identical to the Z waveform at all times while the other outputs are constantly LOW. What circuit faults could cause this malfunction? Assume the circuitry is TTL, and explain each choice.

9-45. The synchronous data transmission system of Figure 9-34 is malfunctioning, and the waveforms are displayed on a high-speed oscilloscope (Fig. 9-56). Note the glitches on the O_1 signal. Consider the two possible faults given below. For each one, explain whether or not it can be the cause of the malfunction.

(a) The connections to the S_1 and S_0 pins of the DEMUX are reversed.

(b) The connections to the Q_1 and Q_0 pins of the receiver's word counter are reversed.

Section 9-12

9-46. Redesign the circuit of Problem 9-14 using a 74LS85 magnitude comparator. Add a "copy overflow" feature that will activate an ALARM output if the OPERATE output fails to stop the machine when the requested number of copies are done.

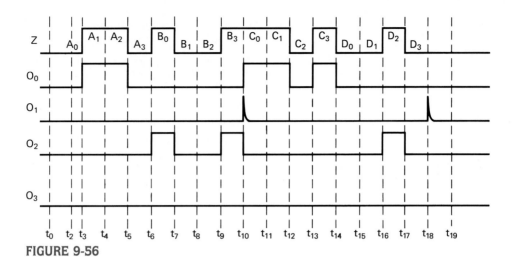

FIGURE 9-56

9-47. Show how to connect 74LS85s to compare two 10-bit numbers.

Sections 9-13 and 9-14

9-48. For the bus arrangement of Figure 9-44, describe the input signal requirements for simultaneously transferring the contents of register C to both of the other registers.

9-49. Assume that the registers in Figure 9-44 are initially $[A] = 1011$, $[B] = 1000$, and $[C] = 0111$. The signals in Figure 9-57 are applied to the register inputs.

    (a) Determine the contents of each register at times t_1, t_2, t_3, and t_4.

    (b) Describe what would happen if $\overline{IE}_A$ were LOW when the third clock pulse occurred.

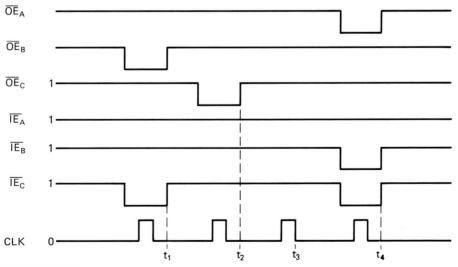

FIGURE 9-57

9-50. Assume the same initial conditions of Problem 9-49 and sketch the signal on DB_3 for the waveforms of Figure 9-57.

9-51. Figure 9-58 shows two more devices that are to be added to the data bus of Figure 9-44. One is a set of buffered switches that can be used to manually enter data into any of the bus registers. The other device is an output register that is used to latch any data that are on the bus during a data transfer operation and display them on a set of LEDs.

  (a) Assume all registers are at 0000. Outline the sequence of operations needed to load the registers with the following data from the switches: $[A] = 1011$, $[B] = 0001$, $[C] = 1110$.

  (b) What will be the state of the LEDs at the end of this sequence?

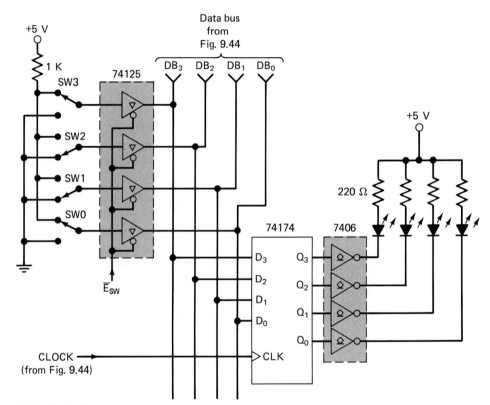

FIGURE 9-58

9-52. Now that the circuitry of Figure 9-58 has been added to Figure 9-44, a total of five devices are connected to the data bus. The circuit in Figure 9-59(a) will now be used to generate the enable signals needed to perform the different data transfers over the data bus. It uses a 74LS139 chip that contains two identical independent 1-of-4 decoders with active LOW enable. The top decoder is used to select the device that will put data on the data bus (output select), and the bottom decoder is used to select the device that is to take the data from the data bus (input select). Assume that the decoder outputs are connected to the corresponding enable inputs of the

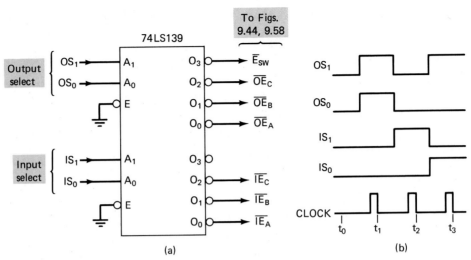

FIGURE 9-59

devices tied to the data bus. Also assume that all registers are initially at 0000 at time t_0, and the switches are in the positions shown in Figure 9-58.

(a) Determine the contents of each register at times t_1, t_2, and t_3 in response to the waveforms in Figure 9-59(b).

(b) Can "bus contention" ever occur with this circuit?

9-53. In theory, the answer to Problem 9-52(b) is "no" because only one decoder output can go LOW at one time, and so only one device's outputs can be enabled. In practice, however, there will be a very short overlap interval where two devices' outputs are enabled as the output select code changes from one code to another. This is usually caused by the fact that the newly enabled device becomes enabled faster than the previously enabled device becomes disabled. What can be done to prevent "bus contention" during the transition from one output select code to another?

INTERFACING WITH THE ANALOG WORLD

10

Upon completion of this chapter, you will be able to:

- Understand the theory of operation and circuit limitations of several types of digital-to-analog converters (DACs).

- Compare different types of multiplying DACs.

- Read and understand the various DAC manufacturer specifications.

- Use different test procedures to troubleshoot DAC circuits.

- Compare the advantages and disadvantages among the digital-ramp analog-to-digital converter (ADC), successive-approximation ADC, and flash ADC.

- Analyze the process by which a computer in conjunction with an ADC digitizes an analog signal and then reconstructs that analog signal from the digital data.

- Describe the basic operation of a digital voltmeter.

- Understand the need for using sample-and-hold circuits in conjunction with ADCs.

- Cite the advantages and limitations of multiplexing ADCs.

10-1 INTERFACING WITH THE ANALOG WORLD

Most physical variables are analog in nature. Recall that this means they can take on any value within a continuous range of values. Examples include temperature, pressure, light intensity, audio signals, position, rotational speed, and flow rate. Digital systems perform all of their internal operations using digital circuitry and digital operations. Any information that has to be input to a digital system must first be put into digital form. Similarly, the outputs from a digital system are always in digital form. When a digital system such as a computer is to be used to monitor and/or control a physical process, we must deal with the difference between the digital nature of the computer, and the analog nature of the process variables. Figure 10-1 illustrates the situation. This diagram shows the five elements that are involved when a computer is monitoring and controlling a physical variable that is assumed to be analog.

1. *Transducer.* The physical variable is normally a nonelectrical quantity. A transducer is a device that converts the physical variable to an electrical vari-

FIGURE 10-1 Analog-to-digital converter (ADC) and digital-to-analog converter (DAC) are used to interface a computer to the analog world so that the computer can monitor and control a physical variable.

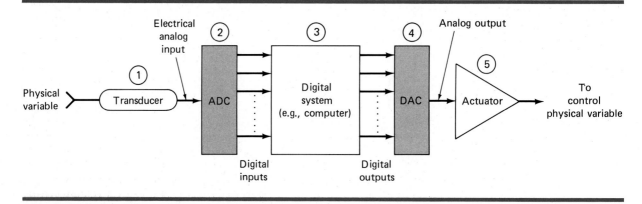

able. Some common transducers include thermistors, photocells, photodiodes, flow meters, pressure transducers, and tachometers. The electrical output of the transducer is an analog current or voltage that is proportional to the physical variable it is monitoring.

For example, the physical variable could be the temperature of water in a large tank that is being filled from cold and hot water pipes. Let's say that the water temperature varies from 80° to 150°F and that a thermistor and its associated circuitry converts this water temperature to a voltage ranging from 800 to 1500 mV. Note that the transducer's output is directly proportional to temperature such that each 1°F produces a 10-mV output. This proportionality factor was chosen for convenience.

2. *Analog-to-digital converter (ADC)*. The transducer's electrical analog output serves as the analog input to the ADC. The ADC converts this analog input to a digital output. This digital output consists of a number of bits that represents the value of the analog input.

For example, the ADC might convert the transducer's 800–1500 mV analog values to binary values ranging from 01010000(80) to 10010110(150). Note that the binary output from the ADC is proportional to the analog input voltage such that each unit of the digital output represents 10 mV.

3. *Computer*. The digital representation of the process variable is transmitted from the ADC to the digital computer, which processes the digital input value according to a program of instructions that it is executing. The program might perform calculations or other operations on this digital representation of temperature to come up with a digital output that will eventually be used to control the temperature.

4. *Digital-to-analog converter (DAC)*. This digital output from the computer is connected to a DAC, which converts it to a proportional analog voltage or current. For example, the computer might produce a digital output ranging from 00000000 to 11111111, which the DAC converts to a voltage ranging from 0 to 10 V.

5. *Actuator*. The analog signal from the DAC is often connected to some device or circuit that serves as an actuator to control the physical variable. For our water temperature example, the actuator might be an electrically controlled valve that regulates the flow of hot water into the tank in accordance with the analog voltage from the DAC. The flow rate would vary in proportion to this analog voltage, with 0 V producing no flow and 10 V producing the maximum flow rate.

Thus we see that ADCs and DACs function as *interfaces* between a completely digital system, like a computer, and the analog world. This function has become increasingly more important as inexpensive microcomputers have moved into areas of process control where computer control was previously not feasible.

REVIEW QUESTIONS

1. What is the function of a transducer?
2. What is the function of an ADC?

3. What does a computer often do with the data it receives from an ADC?

4. What function does a DAC perform?

5. What is the function of an actuator?

10-2 DIGITAL-TO-ANALOG CONVERSION

We will now begin our study of digital-to-analog (D/A) and analog-to-digital (A/D) conversion. Since many A/D conversion methods utilize the D/A conversion process, we will examine D/A conversion first.

Basically, D/A *conversion* is the process of taking a value represented in *digital* code (such as straight binary or BCD) and converting it to a voltage or current which is proportional to the digital value. This voltage or current is an *analog* quantity, since it can take on many different values over a given range. Figure 10-2(a) shows the symbol for a typical 4-bit D/A converter. We will not concern ourselves with the internal circuitry until later. For now, we will examine the various input/output relationships.

The digital inputs *D, C, B,* and *A* are usually derived from the output register of a digital system. The $2^4 = 16$ different binary numbers represented by this 4 bits are listed in Figure 10-2(b). For each input number, the D/A converter output voltage is a unique value. In fact, for this case, the analog output voltage V_{OUT} is equal in volts to the binary number. It could also have been twice the binary number or some other proportionality factor. The same idea would hold true if the D/A output were a current I_{OUT}.

EXAMPLE 10-1

A 5-bit D/A converter has a current output. For a digital input of 10100, an output current of 10 mA is produced. What will I_{OUT} be for a digital input of 11101?

SOLUTION

The digital input 10100 is the binary equivalent of 20_{10}. Since $I_{OUT} = 10$ mA for this case, the proportionality factor is 0.5; that is, $I_{OUT} = 0.5 \times$ binary value. Thus, the binary input 11101 is equivalent to 29_{10}, so $I_{OUT} = 0.5 \times 29 = 14.5$ mA.

Input Weights For the DAC of Figure 10-2 it should be noted that each digital input contributes a different amount to the analog output. This is easily seen if we examine the cases where only one input is HIGH:

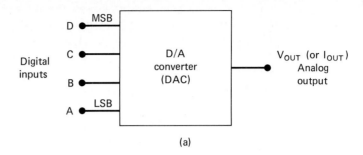

(a)

D	C	B	A	V_{OUT}	
0	0	0	0	0	volts
0	0	0	1	1	
0	0	1	0	2	
0	0	1	1	3	
0	1	0	0	4	
0	1	0	1	5	
0	1	1	0	6	
0	1	1	1	7	
1	0	0	0	8	
1	0	0	1	9	
1	0	1	0	10	
1	0	1	1	11	
1	1	0	0	12	
1	1	0	1	13	
1	1	1	0	14	
1	1	1	1	15	volts

(b)

FIGURE 10-2   Four-bit D/A converter with voltage output.

D	C	B	A		V_{OUT} (V)
0	0	0	1	$\rightarrow$	1
0	0	1	0	$\rightarrow$	2
0	1	0	0	$\rightarrow$	4
1	0	0	0	$\rightarrow$	8

The contributions of each digital input are *weighted* according to their position in the binary number. Thus, *A,* which is the LSB, has a *weight* of 1 V, *B* has a weight of 2 V, *C* has a weight of 4 V, and *D,* the MSB, has the largest weight, 8 V. The weights are successively doubled for each bit, beginning with the LSB. Thus, we can consider V_{OUT} to be the weighted sum of the digital inputs. For instance, to find V_{OUT} for the digital input 0111 we can add the weights of the *C, B,* and *A* bits to obtain 4 V + 2 V + 1 V = 7 V.

EXAMPLE 10-2

A 5-bit D/A converter produces $V_{OUT} = 0.2$ V for a digital input of 00001. Find the value of V_{OUT} for a 11111 input.

SOLUTION
Obviously, 0.2 V is the weight of the LSB. Thus, the weights of the other bits must be 0.4 V, 0.8 V, 1.6 V, and 3.2 V, respectively. For a digital input of 11111, then, the value of V_{OUT} will be 3.2 V + 1.6 V + 0.8 V + 0.4 V + 0.2 V = 6.2 V.

Resolution (Step Size) *Resolution* of a D/A converter is defined as the smallest change that can occur in the analog output as a result of a change in the digital input. Referring to the table in Figure 10-2, we can see that the resolution is 1 V, since V_{OUT} can change by no less than 1 V when the digital input value is changed. The resolution is always equal to the weight of the LSB and is also referred to as the *step size,* since it is the amount that V_{OUT} will change as the digital input value is changed from one step to the next. This is illustrated better in Figure 10-3, where the outputs from a 4-bit binary counter provide the inputs to our DAC. As the counter is being continually cycled through its 16 states by the clock signal, the DAC output is a *staircase* waveform that goes up 1 V per step. When the counter is at 1111, the DAC output is at its maximum value of 15 V; this is its *full-scale output.* When the counter recycles to 0000, the DAC output returns to 0 V. The resolution or step size is the size of the jumps in the staircase waveform; in this case, each step is 1 V.

FIGURE 10-3   Output waveforms of D/A converter as inputs are provided by a binary counter.

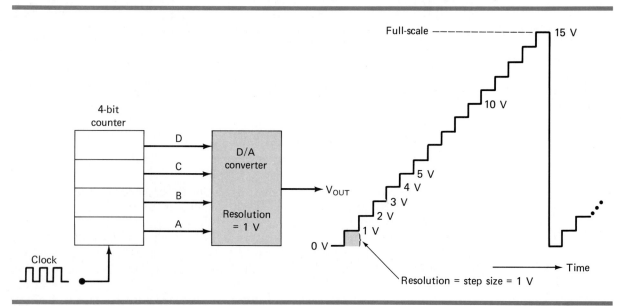

Note that the staircase has 16 levels corresponding to the 16 input states, but there are only 15 steps or jumps between the 0-V level and full-scale. In general, for an N-bit DAC the number of different levels will be 2^N, and the number of steps will be $2^N - 1$.

EXAMPLE 10-3

What is the resolution (step size) of the DAC of Example 10-2? Describe the staircase signal out of this DAC.

SOLUTION
The LSB for this converter has a weight of 0.2 V. This is the resolution or step size. A staircase waveform can be generated by connecting a 5-bit counter to the DAC inputs. The staircase will have 32 levels from 0 V up to a full-scale output of 6.2 V, and 31 steps of 0.2 V each.

Percentage Resolution   Although resolution can be expressed as the amount of voltage or current per step, it is more useful to express it as a percentage of the *full-scale output*. To illustrate, the D/A converter of Figure 10-3 has a maximum full-scale output of 15 V (when the digital input is 1111). The step size is 1 V, which gives a percentage resolution of

$$\% \text{ resolution} = \frac{\text{step size}}{\text{full scale (F.S.)}} \times 100\% \tag{10-1}$$

$$= \frac{1 \text{ V}}{15 \text{ V}} \times 100\% = 6.67\%$$

EXAMPLE 10-4

A 10-bit D/A converter has a step size of 10 mV. Determine the full-scale output voltage and the percentage resolution.

SOLUTION
With 10 bits, there will be $2^{10} - 1 = 1023$ steps of 10 mV each. The full-scale output will therefore be 10 mV $\times$ 1023 = 10.23 V and

$$\% \text{ resolution} = \frac{10 \text{ mV}}{10.23 \text{ V}} \times 100\% \approx 0.1\%$$

Example 10-4 helps to illustrate the fact that the percentage resolution becomes smaller as the number of input bits is increased. In fact, the percentage resolution can also be calculated from

$$\% \text{ resolution} = \frac{1}{\text{total number of steps}} \times 100\% \tag{10-2}$$

For an N-bit binary input code the total number of steps is $2^N - 1$. Thus, for the previous example,

$$\% \text{ resolution } = \frac{1}{2^{10} - 1} \times 100\%$$

$$= \frac{1}{1023} \times 100\%$$

$$\approx 0.1\%$$

This means that it is *only the number of bits* which determines the *percentage* resolution. Increasing the number of bits increases the number of steps to reach full scale, so each step is a smaller part of the full-scale voltage. Most DAC manufacturers specify resolution as the number of bits.

What Does Resolution Mean?

A DAC cannot produce a continuous range of output values, so, strictly speaking, its output is not truly analog. A DAC produces a finite set of output values. In our example of Section 10-1 the computer generates a digital output to provide an analog voltage between 0 and 10 V to an electrically controlled valve. The DAC's resolution (number of bits) determines how many possible voltage values the computer can send to the valve. If a 6-bit DAC is used, there will be 63 possible steps of 0.159 V between 0 and 10 V. When an 8-bit DAC is used, there will be 255 possible steps of 0.039 V between 0 and 10 V. As the number of bits increases, the finer the resolution (the smaller the step size).

The system designer must decide what resolution is needed based on the required system performance. The resolution limits how close the DAC output can come to a given analog value. Generally, the cost of DACs increases with the number of bits, so the designer will use as few bits as necessary.

EXAMPLE 10-5

Figure 10-4 shows a computer controlling the speed of a motor. The 0–2 mA analog current from the DAC is amplified to produce motor speeds from 0 to 1000 rpm (revolutions per minute). How many bits should be used if the computer is to be able to produce a motor speed that is within 2 rpm of the desired speed?

SOLUTION
The motor speed will range from 0 to 1000 rpm as the DAC goes from zero to full-scale. Each step in the DAC output will produce a step in the motor speed.

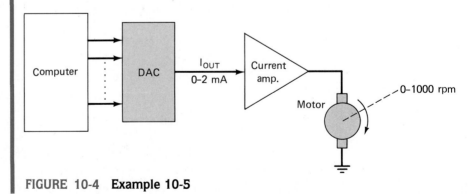

FIGURE 10-4   Example 10-5

We want the step size to be no greater than 2 rpm. Thus, we need at least 500 steps (1000/2). Now we must determine how many bits are required so that there are at least 500 steps from zero to full-scale. We know that the number of steps is $2^N - 1$, so we can say

$$2^N - 1 \geq 500$$

or

$$2^N \geq 501$$

Since $2^8 = 256$ and $2^9 = 512$, the smallest number of bits that will produce at least 500 steps is *nine*. We could use more than 9 bits, but this might add to the cost of the DAC.

EXAMPLE 10-6

Using 9 bits, how close to 326 rpm can the speed be adjusted?

SOLUTION

With 9 bits, there will be 511 steps ($2^9 - 1$). Thus, the motor speed will go up in steps of 1000 rpm/511 = 1.957 rpm. The number of steps needed to reach 326 rpm is 326/1.957 = 166.58. This is not a whole number of steps, so we will round it to 167. The actual motor speed on the 167th step will be 167 × 1.957 = 326.8 rpm. Thus, the computer must output the 9-bit binary equivalent of 167_{10} to produce the desired motor speed within the resolution of the system.

In all of our examples, we have assumed that the DACs have been perfectly accurate in producing an analog output that is directly proportional to the binary input, and that the resolution is the only thing that limits how close we can come to a desired analog value. This, of course, is unrealistic since all devices contain inaccuracies. We will examine the causes and effects of DAC inaccuracy in a later section.

BCD Input Code   The D/A converters we have considered thus far have used a binary input code. Many D/A converters use a BCD input code where 4-bit code groups are used for each decimal digit. Figure 10-5 shows the diagram of an 8-bit

FIGURE 10-5   D/A converter using BCD input code. This converter accepts a 2-digit input and generates 100 possible analog output values.

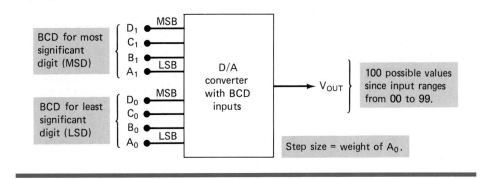

(two-digit) converter of this type. Each 4-bit code group can range from 0000 to 1001, so the BCD inputs can represent any decimal number from 00 to 99. *Within* each code group the weights of the different bits are proportioned the same as in the binary code but the *group* weights are different by a factor of 10. For example, A_0, the LSB of the least significant digit (LSD), could have a weight of 0.1 V. Thus, B_0, C_0, and D_0 would be 0.2, 0.4, and 0.8 V, respectively. The weight of A_1, the LSB of the MSD, would be 1 V (10 times A_0). Similarly, B_1, C_1, and D_1 would be 2, 4, and 8 V, respectively.

EXAMPLE 10-7

If the weight of A_0 is 0.1 V in Figure 10-5, find the following values.

(a) Step size.

(b) Full-scale output and percentage resolution.

(c) V_{OUT} for $D_0C_0B_0A_0 = 1000$ and $D_1C_1B_1A_1 = 0101$.

SOLUTION

(a) Step size is the weight of the LSB of the LSD, 0.1 V.

(b) There are 99 steps since there are two BCD digits. Thus, full-scale output is $99 \times 0.1 = 9.9$ V. The resolution is [using equation (10.1)]

$$\frac{\text{step size}}{\text{F.S.}} \times 100\% = \frac{0.1}{9.9} \times 100\% \approx 1\%$$

We could also have used equation (10.2) to calculate percentage resolution, since the total number of steps is 99.

(c) To find V_{OUT}, add the weights of all the bits that are 1s. Thus,

$$D_0 \qquad C_1 \qquad A_1$$

$$V_{OUT} = 0.8\text{ V} + 4\text{ V} + 1\text{ V} = 5.8\text{ V}$$

Alternatively, $D_1C_1B_1A_1 = 5_{10}$ and $D_0C_0B_0A_0 = 8_{10}$, so the input code is 58, multiplied by 0.1 V per step, to again give 5.8 V.

Bipolar DACs Up to this point we have assumed that the binary input to a DAC has been an unsigned number, and the DAC output has been a positive voltage or current. Some DACs are designed to produce both positive and negative values, such as -10 to $+10$ V. This is generally done by using the binary input as a signed number with the MSB as the sign bit (0 for $+$ and 1 for $-$). Negative input values are often represented in 2's-complement form, although the true-magnitude form is also used by some DACs. For example, suppose that we have a 6-bit bipolar DAC that uses the 2's-complement system and has a resolution of 0.2 V. The binary input values range from 100000 (-32) to 011111 ($+31$) to produce analog outputs in the range from -6.4 to $+6.2$ V. There are 63 steps ($2^6 - 1$) of 0.2 V between these negative and positive limits.

REVIEW QUESTIONS

1. An 8-bit DAC has an output of 3.92 mA for an input of 01100010. What is the DAC's resolution and full-scale output? (*Ans.* 40 μa; 10.2 mA)

2. What is the weight of the MSB of the DAC of question 1? (*Ans.* 5.12 mA)

3. What is the percentage resolution of an 8-bit DAC? (*Ans.* 0.39%)

4. How many different output voltages can a 12-bit DAC produce? (*Ans.* 4096)

5. For the system of Figure 10-4, how many bits should be used if the computer is to control the motor speed within 0.4 rpm? (*Ans.* twelve)

6. Consider a 12-bit DAC with BCD inputs and a resolution of 10 mV. What will be its output for an input of 100001110011? (*Ans.* 8.73 V).

7. *True or false:* The percentage resolution of a DAC depends *only* on the number of bits.

10-3 D/A-CONVERTER CIRCUITRY

There are several methods and circuits for producing the D/A operation which has been described. We shall examine several of the basic schemes, to gain an insight into the ideas used. It is not important to be familiar with all the various circuit schemes because D/A converters are available as ICs or as encapsulated packages that do not require any circuit knowledge. Instead, it is important to know the significant performance characteristics of D/A converters, in general, so that they can be used intelligently. These will be covered in Section 10-4.

Figure 10-6(a) shows the basic circuit for one type of 4-bit D/A converter. The inputs A, B, C, and D are binary inputs which are assumed to have values of either 0 V or 5 V. The *operational amplifier* is employed as a summing amplifier, which produces the weighted sum of these input voltages. It may be recalled that the summing amplifier multiplies each input voltage by the ratio of the feedback resistor R_F to the corresponding input resistor R_{IN}. In this circuit $R_F = 1$ kΩ and the input resistors range from 1 to 8 kΩ. The D input has $R_{IN} = 1$ kΩ, so the summing amplifier passes the voltage at D with no attenuation. The C input has $R_{IN} = 2$ kΩ, so it will be attenuated by 0.5. Similarly, the B input will be attenuated by 0.25 and the A input by 0.125. The amplifier output can thus be expressed as

$$V_{OUT} = -(V_D + 0.5V_C + 0.25V_B + 0.125V_A) \qquad (10\text{-}3)$$

The negative sign is present because the summing amplifier is a polarity-inverting amplifier, but it will not concern us here.

Clearly, the summing amplifier output is an analog voltage which represents a weighted sum of the digital inputs, as shown by the table in Figure 10-6(b). This table lists all the possible input conditions and the resultant amplifier output voltage. The output is evaluated for any input condition by setting the appropriate inputs to

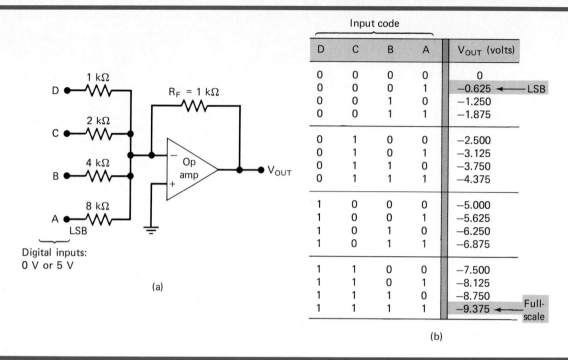

Input code				
D	C	B	A	V_{OUT} (volts)
0	0	0	0	0
0	0	0	1	−0.625 ⟵ LSB
0	0	1	0	−1.250
0	0	1	1	−1.875
0	1	0	0	−2.500
0	1	0	1	−3.125
0	1	1	0	−3.750
0	1	1	1	−4.375
1	0	0	0	−5.000
1	0	0	1	−5.625
1	0	1	0	−6.250
1	0	1	1	−6.875
1	1	0	0	−7.500
1	1	0	1	−8.125
1	1	1	0	−8.750
1	1	1	1	−9.375 ⟵ Full-scale

(b)

FIGURE 10-6   Simple D/A converter using op-amp summing amplifier with binary-weighted resistors.

either 0 V or 5 V. For example, if the digital input is 1010, then $V_D = V_B = 5$ V and $V_C = V_A = 0$ V. Thus, using (10-3),

$$V_{OUT} = -(5 \text{ V} + 0 \text{ V} + 0.25 \times 5 \text{ V} + 0 \text{ V})$$

$$= -6.25 \text{ V}$$

The resolution of this D/A converter is equal to the weighting of the LSB, which is 0.125×5 V = 0.625 V. As shown in the table, the analog output increases by 0.625 V as the binary input number advances one step.

EXAMPLE 10-8

(a)   Determine the weights of each input bit of Figure 10-6(a).

(b)   Change R_F to 250 Ω and determine the full-scale output.

SOLUTION

(a)   The MSB passes with gain = 1, so its weight in the output is 5 V. Thus,

$$\text{MSB} \to 5 \text{ V}$$
$$\text{2nd MSB} \to 2.5 \text{ V}$$
$$\text{3rd MSB} \to 1.25 \text{ V}$$
$$\text{4th MSB} = \text{LSB} \to 0.625 \text{ V}$$

(b)   If R_F is reduced by a factor of 4, to 250 Ω, each input weight will be four times *smaller* than the values above. Thus, the full-scale output will be reduced by this same factor and becomes $-9.375/4 = -2.344$ V.

If we look at the input resistor values in Figure 10-6, it should come as no surprise that they are *binarily weighted*. In other words, starting with the MSB resistor, the resistor values increase by a factor of 2. This of course produces the desired weighting in the voltage output.

Conversion Accuracy   The table in Figure 10-6(b) gives the *ideal* values of V_{OUT} for the various input cases. How close the circuit comes to producing these values depends primarily on two factors: (1) the precision of the input and feedback resistors, and (2) the precision of the input voltage levels. The resistors can be made very accurate (within 0.01 percent of the desired values) by trimming, but the input voltage levels must be handled differently. It should be clear that the digital inputs cannot be taken directly from the outputs of FFs or logic gates because the output logic levels of these devices are not precise values like 0 V and 5 V but vary over a given range. For this reason, it is necessary to insert a *precision-level amplifier* in between each logical input and its input resistor to the summing amplifier. This is shown in Figure 10-7.

The level amplifiers produce precise output levels of 5 V and 0 V, depending on whether the digital inputs are HIGH or LOW. A very stable, precise 5-V *reference* supply is required to produce the accurate 5-V level.

FIGURE 10-7   Complete 4-bit D/A converter including precision-level amplifiers.

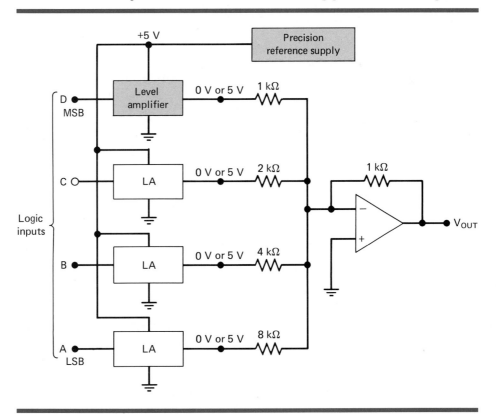

DAC with Current Output Figure 10-8(a) shows one basic scheme for generating an analog output current proportional to a binary input. The circuit shown is a 4-bit DAC using binarily-weighted resistors. The circuit uses four parallel current paths, each controlled by a semiconductor switch such as the CMOS transmission gate (Chapter 8). The state of each switch is controlled by logic levels at the binary inputs. The current through each path is determined by an accurate reference voltage, V_{REF}, and a precision resistor in the path. The resistors are binarily weighted so that the various currents will be binarily weighted, and the total current, I_{OUT}, will be the sum of the individual currents. The MSB path has the smallest resistor, R; the next path has a resistor of twice the value, and so on. The output current can be made to flow through a load R_L which is much smaller than R, so that it has no effect on the value of current. Ideally, R_L should be a short to ground.

FIGURE 10-8 (a) Basic current-output DAC; (b) connected to an op-amp current-to-voltage converter.

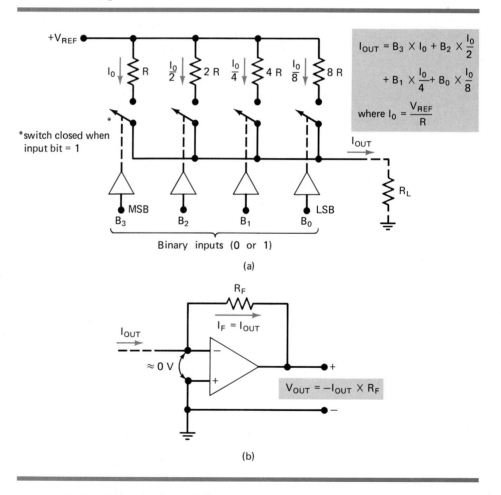

$$I_{OUT} = B_3 \times I_0 + B_2 \times \frac{I_0}{2}$$
$$+ B_1 \times \frac{I_0}{4} + B_0 \times \frac{I_0}{8}$$
$$\text{where } I_0 = \frac{V_{REF}}{R}$$

*switch closed when input bit = 1

Binary inputs (0 or 1)

(a)

$I_F = I_{OUT}$

$V_{OUT} = -I_{OUT} \times R_F$

(b)

EXAMPLE 10-9

Assume $V_{REF} = 10$ V and $R = 10$ kΩ. Determine the resolution and full-scale output for this DAC. Assume that $R_L \ll R$.

SOLUTION

$I_0 = V_{REF}/R = 1$ mA. This is the weight of the MSB. The other three currents wll be 0.5, 0.25, and 0.125 mA. The LSB is 0.125 mA, which is also the resolution.

The full-scale output will occur when the binary inputs are all HIGH so that each current switch is closed and

$$I_{OUT} = 1 + 0.5 + 0.25 + 0.125 = 1.875 \text{ mA}$$

Note that the output current is proportional to V_{REF}. If V_{REF} is increased or decreased, I_{OUT} will change proportionally.

In order for I_{OUT} to be accurate, R_L should be a short to ground. One common way to accomplish this is to use an op-amp as a current-to-voltage converter, as shown in Figure 10-8(b). Here the I_{OUT} from the DAC is connected to the op-amp's "$-$" input, which is virtually at ground. The op-amp negative feedback forces a current equal to I_{OUT} to flow through R_F to produce $V_{OUT} = -I_{OUT} \times R_F$. Thus, V_{OUT} will be an analog voltage that is proportional to the binary input to the DAC. This analog output can drive a wide range of loads without being loaded down.

R/2R Ladder   The DAC circuits we have looked at thus far use binary-weighted resistors to produce the proper weighting of each bit. While this method works in theory, it has some practical limitations. The biggest problem is the large difference in resistor values between the LSB and MSB, especially in high-resolution DACs (i.e., many bits). For example, if the MSB resistor is 1 kΩ in a 12-bit DAC, the LSB resistor will be over 2 MΩ. With the current IC fabrication technology, it is very difficult to produce resistance values over a wide resistance range that maintain an accurate ratio especially with variations in temperature.

For this reason it is preferable to have a circuit that uses resistances that are fairly close in value. One of the most widely used DAC circuits that satisfies this requirement is the *R/2R ladder* network, where the resistance values only span a range of 2 to 1. One such DAC is shown in Figure 10-9.

Note how the resistors are arranged, and especially note that only two different values are used, R and $2R$. The current I_{OUT} depends on the positions of the four switches, and the binary inputs, $B_3B_2B_1B_0$ control the states of the switches. This current is allowed to flow through an op-amp current-to-voltage converter to develop V_{OUT}. We will not perform a detailed analysis of this circuit here, but it can be shown that the value of V_{OUT} is given by the expression

$$V_{OUT} = \frac{-V_{REF}}{8} \times B \tag{10-4}$$

where B is the value of the binary input which can range from 0000(0) to 1111(15). The resolution can be found by setting $B = 0001 = 1$; it is seen to be equal to $V_{REF}/8$. The full-scale output can be determined by setting $B = 1111 = 15$ and evaluating V_{OUT}; the result is $1.875V_{REF}$.

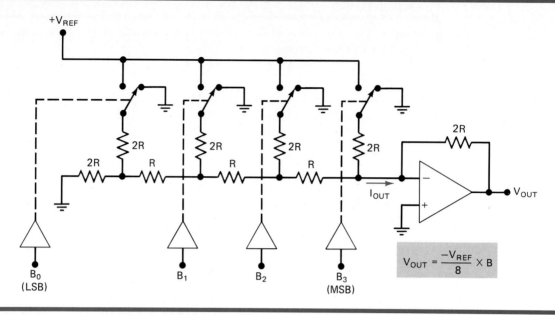

$$V_{OUT} = \frac{-V_{REF}}{8} \times B$$

FIGURE 10-9 Basic R/2R ladder DAC.

Multiplying DACs   Many DACs require a reference voltage source that is used internally to help generate the analog output. In some DACs this reference voltage has to be a fixed, precise value in order to produce an accurate output. In other types the reference voltage can actually be variable and even change polarities. This latter type of DAC is called a *multiplying DAC* because the analog output is the product of the binary input and the reference voltage.

For example, we saw that the output current from the DAC of Figure 10-8(a) is proportional to V_{REF}. That is,

$$I_{OUT} = k(V_{REF} \times B) \qquad (10\text{-}5)$$

where B represents the value of the binary input, and k is the proportionality constant. We can determine the value of k from the results of example 10-9, where we found that $I_{OUT} = 1.875$ mA for $V_{REF} = 10$ V and $B = 1111_2 = 15_{10}$. Thus, we have

$$1.875 \text{ mA} = k(10 \text{ V} \times 15)$$

or

$$k = 1.875 \text{ mA}/150 \text{ V} = 0.0125 \text{ mA/V}$$

Once k is known, I_{OUT} can be found for any values of V_{REF} and binary input.

A multiplying DAC can be used to multiply an analog voltage (V_{REF}) by a binary value to produce an analog output. Some multiplying DACs allow V_{REF} to be of either polarity so that the output product can have either polarity. These are called *two-quadrant* multiplying DACs. Some DACs also allow the binary input to be of either polarity by using the MSB as a sign bit. These are called *four-quadrant* multiplying DACs.

EXAMPLE 10-10

Find I_{OUT} for the DAC of Figure 10-8(a) when $V_{REF} = 3.8$ V and the binary input is 0110 $= 6_{10}$.

SOLUTION
Using equation (10-5) with $k = 0.0125$ mA/V

$$I_{OUT} = (0.0125 \text{ mA/V})(3.8 \text{ V} \times 6)$$
$$= 0.285 \text{ mA}$$

REVIEW QUESTONS

1. What is the advantage of $R/2R$ ladder DACs over those that use binary-weighted resistors?
2. A certain 6-bit DAC uses binary-weighted resistors. If the MSB resistor is 20 kΩ, what is the LSB resistor? (*Ans.* 640 kΩ)
3. What will be the resolution if the value of R_F in Figure 10-6 is changed to 800 Ω? (*Ans.* 0.5 V)
4. What is a multiplying DAC?

10-4 DAC SPECIFICATIONS

A wide variety of DACs are currently available as ICs or as self-contained encapsulated packages. One should be familiar with the more important manufacturers' specifications in order to evelute a DAC for a particular application.

Resolution As mentioned earlier, the percentage resolution of a DAC is dependent solely on the number of bits. For this reason, manufacturers usually specify a DAC resolution as the number of bits. A 10-bit DAC has a finer (smaller) resolution than an 8-bit DAC.

Accuracy DAC manufacturers have several ways of specifying accuracy. The two most common are called *full-scale error* and *linearity error,* which are normally expressed as a percentage of the converter's full-scale output (%F.S.)

Full-scale error is the maximum deviation of the DAC's output from its expected (ideal) value. For example, assume that the DAC of Figure 10-6 has an accuracy of $\pm 0.01\%$ F.S. Since this converter has a full-scale output of 9.375 V, this percentage converts to

$$\pm 0.01\% \times 9.375 \text{ V} = \pm 0.9375 \text{ mV}$$

This means that the output of this DAC can, at any time, be off by as much as 0.9375 mV from its expected value.

Linearity error is the maximum deviation in step size from the ideal step size. For example, the DAC of Figure 10-6 has an expected step size of 0.625 V. If this converter has a linearity error of $\pm 0.01\%$ F.S., this would mean that the actual *step size* could be off by as much as 0.9375 mV.

Some of the more expensive DACs have full-scale and linearity errors as low as 0.001% F.S. General-purpose DACs usually have accuracies in the 0.01–0.1% range.

It is important to understand that accuracy and resolution of a D/A converter must be compatible. It is illogical to have a resolution of, say, 1 percent and an accuracy of 0.1 percent, or vice versa. To illustrate, a D/A converter with a resolution of 1 percent and a F.S. output of 10 V can produce an output analog voltage within 0.1 V of any desired value, assuming perfect accuracy. It makes no sense to have a costly accuracy of 0.01 percent of F.S. (or 1 mV) if the resolution already limits the closeness of the desired value to 0.1 V. The same can be said for having a resolution that is very small (many bits) while the accuracy is poor; it is a waste of input bits.

EXAMPLE 10-11

A certain 8-bit DAC has a full-scale output of 2 mA and an accuracy of $\pm 0.5\%$ F.S. What is the range of possible outputs for an input of 10000000?

SOLUTION
The step size is 2 mA/255 = 7.84 μA. Since 10000000 = 128_{10}, the ideal output should be 128 $\times$ 7.84 μA = 1004 μA. The error can be as much as

$$\pm 0.5\% \times 2 \text{ mA} = \pm 10 \text{ } \mu A$$

Thus, the actual output can deviate by this amount from the ideal 1004 μA, so the actual output can be anywhere from 994 to 1014 μA.

Settling Time   The operating speed of a DAC is usually specified by giving its *settling time,* which is the time required for the DAC output to go from zero to full-scale as the binary input is changed from all 0s to all 1s. Actually, the settling time is measured as the time for the DAC output to settle within $\pm 1/2$ step size (i.e., the resolution) of its final value. For example, if a DAC has a resolution of 10 mV, settling time is measured as the time it takes the output to settle within 5 mV of its full-scale value.

Typical values for settling time range from 50 ns to 10 μs. Generally speaking, DACs with a current output will have shorter settling times than those with voltage outputs. For instance, the DAC1280 can operate as either current output or voltage output. In the current output mode, its settling time is 300 ns; in the voltage output mode, its settling time is 2.5 μs. The main reason for this difference is the response time of the op-amp that is used as the current-to-voltage converter.

Offset Voltage   Ideally, the output of a DAC will be zero volts when the binary input is all 0s. In practice, however, there will be a very small output voltage for this situation; this is called *offset error*. This offset error, if not corrected, will be added to the expected DAC output for *all* input cases. For example, let's say that a 4-bit DAC has an offset error of $+2$ mV and a *perfect* step size of 100 mV. The table below shows the ideal and actual DAC output for several input cases. Note that the actual output is 2 mV greater than expected; this is due to the offset error.

INPUT CODE	IDEAL OUTPUT (mV)	ACTUAL OUTPUT (mV)
0000	0	2
0001	100	102
1000	800	802
1111	1500	1502

Many DACs will have an external offset adjustment that allows you to zero the offset. This is usually accomplished by applying all 0s to the DAC input and monitoring the output while an *offset adjustment potentiometer* is adjusted until the output is as close to 0 V as required.

Monotonicity   A DAC is monotonic if its output either increases or stays the same as the binary input is incremented from one value to the next. Another way to describe this is that the staircase output will have no downward steps as the binary input is incremented from zero to full-scale.

REVIEW QUESTIONS

1. Define full-scale error.
2. What is the meaning of settling time?
3. Describe offset error and its effect on a DAC output.
4. Why are voltage DACs generally slower than current DACs?

10-5   DAC APPLICATIONS

DACs are used whenever the output of a digital circuit has to provide an analog voltage or current to drive an analog device. Some of the most common applications are described in the following paragraphs.

Control   The digital output from a computer can be converted to an analog control signal to adjust the speed of a motor, the temperature of a furnace, or to control almost any physical variable.

Automatic Testing   Computers can be programmed to generate the analog signals (through a DAC) needed to test analog circuitry. The test circuit's analog output response will normally be converted to a digital value by an ADC and fed into the computer to be stored, displayed, and sometimes analyzed.

Digital Amplitude Control   A multiplying DAC can be used to digitally adjust the amplitude of an analog signal. Recall that a multiplying DAC produces an output that is the product of a reference voltage and the binary input. If the reference voltage is a time-varying signal, the DAC output will follow this signal but with an amplitude determined by the binary input code. A typical application of this is digital "volume control," where the output of a digital circuit or computer can adjust the amplitude of an audio signal. We will examine this in Problem 10-13.

A/D Conversion   Several types of ADCs use DACs as part of their circuitry.

10-6 TROUBLESHOOTING DACs

DACs are both digital and analog. Logic probes and pulsers can be used on the digital inputs, but a meter or oscilloscope must be used on the analog output. There are basically two ways to test a DAC's operation: a *static accuracy test* and a *staircase test*.

The static test involves setting the binary input to a fixed value and measuring the analog output with a high-accuracy meter. This test is used to check that the output value falls within the expected range consistent with the DAC's specified accuracy. If it does not, there can be several possible causes. Here are some of them:

- Drift in the DAC's internal component values (e.g., resistor values) caused by temperature, aging, or some other factors. This can easily produce output values outside the expected accuracy range.

- Opens or shorts in any of the binary inputs. This could either prevent an input from adding its weight to the analog output, or cause its weight to be permanently present in the output. This is especially hard to detect when the fault is in the less significant inputs.

- A faulty voltage reference. Since the analog output is directly dependent on V_{REF}, this could produce results that are way off. For DACs that use external reference sources, the reference voltage can easily be checked with a digital voltmeter. Many DACs have internal reference voltages which cannot be checked, except on some DACs that bring the reference voltage out to a pin of the IC.

- Excessive offset error caused by component aging or temperature. This would produce outputs that are off by a fixed amount. If the DAC has an external offset adjustment capability, this type of error can initially be zeroed out, but changes in operating temperature can cause the offset error to reappear.

The staircase test is used to check the monotonicity of the DAC; that is, it checks to see that the output increases step by step as the binary input is incremented as in Figure 10-3. The steps on the staircase must be of the same size, and there should be no missing steps or downward steps until full-scale is reached. This test can help detect internal or external faults that cause an input to either have no contribution or a permanent contribution to the analog output. The following example will illustrate.

EXAMPLE 10-12

How would the staircase waveform appear if the C input to the DAC of Figure 10-3 is open? Assume that the DAC inputs are TTL compatible.

SOLUTION

An open at C will be interpreted as a constant logic 1 by the DAC. Thus, this will contribute a constant 4 V to the DAC output so that the DAC output waveform will appear as shown in Figure 10-10. The dotted lines are the staircase as it would appear if the DAC were working correctly. Note that the faulty output waveform matches the correct one during those times when the bit C input would normally be HIGH.

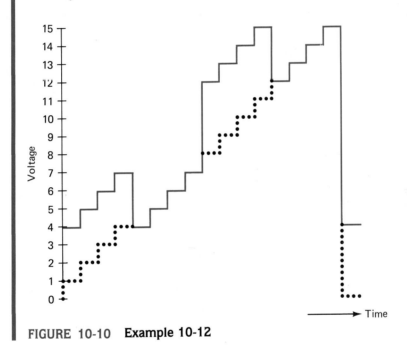

FIGURE 10-10   Example 10-12

10-7   ANALOG-TO-DIGITAL CONVERSION

An A/D *converter* takes an analog input voltage and after a certain amount of time produces a digital output code which represents the analog input. The A/D conversion process is generally more complex and time consuming than the D/A process, and many different methods have been developed and used. We shall examine several of these methods in detail, even though it may never be necessary to design or construct A/D converters (they are available as completely packaged units). However, the techniques that are used provide an insight into what factors determine an A/D converter's performance.

Several important types of ADC utilize a D/A converter as part of their circuitry. Figure 10-11 is a general block diagram for this class of ADC. The timing for the operation is provided by the input clock signal. The control unit contains the logic circuitry for generating the proper sequence of operations in response to the START COMMAND, which initiates the conversion process. The op-amp comparator has two *analog* inputs and a *digital* output that switches states, depending on which analog input is greater.

The basic operation of A/D converters of this type consists of the following steps:

1.   The START COMMAND pulse initiates the operation.
2.   At a rate determined by the clock, the control unit continually modifies the binary number that is stored in the register.
3.   The binary number in the register is converted to an analog voltage $V_{A'}$, by the D/A converter.
4.   The comparator compares $V_{A'}$ with the analog input V_A. As long as $V_{A'} < V_A$, the comparator output stays HIGH. When $V_{A'}$ exceeds V_A by at least an amount $= V_T$ (threshold voltage), the comparator output goes LOW and stops the process of modifying the register number. At this point, $V_{A'}$ is a close approximation to V_A. The digital number in the register, which is the digital equivalent of $V_{A'}$, is also the approximate digital equivalent of V_A, within the resolution and accuracy of the system.

FIGURE 10-11   **General diagram of one class of A/D converters.**

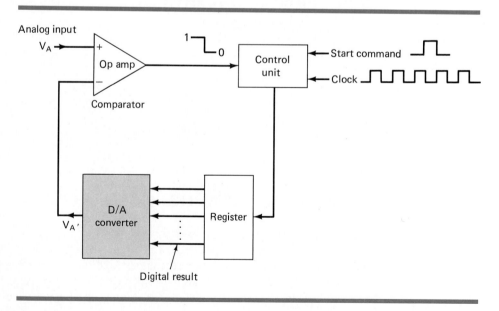

The several variations of this A/D conversion scheme differ mainly in the manner in which the control section continually modifies the numbers in the register. Otherwise, the basic idea is the same, with the register holding the required digital output when the conversion process is complete.

10-8 DIGITAL-RAMP A/D CONVERTER

One of the simplest versions of the general A/D converter of Figure 10-11 uses a binary counter as the register and allows the clock to increment the counter one step at a time until $V_{A'} \geq V_A$. It is called a *digital-ramp A/D converter* because the waveform at $V_{A'}$ is a step-by-step ramp (actually a staircase) like the one shown in Figure 10-3.

Figure 10-12 shows the complete diagram for a digital-ramp A/D converter. If V_A is a positive voltage, the operation proceeds as follows:

1. A positive START pulse is applied, which resets the counter to zero. It also inhibits the AND gate so that no clock pulses get through to the counter while the START pulse is HIGH.

2. With the counter at zero, $V_{A'} = 0$, so the comparator output is HIGH.

FIGURE 10-12   Digital-ramp A/D converter.

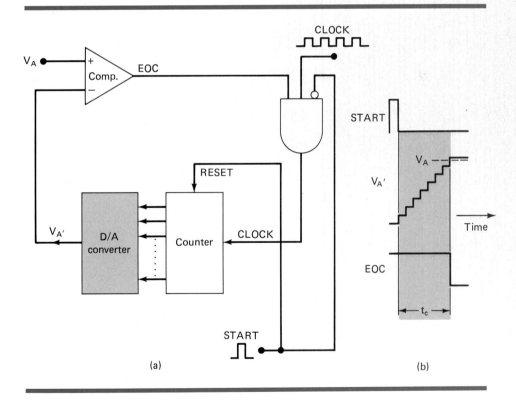

(a)

(b)

3. When the START pulse returns to LOW, the AND gate is enabled and pulses are allowed into the counter.

4. As the counter advances, the DAC output $V_{A'}$ increases in steps of voltage equal to its resolution.

5. This continues until $V_{A'}$ reaches a step that exceeds V_A by an amount equal to V_T or greater.* At this point, the comparator output goes LOW, inhibiting the pulses to the counter so that the counter has stopped at a count that is the desired digital representation of V_A. The conversion process is now complete. The HIGH-to-LOW transition at the comparator output signals the end of the conversion. This signal is often called EOC (end of conversion).

EXAMPLE 10-13

Assume the following values for the A/D converter of Figure 10-12: clock frequency = 1 MHz; V_T = 0.1 mV; D/A converter has F.S. output = 10.23 V and a 10-bit input. Determine the following values.

(a) The digital equivalent obtained for V_A = 3.728 V.

(b) The conversion time.

(c) The resolution of this converter.

SOLUTION

(a) The DAC has a 10-bit input and a 10.23-V F.S. output. Thus, the number of total possible steps is $2^{10} - 1 = 1023$, so the step size is

$$\frac{10.23 \text{ V}}{1023} = 10 \text{ mV}$$

This means that $V_{A'}$ increases in steps of 10 mV as the counter counts up from zero. Since V_A = 3.728 V and V_T = 0.1 mV, then $V_{A'}$ has to reach 3.7281 V or more before the comparator switches LOW. This will require

$$\frac{3.7281 \text{ V}}{10 \text{ mV}} = 372.81 = 373 \text{ steps}$$

At the end of the conversion, then, the counter will hold the binary equivalent of 373, whch is 0101110101. This is the desired digital equivalent of V_A = 3.728 V, as produced by this A/D converter.

(b) Three hundred seventy-three steps were required to complete the conversion. Thus, 373 clock pulses occurred at the rate of one per microsecond. This gives a total conversion time of 373 μs.

(c) The resolution of this converter is equal to the step size of the D/A converter, which is 10 mV. In percent it is $1/1023 \times 100\% \approx 0.1\%$.

A/D Resolution and Accuracy As pointed out in Example 10-13 the resolution of the A/D converter is equal to the resolution of the D/A converter which it contains. The DAC output voltage $V_{A'}$ is a staircase waveform that goes up in discrete steps until it exceeds V_A. Thus, $V_{A'}$ is an approximation to the value of V_A,

---

*IC comparators typically have threshold sensitivities of 10–100 μV.

can expect is that $V_{A'}$ is within 10 mV of V_A if the resolution (step
We can think of the resolution as being an inherent error which is
as *quantization error*. This quantization error, which can be re-
sing the number of bits in the counter and D/A converter, is some-
as an error of $+1$ LSB, indicating that the result could be off by
the finite (nonzero) step size. In Problem 10-24 we will see how
error can be modified so that it is $\pm\frac{1}{2}$ LSB, which is a more
n.

it from a different aspect, the input V_A can take on an *infinite*
from 0 V to F.S. The approximation $V_{A'}$, however, can take on
ber of discrete values. This means that a small range of V_A values
e digital representation. To illustrate, in Example 10-13 any value
V to 3.7299 V will require 373 steps, thereby resulting in the
sentation. In other words, V_A must change by 10 mV (the resolu-
change in digital output.

/A converter, *accuracy* is not related to the resolution but is de-
ccuracy of the circuit components, such as the comparator, the
resistors and level amplifiers, the reference supplies, and so on.
tion of 0.01 percent F.S. indicates that the A/D converter result
1 present of F.S., owing to nonideal components. This error is *in*
ntization error due to the resolution. These two sources of error
ame order of magnitude for a given ADC.

DC has a full-scale input of 2.55 V (i.e., $V_A = 2.55$ V produces
al output). It has a specified error of 0.1% F.S. Determine the
it by which the digital output can differ from the analog input.

The step size is 2.55 V/$(2^8 - 1)$, which is exactly 10 mV. This means that even
if the DAC has no inaccuracies, the digital output could be off by as much as 10
mV because $V_{A'}$ can change only in 10-mV steps; this is the quantization error.
The specified error of 0.1% F.S. is 0.1% $\times$ 2.55 V $=$ 2.55 mV. This means
that the $V_{A'}$ value can be off by as much as 2.55 mV because of component
inaccuracies. Thus, the total possible error could be as much as 10 mV $+$ 2.55
mV $=$ 12.55 mV.

Conversion Time, t_C In the digital-ramp converter, the counter starts at zero
and counts up until $V_{A'} > V_A$. Clearly, then, the time it takes to complete the conversion will depend on the value of V_A. A larger value of V_A will require more steps
before the staircase voltage exceeds V_A. The maximum conversion time occurs when
V_A is slightly lower than full-scale, so the staircase has to reach the full-scale step
in order for $V_{A'} > V_A$. Thus,

$$t_C(\text{max}) = 2^N - 1 \text{ clock cycles}$$

For example, the ADC in Example 10-13 would have a maximum conversion
time of

$$t_C(\text{max}) = (2^{10} - 1) \times 1 \text{ μs} = 1023 \text{ μs}$$

Sometimes, average conversion time is specified; it is half of the maximum conversion time. For the digital-ramp converter, this would be

$$t_C(avg) = \frac{t_C(max)}{2} \approx 2^{N-1} \text{ clock cycles}$$

The major disadvantage of the digital-ramp method is that conversion time essentially doubles for each bit that is added to the counter, so that resolution can be improved only at the cost of a longer t_C. This makes this type of ADC unsuitable for applications where repetitive A/D conversions of a fast-changing analog signal have to be made. For low-speed applications, however, the relative simplicity of the digital-ramp converter is an advantage over the more complex, higher-speed ADCs.

EXAMPLE 10-15

What will happen to the operation of a digital-ramp ADC if the analog input V_A is greater than the full-scale value?

SOLUTION
Referring to Figure 10-12, it should be clear that the comparator output will never go LOW since the staircase voltage can never exceed V_A. Thus, pulses will be continually applied to the counter, so the counter will repetitively count up from zero to maximum, recycle back to zero, count up, and so on. This will produce repetitive staircase waveforms going from zero to full-scale, and this will continue until V_A is decreased below full-scale.

REVIEW QUESTIONS

1. Describe the basic operation of the digital-ramp ADC.
2. Explain what is meant by "quantization error."
3. Why does conversion time increase with the value of the analog input voltage?
4. *True or false:* Everything else being equal, an 8-bit digital-ramp ADC will have a better resolution, but a longer conversion time, than a 10-bit ADC.
5. Give one advantage and one disadvantage of a digital-ramp ADC.
6. For the converter of Example 10-13, determine the digital output for V_A = 1.345 V. Repeat for V_A = 1.342 V. (*Ans.* For both cases, digital result is $0010000111_2 = 135_{10}$)

10-9 DATA ACQUISITION

There are many applications in which analog data have to be *digitized* (converted to digital) and transferred into a computer's memory. The process by which the computer acquires these digitized analog data is referred to as *data acquisition*. The computer can do several different things with the data, depending on the application. In an analog storage application, such as a digital oscilloscope, the computer will store the data and then transfer them to a DAC at a later time to reproduce the analog data. In a process control application, the computer can examine the data or perform computations on them to determine what control outputs to generate.

Figure 10-13(a) shows how a microcomputer is connected to a digital-ramp ADC for the purpose of data acquisition. The computer generates the START pulses that initiate each new A/D conversion. The EOC (end-of-conversion) signal from the ADC is fed to the computer. The computer examines EOC to find out when the current A/D conversion is complete; then it transfers the digital data from the ADC output into its memory.

The waveforms in Figure 10-13(b) illustrate how the computer acquires a digital version of the analog signal, V_A. The $V_{A'}$ staircase waveform that is generated internal to the ADC is shown superimposed on the V_A waveform for purposes of illustration. The process begins at t_0 when the computer generates a START pulse to start an A/D conversion cycle. The conversion is completed at t_1 when the staircase first exceeds V_A, and EOC goes LOW. This NGT at EOC signals the computer that the ADC has a digital output that now represents the value of V_A at point *a*, and the computer will load these data into its memory.

The computer generates a new START pulse shortly after t_1 to initiate a second conversion cycle. Note that this resets the staircase to zero and EOC back HIGH because the START pulse resets the counter in the ADC. The second conversion ends at t_2 when the staircase again exceeds V_A. The computer then loads the digital data corresponding to point *b* into its memory. These steps are repeated at t_3, t_4, and so on.

The process whereby the computer generates a START pulse, examines EOC, and loads ADC data into memory is done under the control of a program that the computer is executing. This data acquisition program will determine how many data points from the analog signal will be stored in the computer memory.

Reconstructing a Digitized Signal

In Figure 10-13(b) the ADC is operating at its maximum speed since a START pulse is generated immediately after the computer acquires the ADC output data from the previous conversion. Note that the conversion times are not constant because the analog input value is changing. The computer will store the digital data obtained from the various conversions so that it has a digitized version of the analog signal in its memory. For example, the digital data for the points *a, b,* and *c* might look like this:

POINT	ACTUAL VOLTAGE (V)	DIGITAL EQUIVALENT
a	1.74	10101110
b	1.47	10010011
c	1.22	01111010

This digital data is often used to later reconstruct the analog signal. For example, in some digital oscilloscopes the analog signal is reconstructed by drawing straight lines from each digitized point to the next. This is illustrated in Figure 10-14.

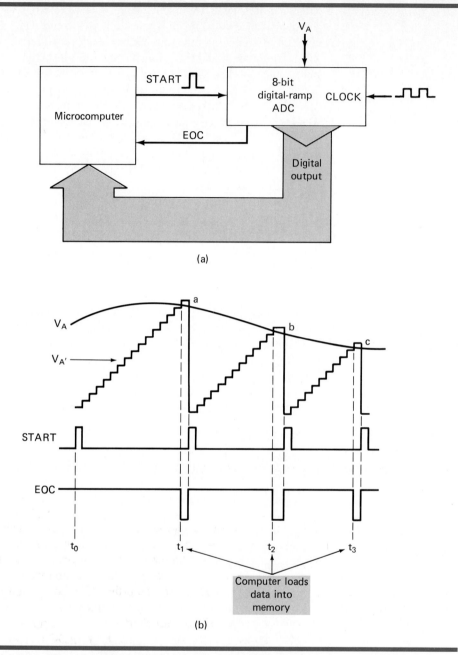

FIGURE 10-13 **(a) Typical computer data acquisition system; (b) waveforms showing how computer initiates each new conversion cycle and then loads digital data into memory at end of conversion.**

In Figure 10-14(a) we see how the ADC continually performs conversions to digitize the analog signal at points *a, b, c, d,* etc. If this digital data is used to reconstruct the signal, the result will look like that in Figure 10-14(b). We can see

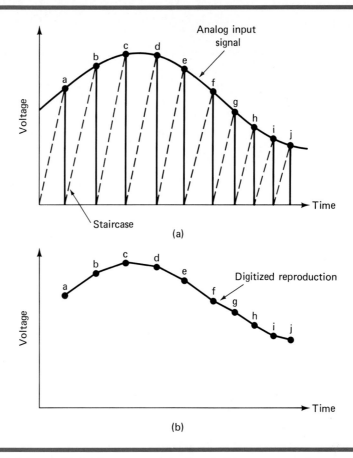

FIGURE 10-14 **(a) Digitizing an analog signal; (b) reconstructing the signal from the digital data.**

that it is a fairly good reproduction of the original analog signal. This is because the analog signal does not make any rapid changes between digitized points. If the analog signal contained higher-frequency variations, the ADC would not be able to follow the variations, and the reproduced version would be much less accurate. For this reason it is important to keep the ADC conversion time short enough so that the analog signal does not change significantly between conversions. This emphasizes the need for ADCs with shorter conversion times than the simple digital-ramp converters. We will examine a much faster ADC method in the next section.

REVIEW QUESTIONS

1. What is the meaning of the term ''digitize''?
2. Describe the steps in a computer data acquisition process.

10-10 SUCCESSIVE-APPROXIMATION ADC

This is one of the most widely used types of ADC. It has more complex circuitry than the digital-ramp ADC but a much shorter conversion time. In addition, successive-approximation converters (SAC) have a fixed value of conversion time that is not dependent on the value of the analog input.

The basic arrangement, as shown in Figure 10-15, is similar to the digital-ramp ADC. The SAC, however, does not use a counter to provide the input to the D/A converter block but uses a register instead. The control logic modifies the contents of the register bit-by-bit until the register data are the digital equivalent of the analog input V_A (within the resolution of the converter). The process takes place as follows:

1. The control logic sets the MSB of the register HIGH and all other bits LOW. This produces a value of $V_{A'}$ at the DAC output equal to the weight of the MSB. If $V_{A'}$ is now greater than V_A, the comparator output, COMP, goes LOW and causes the control logic to clear the MSB back to LOW. Otherwise, the MSB is kept HIGH.

2. The control logic sets the next bit of the register to 1. This produces a new value of $V_{A'}$. If this value is greater than V_A, COMP goes LOW to tell the control logic to clear the bit back to 0. Otherwise, the bit is kept at 1.

3. This process is continued for each of the bits in the register. This trial-and-error process requires one clock cycle per bit. After all bits have been tried, the register holds the digital equivalent of V_A.

We can better illustrate the successive approximation process with a specific example. For simplicity, we will use a 4-bit converter with a step size of 1 V. Table 10-1 shows the sequence of steps for converting $V_A = 10.4$ V. There are

TABLE 10-1

	STEPS	REGISTER	$V_{A'}(V)$	$V_A(V)$	COMPARATOR
	Initial status	0000	0	10.4	HIGH
I. A.	Set MSB to 1	1000	8	10.4	HIGH
B.	Leave it at 1 since $V_{A'} < V_A$.				
II. A.	Set second MSB to 1	1100	12	10.4	LOW
B.	Reset it to 0 since $V_{A'} > V_A$.	1000	8	10.4	HIGH
III. A.	Set third MSB to 1	1010	10	10.4	HIGH
B.	Leave it at 1 since $V_{A'} < V_A$.				
IV. A.	Set LSB to 1.	1011	11	10.4	LOW
B.	Reset it to 0 since $V_{A'} > V_A$.	1010	10	10.4	HIGH
	Digital number now in the REGISTER is the final result.	1010			

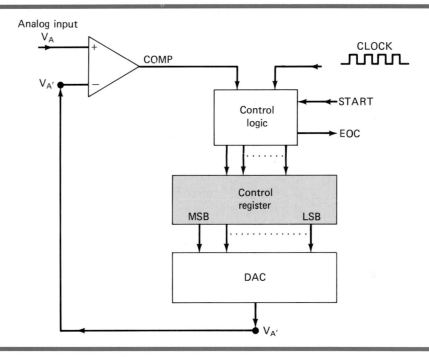

FIGURE 10-15   **Block diagram of a successive-approximation ADC.**

four steps, one per clock cycle. Note how each step generates a better approximation to V_A. At the end of the fourth step the conversion is complete and the required digital value 1010 is in the register. This value is equivalent to 10 V and is *less than V_A;* this is a characteristic of the successive-approximation method.

Conversion Time   The same operation occurs for more practical SACs that use a greater number of bits for better resolution. In general, an *N*-bit SAC will require *N* clock cycles to perform a conversion *regardless of the value of V_A.* This is because the control circuitry has to try a 1 in each bit position to see whether it is needed or not.

EXAMPLE 10-16

Compare the maximum conversion times of a 10-bit digital-ramp A/D converter and a 10-bit successive approximation A/D converter if both utilize a 500-kHz clock frequency.

SOLUTION
For the digital-ramp converter, the maximum conversion time is

$$(2^N - 1) \times (1 \text{ clock cycle}) = 1023 \times 2 \text{ μs} = 2046 \text{ μs}$$

For a 10-bit successive approximation converter, the conversion time is always 10 clock periods or

$$10 \times 2 \text{ μs} = 20 \text{ μs}$$

Thus, it is about 100 times faster than the digital-ramp converter.

Since SACs have relatively fast conversion times, their use in data acquisition applications will permit more data values to be acquired in a given time interval. This can be very important when the analog data are changing at a relatively fast rate.

Because many SACs are available as ICs, it is rarely necessary to design the control logic circuitry, so we will not cover it here. For those who are interested in the details of the control logic, many manufacturers' data books should provide sufficient detail.

An Actual IC: The ADC0801 Successive-Approximation ADC

ADCs are available from several IC manufacturers with a wide range of operating characteristics and features. We will take a look at one of the more popular devices to get an idea of what is actually used in system applications. Figure 10-16 is the pin layout for the ADC0801, which is a 20-pin IC that performs A/D conversion using the successive-approximation method. Some of its important characteristis include:

- It has two analog inputs: $V_{in}(+)$ and $V_{in}(-)$ to allow *differential* inputs. In other words, the actual analog input, V_{in}, is the difference in the voltages applied to these pins [analog $V_{in} = V_{in}(+) - V_{in}(-)$]. In single-ended measurements, the analog input is applied to $V_{in}(+)$, while $V_{in}(-)$ is connected to analog ground. During normal operation, the converter uses $V_{CC} = +5$ V as its reference voltage, and the analog input can range from 0 to 5 V.

FIGURE 10-16 ADC0801 8-bit successive-approximation ADC with tristate outputs. The numbers in parentheses are the IC's pin numbers.

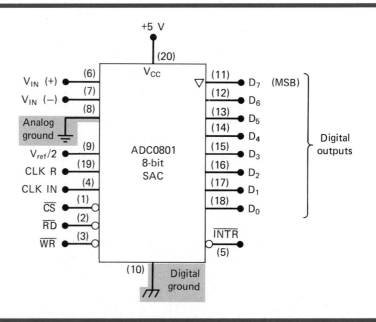

- It converts the analog input voltage to an 8-bit digital output. The digital outputs are tristate buffered so that they can be easily connected in a data bus arrangement. With 8 bits, the resolution is 5 V/255 = 19.6 mV.

- It has an internal clock generator circuit that produces a frequency of $f = 1/(1.1RC)$, where R and C are externally connected components. A typical clock frequency is 606 kHz using $R = 10$ kΩ and $C = 150$ pF. An external clock signal can be used, if desired, by connecting it to the CLK IN pin.

- Using a 606-kHZ clock frequency, the conversion time is approximately 100 μs.

- It has separate ground connections for digital and analog voltages. Pin 8 is the analog ground that is connected to the common reference point of the analog circuit that is generating the analog voltage. Pin 10 is the digital ground that is the one used by all the digital devices in the system. (Note the different symbols used for the different grounds.) The digital ground is inherently noisy because of the rapid current changes that occur as digital devices change states. Although it is not necessary to use a separate analog ground, doing so ensures that the noise from digital ground is prevented from causing premature switching of the analog comparator inside the ADC.

This IC is designed to be easily interfaced to a microprocessor data bus. As such, the names of some of the ADC0801 inputs and outputs are based on functions that are common to microprocessor-based systems. The functions of these inputs and outputs are defined as follows:

$\overline{CS}$ **(Chip Select)** This input has to be in its active-LOW state for the $\overline{RD}$ or $\overline{WR}$ inputs to have any effect. With $\overline{CS}$ HIGH, the digital outputs are in the Hi-Z state, and no conversions can take place.

$\overline{RD}$ **(Output Enable)** This input is used to enable the digital output buffers. With $\overline{CS} = \overline{RD} = $ LOW, the digital output pins will have logic levels representing the results of the *last* A/D conversion.

$\overline{WR}$ **(Start Conversion)** A LOW pulse is applied to this input to signal the start of a new conversion.

$\overline{INTR}$ **(End of Conversion)** This output signal will go HIGH at the start of a conversion, and will return LOW to signal the end of conversion.

$V_{ref}/2$ This is an optional input that can be used to reduce the internal reference voltage and thereby change the analog input range that the converter can handle. When this input is unconnected, it sits at 2.5 V ($V_{CC}/2$) since V_{CC} is being used as the reference voltage. By connecting an external voltage to this pin, the internal reference is changed to twice that voltage, and the analog input range is changed accordingly. The table below shows this.

$V_{ref}/2$	ANALOG INPUT RANGE (V)	RESOLUTION (mV)
Open	0–5	19.6
2.25	0–4.5	17.6
2.0	0–4	15.7
1.5	0–3	11.8

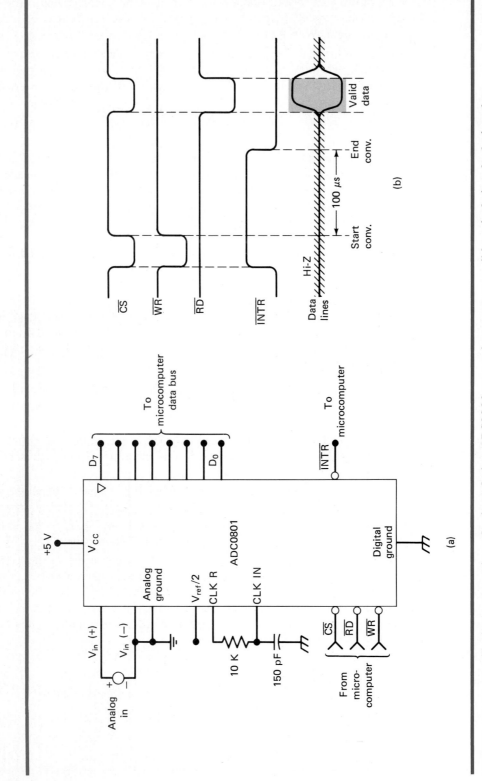

FIGURE 10-17 (a) Typical connections for interfacing ADC0801 to a microcomputer; (b) typical signals during data acquisition process.

CLK R A resistor is connected to this pin for the internal clock.

CLK IN Used for external clock input, or for capacitor connection when internal clock is used.

Figure 10-17(a) shows a typical connection of the ADC0801 to a microcomputer in a data acquisition application. The microcomputer controls when a conversion is to take place by generating the $\overline{CS}$ and $\overline{WR}$ signals. It then acquires the ADC output data by generating the $\overline{CS}$ and $\overline{RD}$ signals after detecting a NGT at $\overline{INTR}$ indicating the end of conversion. The waveforms in Figure 10-17(b) show the signal activity during the data acquisition process. Note that $\overline{INTR}$ goes HIGH when $\overline{CS}$ and $\overline{WR}$ go LOW, but the conversion process does not begin until $\overline{CS}$ and $\overline{WR}$ return HIGH. Also note that the ADC output data lines are in their Hi-Z state until the microcomputer activates $\overline{CS}$ and $\overline{RD}$; at that point the ADC's data buffers are enabled so that the ADC data is sent to the microcomputer over the data bus. The data lines return to the Hi-Z state when $\overline{CS}$ and $\overline{RD}$ are returned HIGH.

REVIEW QUESTIONS

1. What is the main advantage of a successive-approximation converter?
2. What is its principal disadvantage compared to the digital-ramp converter?
3. *True or false:* The conversion time for an SAC increases as the analog voltage increases.
4. Answer the following concerning the ADC0801.
   (a) What is its resolution in bits?
   (b) What is the normal analog input voltage range?
   (c) Describe the functions of the $\overline{CS}$, $\overline{WR}$, and $\overline{RD}$ inputs.
   (d) What is the function of the $\overline{INTR}$ output?
   (e) Why does it have two separate grounds?

10-11 FLASH ADCs

This type is the highest-speed ADC available, but it requires much more circuitry than the other types. For example, a 6-bit flash ADC requires 63 analog comparators, while an 8-bit unit requires 255 comparators, and a 10-bit converter requires 1023 comparators. The large number of comparators needed effectively limits *discrete* flash converters to the 6-bit level. IC flash converters are currently available in 8-bit units, and most manufacturers predict that 9- and 10-bit units will hit the market in the near future.

The principle of operation will be described for a 3-bit flash converter in order to keep the circuitry at a workable level. Once the 3-bit converter is understood, it should be easy to extend the basic idea to higher-bit flash converters.

The flash converter in Figure 10-18(a) has a 3-bit resolution and a step size

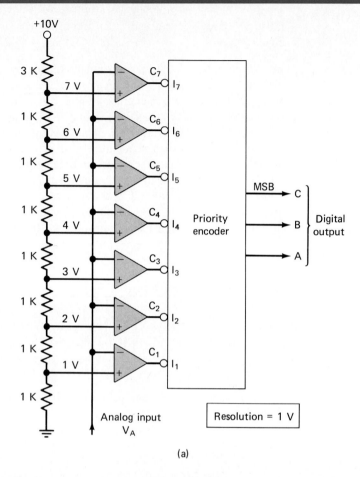

(a)

Analog in	Comparator outputs							Digital output		
V_A	C_1	C_2	C_3	C_4	C_5	C_6	C_7	C	B	A
<1 V	1	1	1	1	1	1	1	0	0	0
>1 V, <2 V	0	1	1	1	1	1	1	0	0	1
>2 V, <3 V	0	0	1	1	1	1	1	0	1	0
>3 V, <4 V	0	0	0	1	1	1	1	0	1	1
>4 V, <5 V	0	0	0	0	1	1	1	1	0	0
>5 V, <6 V	0	0	0	0	0	1	1	1	0	1
>6 V, <7 V	0	0	0	0	0	0	1	1	1	0
>7 V	0	0	0	0	0	0	0	1	1	1

(b)

FIGURE 10-18 (a) Three-bit flash ADC; (b) truth table.

of 1 V. The voltage divider sets up reference levels for each comparator so that there are seven levels corresponding to 1 V (weight of LSB), 2 V, 3 V, . . . , and 7 V (full-scale). The analog input, V_A, is connected to the other input of each comparator.

With $V_A < 1$ V, all the comparator outputs C_1–C_7 will be HIGH. With $V_A > 1$ V, one or more of the comparator outputs will be LOW. The comparator outputs are fed into an active LOW priority encoder that generates a binary output corresponding to the highest-numbered comparator output that is LOW. For example, when V_A is between 3 and 4 V, outputs C_1, C_2, and C_3 will be LOW and all others will be HIGH. The priority encoder will respond only to the LOW at C_3 and will produce a binary output $CBA = 011$, which represents the digital equivalent of V_A, within the resolution of 1 V. When V_A is greater than 7 V, C_1–C_7 will all be LOW, and the encoder will produce $CBA = 111$ as the digital equivalent of V_A. The table in Figure 10-18(b) shows the responses for all possible values of analog input.

The flash converter uses no clock signal because there is no timing or sequencing required. The conversion process takes place as soon as V_A is applied, and the conversion time depends only on the propagation delays of the comparators and the encoder. A typical flash converter can have a conversion time of 50 ns.

It should now be easy to see why the flash-converter circuitry increases dramatically as the number of desired bits increases. The 3-bit converter in Figure 10-18 requires seven comparators because there are $2^3 = 8$ possible voltage levels 0 V, 1 V, 2 V, . . . , 7 V. There is no comparator required for the 0-V level. For an 8-bit flash converter there would be $2^8 = 256$ voltage levels including 0 V, so 255 comparators would be required. In general, then, an N-bit flash converter requires $2^N - 1$ comparators.

Clearly, the major advantage of the flash converter is its operating speed, but it achieves this high-speed operation at the expense of increased circuit complexity. The high cost of these ADCs restricts their use to those applications where high speed is a prime requirement.

REVIEW QUESTIONS

1. *True or false:* A flash ADC does not contain a DAC.
2. How many comparators would a 12-bit flash converter contain? (*Ans.* 4095)
3. State the major advantage and disadvantage of a flash converter.

10-12 DIGITAL VOLTMETER

A digital voltmeter (DVM) converts an analog voltage to its BCD-code representation, which is then decoded and displayed on some type of readout. Figure 10-19 shows a 3-digit DVM circuit that uses a digital-ramp ADC (shown inside the dashed lines). Three cascaded BCD counters provide the inputs to a 3-digit BCD/A converter that has a step size of 10 mV and a full-scale output of 9.99 V. Each BCD counter stage also drives a 4-bit register that feeds a decoder/driver and display. The contents of the BCD counters is transferred to the registers at the end of each

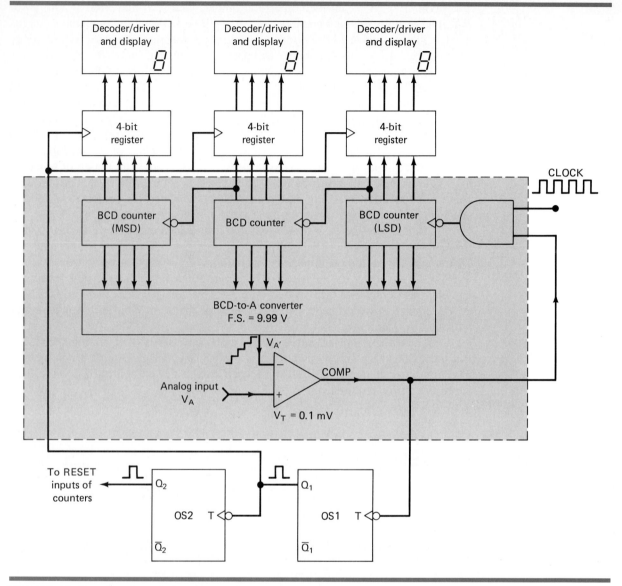

FIGURE 10-19 Continuous-conversion DVM using a digital-ramp ADC.

conversion cycle, so that the displays do not show the counters resetting and counting, but only display the final count that represents the unknown voltage.

The clock pulses are gated into the counters with the comparator output. As long as $V_A > V_{A'}$, COMP = 1 and the counter will receive pulses. As the counter advances, the $V_{A'}$ waveform goes up 10 mV per step until $V_{A'}$ exceeds V_A by 0.1 mV. At that point COMP goes LOW and disables the AND gate so that the counter will no longer advance. The negative transition at COMP also triggers one-shot OS1 which produces 1-μs pulse at Q_1. The PGT of Q_1 provides the clock transition that transfers the BCD counter outputs to their respective registers. The NGT of Q_1 triggers a second one-shot Q_2 producing a 1-μs pulse at Q_2 to reset all the counters back to zero. This brings $V_{A'}$ back to 0V and COMP returns HIGH, allowing pulses into the counter to begin a new conversion cycle.

Thus, this DVM will perform one conversion right after another. Of course, the storage registers will keep the displays from showing the conversion process. The display readings will change only if V_A changes, so that a different counter contents is transferred to the registers at the end of the conversion cycle.

A numerical example will help illustrate this circuit's operation. Assume that V_A is 6.372 V. In order for the COMP output to switch LOW, $V_{A'}$ must exceed 6.3721 V. Since the DAC output increases by 10 mV/step, this requires

$$\frac{6.3721 \text{ V}}{10 \text{ mV}} = 637.21 \longrightarrow 638 \text{ steps}$$

Thus, the counters will count up to 638, which will be transferred to the registers and displayed. A small LED can be used to display a decimal point so that the operator sees 6.38 V.

EXAMPLE 10-17

What will happen if V_A is greater than 9.99 V?

SOLUTION
With $V_A > 9.99$ V, the comparator output will stay HIGH, allowing clock pulses into the counter continuously. The counter will repetitively count up to 999 and recycle to 000. One-shot Q_1 *will never* be triggered and the register contents will not change from its previous value. A well-designed DVM would have some means of detecting this *over-range* condition and activating some type of over-range indicator. One possible method would simply add a 13th bit to the counter string. This bit would go HIGH only if the counter recycled from 999 to 000, and would indicate the over-range condition.

The DVM can be modified to read input voltages over several ranges by using a suitable amplifier or attenuator between V_A and the comparator. For example, if V_A were 63.72 V, it could be attenuated by a factor of 10, so the comparator would receive 6.372 V at its + input and the counters would display 638 at the end of the conversion. The decimal-point indicator would be placed in front of the LSD, so the display would read 63.8 V.

REVIEW QUESTIONS

1. Describe the operation of the DVM.
2. What range of input voltages will produce a reading of 4.55 V? (*Ans.* 4.5400–4.5499 V)

10-13 SAMPLE-AND-HOLD CIRCUITS

When an analog voltage is connected directly to the input of an ADC, the conversion process can be adversely affected if the analog voltage is changing during the conversion time. The stability of the conversion process can be improved by using a *sample-and-hold circuit* to hold the analog voltage constant while the A/D conversion is taking place. A simplified diagram of a sample-and-hold (S/H) circuit is shown in Figure 10-20.

The S/H circuit contains a unity-gain buffer amplifier *A1* that presents a high impedance to the analog signal, and has a low output impedance that can rapidly charge the hold capacitor, C_h. The capacitor will be connected to the output of *A1* when the digitally controlled switch is closed. This is called the *sample* operation. The switch will be closed long enough for C_h to charge to the current value of the analog input. For example, if the switch is closed at time t_0, the *A1* output will quickly charge C_h up to a voltage V_0. When the switch opens, C_h will *hold* this voltage so that the output of *A2* will apply this voltage to the ADC. The unity-gain buffer amplifier *A2* presents a high input impedance that will not discharge the capacitor voltage appreciably during the conversion time of the ADC, so the ADC will essentially receive a dc input voltage V_0.

In a computer-controlled data acquisition system such as the one discussed earlier, the sample-and-hold switch would be controlled by a digital signal from the computer. The computer signal would close the switch in order to charge C_h to a new sample of the analog voltage; the amount of time the switch would have to remain closed is called the *acquisition time* and it depends on the value of C_h and the characteristics of the S/H circuit. The LF198 is an integrated S/H circuit that has a typical acquisition time of 4 μs for C_h = 1000 pF, and 20 μs for C_h = 0.01 μF. The computer signal would then open the switch to allow C_h to hold its value

FIGURE 10-20 **Simplified diagram of a sample-and-hold circuit.**

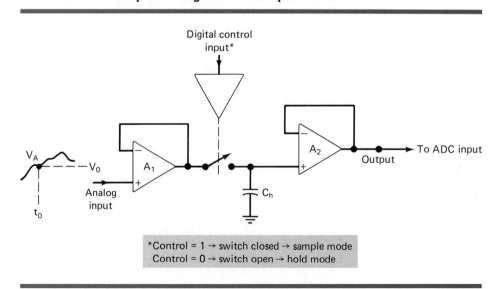

*Control = 1 → switch closed → sample mode
Control = 0 → switch open → hold mode

and provide a relatively constant analog voltage at the *A2* output. For example, with the LF198, the capacitor voltage will typically discharge at the rate of only 30 mV per second for a 1000-pF capacitor.

REVIEW QUESTIONS

1. Describe the function of a sample-and-hold circuit.
2. *True or false:* The amplifiers in the S/H circuit are used to provide voltage amplification.

10-14 MULTIPLEXING

When analog inputs from several sources are to be converted, a multiplexing technique can be used so that one A/D converter may be time-shared. The basic scheme is illustrated in Figure 10-21 for a three-channel acquisition system. Switches S_1, S_2, and S_3 are used to switch each analog signal sequentially to the A/D converter. The control circuitry controls the actuation of these switches, which are usually semiconductor switches, so that only one switch is closed at a time. The control circuitry also generates the START pulse for the ADC. The operation proceeds in the following manner:

1. The control circuit closes S_1, which connects V_{A1} to the ADC input.
2. A START pulse is generated and the ADC converts V_{A1} to its digital equivalent. S_1 stays closed long enough to allow the conversion to be completed.
3. The ADC outputs representing V_{A1} can now be transferred to another location. In many cases this would be to the memory of a computer.
4. The control circuit opens S_1 and closes S_2 to connect V_{A2} to the ADC input.
5. Steps 2 and 3 are repeated.
6. S_2 is open and S_3 is closed to connect V_{A3} to the ADC input.
7. Steps 2 and 3 are repeated.

The multiplexing clock controls the rate at which the analog signals are sequentially switched into the ADC. The maximum rate is determined by the delay time of the switches and the conversion time of the ADC. The switch delay time can be minimized by using semiconductor switches such as the CMOS transmission gates described in Section 8-17. It may be necessary to connect a sample-and-hold circuit at the input of the ADC if the analog inputs will change significantly during the ADC conversion time.

Many integrated ADCs contain the multiplexing circuitry on the same chip as the ADC. The ADC0808, for example, can multiplex eight different analog inputs into one ADC. It uses a 3-bit select input code to determine which analog input is connected to the ADC.

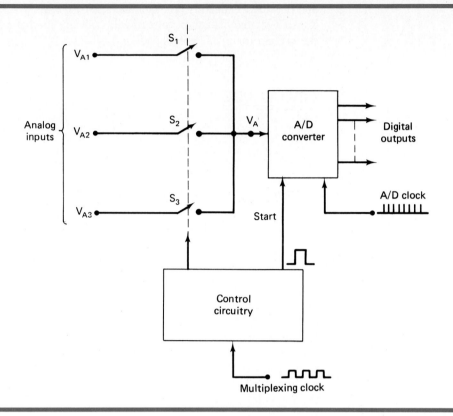

FIGURE 10-21 **Conversion of three analog inputs by multiplexing through one A/D converter.**

Actuator Electrically controlled device that controls a physical variable. [Sec. 10-1]

Analog-to-Digital Converter (ADC) Circuit that converts an analog input to a corresponding digital output. [Sec. 10-1]

Bipolar DAC Digital-to-analog converter that accepts signed binary numbers as input and produces the corresponding positive or negative output value. [Sec. 10-2]

Data Acquisition Process by which a computer acquires digitized analog data. [Sec. 10-9]

Digital-Ramp ADC Type of analog-to-decimal converter in which an internal staircase waveform is generated and utilized for the purpose of accomplishing the conversion. The conversion time for this type of analog-to-decimal converter varies depending on the value of the input analog signal. [Sec. 10-8]

nalog Converter (DAC) Circuit that converts a digital input to a cor-
nalog output. [Sec. 10-1]

cess by which an analog signal is converted to digital data. [Sec.

Type of analog-to-decimal converter that has the highest operating
le. [Sec. 10-11]

nt Multiplying DAC Multiplying digital-to-analog converter which
polarity on its V_{REF} input as well as signed binary numbers on its
[Sec. 10-3]

ror Term used by some digital-to-analog converter manufacturers to
curacy of a digital-to-analog converter. It's defined as the maximum
digital-to-analog converter's output from its expected ideal value.

tput Maximum possible output value of a digital-to-analog conver-
]

or Term used by some digital-to-analog converter manufacturers to
ice's accuracy. It is defined as the maximum deviation in step size
step size. [Sec. 10-4]

A digital-to-analog converter is said to be monotonic when its output
or stays the same as the input is increased. [Sec. 10-4]

AC Type of Digital-to-analog converter in which its output is ob-
multiplication of the analog voltage (V_{REF}) by the digital binary
3]

Under ideal conditions the output of a digital-to-analog converter
volts when the input is all 0s. In reality, there is a very small output
situation. This deviation from the ideal zero volts is called the offset
error. [Sec. 10-4]

Percentage Resolution Ratio of the step size to the full-scale value of a digital-to-analog converter. Percentage resolution can also be defined as the reciprocal of the maximum number of steps of a digital-to-analog converter. [Sec. 10-2]

Quantization Error Caused by the nonzero resolution of an ADC, this is an inherent error of the device. [Sec. 10-8]

R/2R Ladder DAC Type of digital-to-analog converter where its internal resistance values only span a range of 2 to 1. [Sec. 10-3]

Resolution The change that occurs in the analog output of a digital-to-analog converter as a result of the change in the LSB of its digital input. [Secs. 10-2 and 10-4]

Sample-and-Hold Circuit Type of circuit that utilizes a unity-gain buffer amplifier in conjunction with a capacitor to accomplish a more stable analog-to-digital conversion process. [Sec. 10-13]

Settling Time Amount of time that it takes the output of a DAC to go from zero to within ½ step size of its full-scale value as the input is changed from all 0s to all 1s. [Sec. 10-4]

Staircase Test Process by which a digital-to-analog converter's digital input is incremented and its output monitored to determine whether or not it exhibits a staircase format. The staircase waveform should be without any missing steps or any downward steps until it reaches its full-scale value. [Sec. 10-6]

Staircase Waveform Type of waveform generated at the output of a digital-to-analog converter as its digital input signal is incrementally changed. [Sec. 10-2]

Static Accuracy Test When a fixed binary value is applied to the input of a digital-to-analog converter and the analog output is accurately measured. The measured result should fall within the expected range specified by the digital-to-analog converter's manufacturer. [Sec. 10-6]

Step Size *See* Resolution. [Sec. 10-2]

Successive-Approximation ADC Type of analog-to-decimal converter in which an internal parallel register and complex control logic are used to perform the conversion. The conversion time for this type of analog-to-decimal converter is always the same regardless of the value of the input analog signal. [Sec. 10-10]

Transducer Device that converts a physical variable to an electrical variable. [Sec. 10-1]

Two-Quadrant Multiplying DAC Multiplying digital-to-analog converter which accepts either polarity on its V_{REF} input. [Sec. 10-3]

PROBLEMS   Sections 10-1 and 10-2

10-1.   An 8-bit D/A converter produces an output voltage of 2.0 V for an input code of 01100100. What will be the value of V_{OUT} for an input code of 10110011?

10-2.   Determine the weights of each input bit for the D/A converter of Problem 10-1.

10-3.   What is the resolution of the D/A converter of Problem 10-1? Express it in volts and in percent.

10-4.   What is the resolution in volts of a 10-bit D/A converter whose F.S. output is 5 V?

10-5.   How many bits are required for a D/A converter so that its F.S. output is 10 mA and its resolution is less than 40 μA?

10-6.   What is the percentage resolution of the D/A converter of Figure 10-22? What is the step size if the top step is 2 V?

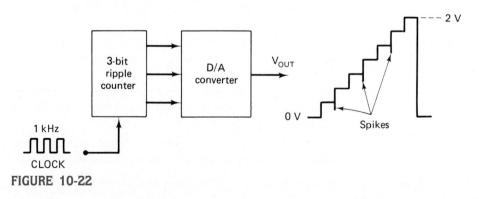

FIGURE 10-22

10-7. What is the cause of the negative-going spikes on the V_{OUT} waveform of Figure 10-22? (*Hint:* Note that the counter is a ripple counter and that the spikes occur on every other step.)

10-8. Assuming a 12-bit DAC with perfect accuracy, how close to 250 rpm can the motor speed be adjusted in Figure 10-4?

10-9. A 12-bit (three-digit) D/A converter which uses the BCD input code has a full-scale output of 9.99 V. Determine the step size, percentage resolution, and the value of V_{OUT} for an input code of 0110 1001 0101.

10-10. Compare the step size and percentage resolution of a DAC with an 8-bit binary input to a DAC with an 8-bit BCD input. Assume 990 mV full-scale for each DAC.

Section 10-3

10-11. The step size of the D/A converter of Figure 10-6 can be changed by changing the value of R_F. Determine the required value of R_F for a step size of 0.5 V. Will the new value of R_F change the percentage resolution?

10-12. Assume that the output of the DAC in Figure 10-8(a) is connected to the op-amp of Figure 10-8(b).
  (a) With $V_{REF} = 5$ V, $R = 20$ kΩ, and $R_F = 10$ kΩ, determine the step size and full-scale voltage at V_{OUT}.
  (b) Change the value of R_F so that the full-scale voltage at V_{OUT} is -2 V.
  (c) Use this new value of R_F, and determine the proportionality factor, k, in the relationship $V_{OUT} = k(V_{REF} \times B)$.

10-13. The DAC from Problem 10-12 can be used as a multiplying DAC, where V_{REF} is a signal and V_{OUT} is an amplified version of V_{REF} with the amplification factor controlled by the binary input to the DAC. This binary input can come from a computer that is being used to control the amplitude of an analog signal.

Assume that V_{REF} is an audio signal with an amplitude of 5 V p-p Determine the amplitude of V_{OUT} for each possible binary input.

Sections 10-4 and 10-5

10-14. An 8-bit D/A converter has a full-scale error of 0.2 percent F.S. If the DAC has a full-scale output of 10 mA, what is the most that it can be in error for any digital input? If the D/A output reads 50 μA for a digital input of 00000001, is this within the specified range of accuracy? (Assume no offset error.)

10-15. The control of a positioning device may be achieved using a *servomotor,* which is a motor designed to drive a mechanical device as long as an error signal exists. Figure 10-23 shows a simple servo-controlled sysem which is controlled by a digital input that could be coming directly from a computer or from an output medium such as magnetic tape. The lever arm is moved vertically by the servomotor. The motor rotates clockwise or counterclockwise, depending on whether the voltage from the power amplifier (PA) is positive or negative. The motor stops when the PA output is zero.

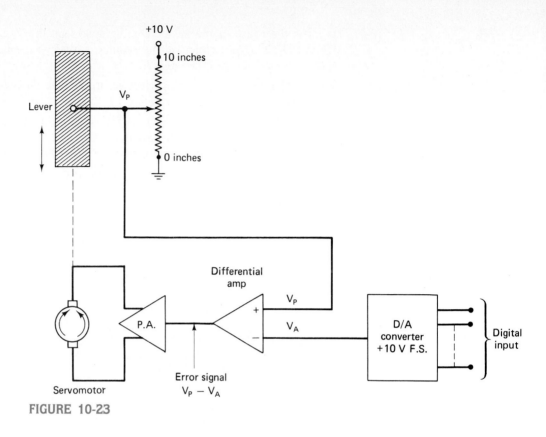

FIGURE 10-23

The mechanical position of the lever is converted to a dc voltage by the potentiometer arrangement shown. When the lever is at its zero reference point, $V_P = 0$ V. The value of V_P increases at the rate of 1 V/inch until the lever is at its highest point (10 inches) and $V_P = 10$ V. The desired position of the lever is provided as a digital code from the computer and is then fed to a D/A converter, producing V_A. The *difference* between V_P and V_A (called *error*) is produced by the *differential* amplifier and amplified by the PA to drive the motor in the direction that causes the error signal to decrease to 0—that is, moves the lever until $V_P = V_A$.

(a) If it is desired to position the lever within a resolution of 0.1 inch, what is the number of bits needed in the digital input code?

(b) In actual operation, the lever arm might oscillate slightly around the desired position, especially if a *wirewound* potentiometer is used. Can you explain why?

10-16. A particular 6-bit DAC has a full-scale output of 1.260 V. Its accuracy is specified as ±0.1% F.S. and it has an offset error of ±1 mV. Assume that the offset error has not been zeroed out. Consider the following measurements made on this DAC, and determine which of them are not within the device's specifications. (*Hint:* The offset error is added to the error caused by component inaccuracies.)

INPUT CODE	OUTPUT (mV)
000010	41.5
000111	140.2
001100	242.5
111111	1.258V

Section 10-6

10-17. A certain DAC has the following specifications: 8-bit resolution, full-scale = 2.55 V, offset $\leq$ 2 mV; accuracy = $\pm0.1\%$ F.S. A static test on this DAC produces the results shown below. What is the probable cause of the malfunction?

INPUT CODE	OUTPUT	
00000000	8	mV
00000001	18.2	mV
00000010	28.5	mV
00000100	48.3	mV
00001111	158.3	mV
10000000	1.289 V	

10-18. Repeat Problem 10-17 using the following measured data.

INPUT CODE	OUTPUT	
00000000	20.5	mV
00000001	30.5	mV
00000010	20.5	mV
00000100	60.6	mV
00001111	150.6	mV
10000000	1.300 V	

10-19. A technician connects a counter to the DAC of Figure 10-3 to perform a staircase test using a 1-kHz clock. The result is shown in Figure 10-24. What is the probable cause of the incorrect staircase signal?

Sections 10-7 and 10-8

10-20. An 8-bit digital-ramp A/D converter with a 40-mV resolution uses a clock frequency of 2.5 MHz and a comparator with $V_T = 1$ mV. Determine the following values.
(a) The digital output for $V_A = 6.000$ V.
(b) The digital output for 6.035 V.
(c) The maximum and average conversion times for this ADC.

10-21. Why were the digital outputs the same for parts (a) and (b) of Problem 10-20?

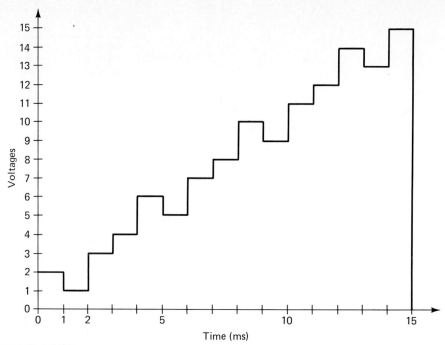

FIGURE 10-24

10-22. What would happen in the A/D converter of Problem 10-20 if an analog voltage of $V_A = 10.853$ V were applied to the input? What waveform would appear at the D/A output? Incorporate the necessary logic in this A/D converter so that an "over-scale" indication will be generated whenever V_A is too large.

10-23. An A/D converter has the following characteristics: resolution, 12 bits; full-scale error, 0.03% F.S.; full-scale output, $+5$ V.

(a) What is the quantization error in volts?

(b) What is the total possible error in volts?

10-24. The quantization error of an A/D converter such as the one in Figure 10-12 is always positive since the $V_{A'}$ value must exceed V_A in order for the COMPARATOR output to switch states. This means the value of $V_{A'}$ could be as much as 1 LSB greater than V_A. This quantization error can be modified so that $V_{A'}$ would be within $\pm\frac{1}{2}$ LSB of V_A. This can be done by adding a fixed voltage equal to $\frac{1}{2}$ LSB ($\frac{1}{2}$ of a step) to the value of $V_{A'}$. Figure 10-25 shows this symbolically for a converter that has a resolution of 10 mV/step. A fixed voltage of $+5$ mV is added to the D/A output in the summing amplifier and the result, $V_{A''}$, is fed to the comparator which has $V_T = 1$ mV.

For this modified converter, determine the digital output for the following V_A values.

(a) $V_A = 5.022$ V

(b) $V_A = 5.028$ V

Determine the quantization error in each case by comparing $V_{A'}$ and V_A. Note that the error is positive in one case and negative in the other.

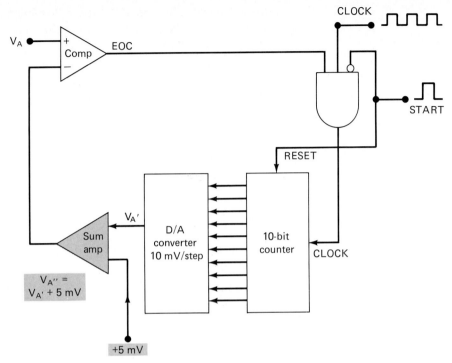

FIGURE 10-25

10-25. For the ADC of Figure 10-25, determine the range of analog input values that will produce a digital output of 0100011100.

Section 10-9

10-26. Assume that the analog signal in Figure 10-26 is to be digitized by performing continuous A/D conversions using an 8-bit digital-ramp converter whose staircase rises at the rate of 1 V every 25 μs. Sketch the recon-

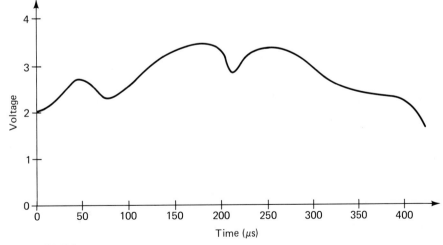

FIGURE 10-26

structed signal using the data obtained during the digitizing process. Compare it to the original signal, and discuss what could be done to make it a more accurate representation.

Section 10-10

10-27. Construct a table such as Table 10-1 showing the sequence of operations for a 6-bit successive-approximation converter which has a resolution of 0.1 V per step and an analog input of $V_A = 6.25$ V. Assume that $V_T = 1$ mV for the comparator.

10-28. A certain 8-bit successive-approximation converter has 2.55 V F.S. The conversion time for $V_A = 1$ V is 80 μs. What will be the conversion time for $V_A = 1.5$ V?

10-29. Figure 10-27 shows the waveform at $V_{A'}$ for a 6-bit SAC with a step size of 40 mV during a complete conversion cycle. Examine this waveform and describe what is occurring at times t_0-t_5. Then determine the resultant digital output.

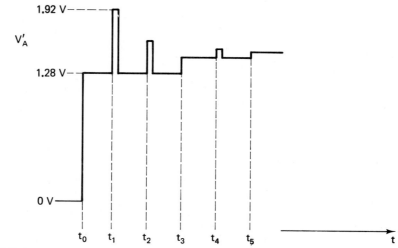

FIGURE 10-27

10-30. Refer to Figure 10-17. What is the approximate value of the analog input, if the microcomputer's data bus is at 10010111 when $\overline{RD}$ is pulsed LOW?

10-31. Connect a 2.0-V reference source to $V_{ref}/2$ and repeat Problem 10-30.

Section 10-11

10-32. Discuss how a flash ADC with a conversion time of 1 μs would work for the situation of Problem 10-26.

10-33. Draw the circuit diagram for a 4-bit flash converter with BCD output and a resolution of 0.1 V. Assume that a +5-V precision supply voltage is available.

10-34. A voltage $V_A = 3.853$ V is applied to the DVM of Figure 10-19. Assume that the counters are initially reset to zero and sketch the waveforms at $V_{A'}$, COMP, Q_1, Q_2, and the AND gate output for *two* complete cycles. Use a 100-kHz clock and assume both one-shots have $t_p = 1$ μs.

10-35. Refer to the DVM circuit in Figure 10-19. Assume that the circuit works correctly for V_A values up to about 7.99 V. However, when V_A is increased above 8 V, the displays do not change from their previous reading, the staircase waveform at $V_{A'}$ steps up to only about 7.99 volts before jumping back down to zero, and COMP remains HIGH. What are some possible causes for this malfunction?

10-36. Add the necessary logic to the DVM of Figure 10-19 to turn ON an over-scale LED whenever the analog input voltage exceeds 9.99 V. The LED should be turned OFF at the start of each new conversion.

Sections 10-13 and 10-14

10-37. Refer to the sample-and-hold circuit of Figure 10-20. What circuit fault would result in V_{OUT} looking exactly like V_A? What fault would cause V_{OUT} to be stuck at zero?

10-38. Use the CMOS 4016 IC (Section 8-17) for S_1, S_2, and S_3 in Figure 10-21 and design the necessary control logic so that each analog input is converted to its digital equivalent in sequence. The A/D converter is a 10-bit successive-approximation type using a 50-kHz clock signal and requires a 10-μs duration start pulse to begin each conversion. The digital outputs are to remain stable for 100 μs after the conversion is complete before switching to the next analog input. Choose the appropriate multiplexing clock frequency.

MEMORY DEVICES

11

OUTLINE

Upon completion of this chapter, you will be able to:

- Understand and correctly use the terminology associated with memory systems.
- Describe the difference between read/write memory and read-only memory.
- Describe the difference between sequential access and random access memory.
- Determine the capacity of a memory device from its inputs and outputs.
- Outline the steps that occur when the CPU reads from or writes to memory.
- Distinguish among the various types of ROMs and cite some common applications.
- Understand and describe the organization and operation of static and dynamic RAMs.
- Combine memory ICs to form larger memory capacities.
- Describe the basic characteristics of various types of magnetic memory devices.
- Use the test results on a RAM or ROM system to determine possible faults in the memory system.

INTRODUCTION

A major advantage of digital over analog systems is the ability to easily store large quantities of digital information and data for short or long periods. This memory capability is what makes digital systems so versatile and adaptable to many situations. For example, in a digital computer the internal main memory stores instructions that tell the computer what to do under *all* possible circumstances so that the computer will do its job with a minimum amount of human intervention.

This chapter is devoted to a study of the most commonly used types of memory devices and systems. We have already become very familiar with the flip-flop, which is an electronic memory device. We have also seen how groups of FFs called registers can be used to store information and how this information can be transferred to other locations. FF registers are high-speed memory elements which are used extensively in the internal operations of a digital computer, where digital information is continually being moved from one location to another. Advances in LSI and VLSI technology have made it possible to obtain large numbers of FFs on a single chip arranged in various memory-array formats. These bipolar and MOS semiconductor memories are the fastest memory devices available, and their cost has been continuously decreasing as LSI technology improves.

Digital data can also be stored as charges on capacitors, and a very important type of semiconductor memory uses this principle to obtain high-density storage at low power-requirement levels. Still another means for storage is provided by magnetic devices. In particular, *magnetic core* memories had been widely used since the late 1950s, although MOS memories have virtually replaced them in most computers.

Semiconductor (and earlier magnetic core) memories are used as the *internal* main memory of a computer (Figure 11-1) where fast operation is important. The internal computer memory is in constant communication with the central processing unit (CPU) as a program of instructions is being executed. The program and any data used by the program are usually stored in the internal memory.

Although semiconductor and magnetic core memories are well suited for high-speed internal memory, their cost per bit of storage prohibits their use as *mass storage* devices. Mass storage refers to memory that is external to the main computer (Figure 11-1) and has the capacity to store millions of bits of data *without* the need for electrical power. Mass memory is normally much slower than

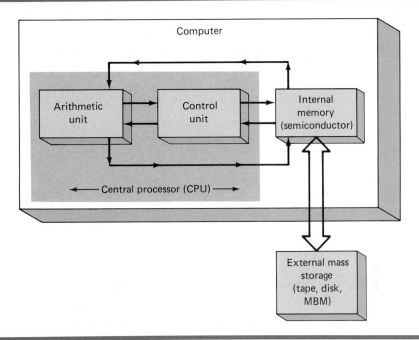

FIGURE 11-1 A computer system normally uses high-speed internal memory and slower external mass memory.

internal memory and is used to store information (programs, data, etc.) not currently being used by the computer. This information is transferred to the internal memory when the computer needs it.

Magnetic tape and magnetic disk are popular mass storage devices that are much less expensive in cost per bit than internal memory devices. A newer entry into the mass memory category is the *magnetic bubble memory* (MBM), a semiconductor device that uses magnetic principles to store millions of bits on one chip. The MBM is relatively slow and cannot be used as internal memory.

We will take a more detailed look at the characteristics of these memory devices after we define some of the more important terms that are common to many memory systems.

11-1 MEMORY TERMINOLOGY

The study of memory devices and systems is filled with terminology that can sometimes be overwhelming to a student. Before we get into any comprehensive discussion of memories, it would be helpful if you had the meaning of some of the more basic terms under your belt. Other new terms will be defined as they appear in the chapter.

Memory Cell A device or electrical circuit used to store a single bit (0 or 1). Examples of memory cells include a flip-flop, a single magnetic core, and a single spot on magnetic tape or disk.

Memory Word A group of bits (cells) in a memory that represents instructions or data of some type. For example, a register consisting of eight FFs can be considered to be a memory that is storing an 8-bit word. Word sizes in modern computers typically range from 4 to 64 bits, depending on the size of the computer.

Byte A special term used for an 8-bit word. A byte always consists of 8 bits, which is the most common word size in microcomputers.

Capacity A way of specifying how many bits can be stored in a particular memory device or complete memory system. To illustrate, suppose that we have a memory which can store 4096 20-bit words. This represents a total capacity of 81,920 bits. We could also express this memory's capacity as 4096 $\times$ 20. When expressed this way, the first number (4096) is the number of words and the second number (20) is the number of bits per word (word size). The number of words in a memory is often a multiple of 1024. It is common to use the designation "1K" to represent 1024 when referring to memory capacity. Thus, a memory that has a storage capacity of 4K $\times$ 20 is actually a 4096 $\times$ 20 memory.

EXAMPLE 11-1

A certain semiconductor memory chip is specified as 2K $\times$ 8. How many words can be stored on this chip? What is the word size? How many total bits can this chip store?

SOLUTION

$$2K = 2 \times 1024 = 2048 \text{ words}$$

Each word is 8 bits (one byte). The total number of bits is therefore

$$2048 \times 8 = 16.384 \text{ bits}$$

Address A number that identifies the location of a word in memory. Each word stored in a memory device or system has a unique address. Addresses are always expressed as a binary number, although octal, hexadecimal, and decimal numbers are often used for convenience. Figure 11-2 illustrates a small memory consisting of eight words. Each of these eight words has a specific address represented as a 3-bit number ranging from 000 to 111. Whenever we refer to a specific word location in memory, we use its address code to identify it.

Read Operation The operation whereby the binary word stored in a specific memory location (address) is sensed and then transferred to another device. For example, if we want to use word 4 of the memory of Figure 11-2 for some purpose, we have to perform a read operation on address 100. The read operation is often called a *fetch* operation, since a word is being fetched from memory. We will use both terms interchangeably.

Write Operation The operation whereby a new word is placed into a particular memory location. It is also referred to as a *store* operation. Whenever a new word is written into a memory location, it replaces the word that was previously stored there.

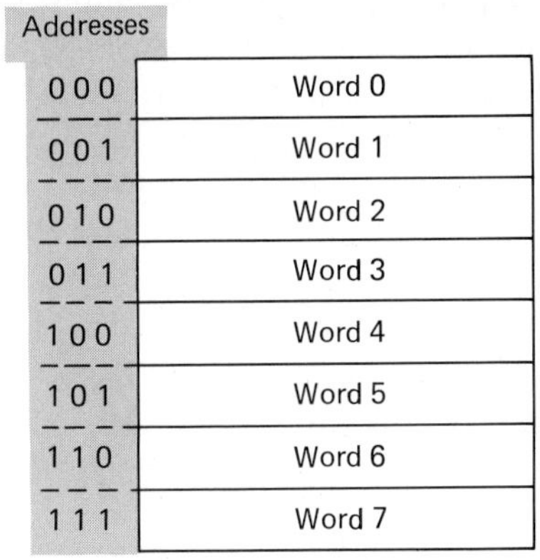

Addresses	
0 0 0	Word 0
0 0 1	Word 1
0 1 0	Word 2
0 1 1	Word 3
1 0 0	Word 4
1 0 1	Word 5
1 1 0	Word 6
1 1 1	Word 7

FIGURE 11-2 **Each word location has a specific binary address.**

Access Time A measure of a memory device's operating speed. It is the amount of time required to perform a read operation. More specifically, it is the time between the memory receiving a new address input and the data becoming available at the memory output. The symbol t_{ACC} is used for access time.

Volatile Memory Any type of memory that requires the application of electrical power in order to store information. If the electrical power is removed, all information stored in the memory will be lost. Many semiconductor memories are volatile, while all magnetic memories are nonvolatile.

Random Access Memory (RAM) Memory in which the actual physical location of a memory word has no effect on how long it takes to read from or write into that location. In other words, the access time is the same for any address in memory. Most semiconductor memories and magnetic core memories are RAMs.

Sequential Access Memory (SAM) A type of memory in which the access time is not constant, but varies depending on the address location. A particular stored word is found by sequencing through all address locations until the desired address is reached. This produces access times which are much longer than those of random access memories. Examples of sequential access memory devices include magnetic tape, disk, and magnetic bubble memory. To illustrate the difference between SAM and RAM, consider the situation where you have recorded 60 minutes of songs on an audio tape cassette. When you want to get to a particular song, you have to rewind or fast-forward the tape until you find it. The process is relatively slow, and the amount of time required depends on where on the tape the desired song is recorded. This is SAM since you have to sequence through all intervening information until you find what you are looking for. The RAM counterpart to this would be a jukebox, where you can quickly select any song by punching in the appropriate code, and it takes the same time no matter what song you select.

Read/Write Memory (RWM) Any memory that can be read from or written into with equal ease.

Read-only Memory (ROM) A broad class of semiconductor memories designed for applications where the ratio of read operations to write operations is very high.

Technically, a ROM can be written into (programmed) only once, and this operation is normally performed at the factory. Thereafter information can only be read from the memory. Other types of ROM are actually read-mostly memories (RMM) which can be written into more than once, but the write operation is more complicated than the read operation and it is not performed very often. The various types of ROM will be discussed later.

Static Memory Devices Semiconductor memory devices in which the stored data will remain permanently stored as long as power is applied, without the need for periodically rewriting the data into memory.

Dynamic Memory Devices Semiconductor memory devices in which the stored data *will not* remain permanently stored, even with power applied, unless the data are periodically rewritten into memory. The latter operation is called a *refresh* operation.

Internal Memory This is also referred to as the computer's *main* memory. It stores the instructions and data that the CPU is currently working on. It is the fastest memory in the computer system, and usually consists of semiconductor memory devices.

Mass Memory This type of memory is also referred to as *auxiliary* memory. It stores large amounts of information external to the computer's internal memory. It is generally slower than internal memory and is always nonvolatile.

REVIEW QUESTIONS

1. Define the following terms: (a) memory cell, (b) memory word, (c) address, (d) byte, (e) access time.
2. A certain memory has a capacity of 8K × 16. How many bits are in each word? How many words are being stored? How many memory cells does this memory contain? (Ans. 16, 8192, 131,072)
3. Explain the difference between the read (fetch) and write (store) operations.
4. *True or false:* A volatile memory will lose its stored data when electrical power is interrupted.
5. Explain the difference between SAM and RAM.
6. Explain the difference between RWM and ROM.
7. *True or false:* A dynamic memory will hold its data as long as electrical power is applied.

11-2 GENERAL MEMORY OPERATION

Although each type of memory is different in its internal operation, certain basic operating principles are the same for all memory systems. An understanding of these basic ideas will help in our study of individual memory devices.

Every memory system requires several different types of input and output lines to perform the following functions:

1. Select the address in memory that is being accessed for a read or write operation.
2. Select either a read or write operation to be performed.
3. Supply the input data to be stored in memory during a write operation.
4. Hold the output data coming from memory during a read operation.
5. Enable (or disable) the memory so that it will (or will not) respond to the address inputs and read/write command.

Figure 11-3(a) illustrates these basic functions in a simplified diagram of a 32×4 memory that stores 32 4-bit words. Since the word size is 4 bits, there are four data input lines I_0–I_3 and four data output lines O_0–O_3. During a write operation the data to be stored into memory have to be applied to the data input lines. During a read operation the word being read from memory appears at the data output lines.

Address Inputs   Since this memory stores 32 words, it has 32 different storage locations and therefore 32 different binary addresses ranging from 00000 to 11111 (0 to 31 in decimal). Thus there are five address inputs, A_0–A_4. To access one of the memory locations for a read or write operation, the 5-bit address code for that particular location is applied to the address inputs. In general, N address inputs are required for a memory that has a capacity of 2^N words.

FIGURE 11-3   **(a) Diagram of a 32 × 4 memory; (b) virtual arrangement of memory cells into 32 4-bit words.**

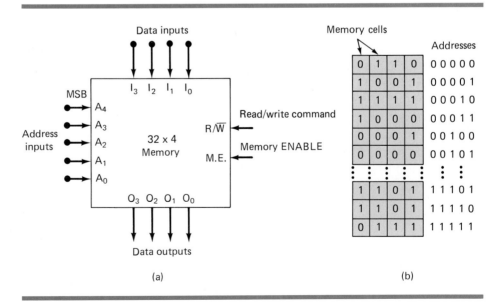

We can visualize the memory of Figure 11-3(a) as an arrangement of 32 registers, with each register holding a 4-bit word as illustrated in Figure 11-3(b). Each address location is shown containing four memory cells that hold 1s and 0s that make up the data word stored at that location. For example, the data word 0110 is stored at address 00000, the data word 1001 is stored at address 00001, and so on.

The R/$\overline{W}$ Input The read/write ($R/\overline{W}$) input line determines which memory operation is to take place. Some memory systems use two separate inputs, one for read and one for write. When a single $R/\overline{W}$ input is used, the read operation takes place for $R/\overline{W} = 1$, and the write operation takes place for $R/\overline{W} = 0$.

A simplified illustration of the read and write operations is shown in Figure 11-4. Part (a) shows the data word 0100 being written into the memory register at address location 00011. This data word would have been applied to the memory's data input lines, and replaces the data that was previously stored at address 00011. Part (b) shows the data word 1101 being read from address 11110. This output data would appear at the memory's data output lines. After the read operation, the data word 1101 is still stored in address 11110. In other words, the read operation does not change the stored data.

MEMORY ENABLE Many memory systems have some means for completely disabling all or part of the memory so that it will not respond to the other inputs. This is represented in Figure 11-3 as the MEMORY ENABLE input, although it can have different names in the various memory systems. Here it is shown as an active-HIGH input that enables the memory to operate normally when it is kept

FIGURE 11-4 **Simplified illustration of the READ and WRITE operations on the 32 × 4 memory: (a) WRITING the data word 0100 into memory location 00011; (b) READING the data word 1101 from memory location 11110.**

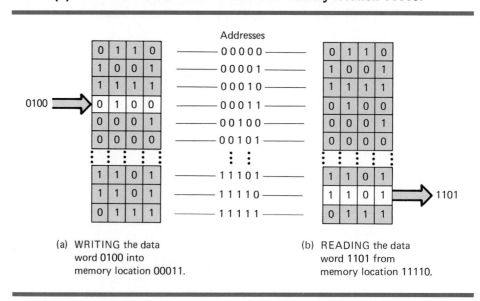

(a) WRITING the data word 0100 into memory location 00011.

(b) READING the data word 1101 from memory location 11110.

HIGH. A LOW on this input disables the memory so that it will not respond to the address and $R/\overline{W}$ inputs. This type of input is useful when several memory modules are combined to form a larger memory. We will examine this idea later.

EXAMPLE 11-2

Describe the conditions at each input and output when the contents of address location 00100 are to be read.

SOLUTION

Address inputs: 00100
Data inputs: xxxx (not used)
$R/\overline{W}$: HIGH
MEMORY ENABLE: HIGH
Data outputs: 0001

EXAMPLE 11-3

Describe the conditions at each input and output when the data word 1110 is to be written into address location 01101.

SOLUTION

Address inputs: 01101
Data inputs: 1110
$R/\overline{W}$: LOW
MEMORY ENABLE: HIGH
Data outputs: xxxx (not used; usually Hi-Z)

EXAMPLE 11-4

A certain memory has a capacity of 4K × 8.

(a) How many data input and data output lines does it have?
(b) How many address lines does it have?
(c) What is its capacity in bytes?

SOLUTION
(a) Eight of each, since the word size is eight.
(b) The memory stores 4K = 4 × 1024 = 4096 words. Thus, there are 4096 memory addresses. Since $4096 = 2^{12}$, it requires a 12-bit address code to specify one of 4096 addresses.
(c) A byte is 8 bits. This memory has a capacity of 4096 bytes.

The example memory in Figure 11-3 illustrates the important input and output functions common to most memory systems. Of course, each type of memory may have other input and output lines that are peculiar to that memory. These will be described as we discuss the individual memory types.

REVIEW QUESTIONS

1. How many address inputs, data inputs, and data outputs are required for a 16K × 12 memory? (*Ans.* 14, 12, 12)
2. What is the function of the $R/\overline{W}$ input?
3. What is the function of the MEMORY ENABLE input?

11-3 CPU–MEMORY CONNECTIONS

A major part of this chapter is devoted to semiconductor memory, which, as pointed out earlier, makes up the internal memory of most modern computers. Remember, this internal memory is in constant communication with the CPU (central processing unit). It is not necessary to be familiar with the detailed operation of a CPU at this point, so the following simplified treatment of the CPU–memory interface will provide the perspective needed to make our study of memory devices more meaningful.

A computer's internal memory is made up of RAM and ROM ICs that are interfaced to the CPU over three groups of signal lines or buses. These are shown in Figure 11-5 as the address lines or address bus, data lines or data bus, and the

FIGURE 11-5 **Three groups of lines (buses) connect the internal memory ICs to the CPU.**

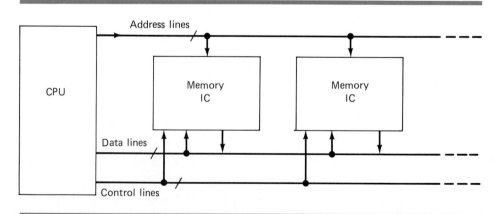

control lines or control bus. Each of these buses consists of several lines (note how they are represented by a single line with a slash), and the number of lines in each bus will vary from one computer to the next. The three buses play a necessary part in allowing the CPU to write data into memory and to retrieve data from memory.

When a computer is executing a program of instructions, the CPU continually fetches (reads) information from those locations in memory that contain (1) the program codes representing the operations to be performed, and (2) the data to be operated upon. The CPU will also store (write) data into memory locations as dictated by the program instructions. Whenever the CPU wants to write data to a particular memory location, the following steps must occur:

WRITE operation

1. The CPU supplies the binary address of the memory location where the data is to be stored. It places this address on the address bus lines.
2. The CPU places the data to be stored on the data bus lines.
3. The CPU activates the appropriate control signal lines for the memory write operation.
4. The memory ICs decode the binary address to determine which location is being selected for the store operation.
5. The data on the data bus is transferred to the selected memory location.

Whenever the CPU wants to read data from a specific memory location, the following steps must occur:

READ operation

1. The CPU supplies the binary address of the memory location from which data is to be retrieved. It places this address on the address bus lines.
2. The CPU activates the appropriate control signal lines for the memory read operation.
3. The memory ICs decode the binary address to determine which location is being selected for the read operation.
4. The memory ICs place data from the selected memory location on to the data bus, from where it is transferred to the CPU.

The steps above should make clear the function of each of the system buses:

Address Bus This is a *unidirectional* bus that carries the binary address outputs from the CPU to the memory ICs.

Data Bus This is a *bidirectional* bus that carries data between the CPU and the memory ICs.

Control Bus This bus carries control signals (such as the $R/\overline{W}$ signal) from the CPU to the memory ICs.

As we get into discussions of actual memory ICs, we will examine the signal activity that appears on these buses for the read and write operations.

REVIEW QUESTIONS

1. Name the three groups of lines that connect the CPU and internal memory.
2. Outline the steps that take place when the CPU reads from memory.
3. Outline the steps that occur when the CPU writes to memory.

11-4 SEMICONDUCTOR MEMORY TECHNOLOGIES

On the basis of their manufacturing process or *technology,* semiconductor memories are divided into two categories: *bipolar memories* (made with bipolar transistors) and *MOS memories* (made with MOSFETs).

The memory devices that are manufactured using the various technologies will generally differ in their electrical and physical characteristics, and they are usually compared acccording to the following criteria:

1. Density (i.e., capacity in bits per chip).
2. Speed (i.e., access time).
3. Power requirements (i.e., μW or mW per bit).
4. Cost per bit.
5. Noise immunity.

Both bipolar and MOS memory technologies have their particular strengths and weaknesses based on the above criteria. In general, though, the bipolar devices have the advantage of higher speed, while MOS memory devices are superior in all the other areas.

Bipolar Memories   All the bipolar technologies use resistors, diodes, and bipolar transistors for the memory circuitry on the chip. The relative complexity of the bipolar circuitry imposes a limitation on the memory capacity that can be placed on a single silicon chip, and bipolar memories have the lowest capacity of any technology.

Standard TTL and Schottky TTL technologies account for the major portion of bipolar memory devices in use today. These technologies use the same basic structures that were discussed in Chapter 8. The basic characteristics of standard TTL technology memories are high speed, moderate capacity, high power consumption, relatively low noise immunity, and relatively high cost. Schottky TTL memories share these same characteristics but are faster than standard TTL memories. TTL memories are suited for applications that demand relatively high speed and medium capacity and where power consumption is not a critical factor.

ECL memories are the fastest memory devices available today because they use the nonsaturating bipolar transistor operation described in Chapter 8. Unfortu-

TABLE 11-1

TECHNOLOGY	SPEED	POWER REQUIRED	CAPACITY	NOISE IMMUNITY	COST
TTL/S-TTL	Fast	High	Low	Low	High
ECL	Fastest	Highest	Lowest	Lowest	Highest
NMOS	Medium (approaching TTL)	Low	Highest	High	Lowest
CMOS/SOS	Slowest	Lowest	High	Highest	Low

nately, high speed is the only advantage it has over the other technologies, and it is used only where the utmost in operating speed is required without regard to cost or power consumption.

MOS Memories

The workhorse of memory technologies is *N*-channel MOS technology. NMOS technology currently provides the highest capacity of all memory technologies at the lowest cost per bit, and at power levels that are much lower than bipolar memories. Their only disadvantage is their slower speed compared to bipolar. The refinements in the NMOS fabrication process, however, have not yet reached their limits in terms of speed and density, and NMOS memory devices are gradually approaching the operating speed of some of the lower-speed bipolar memories. *VMOS* (*vertical MOS*) and *HMOS* (*high-performance MOS*) are two of the newer MOS processes that appear to be approaching the performance of bipolar technologies.

CMOS memory devices provide the lowest power consumption of all the memory technologies. CMOS memories also have the highest noise immunity of any technology. In capacity, on the other hand, CMOS is somewhat lower than NMOS, although it is higher than bipolar, and CMOS operating speed is the slowest of any technology. An improved CMOS process uses a *silicon-on-sapphire* (SOS) substrate instead of pure silicon to achieve faster speeds and higher densities. CMOS and SOS are best suited for applications in high-noise environments or where power consumption is extremely critical (i.e., battery-operated systems).

Table 11-1 is a comparison of the relative characteristics of the various memory technologies.

REVIEW QUESTIONS

1. Which memory technology has the greatest capacity per chip?
2. Which memory technology would be best to use in a hand calculator?
3. Which memory technology would be best to use in a very high speed computer?
4. Which memory technology generally has the lowest cost per bit?
5. Which memory technology would be best to use in a high-noise industrial environment?

This type of semiconductor memory is designed to hold data that either are permanent or will not change frequently. During normal operation, no new data can be written into a ROM but data can be read from ROM. For some ROMs the data that are stored have to be built in during the manufacturing process; for other ROMs the data can be entered electrically. The process of entering data is called *programming* or *burning* the ROM. Some ROMs cannot have their data changed once they have been programmed; others can be *erased* and reprogrammed as often as desired. We will take a detailed look later at these various types of ROMs. For now, we will assume that the ROMs have been programmed and are holding data.

ROMs are used to store data and information that is not to change during the operation of a system. A major use for ROMs is in the storage of programs in microcomputers. Since all ROMs are *nonvolatile,* these programs are not lost when the microcomputer is turned off. When the microcomputer is turned on, it can immediately begin executing the program stored in ROM. ROMs are also used for program and data storage in microprocessor-controlled equipment such as sophisticated electronic cash registers.

ROM Block Diagram

A typical block diagram for a ROM is shown in Figure 11-6(a). It has three sets of signals: address inputs, control input(s), and data outputs. From our previous discussions we can determine that this ROM is storing 16 words, since it has $2^4 = 16$ possible addresses, and each word contains 8 bits, since there are eight data outputs. Thus, this is a 16×8 ROM. Another way to describe this ROM's capacity is to say that it stores 16 bytes of data.

The data outputs of most ROM ICs are either open-collector or tristate outputs to permit the connection of many ROM chips to the same data bus for memory expansion. The most common numbers of data outputs for ROMs are 4 bits and 8 bits, with 8-bit words being the most common.

The control input *CS* stands for *chip select.* This is essentially an enable input that enables or disables the ROM outputs. Some manufacturers use different labels for the control input, such as *CE* (chip enable) or *OE* (output enable). Many ROMs have more than two or more control inputs that have to be activated in order to enable the data outputs. The *CS* input shown in Figure 11-6(a) is active HIGH. If it had a small circle or an inversion symbol ($\overline{CS}$), it would be active LOW. Note that there is no *R/W* (read/write) input because the ROM cannot be written into under normal operating circumstances.

The READ Operation

Let's assume that the ROM has been programmed with the data shown in the table of Figure 11-6(b). Sixteen different data words are stored at the 16 different address locations. For example, the data word stored at location 0011 is 10101111. Of course, the data are stored in binary inside the ROM, but very often we use hexadecimal notation to efficiently show the programmed data. This is done in Figure 11-6(c).

In order to read a data word from ROM, we need to do two things: apply the appropriate address inputs, and then activate the control inputs. For example, if we want to read the data stored at location 0111 of the ROM in Figure 11-6, we have

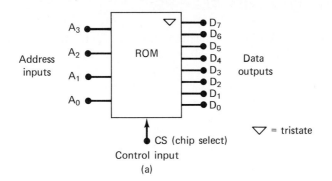

FIGURE 11-6   (a) ROM block diagram; (b) programmed data; (c) hex data.

Word	A_3	A_2	A_1	A_0	D_7	D_6	D_5	D_4	D_3	D_2	D_1	D_0
0	0	0	0	0	1	1	0	1	1	1	1	0
1	0	0	0	1	0	0	1	1	1	0	1	0
2	0	0	1	0	1	0	0	0	0	1	0	1
3	0	0	1	1	1	0	1	0	1	1	1	1
4	0	1	0	0	0	0	0	1	1	0	0	1
5	0	1	0	1	0	1	1	1	1	0	1	1
6	0	1	1	0	0	0	0	0	0	0	0	0
7	0	1	1	1	1	1	1	0	1	1	0	1
8	1	0	0	0	0	0	1	1	1	1	0	0
9	1	0	0	1	1	1	1	1	1	1	1	1
10	1	0	1	0	1	0	1	1	1	0	0	0
11	1	0	1	1	1	1	0	0	0	1	1	1
12	1	1	0	0	0	0	1	0	0	1	1	1
13	1	1	0	1	0	1	1	0	1	0	1	0
14	1	1	1	0	1	1	0	1	0	0	1	0
15	1	1	1	1	0	1	0	1	1	0	1	1

(b)

Word	A_3 A_2 A_1 A_0	D_7-D_0
0	0	DE
1	1	3A
2	2	85
3	3	AF
4	4	19
5	5	7B
6	6	00
7	7	ED
8	8	3C
9	9	FF
10	A	B8
11	B	C7
12	C	27
13	D	6A
14	E	D2
15	F	5B

(c)

to apply $A_3A_2A_1A_0 = 0111$ to the address inputs, and then apply a HIGH to CS. The address inputs will be decoded inside the ROM to select the correct data word, 11101101, that will appear at the D_7–D_0 outputs. If CS is kept LOW, the ROM outputs will be disabled and will be in the Hi-Z state.

REVIEW QUESTIONS

1.   *True or false:* All ROMs are nonvolatile.
2.   Describe the procedure for reading from ROM.

11-6 ROM ARCHITECTURE

The internal architecture (structure) of a ROM IC is very complex, and we need not be familiar with all its detail. It is instructive, however, to look at a simplified diagram of the internal architecture, such as that shown in Figure 11-7 for the 16×8 ROM. There are four basic parts: *row decoder, column decoder, register array, output buffers*.

Register Array The register array stores the data that have been programmed into the ROM. Each register contains a number of memory cells equal to the word

FIGURE 11-7 **Architecture of 16 × 8 ROM.**

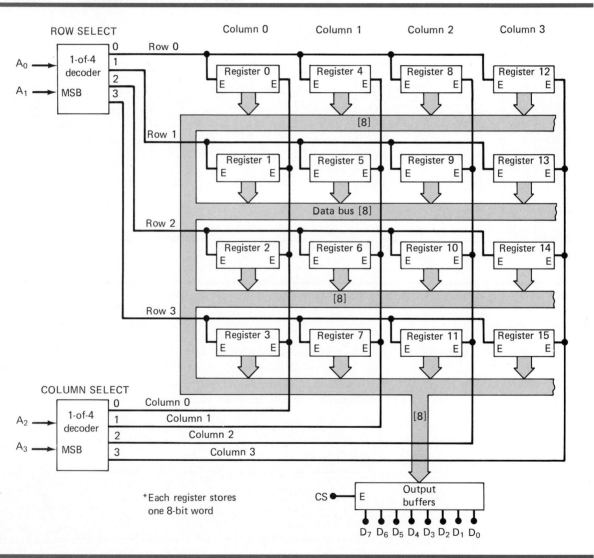

size. In this case, each register stores an 8-bit word. The registers are arranged in a square matrix array that is common to many semiconductor memory chips. We can specify the position of each register as being in a specific row and specific column. For example, register 0 is in row 0/column 0, and register 9 is in row 1/column 2.

The eight data outputs of each register are connected to an internal data bus that runs through the entire circuit. Each register has two enable inputs (E); both have to be HIGH in order for the register's data to be placed on the bus.

Address Decoders The applied address code $A_3A_2A_1A_0$ determines which register in the array will be enabled to place its 8-bit data word onto the bus. Address bits A_1A_0 are fed to a 1-of-4 decoder which activates one row-select line, and address bits A_3A_2 are fed to a second 1-of-4 decoder which activates one column-select line. Only one register will be in both the row and the column selected by the address inputs, and this one will be enabled.

EXAMPLE 11-5

Which register will be enabled by input address 1101?

SOLUTION

$A_3A_2 = 11$ will cause the column decoder to activate the column 3 select line, and $A_1A_0 = 01$ will cause the row decoder to activate the row 1 select line. This will place HIGHs at both enable inputs of register 13, thereby causing its data outputs to be placed on the bus. Note that the other registers in column 3 will only have one enable input activated; likewise for the other row 1 registers.

EXAMPLE 11-6

What input address will enable register 7?

SOLUTION

The enable inputs of this register are connected to the row 3 and column 1 select lines, respectively. To select row 3, the A_1A_0 inputs have to be at 11, and to select column 1, the A_3A_2 inputs have to be at 01. Thus, the required address will be $A_3A_2A_1A_0 = 0111$.

Output Buffers The register that is enabled by the address inputs will place its data on the data bus. These data feed into the output buffers, which will pass the data to the external data outputs, provided that CS is HIGH. If CS is LOW, the output buffers are in the Hi-Z state, and D_7–D_0 will be floating.

The architecture shown in Figure 11-7 is similar to that of most IC ROMs. Depending on the number of stored data words, the registers in some ROMs will not be arranged in a square array. For example, the Intel 2708 is a MOS ROM that stores 1024 8-bit words. Its 1024 registers are arranged in a 64-by-16 array. Typically, ROM capacities range from 32×8 to $512K \times 8$.

EXAMPLE 11-7

Describe the internal architecture of a ROM that stores 4K bytes and uses a square register array.

SOLUTION

4K is actually $4 \times 1024 = 4096$, so this ROM holds 4096 8-bit words. Each word can be thought of as being stored in an 8-bit register, and there are 4096 registers connected to a common data bus internal to the chip. Since $4096 = 64^2$, the registers are arranged in a 64-by-64 array; that is, there are 64 rows and 64 columns. This requires a 1-of-64 decoder to decode six address inputs for the row select, and a second 1-of-64 decoder to decode six other address inputs for the column select. Thus, a total of 12 address inputs are required. This makes sense, since $2^{12} = 4096$, and there are 4096 different addresses.

REVIEW QUESTIONS

1. What address inputs are required if we want to read the data from register 9 in Figure 11-7? (*Ans.* $A_3A_2A_1A_0 = 1001$)
2. Describe the function of the row-select decoder, column-select decoder, and output buffers in the ROM architecture.

11-7 ROM TIMING

There will be a propagation delay between the application of a ROM's inputs and the appearance of the data outputs during a READ operation. This time delay, called *access time*, t_{ACC}, is a measure of the ROM's operating speed. Access time is described graphically by the waveforms in Figure 11-8.

The top waveform represents the address inputs, the middle waveform is an active LOW chip select, $\overline{CS}$, and the bottom waveform represents the data outputs. At time t_0 the address inputs are all at some specific level, some HIGH, and some LOW. $\overline{CS}$ is HIGH, so that the ROM data outputs are in their Hi-Z state (represented by hashed line).

Just prior to t_1, the address inputs are changing to a new address for a new READ operation. At t_1 the new address is valid; that is, each address input is at a valid logic level. At this point the internal ROM circuitry begins to decode the new address inputs to select the register which is to send its data to the output buffers. At t_2 the $\overline{CS}$ input is activated to enable the output buffers. Finally, at t_3, the outputs change from the Hi-Z state to the valid data that represent the data stored at the specified address.

The time delay between t_1, when the new address becomes valid, and t_3, when the data outputs become valid, is the access time t_{ACC}. Typical bipolar ROMS will

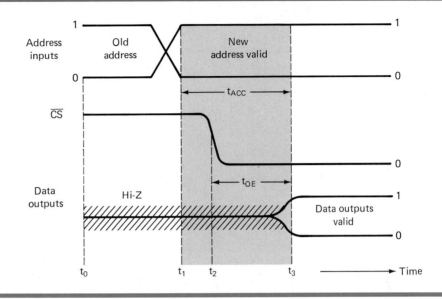

FIGURE 11-8 Typical timing for a ROM READ operation.

have access times in the range 30 to 90 ns, while MOS ROMs will range from 200 to 900 ns.

Another important timing parameter is the *output enable time, t_{OE},* which is the delay between the $\overline{CS}$ input and the valid data output. Typical values of t_{OE} are 20 ns for bipolar ROMs and 100 ns for MOS ROMs. This timing parameter is important in situations where the address inputs are already set to their new values, but the ROM outputs have not yet been enabled. When $\overline{CS}$ goes LOW to enable the outputs, the delay will be t_{OE}.

11-8 TYPES OF ROMs

Now that we have a general understanding of the internal architecture and external operation of ROM devices, we will take a brief look at the various types of ROMs to see how they differ in the way they are programmed and in their ability to be erased and reprogrammed.

Mask-Programmed ROM This type of ROM has its storage locations written into (programmed) by the manufacturer according to the customer's specifications. A photographic negative called a *mask* is used to control the electrical interconnections on the chip. A special mask is required for each different set of information to be stored in the ROM. Since these masks are expensive, this type of ROM is economical only if you need a large quantity of the same ROM. Some ROMs of this type are available as off-the-shelf devices preprogrammed with commonly used

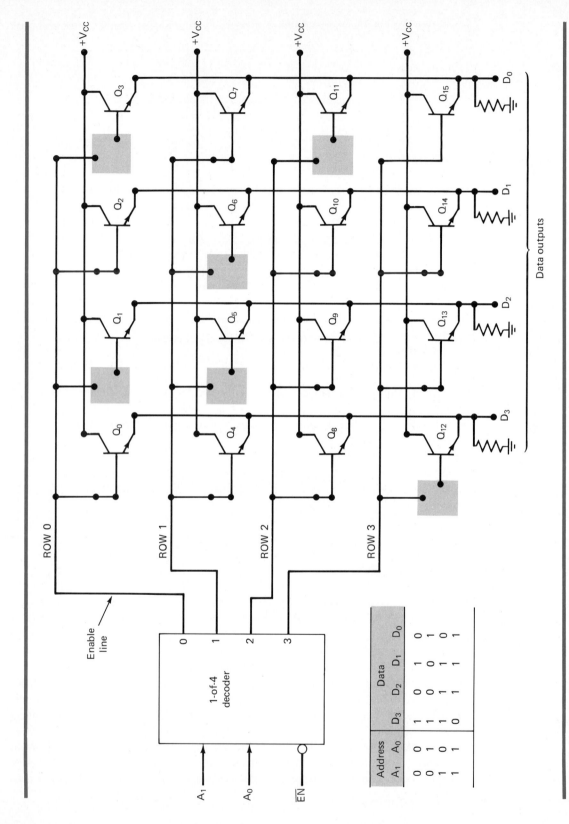

FIGURE 11-9 Structure of a bipolar MROM shows one bipolar transistor used for each memory cell. An open base connection stores a "0"; a closed base connection stores a "1."

information or data such as certain mathematical tables and character generator codes for CRT displays. A major disadvantage of this type of ROM is that it cannot be reprogrammed in the event of a design change requiring a modification of the stored data. The ROM would have to be replaced by a new one with the desired data written into it. Several types of user-programmable ROMs have been developed to overcome this disadvantage. Mask-programmed ROMs, however, still represent the most economical approach when a large quantity of identically programmed ROMs are needed.

Mask-programmed ROMs are commonly referred to as just ROMs, but this can be confusing since the term ROM actually represents the broad category of devices that, during normal operation, are only read from. We will use the mnemonic MROM whenever we refer to mask-programmed ROMs.

Figure 11-9 shows the structure of a small bipolar MROM. It consists of 16 memory cells arranged in four rows of four cells. Each cell is an NPN bipolar transistor connected in the common-collector configuration (input at base, output at emitter). The top row of cells (ROW 0) constitutes a 4-bit register. Note how some of the transistors (Q_0 and Q_2) have their bases connected to the ROW 0 enable line, while others (Q_1 and Q_3) do not. The same is true of the cells in each of the other rows. The presence or absence of these base connections determines whether a cell is storing a 1 or 0, respectively. The condition of each base connection is controlled by the photographic mask during production based on the customer supplied data.

Note that cells that are in corresponding positions in each row (register) have their emitters connected to a common output. For instance, the emitters of Q_0, Q_4, Q_8, and Q_{12} are connected together as data output D_3. As we shall see, this will present no problem because only one row of cells will be activated at one time.

The 1-of-4 decoder is used to decode the address inputs A_1A_0 to select which row (register) is to have its data read. The decoder's active-HIGH outputs provide the ROW enable lines that are the base inputs for the various rows of cells. If the decoder's enable input, $\overline{EN}$, is held HIGH, all of the decoder outputs will be in their inactive-LOW state, and all of the transistors in the array will be OFF because of the absence of any base voltage. For this situation, the data outputs will all be in the LOW state.

When $\overline{EN}$ is in its active-LOW state, the conditions at the address inputs determine which row (register) will be enabled so that its data can be read at the data outputs. For example, to read ROW 0, the A_1A_0 inputs are set to 00. This places a HIGH at the ROW 0 line; all other row lines are at 0 V. This HIGH at ROW 0 turns ON transistors Q_0 and Q_2, but not Q_1 and Q_3. With Q_0 and Q_2 conducting, the data outputs D_3 and D_1 will be HIGH; outputs D_2 and D_0 are still LOW. In a similar manner, application of the other address codes will produce data outputs from the corresponding register. The table in Figure 11-9 shows the data for each address. You should verify how this correlates with the base connections to the various cells.

EXAMPLE 11-8

MROMs are often used to store tables of mathematical functions. Show how the MROM in Figure 11-9 can be used to store the function $y = x^2 + 3$, where the input address supplies the value for x and the value of the output data is y.

SOLUTION

The first step is to set up a table showing the desired output for each set of inputs. The input binary number, x, is represented by the address A_1A_0. The output binary number is the desired value of y. For example, when $x = A_1A_0 = 10_2 = 2_{10}$, the output should be $2^2 + 3 = 7_{10} = 0111_2$. The complete table is shown below.

x		$y = x^2 + 3$			
A_1	A_0	D_3	D_2	D_1	D_0
0	0	0	0	1	1
0	1	0	1	0	0
1	0	0	1	1	1
1	1	1	1	0	0

This table is supplied to the MROM manufacturer for developing the mask that will make the appropriate connections to the bases of the memory cells during the fabrication process. For instance, the first row in the table indicates that the connections to the bases of Q_0 and Q_1 will be left unconnected, while the connections to Q_2 and Q_3 will be made.

Bipolar MROMs are available in several capacities. One of the more popular ones is the 74187 whose logic symbol is shown in Figure 11-10. It is organized as a 256×4 memory, and has an access time of 40 ns. Its outputs are open-collector types which require external pull-up resistors. Another bipolar ROM is the 7488A, which has a capacity of 32×8 and an access time of 45 ns.

FIGURE 11-10   Logic symbol for 74187 bipolar MROM.

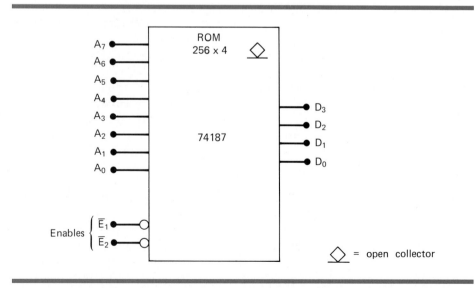

MOS MROMs have an internal structure similar to Figure 11-9 except that the cells are MOSFETs rather than bipolar transistors. The TMS47256 is an NMOS version that has a capacity of 32K × 8. Its symbol is shown in Figure 11-11. Note that it has tristate outputs to permit easy interfacing to a computer data bus. In addition to the fourteen address inputs, it has two enable inputs, $\overline{E}$ and $\overline{S}$. Both of these have to be LOW to enable the MROM outputs. The $\overline{E}$ input also performs a *power-down* function. When $\overline{E}$ is kept HIGH, the chip's internal circuitry is put into a low-power standby state where it draws about one fourth of the normal supply current. The TMS47256 has an access time of 200 ns and a standby power of 82.5 mW. The CMOS version, the TMS47C256, has an access time of 150 ns and a standby power of only 2.8 mW.

Programmable ROMs (PROMs)

A mask-programmable ROM is very expensive and would not be used except in high-volume applications, where the cost would be spread out over many units. For lower-volume applications, manufacturers have developed *fusible-link* PROMs that are user-programmable; that is, they are not programmed during the manufacturing process but are custom programmed by the user. Once programmed, however, a PROM is like a MROM in that it cannot be erased and reprogrammed. Thus, if the program in the PROM is faulty or has to be changed, the PROM has to be thrown away.

The fusible-link PROM structure is very similar to the MROM in that certain connections either are left intact or are opened in order to program a memory cell as a 1 or 0. In the MROM of Figure 11-9 these connections were from the enable lines to the transistor bases. In a PROM each of these connections is made with a

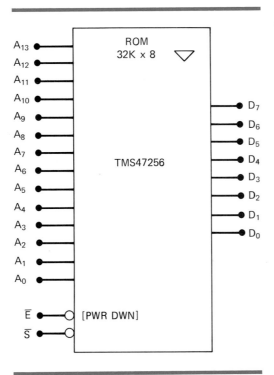

FIGURE 11-11 Logic symbol for the TMS47256 MROM made using NMOS technology.

thin fuse link that comes intact from the manufacturer (see Figure 11-12). The user can selectively *blow* any of these fuse links to produce the desired stored memory data. Typically this is accomplished by applying a carefully controlled voltage to the device to produce a current flow that will cause the fuse link to open in a manner similar to blowing a fuse. Once a fuse link is broken, it cannot be reconnected.

The process of programming a PROM and then verifying the programmed data can be extremely time-consuming and tedious when done manually. A number of commercial PROM programmers are available for several hundred dollars that permit keyboard entry of the program into RWM (read/write memory) and then perform the fuse blowing and verification automatically without user intervention.

A popular bipolar PROM IC is the 74186, which is organized as 64 8-bit words and has a typical access time of 50 ns. Another one is the TBP28S166, which is a 2K × 8 chip. MOS PROMs are available with much greater capacities than bipolar. For example, the TMS27P64 is an NMOS PROM with a capacity of 8K × 8 and an access time of 250 ns.

Erasable Programmable ROM (EPROM)

An EPROM can be programmed by the user and it can also be *erased* and reprogrammed as often as desired. Once programmed, the EPROM is a *nonvolatile* memory that will hold its stored data indefinitely. The process for programming an EPROM involves the application of special voltage levels (typically in the 25- to 50-V range) to the appropriate chip inputs for a specified amount of time (typically 50 ms per address location). The programming process is usually performed by a special programming circuit that is separate from the circuit in which the EPROM will eventually be working. The complete programming process can take up to several minutes for one EPROM chip.

The storage cells in an EPROM are field-effect transistors with a silicon gate that has no electrical connections (i.e., a *floating gate*). By applying a special high-

FIGURE 11-12 PROMs use fusible links that can be selectively blown open by the user to program a logic 0 into a cell.

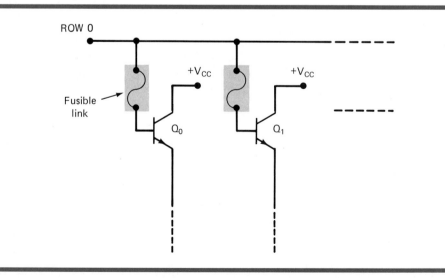

voltage programming pulse to the device, high-energy electrons are injected into the floating-gate region, thus turning the transistor "on." These electrons remain trapped in this region once the pulse is terminated, since there is no discharge path. During this programming process, the chip's address and data pins are used to determine which memory cells will be affected by the programming pulse.

Once a memory cell has been programmed, it can be erased only by exposing it to ultraviolet (UV) light applied through a window on the chip. The UV light causes a flow of photocurrent from the floating gate back to the silicon substrate, thereby restoring the gate to its initial condition. Note that there is no way to expose just a single cell to the UV light without exposing all the cells. Thus, the UV will erase the entire memory. The erasure process typically requires 15 to 30 minutes of exposure to the UV rays.

EPROMs are available in a wide selection with capacities up to 32K × 8 and access times down to 150 ns. We will use the popular Intel 2716 to illustrate typical EPROM operation. The 2716 is a 2K × 8 NMOS EPROM that operates from a single +5-V supply, unlike earlier EPROMs such as the 2708 which required three different supply voltages. The block symbol for the 2716 is shown in Figure 11-13. Note that it has eleven address inputs, since $2^{11} = 2048$, and eight data outputs. It has two power-supply inputs, V_{cc} and V_{PP}, both of which require +5 V during normal operation. V_{PP}, however, has to be set at +25 V during the programming process.

FIGURE 11-13 (a) Logic symbol for 2716 EPROM; (b) typical EPROM package showing ultraviolet window.

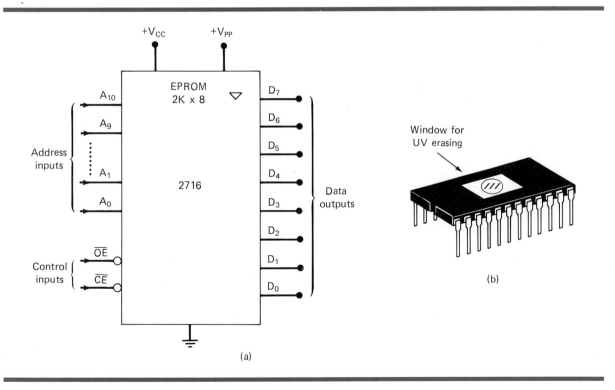

There are two control inputs. $\overline{OE}$ is the output enable which controls the output buffers to determine whether or not EPROM data appears at the output terminals. $\overline{CE}$ is a chip enable input that has two different functions. During normal operation, it acts as a power-control input that determines whether or not the chip will operate at all. To read from the EPROM, $\overline{CE}$ has to be LOW in order for the internal circuitry to select the data and route it to the output buffers, and $\overline{OE}$ has to be LOW for the data to appear at the output pins. When $\overline{CE}$ is HIGH, the chip is in a *standby* mode where it consumes much less power (132 mW) than it does in the active mode (525 mW).

$\overline{CE}$ is also used during the programming process where it has to be pulsed from LOW to HIGH for 50 ms each time a new word is to be programmed into one of the 2048 address locations. Figure 11-14 shows the conditions needed to program a data word into one location. Assume that the EPROM has previously been erased with UV light so that it is a "clean" EPROM (all cells are 1s). Note that V_{PP} is connected to +25 V. The steps required to program any address location are as follows:

1. Apply the desired address to the address inputs.

2. Apply the desired 8-bit data word to the data pins D_7–D_0. These data pins will function as inputs in the program mode (note $\overline{OE}$ is HIGH).

3. Apply a 50-ms LOW-to-HIGH pulse to $\overline{CE}$. At the termination of this pulse, the selected address location should be storing the applied data word.

FIGURE 11-14 To program a 2716, the desired data are applied to the data pins, and a program pulse is applied to $\overline{CE}$.

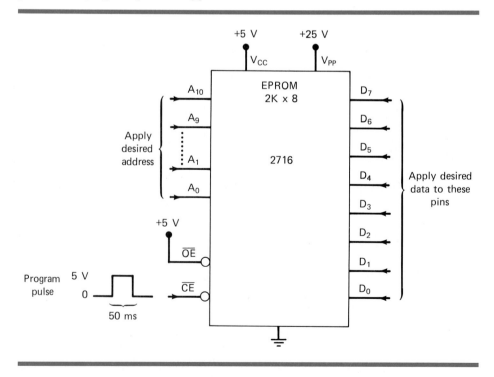

4. In order to verify that the data word has been programmed properly, the address location should be read. This is done by applying LOWs to $\overline{CE}$ and $\overline{OE}$ and reading the levels at the data output pins. *Note:* BEFORE THIS VERIFY STEP, THE DATA INPUTS THAT WERE APPLIED DURING THE PROGRAMMING STEPS MUST BE DISCONNECTED FROM THE DATA OUTPUT PINS.

The programming process when done manually can take several hours. Numerous commercial EPROM programmers are available that can program and verify an entire 2716 in less than two minutes, once the user has entered the data into the EPROM programmer's RWM from a keyboard. Some EPROM programmers allow data to be transferred into the RWM from a computer. This process, called *downloading,* allows the user to first develop his program on a computer and then, when it is finished and working, transfer it from the computer's memory to the EPROM programmer, which will "burn it" into the EPROM.

Electrically Erasable PROM (EEPROM)

As noted above, EPROMs have two major disadvantages. First, they have to be removed from their sockets in order to be erased and reprogrammed. Second, erasure removes the complete memory contents; this necessitates complete reprogramming even when only one memory word has to be changed. The *electrically erasable PROM* (EEPROM) was developed around 1980 as an improvement to the EPROM.

The EEPROM takes advantage of the same floating-gate structure as the EPROM. It adds the feature of *electrical* erasability through the addition of a thin oxide region above the drain of the MOSFET memory cell. By applying a HIGH voltage (21 V) between the MOSFET's gate and drain, a charge can be induced onto the floating gate, where it will remain even when power is removed. Reversal of the same voltage causes a removal of the trapped charges from the floating gate and erases the cell. Since this charge-transport mechanism requires very low currents, programming and erasing of an EEPROM can usually be done *in-circuit* (i.e., without a UV light source and PROM programmer unit).

A major advantage offered by EEPROMs over EPROMs is the ability to electrically erase and reprogram *individual* words in the memory array. Another advantage is that a complete EEPROM can be erased in about 10 ms (in circuit) versus about 30 minutes for an EPROM in external UV light. An EEPROM can also be programmed more rapidly; it requires only a 10-ms programming pulse for each data word, as compared to 50 ms for an EPROM.

Because the EEPROM can be erased and reprogrammed by applying appropriate voltages, we need not remove it from the circuit it is part of, provided that additional support components are also part of the circuitry. This support circuitry includes the 21-V programming voltage (V_{PP}), which is usually generated from the +5-V supply through a DC-to-DC converter, and circuitry to control the 10-ms timing and sequencing for the erasure and programming operations.

The Intel 2816 was the original EEPROM introduced in 1981. It has a 2K $\times$ 8 capacity, an access time of 250 ns, and the characteristics described above. The 2816 has the same pin layout as the 2716 EPROM and the same logic symbol (Figure 11-13), and they can be used interchangeably once they are programmed. The requirements for erasing and programming the 2816 are described in Intel's *Memory Components Handbook*.

Improved versions of the 2816 are currently available from Intel. The 2816A, 2817, and 2817A offer many improvements, the main one being the on-chip inclusion of the support circuitry needed for erasing and programming. This greatly reduces the amount of external circuitry needed to incorporate an EEPROM into a system.

REVIEW QUESTIONS

1. *True or false:* A MROM can be programmed by the user.
2. How does a PROM differ from an MROM? Can it be erased and reprogrammed?
3. *True or false:* PROMs are available in both bipolar and MOS versions.
4. How is an EPROM erased?
5. *True or false:* There is no way to erase only a portion of an EPROM's memory.
6. Outline the steps in programming and verifying a 2716 EPROM.
7. What are the advantages of an EEPROM over an EPROM?

11-9 ROM APPLICATIONS

ROMs can be used in any application requiring nonvolatile data storage where the data rarely or never have to change. We will briefly describe some of the most common application areas.

Microcomputer Program Storage (Firmware) At present, this is the most widespread application of ROMs. Personal and business microcomputers use ROMs to store their operating system programs and their language interpreters (e.g., BASIC) so that the computer can be used immediately after power is turned on. Products that include a microcomputer to control their operation use ROMs to store the control programs. Examples are electronic games, electronic cash registers, electronic scales, and microcomputer-controlled automobile fuel injection.

The microcomputer programs that are stored in ROM are referred to as *firmware* because they are not subject to change; programs that are stored in RWM are referred to as *software* because they can be easily altered.

Bootstrap Memory Some microcomputers and most larger computers do not have their operating system programs stored in ROM. Instead, these programs are stored in external mass memory, usually magnetic disk. How, then, do these computers know what to do when they are powered on? A relatively small program, called a *bootstrap program,* is stored in ROM. (The term "bootstrap" comes from the idea of pulling oneself up by one's own boot straps.) When the computer is powered on, it will execute the instructions that are in this bootstrap program. These instructions typically cause the CPU to initialize the system hardware. The bootstrap program then reads the operating system programs from mass storage into its main internal memory. At that point the computer begins executing the operating system program and is ready to respond to the user commands. This startup process is often called ''booting up the system.''

Implementing Combinatorial Logic   Once a ROM is programmed, the relationship between the address inputs and data outputs is fixed and will not change. This relationship can be described in a truth table like the one in Figure 11-6, which could just as easily represent the truth table for a combinatorial logic circuit. In other words, we can think of a ROM as acting like a combinatorial logic circuit that has a number of inputs equal to its address inputs and a number of outputs equal to its data outputs. The ROM can be programmed so that each data output represents a specific function of the inputs.

When used in this way, a ROM performs the same function as a PLA* with the same advantages of reduced chip count, board size, and improved reliability. A ROM is generally slower than a PLA, however, and this must be taken into account when deciding which one to use.

Data Tables   ROMs are often used to store tables of data that do not change. Some examples are the trigonometric tables (i.e., sine, cosine, etc.) and code-conversion tables.

Several standard ROM ''look-up'' tables are available with trig functions. One, the National Semiconductor MM4220BM, stores the sine function for angles between 0 and 90 degrees. This ROM is organized as a 128×8, with seven address inputs and eight data outputs. The address inputs represent the angle in increments of approximately 0.7°. For example, address 0000000 is 0°, address 0000001 is 0.7°, address 0000010 is 1.41°, and so on up to address 1111111, which is 89.3°. When an address is applied to the ROM, the data outputs will represent the approximate sine of the angle. For example, with input address 1000000 (representing approximately 45°) the data outputs will be 10110101. Since the sine is less than or equal to one, these data are interpreted as a fraction; that is, .10110101, which when converted to decimal equals .707 (the sine of 45°).

A *code converter* is a circuit that takes data expressed in one type of code and produces an output expressed in another type. Code conversion is needed, for example, when a computer is outputting data in straight binary code, and we want to convert it to BCD in order to display it on 7-segment LED readouts.

One of the easiest methods of code conversion uses a ROM programmed so that the application of a particular address (the old code) produces a data output that represents the equivalent in the new code. The 74185 is a TTL ROM that stores the binary-to-BCD code conversion for 6-bit binary input. To illustrate, a binary address input of 100110, (decimal 38) will produce a data output of 00111000, which is the BCD code for decimal 38. Problem 11.15 will deal with this ROM.

Character Generators   If you have ever looked closely at the alphanumeric characters (letters, numbers, etc.) printed on a computer's video display screen, you have noticed that each is generally made up of a group of dots. Depending on the

---

*Programmable logic array—to be presented in the next section.

character being displayed, some dot positions are made bright while others are dark. Each character is made to fit into a pattern of dot positions, usually arranged as a 5 × 7 or 7 × 9 matrix. The pattern of dots for each character can be represented as a binary code (i.e., bright dot = 1, dark dot = 0).

A *character-generator ROM* stores the dot pattern codes for each character at an address corresponding to the ASCII code for that character. For example, the dot pattern for the letter "A" would be stored at address 1000001, where 1000001 is the ASCII code for uppercase A. Character-generator ROMs are used extensively in any application that displays or prints out alphanumeric characters.

REVIEW QUESTIONS

1. What is firmware?
2. Describe how a computer uses a bootstrap program.
3. What is a code converter?
4. What kind of information is stored in a character-generator ROM?

11-10 PROGRAMMABLE LOGIC DEVICES

Logic designers have a wide range of standard ICs available to them with numerous logic functions and logic circuit arrangements on a chip. In addition, these ICs are available from many manufacturers and at a reasonably low cost. For these reasons, designers have been interconnecting standard ICs to form an almost endless variety of different circuits and systems and will continue to do so in the indefinite future.

However, there are some problems with circuit and system designs that use only standard ICs. Some system designs might require hundreds or thousands of these ICs. This large number of ICs requires a considerable amount of circuit board space, and a great deal of time and cost in inserting, soldering, and testing of the ICs. The equipment manufacturer must keep a significant inventory of all the different types of ICs that are used in the design. Finally, equipment that uses only standard ICs becomes relatively easy for competitors to copy.

The development of *programmable logic devices (PLDs)* have presented logic designers with an alternative to using standard ICs to overcome some of the above problems. A PLD is an IC that contains large numbers of gates, FFs, registers, and other logic functions that are all interconnected on the chip. (A typical PLD IC may have hundreds or thousands of gates.) Many of the interconnections are fusible links similar to those used in the PROMs discussed earlier. The IC is said to be *programmable* because the specific function of the IC for a given application is determined by the selective breaking of some of the interconnections while leaving others intact. The "fuse blowing" process can be done by the manufacturer in accordance with the customer's instructions, or it can be done by the customer. This process is called *programming* because it produces the desired circuit pattern interconnecting the gates, FFs, registers, etc.

PLAs One of the most common types of PLD is an IC that contains arrays of AND gates and OR gates that can be interconnected to produce the desired logic functions of several input variables. This type of IC is called a *programmable logic array (PLA)*. Figure 11-15 shows an example of a small PLA. In the interest of clarity, this example uses only a small number of inputs, outputs, and gates. It should be easy to see, however, how the basic structure can be expanded to a much larger scale.

This particular PLA has three input variables, A, B, and C, and three outputs,

FIGURE 11-15 Scaled-down version of a typical PLA structure.

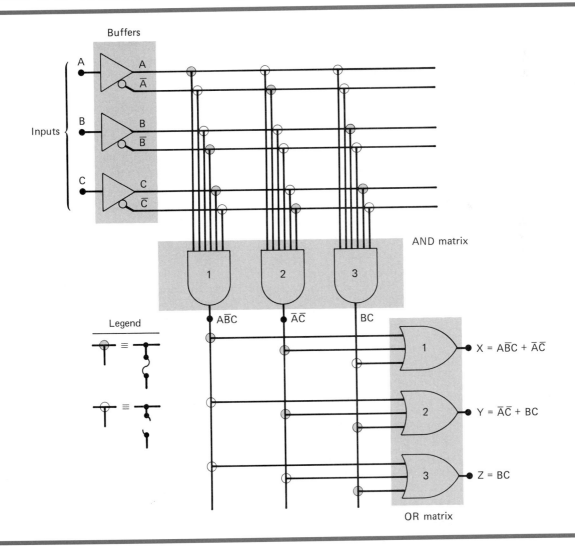

X, Y, and Z. The inputs are connected to buffers that provide both the normal and inverted forms of the variables for connection to a matrix of six-input AND gates. Each buffer output is originally connected to each AND gate; some of these connections, of course, will be removed during the programming process. For purposes of illustration, we have circled each of these connections, using open circles to indicate connections that have been opened during programming. Closed circles indicate connections that have been left intact. For example, the diagram shows that the $\overline{A}$, B, and $\overline{C}$ lines have been disconnected from AND-gate 1.

The AND outputs are connected to a matrix of three-input OR gates. Again, each AND output comes originally connected to each OR gate, and some of these interconnections will be opened during programming. In this illustration, for example, the connections from AND-gate 1 to OR-gates 2 and 3 have been opened. The results of selectively breaking some of the connections to the AND and OR gates are three outputs which are the desired functions of the input variables.

The example PLA in Figure 11-15 is simple compared to the many PLDs that are either currently available or under development, but even this simple example shows how PLDs can prove advantageous over standard ICs. The logic functions performed by the PLA in Figure 11-15 would require three standard ICs (such as 7400 NAND-gate chips).

An example of a commercially available PLA is the PLS100 from Signetics Corporation. It is a 28-pin chip that has 16 input variables and 8 output functions. It contains 144 AND gates and 128 OR gates, and when fully utilized, it replaces well over 100 standard ICs. The PLS100 also has the capability of selectively inverting any of the outputs by using EX-OR gates which can be programmed to invert or not invert. Another example is the PLS105 from Signetics. This IC is actually called a *programmable logic sequencer (PLS)* because it contains 14 edge-triggered FFs, as well as hundreds of AND and OR gates, that can be interconnected to produce a wide range of sequential logic circuits.

REVIEW QUESTIONS

1. What are some of the drawbacks in using only standard ICs in a complex design?
2. What is the basic idea of a PLD?

11-11 SEMICONDUCTOR RAMS

Recall that the term RAM stands for random-access memory, meaning that any memory address location is as easily accessible as any other. Many types of memory can be classified as having random access, but when the term RAM is used with semiconductor memories it is usually taken to mean read/write memory (RWM) as opposed to ROM. Since it is common practice to use RAM to mean semiconductor RWM, we will do so throughout the following discussions.

RAMs are used in computers for the *temporary* storage of programs and data. The contents of many RAM address locations will be changing continually as the computer executes a program. This requires fast read and write cycle times for the RAM so as not to slow down the computer operation.

A major disadvantage of RAMs is that they are volatile and will lose all stored information if power is interrupted or turned off. Some RAMs, however, use such small amounts of power in the standby mode (no read or write operations taking place) that they can be powered from batteries whenever the main power is interrupted. Of course, the main advantage of RAMs is that they can be written into and read from rapidly with equal ease.

The following discussion of RAMs will draw on some of the material covered in our treatment of ROMs, since many of the basic concepts are common to both types of memory.

11-12 RAM ARCHITECTURE

As with the ROM, it is helpful to think of the RAM as consisting of a number of registers, each storing a single data word, and each having a unique address. RAMs typically come with word capacities of 1K, 4K, 8K, 16K, 64K, 128K, 256K, and 1024K, and word sizes of 1, 4, or 8 bits. As we will see later, the word capacity and word size can be expanded by combining memory chips.

Figure 11-16 shows the simplified architecture of a RAM which stores 64 words of 4 bits each (i.e., a 64 $\times$ 4 memory). These words have addresses ranging from 0 to 63_{10}. In order to select one of the 64 address locations for reading or writing, a binary address code is applied to a decoder circuit. Since $64 = 2^6$, the decoder requires a 6-bit input code. Each address code activates one particular decoder output, which, in turn, enables its corresponding register. For example, assume an applied address code of

$$A_5A_4A_3A_2A_1A_0 = 011010$$

Since $011010_2 = 26_{10}$, decoder output 26 will go high, selecting register 26 for either a read or a write operation.

Read Operation   The address code picks out one register in the memory chip for reading or writing. In order to *read* the contents of the selected register, the READ/WRITE $(R/\overline{W})$* input must be a 1. In addition, the CHIP SELECT (CS) input must be activated (a 1 in this case). The combination of $R/\overline{W} = 1$ and $CS = 1$ enables the output buffers so that the contents of the selected register will appear at the four data outputs. $R/\overline{W} = 1$ also *disables* the input buffers so that the data inputs do not affect the memory during a read operation.

Write Operation   To write a new 4-bit word into the selected register requires $R/\overline{W} = 0$ and $CS = 1$. This combination *enables* the input buffers so that the 4-bit word applied to the data inputs will be loaded into the selected register. The $R/\overline{W} = 0$ also *disables* the output buffers, which are tristate, so that the data outputs are

---

*Some IC manufacturers use the symbol $\overline{WE}$ (write enable) instead of $R/\overline{W}$. In either case the operation is the same.

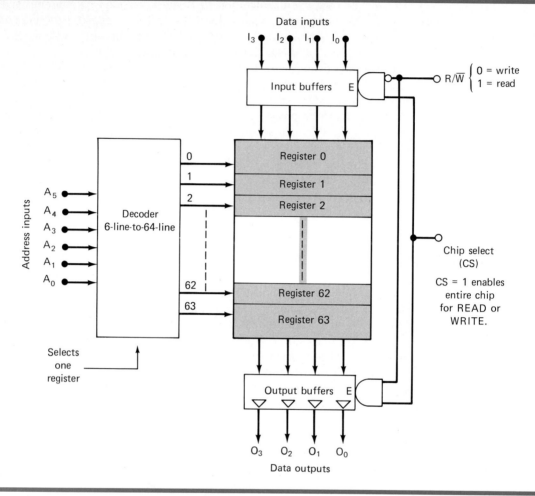

Data inputs

I_3 I_2 I_1 I_0

Input buffers   E

$R/\overline{W}$ $\begin{cases} 0 = \text{write} \\ 1 = \text{read} \end{cases}$

Address inputs

A_5
A_4
A_3
A_2
A_1
A_0

Decoder
6-line-to-64-line

0
1
2
62
63

Register 0
Register 1
Register 2
Register 62
Register 63

Chip select
(CS)

CS = 1 enables
entire chip
for READ or
WRITE.

Selects
one
register

Output buffers   E

O_3 O_2 O_1 O_0

Data outputs

FIGURE 11-16   **Internal organization of a 64 × 4 RAM.**

in their Hi-Z state during a write operation. The write operation, of course, destroys the word that was previously stored at that address.

Chip Select   Most memory chips have one or more *CS* inputs which are used to enable the entire chip or disable it completely. In the disabled mode all data inputs and data outputs are disabled (Hi-Z) so that neither a read or write operation can take place. In this mode the contents of the memory are unaffected. The reason for having *CS* inputs will become clear when we combine memory chips to obtain larger memories. Note that many manufacturers call these inputs CHIP ENABLE (*CE*) rather than *CS*.

Common Input/Output Pins   In order to conserve pins on an IC package, manufacturers often combine the data input and data output functions using common input/output pins. The $R/\overline{W}$ input controls the function of these I/O pins. During a read operation, the I/O pins act as data outputs which reproduce the contents of the selected address location. During a write operation, the I/O pins act as data inputs.

We can see why this is done by considering the chip in Figure 11-16. With separate input and output pins, a total of 18 pins is required (including ground and power supply). With four common I/O pins, only 14 pins are required. The pin savings becomes even more significant for chips with larger word size.

EXAMPLE 11-9

The 2125A is an NMOS RAM that is organized as a 1K $\times$ 1 with separate data input and output, and a single active-LOW chip-select input. Draw the logic symbol for this chip, showing all pin functions.

SOLUTION
The logic symbol is shown in Figure 11-17(a).

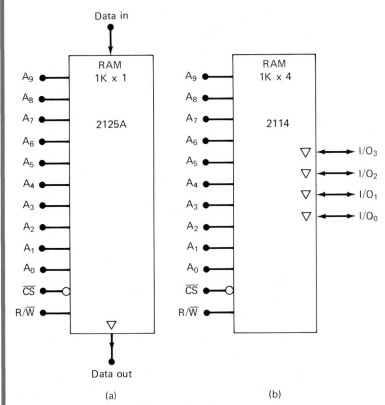

FIGURE 11-17 (a) Logic symbol for 2125 RAM; (b) for 2114 RAM.

EXAMPLE 11-10

The 2114 is an NMOS RAM organized as a 1K $\times$ 4 with common input/output pins and an active-LOW chip-select. Draw its logic symbol.

SOLUTION
The logic symbol is shown in Figure 11-17(b).

The architecture illustrated in Figure 11-16 for a 64 $\times$ 4 RAM will be somewhat different for larger-capacity RAMs. The registers will be arranged in a matrix such as shown for the ROM architecture in Figure 11.6. The address decoders will select the row and column of the register that is being acccessed for a read or write operation. This matrix architecture reduces the size of the required decoding circuitry.

REVIEW QUESTIONS

1. Describe the input conditions needed to read a word from a specific RAM address location.
2. Why do some RAM chips have common input/output pins?
3. How many pins are required for a 256 $\times$ 8 RAM with common I/O and one CS input? (*Ans.* 20, including V_{CC} and ground)

11-13 STATIC RAM

The RAM operation that we have been discussing up to this point applies to a *static* RAM—one that can store data as long as power is applied to the chip. Static-RAM memory cells are essentially flip-flops that will stay in a given state (store a bit) indefinitely, provided that power to the circuit is not interrupted. Later we will describe *dynamic* RAMs, which store data as charges on capacitors. With dynamic RAMs the stored data will gradually disappear because of capacitor discharge, so it is necessary to periodically refresh the data (i.e., recharge the capacitors).

Static RAMs are available in bipolar and MOS technologies, although the vast majority of applications use NMOS or CMOS RAMs. As stated earlier, the bipolars have the advantage in speed (though NMOS is gradually closing the gap), and MOS devices have much greater capacities. Figure 11-18 shows for comparison a typical bipolar static memory cell and a typical NMOS static memory cell. The bipolar cell contains two bipolar transistors and two resistors, while the NMOS cell contains four N-channel MOSFETs. The bipolar cell requires more chip area than the MOS cell because a bipolar transistor is more complex than a MOSFET, and because the bipolar cell requires separate resistors while the MOS cell uses MOSFETS as resistors (Q_3 and Q_4). A CMOS memory cell would be similar to the NMOS cell except that it would use P-channel *MOSFETs in place of* Q_3 and Q_4. This results in lower power consumption but increases the chip complexity.

Table 11-2 lists some of the standard static RAM ICs along with their major characteristics. Note that the ECL device has the shortest access time. Also note that the bipolar devices (TTL, ECL) are generally of lower capacity than the MOS devices. The CMOS RAMs have the lowest power requirements, and ECL has the highest. Some memory ICs have a standby (power-down) mode when they are not

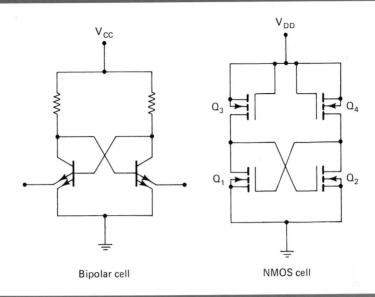

FIGURE 11-18 Typical bipolar and NMOS static RAM cells.

to be read from or written into. The numbers in parentheses represent the power consumption when a device is in its standby power-down mode.

Static RAM Timing RAM ICs are most often used as the internal memory of a computer. The CPU (central processing unit) continually performs read and write operations on this memory at a very fast rate determined by the limitations of the CPU. The memory chips that are interfaced to the CPU have to be fast enough to respond to the CPU read and write commands, and a computer designer has to be concerned with the RAM's various timing characteristics.

Not all RAMs have the same timing characteristics, but most of them are similar, so we will use a typical set of characteristics for illustrative purposes. The nomenclature for the different timing parameters will vary from one manufacturer to another, but the meaning of each parameter is usually easy to determine from the

TABLE 11-2 Standard Static Rams

DEVICE	TECHNOLOGY	ORGANIZATION	ACCESS TIME (μS)	POWER/BIT (μW)
10425	ECL	1K × 1	12	525
74LS89	TTL	16 × 4	37	3000
93425A	TTL	1K × 1	30	475
6116	CMOS	2K × 8	120	11 (.5)
51C69	CMOS	4K × 4	25	28
2114A	NMOS	1K × 4	100	50
2147H-1	NMOS	4K × 1	35	220 (35)
2149H	NMOS	1K × 4	35	180

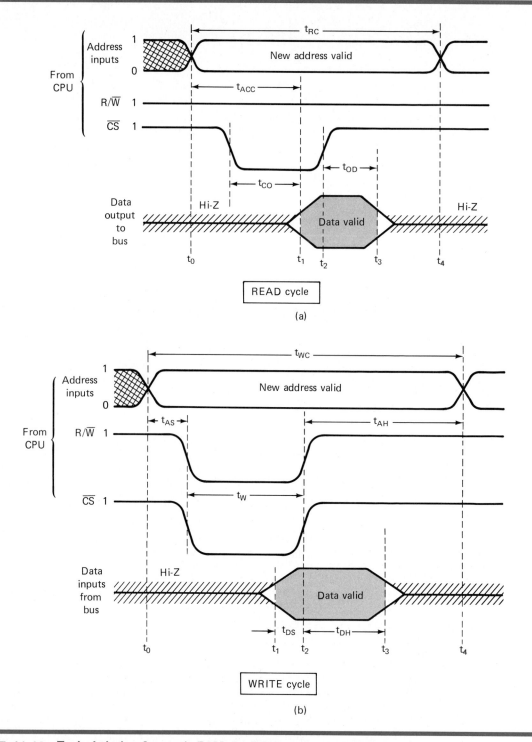

READ cycle

(a)

WRITE cycle

(b)

FIGURE 11-19 Typical timing for static RAM: (a) READ cycle; (b) WRITE cycle.

memory timing diagrams on the RAM data sheets. Figure 11-19 shows the timing diagrams for a complete read cycle and complete write cycle for a typical RAM chip.

Read Cycle   The waveforms in Figure 11-19(a) show how the address, $R/\overline{W}$, and chip-select inputs behave during a memory read cycle. As noted, the CPU supplies these input signals to the RAM when it wants to read data from a specific RAM address location. Although a RAM may have many address inputs coming from the CPU's address bus, for clarity the diagram shows only two. The RAM's data output is also shown; we will assume that this particular RAM has one data output. Recall that the RAM's data output is connected to the CPU data bus (Figure 11-5).

The read cycle begins at time t_0. Prior to that time, the address inputs will be whatever address is on the address bus from the preceding operation. Since the RAM's chip-select is not active, it will not respond to this "old" address. Note that the $R/\overline{W}$ line is HIGH prior to t_0 and stays HIGH throughout the read cycle. In most memory systems, $R/\overline{W}$ is normally kept in the HIGH state except when it is driven LOW during a write cycle. The RAM's data output is in its Hi-Z state since $\overline{CS} = 1$.

At t_0, the CPU applies a new address to the RAM inputs; this is the address of the location to be read. After allowing time for the address signals to stabilize, the $\overline{CS}$ line is activated. The RAM responds by placing the data from the addressed location onto the data output line at t_1. The time between t_0 and t_1 is the RAM's access time, t_{ACC}, and is the time between the application of the new address, and the appearance of valid output data. The timing parameter, t_{CO}, is the time it takes for the RAM output to go from Hi-Z to a valid data level once $\overline{CS}$ is activated.

At time t_2, the $\overline{CS}$ is returned HIGH, and the RAM output returns to its Hi-Z state after a time interval, t_{OD}. Thus, the RAM data will be on the data bus between t_0 and t_3. The device reading this data can take it from the data bus at any point during this interval. In most computers, the CPU will use the PGT of the $\overline{CS}$ signal at t_2 to latch this data into one of its internal registers.

The complete read cycle time, t_{RC}, extends from t_0 to t_4 when the CPU changes the address inputs to a different address for the next read or write cycle.

Write Cycle   Figure 11-19(b) shows the signal activity for a write cycle that begins when the CPU supplies a new address to the RAM at a time t_0. The CPU drives the $R/\overline{W}$ and $\overline{CS}$ lines LOW after waiting for a time interval t_{AS}, called the address setup time. This gives the RAM's address decoders time to respond to the new address. $R/\overline{W}$ and $\overline{CS}$ are held LOW for a time interval t_W, called the write-time interval.

During this write-time interval at time t_1 the CPU applies valid data to the data bus to be written into the RAM. This data has to be held at the RAM input for at least a time interval t_{DS} prior to, and for at least a time interval t_{DH} after, the deactivation of $R/\overline{W}$ and $\overline{CS}$ at t_2. The t_{DS} interval is called the data setup time, and t_{DH} is called the data hold time. Similarly, the address inputs have to remain stable for the address hold-time interval, t_{AH}, after t_2. If any of these setup or hold-time requirements are not met, the write operation will not take place reliably.

The complete write-cycle time, t_{WC}, extends from t_0 to t_4 when the CPU changes the address lines to a new address for the next read or write cycle.

REVIEW QUESTIONS

1. How does a static RAM differ from a dynamic RAM?
2. Which memory technology generally uses the least amount of power?
3. What device places data on the data bus during a read cycle?
4. What device places data on the data bus during a write cycle?
5. Describe the signal activity during a read operation.
6. Describe the signal activity during a write operation.

11-14 DYNAMIC RAM

Dynamic RAMs are fabricated using MOS technology and are noted for their high capacity, low power requirement and moderate operating speed. As we stated earlier, unlike static RAMs, which store information in FFs, dynamic RAMs store 1s and 0s as charges on a small MOS capacitor (typically a few picofarads). Because of the tendency for these charges to leak off after a period of time, dynamic RAMs require periodic recharging of the memory cells; this is called *refreshing* the dynamic RAM. Typically, each cell must be refreshed at least every 2 ms or its data will be lost.

The need for refreshing is a disadvantage of dynamic versus static RAMs because it adds a requirement to the memory system design. Up until recently, system designers had to include added circuitry to implement the memory refresh operation during time intervals when the memory was not being accessed for a read or write operation. Now there are two alternatives available to designers to help neutralize this disadvantage. For relatively small memories (<64K words) the *integrated RAM (iRAM)* provides a solution. An iRAM is an IC that includes the refresh circuitry on the same chip with the memory cell array. The result is a chip that externally operates just like a static RAM chip—you apply the address and collect the data—but internally uses the high-density dynamic RAM structure. The designer does not have to be concerned with the memory refresh operation, it is done internally and automatically.

For larger memory systems (>64K), a more cost-effective approach uses LSI chips called *dynamic memory controllers* which contain all of the necessary logic for refreshing the dynamic RAM chips that make up the system. This greatly reduces the amount of extra circuitry in a dynamic RAM system.

For applications where speed and reduced complexity are more critical than space and power considerations, static RAMs are still the best. They are generally faster than dynamic RAMs and require no refresh operation. They are simpler to design with, but they cannot compete with the higher capacity and lower power requirement of dynamic RAMs.

Because of their simple cell structure, dynamic RAMs typically have four times the density of static RAMs. This allows four times as much memory capacity to be placed on a single board, or alternatively, requires one-fourth as much board space for the same amount of memory. The cost per bit of dynamic RAM storage

TABLE 11-3

	2164 (DYNAMIC)	51C67 (STATIC)
Capacity (bits)	64K	16K
Access time (ns)	150	30
Active power (mW)	135	300
Standby power (mW)	17.5	15
Cost per bit (millicents)	45	250

is typically four or five times less than for static RAMs. A further cost saving is realized because the lower power requirements of a dynamic RAM, typically two to six times less than those of a static RAM, allow the use of smaller, less expensive power supplies. These advantages of dynamic RAMs are illustrated in Table 11-3, which compares two of the newer RAM chips.

REVIEW QUESTIONS

1. What is the main drawback of dynamic RAM compared to static?
2. List the advantages of dynamic versus static RAM.
3. What are the two means available to dynamic memory system designers for handling the refresh operation without the need for a lot of extra circuitry?

11-15   DYNAMIC RAM STRUCTURE AND OPERATION

Most dynamic RAMs (DRAMs) can be visualized as an array of single-bit cells as illustrated in Figure 11-20. Here there are 16,384 cells arranged in a 128×128 array. Each cell occupies a unique row and column position within the array. Fourteen address inputs are needed to select one of the cells ($2^{14} = 16,384$); the lower address bits, A_0–A_6, select the row, and the higher-order bits, A_7–A_{13}, select the column. Each 14-bit address selects a unique cell to be written into or read from. The structure in Figure 11-20 is a $16K \times 1$ DRAM chip. DRAM chips are currently available in capacities up to $1024K \times 1$. Since most DRAMs have a 1-bit or 4-bit word size, several chips have to be combined to produce larger word sizes. We will see how to do this later.

Figure 11-21 is a symbolic representation of a dynamic memory cell and its associated circuitry. Many of the circuit details are not shown, but this simplified diagram can be used to describe the essential ideas involved in writing to and reading from a DRAM. The switches S1 through S4 are actually MOSFETs that are controlled by various address decoder outputs and the $R/\overline{W}$ signal. The capacitor, of course, is the actual storage cell.

To write data to the cell, signals from the address decoding and read/write logic will close switches S1 and S2, while keeping S3 and S4 open. This connects

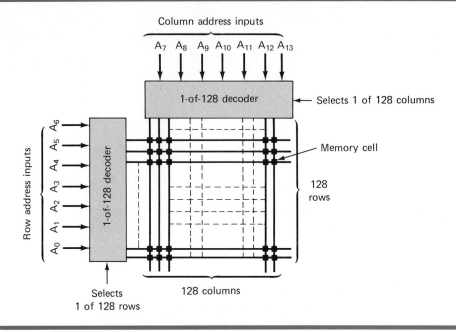

Column address inputs

A_7 A_8 A_9 A_{10} A_{11} A_{12} A_{13}

1-of-128 decoder

Selects 1 of 128 columns

Memory cell

128 rows

Row address inputs

A_6 A_5 A_4 A_3 A_2 A_1 A_0

1-of-128 decoder

Selects 1 of 128 rows

128 columns

FIGURE 11-20   Cell arrangement in 16K × 1 dynamic RAM.

the input data to C. A logic 1 at the data input charges C, and a logic 0 discharges it. Then the switches are open so that C is disconnected from the rest of the circuit. Ideally, C would retain its charge indefinitely, but there is always some leakage path through the off switches, so C will gradually lose its charge.

To read data from the cell, switches S2, S3, and S4 are closed and S1 is kept open. This connects the stored capacitor voltage to the *sense amplifier*. The sense amplifier compares the voltage to some reference value to determine if it is a logic 0 or 1, and produces a solid 0 or 1 for the data output. This data output is also connected to C (S2 and S4 are closed) and refreshes the capacitor voltage by recharging or discharging. In other words, the data in a memory cell is refreshed each time it is read.

FIGURE 11-21   Symbolic representation of a dynamic memory cell. During a WRITE operation, semiconductor switches S1 and S2 are closed. During a READ operation, all switches are closed except S1.

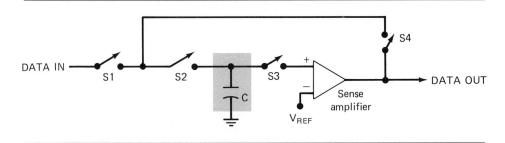

Address Multiplexing The 16K $\times$ 1 DRAM array depicted in Figure 11-20 has 14 address inputs. A 64K $\times$ 1 DRAM array would have 16 address inputs. High-capacity memory chips such as these would require many pins if each address input required a separate pin. In order to reduce the number of pins on their high-capacity DRAM chips, manufacturers utilize *address multiplexing* whereby each address input pin can accommodate two different address bits. The savings in pin count translates to a significant decrease in the size of the IC packages. This is very important in large-capacity memory boards, where you want to maximize the amount of memory that can fit on one board.

We will use the 2118 16K $\times$ 1 DRAM to illustrate the address multiplexing idea. A simplified diagram of this chip's internal architecture is shown in Figure 11-22(a). It contains an array of cells arranged as 128 rows by 128 columns. There is a single data input line, a single data output line, and a $R/\overline{W}$ input. There are seven address inputs, and each one has a dual function (e.g., A_0/A_7 will function as both A_0 and A_7). Two address *strobe* inputs are included for clocking the row and column addresses into their respective on-chip registers. The *row address strobe* $\overline{RAS}$ clocks the 7-bit row address register, and the *column address latch* $\overline{CAS}$ clocks the 7-bit column address register.

A 14-bit address is applied to this DRAM in two steps using $\overline{RAS}$ and $\overline{CAS}$. The timing is shown in Figure 11-22(b). Initially, $\overline{RAS}$ and $\overline{CAS}$ are both HIGH. At time t_0, the 7-bit row address (A_0–A_6) is applied to the address inputs. After allowing time for the setup time requirement (t_{RS}) of the row address register, the $\overline{RAS}$ input is driven LOW at t_1. This NGT loads the row address into the row address register so that A_0–A_6 now appear at the row decoder inputs. The LOW at $\overline{RAS}$ also enables this decoder so it can decode the row address and select one row of the array.

At time t_2, the 7-bit column address (A_7–A_{13}) is applied to the address inputs. At t_3, the $\overline{CAS}$ input is driven LOW to clock the column address into the column address register. $\overline{CAS}$ also enables the column decoder so that it can decode the column address and select one column of the array.

At this point the two parts of the address are in their respective registers, the decoders have decoded them to select the one cell corresponding to the row and column address, and a read or a write operation can be performed on that cell just as in a static RAM.

You may have noticed that this DRAM does not have a chip-select (CS) input. The $\overline{RAS}$ and $\overline{CAS}$ signals perform the chip-select function since they must both be LOW for the decoders to select a cell for reading or writing.

EXAMPLE 11-11

How many pins are saved by using the address multiplexing for the 16K $\times$ 1 DRAM?

SOLUTION
Seven address inputs are used instead of 14; $\overline{RAS}$ and $\overline{CAS}$ are added; no CS is required. Thus, there is a net savings of *six* pins.

In a typical computer system, the address inputs to the memory system come from the central processing unit (CPU). When the CPU wants to access a particular

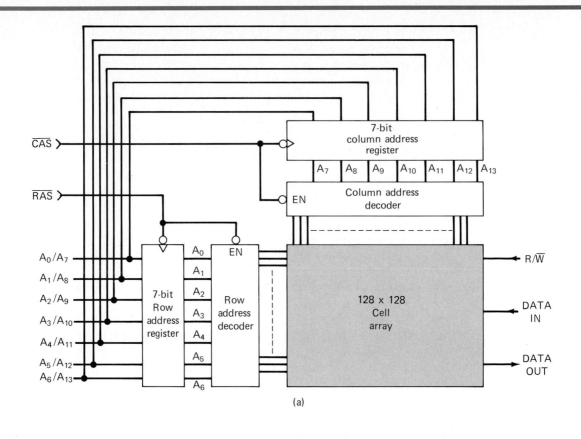

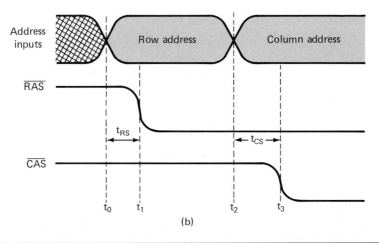

FIGURE 11-22 (a) Simplified architecture of the 2118, a 16K × 1 DRAM; (b) $\overline{RAS}/\overline{CAS}$ timing.

memory location, it generates the complete address and places it on address lines that make up an *address bus*. Figure 11-23 (a) shows this for a memory that has a capacity of 16K words and therefore requires a 14-line address bus going directly from the CPU to the memory.

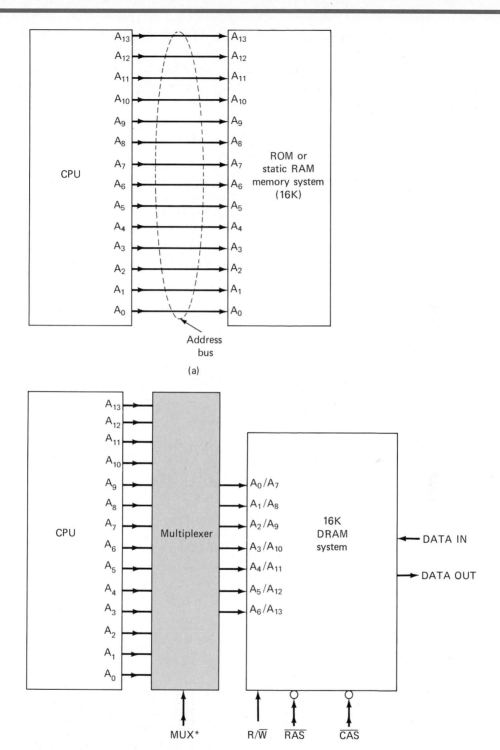

FIGURE 11-23 (a) CPU address bus driving ROM or static RAM memory; (b) CPU addresses driving
DRAM with multiplexer used to multiplex the CPU address lines.

This arrangement works for ROM or for static RAM, but it has to be modified for DRAM that uses multiplexed addressing. If the memory is DRAM, it will have only 7 address inputs. This means that the 14 address lines from the CPU address bus have to be fed into a multiplexer circuit that will transmit 7 address bits at a time to the memory address inputs. This is shown symbolically in Figure 11-23(b). The multiplexer select input, labeled *MUX,* controls whether CPU address lines A_0–A_6 or address lines A_7–A_{13} will be present at the DRAM address inputs.

The timing of the *MUX* signal has to be synchronized with the $\overline{RAS}$ and $\overline{CAS}$ signals that clock the addresses into the DRAM. This is shown in Figure 11-24. *MUX* has to be LOW when $\overline{RAS}$ is pulsed LOW so that address lines A_0–A_6 from the CPU will reach the DRAM address inputs to be latched on the NGT of $\overline{RAS}$. Likewise, *MUX* has to be HIGH when $\overline{CAS}$ is pulsed LOW so that A_7–A_{13} from the CPU will be present at the DRAM inputs to be latched on the NGT of $\overline{CAS}$.

The actual multiplexing and timing circuitry will not be shown here but will be left to the end-of-chapter problems (Problems 11-20 and 11-21).

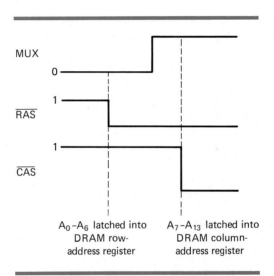

FIGURE 11-24 **Timing required for address multiplexing.**

REVIEW QUESTIONS

1. Describe the array structure of a 64K $\times$ 1 DRAM. (*Ans.* 256 rows by 256 columns)
2. What is the benefit of address multiplexing?
3. How many address inputs would there be on a 256K $\times$ 1 DRAM chip? (*Ans.* Nine)
4. What are the functions of the *RAS* and *CAS* signals?
5. What is the function of the *MUX* signal?

11-16 DRAM READ/WRITE CYCLES

The timing of the read and write operations of a DRAM is much more complex than for a static RAM, and there are many critical timing requirements that the DRAM memory designer has to consider. At this point, a detailed discussion of these requirements would probably cause more confusion than enlightenment. We will concentrate on the basic timing sequence for the read and write operations for a typical DRAM system like that of Figure 11-23(b).

DRAM Read Cycle   Figure 11-25 shows typical signal activity during the read operation. It is assumed that $R/\overline{W}$ is in its HIGH state throughout the operation. The following is a step-by-step description of the events that occur at the times indicated on the diagram.

t_0: MUX is driven LOW to apply the row address bits (A_0–A_6) to the DRAM address inputs.

t_1: $\overline{RAS}$ is driven LOW to latch the row address into the DRAM.

FIGURE 11-25   Signal activity for a read operation on a dynamic RAM. The $R/\overline{W}$ input (not shown) is assumed to be HIGH.

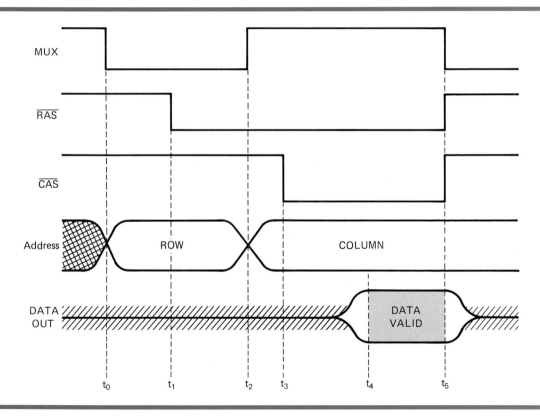

t₂: *MUX* goes HIGH to place the column address (A_7–A_{13}) to the DRAM address inputs.

t₃: $\overline{CAS}$ goes LOW to latch the column address into the DRAM.

t₄: The DRAM responds by placing valid data from the selected memory cell onto the DATA OUT line.

t₅: *MUX*, $\overline{RAS}$, $\overline{CAS}$, and DATA OUT return to their initial states.

DRAM Write Cycle   Figure 11-26 shows typical signal activity during a DRAM write operation. Here is a description of the sequence of events.

t₀: The LOW at *MUX* places the row address at the DRAM inputs.

t₁: The NGT at $\overline{RAS}$ clocks the row address into the DRAM.

t₂: *MUX* goes HIGH to place the column address at the DRAM inputs.

t₃: The NGT at $\overline{CAS}$ latches the column address into the DRAM.

FIGURE 11-26   Signal activity for a write operation on a dynamic RAM.

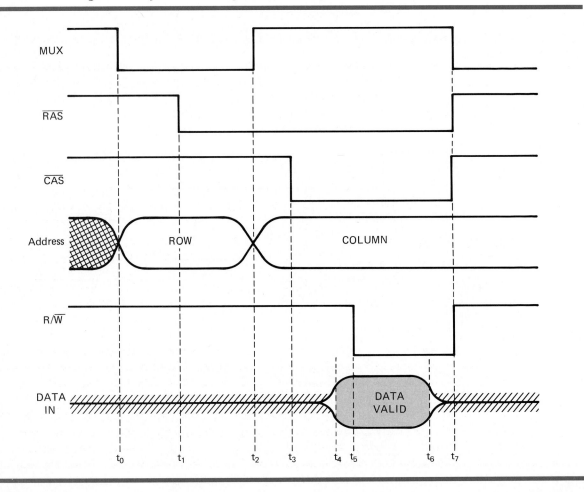

t₄: Data to be written is placed on the DATA IN line.

t₅: $R/\overline{W}$ is pulsed LOW to write the data into the selected cell.

t₆: Input data is removed from DATA IN.

t₇: MUX, $\overline{RAS}$, $\overline{CAS}$, and $R/\overline{W}$ are returned to their initial states.

REVIEW QUESTIONS

1. List the steps in a read operation from DRAM.
2. List the steps in a write operation from DRAM.

11-17 DRAM REFRESHING

DRAM cells have to be refreshed periodically (typically, every 2 ms) or the stored data will be lost. As we saw in our discussion of the DRAM cell of Figure 11-21, a DRAM cell is refreshed every time a read operation is performed on that cell. It would appear then that for a 16K × 1 DRAM it would be necessary to perform read operations at the rate of one every 122 ns (2 ms/16384 = 122 ns). This is much too fast for most currently available DRAMs, and even if the DRAMs were fast enough, it is highly unlikely that during normal operation every cell will be read from. For this reason, manufacturers have designed their DRAM chips so that

whenever a read operation is performed on a cell, all the cells in that same row will be refreshed.

This greatly reduces the number of read operations that must take place in order to refresh the complete memory; it is only necessary to read from each of the 128 rows once every 2 ms. It is still highly unlikely that during normal operation, each of the 128 rows will be read from, so the DRAM refresh operation has to be performed by some other means.

The most common means is provided by a 7-bit *refresh counter* which is used to cycle through the 128 different row addresses. The counter will start at 0000000 corresponding to row 0. This address is applied to the DRAM address inputs (MUX = 0), and $\overline{RAS}$ is pulsed LOW (see Figure 11-27) while $R/\overline{W}$ and $\overline{CAS}$ are kept HIGH. This refreshes row 0. The counter is incremented and the sequence is repeated up to row 127. This complete refresh process can be done in about 50 μs.

While the refresh counter idea seems easy enough, we must realize that the row addresses from the refresh counter cannot interfere with the addresses coming from the CPU during normal read/write operations. For this reason, the refresh counter addresses have to be multiplexed with the CPU addresses, so that the proper source of DRAM addresses is activated at the proper times.

Most manufacturers of dynamic memory ICs have developed special ICs to handle the refresh operation as well as the address multiplexing needed by DRAM systems. These ICs are called *dynamic RAM controllers*. We will look at a typical

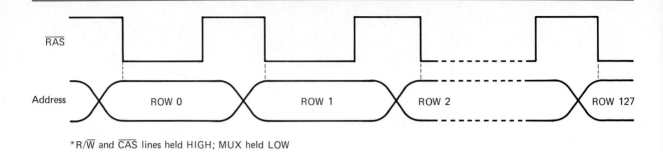

RAS

Address

ROW 0   ROW 1   ROW 2   ROW 127

*R/$\overline{W}$ and $\overline{CAS}$ lines held HIGH; MUX held LOW

FIGURE 11-27   A 16K DRAM can be refreshed by sequential latching of row addresses using $\overline{RAS}$.

IC of this type, the Intel 3242, which is designed for use with 16K DRAMs. The block diagram in Figure 11-28 shows the 3242 used with a 16K × 1 DRAM.

The 3242 outputs a 7-bit multiplexed address to the DRAM's address inputs. There are three possible sources for this address. The 7-bit refresh counter is incremented by an external clock signal at its $\overline{COUNT}$ input. This counter supplies the row addresses to the DRAM during a refresh operation. The 3242 also takes the 14-bit address from the CPU address bus and multiplexes it into the row and column addresses that are used when the CPU performs a read or write operation on the DRAM. The logic levels applied to the REFRESH ENABLE and ROW ENABLE inputs determine which 7-bit address will appear at the controller outputs according to the table given in the diagram.

REVIEW QUESTIONS

1. What is the function of a refresh counter?
2. What functions does a DRAM controller perform?

11-18 EXPANDING WORD SIZE AND CAPACITY

In most IC memory applications the required memory capacity or word size cannot be satisfied by one memory chip. Instead, several memory chips have to be combined to provide the desired capacity and word size. We will see how this is done through several examples that illustrate all the important concepts that will be needed when we interface memory chips to a microprocessor.

Expanding Word Size   Suppose we need a memory that can store sixteen 8-bit words and all we have are RAM chips which are arranged as 16 × 4 memories with common I/O lines. We can combine two of these 16 × 4 chips to produce the desired memory. The configuration for doing so is shown in Figure 11-29. Examine this diagram carefully and see what you can find out from it before reading on.

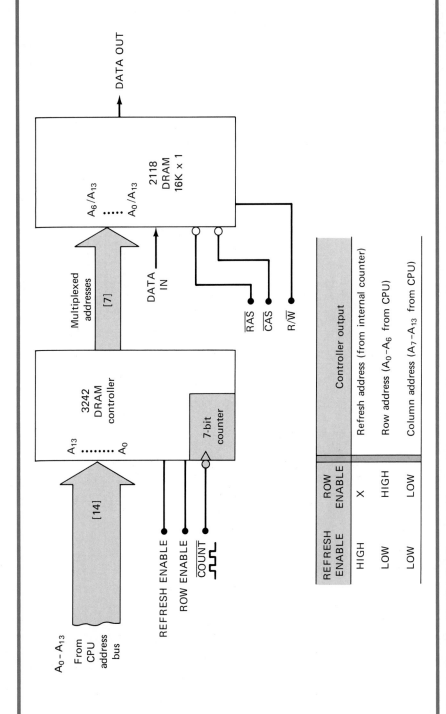

FIGURE 11-28 The 3242 DRAM controller performs the address multiplexing and refresh address counting for a 16K DRAM.

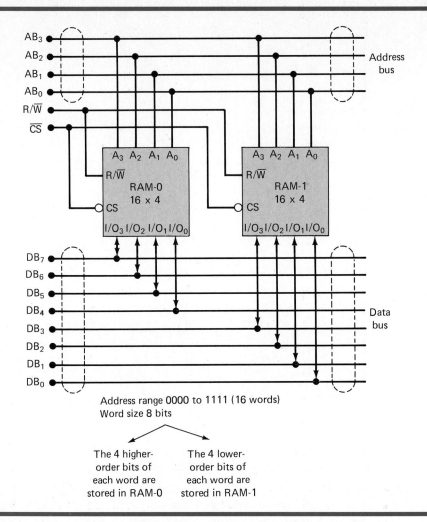

FIGURE 11-29 **Combining two 16 × 4 RAMs for a 16 × 8 module.**

Since each chip can store sixteen 4-bit words and we want to store sixteen 8-bit words, we are using each chip to store *half* of each word. In other words, RAM-0 stores the four *higher*-order bits of each of the 16 words, and RAM-1 stores the four *lower*-order bits of each of the 16 words. A full 8-bit word is available at the RAM outputs connected to the data bus.

Any one of the 16 words is selected by applying the appropriate address code to the four-line *address bus* (AB_3, AB_2, AB_1, AB_0). The address lines typically originate at the CPU. Note that each address bus line is connected to the corresponding address input of each chip. This means that once an address code is placed on the address bus, this same address code is applied to both chips so that the same location in each chip is accessed at the same time.

Once the address is selected, we can read or write at this address under control of the common $R/\overline{W}$ and $\overline{CS}$ line. To read, $R/\overline{W}$ must be high and $\overline{CS}$ must be low. This causes the RAM I/O lines to act as *outputs*. RAM-0 places its selected 4-bit word on the upper four data bus lines and RAM-1 places its selected 4-bit word on the lower four data bus lines. The data bus then contains the full selected 8-bit word, which can now be transmitted to some other device (usually a register in the CPU).

To write, $R/\overline{W} = 0$ and $\overline{CS} = 0$ causes the RAM I/O lines to act as *inputs*. The 8-bit word to be written is placed on the data bus (usually by the CPU). The higher 4 bits will be written into the selected location of RAM-0 and the lower 4 bits will be written into RAM-1.

In essence, the combination of the two RAM chips acts like a single 16×8 memory chip. We would refer to this combination as a 16×8 memory module.

The same basic idea for expanding word size will work for many different situations. Read the following example and draw a rough diagram for what the system will look like before looking at the solution.

EXAMPLE 11-12

Show how to combine several 2125A static RAM ICs (Figure 11-17) to form a $1K \times 8$ module.

SOLUTION

The arrangement is shown in Figure 11-30 where eight 2125A chips are used (see page 666). Each chip stores one of the bits of each of the 1024 8-bit words. Note that all of the $R/\overline{W}$ and $\overline{CS}$ inputs are wired together, and the 10-line address bus is connected to the address inputs of each chip. Also note that since the 2125A has separate data in and data out pins, both of these pins of each chip are tied to the same data bus line.

Expanding Capacity   Suppose we need a memory that can store 32 4-bit words and all we have are the 16×4 chips. By combining two 16×4 chips as shown in Figure 11-31, we can produce the desired memory. Once again, examine this diagram and see what you can determine from it before reading on.

Each RAM is used to store sixteen 4-bit words. The data I/O pins of each RAM are connected to a common four-line data bus. Only one of the RAM chips can be selected (enabled) at one time so that there will be no bus-contention problems. This is ensured by driving the respective $\overline{CS}$ inputs from different logic signals.

Since the total capacity of this memory module is 32×4, there have to be 32 different addresses. This requires *five* address bus lines. The upper address line AB_4 is used to select one RAM or the other (via the $\overline{CS}$ inputs) as the one that will be read from or written into. The other four address lines AB_0–AB_3 are used to select the one memory location out of 16 from the selected RAM chip.

To illustrate, when $AB_4 = 0$, the $\overline{CS}$ of RAM-0 enables this chip for read or write. Then, any address location in RAM-0 can be accessed by AB_3–AB_0. The latter four address lines can range from 0000 to 1111 to select the desired location. Thus, the range of addresses representing locations in RAM-0 are

$$AB_4AB_3AB_2AB_1AB_0 = 00000 \text{ to } 01111$$

Note that when $AB_4 = 0$, the $\overline{CS}$ of RAM-1 is high, so that its I/O lines are disabled and cannot communicate (give or take data) with the data bus.

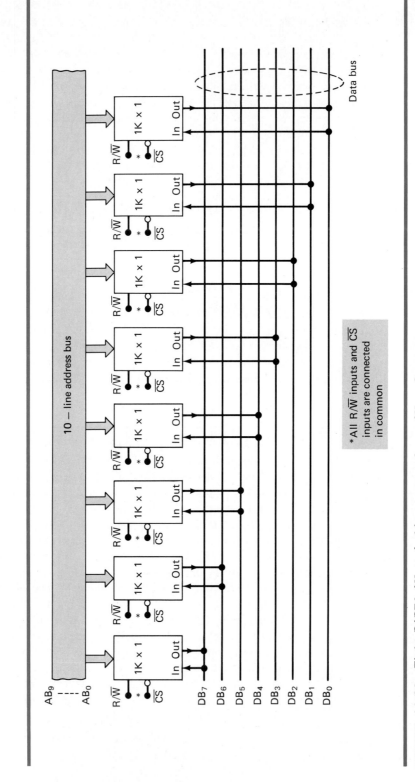

FIGURE 11-30 Eight 2125A 1K × 1 chips arranged as a 1K × 8 memory.

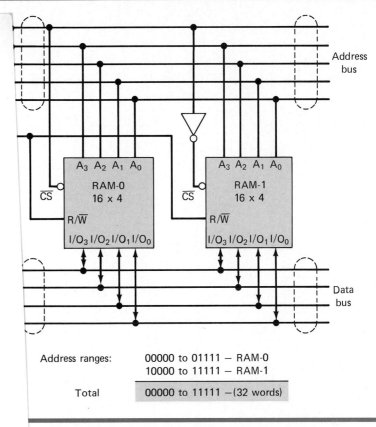

Address ranges:	00000 to 01111 — RAM-0
	10000 to 11111 — RAM-1
Total	00000 to 11111 —(32 words)

Combining two 16 × 4 chips for a 32 × 4 memory.

be clear that when $AB_4 = 1$, the roles of RAM-0 and RAM-1 are
-1 is now enabled and the AB_3–AB_0 lines select one of its locations.
Thus, the range of addresses located in RAM-1 is

$$AB_4 AB_3 AB_2 AB_1 AB_0 = 10000 \text{ to } 11111$$

EXAMPLE 11-13

It is desired to combine several 2K × 8 PROMs to produce a total capacity of 8K × 8. How many PROM chips are needed? How many address bus lines are required?

SOLUTION

Four PROM chips are required, with each one storing 2K of the 8K words. Since $8K = 8 \times 1024 = 8192 = 2^{13}$, *thirteen* address lines are needed.

The configuration for the memory of Example 11-13 is similar to the 32 × 4 memory of Figure 11-31. However, it is slightly more complex, because it requires a decoder circuit for generating the $\overline{CS}$ input signals. The complete diagram for this 8192 × 8 memory is shown in Figure 11-32.

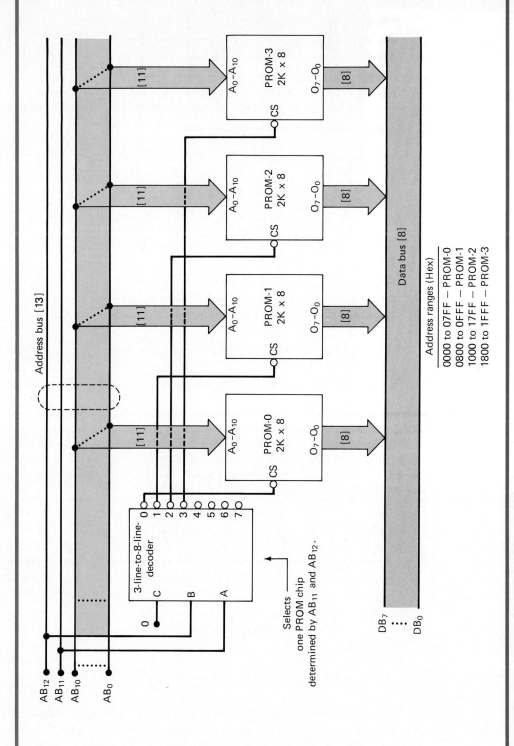

FIGURE 11-32 Four 2K × 8 PROMs arranged to form a total capacity of 8K × 8.

Since the total capacity is 8192 words, 13 address bus lines are required. The two highest-order lines, AB_{11} and AB_{12} are used to select *one* of the PROM chips; the other eight address bus lines go to each PROM to select the desired location within the selected PROM. The PROM selection is accomplished by feeding AB_{11} and AB_{12} into the decoder circuit. The four possible combinations are decoded to generate active-LOW signals, which are applied to the $\overline{CS}$ inputs. For example, when $AB_{11} = AB_{12} = 0$, the 0 output of the decoder goes low (all others are high) and enables PROM-0. This causes the PROM-0 outputs to generate the data word internally stored at the address determined by AB_0–AB_{10}. All other PROMs are disabled, so there is no bus contention.

While $AB_{12} = AB_{11} = 0$, the values of AB_{10}–AB_0 can range from all 0s to all 1s. Thus, PROM-0 will respond to the following range of 13-bit addresses

$$AB_{12}\text{–}AB_0 = 0000000000000 \text{ to } 0011111111111$$

For convenience, these addresses can be more easily expressed in hexadecimal code to give a range of 0000–07FF.

Similarly, when $AB_{12} = 0$ and $AB_{11} = 1$, the decoder selects PROM-1, which then responds by outputting the data word it has internally stored at the address AB_{10}–AB_0. Thus, PROM-1 responds to the following range of addresses:

$$0100000000000 \text{ to } 0111111111111 \text{ (binary)}$$

or

$$0800 \text{ to } 0FFF \text{ (hex)}$$

You should verify the PROM-2 and PROM-3 address ranges given in the figure.

Clearly, address lines AB_{12} and AB_1, are used to select one of the four PROM chips, while AB_{10}–AB_0 selects the word stored in the selected PROM.

EXAMPLE 11-14

What size decoder would be needed to expand the memory of Figure 11-32 to 32K $\times$ 8? Describe what address lines are used.

SOLUTION
A 32K capacity will require 16 PROM chips. To select one of the 16 PROMs will require a 4-line-to-16-line decoder. Four address lines (AB_{14}, AB_{13}, AB_{12}, AB_{11}) will be connected as inputs to this decoder. The address lines AB_{10} to AB_0 are connected to the address inputs of each of the 16 PROMs. Thus, a total of 15 address lines are used. This agrees with the fact that $2^{15} = 32,768 = 32$K.

Combining DRAMs   As we stated earlier, DRAM ICs usually come with word sizes of 1 or 4 bits. In order to use these ICs in computer systems requiring word sizes of 8 or 16 bits, it is necessary to combine several of them in a similar manner to what we have done with the static RAMs and ROMs. Figure 11-33 shows how to combine eight 16K $\times$ 1 DRAMs to form a 16K $\times$ 8 memory module. The 3242 DRAM controller is used to provide the address multiplexing and refresh functions for each chip.

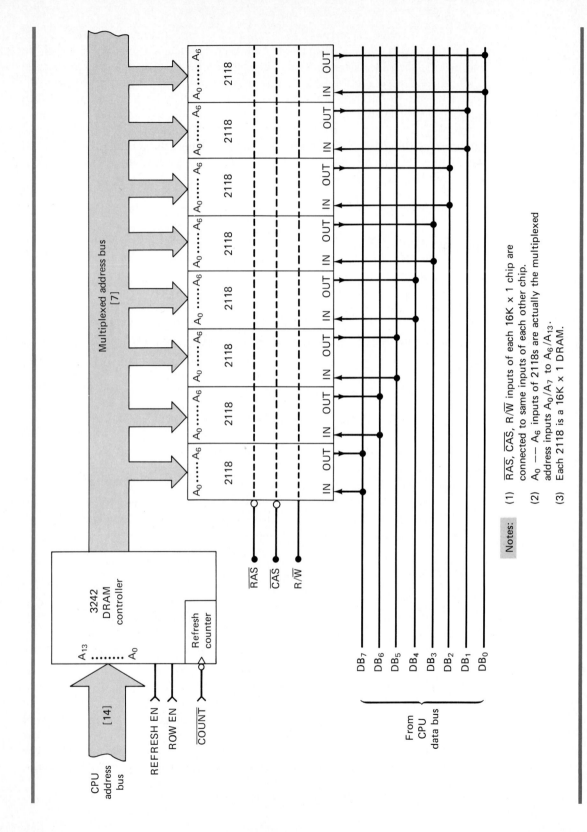

Notes:

(1) $\overline{RAS}$, $\overline{CAS}$, R/$\overline{W}$ inputs of each 16K × 1 chip are connected to same inputs of each other chip.

(2) A_0 — A_6 inputs of 2118s are actually the multiplexed address inputs A_0/A_7 to A_6/A_{13}.

(3) Each 2118 is a 16K × 1 DRAM.

FIGURE 11-33 Eight 16K × 1 DRAM chips combined to form a 16K × 8 memory module.

REVIEW QUESTIONS

1. The 2114 is a 1K $\times$ 4 static RAM chip. How many of these are needed to form a 16K $\times$ 4 module? (*Ans.* 16)
2. How many 2114s are needed for a 1K $\times$ 16 module? (*Ans.* Four)
3. *True or false:* To increase memory word size, the ICs are connected so that each one is connected to different data bus lines.

11-19 SEQUENTIAL MEMORIES

The semiconductor memories that we have discussed are random access memories because the data from any memory location can be obtained quickly without sequencing through other memory locations. The high-speed operation of random-access devices makes them suitable for use as the internal memory of a computer. Sequential access semiconductor memories utilize shift registers to store data that can be accessed in a sequential manner. Although they are not useful as internal computer memory because of their relatively slow speed, shift register memories find application in areas where sequential, repetitive data is required. A primary example is the storage and sequential transmission of ASCII-coded data for the characters on a video display. This data has to be supplied to the video display circuits periodically in order to refresh the displayed image on the screen. By using shift registers, the stored data can be *recirculated* to refresh the screen image periodically.

Recirculating Shift Register Figure 11-34(a) shows the block diagram for a recirculating shift register. There can be any number of FFs in the shift register, and we will use four FFs for our illustrations. Data enters the shift register from the serial input D_S, which shifts into Q_3, Q_3 shifts into Q_2, Q_2 into Q_1, and Q_1 into Q_0. The Q_0 output is recirculated back to the serial input through some control logic. This logic provides two modes of operation that are controlled by the *recirculation* input REC. The level at REC determines the source of the data that will reach the serial input.

Recirculate mode (REC = 1) In this mode, the upper AND gate is enabled, and the Q_0 output (DATA OUT) is applied to D_S. As clock pulses are applied, the data in the shift register will recirculate as shown below.

$$Q_3 \rightarrow Q_2 \rightarrow Q_1 \rightarrow Q_0 \rightarrow \text{DATA OUT}$$

As it recirculates in the register, the data also appears at DATA OUT, one bit at a time. In this mode, DATA IN is inhibited and has no effect on the register data.

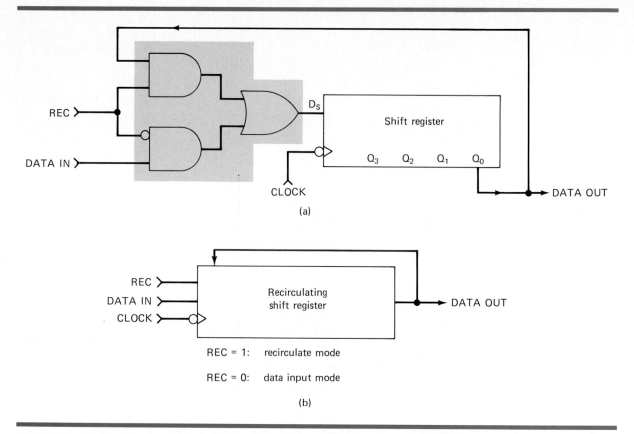

FIGURE 11-34   (a) Recirculating shift register; (b) simplified representation.

Data Input Mode (REC = 0)   In this mode, the lower AND gate is enabled, and the DATA IN signal is applied to D_S. As clock pulses are applied, the data will shift as shown below:

$$\text{DATA IN} \rightarrow Q_3 \rightarrow Q_2 \rightarrow Q_1 \rightarrow Q_0 \rightarrow \text{DATA OUT}$$

There is no recirculation of data since the upper AND gate is inhibited by $REC = 0$. This mode is used to enter new data at DATA IN for storage in the register.

Figure 11-34(b) is a simplified symbol that we will use for the recirculating shift register. The control logic is understood to be built into the symbol.

EXAMPLE 11-15

Assume the shift register data $Q_3Q_2Q_1Q_0 = 1011$, $REC = 1$, and DATA IN $= 0$. What will be the level at DATA OUT after the application of five clock pulses?

SOLUTION

In the recirculate mode, DATA IN will have no effect. The table below shows how the register FFs change as clock pulses are applied.

Q_3	Q_2	Q_1	Q_0	
1	0	1	1	Before 1st clock pulse
1	1	0	1	After 1st clock pulse
1	1	1	0	After 2nd clock pulse
0	1	1	1	After 3rd clock pulse
1	0	1	1	After 4th clock pulse
1	1	0	1	After 5th clock pulse

After the fifth pulse Q_0 is a 1, so DATA OUT $= 1$.

EXAMPLE 11-16

Figure 11-35 shows typical input and output waveforms for the recirculating shift register. It is assumed that the register was initially reset to all 0s. Describe the sequence of events depicted by these waveforms.

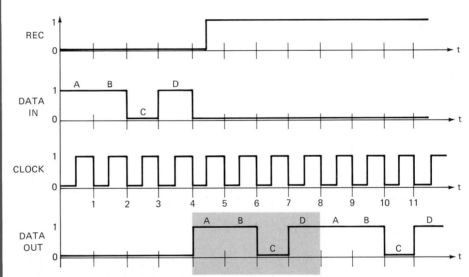

FIGURE 11-35   Input and output waveforms for the two operating modes of the recirculating shift register.

SOLUTION

During the first four clock cycles $REC = 0$ selects the data input mode. During this interval, data from the DATA IN line is shifted in one bit at a time (A, B, C, then D) on the NGTs of CLOCK. DATE OUT will stay at 0 until the fourth NGT shifts bit A into Q_0. Before the fifth NGT of CLOCK, REC goes HIGH to begin the recirculate mode. For all subsequent clock cycles the data within the register will recirculate so that the sequence of bits A, B, C, and D will appear at DATA OUT repetitively.

Figure 11-36 shows how seven recirculating shift registers are combined to store 7-bit ASCII-coded data for repetitive transmission to the circuits of a video display. The REC and CLOCK inputs of the registers are tied together so that all

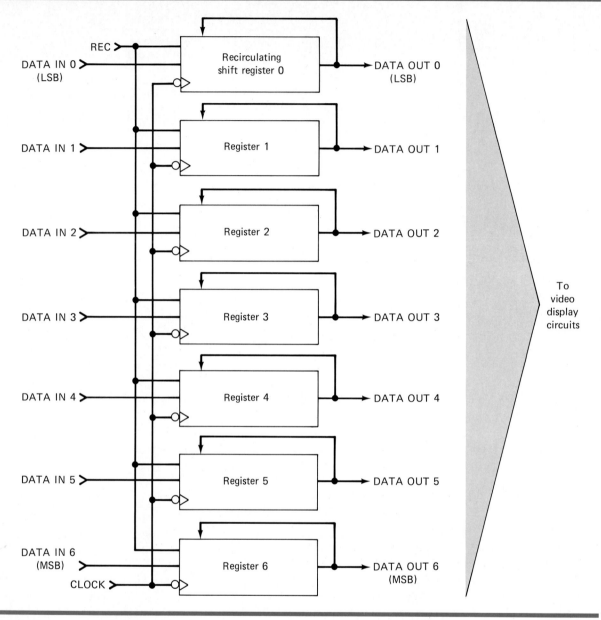

FIGURE 11-36 Combining several recirculating shift registers to store and repetitively transmit 7-bit data words.

registers will be in the same mode and will shift data at the same rate. The seven DATA IN lines represent ASCII data words that are entered into the registers for storage in the data input mode ($REC = 0$). Register 0 gets the LSB of each ASCII data word, register 1 gets the second LSB of each word, and so on. As clock pulses are applied, the seven bits of a given ASCII data word shift along through the seven registers so that they are always at the same relative position in their respective registers. The seven DATA OUT lines will reflect the ASCII data words as they are shifted out of the registers and sent to refresh the video display as well as recirculated back to the input.

EXAMPLE 11-17

A certain video display system displays 24 lines of 80 characters. How does this affect the arrangement of Figure 11-36?

SOLUTION
A total of $24 \times 80 = 1920$ characters are to be displayed on the screen. This means that 1920 ASCII codes are to be stored and recirculated by the sequential memory. Thus, each shift register has to have a length of 1920 bits.

EXAMPLE 11-18

What is the storage capacity of a shift register memory that uses eight 512-bit registers?

SOLUTION
The number of registers is the word size. The number of bits per register is the number of words. Thus, the total capacity is 512×8, or 512 8-bit words.

First-In First-Out Memories (FIFOs) The FIFO is another type of sequential access memory that uses shift registers. It is similar to the recirculating shift register memory of Figure 11-36 in that the order in which the data words are entered at DATA IN is the same order in which they are read out at DATA OUT. In other words, the first word that is written in is the first word that is read out— hence the name FIFO; There are two important differences, however, between FIFOs and the recirculating shift register memory. First, in a FIFO the output data is not recirculated; once it is shifted out, it is lost. Second, in a FIFO the operation of shifting data into the memory is completely independent from the operation of shifting data out of the memory. In fact, the rate at which data is shifted in is usually much different than the rate at which it is shifted out. This makes it particularly suitable for the transfer of data between systems that are operating at widely different speeds.

One common example is the data transfer from a computer to a printer (Figure 11-37). The computer can send data to the printer at a much more rapid rate than the printer can accept it and print it out. A FIFO can act as a *data-rate buffer* between the computer and the printer by accepting data from the computer at a fast rate (say, one word every 10 μs) and storing it; the data is then shifted out to the printer at a slow rate (say, one word every 10 ms). The FIFO can also be used as a data-rate buffer for the transmission of data from a relatively slow device like a keyboard to a much faster device like a computer. For this situation the FIFO accepts data from the keyboard at a slow rate and stores it; the data is then shifted out to the computer at a fast rate. In this way, the computer can be performing other tasks while the FIFO is slowly being filled with data.

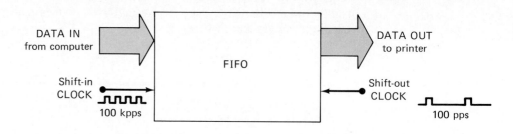

FIGURE 11-37 A FIFO is often used as a data-rate buffer between a fast device such as a computer, and a slower device like a printer.

REVIEW QUESTIONS

1. What are the operating modes of a recirculating shift register?
2. For the shift register of Figure 11-34 determine the level at DATA OUT after six clock pulses if $REC = 0$, DATA IN = 1, and the register FFs are initially all 0. (*Ans.* 1)
3. Repeat the last question with $REC = 1$. (*Ans.* 0)
4. Describe the structure of a shift register memory that is to store 64 9-bit words. (*Ans.* It contains nine 64-bit recirculating shift registers connected as in Figure 11-36)
5. How does a FIFO differ from a recirculating shift register memory?
6. What is a data-rate buffer?

11-20 MAGNETIC BUBBLE MEMORY

Magnetic bubble memory (MBM) is a solid-state memory that stores binary data in the form of tiny cylindrical domains (bubbles) in a thin film of magnetic material. The presence or absence of a bubble in a specific position is interpreted as a 1 or 0, respectively. Continuously changing magnetic fields are used to move the bubbles around in loops inside the magnetic material much like recirculating shift registers. The data circulates past a pickup point where it is available to the outside world at the rate of typically 50000 bits per second. Clearly, MBMs are sequential access devices.

The major advantage of MBMs is their nonvolatility; if electrical power is removed, the stored data is not lost since the bubbles simply remain in their fixed positions. When power is restored, the bubbles resume circulating around their loops. Unlike other nonvolatile semiconductor memories such as ROMs, PROMs, EPROMs, and EEPROMs, the MBM can be written into and read from with equal ease. Compared to nonvolatile tape and disk memories, an MBM system has no moving parts and is therefore quieter and more reliable.

MBMs are compact and dissipate very little power (typically 1 μW per bit). A typical MBM device can store a million or more bits, which is much more than any other semiconductor memory at this time. With all these advantages, the MBM does not possess the speed needed for it to function as the main internal memory of a computer; typical access times are in the range 1 to 20 ms. For this reason MBMs are better suited for the external mass storage functions that have been previously dominated by magnetic tape and disk storage. At present, MBMs are about 100 times faster than floppy disks but somewhat more costly because of the required support circuitry. As their cost comes down, we should see more and more MBM systems being used in place of the slower, less reliable floppy disk systems.

MBM Unit and Support Chips

A typical MBM unit (Figure 11-38) contains the bubble chip, two coils that generate the changing magnetic field, two permanent magnets that produce a steady bias field, and a magnetic shield that prevents disturbances from external magnetic fields. This is the actual unit that stores the binary data, and it may typically store one million bits (i.e., one megabit). This MBM unit requires several support chips in order to function as a complete memory system interfaced to a computer.

Figure 11-39 shows a typical bubble memory system configuration using Intel's 7110 MBM unit and support chips. The shaded portion is Intel's BPK70

FIGURE 11-38 Exploded view of an MBM unit. (Courtesy of Intel Corporation)

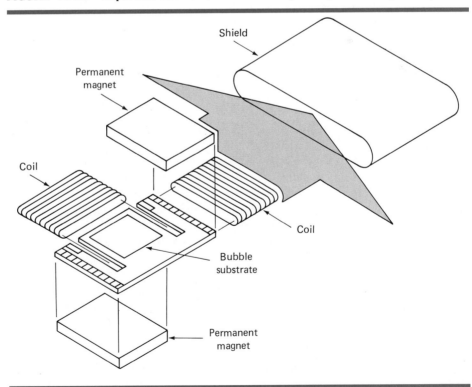

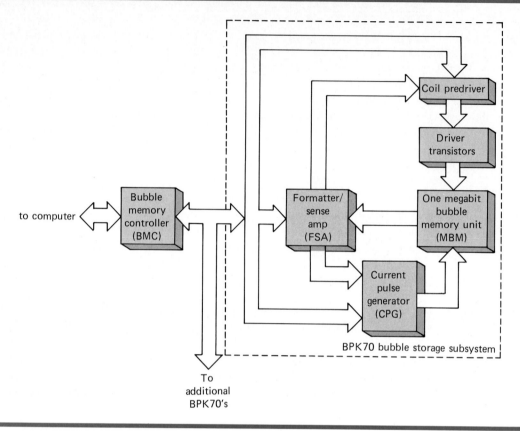

FIGURE 11-39 Intel MBM system. (Courtesy of Intel Corporation)

Bubble Storage Subsystem. It contains a 1-megabit MBM chip and four support chips. External to the subsystem we have the bubble memory controller (BMC), which is the interface between the memory system and the equipment it serves (i.e., computer). The BMC generates all the timing signals required by the other support chips. It also performs the parallel-to-serial and serial-to-parallel conversions needed because the bubble memory subsystem uses serial data, while the computer uses parallel data. The BMC also serves as a speed-matching device between the faster computer and the slower bubble memory subsystem. The BMC can serve up to eight bubble subsystems for a total of 8 megabits.

Serial data that are to be stored in or read from the bubble memory must pass through the formatter/sense amplifier (FSA). The FSA has many functions, including shifting serial data to and from the correct magnetic bubble loops in the MBM. The coil predriver chip takes timing signals from the BMC and converts them to pulsed outputs that are amplified by the driver transistor chip to drive the coils that surround the bubble chip. The current pulse generator (CPG) generates the current pulse needed to produce the magnetic bubble when a logic 1 is to be stored.

REVIEW QUESTIONS

1. What is the principal advantage of the MBM compared to semiconductor RAM? What is its principal disadvantage?

2. What are its advantages compared to magnetic tape and disk memory? What is its disadvantage?

11-21 MAGNETIC CORE MEMORY

The basic memory cell in magnetic core memories is a small toroidal (ring-shaped) piece of ferromagnetic material called a *ferrite core*. These cores are typically 50 mils in diameter, although more recently 18-mil cores have been introduced. Ferrite cores have a very low reluctance, which means that they are easily magnetized, and they have a high retentivity, which means that they will stay magnetized indefinitely when not disturbed.

Figure 11-40 shows a ferrite core, many times its actual size, with a wire threaded through the core. If a current pulse is passed through the wire, magnetic flux will be produced around the wire in a direction dependent on the direction of current.* Because of the ferrite core's low magnetic reluctance, these lines of flux will enter the core and magnetize the core in either a clockwise or counterclockwise direction. When the current in the wire is reduced to zero, the retentivity of the ferrite core results in the core retaining a large portion of the flux, *provided that the wire current exceeded the threshold magnetizing current value, I_M.* The two possible states of magnetization of the ferrite core act just like the logic levels of a flip-flop and can be assigned logic values. The usual convention is:

$$CW \text{ flux} = \text{logical 1 state}$$

$$CCW \text{ flux} = \text{logical 0 state}$$

FIGURE 11-40 **Magnetic core.**

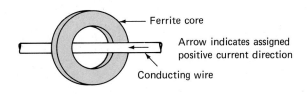

*The right-hand rule can be used to determine the flux direction as follows: Grasp the wire in the right hand with the thumb pointing in the direction of *conventional* current. The fingers then wrap around the wire in the direction of the flux lines.

Sensing a Core   Once a core has been pulsed to a given logic state, it can retain that logical state indefinitely without the need for any electrical power. Each core is a memory cell storing 0 or 1. Since a core's logic state is magnetic rather than electrical, a somewhat unusual technique is required to convert the core's stored information to an electrical signal to determine whether the core contains a 0 or 1. This technique of *sensing* or *reading* the core requires a second wire, called a *sense line,* threaded through the core as illustrated in Figure 11-41.

To sense or read the logic state of the core, a negative current pulse is applied to the input line to bring the core to the 0 state (CCW). If the core was previously in the 0 state, there will be only a very small flux change in the core which will produce a small voltage pulse across the sense line. If the core was in the 1 state, the flux change will be much greater and will produce a large voltage pulse across the sense line.

As Figure 11-41(b) shows, the 1-state output waveform has a much larger amplitude and pulse width than the 0-state output waveform. Thus, a 1 can be detected by using an amplitude-discriminating amplifier. The amplifier will amplify the 1 output pulse to several volts so that it can drive logic circuitry, but it will not amplify the 0 pulse.

FIGURE 11-41   Sensing the state of a core.

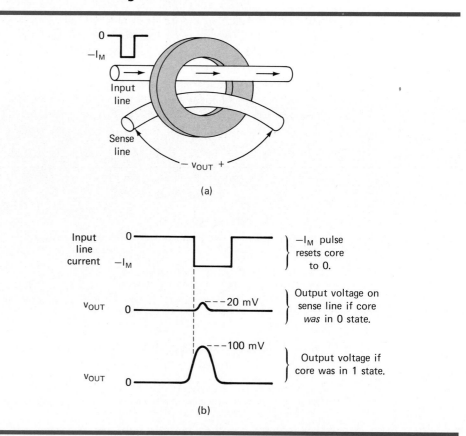

Destructive Readout   The main drawback of this technique for sensing the states of magnetic cores is that all the cores end up reset to the 0 state. This means that the information previously contained in the cores is no longer there. This sensing technique is therefore referred to as *destructive readout*. Generally speaking, it is necessary to retain the information in the core memory after it has been read. This requires that the information be written back into the memory after each read operation.

Magnetic Core Memory Characteristics   Individual core memory cells are usually arranged in core planes, typically containing 4096 cores per plane. This is illustrated symbolically in Figure 11-42 on a smaller scale. Each of the four planes contains 16 cores arranged in a 4×4 matrix. A 4-bit data word is stored vertically with 1 bit in each plane. For example, the cores labeled *a, b, c, d* represent a single 4-bit word.

A typical magnetic core memory system would contain sixteen 64×64 core planes to store 4096 16-bit words. Larger word sizes are obtained by increasing the number of planes; larger capacity is obtained by increasing the size of each plane.

Prior to the development of semiconductor memories, core memory systems were the dominant technology used in internal computer memories. At present, cores cannot compete with semiconductor RAM in the areas of speed, size, and cost. Typical core memory access times are in the 100–500 ns range. A core memory system requires an extensive amount of external support circuitry in addition to the cores themselves. This circuitry includes (1) address decoders to select a given data word location; (2) current generators to produce the relatively large current pulses (100 mA to 1 A) needed to magnetize the cores; (3) sense amplifiers to

FIGURE 11-42   **Core memory is arranged as a stack of core planes where each plane contains a matrix of cores.**

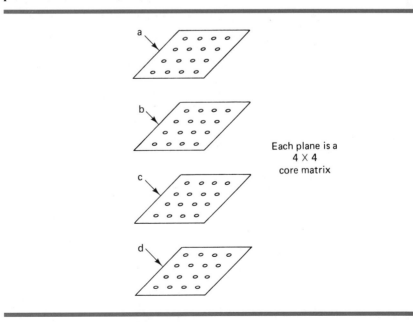

Each plane is a
4×4
core matrix

amplify the small sense line outputs. This circuitry makes a core memory system larger and more expensive than a typical semiconductor memory system, especially when large capacities are needed.

The major advantage of core memory is its nonvolatility. In applications where this is critical, cores offer a reasonably fast form of random-access memory.

REVIEW QUESTIONS

1. How are 1s and 0s written into a magnetic core?
2. How is the logic state of a memory cell sensed?
3. What is the principal advantage of core memory over semiconductor memory?

11-22 MAGNETIC TAPE AND DISK MEMORY

Magnetic tape and disks are memory devices that involve recording and reading magnetic spots on a moving surface of magnetic material. For each of these devices, a thin coating of magnetic material is applied to a smooth nonmagnetic surface. Magnetic tapes, for example, consist of a layer of magnetic material deposited on plastic tape. Disks have the magnetic material coated on both sides of a flat disk that resembles a phonograph record.

Recording and reading of binary information on tapes and disks uses the same basic principles. Figure 11-43 illustrates the fundamental concept of recording on a moving magnetic surface. The read/write head is a core of high-permeability soft

FIGURE 11-43 Fundamental parts for recording on moving magnetic surface.

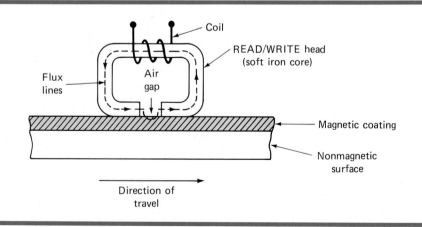

iron with a coil wound around it and a small air gap (typically 0.001 inch wide). In writing, a current is driven through the coil, establishing magnetic lines of flux in the core. These flux lines remain in the core until they encounter the air gap, which has a very high reluctance to magnetic flux. This causes the flux lines to deviate and travel through the magnetic coating on the moving surface. Thus, pulses of current in the coil result in spots of magnetism on the moving surface. These spots remain magnetized after they pass the read/write head.

The read operation is just the opposite of the write operation. During the read operation the coil is used as a sense line. As the magnetic surface is moved under the read/write head, the spots that have been magnetized produce a flux up through the air gap and into the core. This change in the core's flux induces a voltage signal in the sense winding which is then amplified and interpreted.

The coil around the core usually has a center tap, so that half the coil is used as a sense winding and the other half is heavier-gauge wire used for supplying current in the write operation.

Disk Memory   Magnetic-disk memories provide a very large storage capacity at moderate operating speeds. The magnetic disk memory costs somewhat more than magnetic tape. In this method of storage, data are recorded on a layer of magnetic material that is coated on hard flat disks resembling phonograph records. A number of rotating disks are stacked with space between each disk (refer to Figure 11-44). Information is recorded *serially* on concentric bands or tracks on both surfaces of each disk by movable read/write heads. The packing density along a given track can be greater than 5000 bits per inch. Typical rotation speeds can range up to 6000 rpm. A single disk surface usually has from 200 to 500 tracks and can store up to 250 million bits.

FIGURE 11-44   **Typical hard disk arrangement.**

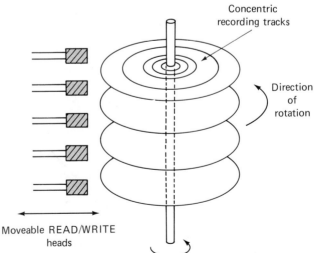

The read/write heads move radially in and out between disks in the space provided. The access time depends on rotation speed and disk diameter. For faster access time, additional heads are used to serve the same disks, some serving the inner tracks and others the outer tracks. The heads do not contact the disk surface, because this would cause excessive wear. Instead, most manufacturers use a *flying head*, which is shaped to ride along on a thin layer of air that adheres to the rotating disk. In most fast-acting mechanisms, a compressed-air piston is used to force the head toward the disk into or out of the air cushion.

Hard disk memories have typical access times of 10–40 ms; basically, this is the time required for the read/write head to become positioned on the desired track. Once the head is in position, data can be transferred to or from the disk at a typical rate of 2.5 megabits per second.

In a hard disk system, the flying head is only a few millionths of an inch above the disk surface; this is smaller than even a particle of smoke. Since the air is full of particles that could interfere with the operation of the read/write head, a hard disk unit requires that the head and disk be situated in an environmentally protected area.

A *Winchester* disk system houses the disk and head in an airtight container that is filled with filtered air or an inert gas. This provides excellent protection against environmental contaminants.

Floppy Disks   A relatively recent innovation in disk storage, which was originally developed at IBM, uses a small flexible disk with a plastic base in place of the more conventional metal disk. A floppy disk (also called a *diskette*) is a flexible disk that looks like a 45-rpm phonograph record without grooves. Because they can be bent easily, floppy disks are enclosed in a protective cardboard envelope. This keeps them rigid as they are spun around by the disk drive unit in much the same way as phonograph records.

Figure 11-45 shows the standard floppy disk, which has a diameter slightly less than 8 inches.* The protective envelope has cutouts for the drive spindle, the read/record head, and an index position hole. The access slot allows the read/record head to make contact with the surface of the disk as it spins inside the envelope. The index hole allows a photoelectric sensor to be used to determine a reference point for all the tracks on the disk.

Floppy disks are normally rotated at a speed of 360 rpm, which corresponds to one rotation in about 167 ms. As the disk rotates, the read/record head makes contact with it through the access slot so that it can write or sense magnetic pulses on the disk surface. The standard 8-inch disk surface is divided into 77 concentric tracks numbered 0 to 76 [see Figure 11.34(a)], with track number 0 being the outside track. These tracks are normally spaced about 0.02 inch apart, with track 0 having a total length of about 25 inches and track 77 a total length of 15 inches. This means that the longer outside tracks have space for more information than do the shorter inside tracks. For convenience in accessing a given point on the disk, the disk circumference is divided into 26 equal-size *sectors* [see Figure 11-46(b)] in which information can be stored. This means that the same amount of storage space is used on each track irrespective of the length of the track; thus, more space is wasted (not used) on the longer tracks.

---

*A "mini-floppy" has a diameter of 5 1/4 inches. This is the most popular size for personal computers.

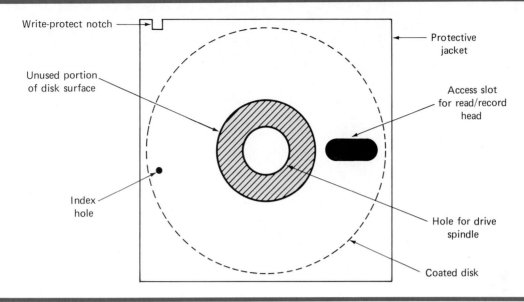

FIGURE 11-45   Floppy disk.

The index hole is placed between the last sector (26) and the first sector (1). Using a light source/photodetector combination, the disk-drive circuitry will know when the disk is in its reference position with sector 1 next to the index hole. By counting each sector as it passes under the read/record head, the drive circuitry will always know which sector is passing under the read/record head.

FIGURE 11-46   (a) Disk is divided into 77 concentric tracks; (b) each track is divided into 26 equal sectors.

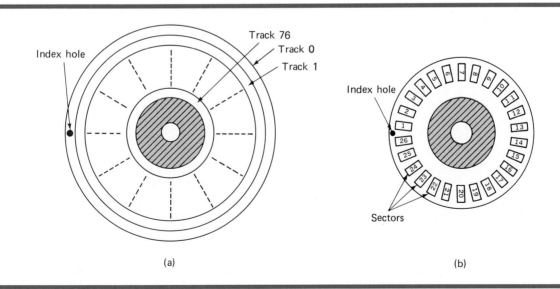

(a)   (b)

Each sector usually stores 128 bytes of information, so that the total storage capacity of a single floppy disk surface will be

$$128\frac{\text{bytes}}{\text{sector}} \times 26\frac{\text{sectors}}{\text{track}} \times 77\frac{\text{tracks}}{\text{disk}} \approx 256{,}000\frac{\text{bytes}}{\text{disk}} \approx 2 \times 10^6\frac{\text{bits}}{\text{disk}}$$

Some manufacturers use "double-density" floppy disks, which store 256 bytes into each sector.

Access Time

Data are recorded or read from the floppy disk one bit at a time at a nominal rate of 1 bit every 4 μs* or 1 byte every 32 μs. This rate, however, occurs after the read/record head has been positioned on the desired sector of the desired track. It takes approximately 6 ms for the head-positioning mechanism to move the head from one track to the next and 16 ms more to lower the head onto the track. With 77 tracks, it will take about 1/2 s in the worst case for the head to move from its present track to the track that is to be accessed for reading or writing. Once the head is on the desired track, it has to find the desired sector on that track. Since the disk makes one revolution in 167 ms, it will take 167 ms to find the right sector in the worst case. Thus, in the worst case it takes a little over 0.6 s to find the desired record on the disk. On the average, this access time will be more like 0.3 s. This is much faster than the time required to randomly access a record on even the fastest magnetic tape system.

Magnetic Tape

Currently, magnetic tape is the most popular medium for storing very large quantities of digital information. Most medium-size and large-size computers have one or more magnetic tape units. Magnetic tape represents the least expensive type of storage; the cost averages about 0.0001 cent per bit. Tapes are usually 1/2–1 inch wide and have 7–14 tracks across the width of the tape. Packing density usually ranges from 200 to 1600 bits per inch along each track. A typical tape reel is about 2400 ft and can store more than 10^8 bits of information.

A major difference between tape and disks is that the tape is not kept in continuous motion but is started and stopped each time it is used. The tape transports have the ability to start and stop very quickly. The necessity to accelerate and decelerate the tape quickly is due to two reasons. First, since the reading and writing processes cannot begin until the tape has reached sufficient speed, a time delay occurs as the tape comes up to speed. Second, the tape which passes under the heads during the acceleration and deceleration processes is wasted, so that fast starting and stopping conserves tape space. Typical start and stop times are of the order of 1 msec.

Most modern tape units use one READ/WRITE head per track and usually these heads have *two* air gaps, one for writing and one for reading [refer to Figure 11-47(a)]. The read gap is positioned after the write gap so that the information written on the tape can be immediately read out and compared with the input information (usually in a register) to see that it is correct.

---

*This is a data transfer rate of 250 kbits/sec, about 10 times slower than conventional hard disk systems.

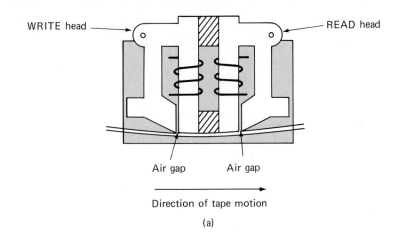

Direction of tape motion

(a)

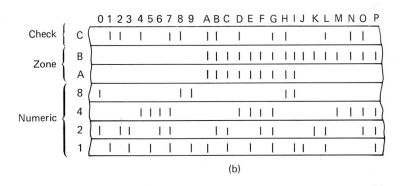

(b)

FIGURE 11-47 **(a) Two-gap READ/WRITE head; (b) seven-track tape format.**

These are several formats for recording data on magnetic tape. Typically, one character (number, letter, symbol) is recorded on a given row along the tape. Figure 11-47(b) shows a seven-track format where one of the tracks is for a parity bit. Data are recorded on tape in blocks called *records* with spaces between records to allow for starting and stopping of the tape.

Magnetic tape has become a very reliable storage medium, and its basic limitation continues to be its relatively long access time. However, despite this limitation, it continues to be an economical and convenient medium for bulk storage of large quantities of information.

Tape Cassette Storage Microcomputers often use a low-cost type of magnetic tape storage utilizing standard audio cassettes. A typical cassette storage system records different audio tones to represent digital data. These systems are relatively slow, having access times that can be several minutes, and data transfer rates of 300–1200 bits per second. Their advantages are simplicity and low cost.

REVIEW QUESTIONS

1. *True or false:* Floppy disks have a greater data transfer rate than hard disks.
2. *True or false:* Magnetic tape is slower than disk.
3. How does a Winchester disk unit differ from a standard hard disk?

11-23 TROUBLESHOOTING RAM SYSTEMS

All computers use RAM. Many general-purpose computers and most special-purpose computers (such as microprocessor-based controllers and process-control computers) also use some form of ROM. Each RAM and ROM IC that is part of a computer's internal memory typically contains thousands of memory cells. A single faulty memory cell can cause a complete system failure (commonly referred to as a "system crash") or, at the least, unreliable system operation. The testing and troubleshooting of memory systems involves the use of techniques that are not often used on other parts of the digital system. Because memory consists of thousands of identical circuits acting as storage locations, any tests of its operation must involve checking to see exactly which locations are working and which are not. Then, by looking at the pattern of good and bad locations along with the organization of the memory circuitry, the possible causes of the memory malfunction can be determined. The problem typically can be traced to a bad memory IC, a bad decoder IC, logic gate or signal buffer, or a problem in the circuit connections (i.e., shorts, opens).

Because RAM has to be written to and read from, testing RAM is generally more complex than testing ROM. In this section we will look at some common procedures for testing the RAM portion of memory and interpreting the test results. We will examine ROM testing in the next section.

Know the Operation The RAM memory system shown in Figure 11-48 will be used in our examples. As we emphasized in earlier discussions, successful troubleshooting of a relatively complex circuit or system begins with a thorough knowledge of its operation. Before we can discuss the testing of this RAM system, we should first analyze it carefully so that we fully understand its operation.

The total capacity is 4K $\times$ 8 and is made up of four 1K $\times$ 8 RAM modules. A module may be just a single IC, or it may consist of several ICs (e.g., two 1K $\times$ 4 chips). Each module is connected to the CPU through the address and data buses, and the $R/\overline{W}$ control line. The modules have common I/O data lines. During a read operation these lines become data output lines through which the selected module places its data on the bus for the CPU to read. During a write operation, these lines act as input lines to accept CPU-generated data from the data bus for writing into the selected location.

The 74LS138 decoder and the four-input OR gate combine to decode the six high-order address lines to generate the chip-select signals $\overline{K}0$, $\overline{K}1$, $\overline{K}2$, and $\overline{K}3$.

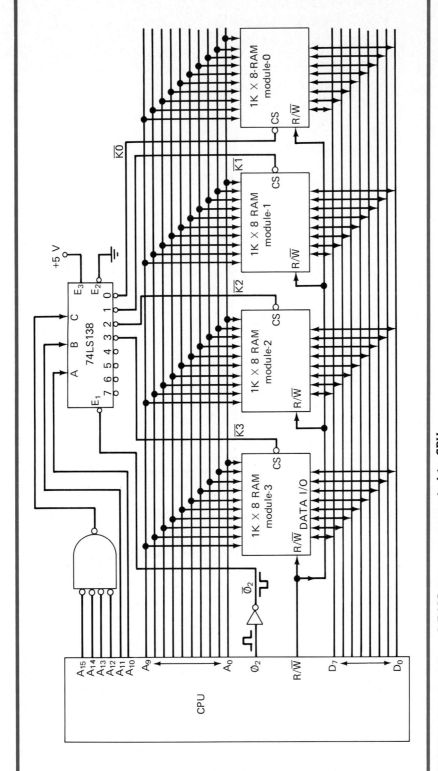

FIGURE 11-48 4K × 8 RAM memory connected to CPU.

These signals enable a specific RAM module for a read or write operation. The INVERTER is used to invert the CPU output clock signal ϕ_2 so that the decoder is enabled only while ϕ_2 is HIGH. The ϕ_2 pulse occurs only after allowing enough time for the address lines to stabilize following the application of a new address on the address bus. ϕ_2 will be LOW while the address and $R/\overline{W}$ lines are changing; this prevents the occurrence of decoder output glitches that could erroneously activate a memory chip and possibly cause invalid data to be stored.

Each RAM module has its address inputs connected to the CPU address bus lines A_0–A_9. The high-order address lines A_{10}–A_{15} select one of the RAM modules. The selected module decodes address lines A_0–A_9 to find the word location that is being addressed. The following examples will show how to determine the addresses that correspond to each module.

EXAMPLE 11-19

Assume that the CPU is performing a read operation from address 06A3(hex). Which RAM module, if any, is being read from?

SOLUTION
First write out the address in binary.

$$A_{15}\,A_{14}\,A_{13}A_{12}A_{11}A_{10}A_9A_8A_7A_6A_5A_4A_3A_2A_1A_0$$

$$0\;\;0\;\;0\;\;0\;\;0\;\;1\;\;1\;0\;1\;0\;1\;0\;0\;0\;1\;1$$

You should be able to verify that the A_{15}–A_{10} levels will activate decoder output $\overline{K1}$ to select RAM module-1. This module internally decodes the A_9–A_0 address lines to select the location whose data is to be placed on the data bus.

EXAMPLE 11-20

Which RAM module will have data written into it when the CPU executes a write operation to address 1C65?

SOLUTION
Writing out the address in binary, we can see that $A_{15} = 1$. This produces a HIGH out of the OR gate and at the C input of the decoder. With $A_{11} = A_{10} = 1$, the decoder inputs are 111, which activates output 7. Outputs $\overline{K0}$–$\overline{K3}$ will be inactive, so none of the RAM modules will be enabled. In other words, the data placed on the data bus by the CPU will not be accepted by any of the RAMs.

EXAMPLE 11-21

Determine the range of addresses for each module in Figure 11-48.

SOLUTION
Each module stores 1024 8-bit words. To determine the addresses of the words stored in any module, we start by determining the address bus conditions that

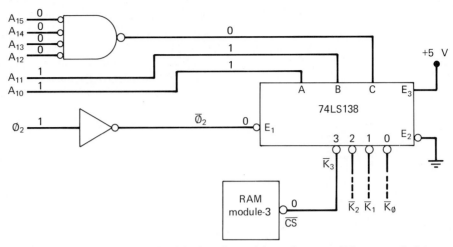

FIGURE 11-49 Example 11-21 showing address bus conditions needed to select RAM module-3.

activate that module's chip-select input. For example, module-3 will be selected when decoder output $\overline{K3}$ is LOW (Figure 11-49). $\overline{K3}$ will be LOW for $CBA = 011$. Working back to the CPU address lines A_{15}–A_{10}, we see that module-3 will be enabled when the following address is placed on the address bus:

$A_{15}\ A_{14}\ A_{13}\ A_{12}\ A_{11}\ A_{10}\ A_9\ A_8\ A_7\ A_6\ A_5\ A_4\ A_3\ A_2\ A_1\ A_0$

0   0   0   0   1   1   x   x   x   x   x   x   x   x   x   x -Selects module-3

The x's under A_9–A_0 indicate "don't care" because these address lines are not used by the decoder to select module-3. A_9–A_0 can be any combination ranging from 0000000000 to 1111111111, depending on which word in module-3 is being accessed. Thus, the complete range of addresses for module-3 is determined by using all 0s, and then all 1s for the x's.

$A_{15}\ A_{14}\ A_{13}\ A_{12}\ A_{11}\ A_{10}\ A_9\ A_8\ A_7\ A_6\ A_5\ A_4\ A_3\ A_2\ A_1\ A_0$

0   0   0   0   1   1   0   0   0   0   0   0   0   0   0   0 $\longrightarrow$ 0C00$_{16}$

0   0   0   0   1   1   1   1   1   1   1   1   1   1   1   1 $\longrightarrow$ 0FFF$_{16}$

Finally, this give us 0C00 to 0FFF as the range of hex addresses stored in module-3. When the CPU places any address in this range onto the address bus, module-3 will be enabled for either a read or write, depending on the state of $R/\overline{W}$.

A similar analysis can be used to determine the address ranges for each of the other RAM modules. The results are:

Module-0: 0000-03FF

Module-1: 0400-07FF

Module-2: 0800-0BFF

Module-3: 0C00-0FFF

Note that the four modules combine for a total address range of 0000 to 0FFF.

Testing the Decoding Logic
In some situations, the decoding logic portion of the RAM circuit (Figure 11-49) can be tested using the various techniques we have applied to combinatorial circuits. It can be tested by applying signals to the six most significant address lines and $\emptyset_2$, and monitoring the decoder outputs. To do this, it has to be possible to easily disconnect the CPU from these signal lines. If the CPU is a microprocessor chip in a socket, it can simply be removed from its socket.

Once the CPU is disconnected, you can supply the A_{10}–A_{15} and $\emptyset_2$ signals from an external test circuit to perform a static test, using manually operated switches for each signal, or a dynamic test, using some type of counter to cycle through the various address codes. With these test signals applied, the decoder output lines can be checked for the proper response. Standard signal tracing techniques can be used to isolate any faults in the decoding logic.

EXAMPLE 11-22

A dynamic test is performed on the decoding logic of Figure 11-49 by keeping $\emptyset_2 = 1$ and connecting the outputs of a 6-bit counter to the address inputs A_{10}–A_{15}. The decoder outputs are monitored as the counter repetitively cycles through all 6-bit codes. A logic probe check on the decoder outputs shows pulses at $\overline{K1}$ and $\overline{K3}$, but shows $\overline{K0}$ and $\overline{K2}$ remaining HIGH. What are the most probable faults?

SOLUTION
It is possible, but highly unlikely, that $\overline{K0}$ and $\overline{K2}$ could both be stuck HIGH due to either an internal or external short to V_{CC}. A more likely fault is an open between A_{10} and the A input of the decoder since this would act as a logic HIGH and prevent any even-numbered decoder output from being activated. It is also possible that the decoder's A input is shorted to V_{CC}, but this is also unlikely because this short would most probably affect the operation of the counter that is supplying the address inputs.

Testing the Complete RAM System
Testing and troubleshooting the decoding logic will not reveal problems with the memory chips and their connections to the CPU buses. The most common methods for testing the operation of the *complete* RAM system involve writing known patterns of 1's and 0's to each memory location and reading them back to verify that the location stored the pattern properly. While there are many different patterns that can be used, one of the most widely used is the "checkerboard pattern." In this pattern, 1s and 0s are alternated as in 01010101. Once all locations have been tested using this pattern, the pattern is reversed (i.e., 10101010) and each location is tested again. Note that this sequence of tests will check each cell for the ability to store and read both a 1 and a 0. Because it alternates 1s and 0s, the checkerboard pattern will also detect any interactions or shorts between adjacent cells. Many other patterns can be used to detect various failure modes within RAM chips.

No memory test can catch all possible RAM faults with 100 percent accuracy, even though it may show that each cell can store and read a 0 or 1. Some faulty RAMs can be pattern sensitive. For instance, it may be able to store and read 01010101 and 10101010, but it may fail when 11100011 is stored. Even for a small

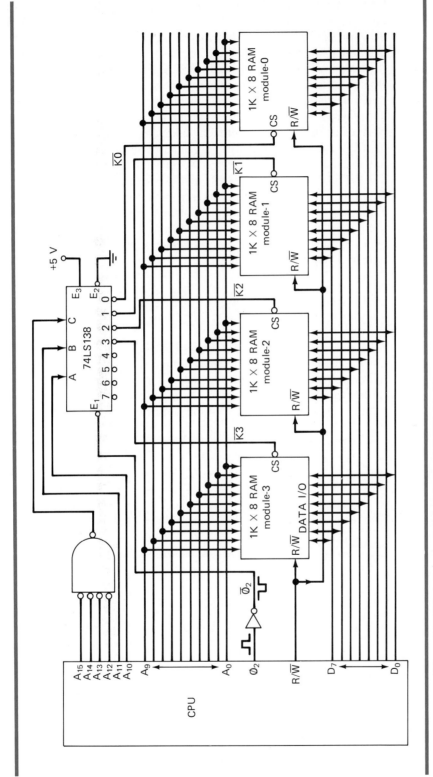

FIGURE 11-50 4K × 8 RAM system repeated from FIGURE 11-48.

RAM system, it would take a prohibitively long time to try storing and reading every possible pattern in each location. For this reason, if a RAM system passes the checkerboard test, you can conclude that it is *probably* good; if it fails the test, then it *definitely* contains a fault.

Manually testing thousands of RAM locations by storing and reading checkerboard patterns would take hundreds of hours, and is obviously not feasible. RAM pattern testing is usually done automatically either by having the CPU run a memory test program or by connecting a special test instrument to the RAM system buses in place of the CPU. In fact, in many computers and microprocessor-based equipment the CPU will automatically run a memory test program every time it is powered up; this is called an *automatic power-up self-test*. The self-test routine (we will call it SELF-TEST) is stored in ROM, and it is executed whenever the system is turned on, or when the operator requests it from the keyboard. As the CPU executes SELF-TEST, it will write and read test patterns to each RAM location, and will display some type of message to the user. It may be something as simple as an LED to indicate faulty memory, or it may be a descriptive message printed on the screen or printer. Typical messages might be:

```
RAM module-3 test OK
All RAM working properly
Location 027F faulty in bit positions 6 and 7
```

With messages like these and a knowledge of the RAM system operation, the troubleshooter can determine what further action is needed to isolate the fault. The following two examples will illustrate the procedure for the 4K × 8 RAM system; its diagram is repeated in Figure 11-50 for convenience (see page 693). Remember from our previous examples that the addresses for each RAM module are

Module-0: 0000–03FF

Module-1: 0400–07FF

Module-2: 0800–0BFF

Module-3: 0C00–0FFF

EXAMPLE 11-23

The following messages are printed out after the SELF-TEST is run:

```
module-0 test OK
module-1 test OK
address 09A7 faulty at bit 6
module-3 test OK
```

Examine these messages and decide what to do next.

SOLUTION

The faulty address location is in module-2. There is no circuit fault external to module-2 that could cause only that single address to be bad and no others. Thus, it appears that there is an internal fault in module-2. Before we reach this conclusion, however, the SELF-TEST program should be run several more times to see if the same messages appear. The original fault message may have been the result of noise rather than a bad memory cell; if so, further tests may produce no fault messages or different fault messages. This would point up the need to check for sources of noise in the system such as nonexistent or nonfunctioning power-supply decoupling capacitors, or crosstalk between closely spaced signal lines.

EXAMPLE 11-24

Consider the following messages from SELF-TEST:

```
module-0 test OK
address 0400 faulty at bits 0-7
address 0401 faulty at bits 0-7
address 0402 faulty at bits 0-7
         .    .     .      .    .
         .    .     .      .    .
         .    .     .      .    .
         .    .     .      .    .
address 07FE faulty at bits 0-7
address 07FF faulty at bits 0-7
module-2 test OK
module-3 test OK
```

Examine these messages, list the possible faults, and describe what to do next.

SOLUTION
Clearly, every address in module-1 is listed as being faulty, so module-1 is not working at all. Here are several possible faults that could cause this:

- An open at the $R/\overline{W}$ input to module-1.
- An open between $\overline{K1}$ and the module-1 chip-select.
- A faulty decoder output $\overline{K1}$ (stays HIGH).
- The $\overline{K1}/\overline{CS}$ node is externally shorted to V_{cc}.
- A faulty RAM module-1 (does not respond to $\overline{CS}$).
- An unconnected V_{CC} or ground on module-1.

Standard troubleshooting techniques can now be used to isolate the fault.

As these examples have shown, the memory test program is useful in narrowing the problem to a specific area of the circuit. The troubleshooter uses the information provided from the memory test to determine possible faults and then proceeds to locate the actual fault.

It should be pointed out that there are some RAM circuit faults that will prevent the CPU from executing the SELF-TEST program on power-up. For example, any RAM circuit fault that causes a data or address line to be stuck LOW or HIGH will cause erroneous operation when the CPU tries to execute the SELF-TEST program in ROM (remember that the ROM shares the address and data buses with the RAM). When this happens, you should check for stuck bus lines and, if one is found, use the standard techniques for determining the exact cause of the stuck line.

REVIEW QUESTIONS

1. What is $\emptyset_2$'s function in the RAM circuit of Figure 11-48?
2. What is the checkerboard test? Why is it used?
3. What is a power-up self-test?

11-24 TESTING ROM

The ROM circuitry in a computer is very similar to the RAM circuitry (compare Figures 11-32 and 11-48). The ROM decoding logic can be tested in the same manner described in the preceding section for the RAM system. The ROM chips, however, must be tested differently than RAM chips, because we cannot write patterns into ROM and read them back as we can for RAM. There are several methods used to check the contents of a ROM IC.

In one approach the ROM is placed in a socket in a special test instrument which is typically microprocessor-controlled. The special test instrument can be programmed to read every location in the test ROM and print out a listing of the contents of each location. The listing can then be compared to what the ROM is supposed to contain. Except for low-capacity ROM chips, this process can be very time consuming.

In a more efficient approach, the test instrument would have the correct data stored in its own *reference ROM* chip. The test instrument is then programmed to read the contents of each location of the test ROM and compare it to the contents of the reference ROM. This approach, of course, requires the availability of a pre-progammed reference ROM.

A third approach uses a *checksum*. This is a special code placed in the last one or two locations of the ROM chip when it is programmed. This code is derived by adding up the data words to be stored in all of the ROM locations (excluding those containing the checksum). As the test instrument reads the data from each test ROM location, it will add them up and develop its own checksum. It then compares its calculated checksum with that stored in the last ROM locations, and they should agree. If so, there is a good probability that the ROM is good (there is a finite chance that a combination of errors in the test ROM data could still produce the same checksum value). If they do not agree, then there is a definite problem in the test ROM.

The checksum idea is illustrated in Figure 11-51(a) for a very small ROM. The data word stored in the last address is the 8-bit sum of the other seven data words (ignoring carries from the MSB). When this ROM is programmed, the checksum is placed in the last location. Figure 11-51(b) shows the data that might actually be read from a faulty ROM that was originally programmed with the data in (a). Note the error in the word at address 011. As the test instrument reads the data from each location of the faulty ROM, it calculates its own checksum from that data. Because of the error, the calculated checksum will be 10010011. When the test instrument compares this with the checksum value stored at ROM location 111,

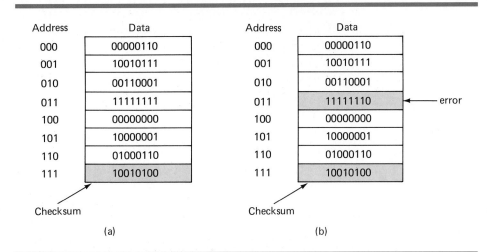

Address	Data
000	00000110
001	10010111
010	00110001
011	11111111
100	00000000
101	10000001
110	01000110
111	10010100

Checksum

(a)

Address	Data	
000	00000110	
001	10010111	
010	00110001	
011	11111110	← error
100	00000000	
101	10000001	
110	01000110	
111	10010100	

Checksum

(b)

FIGURE 11-51 Checksum method for an 8 × 8 ROM: (a) ROM with correct data; (b) ROM with error in its data.

it sees that they disagree, and a ROM error is indicated. Of course, the exact location of the error is not determined.

The checksum method can also be used by a computer or microprocessor-based equipment during an automatic power-up self-test to checkout the contents of the system ROMs. Again, as in the self-test used for RAM, the CPU would execute a program on power-up that would do a checksum test on each ROM chip and would print out some type of status message. The self-test program itself will be located in a ROM, so any error in that ROM could prevent the successful execution of the checksum tests.

REVIEW QUESTIONS

1. Describe the various approaches for testing ROM.
2. What is a checksum and how is it used?

Address Multiplexing Used in dynamic RAMs to save IC pins, it involves latching the two halves of a complete address into the IC in separate steps. [Sec. 11-15]

Bootstrap Program Program, stored in ROM, that a computer executes on power-up. [Sec. 11-9]

Byte 8-bit word. [Sec. 11-1]

Capacity Amount of storage space in a memory expressed as number of bits or number of words. [Sec. 11-1]

Checksum Special data word stored in last ROM location. It is derived from the addition of all other data words in the ROM, and is used for error-checking purposes. [Sec. 11-24]

Code Converter Circuit that takes data expressed in one type of code and produces an output in another type of code. [Sec. 11-9]

Column Address Strobe (CAS) Signal used to latch the column address into a DRAM chip. [Sec. 11-15]

Control Bus Bus carrying control signals from CPU to memory. [Sec. 11-3]

Data Bus Bidirectional bus that carries data between CPU and memory. [Sec. 11-3]

Data Rate Buffer Application of FIFOs where sequential data is written into the FIFO at one rate, and read out at a different rate. [Sec. 11-19]

DRAM Controller IC used to handle refresh and address multiplexing operations needed by DRAM systems. [Sec. 11-17]

Dynamic RAM (DRAM) Type of semiconductor memory that stores data as capacitor charges that need to be refreshed periodically. [Sec. 11-14]

Electrical Erasable Programmable ROM (EEPROM) ROM that can be electrically programmed, erased, and reprogrammed. [Sec. 11-8]

Erasable Programmable ROM (EPROM) ROM that can be electrically programmed by the user. It can be erased (usually with ultraviolet light) and reprogrammed as often as desired.

Firmware Computer programs stored in ROM. [Secs. 11-8 and 11-9]

First-In First-Out Memory (FIFO) Semiconductor sequential access memory in which data words are read out in the same order that they were written in. [Sec. 11-19]

Magnetic Bubble Memory (MBM) Solid-state, nonvolatile, sequential access, mass storage memory device consisting of tiny magnetic domains (bubbles). [Sec. 11-20]

Magnetic Core Memory (MCM) Nonvolatile random access memory made up of small ferrite cores. [Sec. 11-21]

Magnetic Disk Memory Mass storage memory that stores data as magnetized spots on a rotating flat disk surface. [Sec. 11-22]

Magnetic Tape Memory Mass storage memory that stores data as magnetized spots on a magnetically coated plastic tape. [Sec. 11-22]

Mask-Programmed ROM (MROM) ROM that is programmed by the manufacturer according to the customer's specifications. It cannot be erased or reprogrammed. [Sec. 11-8]

Memory Cell Device that stores a single bit. [Sec. 11-1]

Memory Word Group of bits in memory that represents instructions or data of some type. [Sec. 11-1]

Nonvolatile Memory Memory that will keep storing its information without the need for electrical power. [Sec. 11-5]

Power-Up Self-Test Program stored in ROM and executed by the CPU on power-up to test RAM and/or ROM portions of the computer circuitry. [Sec. 11-23]

Programmable Logic Array (PLA) Type of programmable logic device that contains an array of AND gates and OR gates. [Sec. 11-10]

Programmable Logic Device (PLD) IC that contains a large number of interconnected logic functions. The user can program the IC for a specific function by selectively breaking the appropriate interconnections. [Sec. 11-10]

Programmable ROM (PROM) ROM that can be electrically programmed by the user. It cannnot be erased and reprogrammed. [Sec. 11-8]

Random Access Memory (RAM) Memory in which the access time is the same for any location. [Sec. 11-1]

Read Operation Word in a specific memory location is sensed and possibly transferred to another device. [Sec. 11-1]

Read-Only Memory (ROM) Memory devices that are designed for applications where the ratio of read operations to write operations is very high. [Sec. 11-1]

Read/Write Memory (RWM) Any memory that can be read from and written into with equal ease. [Sec. 11.1]

Refreshing Process of recharging the cells of a dynamic memory. [Sec. 11-14]

Row Address Strobe (RAS) Signal used to latch row address into dynamic RAM chip. [Sec. 11-15]

Sequential Access Memory (SAM) Memory in which the access time will vary depending on where the data is stored. [Sec. 11-1]

Static RAM Semiconductor RAM that stores information in flip-flop cells that do not have to be periodically refreshed. [Sec. 11-13]

Volatile Memory Requires electrical power to keep information stored. [Sec. 11-1]

Write Operation A new word is placed into a specific memory location. [Sec. 11-1]

PROBLEMS   Sections 11-1 and 11-2

11-1.   A certain memory has a capacity of 16K $\times$ 32. How many words does it store? What is the number of bits per word? How many memory cells does it contain?

11-2.   How many different addresses are required by the memory of Problem 11-1?

11-3.   What is the capacity of a memory that has 16 address inputs, four data inputs, and four data outputs?

11-4. A certain memory stores 8K 16-bit words. How many data input and data output lines does it have? How many address lines does it have? What is its capacity in bytes?

Sections 11-5 and 11-6

11-5. Refer to Figure 11-6. Determine the data outputs for each of the following input conditions.
(a) $[A] = 1011$; $CS = 0$
(b) $[A] = 0111$; $CS = 1$

11-6. Refer to Figure 11-7.
(a) What register is enabled by input address 1011?
(b) What input address code selects register 4?

11-7. A certain ROM has a capacity of 16K $\times$ 4 and an internal structure like that shown in Figure 11-7.
(a) How many registers are in the array?
(b) How many bits are there per register?
(c) What size decoders does it require?

Section 11-7

11-8. Figure 11-52 shows how data from a ROM can be transferred to an external register. The ROM has the following timing parameters: $t_{ACC} = 250$ ns and $t_{OE} = 120$ ns. Assume that the new address inputs have been applied to the ROM 500 ns before the occurrence of the TRANSFER pulse. Determine the minimum duration of the TRANSFER pulse for reliable transfer of data.

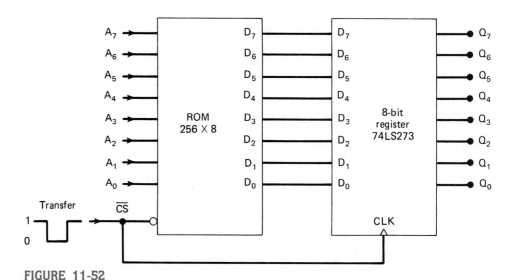

FIGURE 11-52

11-9. Repeat Problem 11-8 if the address inputs are changed 70 ns prior to the TRANSFER pulse.

11-10. Which transistors in Figure 11-9 will be conducting when $A_1 = A_0 = 1$ and $\overline{EN} = 0$?

11-11. Change the MROM connections in Figure 11-9 so that the MROM stores the function $y = 3x + 5$.

11-12. Figure 11-53 shows a simple circuit for manually programming a 2716 EPROM. Each EPROM data pin is connected to a switch that can be set at a 1 or 0 level. The address inputs are driven by an 11-bit counter. The 50-ms programming pulse comes from a one-shot each time the PROGRAM pushbutton is actuated.

   (a) Explain how this circuit can be used to sequentially program the EPROM memory locations with the desired data.

   (b) Show how 74293s and a 74121 can be used to implement this circuit.

   (c) Should switch bounce have any effect on the circuit operation?

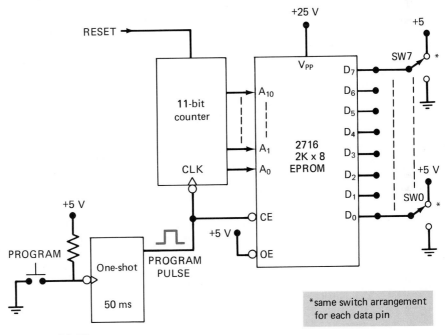

FIGURE 11-53

11-13. Another ROM application is the generation of timing and control signals. Figure 11-54 shows a 16 × 8 ROM with its address inputs driven by a MOD-16 counter so that the ROM addresses are incremented with each input pulse. Assume the ROM is programmed as in Figure 11-6, and sketch the waveforms at each ROM output as the pulses are applied. Ignore ROM delay times.

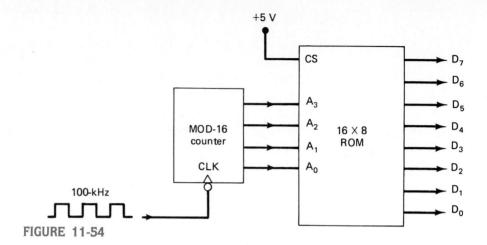

FIGURE 11-54

11-14. Change the program stored in the ROM of Problem 11-13 to generate the D_7 waveform of Figure 11-55.

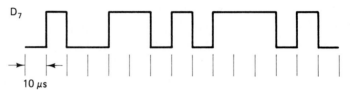

FIGURE 11-55

11-15. Figure 11-56(a) shows the logic symbol for the 74185 IC, which is an "off the shelf" ROM that is programmed as a *BCD-to-binary converter*. It converts a 6-bit binary input to a two-digit BCD output. Note how the LSB of the 6-bit binary input is not connected to the ROM as an address input; instead, it is simply routed to the output and becomes the LSB of the LSD of the output. Figure 11-56(b) shows part of the truth table showing the contents of some of the locations in this ROM. Examine it to verify how the conversion from binary to BCD is taking place. Then fill in the missing entries in the table.

Section 11-10

11-16. Change the arrangement of the PLA in Figure 11-15 to generate the following functions:

$$X = A\overline{C} + AB$$

$$Y = A\overline{B}C$$

$$Z = AB + \overline{A}\,\overline{B}$$

Section 11-12

11-17. (a) Draw the logic symbol for a 51C66 which is a CMOS static RAM organized as a 16K $\times$ 1 with separate data in and data out, and an active-LOW chip enable.

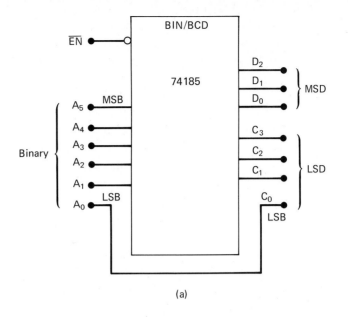

(a)

Binary						BCD						
A_5	A_4	A_3	A_2	A_1	A_0	D_2	D_1	D_0	C_3	C_2	C_1	C_0
1	0	0	1	0	1	0	1	1	0	1	1	1
0	0	1	1	1	0	0	0	1	0	1	0	0
1	1	1	1	1	0	1	1	0	0	0	1	0
0	1	1	1	0	0	0	1	0	1	0	0	0
1	1	0	1	0	1	___	___	___	___	___	___	___
0	0	0	1	1	0	___	___	___	___	___	___	___
___	___	___	___	___	___	0	0	1	1	0	0	1
___	___	___	___	___	___	1	0	0	1	0	0	0

(b)

FIGURE 11-56

(b) Draw the logic symbol for a 51C68, which is a CMOS static RAM organized as a 4K × 4 with common I/O and an active-LOW chip enable.

Section 11-13

11-18. A certain static RAM has the following timing parameters (in nanoseconds):

$$t_{RC} = 100 \qquad t_{AS} = 20$$

$$t_{ACC} = 100 \qquad t_{AH} = \text{not given}$$

$$t_{CO} = 70 \qquad t_{w} = 40$$

$$t_{OD} = 30 \qquad t_{DS} = 10$$

$$t_{WC} = 100 \qquad t_{DH} = 20$$

(a) How long after the address lines stabilize will valid data appear at the outputs during a read cycle?

(b) How long will output data remain valid after $\overline{CS}$ returns HIGH?

(c) How many read operations can be performed per second?

(d) How long should $R/\overline{W}$ and $\overline{CS}$ be kept HIGH after the new address stabilizes during a write cycle?

(e) What is the minimum time that input data have to remain valid for a reliable write operation to occur?

(f) How long must the address inputs remain stable after $R/\overline{W}$ and $\overline{CS}$ return HIGH?

(g) How many write operations can be performed per second?

Sections 11-14–11-17

11-19. Draw the logic symbol for the TMS4256, which is a 256K $\times$ 1 DRAM. How many pins are saved by using address multiplexing for this DRAM?

11-20. Figure 11-57(a) shows a circuit that generates the $\overline{RAS}$, $\overline{CAS}$, and *MUX* signals needed for proper operation of the circuit of Figure 11-23(b). The 10-MHz master clock signal provides the basic timing for the computer. The memory request signal (*MEMR*) is generated by the CPU in synchronism with the master clock as shown in part (b) of the figure. *MEMR* is normally LOW and is driven HIGH whenever the CPU wants to access memory for a read or write operation. Determine the waveforms at Q_0, $\overline{Q}_1$, and Q_2 and compare them to the desired waveforms of Figure 11-24.

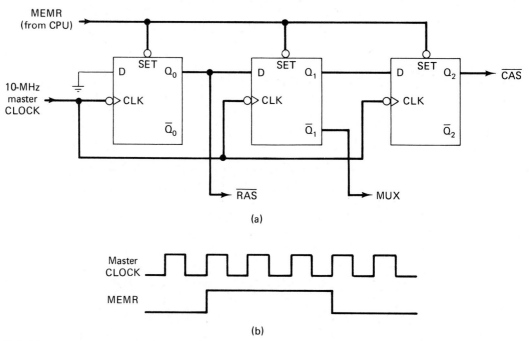

(a)

(b)

FIGURE 11-57

11-21. Show how to connect two 74157 multiplexers to provide the multiplexing function required in Figure 11-23(b).

11-22. Refer to the signals in Figure 11-25. Describe what occurs at each of the labeled time points.

11-23. Repeat for Figure 11-26.

11-24. Show how the outputs from the circuit of Figure 11-57 can be used in the circuit of Figure 11-28. Pay careful attention to the 3242's truth table.

Section 11-18

11-25. Show how to combine two 2114 RAM chips (Figure 11-17) to produce a 1K × 8 module.

11-26. Show how to connect two of the ROM chips symbolized in Figure 11-58 to produce a 2K × 8 ROM module. The circuit should use no other logic.

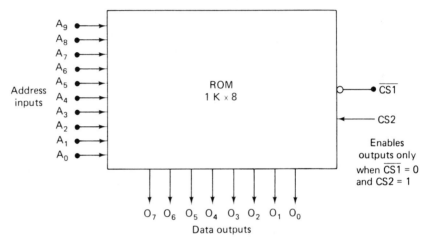

FIGURE 11-58

11-27. Describe how to modify the circuit of Figure 11-32 so that it has a total capacity of 16K × 8. Use the same type of PROM chips.

11-28. Modify the decoding circuit of Figure 11-32 to operate from a 16-line address bus (i.e., add A_{13}, A_{14}, and A_{15}). The four PROMs are to maintain the same hex address ranges.

11-29. Examine the memory circuit of Figure 11-59.
  (a) Determine the total capacity and word size.
  (b) Note that there are more address bus lines than are necessary to select one of the memory locations. This is not an unusual situation, especially in small computer systems, where the actual amount of memory circuitry is much less than the maximum which the computer address bus can handle. Which RAMs will put data on the data bus when $R/\overline{W}$ = 1 and the address bus is at 00010110?
  (c) Determine the range of addresses stored in the RAM-0/RAM-1 combination. Repeat for the RAM-2/RAM-3 combination.

11-30. Draw the complete diagram for a 4K × 4 memory that uses static RAM chips with the following specifications: 1K × 1 capacity, common input/output line, and two active-LOW chip-select inputs. [*Hint:* The circuit can be designed using only two inverters (plus memory chips).]

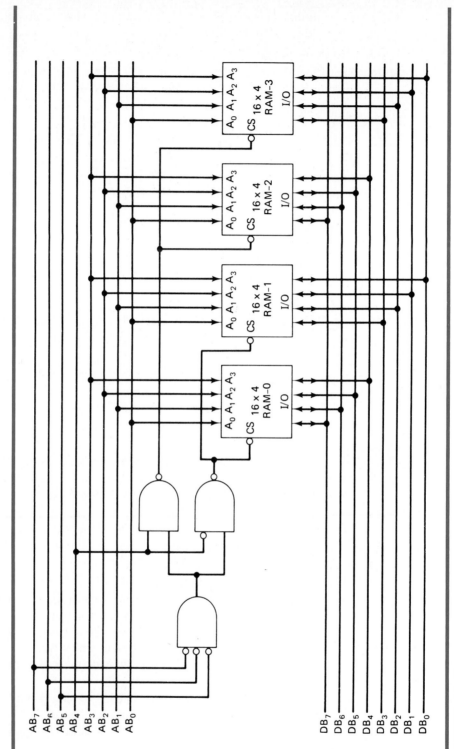

FIGURE 11-59

11-31. Refer to the recirculating shift register of Figure 11-34. The control logic that selects the source of data that will be shifted into D_S is really just a 2-input MUX. Show how to combine four 74164 shift registers (Section 7-21) with a 74157 MUX (Section 9-7) to produce an 8×4 shift register memory.

11-32. Refer to the shift register memory of Figure 11-36. Each register is 8 bits wide. The initial contents of each register are as follows:

> *Register 0:* 01111000
> *Register 1:* 00100001
> *Register 2:* 01001000
> *Register 3:* 00000001
> *Register 4:* 00000100
> *Register 5:* 10000011
> *Register 6:* 01111100

What is the ASCII-code message that will be transmitted to the video display as clock pulses are applied?

11-33. Modify the RAM circuit of Figure 11.48 as follows: change the OR gate to an AND gate and disconnect its output from C; connect the AND output to E_3; connect C to ground. Determine the address range for each RAM module.

11-34. Show how to expand the system of Figure 11-48 to an $8K \times 8$ with addresses ranging from 0000 to 1FFF. (Hint: This can be done by adding the necessary memory modules and modifying the existing decoding logic.)

11-35. A dynamic test is performed on the decoding logic of Figure 11-48 by keeping $\emptyset_2 = 1$ and connecting the outputs of a 6-bit counter to address inputs $A_{10}-A_{15}$. The decoder outputs are monitored with an oscilloscope (or logic analyzer) as the counter is continuously pulsed by a 1-MHz clock. Figure 11.60(a) shows the displayed signals. What are the most probable faults?

11-36. Repeat Problem 11-35 for the decoder outputs shown in Figure 11-60(b).

11-37. Consider the RAM system of Figure 11-48. The checkerboard pattern test will not be able to detect certain types of faults. For instance, assume that there is a break in the connection to the A input to the decoder. If a checkerboard pattern SELF-TEST is performed on this circuit, the displayed messages will state that the memory is OK.
(a) Explain why the circuit fault was not detected.
(b) How would you modify the SELF-TEST so that faults such as this will be detected?

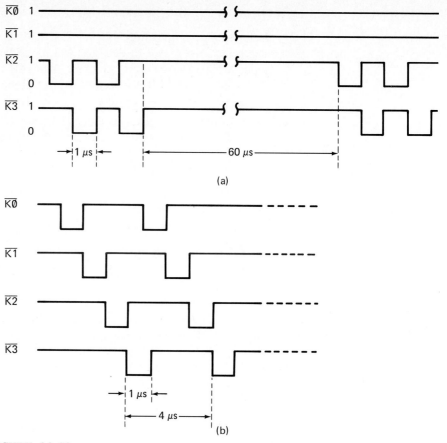

FIGURE 11-60

11-38. Assume that the 1K × 8 modules used in Figure 11-48 are formed from two 1K × 4 RAM chips (as in Problem 11-25). The following messages are printed out when the power-up self-test is performed on this RAM system.

```
module-0 test OK
module-1 test OK
address 0800 faulty at bits 4-7
address 0801 faulty at bits 4-7
address 0802 faulty at bits 4-7
       .      .     .     .     .
       .      .     .     .     .
       .      .     .     .     .
       .      .     .     .     .
address 0BFE faulty at bits 4-7
address 0BFF faulty at bits 4-7
module-3 test OK
```

Examine these messages and list the possible faults.

11-39. The following messages are printed out when the power-up self-test is performed on the RAM system of Figure 11-50.

```
module-0 test OK
module-1 test OK
module-2 test OK
address 0C00 faulty at bit 7
address 0C01 faulty at bit 7
address 0C02 faulty at bit 7
    .       .       .       .
    .       .       .       .
    .       .       .       .
    .       .       .       .
address 0FFE faulty at bit 7
address 0FFF faulty at bit 7
```

Examine these messages and list the possible faults.

11-40. What messages would be printed out when a power-up self-test is performed on the RAM system of Figure 11.48 if there is a short between the decoder's $\overline{K2}$ and $\overline{K3}$ outputs?

Section 11-24

11-41. Consider the 16×8 ROM in Figure 11-6. Replace the data word stored at address location 1111 with a checksum calculated from the other 15 data words.

INTRODUCTION TO THE MICROPROCESSOR AND MICROCOMPUTER

Upon completion of this chapter, you will be able to:

- Describe the function and operation of each one of the five basic elements of any computer organization.

- Evaluate the advantages and disadvantages among the various types of computer word formats.

- Understand the difference between a microprocessor and a microcomputer.

- Analyze the fetch and execute cycles during the execution of a machine-language program.

- Understand the operational role of the different types of buses and their signals in a microcomputer.

- Determine the number of READ and WRITE cycles used by a μC during the execution of a program.

- Understand how and when the various CPU internal registers are used.

INTRODUCTION

It is no exaggeration to say that the microprocessor and microcomputer have revolutionized the electronics industry and have had a tremendous impact on many aspects of our lives. Large-scale integration (LSI) has so sharply reduced the size and cost of computers that designers routinely consider using the power and versatility of the microprocessor and microcomputer in a wide variety of applications.

In this chapter we will study the basic principles of operation of microcomputers. Since complete textbooks are devoted to the study of computers, our main objective here is only to provide a firm foundation for further study. Although we will concentrate on the microcomputer, most of the concepts and principles apply equally to computers of all sizes.

12-1   WHAT IS A DIGITAL COMPUTER?

A digital computer is a combination of digital devices and circuits that can perform a *programmed* sequence of operations with a minimum of human intervention. The sequence of operations is called a *program*. The program is a set of coded instructions that is stored in the computer's internal memory along with all the data that the program requires. When the computer is commanded to execute the program, it performs the instructions in the order that they are stored in memory until the program is completed. It does this at extremely high speeds without making any errors.

12-2   HOW DO COMPUTERS THINK?

Computers do not think! The computer *programmer* provides a *program* of instructions and data which specifies every detail of what to do, what to do it to, and when to do it. The computer is simply a high-speed machine that can manipulate data, solve problems, and make decisions, all under the control of the program. If the programmer makes a mistake in the program or puts in the wrong data, the computer will produce wrong results. A popular saying in the computer field is "garbage in gives you garbage out."

Perhaps a better question to ask at this point is: How does a computer go about executing a program of instructions? Typically, this question is answered by showing a diagram of a computer's architecture (arrangement of its various elements) and then going through the step-by-step process which the computer follows in executing the program. We will do this—but not yet. First, we will look at a somewhat far-fetched analogy that contains many of the concepts involved in computer operation.

12-3 SECRET AGENT 89

Secret Agent 89 is trying to find out how many days before a certain world leader is to be assassinated. His contact tells him that this information is located in a series of post office boxes. To ensure that no one else gets the information, it is spread through 10 different boxes. His contact gives him 10 keys along with the following instructions:

1. The information in each box is written in code.
2. Open box 1 first and execute the instruction located there.
3. Continue through the rest of the boxes in sequence unless instructed to do otherwise.
4. One of the boxes is wired to explode upon opening.

Agent 89 takes the 10 keys and proceeds to the post office, code book in hand.

Figure 12-1 shows the contents of the 10 post office boxes after having been decoded. Assume that you are Agent 89; begin at box 1 and go through the sequence of operations to find the number of days before the assassination attempt. Of course, it should not be as much work for you as it was for Agent 89 because you don't have to decode the messages. The answer is given in the next paragraph.

If you have proceeded correctly, you should have ended up at box 6 with an answer of 17. If you made a mistake, you might have opened box 7, in which case you are no longer with us. As you went through the sequence of operations, you essentially duplicated the types of operations and encountered many of the concepts that are part of a computer. We will now discuss these operations and concepts in the context of the secret-agent analogy and see how they are related to actual computers.

In case you have not already guessed, the post office boxes are like the *memory* in a computer, where *instructions* and *data* are stored. Post office boxes 1–6 contain instructions to be executed by the secret agent and boxes 8–10 contain the data called for by the instructions. (The contents of box 7, to our knowledge, has no counterpart in computers.) The numbers on each box are like the *addresses* of the locations in memory.

① Add the number stored in box ⑨ to your secret agent code number.	**②** Divide the previous result by the number stored in box ⑩.
③ Subtract the number stored in box⑧	**④** If the previous result is not equal to 30, go to box ⑦. Otherwise continue to next box.
⑤ Subtract 13 from the previous result.	**⑥** HALT. You now have the answer
⑦ BOMB! (too bad)	**⑧** 20
⑨ 11	**⑩** 2

FIGURE 12-1 **Ten post office boxes with coded message for Agent 89.**

Three different classes of instructions are present in boxes 1–6. Boxes 1, 2, 3, and 5 are instructions that call for *arithmetic operations*. Box 4 contains a *decision-making* instruction called a *conditional jump* or *conditional branch*. This instruction calls for the agent (or computer) to decide whether to jump to address 7 or to continue to address 5, depending on the result of the previous arithmetic operation. Box 6 contains a simple control instruction that requires no data or refers to no other address (box number). This *halt* instruction tells the agent that the problem is finished (program is completed) and to go no further.

Each of the arithmetic and conditional jump instructions consists of two parts—an *operation* and an *address*. For example, the first part of the first instruction specifies the operation of addition. The second part gives the address (box 9) of the data to be used in the addition. These data are usually called the *operand* and their address the *operand address*. The instruction in box 5 is a special case in which no operand address is specified. Instead, the operand (data) to be used in the subtraction operation is included as part of the instruction.

A computer, like the secret agent, decodes and then executes the instructions stored in memory *sequentially,* beginning with the first location. The instructions are executed in order unless some type of *branch* instruction (such as box 4) causes the operation to branch or jump to a new address location to obtain the next instruction. Once the branching occurs, instructions are executed sequentially beginning at the new address.

This is about as much information as we can extract from the secret-agent analogy. Each of the concepts we encountered will be encountered again in subsequent material. Hopefully, the analogy has furnished insights that should prove useful as we begin a more technical study of computers.

12-4 BASIC COMPUTER SYSTEM ORGANIZATION

Every computer contains five essential elements or units: the *arithmetic logic unit* (ALU), the *memory unit,* the *control unit,* the *input unit,* and the *output unit.* The basic interconnection of these units is shown in Figure 12-2. The arrows in this diagram indicate the direction in which data, information, or control signals are flowing. Two different-size arrows are used; the larger arrows represent data or information that actually consists of a relatively large number of parallel lines, and the smaller arrows represent control signals that are normally only one or a few lines. The various arrows are also numbered to allow easy reference to them in the following descriptions.

FIGURE 12-2   **Basic computer organization.**

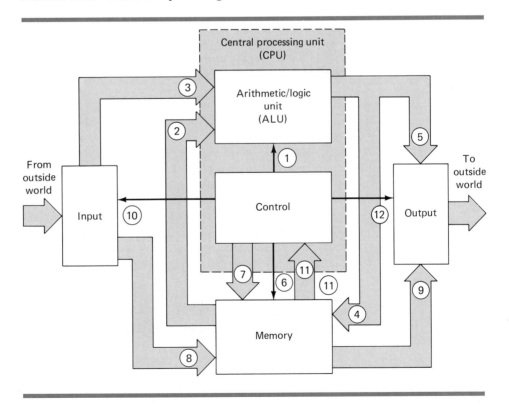

Arithmetic/Logic Unit   The ALU is the area of the computer in which arithmetic and logic operations are performed on data. The type of operation that is to be performed is determined by signals from the control unit (arrow 1). The data that are to be operated on by the ALU can come from either the memory unit (arrow 2) or the input unit (arrow 3). Results of operations performed in the ALU can be transferred to either the memory unit for storage (arrow 4) or to the output unit (arrow 5).

Memory Unit   The memory stores groups of binary digits (words) that can represent instructions (program) that the computer is to perform and the data that are to be operated on by the program. The memory also serves as storage for intermediate and final results of arithmetic operations (arrow 4). Operation of the memory is controlled by the control unit (arrow 6), which signals for either a read or a write operation. A given location in memory is accessed by the control unit, providing the appropriate address code (arrow 7). Information can be written into the memory from the ALU or the input unit (arrow 8), again under control of the control unit. Information can be read from memory into the ALU (arrow 2) or into the output unit (arrow 9).

Input Unit   This unit consists of all of the devices used to take information and data that are external to the computer and put them into the memory unit (arrow 8) or the ALU (arrow 3). The control unit determines where the input information is sent (arrow 10). The input unit is used to enter the program and data into the memory unit prior to starting the computer. This unit is also used to enter data into the ALU from an external device during the execution of a program. Some of the common input devices are keyboards, toggle switches, teletypewriters, punched-card readers, magnetic disk units, magnetic tape units, and analog-to-digital converters (ADC).

Output Unit   This unit consists of the devices used to transfer data and information from the computer to the "outside world." The output devices are directed by the control unit (arrow 12) and can receive data from memory (arrow 9) or the ALU (arrow 5), which is then put into appropriate form for external use. Examples of common output devices are LED readouts, indicator lights, printers, disk or tape units, cathode-ray-tube displays, and digital-to-analog converters (DAC).

As the computer executes its program, it usually has results or control signals that it must present to the external world. For example, a large computer system might have a line printer as an output device. Here, the computer sends out signals to print out the results on paper. A microcomputer might display its results on indicator lights or on LED displays.

Interfacing   The most important aspect of the I/O units involves *interfacing,* which can be defined as the joining of dissimilar devices in such a way that they are able to function in a compatible and coordinated manner. *Computer interfacing* is more specifically defined as the synchronization of digital information transmission between the computer and external input/output devices.

Many input/output devices are not directly compatible with the computer because of differences in such characteristics as operating speed, data format (e.g., hex, ASCII, binary), data transmission mode (e.g., serial, parallel), and logic signal level. Such I/O devices require special interface circuits which allow them to communicate with the CONTROL, MEMORY, and ALU portions of the computer system. A common example is the teletypewriter (abbreviated TTY), which can operate both as an input and an output device. The TTY transmits and receives data serially (one bit at a time) while most computers handle data in parallel form. Thus, a TTY requires interface circuitry in order to send data to or receive data from a computer.

Control Unit   The function of the control unit should now be obvious. It directs the operation of all the other units by providing timing and control signals. In a sense, the control unit is like the conductor of an orchestra, who is responsible for keeping each of the orchestra members in proper synchronization. This unit contains logic and timing circuits that generate the proper signals necessary to execute each instruction in a program.

The control unit *fetches* an instruction from memory by sending an address (arrow 7) and a read command (arrow 6) to the memory unit. The instruction word stored at the memory location is then transferred to the control unit (arrow 11). This instruction word, which is in some form of binary code, is then decoded by logic circuitry in the control unit to determine which instruction is being called for. The control unit uses this information to generate the necessary signals for *executing* the instruction.

Central Processing Unit (CPU)   In Figure 12-2, the ALU and control units are shown combined into one unit called the central processing unit (CPU). This is commonly done to separate the actual ''brains'' of the computer from the other units. We will use the CPU designation often in our work on microcomputers because, as we shall see, in microcomputers the CPU is often contained in a single LSI chip called the *microprocessor* chip.

REVIEW QUESTIONS

1. Name the five basic units of a computer and describe the major functions of each.
2. What is the CPU?
3. What is meant by interfacing in a computer system?

12-5   BASIC μC ELEMENTS

It is important that we understand the distinction between the microcomputer (μC) and the microprocessor (μP). A μC contains many elements, one of which is the μP. The μP is the central processing unit (CPU) portion of the μC. This is illustrated in Figure 12-3, where the basic elements of a μC are shown. The μP is typically a single LSI chip that contains all the control and arithmetic circuits of the

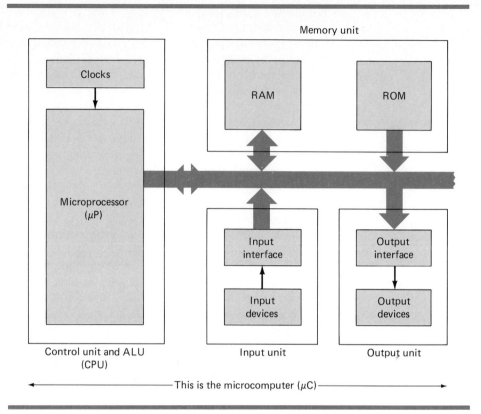

Memory unit

Clocks

RAM

ROM

Microprocessor
(μP)

Input
interface

Output
interface

Input
devices

Output
devices

Control unit and ALU
(CPU)

Input unit

Output unit

This is the microcomputer (μC)

FIGURE 12-3 **Basic elements of a μC.**

μC. The μP may consist of more than one chip. This is true, for example, of bipolar μPs (TTL, Shottky TTL, ECL) which do not have the high packing densities of the MOS devices, and so require two or more chips to produce a μP with appropriate word size.

The memory unit shows both RAM and ROM devices, typical of most μCs, although one or the other might not be present in certain applications. The RAM section consists of one or more LSI chips arranged to provide the designed memory capacity. This section of memory is used to store programs and data, which will change often during the course of operation. It is also used as storage for intermediate and final results of operations performed during execution of a program.

The ROM section consists of one or more LSI chips to store instructions and data that do not change. For example, it might store a program that causes the μC to continually monitor a keyboard, or it might store a table of ASCII codes needed for outputting information to a teletype unit.

The input and output sections contain the interface circuits needed to allow the I/O devices to properly communicate with the rest of the computer. In some cases these interface circuits are LSI chips designed by the μP manufacturer to interface the μP to a variety of I/O devices. In other cases the interface circuits may be as simple as a buffer register.

12-6 COMPUTER WORDS

The preceding description of how the various units in a computer interact has been, by necessity, somewhat oversimplified. To proceed in more detail, we must define the various forms of information that are continually being transferred and manipulated within the computer.

In a computer, the most elementary unit of information is the binary digit (bit). A single bit, however, can impart very little information. For this reason, the primary unit of information in a computer is a group of bits referred to as the *computer word*. Word size is so important that it is often used in describing a computer. For example, a 16-bit computer is a computer in which data and instructions are processed in 16-bit units. Of course, the word size also indicates the word size of the memory unit. Thus, a 16-bit computer has a memory unit that stores a certain number of 16-bit words.

A large variety of word sizes have been used by computer manufacturers. The larger (maxi) computers have word sizes that range from 16 to 64 bits, with 32 bits being the most common. Minicomputer word sizes run from 8 to 32, with 16 bits representing the overwhelming majority. Most microcomputers use an 8-bit word size. There are several 4-bit microcomputers which are designed for replacing digital logic circuits, some 16-bit and a few 32-bit microcomputers which are aimed at competing with minicomputers.

The Byte   A group of 8 bits is called a *byte* and represents a universally used unit in the computer industry. For example, a microcomputer with an 8-bit word size is said to have a word size of one byte. A 16-bit computer can be said to have a word size of two bytes. When we deal with microcomputers that have an 8-bit word size, we will use the terms "word" and "byte" interchangeably.

The 4-bit microcomputers have a word size of one-half byte. This is commonly referred to as a "nibble." Thus, each word in a 4-bit microcomputer is a nibble and two nibbles constitute a byte.

Types of Computer Words   A word stored in a computer's memory unit can contain several different types of information, depending on what the programmer intended for that particular word. We can classify computer words into three categories: (1) pure binary numerical data, (2) coded data, and (3) instructions. These will now be examined in detail.

12-7 BINARY DATA WORDS

These are words that simply represent a numerical quantity in the binary number system. For example, a certain location in the memory of an 8-bit process control microcomputer might contain the word 01110011, representing the desired process temperature in Fahrenheit degrees. This binary number 01110011 is equivalent to 115_{10}.

Here is an example of a 16-bit data word:

$$1010000101001001$$

which is equivalent to $41,289_{10}$.

Obviously, a wider range of numerical data can be represented with a larger word size. With an 8-bit word size, the largest data word (11111111_2) is equivalent to $2^8 - 1 = 255_{10}$. With a 16-bit word size, the largest data word is equivalent to $2^{16} - 1 = 65,535_{10}$. With 32 bits (four bytes) we can represent numbers greater than 4 billion.

Signed Data Words   A computer would not be too useful if it could only handle positive numbers. For this reason, most computers use the signed 2's-complement system. Recall that the most significant bit (MSB) is used as the *sign* bit (0 is positive and 1 is negative). Here is how the values $+9$ and -9 would be represented in an 8-bit computer:

$$+9 \longrightarrow \underset{\curvearrowright}{00001001}$$

$$+ \qquad \uparrow\text{——binary for } 9_{10}$$

$$-9 \longrightarrow \underset{\curvearrowright}{11110111}$$

$$- \qquad \uparrow\text{———— 2's complement of 0001001}$$

Here, of course, only 7 bits are reserved for the magnitude of the number. Thus, in the signed 2's-complement system, we can only represent numbers from -128_{10} to $+127_{10}$. Similarly, with 16-bit words, we can have a range from $-32,768_{10}$ to $+32,767_{10}$.

Multiword Data Units   Very often a computer needs to process data that extend beyond the range possible with a single word. For such cases, two or more memory words can be used to store the data in parts. For example, the 16-bit data word 1010101100101001_2 can be stored in two consecutive 8-bit memory locations, as shown below:

MEMORY ADDRESS (HEX)	MEMORY CONTENTS		
0030	10101011	→	8 high-order bits of 16-bit number
0031	00101001	→	8 low-order bits of 16-bit number

Here, address location 0030_{16} stores the 8 higher-order bits of the 16-bit data word. This is also called the *high-order byte*. Similarly, address 0031_{16} stores the *low-order byte*. The two bytes combined make up the full data word.

There is no actual limit to the number of memory words that can be combined to store large numbers.

Octal and Hex Data Representation   For purposes of convenience in writing and displaying data words, they can be represented in either octal or hexadecimal codes. For example, the number $+116_{10}$ can be represented in a single byte as 01110100_2. The hex and octal representations are:

$$01110100_2 = 74_{16}$$

$$01110100_2 = 164_8$$

It is important to realize that the use of hex or octal representations is solely for convenience of the computer user; the computer memory still stores the binary numbers (0s and 1s), and these are what the computer processes.

12-8   CODED DATA WORDS

Data processed by a computer do not have to be pure binary numbers. One of the other common data forms uses the BCD code, where each group of 4 bits can represent a single decimal digit. Thus, an 8-bit word can represent two decimal digits, a 16-bit word can represent four decimal digits, and so on. Many computers can perform arithmetic operations on BCD-coded numbers as part of their normal instruction repertoire; others, especially some microcomputers, require special effort on the part of the programmer in order to do BCD arithmetic.

Data words are not restricted to representing only numbers. They are often used to represent alphabetic characters and other special characters or symbols using codes such as the 7-bit ASCII code (Chapter 2). The ASCII code is used by all minicomputer and microcomputer manufacturers. Although the basic ASCII code uses 7 bits, an extra parity bit (Chapter 2) is added to each code word, producing a one-byte ASCII code. The example below shows how a message might be stored in a sequence of memory locations using ASCII code with an *even* parity bit. The contents of each location are also given in hex code. Use Table 2-5 to determine the message. Note that the leftmost bit is the parity bit and the first character is stored in location $012A_{16}$. The decoded message is the familiar electrical Ohm's law, $I = V/R$.

CONTENTS		
ADDRESS LOCATION	**BINARY**	**HEX**
012A	11001001	C9
012B	10111101	BD
012C	01010110	56
012D	10101111	AF
012E	11010010	D2
	ASCII	

Interpretation of Data Words   Suppose you are told that a particular data word in a microcomputer's memory is 01010110. This word can be interpreted in several ways. It could be the binary representation of 86_{10}; it could be the BCD representation of 56_{10}; or it could be the ASCII code for the character V. How should this data word be interpreted? It is up to the programmer, since he or she is the one who places the data in memory along with instructions that make up the program. The programmer knows what type of data word he or she is using and must make sure that the program of instructions executed by the computer interprets the data properly.

12-9   INSTRUCTION WORDS

The format used for *data* words varies only slightly among different computers, especially those with the same word size. This is not true, however, of the format for *instruction* words. These words contain the information necessary for a computer to execute its various operations, and the format and codes for these can vary widely from computer to computer. Depending on the computer, the information contained in an instruction word can be different. But, for most computers, the instruction words carry two basic units of information: the *operation* to be performed and the *address* of the *operand* (data) that is to be operated upon.

Figure 12-4 shows an example of a *single-address instruction word* for a hypothetical 20-bit computer. The 20 bits of the instruction word are divided into two parts. The first part of the word (bits 16–19) contains the *operation code (op code,* for short). The 4-bit op code represents the operation that the computer is being instructed to perform, such as addition or subtraction. The second part (bits 0–15) is the *operand address* which represents the location in memory where the operand is stored.

With 4 bits used for the op code, there are $2^4 = 16$ different possible op codes, with each one indicating a different instruction. This means that a computer using this instruction-word format is limited to 16 different possible instructions which it can perform. A more versatile computer would have a greater number of instructions and would therefore require more bits in its op code. In any case, each instruction that a computer can perform has a specific op code which the computer (control unit) must interpret (decode).

The instruction word of Figure 12-4 has 16 bits reserved for the operand ad-

FIGURE 12-4   **Typical single-address instruction word.**

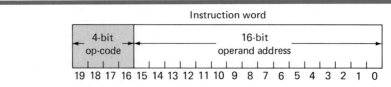

dress code. With 16 bits, there are $2^{16} = 65{,}536$ different possible addresses. Thus, this instruction word can specify 16 different instructions and 65,536 operand addresses. As an example, a 20-bit instruction word might be

Op code

Address code

The op code 0100 represents one of 16 possible operations; let's assume that it is the code for *addition* (ADD). The address code is 0101101001110010 or, more conveniently, 5A72 in hexadecimal. In fact, this complete instruction word can be expressed in hexadecimal as

$$4 \quad 5 \quad A \quad 7 \quad 2$$

Op code

Address

This complete instruction word, then, tells the computer to do the following:

Fetch the data word stored in address location 5A72, send it to the ALU and **add** *it to the number in the accumulator register. The sum will then be stored in the accumulator. (Previous contents of accumulator is lost.)*

We will examine this and other instructions more thoroughly later.

Multiple-Address Instructions   The single-address instruction described above is the basic type used in small computers and was once the principal type used in larger computers. The larger computers, however, have begun to use several other instruction formats which provide more information per instruction word.

Figure 12-5 shows two instruction-word formats that contain more than one address. The two-address instruction has the op code plus the addresses of *both* operands which are to take part in the specified operation. The three-address instruction has the addresses of both operands plus the address in memory where the result is to be stored.

These multiple-address instruction words have the obvious advantage that they contain more information than does a single-address instruction. This means that a computer using multiple-address instructions will require fewer instructions to execute a particular program. Of course, the longer instruction words require a memory unit with a larger word size. We will not concern ourselves further with multiple-address instructions since they are not used in microcomputers.

Multibyte Instructions   We have looked at instruction-word formats that contain op code and operand address information in a *single* word. In other words, a complete instruction such as those in Figure 12-4 or Figure 12-5 is stored in a *single* memory location. This is typical of computers with relatively large word sizes. For most microcomputers and many minicomputers, the smaller word size makes it impossible to provide the op code and operand address in a single word.

Op code	Address of operand 1	Address of operand 2

Two-address instruction

(a)

Op code	Address of operand 1	Address of operand 2	Address of where to store result

Three-address instruction

(b)

FIGURE 12-5 **Multiple-address instruction formats.**

Since the vast majority of microcomputers use an 8-bit (one-byte) word length, we will describe the instruction formats used in 8-bit computers. With a one-byte word size there are *three* basic instruction formats: single-byte, two-byte, and three-byte instructions. These are illustrated in Figure 12-6.

The single-byte instruction contains only an 8-bit op code, with no address portion. Clearly, this type of instruction does not specify any data from memory to be operated on. As such, single-byte instructions are used for operations that do not require memory data. An example would be the instruction *clear the accumulator register to zero* (CLA), which instructs the computer to clear all the FFs in the ALU's accumulator.

The first byte of the two-byte instruction is an op code and the second byte is an 8-bit address code specifying the memory location of the operand. In the three-byte instruction, the second and third bytes form a 16-bit operand address. For these multibyte instructions, the two or three bytes making up the complete instruction have to be stored in successive memory locations. This is illustrated in Table 12-1

TABLE 12-1

MEMORY ADDRESS (HEX)	MEMORY WORD		DESCRIPTION
	BINARY	HEX	
0020	01001010	4A	Op code for ADD
0021	00110101	F6	Low-order address bits (LO)
0022	11110110	35	High-order address bits (HI)
.	.	.	
.	.	.	
.	.	.	
.	.	.	
.	.	.	
35F6	01111100	7C	Operand

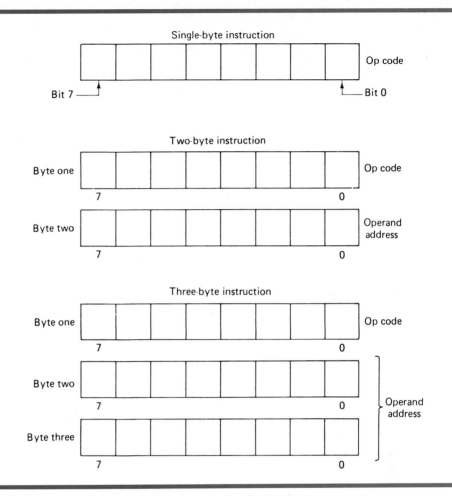

FIGURE 12-6 **Instruction formats used in 8-bit microcomputers.**

for a three-byte instruction. The left-hand column lists the address locations in memory where each byte (word) is stored. These addresses are given in hexadecimal code. The second column gives the binary word as it is actually stored in memory; the third column is the hex equivalent of this word. Examine this complete example before reading further and try to figure out what it represents.

The three bytes stored in locations 0020, 0021, and 0022 constitute the complete instruction for adding the data word stored in address location 35F6 to the accumulator. The second and third bytes hold the 8 low-order bits (LO), and 8 high-order bits (HI), respectively, of the operand address. Some microcomputers use the reverse order, with HI stored in the second byte and LO in the third byte of the instruction sequence. Memory location 35F6 is also shown; its contents is the data word (7C) which the control unit of the computer will fetch and send to the ALU for addition.

REVIEW QUESTIONS

1. Explain the difference between a data word and an instruction word.
2. What information is contained in a single-address instruction word?
3. How is this information represented in a three-byte instruction?

12-10 EXECUTING A MACHINE-LANGUAGE PROGRAM

The instruction words we have been describing are called *machine-language* instructions because they are represented by 1s and 0s, the only language the machine (computer) understands. Many other languages can be used to program a computer, and you may be familiar with some of them, such as BASIC or PASCAL. These *high-level languages* are designed to make it easy to write a program. It is important to understand that these high-level programs have to be converted to machine-language instructions and placed in the computer's internal memory before the computer can execute them. In a typical microcomputer, the conversion from BASIC to machine language is accomplished through a special machine-language program stored in ROM.

In order to illustrate how a microcomputer executes a machine-language program, we will use the instructions described in Table 12-2. These instructions do not pertain to any particular microcomputer but they are typical of the kinds of

TABLE 12-2

MNEMONIC	OP CODE, BINARY/HEX	DESCRIPTION OF OPERATION
LDA	01011011/5B	*Load accumulator:* the data stored at the operand address are loaded into the accumulator register.
ADD	01101101/6D	*Add:* the operand is added to the number stored in the accumulator and the resultant sum is stored in the accumulator.
SUB	10100010/A2	*Subtract:* the operand is subtracted from the contents of the accumulator and the result is stored in the accumulator.
STA	10001010/8A	*Store accumulator:* the contents of the accumulator is stored in memory at the location specified by the operand address.
JMP	01001100/4C	*Jump (unconditonally):* the next instruction is taken from the location specified by the operand address instead of in sequence.
JPZ	01110110/76	*Jump on zero:* the next instruction is taken from operand address *if* the accumulator contents is *zero.* Otherwise, the next instruction is taken in sequence.
HLT	00111111/3F	*Halt:* the computer operation is halted. No further instructions are executed.

instructions that most microcomputers can execute. Each instruction is accompanied by a *mnemonic* or abbreviation that is easier to remember than the op code. Read the description of each instruction carefully and keep in mind that the operand that is being referred to is the data word stored at the operand address location.

We will use some of the instructions listed in Table 12-2 to write a machine-language program that starts at address 0000 and does the following:

1. Subtracts an 8-bit data word (*X*) from another 8-bit data word (*Y*). The values of *X* and *Y* are stored in hex address locations 0300 and 0301, respectively.

2. Stores the difference (DIF) in memory location 0302.

3. If the value of DIF is zero, the program jumps to address 0400 for its next instruction. Otherwise, the program halts.

Table 12-3 shows the complete program as it would be entered into the computer's memory. Actually, the first two columns are the machine-language program; the other information is included for descriptive purposes. The first column lists the hex address of each memory location being used by the program. The second column gives the hex equivalent of the word stored in each memory location. Remember, these hex values represent the actual binary addresses and instruction codes that the computer understands.

The third column gives the mnemonic and operand address (if any) associated with each instruction. The last column describes the operation performed by each

TABLE 12-3

MEMORY ADDRESS (HEX)	MEMORY WORD (HEX)	MNEMONIC	DESCRIPTION
0000	5B	LDA$0300	Load data (*Y*) from address 0300 into accumulator
0001	00⎱		
0002	03⎰		
0003	A2	SUB$0301	Subtract data (*X*) at address 0301 from accumulator
0004	01⎱		
0005	03⎰		
0006	8A	STA$0302	Store difference in address 0302
0007	02⎱		
0008	03⎰		
0009	76	JPZ$0400	If difference (DIF) is zero, jump to 0400 for next instruction
000A	00⎱		
000B	04⎰		
000C	3F	HLT	If DIF is *not* zero, halt program.
.	.	.	.
.	.	.	.
.	.	.	.
0300	??	*Y*	Address of data *Y*
0301	??	*X*	Address of data *X*
0302	??	DIF	Address where difference is to be stored

instruction. For example, the first instruction in the program is a three-byte instruction. The first byte stored in memory address 0000 is the op code 5B. The second and third bytes stored in addresses 0001 and 0002 represent the operand address 0300. The mnemonic for this instruction is LDA $0300. The LDA is the abbreviation for the load accumulator operation, and the $0300 is the operand address. The $ is often used to indicate that the address is represented in hex. When the computer executes this instruction, it will read the data Y that are stored in location 0300 and will load them into the accumulator.

The program starts at address 0000 and ends at 000C. Address locations 0300–0302 are used for data storage. The contents of these latter locations are not given because they will generally be variables. Address locations between 000C and 0300 are not being used.

Program Execution

We will now proceed through the complete execution of this program and describe what the computer does at each step. Our intent here is to outline only the main operations without getting bogged down in the details of all the activity taking place inside the computer. In particular, we will see that the computer is always in one of two kinds of operating cycles: (1) a *fetch cycle* during which the CONTROL unit fetches the instruction codes (op code and operand address) from memory and (2) an *execute cycle* during which the CONTROL unit performs the operation called for by the op code.

The operation starts when the operator activates a START or RUN switch. This will initialize a PROGRAM COUNTER (PC) to a starting count of 0000. The PC is a counter within the CONTROL unit that keeps track of the program addresses as the computer sequences through them.

1. The CONTROL unit fetches the first byte from address 0000 as determined by the PC. This byte is 5B, which is the op code for the first instruction. Circuitry within the CONTROL unit determines that this op code calls for the LDA (load accumulator) operation and that an operand address follows the op code.

2. The CONTROL unit increments the PC to 0001 and then fetches the byte stored at this address. This byte is 00 and represents the low-order byte of the operand address. The CONTROL unit then increments PC again to 0002 and fetches the byte that is stored at this address. This byte is 03 and represents the high-order byte of the operand address.

3. This completes the first fetch cycle, and the CONTROL unit now holds the op code and operand address needed to execute the LDA instruction.

4. The CONTROL unit executes this instruction by reading the data word from address 0300 and loading it into the accumulator register in the ALU. This completes the execute cycle.

5. The PC is incremented to 0003 and the CONTROL unit begins a new fetch cycle by fetching the op code stored at this address. The CONTROL unit recognizes the op code A2 as calling for the SUB operation which requires an operand address that is stored in the next two bytes of the program.

6. The CONTROL unit fetches the operand address 0301 in the same manner as in step 2 above. The PC is now at 0005.

7. This completes the second fetch cycle, and the CONTROL unit now holds the op code and operand address needed to perform the SUB instruction. The CONTROL unit executes this instruction by reading the data word from address 0301 and sending it to the ALU to be subtracted from the contents of the accumulator.

8. The PC is incremented to 0006 to initiate a new fetch cycle. The CONTROL unit fetches the op code (8A) stored at this address, recognizes it as calling for the STA operation, and knows that the operand address is stored in the next two bytes of the program.

9. The CONTROL unit fetches the operand address 0302. The PC is now at 0008. The contents of the accumulator is stored in 0302.

10. The PC is incremented to 0009 to initiate a new fetch cycle, and the CONTROL unit fetches the op code (76) stored at this address. The CONTROL unit recognizes the JPZ instruction and knows that it has to make a decision based on the current contents of the accumulator.

11. If the data in the accumulator are exactly equal to zero, the CONTROL unit fetches the operand address 0400 from the next two memory addresses (000A and 000B). This operand address is loaded into the PC so that the CONTROL unit will take its next instruction from address 0400 and will continue executing instructions from that point.

12. If the data in the accumulator are *not* equal to zero, the CONTROL unit increments the PC three times so that PC = 000C. The next instruction is fetched from this address, and it is recognized as the HLT instruction which causes the CONTROL unit to stop fetching and executing instructions.

This simple program illustrates many of the kinds of things that take place as a computer executes a machine-language program, but it does not even begin to show the capabilities and versatility of a computer. It is important to understand that the program execution is performed in a step-by-step sequential manner starting at the first address in the program (0000 in our example). The program's address sequence can change when either a JMP or JPZ instruction is encountered (e.g., steps 10 and 11 above). Of course, even though the computer performs one step at a time, each step can be done so rapidly that the overall execution time for a program can be very short. For example, this short program would take less than 20 μs on even the slowest microcomputer.

REVIEW QUESTIONS

1. What is the difference between a machine-language program and a high-level-language program?
2. What is an instruction mnemonic?
3. What generally takes place during a fetch cycle? During an execute cycle?
4. What is the function of the PROGRAM COUNTER?

12-11 TYPICAL μC STRUCTURE

We are now prepared to take a more·detailed look at μC organization. The many possible μC structures are essentially the same in principle, although they vary as to the size of the data and address buses, and the types of control signals they use. In order to provide the clearest means for learning the principles of μC operation, it is necessary to choose a single type of μC structure and study it in detail. Once a solid understanding of this typical μC is obtained, it will be relatively easy to learn about any other type. The μC structure we have chosen to present here represents a common one in use today and is shown in Figure 12-7.

This diagram shows the basic elements of an 8-bit microcomputer system and the various buses that connect them together. Although this diagram looks somewhat complex, it still does not show all the details of the μC system. For our purposes, however, it will be sufficient.

FIGURE 12-7 **Typical 8-bit μC structure.**

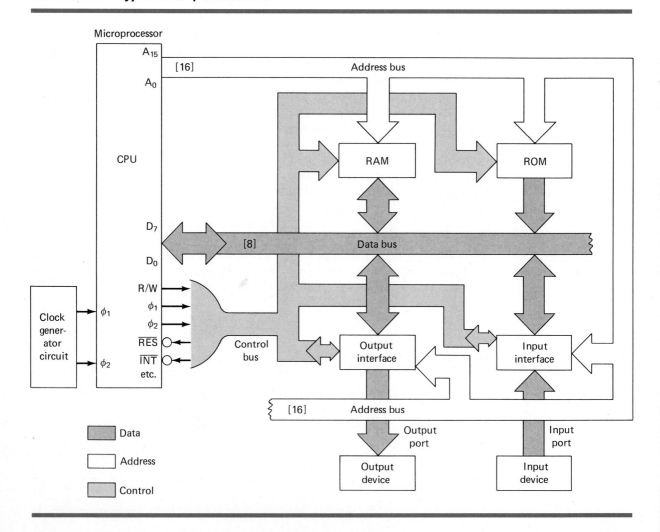

The Bus System   The μC has three buses which carry all the information and signals involved in the system operation. These buses connect the microprocessor (CPU) to each of the memory and I/O elements so that data and information can flow between the μP and any of these other elements. In other words, the CPU is continually involved in sending or receiving information to or from a location in memory, an input device, or an output device.*

In the μC, all information transfers are referenced to the CPU. When the CPU is sending data to another computer element, it is called a WRITE operation and the CPU is WRITING into the selected element. When the CPU is receiving data from another element, it is called a read operation and the CPU is READING from the selected element. It is very important to realize that the terms READ and WRITE always refer to operations performed by the CPU.

The buses involved in all the data transfers have functions that are described as follows:

Address bus: This is a *undirectional* bus, because information flows over it in only one direction, from the CPU to the memory or I/O elements. The CPU alone can place logic levels on the line of the address bus, thereby generating $2^{16} = 65{,}536$ different possible addresses. Each of these addresses corresponds to one memory location or one I/O element. For example, address $20A0_{16}$ might be a location in RAM or ROM where an 8-bit word is stored, or it might be an 8-bit buffer register that is part of the interface circuitry for a keyboard input device.

When the CPU wants to communicate (READ or WRITE) with a certain memory location or I/O device, it places the appropriate 16-bit address code on its 16 address pin outputs, $A_0 - A_{15}$, and onto the address bus. These address bits are then *decoded* to select the desired memory location or I/O device. This decoding process usually requires decoder circuitry not shown on this diagram.

Data bus: This is a *bidirectional* bus, because data can flow to or from the CPU. The CPU's eight data pins, $D_0 - D_7$, can be either inputs or outputs, depending on whether the CPU is performing a READ or a WRITE operation. During a READ operation they act as inputs and receive data that have been placed on the data bus by the memory or I/O element selected by the address code on the address bus. During a WRITE operation the CPU's data pins act as outputs and place data on the data bus, which are then sent to the selected memory or I/O element. In all cases, the transmitted data words are 8 bits long because the CPU handles 8-bit data words, making this an 8-bit μC.

In some microprocessors, the data pins are used to transmit other information in addition to data (e.g., address bits or CPU status information). That is, the data pins are time-shared or *multiplexed,* which means that special control signals must be generated by the CPU to tell the other elements exactly what is on the data bus at a particular time. We will not concern ourselves with this type of operation.

Control bus: This is the set of signals that is used to synchronize the activities of the separate μC elements. Some of these control signals, such as $R/\overline{W}$, are signals the CPU sends to the other elements to tell them what type of operation is currently in progress. The I/O elements can send control signals to the CPU. An example is the reset input $(\overline{RES})$ of the CPU, which, when driven LOW, causes the CPU to reset to a particular starting state. Another example is the CPU's interrupt input

---

*In some μC systems, it is possible for I/O devices to send data directly to or receive data directly from memory without the CPU being involved. This type of operation is called *direct memory access* (DMA).

(INT), used by I/O devices to get the attention of the CPU when it is performing other tasks.

The control bus signals will vary widely from one μC to another. There are certain control signals that all μCs use, but there are also many control signals that are peculiar to the μP upon which the μC is based. We will include only the essential control signals in this discussion.

Timing signals   The most important signals on the control bus are the system clock signals that generate the time intervals during which all system operations take place. Different μCs use different kinds of clock signals, depending on the type of μP being used. Some μPs, like the 8085, the 6502, and the Z-80, do not require an external clock-generating circuit. A crystal or RC network connected to the appropriate μP pins sets the operating frequency for the clock signals which are generated on the μP chip. Other μPs, such as the 8080A and 6800, require an external circuit to generate the clock signals needed by the CPU and the other μC elements. These manufacturers often provide a special clock-generator chip designed to be used with their μP.

Many of the currently popular μPs (8085, 6802, 6502) use a two-phase clock system with nonoverlapping pulses such as those shown in Figure 12-8. Other widely used μPs such as the Z-80 operate from a single clock signal. In our subsequent discussions we will use the two-phase clock system. The two clock phases, ϕ_1 and ϕ_2, are always part of the control bus. Other timing signals, derived from ϕ_1 and ϕ_2, are sometimes generated by the CPU and become part of the control bus.

I/O Ports   During the execution of a program, the CPU is constantly READING or WRITING into memory. The program may also call on the CPU to READ from one of the input devices or WRITE into one of the output devices. Although the diagram of the 8-bit μC (repeated in Figure 12-9) only shows one input and one output device there can be any number of each tied to the μC bus system. Each I/O device is normally connected to the μC bus system through some type of interface circuit. The function of the interface is to make the μC and the device compatible so that data can be easily passed between them. The interface is needed whenever the I/O device uses different signal levels, signal timing, or signal format than the μC.

For example, a typical I/O device is the standard teletype unit (abbreviated

FIGURE 12-8 Two-phase clock system.

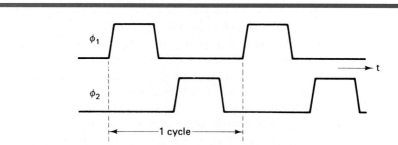

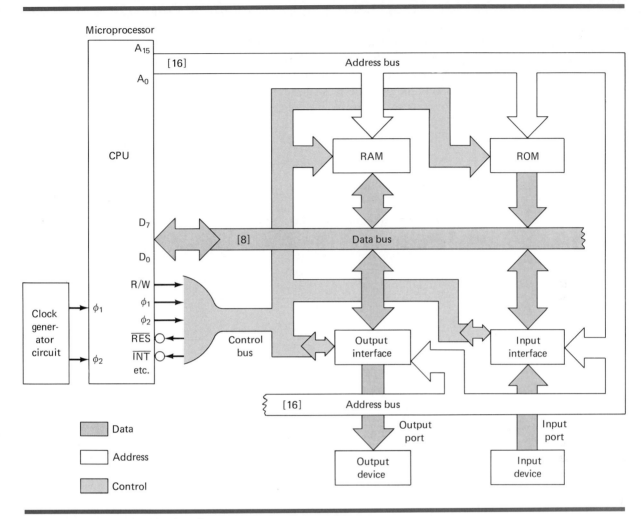

FIGURE 12-9 Typical 8-bit μC structure.

TTY), which sends ASCII-coded information to the computer in *serial* fashion (1 bit at a time over a single line). The μC, however, accepts data from the data bus as 8 *parallel* bits. Thus, an interface circuit is used to convert the TTY's serial signal to an 8-bit parallel data word, and another to convert the μC's parallel output data to a serial signal for the TTY.

It was mentioned during the discussion of the address bus that the CPU places a 16-bit address on this bus to select a certain memory location or a certain I/O device. This means that each I/O device has a specific address just like any location in memory. In many μCs, the CPU does not distinguish between memory and I/O, and it communicates with both in the same way using the same control signals. This method is called *memory-mapped I/O*. Other μCs use separate control signals and separate address decoders for I/O. This is called *isolated I/O*. We will concentrate mainly on the memory-mapped I/O technique, since it is the most common and has several advantages over the isolated I/O technique.

Although I/O devices are treated like memory locations, they are significantly different from memory in some respects. One big difference is that I/O devices can have the capability to *interrupt* the μC while it is executing a program. What this means is that an I/O device can send a signal to the μP chip's interrupt ($\overline{INT}$) input to tell the CPU that it wishes to communicate with it. The CPU will then suspend execution of the program it is currently working on and will perform the appropriate operation with the interrupting I/O device. RAM and ROM do not have interrupting capability.

REVIEW QUESTIONS

1. Describe the functions of the three buses that are part of a typical microcomputer system.
2. Which bus is unidirectional?
3. How does memory-mapped I/O differ from isolated I/O?

12-12 READ AND WRITE OPERATIONS

We are now ready to take a more detailed look at how the μP communicates with the other μC elements. Remember, the **μP is the CPU** and contains all the control and arithmetic/logic circuitry needed to execute a program of instructions stored in RAM or ROM. The CPU is continually performing READ and WRITE operations as it executes a program. It fetches each instruction from memory with a READ operation. After interpreting the instruction, it may have to perform a READ operation to obtain the operand from memory, or it may have to WRITE data into memory. In some cases, the instruction may call for the CPU to READ data from an input device (such as a keyboard or TTY) or to WRITE data into an output device (like an LED display or a magnetic tape cassette).

The READ Operation   The following steps take place during a READ operation:

1. The CPU generates the proper logic level on its $R/\overline{W}$ line for initiating a READ operation. Normally, $R/\overline{W} = 1$ for READ. The $R/\overline{W}$ line is part of the control bus and goes to all the memory and I/O elements.
2. Simultaneously, the CPU places the 16-bit address code onto the address bus to select the particular memory location or I/O device from which the CPU wants to receive data.
3. The selected memory or I/O element places an 8-bit word on the data bus. All nonselected memory and I/O elements will not affect the data bus because their *tristate* outputs will be in the disabled (Hi-Z) condition.

4. The CPU receives the 8-bit word from the data bus on its data pins, D_0–D_7. These data pins act as inputs whenever $R/\overline{W} = 1$. This 8-bit word is then latched into one of the CPU's internal registers, such as the accumulator.

This sequence can be better understood with the help of a timing diagram showing the interrelationship between the signals on the various buses (see Figure 12-10). Everything is referenced to the ϕ_1 and ϕ_2 clock signals. The complete READ operation occurs in one clock cycle. This is typically 1 μs for MOS microprocessors. The leading edge of ϕ_1 initiates the CPU's generation of the proper $R/\overline{W}$ and address signals. After a short delay, typically 100 ns for a MOS μP, the $R/\overline{W}$ line goes HIGH and the address bus holds the new address code (point 1 on timing diagram). Note that the address-bus waveform shows both possible transitions (LOW-to-HIGH and HIGH-to-LOW), because some of the 16 address lines

FIGURE 12-10   Typical μC timing for a READ operation.

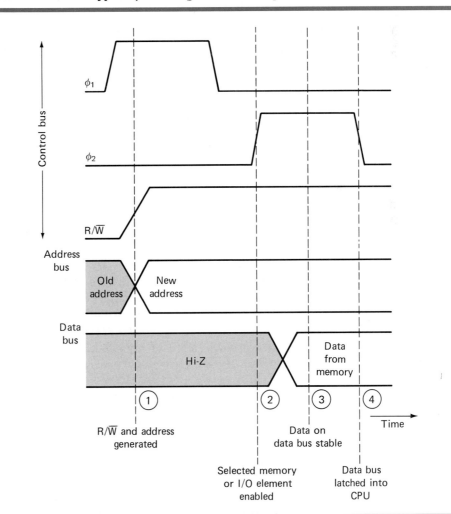

will be changing in one direction while others will be changing in the opposite direction.

During the ϕ_2 pulse, the selected memory or I/O device is enabled (point 2) and it proceeds to put its data word on the data bus. Prior to this, the data bus is in its Hi-Z state, since no device connected to it has been enabled. At some point during the ϕ_2 pulse, the data on the data bus become stable (point 3). Again, both possible data-line transitions are shown on the diagram. The delay between the start of the ϕ_2 pulse and the data-bus stabilizing depends on the speed of the memory and I/O elements. For memory this delay would be its *access time*. On the falling edge of ϕ_2, the data on the data bus are latched into the CPU (point 4). Clearly, then, the memory and I/O devices must be capable of putting data on the bus prior to the falling edge of ϕ_2, or proper transfer to the CPU will not occur. Thus, it is necessary to ensure that these devices have a speed compatible with the μC clock frequency.

EXAMPLE 12-1

A certain type of PROM has an access time specified as 750 ns (typical) and 1 μs (maximum). Can it be used with a μC that has a clock frequency of 1 MHz?

SOLUTION

No, with a clock frequency of 1 MHz the ϕ_2 pulse duration will be less than 500 ns. Thus, the PROM would have to have an access time of less than 500 ns for proper data transfer to the CPU.

The WRITE Operation   The following steps occur during a WRITE operation:

1. The CPU generates the proper logic level on the $R/\overline{W}$ line for initiating a WRITE operation. Normally, $R/\overline{W} = 0$ for WRITE.
2. Simultaneously, the CPU places the 16-bit address code onto the address bus.
3. The CPU then places an 8-bit word on the data bus via its data pins D_0–D_7, which are now acting as outputs. This 8-bit word typically comes from an internal CPU register, such as the accumulator. All other devices connected to the data bus have their outputs disabled.
4. The selected memory or I/O element takes the data from the data bus. All nonselected memory and I/O elements will not have their inputs enabled.

This sequence has the timing diagram shown in Figure 12-11. Once again, the leading edge of ϕ_1 initiates the $R/\overline{W}$ and address-bus signals (point 1). During the ϕ_2 pulse, the selected memory or I/O device is enabled (point 2) and the CPU places its data on the data bus. The data-bus levels become stabilized a short time into the ϕ_2 pulse (typically 100 ns). These data are then written into the selected memory location while ϕ_2 is high. If an I/O device has been selected, it usually latches the data from the data bus on the falling edge of ϕ_2 (point 4).

The READ and WRITE operations encompass most of the μC activity that takes place outside the CPU. The following example illustrates.

EXAMPLE 12-2

Below is a short program that is stored in memory locations 0020_{16}–0029_{16} of an 8-bit μC. Note that since each word is one byte (8 bits), it takes two successive bytes to represent a 16-bit operand address code. Determine the total number of READ and WRITE operations that the μC will perform as it executes this program.

MEMORY ADDRESS (HEX)	MEMORY WORD (HEX)	MNEMONIC	DESCRIPTION
0020	5B	LDA$5001	Load accumulator (ACC) with X.
0021	01⎫		⎡Address of⎤
0022	50⎭		⎣operand X⎦
0023	6D	ADD$5002	Add Y to contents of ACC.
0024	02⎫		⎡Address of⎤
0025	50⎭		⎣operand Y⎦
0026	8A	STA$5003	Store ACC contents.
0027	03⎫		⎡Address where ACC⎤
0028	50⎭		⎣will be stored⎦
0029	3F	HLT	Halt operation.

SOLUTION

The CPU begins executing the program by READING the contents of memory location 0020. The word stored there (5B) is taken into the CPU and is interpreted as an instruction op code (In other words, the CPU *always* interprets the first word of a program as an op code and the programmer must *always* adhere to this format.) The CPU decodes this op code to determine the operation to be performed and to determine if an operand address follows the op code. In this case, an operand address is required and is stored in the next two successive bytes in the program. Thus, the CPU must READ locations 0021 and 0022 to obtain the address of data X. Once this address (5001) has been read into the CPU, the CPU proceeds to execute the LDA instruction by READING memory location 5001 and putting its contents into the accumulator. Thus, the complete execution of this first instruction requires *four* separate READ operations; one for the op code, two for the address, and one for the LDA operation.

Similarly, *four* READ operations are needed for the second instruction, which begins at 0023. This instruction also uses a two-byte operand address and requires reading the contents of this address for transfer to the CPU's arithmetic unit.

The third instruction op code at 0026 uses a two-byte operand address. The CPU uses *three* READ operations to obtain these. However, the instruction STA calls for a WRITE operation whereby the CPU transfers the contents of the accumulator to memory location 5003.

The final instruction at 0029 is simply an op code with no operand address. The CPU READS this op code (3F), decodes it, and then halts further operations.

The total number of READ operations, then, is *twelve* and the total number of WRITE operations is *one*.

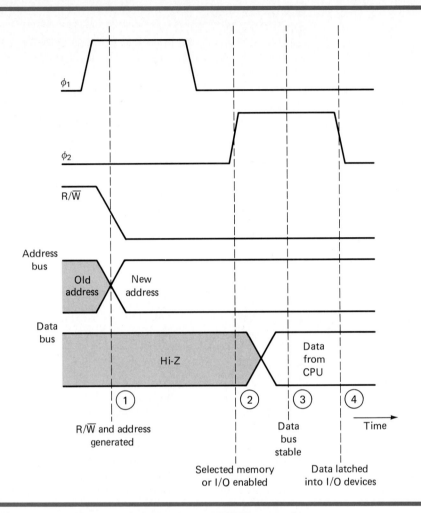

ϕ_1

ϕ_2

R/$\overline{\text{W}}$

Address bus

Old address   New address

Data bus

Hi-Z   Data from CPU

① ② ③ ④   Time

R/$\overline{\text{W}}$ and address generated

Data bus stable

Selected memory or I/O enabled

Data latched into I/O devices

FIGURE 12-11   Typical μC timing for a WRITE operation.

REVIEW QUESTIONS

1. Describe the steps that take place when the CPU performs a READ operation.
2. Repeat for a WRITE operation.
3. What device puts data on the data bus during a WRITE operation?

12-13 THE MICROPROCESSOR

All microcomputers, although they vary in their architecture, have one element in common—the microprocessor chip. As we know, the μP functions as the central processing unit of the μC. In essence, the μP is the heart of the μC because its capabilities determine the capabilities of the μC. Its speed determines the maximum speed of the μC, its address and data pins determine the μC's memory capacity and word size, and its control pins determine the type of I/O interfacing that must be used.

The μP performs a large number of functions, including:

1. Providing timing and control signals for all elements of the μC.
2. Fetching instructions and data from memory.
3. Transferring data to and from I/O devices.
4. Decoding instructions.
5. Performing arithmetic and logic operations called for by instructions.
6. Responding to I/O-generated control signals such as RESET and INTER-RUPT.

The μP contains all the logic circuitry for performing these functions, but it should be kept in mind that a great deal of the μP's internal logic is not externally accessible. For example, we cannot apply an external signal to the μP chip to increment the program counter (PC). Instead, the μP elements are *software*-accessible. This means that we can affect the internal μP circuitry only by the *program* we put in memory for the μP to execute. This is what makes the μP so versatile and flexible—when we want to change the μP's operation, we simply change the program (e.g., by changing the ROMs that store the program). This is generally easier than rewiring hardware.

Since we must construct the programs that tell the internal μP logic what to do, we need to become familiar with the internal μP structure, its characteristics, and its capabilities. The μP logic is extremely complex, but we can divide it into three areas: the register section, the control and timing section, and the ALU (see Figure 12-12). Although there is some overlap among these three areas, for clarity we will consider them separately.

FIGURE 12-12   Major functional areas of a μP chip.

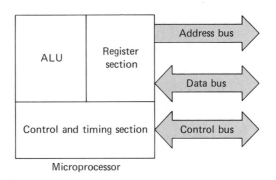

12-14 TIMING AND CONTROL SECTION

We will not discuss this part of the μP in too much detail for two reasons. First, it is the one area on the μP chip over which we have very little control. Second, we do not have to know the detailed structure of the timing and control section to be able to develop useful programs for the μP.

The major function of this μP section is to fetch and decode instructions from program memory, and then to generate the necessary control signals required by the ALU and register section for executing these instructions. The fetching and decoding functions correspond to the *fetch cycle,* and the control-signal generation takes place during the *execute cycle.* Both of these cycles were discussed earlier and we need not elaborate on them further.

Control-Bus Signals   The control section also generates *external* control signals that are sent to other μC elements as part of the system's control bus. We used some of these control-bus signals in our description of the μC, namely $R/\overline{W}$ and ϕ_2. In addition to generating output signals for the control bus, the μP control section also responds to control-bus signals that are sent from other μC elements to the μP chip. The RESET and INT (interrupt) mentioned earlier are examples.

Each μP has its own unique set of input and output control signals that are described in detail in the manufacturers' operation manuals. We will not attempt to define all of these here. Instead, we will describe some of the control-bus signals that are common to several different microprocessors.

RESET: All μPs have this input. When this input is activated, most of the μP's internal registers are reset to 0. In many μPs, the program counter (PC) is reset to 0 so that the instruction stored at memory location 0000_{16} is the first to be executed. In some μPs, activating the RESET input does not clear the PC. Instead, the PC is loaded from two specific memory locations (such as FFFE and FFFF). In other words, upon RESET the address of the first instruction is taken from these memory locations (each one stores one byte of the 16-bit address). Usually, this starting address is stored in ROM and is often referred to as an *address vector.*

$R/\overline{W}$: This μP output line informs the rest of the μC as to whether the μP is in a READ or WRITE operation. Some μPs use separate control lines, *RD* to indicate a READ operation and *WR* to indicate a WRITE operation.

MREQ (memory request): This μP output indicates that a memory access is in progress.

IORQ(I/O request): A μP output which indicates that an I/O device is being accessed. Some μPs use this signal along with MREQ to distinguish between memory and I/O operations. This allows memory and I/O to use the same addresses because the IORQ and MREQ signals determine which one (I/O or memory) is enabled. This technique of treating I/O separately from memory is called *isolated I/O.*

READY: This μP input is used by slow memory or I/O devices which cannot respond to a μP access request within one μP clock cycle. When the slow device is selected by the address decoding circuitry, it immediately sends a READY signal to the μP. In response, the μP suspends all its internal operations and enters what is called a WAIT state. It remains there until the device is ready to send or receive data, which the device indicates by removing the READY signal.

INT or IRQ (interrupt request): This is a µP input used by I/O devices to interrupt the execution of the current program and cause the µP to jump to a special program, called the *interrupt service routine*. The µP executes this special program, which normally involves servicing the interrupting device. When this execution is complete, the µP resumes execution of the program it had been working on when it was interrupted.

INTE (interrupt enable): This is a µP output that indicates to external devices whether or not the internal µP interrupt logic is enabled or disabled. If enabled (*INTE* = 1), the µP can be interrupted as described above. If disabled (*INTE* = 0), the µP will not respond to the *INT* or *IRQ* inputs. The state of *INTE* can be software-controlled. For example, a program can contain an instruction that makes *INTE* = 0 so that the interrupt operation is disabled.

NMI (nonmaskable interrupt): This is another µP interrupt input, but it differs from *INT* or *IRQ* in that its effect cannot be disabled. In other words, the proper signal on *NMI* will always interrupt the µP, regardless of the interrupt enable status.

REVIEW QUESTIONS

1. What are the major sections of a microprocessor?
2. What are the main functions of the timing and control section?

12-15 REGISTER SECTION

The most common operation that takes place *inside* the µP chip is the transfer of binary information from one register to another. The number and types of registers that a µP contains is a key part of its architecture, and it has a major effect on the programming effort required in a given application. The register structure of different µPs varies considerably from manufacturer to manufacturer. However, the basic functions performed by the various registers is essentially the same in all µPs. They are used to store data, addresses, instruction codes, and information on the status of various µP operations. Some are used as counters which can be controlled by software (program instructions) to keep track of such things as the number of times a particular sequence of instructions has been executed or sequential memory locations from which data are to be taken.

We will briefly describe the most common types of registers and their functions.

The Instruction Register (IR) This register is used to store the op code of the current instruction that is being fetched and executed. When the µP fetches the op code from memory, it stores it in the IR while the decoding circuitry determines which operation is to be performed. The IR is automatically used by the CPU during each instruction cycle and the programmer never needs to access this register. The size of the IR will be the same as the word size. For an 8-bit μP, the IR is 8 bits long.

The Program Counter (PC)

This register has also been discussed previously. The PC always contains the address in memory of the next instruction (or portion of an instruction) that the CPU is to fetch. When the μP RESET input is activated, the PC is set to the address of the first instruction to be executed. The μP places the contents of the PC on the address bus and fetches the first byte of the instruction from that memory location. (Recall that in 8-bit μCs one byte is used for the op code and the following one or two bytes for the operand address.) The μC automatically increments the PC after each use and in this way executes the stored program sequentially unless the program contains an instruction which alters the sequence (e.g., a JUMP instruction).

The size of the PC depends on the number of address bits the μP can handle. Most of the more common μPs use 16-bit addresses, but some use 12 bits. In either case, the PC will have the same number of bits as the address bus. In many μPs, the PC is divided into two smaller registers, PCH and PCL, each of which holds half an address. This is illustrated in Figure 12-13. PCH holds the 8 high-order bits of the 16-bit address and PCL holds the 8 low-order bits. The reason for using PCH and PCL is that it is often necessary to store the contents of PC in memory. Since the memory stores 8-bit words, the 16-bit PC must be broken into two halves and stored in two succcessive memory locations.

The programmer has no direct access to the PC. That is, there are no direct instructions to load the PC from memory or to store the PC contents in memory. However, there are many instructions that cause the PC to take on a value other than its normal sequential value. JMP (jump) and JPZ (jump on zero) are examples of instructions that change the PC so that the CPU executes instructions out of their normal sequential order.

Memory Address Register (MAR)

This is sometimes called a *storage address register* or an *address latching register*. It is used to hold the address of *data* the CPU is reading from or writing into memory. For example, when executing an ADD instruction, the CPU places the operand address portion of the ADD instruc-

FIGURE 12-13   **16-bit PC broken into two 8-bit units.**

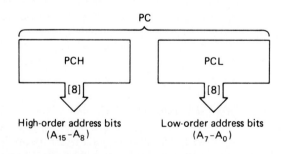

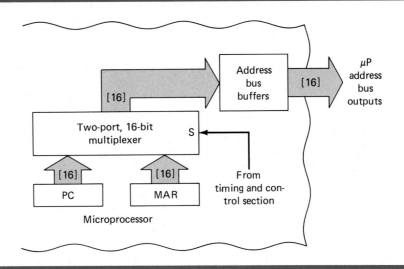

FIGURE 12-14 Portion of the μP showing how address source is selected for the address bus.

tion into the MAR. The contents of the MAR are then placed on the address bus so that the CPU can fetch the data during the next clock cycle.

It should now be apparent that there are two sources of addresses for the μP's address bus, the PC and the MAR. The PC is used for *instruction* addresses and the MAR is used for *data* addresses. A multiplexer is used to switch either the PC or MAR onto the address bus, depending on whether the CPU is an instruction cycle or an execution cycle. This is illustrated in Figure 12-14. Here the PC and MAR outputs are fed to a two-port multiplexer. The multiplexer's single output port is connected to the μP's address bus via the address bus buffers. The multiplexer's select input S is controlled by a signal from the μP's control section. One value of S selects the PC to be placed onto the address bus, and the other value of S selects the MAR.

Accumulator   The accumulator is a register that takes part in most of the operations performed by the ALU. It is also the register in which the results are placed after most ALU operations. In many ALU instructions, the accumulator is the source of one of the operands and the destination of the result.

In addition to its use in ALU instructions, the accumulator has other uses; for instance, it can be used as a storage register for data that are being sent to an output device or as a receiving register for data that are being read from an input device. Some μPs have more than one accumulator. In these μPs, the instruction op code specifies which accumulator is to be used.

The accumulator generally has the same number of bits as the μP's word size. Thus, an 8-bit μP will have an 8-bit accumulator. Some μPs also have an *extension register,* which is used in conjunction with the accumulator for handling binary numbers with more than 8 bits. For example, the μP has an arithmetic mode called "double-precision arithmetic," where each number is 16 bits long and is stored in two words of memory. When these 16-bit data words are sent to the CPU, the extension register stores the 8 least significant bits and the accumulator stores the 8 most significant bits. These two registers can be considered to be one 16-bit register.

General-Purpose Registers

These registers are used for many of the temporary storage functions required inside the CPU. They can be used to store data that are used frequently during a program, thereby speeding up the program execution, since the CPU does not have to perform a memory READ operation each time the data are needed. These registers are often used as counters which are used to keep track of the number of times a particular instruction sequence in a program has been executed. In some μPs, the general-purpose registers can also be used as index registers (described below) or as accumulators. The number of general-purpose registers will vary from μP to μP. Some μPs have none, while others may have 12 or more.

Index Registers

An index register, like general-purpose registers, can be used for general CPU storage functions and as a counter. In addition, it has a special function which is of great use in a program where tables or arrays of data must be handled. In this function, the index register takes part in determining the addresses of data that the CPU is accessing. This operation is called *indexed addressing* and is a special form of memory addressing available to the programmer. There are several different forms of indexed addressing, but the basic idea of indexed addressing is that the actual or *effective* operand address called for by an instruction is the sum of the operand address portion of the instruction *plus* the contents of the index register.

The Status Register

Also referred to as a *process status register* or *condition register,* the status register consists of individual bits with different meanings assigned by the μP manufacturer. These bits are called *flags,* and each flag is used to indicate the status of a particular μP condition. The value of some of the flags can be examined under program control to determine what sequence of instructions to follow. Two of the most common flags are the ZERO flag, Z, and the CARRY flag, C. The value of Z will always indicate whether or not the previous data-manipulation instruction produced a result of zero. Normally, a μP control signal sets Z to the HIGH state when the result of an instruction is zero and clears Z to a LOW when the result of an instruction is not zero. The value of C always indicates whether the previous instruction produced a result that exceeded the μP word size. For example, whenever the addition of two 8-bit data words produces a sum that exceeds 8 bits (i.e., a carry-out of the eighth position), the C flag will be set to 1. If the addition produces no carry, C will be 0. The C flag can be thought of as the ninth bit of any arithmetic result.

Most μP instruction sets contain several *conditional branch* instructions which determine the sequence of instructions to be executed, depending on the flag values. The jump-on-zero (JPZ) instruction is a prime example. When the CPU executes the JPZ instruction, it examines the value of the ZERO flag. If the ZERO flag is LOW, indicating that the previous instruction produced a nonzero result, the next instruction to be executed will be taken in normal sequence. If the ZERO flag is HIGH, indicating a zero result, the program will jump to the operand address for its next instruction and will continue in sequence from there. Other examples are jump-on-carry-set (JCS) and jump-on-carry-cleared (JCC), both of which examine the CARRY flag.

The Stack Pointer Register (SP) Before defining the function of this register, we must first define the *stack*. The stack is a portion of RAM reserved for the temporary storage and retrieval of information, typically the contents of the μP's internal registers. The *stack pointer* register acts as a special memory address register used only for the stack portion of RAM. Whenever a word is to be stored on the stack, it is stored at the address contained in the stack pointer. Similarly, whenever a word is to be read from the stack, it is read from the address specified by the stack pointer. The contents of the stack pointer is initialized by the progammer at the beginning of the program. Thereafter, the SP is automatically decremented *after* a word is stored on the stack and incremented *before* a word is read from the stack. (The incrementing and decrementing is done automatically by the μP control section.)

REVIEW QUESTION

1. For each statement below, indicate which μP register is being described.
    (a) Holds data address.
    (b) Results of arithmetic operations are placed here.
    (c) Is used in a special type of memory addressing.
    (d) Contains flags that indicate various conditions in the μP.
    (e) Holds instruction addresses.

12-16 ARITHMETIC/LOGIC UNIT

Most modern μPs have ALUs capable of performing a wide variety of arithmetic and logical operations. These operations can involve two operands, such as the accumulator and a data word from memory, or the accumulator and another μP internal register. Some of the operations involve only a single operand, such as the accumulator, a register, or a word from memory.

A simplified diagram of a typical ALU is shown in Figure 12-15. The ALU block represents all the logic circuitry used to perform arithmetic, logic, and manipulation operations on the operand inputs. Two 8-bit operands are shown as inputs to the ALU, although frequently only one operand is used. Also shown as an input to the ALU is the carry flag, C, from the μP status register. The reason for this will be explained later. The functions that the ALU will perform are determined by the control-signal inputs from the μP control section. The number of these control inputs varies from μP to μP.

The ALU produces two sets of outputs. One set is an 8-bit output, representing the results of the operation performed on the operands. The other is a set of

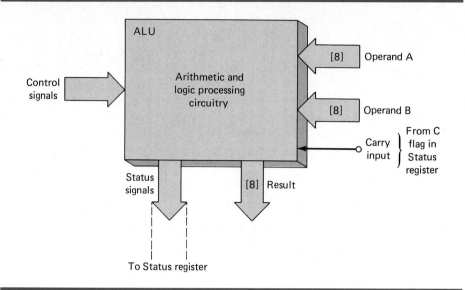

FIGURE 12-15 Simplified diagram of a typical μP ALU.

status signals which are sent to the μP status register to set or clear various flag bits. For example, if the result of an ALU operation is exactly zero, a signal is sent to the Z flag in the status register to set it to the HIGH state. If the result of the ALU operation produces a carry out of the MSB position, a signal is sent to set the C flag in the status register. Other signals from the ALU will set or clear other flags in the status register.

The operand inputs *A* and *B* can come from several sources. When two operands are to be operated on by the ALU, one of the operands comes from the accumulator and the other operand comes from a data word fetched from memory which is stored in a data buffer register within the μP. In some μPs, the second operand can also be the contents of one of the general-purpose registers. The result of the ALU operation performed on the two operands is normally sent to the accumulator.

When only one operand is to be operated on by the ALU, the operand can be the contents of the accumulator, a general-purpose register, an index register, or a memory data word. The result of the ALU operation is then sent back to the source of the operand. When the single operand is a memory word, the result is sent to a data buffer register, from where the CPU writes it back into the memory location of the operand.

Single-Operand Operations We will now describe some of the common ALU operations performed on a single operand.

1. Clear: All bits of the operand are cleared to 0. If the operand comes from the accumulator, for example, we can symbolically represent this operation as $0 \rightarrow [A]$.

2. Complement (or invert): All bits are changed to their opposite logic level. If the operand comes from register X, for example, this operation is represented as $[\overline{X}] \rightarrow [X]$.

3. Increment: The operand is increased by 1. For example, if the operand is 11010011, it will have 00000001 added to it in the ALU, to produce a result of 11010100. This instruction is very useful when the program is using one of the μP registers as a counter. Symbolically, this is represented as $[X + 1] \to [X]$.

4. Decrement: The operand is decreased by 1. In other words, the number 00000001 is subtracted from the operand. This instruction is useful in programs where a register is used to count down from an initial value. Symbolically, this operation is represented as $[X - 1] \to [X]$.

5. Shift: The bits of the operand are shifted to the left or to the right one place and the empty bit is made a 0. This process is illustrated below for a *shift-left* operation.

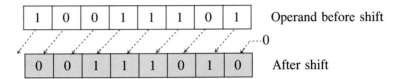

Note that a 0 is shifted into the rightmost bit. Also note that the original leftmost bit is shifted out and is lost. In most μPs the bit that is shifted out of the operand is not lost; instead, it is shifted into the carry flag bit, C, of the status register. An illustration of a *shift-right* operation is given below:

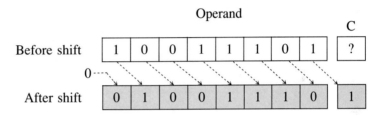

Note that a 0 is shifted into the leftmost bit and also note that the original rightmost bit is shifted into C (the original value of C is lost). For the corresponding shift-left operation, the leftmost bit would be shifted into C. This type of operation is used by the programmer to test the value of a specific bit in an operand. He does this by shifting the operand the required number of times until the bit value is in the C flag. A conditional jump instruction is then used to test the C flag to determine what instruction to execute next.

6. Rotate: This is a modified shift operation in which the C flag becomes part of a circulating shift register along with the operand; that is, the value shifted out of the operand is shifted into C and the previous value of C is shifted into the empty bit of the operand. This is illustrated below for a "rotate-right" operation.

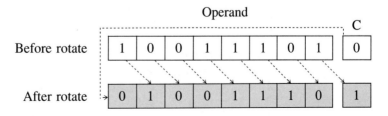

Note that the original C bit is shifted into the leftmost bit of the operand and that the rightmost bit of the operand is shifted into C. A rotate-left operates in the same manner except in the opposite direction.

Two-Operand ALU Operations

We will now describe some of the ALU operations performed on two operands.

1. Add: The ALU produces the binary sum of the two operands. Generally, one of the operands comes from the accumulator, the other from memory, and the result is sent to the accumulator. Symbolically, this operation is written $[A] + [M] \rightarrow [A]$. If the addition of the two operands produces a carry-out of the MSB position, the carry flag, C, in the status register is set to 1. Otherwise, C is cleared to 0. In other words, C serves as the ninth bit of the result.

2. Subtract: The ALU subtracts one operand (obtained from memory) from the second operand (the accumulator) and places the result in the accumulator. Symbolically, this is written $[A] - [M] \rightarrow [A]$. Most μPs use the 2's-complement method of subtraction (Chapter 6), whereby the operand from memory is 2's-complemented and then added to the operand from the accumulator. Once again, the carry bit generated by this operation is stored in the C flag.

3. Compare: This operation is the same as subtraction except that the result is not placed in the accumulator. Symbolically, this is written $[A] - [M]$. The subtraction is performed solely as a means for determining which operand is larger without affecting the contents of the accumulator. Depending on whether the result is positive, negative, or zero, various condition flags will be affected. The programmer can then use conditional jump instructions to test these flags.

4. Logical AND: The corresponding bits of the two operands are ANDed and the result is placed in the accumulator. Symbolically, this is written $[A] \cdot [M] \rightarrow [A]$. One of the operands is always the accumulator and the other comes from memory. As an example, let's assume that $[A] = 10110101$ and $[M] = 01100001$. The AND operation is performed as follows:

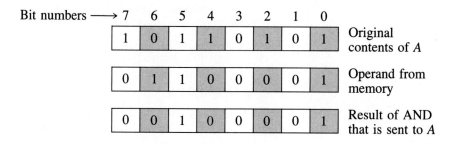

Note that each bit of the result is obtained by ANDing the corresponding bits of the operands. For example, bit 7 of the accumulator is a 1 and bit 7 of the memory word is a 0. Thus, bit 7 of the result is $1 \cdot 0 = 0$. Similarly, bit 5 of each operand is a 1, so that bit 5 of the result is $1 \cdot 1 = 1$.

5. Logical OR: The corresponding bits of the two operands are ORed and the result placed in the accumulator. Symbolically, this is shown as $[A]$ OR $[M] \rightarrow [A]$. A plus sign $(+)$ can be used in place of OR but it might cause confusion with the binary addition operation. Using the same two operands used in the illustration of the AND operation above, the result of the OR operation will be 11110101.

6. Exclusive-OR: The corresponding bits of the two operands are exclusive-ORed and the result is placed in the accumulator. Symbolically, this is written $[A] \oplus [M] \rightarrow [A]$. Using the same two operands used in the prior illustrations, the result of the EX-OR operation will be 11010100. The logic AND, OR, and EX-OR operations are very useful to a programmer.

7. Decimal (BCD) arithmetic: Many μPs have some provision for performing addition and subtraction in the BCD system whereby an 8-bit data word is treated as two BCD-coded decimal digits. As we pointed out in Chapter 6, arithmetic operations on BCD numbers require extra steps to obtain the correct results. While in some μPs these extra steps are performed automatically whenever BCD arithmetic instructions are executed, other μPs require the programmer to insert special instructions to correct the results.

This list of arithmetic/logic operations does not include the more complex operations of multiplication, division, square roots, and so on. These operations are not explicitly performed by most currently available μPs because of the extra circuitry they would require. To perform such operations, the programmer can instruct the μP to execute an appropriate series of simple arithmetic operations. For example, the multiplication operation can be obtained through a series of shifting and adding operations, and the division operation by a series of shifting and subtracting operations.

The sequences of instructions that the programmer develops for performing such complex arithmetic operations are called *subroutines*. A multiplication subroutine for obtaining the product of two 8-bit numbers might consist of as many as 20 or 30 instruction steps requiring 200 μs to execute. If this long execution time is undesirable, it is possible to use an external LSI chip containing a high-speed multiplier. Such LSI multipliers are available with execution times of only 100 ns and can be connected to the μP as I/O devices. Another alternative to slow subroutines is to use a ROM as a storage table which can store such data as multiplication tables, trigonometric tables, and logarithmic tables.

Signed Numbers   The circuitry in the ALU performs the add and subtract operations in the binary number system and always treats the two operands as 8-bit binary numbers. This is true even when the program calls for BCD arithmetic, the only difference being the extra steps needed to correct the BCD result. With an 8-bit data word, decimal numbers from 0 to 255 can be represented in binary code, assuming that all 8 bits are used for the numerical value. However, if the programming wishes to use both positive and negative numbers, the MSB of each data word is used as the *sign bit* with 0 for + and 1 for −. The other 7 bits of each data word represent the magnitude; for positive numbers the magnitude is in true binary form, while for negative numbers the magnitude is in 2's-complement form.

The beauty of this method for representing signed numbers is that it requires no special operations by the ALU. The circuitry in the ALU will perform addition and subtraction on the two operands in the same manner, regardless of whether the operands are unsigned 8-bit data words or 7-bit data words plus a sign bit. As was shown in Chapter 6, the sign bit participates in the add and subtract operations just like the rest of the bits. What this means is that the μP really does not know or care whether the data it is processing are signed or unsigned. Only the programmer is concerned about the distinction, and he must at all times know the format of the data he is using in his program.

As an aid to the programmer, almost all μPs transfer the value of bit 7 of the ALU result to a flag bit in the status register. This flag bit, sometimes called the SIGN flag, S, or the NEGATIVE flag, N, will be set to 1 if the ALU result has a 1 in bit 7 and will be cleared to 0 if bit 7 is 0. The program can then test to see if the result was positive or negative by performing a conditional jump instruction based on the SIGN flag. For example, a jump-on-negative instruction (JPN) will cause the CPU to examine the SIGN flag in the status register to determine what sequence of instructions to follow next.

REVIEW QUESTIONS

1. Indicate whether one or two operands are required for each of the ALU operations listed below:
   (a) Shift.
   (b) Decrement.
   (c) Exclusive-OR.
   (d) Rotate.
   (e) Complement.
   (f) Compare.

2. Assume the data word in the accumulator is 5B (hex), and a data word in a certain memory location is 16 (hex). Determine the result in the accumulator after these two words are ANDed. (*Ans.* $00010010_2 = 12_{16}$)

3. Repeat for the EX-OR operation. (Ans. $01001101_2 = 4D_{16}$)

4. Assume [A] = 36 (hex). What will be the accumulator data after two increments and a shift-right operation? (*Ans.* 1C)

12-17 FINAL COMMENTS

Our discussion has been, by necessity, only a brief introduction to the basic principles, terminology, and operations common to most microprocessors and microcomputers. We have not even begun to gain an insight into the capabilities and applications of these devices. Almost all areas of technology have started taking advantage of the inexpensive computer control which microprocessors can provide. Some typical applications include microwave ovens, traffic controllers, home computers electronic measuring instruments, industrial process control, electronic games, automobile emission control, and a rapidly growing number of new products.

With the revolutionary impact that microprocessors have had on the electronics industry, it is not unreasonable to expect that everyone working in electronics and related areas will have to become knowledgeable in the operation of microprocessors. Hopefully, our introduction will serve as a firm foundation for further study in this important area.

GLOSSARY

Access Time Time from when a memory location is selected to when it has stable output data. [Sec. 12-12]

Accumulator Principal register of an arithmetic/logic unit (ALU). [Sec. 12-15]

Address Bus Unidirectional lines that carry the address code from the CPU to memory and I/O devices. [Sec. 12-11]

Address Latching Register *See* Memory Address Register (MAR). [Sec. 12-15]

Arithmetic/Logic Unit (ALU) A digital circuit used in computers to perform various arithmetic and logic operations. [Secs. 12-4 and 12-16]

BASIC High-level language (Beginner's All-purpose Symbolic Instruction Code). [Sec. 12-10]

Byte Group of 8 bits. [Sec. 12-6]

Central Processing Unit (CPU) The part of a computer which is composed of the arithmetic/logic unit (ALU) and the control unit. [Sec. 12-4]

Computer Word Group of binary bits which is the primary unit of information in a computer. [Sec. 12-6]

Condition Register *See* Status Register. [Sec. 12-15]

Conditional Branch *See* Conditional Jump. [Sec. 12-3]

Conditional Jump Computer decision-making instruction which causes the CPU to jump to an address other than the next one in sequence *provided that a specific condition has been met*. [Sec. 12-10]

Control Bus Set of signal lines that are used to synchronize the activities of the CPU and the separate μC elements. [Sec. 12-11]

Control Unit That part of a computer which provides decoding of program instructions and the necessary timing and control signals for the execution of such instructions. [Sec. 12-4]

Data Bus Bidirectional lines that carry data between the CPU and memory, or between the CPU and I/O devices. [Sec. 12-11]

Execute Cycle Period during which a computer's control unit performs the operation specified by the fetched op code. [Sec. 12-10]

Fetch Cycle Period during which a computer's control unit obtains instruction codes from memory. [Sec. 12-10]

General-Purpose Register CPU register that functions as temporary storage of frequently used data during the execution of a program. [Sec. 12-15]

High-Level Language Computer programming language that utilizes the English language in order to facilitate the writing of a computer program. [Sec. 12-10]

High-Order Byte Eight higher-order bits of a 16-bit data word. [Sec. 12-7]

Index Register CPU register that functions both as general-purpose register and as special type of register used by the indexed addressing operations. [Sec. 12-15]

Indexed Addressing Special form of memory addressing where the effective address is determined by the addition of the operand address portion of the instruction to the contents of the index register. [Sec. 12-15]

Input Unit That part of a computer which facilitates the feeding of external information into the computer's memory unit or ALU. [Sec. 12-4]

Instruction Register (IR) CPU register where the op code of the current instruction is stored. [Sec. 12-15]

Interfacing Joining of dissimilar devices in such a way that they are able to function in a compatible and coordinated manner. [Sec. 12-4]

Isolated I/O When the CPU distinguishes between memory and I/O, and communicates with both in a different way using different control signals. [Sec. 12-11]

Low-Order Byte Eight lower-order bits of a 16-bit data word. [Sec. 12-7]

Machine Language Computer programming language in which groups of 1s and 0s are used to represent instructions. Machine language is also the only language that the computer circuitry understands. [Sec. 12-10]

Memory Place within the architecture of a computer where programming instructions and data reside. [Sec. 12-3]

Memory Address Register (MAR) CPU register where the address of data to be read from or written into memory is temporarily stored. [Sec. 12-15]

Memory-Mapped I/O When the CPU does not distinguish between memory and I/O, and communicates with both in the same way using the same control signals. [Sec. 12-11]

Memory Unit Part of a computer which stores instructions and data received from the input unit, as well as results from the arithmetic/logic unit. [Sec. 12-4]

Microprocessor (MPU) LSI computer chip that contains the central processing unit (CPU). [Sec. 12-4]

Mnemonic Abbreviation that represents the op code of a computer instruction. [Sec. 12-10]

Multibyte Instruction Computer instruction which is represented by more than one byte. [Sec. 12-9]

Multiple-Address Instruction Computer instruction word that contains more than one address. [Sec. 12-9]

Nibble Group of 4 bits. [Sec. 12-6]

Op Code That part of a computer instruction that defines what type of operation the computer is to execute on specified data. [Sec. 12-9]

Operand Data that are operated on by the computer as it executes an instruction. [Sec. 12-3]

Operand Address Address in memory where operand is currently stored or is to be stored. [Secs. 12-3 and 12-9]

Output Unit That part of a computer which receives data from the memory unit or ALU and presents it to the outside world. [Sec. 12-4]

Process Status Register *See* Status Register. [Sec. 12-15]

Program Sequence of binary-coded instructions designed to accomplish a particular task by a computer. [Sec. 12-2]

Program Counter (PC) CPU register where the address of the next instruction to be fetched is stored. [Secs. 12-10 and 12-15]

RAM One or more LSI memory chips which are used to store programs and data that will change often during the course of computer operation. [Sec. 12-5]

READ Term used to describe the condition when the CPU is receiving data from another element. [Sec. 12-11]

ROM One or more LSI memory chips which are used to store programs and data that do not change during the course of computer operation. [Sec. 12-5]

Signed Data Word Computer data word in which the MSB indicates whether that word represents a positive or negative quantity. [Sec. 12-7]

Stack Portion of RAM dedicated to the temporary storage and retrieval of data. [Sec. 12-15]

Stack Pointer Register (SP) Memory address register used to keep track of the next available stack location. [Sec. 12-15]

Status Register CPU register which consists of individual bits (flags) with different meanings assigned by the µP manufacturer. [Sec. 12-15]

Storage Address Register *See* Memory Address Register (MAR). [Sec. 12-15]

WRITE Term used to describe the condition when the CPU is sending data to another element. [Sec. 12-11]

APPENDIX I: POWERS OF 2

2^n	n	2^{-n}
1	0	1.0
2	1	0.5
4	2	0.25
8	3	0.125
16	4	0.062 5
32	5	0.031 25
64	6	0.015 625
128	7	0.070 812 5
256	8	0.003 906 25
512	9	0.001 953 125
1 024	10	0.000 976 562 5
2 048	11	0.000 488 281 25
4 096	12	0.000 244 140 625
8 192	13	0.000 122 070 312 5
16 384	14	0.000 061 035 156 25
32 768	15	0.000 030 517 578 125
65 536	16	0.000 015 258 789 062 5
131 072	17	0.000 007 629 394 531 25
262 144	18	0.000 003 814 697 265 625
524 288	19	0.000 001 907 348 632 812 5
1 048 576	20	0.000 000 953 674 316 406 25
2 097 152	21	0.000 000 476 837 158 203 125
4 194 304	22	0.000 000 238 418 579 101 562 5
8 388 608	23	0.000 000 119 209 289 550 781 25
16 777 216	24	0.000 000 059 604 644 775 390 625
33 554 432	25	0.000 000 029 802 322 387 695 312 5
67 108 864	26	0.000 000 014 901 161 193 847 656 25
134 217 728	27	0.000 000 007 450 580 596 923 828 125

APPENDIX II: MANUFACTURERS' IC DATA SHEETS

These data sheets are presented through the courtesy of Fairchild, a Schlumberger Company, and Texas Instruments.

CONNECTION DIAGRAMS
PINOUT A

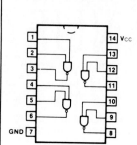

54/7400
54H/74H00
54S/74S00
54LS/74LS00
QUAD 2-INPUT NAND GATE

ORDERING CODE: See Section 9

PKGS	PIN OUT	COMMERCIAL GRADE V_{CC} = +5.0 V ±5%, T_A = 0° C to +70° C	MILITARY GRADE V_{CC} = +5.0 V ±10%, T_A = -55° C to +125° C	PKG TYPE
Plastic DIP (P)	A	7400PC, 74H00PC 74LS00PC, 74S00PC		9A
Ceramic DIP (D)	A	7400DC, 74H00DC 74LS00DC, 74S00DC	5400DM, 54H00DM 54LS00DM, 54S00DM	6A
Flatpak (F)	A	74LS00FC, 74S00FC	54LS00FM, 54S00FM	3I
	B	7400FC, 74H00FC	5400FM, 54H00FM	

PINOUT B

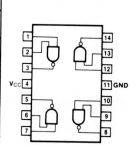

INPUT LOADING/FAN-OUT: See Section 3 for U.L. definitions

PINS	54/74 (U.L.) HIGH/LOW	54/74H (U.L.) HIGH/LOW	54/74S (U.L.) HIGH/LOW	54/74LS (U.L.) HIGH/LOW
Inputs	1.0/1.0	1.25/1.25	1.25/1.25	0.5/0.25
Outputs	20/10	12.5/12.5	25/12.5	10/5.0 (2.5)

DC AND AC CHARACTERISTICS: See Section 3*

SYMBOL	PARAMETER	54/74 Min	54/74 Max	54/74H Min	54/74H Max	54/74S Min	54/74S Max	54/74LS Min	54/74LS Max	UNITS	CONDITIONS
I_{CCH}	Power Supply		8.0		16.8		16		1.6	mA	V_{IN} = Gnd
I_{CCL}	Current		22		40		36		4.4		V_{IN} = Open, V_{CC} = Max
t_{PLH}	Propagation Delay		22		10	2.0	4.5		10	ns	Figs. 3-1, 3-4
t_{PHL}			15		10	2.0	5.0		10		

*DC limits apply over operating temperature range; AC limits apply at T_A = +25°C and V_{CC} = +5.0 V.

54/7404
54H/74H04
54S/74S04
54S/74S04A
54LS/74LS04

HEX INVERTER

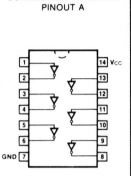

ORDERING CODE: See Section 9

PKGS	PIN OUT	COMMERCIAL GRADE $V_{CC} = +5.0$ V ±5%, $T_A = 0°$C to +70°C	MILITARY GRADE $V_{CC} = +5.0$ V ±10%, $T_A = -55°$C to +125°C	PKG TYPE
Plastic DIP (P)	A	7404PC, 74H04PC 74S04PC, 74S04APC 74LS04PC		9A
Ceramic DIP (D)	A	7404DC, 74H04DC 74S04DC, 74S04ADC 74LS04DC	5404DM, 54H04DM 54S04DM, 54S04ADM 54LS04DM	6A
Flatpak (F)	A	74S04FC, 74S04AFC 74LS04FC	54S04FM, 54S04AFM 54LS04FM	3I
	B	7404FC, 74H04FC	5404FM, 54H04FM	

PINOUT B

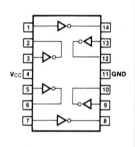

INPUT LOADING/FAN-OUT: See Section 3 for U.L. definitions

PINS	54/74 (U.L.) HIGH/LOW	54/74H (U.L.) HIGH/LOW	54/74S (U.L.) HIGH/LOW	54/74LS (U.L.) HIGH/LOW
Inputs	1.0/1.0	1.25/1.25	1.25/1.25	0.5/0.25
Outputs	20/10	12.5/12.5	25/12.5	10/5.0 (2.5)

DC AND AC CHARACTERISTICS: See Section 3*

SYMBOL	PARAMETER	54/74 Min	54/74 Max	54/74H Min	54/74H Max	54/74S Min	54/74S Max	54/74LS Min	54/74LS Max	UNITS	CONDITIONS
I_{CCH}	Power Supply		12		26		24		2.4	mA	V_{IN} = Gnd, V_{CC} = Max
I_{CCL}	Current		33		58		54		6.6		V_{IN} = Open
t_{PLH}	Propagation Delay		22		10	2.0	4.5		10	ns	Fig. 3-1, 3-4
t_{PHL}			15		10	2.0	5.0		10		
t_{PLH}	Propagation Delay					1.0	3.5			ns	Fig. 3-1, 3-4
t_{PHL}	(54/74S04A only)					1.0	4.0				

*DC limits apply over operating temperature range; AC limits apply at T_A = +25°C and V_{CC} = +5.0 V.

54/7405
54H/74H05
54S/74S05
54S/74S05A
54LS/74LS05

HEX INVERTER
(With Open-Collector Output)

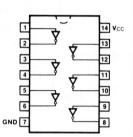

ORDERING CODE: See Section 9

PKGS	PIN OUT	COMMERCIAL GRADE V_{CC} = +5.0 V ±5%, T_A = 0° C to +70° C	MILITARY GRADE V_{CC} = +5.0 V, ±10%, T_A = -55° C to +125° C	PKG TYPE
Plastic DIP (P)	A	7405PC, 74H05PC 74S05PC, 74S05APC 74LS05PC		9A
Ceramic DIP (D)	A	7405DC, 74H05DC 74S05DC, 74S05ADC 74LS05DC	5405DM, 54H05DM 54S05DM, 54S05ADM 54LS05DM	6A
Flatpak (F)	A	74S05FC, 74S05AFC 74LS05FC	54S05FM, 54S05AFM 54LS05FM	3I
	B	7405FC, 74H05FC	5405FM, 54H05FM	

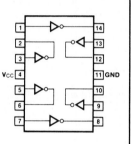

INPUT LOADING/FAN-OUT: See Section 3 for U.L. definitions

PINS	54/74 (U.L.) HIGH/LOW	54/74H (U.L.) HIGH/LOW	54/74S (U.L.) HIGH/LOW	54/74LS (U.L.) HIGH/LOW
Inputs	1.0/1.0	1.25/1.25	1.25/1.25	0.5/0.25
Outputs	OC**/10	OC**/12.5	OC**/12.5	OC**/5.0 (2.5)

DC AND AC CHARACTERISTICS: See Section 3*

SYMBOL	PARAMETER	54/74 Min	54/74 Max	54/74H Min	54/74H Max	54/74S Min	54/74S Max	54/74LS Min	54/74LS Max	UNITS	CONDITIONS
I_{CCH}	Power Supply		12		26		19.8		2.4	mA	V_{IN} = Gnd, V_{CC} = Max
I_{CCL}	Current		33		58		54		6.6		V_{IN} = Open
t_{PLH}	Propagation Delay		55		18	2.0	7.5		22	ns	Fig. 3-2, 3-4
t_{PHL}			15		15	2.0	7.0		18		
t_{PLH}	Propagation Delay					2.0	5.5			ns	Fig. 3-2, 3-4
t_{PHL}	(54S/74S05A only)					1.5	5.0				

*DC limits apply over operating temperature range; AC limits apply at T_A = +25°C and V_{CC} = +5.0 V.
**OC — Collector

54/7406

HEX INVERTER BUFFER/DRIVER
(With Open-Collector High-Voltage Output)

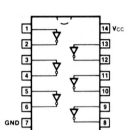

ORDERING CODE: See Section 9

PKGS	PIN OUT	COMMERCIAL GRADE V_{CC} = +5.0 V ±5%, T_A = 0°C to +70°C	MILITARY GRADE V_{CC} = +5.0 V ±10%, T_A = -55°C to +125°C	PKG TYPE
Plastic DIP (P)	A	7406PC		9A
Ceramic DIP (D)	A	7406DC	5406DM	6A
Flatpak (F)	A	7406FC	5406FM	3I

INPUT LOADING/FAN-OUT: See Section 3 for U.L. definitions

PINS	54/74 (U.L.) HIGH/LOW
Inputs	1.0/1.0
Outputs	OC**/10

DC AND AC CHARATERISTICS: See Section 3*

SYMBOL	PARAMETER		54/74 Min	54/74 Max	UNITS	CONDITIONS	
V_{OL}	Output LOW Voltage	XC		0.7	V	I_{OL} = 40 mA	V_{CC} = Min
		XM		0.7		I_{OL} = 30 mA	V_{IN} = V_{IH}
		XC, XM		0.4		I_{OL} = 16 mA	
I_{OH}	Output HIGH Current			0.25	mA	V_{OH} = 30 V, V_{CC} = Min V_{IN} = V_{IL}	
I_{CCH}	Power Supply Current			48	mA	V_{IN} = Gnd	V_{CC} = Max
I_{CCL}				51		V_{IN} = Open	
t_{PLH}	Propagation Delay			15	ns	Fig. 3-2, 3-4	
t_{PHL}				23			

*DC limits apply over operating temperature range; AC limits apply at T_A = +25°C and V_{CC} = +5.0 V.
**OC — Open Collector

54/7414
54LS/74LS14
HEX SCHMITT TRIGGER INVERTER

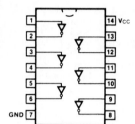

ORDERING CODE: See Section 9

PKGS	PIN OUT	COMMERCIAL GRADE V_{CC} = +5.0 V ±5%, T_A = 0°C to +70°C	MILITARY GRADE V_{CC} = +5.0 V ±10%, T_A = -55°C to +125°C	PKG TYPE
Plastic DIP (P)	A	7414PC, 74LS14PC		9A
Ceramic DIP (D)	A	7414DC, 74LS14DC	5414DM, 54LS14DM	6A
Flatpak (F)	A	7414FC, 74LS14FC	5414FM, 54LS14FM	3I

INPUT LOADING/FAN-OUT: See Section 3 for U.L. definitions

PINS	54/74 (U.L.) HIGH/LOW	54/74LS (U.L.) HIGH/LOW
Inputs	1.0/1.0	0.5/0.25
Outputs	20/10	10/5.0 (2.5)

DC AND AC CHARACTERISTICS: See Section 3*

SYMBOL	PARAMETER	54/74 Min	54/74 Max	54/74LS Min	54/74LS Max	UNITS	CONDITIONS
V_{T+}	Positive-going Threshold Voltage	1.5	2.0	1.5	2.0	V	V_{CC} = +5.0 V
V_{T-}	Negative-going Threshold Voltage	0.6	1.1	0.6	1.1	V	V_{CC} = +5.0 V
$V_{T+} - V_{T-}$	Hysteresis Voltage	0.4		0.4		V	V_{CC} = +5.0 V
I_{T+}	Input Current at Positive-going Threshold	-0.43**		-0.14**		mA	V_{CC} = +5.0 V, V_{IN} = V_{T+}
I_{T-}	Input Current at Negative-going Threshold	-0.56**		-0.18**		mA	V_{CC} = +5.0 V, V_{IN} = V_{T-}
I_{IL}	Input LOW Current		-1.2		-0.4	mA	V_{CC} = Max, V_{IN} = 0.4 V
I_{OS}	Output Short Circuit Current	-18	-55	-20	-100	mA	V_{CC} = Max, V_{OUT} = 0 V
I_{CCH}	Power Supply Current		36		16	mA	V_{IN} = Gnd / V_{CC} = Max
I_{CCL}			60		21		V_{IN} = Open / V_{CC} = Max
t_{PLH} t_{PHL}	Propagation Delay	22 22		22 22		ns	Figs. 3-1, 3-15

*DC limits apply over operating temperature range; AC limits apply at T_A = +25°C and V_{CC} = +5.0 V. **Typical Value

54/7420
54H/74H20
54S/74S20
54LS/74LS20
DUAL 4-INPUT NAND GATE

CONNECTION DIAGRAMS
PINOUT A

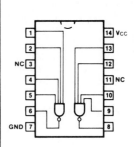

ORDERING CODE: See Section 9

PKGS	PIN OUT	COMMERCIAL GRADE Vcc = +5.0 V ±5%, TA = 0°C to +70°C	MILITARY GRADE Vcc = +5.0 V ±10%, TA = -55°C to +125°C	PKG TYPE
Plastic DIP (P)	A	7420PC, 74H20PC 74S20PC, 74LS20PC		9A
Ceramic DIP (D)	A	7420DC, 74H20DC 74S20DC, 74LS20DC	5420DM, 54H20DM 54S20DM, 54LS20DM	6A
Flatpak (F)	A	74S20FC, 74LS20FC	54S20FM, 54LS20FM	3I
	B	7420FC, 74H20FC	5420FM, 54H20FM	

PINOUT B

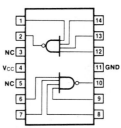

INPUT LOADING/FAN-OUT: See Section 3 for U.L. definitions

PINS	54/74 (U.L.) HIGH/LOW	54/74H (U.L.) HIGH/LOW	54/74S (U.L.) HIGH/LOW	54/74LS (U.L.) HIGH/LOW
Inputs	1.0/1.0	1.25/1.25	1.25/1.25	0.5/0.25
Outputs	20/10	12.5/12.5	25/12.5	10/5.0 (2.5)

DC AND AC CHARACTERISTICS: See Section 3*

SYMBOL	PARAMETER	'54/74 Min Max	54/74H Min Max	54/74S Min Max	54/74LS Min Max	UNITS	CONDITIONS
I_{CCH}	Power Supply	4.0	8.4	8.0	0.8	mA	V_{IN} = Gnd
I_{CCL}	Current	11	20	18	2.2		V_{IN} = Open, V_{CC} = Max
t_{PLH}	Propagation Delay	22	10	2.0 4.5	15	ns	Figs. 3-1, 3-4
t_{PHL}		15	10	2.0 5.0	15		

*DC limits apply over operating temperature range; AC limits apply at T_A = +25°C and V_{CC} = +5.0 V.

D2661, APRIL 1982 — REVISED MAY 1986

- Package Options Include Plastic "Small Outline" Packages, Ceramic Chip Carriers, and Standard Plastic and Ceramic 300-mil DIPs

- Dependable Texas Instruments Quality and Reliability

SN54ALS20A, SN54AS20 . . . J PACKAGE
SN74ALS20A, SN74AS20 . . . D OR N PACKAGE
(TOP VIEW)

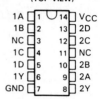

description

These devices contain two independent 4-input NAND gates. They perform the Boolean functions $Y = \overline{A \cdot B \cdot C \cdot D}$ or $Y = \overline{A} + \overline{B} + \overline{C} + \overline{D}$ in positive logic.

The SN54ALS20A and SN54AS20 are characterized for operation over the full military temperature range of $-55\,°C$ to $125\,°C$. The SN74ALS20A and SN74AS20 are characterized for operation from $0\,°C$ to $70\,°C$.

SN54ALS20A, SN54AS20 . . . FK PACKAGE
(TOP VIEW)

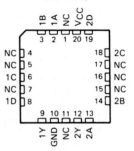

NC — No internal connection

FUNCTION TABLE (each gate)

INPUTS				OUTPUT
A	**B**	**C**	**D**	**Y**
H	H	H	H	L
L	X	X	X	H
X	L	X	X	H
X	X	L	X	H
X	X	X	L	H

logic diagram (positive logic)

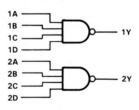

logic symbol[†]

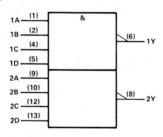

[†] This symbol is in accordance with ANSI/IEEE Std 91-1984 and IEC Publication 617-12.
Pin numbers shown are for D, J, and N packages.

TEXAS INSTRUMENTS
POST OFFICE BOX 655012 • DALLAS, TEXAS 75265

SN54ALS20A, SN74ALS20A
DUAL 4-INPUT POSITIVE-NAND GATES

absolute maximum ratings over operating free-air temperature range (unless otherwise noted)

Supply voltage, V_{CC} . . . . 7 V
Input voltage . . . . 7 V
Operating free-air temperature range: SN54ALS20A . . . . −55°C to 125°C
SN74ALS20A . . . . 0°C to 70°C
Storage temperature range . . . . −65°C to 150°C

recommended operating conditions

		SN54ALS20A			SN74ALS20A			UNIT
		MIN	NOM	MAX	MIN	NOM	MAX	
V_{CC}	Supply voltage	4.5	5	5.5	4.5	5	5.5	V
V_{IH}	High-level input voltage	2			2			V
V_{IL}	Low-level input voltage			0.7			0.8	V
I_{OH}	High-level output current			−0.4			−0.4	mA
I_{OL}	Low-level output current			4			8	mA
T_A	Operating free-air temperature	−55		125	0		70	°C

electrical characteristics over recommended operating-free-air temperature range (unless otherwise noted)

PARAMETER	TEST CONDITIONS		SN54ALS20A			SN74ALS20A			UNIT
			MIN	TYP†	MAX	MIN	TYP†	MAX	
V_{IK}	V_{CC} = 4.5 V,	I_I = −18 mA			−1.5			−1.5	V
V_{OH}	V_{CC} = 4.5 V to 5.5 V,	I_{OH} = −0.4 mA	$V_{CC}-2$			$V_{CC}-2$			V
V_{OL}	V_{CC} = 4.5 V,	I_{OL} = 4 mA		0.25	0.4		0.25	0.4	V
	V_{CC} = 4.5 V,	I_{OL} = 8 mA					0.35	0.5	
I_I	V_{CC} = 5.5 V,	V_I = 7 V			0.1			0.1	mA
I_{IH}	V_{CC} = 5.5 V,	V_I = 2.7 V			20			20	µA
I_{IL}	V_{CC} = 5.5 V,	V_I = 0.4 V			−0.1			−0.1	mA
I_O‡	V_{CC} = 5.5 V,	V_O = 2.25 V	−30		−112	−30		−112	mA
I_{CCH}	V_{CC} = 5.5 V,	V_I = 0 V		0.22	0.4		0.22	0.4	mA
I_{CCL}	V_{CC} = 5.5 V,	V_I = 4.5 V		0.81	1.5		0.81	1.5	mA

† All typical values are at V_{CC} = 5 V, T_A = 25°C.
‡ The output conditions have been chosen to produce a current that closely approximates one half of the true short-circuit output current, I_{OS}.

switching characteristics (see Note 1)

PARAMETER	FROM (INPUT)	TO (OUTPUT)	V_{CC} = 5 V, C_L = 50 pF, R_L = 500 Ω, T_A = 25°C	V_{CC} = 4.5 V to 5.5 V, C_L = 50 pF, R_L = 500 Ω, T_A = MIN to MAX				UNIT
			'ALS20A	SN54ALS20A		SN74ALS20A		
			TYP	MIN	MAX	MIN	MAX	
t_{PLH}	Any	Y	7	1	18	3	11	ns
t_{PHL}	Any	Y	6	1	15	3	10	

NOTE 1: Load circuit and voltage waveforms are shown in Section 1.

TEXAS
INSTRUMENTS
POST OFFICE BOX 655012 • DALLAS, TEXAS 75265

absolute maximum ratings over operating free-air temperature range (unless otherwise noted)

Supply voltage, V_{CC} . . . . 7 V
Input voltage . . . . 7 V
Operating free-air temperature range: SN54AS20 . . . . −55°C to 125°C
SN74AS20 . . . . 0°C to 70°C
Storage temperature range . . . . −65°C to 150°C

recommended operating conditions

		SN54AS20			SN74AS20			UNIT
		MIN	NOM	MAX	MIN	NOM	MAX	
V_{CC}	Supply voltage	4.5	5	5.5	4.5	5	5.5	V
V_{IH}	High-level input voltage	2			2			V
V_{IL}	Low-level input voltage			0.8			0.8	V
I_{OH}	High-level output current			−2			−2	mA
I_{OL}	Low-level output current			20			20	mA
T_A	Operating free-air temperature	−55		125	0		70	°C

electrical characteristics over recommended operating-free-air temperature range (unless otherwise noted)

PARAMETER	TEST CONDITIONS		SN54AS20			SN74AS20			UNIT
			MIN	TYP†	MAX	MIN	TYP†	MAX	
V_{IK}	$V_{CC} = 4.5$ V,	$I_I = -18$ mA			−1.2			−1.2	V
V_{OH}	$V_{CC} = 4.5$ V to 5.5 V,	$I_{OH} = -2$ mA	$V_{CC}-2$			$V_{CC}-2$			V
V_{OL}	$V_{CC} = 4.5$ V,	$I_{OL} = 20$ mA		0.35	0.5		0.35	0.5	V
I_I	$V_{CC} = 5.5$ V,	$V_I = 7$ V			0.1			0.1	mA
I_{IH}	$V_{CC} = 5.5$ V,	$V_I = 2.7$ V			20			20	μA
I_{IL}	$V_{CC} = 5.5$ V,	$V_I = 0.4$ V			−0.5			−0.5	mA
I_O‡	$V_{CC} = 5.5$ V,	$V_O = 2.25$ V	−30		−112	−30		−112	mA
I_{CCH}	$V_{CC} = 5.5$ V,	$V_I = 0$ V		1	1.6		1	1.6	mA
I_{CCL}	$V_{CC} = 5.5$ V,	$V_I = 4.5$ V		5.4	8.7		5.4	8.7	mA

† All typical values are at $V_{CC} = 5$ V, $T_A = 25$°C.
‡ The output conditions have been chosen to produce a current that closely approximates one half of the true short-circuit output current, I_{OS}.

switching characteristics (see Note 1)

PARAMETER	FROM (INPUT)	TO (OUTPUT)	$V_{CC} = 4.5$ V to 5.5 V, $C_L = 50$ pF, $R_L = 500$ Ω, $T_A =$ MIN to MAX				UNIT
			SN54AS20		SN74AS20		
			MIN	MAX	MIN	MAX	
t_{PLH}	Any	Y	1	5.5	1	5	ns
t_{PHL}	Any	Y	1	5	1	4.5	

NOTE 1: Load circuit and voltage waveforms are shown in Section 1.

TYPES SN54LS20, SN74LS20
DUAL 4-INPUT POSITIVE-NAND GATES

recommended operating conditions

		SN54LS20			SN74LS20			UNIT
		MIN	NOM	MAX	MIN	NOM	MAX	
V_{CC}	Supply voltage	4.5	5	5.5	4.75	5	5.25	V
V_{IH}	High-level input voltage	2			2			V
V_{IL}	Low-level input voltage			0.7			0.8	V
I_{OH}	High-level output current			−0.4			−0.4	mA
I_{OL}	Low-level output current			4			8	mA
T_A	Operating free-air temperature	−55		125	0		70	°C

electrical characteristics over recommended operating free-air temperature range (unless otherwise noted)

PARAMETER	TEST CONDITIONS †			SN54LS20			SN74LS20			UNIT
			MIN	TYP‡	MAX	MIN	TYP‡	MAX		
V_{IK}	V_{CC} = MIN,	I_I = −18 mA			−1.5			−1.5		V
V_{OH}	V_{CC} = MIN,	V_{IL} = MAX, I_{OH} = −0.4 mA	2.5	3.4		2.7	3.4			V
V_{OL}	V_{CC} = MIN,	V_{IH} = 2 V, I_{OL} = 4 mA		0.25	0.4			0.4		V
	V_{CC} = MIN,	V_{IH} = 2 V, I_{OL} = 8 mA					0.25	0.5		
I_I	V_{CC} = MAX,	V_I = 7 V			0.1			0.1		mA
I_{IH}	V_{CC} = MAX,	V_I = 2.7 V			20			20		μA
I_{IL}	V_{CC} = MAX,	V_I = 0.4 V			−0.4			−0.4		mA
I_{OS} §	V_{CC} = MAX		−20		−100	−20		−100		mA
I_{CCH}	V_{CC} = MAX,	V_I = 0 V		0.4	0.8		0.4	0.8		mA
I_{CCL}	V_{CC} = MAX,	V_I = 4.5 V		1.2	2.2		1.2	2.2		mA

† For conditions shown as MIN or MAX, use the appropriate value specified under recommended operating conditions.
‡ All typical values are at V_{CC} = 5 V, T_A = 25°C.
§ Not more than one output should be shorted at a time, and the duration of the short-circuit should not exceed one second.

switching characteristics, V_{CC} = 5 V, T_A = 25°C (see note 2)

PARAMETER	FROM (INPUT)	TO (OUTPUT)	TEST CONDITIONS		MIN	TYP	MAX	UNIT
t_{PLH}	Any	Y	R_L = 2 kΩ,	C_L = 15 pF		9	15	ns
t_{PHL}						10	15	ns

NOTE 2: See General Information Section for load circuits and voltage waveforms.

Texas Instruments
POST OFFICE BOX 225012 • DALLAS, TEXAS 75265

D2804, DECEMBER 1982 – REVISED MARCH 1984

absolute maximum ratings over operating free-air temperature range[†]

Supply voltage range, V_{CC} . . . . –0.5 V to 7 V
Input diode current, $I_{IK}(V_I < 0$ or $V_I > V_{CC})$ . . . . ±20 mA
Output diode current, $I_{OK}(V_O < 0$ or $V_O > V_{CC})$ . . . . ±20 mA
Continuous output current, I_O $(V_O = 0$ to $V_{CC})$ . . . . ±25 mA
Continuous current through V_{CC} or GND pins . . . . ±50 mA
Lead temperature 1,6 mm (1/16 inch) from case for 60 seconds: FH, FK, or J package . . . . 300°C
Lead temperature 1,6 mm (1/16 inch) from case for 10 seconds: FN or N package . . . . 260°C
Storage temperature range . . . . –65°C to 150°C

[†] Stresses beyond those listed under ''absolute maximum ratings'' may cause permanent damage to the device. These are stress ratings only and functional operation of the device at these or any other conditions beyond those indicated under ''recommended operating conditions'' is not implied. Exposure to absolute-maximum-rated conditions for extended periods may affect device reliability.

recommended operating conditions

			SN54HC'			SN74HC'			UNIT
			MIN	NOM	MAX	MIN	NOM	MAX	
V_{CC}	Supply voltage		2	5	6	2	5	6	V
V_{IH}	High-level input voltage	V_{CC} = 2 V	1.5			1.5			V
		V_{CC} = 4.5 V	3.15			3.15			
		V_{CC} = 6 V	4.2			4.2			
V_{IL}	Low-level input voltage	V_{CC} = 2 V	0		0.3	0		0.3	V
		V_{CC} = 4.5 V	0		0.9	0		0.9	
		V_{CC} = 6 V	0		1.2	0		1.2	
V_I	Input voltage		0		V_{CC}	0		V_{CC}	V
V_O	Output voltage		0		V_{CC}	0		V_{CC}	V
t_t	Input transition (rise and fall) times (except Schmitt-trigger inputs)	V_{CC} = 2 V	0		1000	0		1000	ns
		V_{CC} = 4.5 V	0		500	0		500	
		V_{CC} = 6 V	0		400	0		400	
T_A	Operating free-air temperature		–55		125	–40		85	°C

electrical characteristics over recommended operating free-air temperature range (unless otherwise noted)

PARAMETER	TEST CONDITIONS	V_{CC}	T_A = 25°C			SN54HC'		SN74HC'		UNIT
			MIN	TYP	MAX	MIN	MAX	MIN	MAX	
V_{OH} (Totem-pole outputs)	V_I = V_{IH} or V_{IL}, I_{OH} = –20 µA	2 V	1.9	1.998		1.9		1.9		V
		4.5 V	4.4	4.499		4.4		4.4		
		6 V	5.9	5.999		5.9		5.9		
	V_I = V_{IH} or V_{IL}, I_{OH} = –4 mA	4.5 V	3.98	4.30		3.7		3.84		
	V_I = V_{IH} or V_{IL}, I_{OH} = –5.2 mA	6 V	5.48	5.80		5.2		5.34		
I_{OH} (Open-drain outputs)	V_I = V_{IH} or V_{IL}, V_O = V_{CC}	6 V		0.01	0.5		10		5	µA
V_{OL}	V_I = V_{IH} or V_{IL}, I_{OL} = 20 µA	2 V		0.002	0.1		0.1		0.1	V
		4.5 V		0.001	0.1		0.1		0.1	
		6 V		0.001	0.1		0.1		0.1	
	V_I = V_{IH} or V_{IL}, I_{OL} = 4 mA	4.5 V		0.17	0.26		0.4		0.33	
	V_I = V_{IH} or V_{IL}, I_{OL} = 5.2 mA	6 V		0.15	0.26		0.4		0.33	
V_{T+}[†]		2 V	0.8	1.2	1.5					V
		4.5 V	2	2.5	3.15					
		6 V	2.5	3.3	4.2					
V_{T-}[†]		2 V	0.3	0.6	0.8					V
		4.5 V	0.9	1.6	2					
		6 V	1.2	2	2.5					
V_{T+} – V_{T-}[†]		2 V	0.2	0.6	1					V
		4.5 V	0.4	0.9	1.4					
		6 V	0.5	1.3	1.7					
I_I	V_I = 0 to V_{CC}	6 V		±0.1	±100		±1000		±1000	nA
I_{CC}	V_I = V_{CC} or 0, I_O = 0	6 V			2		40		20	µA
C_I		2 to 6 V		3	10		10		10	pF

[†]This parameter applies only for Schmitt-trigger inputs.

- Package Options Include Both Plastic and
  Ceramic Chip Carriers in Addition to Plastic
  and Ceramic DIPs

- Dependable Texas Instruments Quality
  and Reliability

**SN54HC20 . . . J PACKAGE
SN74HC20 . . . J OR N PACKAGE
(TOP VIEW)**

```
1A   [1  U 14]  VCC
1B   [2    13]  2D
NC   [3    12]  2C
1C   [4    11]  NC
1D   [5    10]  2B
1Y   [6     9]  2A
GND  [7     8]  2Y
```

description

These devices contain two independent 4-input NAND gates. They perform the Boolean functions $Y = \overline{A \cdot B \cdot C \cdot D}$ or $Y = \overline{A} + \overline{B} + \overline{C} + \overline{D}$ in positive logic.

The SN54HC20 is characterized for operation over the full military temperature range of −55°C to 125°C. The SN74HC20 is characterized for operation from −40°C to 85°C.

**SN54HC20 . . . FH OR FK PACKAGE
SN74HC20 . . . FH OR FN PACKAGE
(TOP VIEW)**

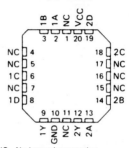

NC—No internal connection

logic symbol

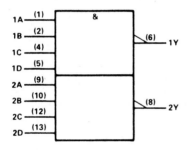

Pin numbers shown are for J and N packages.

FUNCTION TABLE (each gate)

INPUTS				OUTPUT
A	**B**	**C**	**D**	**Y**
H	H	H	H	L
L	X	X	X	H
X	L	X	X	H
X	X	L	X	H
X	X	X	L	H

maximum ratings, recommended operating conditions, and electrical characteristics

See Table I, page 2-4.

switching characteristics over recommended operating free-air temperature range (unless otherwise noted), C_L = 50 pF (see Note 1)

PARAMETER	FROM (INPUT)	TO (OUTPUT)	V_{CC}	T_A = 25°C			SN54HC20		SN74HC20		UNIT
				MIN	TYP	MAX	MIN	MAX	MIN	MAX	
t_{pd}	A, B, C, or D	Y	2 V		45	110		165		140	ns
			4.5 V		14	22		33		28	
			6 V		11	19		28		24	
t_t		Y	2 V		27	75		110		95	ns
			4.5 V		9	15		22		19	
			6 V		7	13		19		16	

C_{pd}	Power dissipation capacitance per gate	No load, T_A = 25°C	25 pF typ

NOTE 1: For load circuit and voltage waveforms, see page 1-14.

**TEXAS
INSTRUMENTS**
POST OFFICE BOX 225012 • DALLAS, TEXAS 75265

54/7432
54S/74S32
54LS/74LS32

QUAD 2-INPUT OR GATE

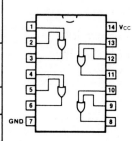

ORDERING CODE: See Section 9

PKGS	PIN OUT	COMMERCIAL GRADE V_{CC} = +5.0 V ±5%, T_A = 0°C to +70°C	MILITARY GRADE V_{CC} = +5.0 V ±10%, T_A = -55°C to +125°C	PKG TYPE
Plastic DIP (P)	A	7432PC, 74S32PC 74LS32PC		9A
Ceramic DIP (D)	A	7432DC, 74S32DC 74LS32DC	5432DM, 54S32DM 54LS32DM	6A
Flatpak (F)	A	7432FC, 74S32FC 74LS32FC	5432FM, 54S32FM 54LS32FM	3I

INPUT LOADING/FAN-OUT: See Section 3 for U.L. definitions

PINS	54/74 (U.L.) HIGH/LOW	54/74S (U.L.) HIGH/LOW	54/74LS (U.L.) HIGH/LOW
Inputs	1.0/1.0	1.25/1.25	0.5/0.25
Outputs	20/10	25/12.5	10/5.0 (2.5)

DC AND AC CHARACTERISTICS: See Section 3 for U.L. definitions

SYMBOL	PARAMETER	54/74 Min	54/74 Max	54/74S Min	54/74S Max	54/74LS Min	54/74LS Max	UNITS	CONDITIONS
I_{CCH}	Power Supply Current		22		32		6.2	mA	V_{IN} = Open, V_{CC} = Max
I_{CCL}			38		68		9.8		V_{IN} = Gnd
t_{PLH}	Propagation Delay		15	2.0	7.0		15	ns	Figs. 3-1, 3-5
t_{PHL}			22	2.0	7.0		15		

*DC limits apply over operating temperature range; AC limits apply at T_A = +25°C and V_{CC} = +5.0 V.

54/7437
54LS/74LS37
QUAD 2-INPUT NAND BUFFER

ORDERING CODE: See Section 9

PKGS	PIN OUT	COMMERCIAL GRADE $V_{CC} = +5.0\ V \pm 5\%$, $T_A = 0°C$ to $+70°C$	MILITARY GRADE $V_{CC} = +5.0\ V \pm 10\%$, $T_A = -55°c$ to $+125°C$	PKG TYPE
Plastic DIP (P)	A	7437PC, 74LS37PC		9A
Ceramic DIP (D)	A	7437DC, 74LS37DC	5437DM, 54LS37DM	6A
Flatpak (F)	A	7437FC, 74LS37FC	5437FM, 54LS37FM	3I

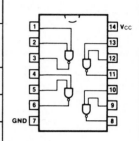

CONNECTION DIAGRAM
PINOUT A

INPUT LOADING/FAN-OUT: See Section 3 for U.L. definitions

PINS	54/74 (U.L.) HIGH/LOW	54/74LS (U.L.) HIGH/LOW
Inputs	1.0/1.0	0.5/0.25
Outputs	30/30	30/15 (7.5)

DC AND AC CHARACTERISTICS: See Section 3*

SYMBOL	PARAMETER		54/74 Min Max	54/74LS Min Max	UNITS	CONDITIONS
V_{OH}	Output HIGH Voltage	XM	2.4	2.5	V	V_{CC} = Max, I_{OH} = -1.2 mA
		XC	2.4	2.7		$V_{IN} = V_{IL}$
V_{OL}	Output LOW Voltage	XM, XC	0.4		V	I_{OL} = 48 mA
		XM		0.4		I_{OL} = 12 mA V_{CC} = Min
		XC		0.5		I_{OL} = 24 mA V_{IN} = 2.0 V
I_{OS}	Output Short Circuit Current	XM	-20 -70	-30 -130	mA	V_{CC} = Min, V_{OUT} = 0 V
		XC	-18 -70	-30 -130		
I_{CCH}	Power Supply Current		15.5	2.0	mA	V_{IN} = Gnd V_{CC} = Max
I_{CCL}			54	12		V_{IN} = Open
t_{PLH}	Propagation Delay		22	20	ns	Figs. 3-1, 3-4
t_{PHL}			15	20		

*DC limits apply over operating temperature range; AC limits apply at T_A = +25°C and V_{CC} = +5.0 V.

54/7486
54S/74S86
54LS/74LS86

QUAD 2-INPUT EXCLUSIVE-OR GATE

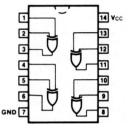

ORDERING CODE: See Section 9

PKGS	PIN OUT	COMMERCIAL GRADE V_{CC} = +5.0 V ±5%, T_A = 0° C to +70° C	MILITARY GRADE V_{CC} = +5.0 V ±10%, T_A = -55° C to +125° C	PKG TYPE
Plastic DIP (P)	A	7486PC, 74S86PC 74LS86PC		9A
Ceramic DIP (D)	A	7486DC, 74S86DC 74LS86DC	5486DM, 54S86DM 54LS86DM	6A
Flatpak (F)	A	7486FC, 74S86FC 74LS86FC	5486FM, 54S86FM 54LS86FM	3I

INPUT LOADING/FAN-OUT: See Section 3 for U.L. definitions

PINS	54/74 (U.L.) HIGH/LOW	54/74S (U.L.) HIGH/LOW	54/74LS (U.L.) HIGH/LOW
Inputs	1.0/1.0	1.25/1.25	1.0/0.375
Outputs	20/10	25/12.5	10/5.0 (2.5)

DC AND AC CHARACTERISTICS: See Section 3*

SYMBOL	PARAMETER		54/74 Min Max	54/74S Min Max	54/74LS Min Max	UNITS	CONDITIONS
I_{CC}	Power Supply Current	XM	43	75	10	mA	V_{CC} = Max, V_{IN} = Gnd
		XC	50	75	10		
t_{PLH}	Propagation Delay		23	3.5 10.5	12	ns	Other Input LOW Figs. 3-1, 3-5
t_{PHL}			17	3.0 10	17		
t_{PLH}	Propagation Delay		30	3.5 10.5	13	ns	Other Input HIGH Figs. 3-1, 3-4
t_{PHL}			22	3.0 10	12		

*DC limits apply over operating temperature range; AC limits apply at T_A = +25° C and V_{CC} = +5.0 V.

54/7490A
54LS/74LS90
DECADE COUNTER

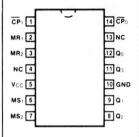

DESCRIPTION — The '90 is a 4-stage ripple counter containing a high speed flip-flop acting as a divide-by-two and three flip-flops connected as a divide-by-five counter. It can be connected to operate with a conventional BCD output pattern or it can be connected to provide a 50% duty cycle output. In the BCD mode, HIGH signals on the Master Set (MS) inputs set the outputs to BCD nine. HIGH signals on the Master Reset (MR) inputs force all outputs LOW. For a similar counter with corner power pins, see the 'LS290; for dual versions, see the 'LS390 and 'LS490.

LOGIC SYMBOL

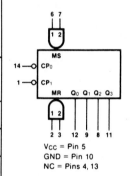

V_{CC} = Pin 5
GND = Pin 10
NC = Pins 4, 13

ORDERING CODE: See Section 9

PKGS	PIN OUT	COMMERCIAL GRADE V_{CC} = +5.0 V ±5%, T_A = 0°C to +70°C	MILITARY GRADE V_{CC} = +5.0 V ±10%, T_A = -55°C to +125°C	PKG TYPE
Plastic DIP (P)	A	7490APC, 74LS90PC		9A
Ceramic DIP (D)	A	7490ADC, 74LS90DC	5490ADM, 54LS90DM	6A
Flatpak (F)	A	7490AFC, 74LS90FC	5490AFM, 54LS90FM	3I

INPUT LOADING/FAN-OUT: See Section 3 for U.L. defintions

PIN NAMES	DESCRIPTION	54/74 (U.L.) HIGH/LOW	54/74LS (U.L.) HIGH/LOW
$\overline{CP_0}$	÷2 Section Clock Input (Active Falling Edge)	2.0/2.0	0.125/1.5
$\overline{CP_1}$	÷5 Section Clock Input (Active Falling Edge)	3.0/3.0	0.250/2.0
MR_1, MR_2	Asynchronous Master Reset Inputs (Active HIGH)	1.0/1.0	0.5/0.25
MS_1, MS_2	Asynchronous Master Set (Preset 9) Inputs (Active HIGH)	1.0/1.0	0.5/0.25
Q_0	÷2 Section Output*	20/10	10/5.0 (2.5)
$Q_1 - Q_3$	÷5 Section Outputs	20/10	10/5.0 (2.5)

*The Q_0 output is guaranteed to drive the full rated fan-out plus the $\overline{CP_1}$ input.

90

FUNCTIONAL DESCRIPTION — The '90 is a 4-bit ripple type decade counter. It consists of four master/slave flip-flops which are internally connected to provide a divide-by-two section and a divide-by-five section. Each section has a separate clock input which initiates state changes of the counter on the HIGH-to-LOW clock transition. State changes of the Q outputs do not occur simultaneously because of internal ripple delays. Therefore, decoded output signals are subject to decoding spikes and should not be used for clocks or strobes. The Q_0 output of each device is designed and specified to drive the rated fan-out plus the $\overline{CP}_1$ input. A gated AND asynchronous Master Reset (MR_1, MR_2) is provided which overrides the clocks and resets (clears) all the flip-flops. A gated AND asynchronous Master Set (MS_1, MS_2) is provided which overrides the clocks and the MR inputs and sets the outputs to nine (HLLH). Since the output from the divide-by-two section is not internally connected to the succeeding stages, the devices may be operated in various counting modes.:

A. BCD Decade (8421) Counter — The $\overline{CP}_1$ input must be externally connected to the Q_0 output. The $\overline{CP}_0$ input receives the incoming count and a BCD count sequence is produced.

B. Symmetrical Bi-quinary Divide-By-Ten Counter — The Q_3 output must be externally connected to the $\overline{CP}_0$ input. The input count is then applied to the $\overline{CP}_1$ input and a divide-by-ten square wave is obtained at output Q_0.

C. Divide-By-Two and Divide-By-Five Counter — No external interconnections are required. The first flip-flop is used as a binary element for the divide-by-two function ($\overline{CP}_0$ as the input and Q_0 as the output). The $\overline{CP}_1$ input is used to obtain binary divide-by-five operation at the Q_3 output.

MODE SELECTION

RESET/SET INPUTS				OUTPUTS			
MR_1	MR_2	MS_1	MS_2	Q_0	Q_1	Q_3	Q_3
H	H	L	X	L	L	L	L
H	H	X	L	L	L	L	L
X	X	H	H	H	L	L	H
L	X	L	X	Count			
X	L	X	L	Count			
L	X	X	L	Count			
X	L	L	X	Count			

H = HIGH Voltage Level
L = LOW Voltage Level
X = Immaterial

BCD COUNT SEQUENCE

COUNT	OUTPUTS			
	Q_0	Q_1	Q_2	Q_3
0	L	L	L	L
1	H	L	L	L
2	L	H	L	L
3	H	H	L	L
4	L	L	H	L
5	H	L	H	L
6	L	H	H	L
7	H	H	H	L
8	L	L	L	H
9	H	L	L	H

NOTE: Output Q_0 is connected to Input $\overline{CP}_1$ for BCD count.

LOGIC DIAGRAM

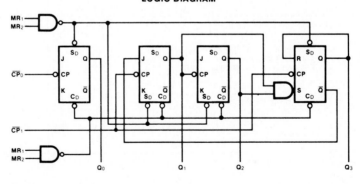

54/74107
54LS/74LS107

DUAL JK FLIP-FLOP
(With Separate Clears and Clocks)

DESCRIPTION — The '107 dual JK master/slave flip-flops have a separate clock for each flip-flop. Inputs to the master section are controlled by the clock pulse. The clock pulse also regulates the state of the coupling transistors which connect the master and slave sections. The sequence of operation is as follows: 1) isolate slave from master; 2) enter information from J and K inputs to master; 3) disable J and K inputs; 4) transfer information from master to slave.

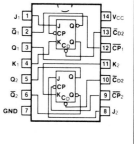

TRUTH TABLE

INPUTS		OUTPUT
@ t_n		@ t_{n+1}
J	K	Q
L	L	Q_n
L	H	L
H	L	H
H	H	$\overline{Q}_n$

H = HIGH Voltage Level
L = LOW Voltage Level
t_n = Bit time before clock pulse.
t_{n+1} = Bit time after clock pulse.

CLOCK WAVEFORM

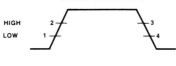

Asynchronous Input:
LOW input to $\overline{C}_D$ sets Q to LOW level
Clear is independent of clock

The 'LS107 offers individual J, K, clear, and clock inputs. These dual flip-flops are designed so that when the clock goes HIGH, the inputs are enabled and data will be accepted. The logic level of the J and K inputs may be allowed to change when the clock is HIGH and the bistable will perform according to the Truth Table as long as minimum setup times are observed. Input data is transferred to the outputs on the negative-going edge of the clock pulse.

ORDERING CODE: See Section 9

PKGS	PIN OUT	COMMERCIAL GRADE V_{CC} = +5.0 V ±5%, T_A = 0°C to +125°C	MILITARY GRADE V_{CC} = +5.0 V ±10%, T_A = -55°C to +125°C	PKG TYPE
Plastic DIP (P)	A	74107PC, 74LS107PC		9A
Ceramic DIP (D)	A	74107DC, 74LS107DC	54107DM, 54LS107DM	6A
Flatpak (F)	A	74107FC, 74LS107FC	54107FM, 54LS107FM	3I

LOGIC SYMBOL

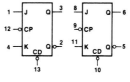

V_{CC} = Pin 14
GND = Pin 7

INPUT LOADING/FAN-OUT: See Section 3 for U.L. definitions

PIN NAMES	DESCRIPTION	54/74 (U.L.) HIGH/LOW	54/74LS (U.L.) HIGH/LOW
J_1, J_2, K_1, K_2	Data Inputs	1.0/1.0	0.5/0.25
$\overline{CP}_1$, $\overline{CP}_2$	Clock Pulse Inputs (Active Falling Edge)	2.0/2.0	2.0/0.5
$\overline{C}_{D1}$, $\overline{C}_{D2}$	Direct Clear Inputs (Active LOW)	2.0/2.0	1.5/0.5
Q_1, Q_2, $\overline{Q}_1$, $\overline{Q}_2$	Outputs	20/10	10/5.0 (2.5)

LOGIC DIAGRAM (one half shown)

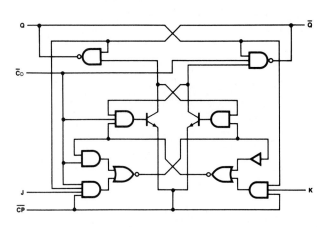

DC CHARACTERISTICS OVER OPERATING TEMPERATURE RANGE (unless otherwise specified)

SYMBOL	PARAMETER	54/74 Min	54/74 Max	54/74LS Min	54/74LS Max	UNITS	CONDITIONS
I_{CC}	Power Supply Current		40		8.0	mA	V_{CC} = Max, V_{CP} = 0 V

AC CHARACTERISTICS: V_{CC} = +5.0 V, T_A = +25° C (See Section 3 for waveforms and load configurations)

SYMBOL	PARAMETER	54/74 C_L − 15 pF R_L = 400 Ω Min	Max	54/74LS C_L − 15 pF Min	Max	UNITS	CONDITIONS
f_{max}	Maximum Clock Frequency	15		30		MHz	Figs. 3-1, 3-9
t_{PLH} t_{PHL}	Propagation Delay $\overline{CP}_n$ to Q_n or $\overline{Q}_n$		25 40		20 30	ns	Figs. 3-1, 3-9
t_{PLH} t_{PHL}	Propagation Delay $\overline{C}_{Dn}$ to Q_n or $\overline{Q}_n$		25 40		20 30	ns	Figs. 3-1, 3-10

107

		AC OPERATING REQUIREMENTS: V_{CC} = +5.0 V, T_A = +25°C					

SYMBOL	PARAMETER	54/74		54/74LS		UNITS	CONDITIONS
		Min	Max	Min	Max		
t_s (H)	Setup Time HIGH J_n or K_n to $\overline{CP}_n$	0		20		ns	Fig. 3-18 ('107) Fig. 3-7 ('LS107)
t_h (H)	Hold Time HIGH J_n or K_n to $\overline{CP}_n$	0		0		ns	
t_s (L)	Setup Time LOW J_n or K_n to $\overline{CP}_n$	0		20		ns	
t_h (L)	Hold Time LOW J_n or K_n to $\overline{CP}_n$	0		0		ns	
t_w (H) t_w (L)	$\overline{CP}_n$ Pulse Width	20 47		13.5 20		ns	Fig. 3-9
t_w (L)	$\overline{C}_{Dn}$ Pulse Width LOW	25		25		ns	Fig. 3-10

54S/74S112
54LS/74LS112
DUAL JK NEGATIVE
EDGE-TRIGGERED FLIP-FLOP

DESCRIPTION — The '112 features individual J, K, Clock and asynchronous Set and Clear inputs to each flip-flop. When the clock goes HIGH, the inputs are enabled and data will be accepted. The logic level of the J and K inputs may change when the clock is HIGH and the bistable will perform according to the Truth Table as long as minimum setup and hold times are observed. Input data is transferred to the outputs on the falling edge of the clock pulse.

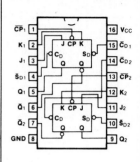

TRUTH TABLE

INPUTS		OUTPUT
@ t_n		@ t_{n+1}
J	K	Q
L	L	Q_n
L	H	L
H	L	H
H	H	$\bar{Q}_n$

Asynchronous Inputs:
LOW input to $\bar{S}_D$ sets Q to HIGH level
LOW input to $\bar{C}_D$ sets Q to LOW level
Clear and Set are independent of clock
Simultaneous LOW on $\bar{C}_D$ and $\bar{S}_D$
makes both Q and $\bar{Q}$ HIGH

t_n = Bit time before clock pulse.
t_{n+1} = Bit time after clock pulse.
H = HIGH Voltage Level
L = LOW Voltage Level

LOGIC SYMBOL

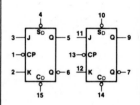

V_{CC} = Pin 16
GND = Pin 8

ORDERING CODE: See Section 9

PKGS	PIN OUT	COMMERCIAL GRADE V_{CC} = +5.0 V ±5%, T_A = 0°C to +70°C	MILITARY GRADE V_{CC} = +5.0 V ±10%, T_A = -55°C to +125°C	PKG TYPE
Plastic DIP (P)	A	74S112PC, 74LS112PC		9B
Ceramic DIP (D)	A	74S112DC, 74LS112DC	54S112DM, 54LS112DM	6B
Flatpak (F)	A	74S112FC, 74LS112FC	54S112FM, 54LS112FM	4L

INPUT LOADING/FAN-OUT: See Section 3 for U.L. definitions

PIN NAMES	DESCRIPTION	54/74S (U.L.) HIGH/LOW	54/74LS (U.L.) HIGH/LOW
J_1, J_2, K_1, K_2	Data Inputs	1.25/1.0	0.5/0.25
$\overline{CP}_1$, $\overline{CP}_2$	Clock Pulse Inputs (Active Falling Edge)	2.5/2.5	2.0/0.5
$\bar{C}_{D1}$, $\bar{C}_{D2}$	Direct Clear Inputs (Active LOW)	2.5/4.375	1.5/0.5
$\bar{S}_{D1}$, $\bar{S}_{D2}$	Direct Set Inputs (Active LOW)	2.5/4.375	1.5/0.5
Q_1, Q_2, $\bar{Q}_1$, $\bar{Q}_2$	Outputs	25/12.5	10/5.0 (2.5)

LOGIC DIAGRAM (one half shown)

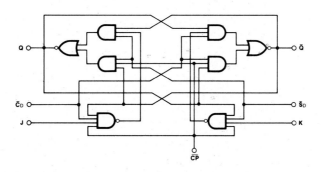

DC CHARACTERISTICS OVER OPERATING TEMPERATURE RANGE (unless otherwise specified)

SYMBOL	PARAMETER	54/74S		54/74LS		UNITS	CONDITIONS
		Min	Max	Min	Max		
I_{CC}	Power Supply Current		50		8.0	mA	V_{CC} = Max, V_{CP} = 0 V

AC CHARACTERISTICS: V_{CC} = +5.0 V, T_A = +25° C (See Section 3 for waveforms and load configurations)

SYMBOL	PARAMETER	54/74S		54/74LS		UNITS	CONDITIONS
		C_L = 15 pF R_L = 280 Ω		C_L = 15 pF			
		Min	Max	Min	Max		
f_{max}	Maximum Clock Frequency	80		30		MHz	Figs. 3-1, 3-9
t_{PLH} t_{PHL}	Propagation Delay $\overline{CP}_n$ to Q_n or $\overline{Q}_n$		7.0 7.0		16 24	ns	Figs. 3-1, 3-9
t_{PLH} t_{PHL}	Propagation Delay $\overline{C}_{Dn}$ or $\overline{S}_{Dn}$ to Q_n or $\overline{Q}_n$		7.0 7.0		16 24	ns	Figs. 3-1, 3-10

AC OPERATING REQUIREMENTS: V_{CC} = +5.0 V, T_A = +25° C

SYMBOL	PARAMETER	54/74S		54/74LS		UNITS	CONDITIONS
		Min	Max	Min	Max		
t_s (H) t_s (L)	Setup Time J_n or K_n to $\overline{CP}_n$	7.0 7.0		20 15		ns	Fig. 3-7
t_h (H) t_h (L)	Hold Time J_n or K_n to $\overline{CP}_n$	0 0		0 0		ns	
t_w (H) t_w (L)	$\overline{CP}_n$ Pulse Width	6.0 6.5		20 15		ns	Fig. 3-9
t_w (L)	$\overline{C}_{Dn}$ or $\overline{S}_{Dn}$ Pulse Width LOW	8.0		15		ns	Fig. 3-10

54/74121
MONOSTABLE MULTIVIBRATOR

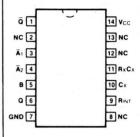

DESCRIPTION — The '121 features positive and negative dc level triggering inputs and complementary outputs. Input pin 5 directly activates a Schmitt circuit which provides temperature compensated level detection, increases immunity to positive-going noise and assures jitter-free response to slowly rising triggers.

When triggering occurs, internal feedback latches the circuit, prevents re-triggering while the output pulse is in progress and increases immunity to negative-going noise. Noise immunity is typically 1.2 V at the inputs and 1.5 V on V_{CC}.

Output pulse width stability is primarily a function of the external R_x and C_x chosen for the application. A 2 kΩ internal resistor is provided for optional use where output pulse width stability requirements are less stringent. Maximum duty cycle capability ranges from 67% with a 2 kΩ resistor to 90% with a 40 kΩ resistor. Duty cycles beyond this range tend to reduce the output pulse width. Otherwise, output pulse width follows the relationship:

$$t_w = 0.69 \, R_x C_x$$

ORDERING CODE: See Section 9

LOGIC SYMBOL

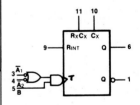

V_{CC} = Pin 14
GND = Pin 7
NC = Pins 2,8,12,13

PKGS	PIN OUT	COMMERCIAL GRADE V_{CC} = +5.0 V ±5%, T_A = 0°C to +70°C	MILITARY GRADE V_{CC} = +5.0 V ±10%, T_A = -55°C to +125°C	PKG TYPE
Plastic DIP (P)	A	74121PC		9A
Ceramic DIP (D)	A	74121DC	54121DM	6A
Flatpak (F)	A	74121FC	54121FM	3I

INPUT LOADING/FAN-OUT: See Section 3 for U.L. definitions

PIN NAMES	DESCRIPTION	54/74 (U.L.) HIGH/LOW
$\overline{A}_1$, $\overline{A}_2$	Trigger Inputs (Active Falling Edge)	1.0/1.0
B	Schmitt Trigger Input (Active Rising Edge)	2.0/2.0
Q, $\overline{Q}$	Outputs	20/10

54/74193
54LS/74LS193

UP/DOWN BINARY COUNTER
(With Separate Up/down Clocks)

DESCRIPTION — The '193 is an up/down modulo-16 binary counter. Separate Count Up and Count Down Clocks are used and in either counting mode the circuits operate synchronously. The outputs change state synchronous with the LOW-to-HIGH transitions on the clock inputs. Separate Terminal Count Up and Terminal Count Down outputs are provided which are used as the clocks for subsequent stages without extra logic, thus simplifying multistage counter designs. Individual preset inputs allow the circuits to be used as programmable counters. Both the Parallel Load ($\overline{PL}$) and the Master Reset (MR) inputs asynchronously override the clocks. For functional description and detail specifications please refer to the '192 data sheet.

CONNECTION DIAGRAM
PINOUT A

P_1 1	16 V_{CC}
Q_1 2	15 P_0
Q_0 3	14 MR
CP_D 4	13 $\overline{TC}_D$
CP_U 5	12 $\overline{TC}_U$
Q_2 6	11 $\overline{PL}$
Q_3 7	10 P_2
GND 8	9 P_3

LOGIC SYMBOL

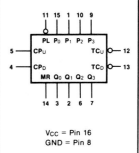

V_{CC} = Pin 16
GND = Pin 8

ORDERING CODE: See Section 9

PKGS	PIN OUT	COMMERCIAL GRADE V_{CC} = +5.0 V ±5%, T_A = 0°C to +70°C	MILITARY GRADE V_{CC} = +5.0 V ±10%, T_A = -55°C to +125°C	PKG TYPE
Plastic DIP (P)	A	74193PC, 74LS193PC		9B
Ceramic DIP (D)	A	74193DC, 74LS193DC	54193DM, 54LS193DM	6B
Flatpak (F)	A	74193FC, 74LS193FC	54193FM, 54LS193FM	4L

INPUT LOADING/FAN-OUT: See Section 3 for U.L. definitions

PIN NAMES	DESCRIPTION	54/74 (U.L.) HIGH/LOW	54/74LS (U.L.) HIGH/LOW
CP_U	Count Up Clock Input (Active Rising Edge)	1.0/1.0	0.5/0.25
CP_D	Count Down Clock Input (Active Rising Edge)	1.0/1.0	0.5/0.25
MR	Asynchronous Master Reset Input (Active HIGH)	1.0/1.0	0.5/0.25
$\overline{PL}$	Asynchronous Parallel Load Input (Active LOW)	1.0/1.0	0.5/0.25
P_0 — P_3	Parallel Data Inputs	1.0/1.0	0.5/0.25
Q_0 — Q_3	Flip-flop Outputs	20/10	10/5.0 (2.5)
$\overline{TC}_D$	Terminal Count Down (Borrow) Output (Active LOW)	20/10	10/5.0 (2.5)
$\overline{TC}_U$	Terminal Count Up (Carry) Output (Active LOW)	20/10	10/5.0 (2.5)

193

MODE SELECT TABLE

MR	$\overline{PL}$	CP_U	CP_D	MODE
H	X	X	X	Reset (Asyn.)
L	L	X	X	Preset (Asyn.)
L	H	H	H	No Change
L	H	⎍	H	Count Up
L	H	H	⎍	Count Down

H = HIGH Voltage Level
L = LOW Voltage Level
X = Immaterial
Z = High Impedance

STATE DIAGRAM

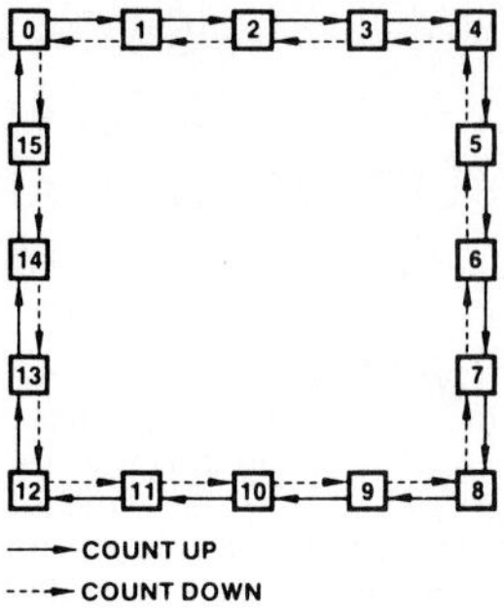

──▶ COUNT UP
----▶ COUNT DOWN

**LOGIC EQUATIONS
FOR TERMINAL COUNT**

$$\overline{TC}_U = Q_0 \bullet Q_1 \bullet Q_2 \bullet Q_3 \bullet \overline{CP}_U$$
$$\overline{TC}_D = \overline{Q}_0 \bullet \overline{Q}_1 \bullet \overline{Q}_2 \bullet \overline{Q}_3 \bullet \overline{CP}_D$$

LOGIC DIAGRAM

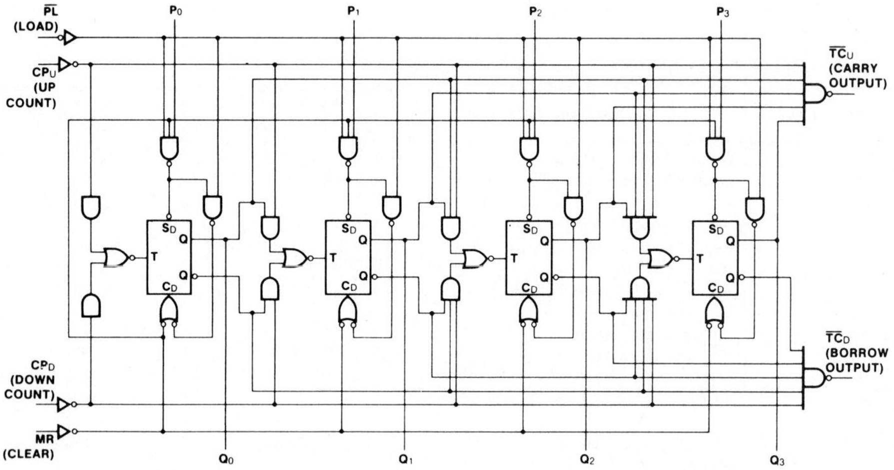

4001B   4002B
QUAD 2-INPUT NOR GATE ● DUAL 4-INPUT NOR GATE

DESCRIPTION — These CMOS logic elements provide the positive input NOR function. The outputs are fully buffered for highest noise immunity and pattern insensitivity of output impedance.

4001B
LOGIC AND CONNECTION DIAGRAM
DIP (TOP VIEW)

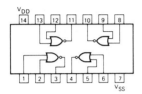

NOTE:
The Flatpak versions have the same pinouts (Connection Diagram) as the Dual In-line Package.

4002B
LOGIC AND CONNECTION DIAGRAM
DIP (TOP VIEW)

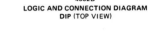
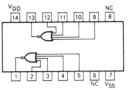

DC CHARACTERISTICS: V_{DD} as shown, V_{SS} = 0 V

SYMBOL	PARAMETER		LIMITS									UNITS	TEMP	TEST CONDITIONS See Note 1
			V_{DD} = 5 V			V_{DD} = 10 V			V_{DD} = 15 V					
			MIN	TYP	MAX	MIN	TYP	MAX	MIN	TYP	MAX			
I_{DD}	Quiescent Power Supply Current	XC			1			2			4	μA	MIN, 25°C	All inputs at 0 V or V_{DD}
					7.5			15			30		MAX	
		XM			0.25			0.5			1	μA	MIN, 25°C	
					7.5			15			30		MAX	

AC CHARACTERISTICS: V_{DD} as shown, V_{SS} = 0 V, T_A = 25°C, 4001B only (See Note 2)

SYMBOL	PARAMETER		LIMITS									UNITS	TEST CONDITIONS See Note 2
			V_{DD} = 5 V			V_{DD} = 10 V			V_{DD} = 15V				
			MIN	TYP	MAX	MIN	TYP	MAX	MIN	TYP	MAX		
t_{PLH}	Propagation Delay			60	110		25	60		20	48	ns	C_L = 50 pF,
t_{PHL}				60	110		25	60		20	48	ns	R_L = 200 kΩ
t_{TLH}	Output Transition Time			60	135		30	70		20	45	ns	Input Transition
t_{THL}				60	135		30	70		20	45	ns	Times ⩽ 20 ns

AC CHARACTERISTICS: V_{DD} as shown, V_{SS} = 0 V, T_A = 25°C, 4002B only

SYMBOL	PARAMETER		LIMITS									UNITS	TEST CONDITIONS See Note 2
			V_{DD} = 5 V			V_{DD} = 10 V			V_{DD} = 15V				
			MIN	TYP	MAX	MIN	TYP	MAX	MIN	TYP	MAX		
t_{PLH}	Propagation Delay			65	110		30	60		20	48	ns	C_L = 50 pF,
t_{PHL}				70	110		30	60		23	48	ns	R_L = 200 kΩ
t_{TLH}	Output Transition Time			75	135		40	70		30	45	ns	Input Transition
t_{THL}				60	135		23	70		15	45	ns	Times ⩽ 20 ns

NOTES:
1. Additional DC Characteristics are listed in this section under 4000B Series CMOS Family Characteristics.
2. Propagation Delays and Output Transition Times are graphically described in this section under 4000B Series CMOS Family Characteristics.

4001B

FAIRCHILD 4000B SERIES CMOS FAMILY CHARACTERISTICS

DC CHARACTERISTICS FOR THE 4000B SERIES CMOS FAMILY — Parametric Limits listed below are guaranteed for the entire Fairchild CMOS Family unless otherwise specified on the individual data sheets.

DC CHARACTERISTICS: V_{DD} = 5 V, V_{SS} = 0 V

SYMBOL	PARAMETER			LIMITS		UNITS	TEMP	TEST CONDITIONS	
			MIN	TYP	MAX				
V_{IH}	Input HIGH Voltage		3.5			V	All	Guaranteed Input HIGH Voltage	
V_{IL}	Input LOW Voltage				1.5	V	All	Guaranteed Input Low Voltage	
V_{OH}	Output HIGH Voltage		4.95			V	Min, 25°C	$I_{OH} < 1\,\mu A$, Inputs at 0 or 5 V per	
			4.95				MAX	the Logic Function or Truth Table	
			4.5			V	All	$I_{OH} < 1\,\mu A$, Inputs at 1.5 or 3.5 V	
V_{OL}	Output LOW Voltage				0.05	V	MIN, 25°C	$I_{OL} < 1\,\mu A$, Inputs at 0 or 5 V per	
					0.05		MAX	the Logic Function or Truth Table	
					0.5	V	All	$I_{OL} < 1\,\mu A$, Inputs at 1.5 or 3.5 V	
I_{OH}	Output HIGH Current		-0.63			mA	MIN, 25°C	V_{OUT} = 4.6 V	Inputs at 0 or 5 V per the Logic Function or Truth Table
			-0.36				MAX		
I_{OL}	Output LOW Current		1			mA	MIN, 25°C	V_{OUT} = 0.4 V	
			0.8						
			0.4				MAX		
C_{IN}	Input Capacitance Per Unit Load				7.5	pF	25°C	Any Input	
I_{DD}	Quiescent Power Supply Current	Gates	XC			1	μA	MIN, 25°C	All Inputs at 0 V or V_{DD} for all Valid Input Combinations
						7.5		MAX	
			XM			0.25	μA	MIN, 25°C	
						7.5		MAX	
		Buffers and Flip-Flops	XC			4	μA	MIN, 25°C	
						30		MAX	
			XM			1	μA	MIN, 25°C	
						30		MAX	
		MSI	XC			20	μA	MIN, 25°C	
						150		MAX	
			XM			5	μA	MIN, 25°C	
						150		MAX	

4049B • 4050B
4049B HEX INVERTING BUFFER • 4050B HEX NON-INVERTING BUFFER

DESCRIPTION — These CMOS buffers provide high current output capability suitable for driving TTL or high capacitance loads. Since input voltages in excess of the buffers' supply voltage are permitted, these buffers may also be used to convert logic levels of up to 15 V to standard TTL levels. The 4049B provides six inverting buffers, the 4050B six non-inverting buffers. Their guaranteed fan out into common bipolar logic elements is shown in Table 1.

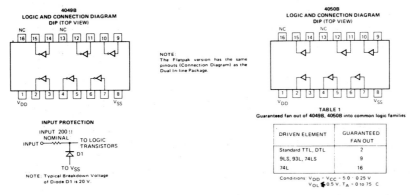

4049B LOGIC AND CONNECTION DIAGRAM DIP (TOP VIEW)

NOTE:
The Flatpak version has the same pinouts (Connection Diagram) as the Dual In-line Package.

4050B LOGIC AND CONNECTION DIAGRAM DIP (TOP VIEW)

INPUT PROTECTION

INPUT 200 Ω NOMINAL
INPUT ○——— TO LOGIC TRANSISTORS
                D1
          TO V_SS

NOTE: Typical Breakdown Voltage of Diode D1 is 20 V.

TABLE 1
Guaranteed fan out of 4049B, 4050B into common logic families

DRIVEN ELEMENT	GUARANTEED FAN OUT
Standard TTL, DTL	2
9LS, 93L, 74LS	9
74L	16

Conditions: $V_{DD} = V_{CC} = 5.0 \pm 0.25$ V
$V_{OL} \leq 0.5$ V, $T_A = 0$ to 75°C

DC CHARACTERISTICS: V_{DD} as shown, $V_{SS} = 0$ V, 4049BXM and 4050BXM (See Note 1)

SYMBOL	PARAMETER	$V_{DD} = 5$ V			$V_{DD} = 10$ V			$V_{DD} = 15$ V			UNITS	TEMP	TEST CONDITIONS
		MIN	TYP	MAX	MIN	TYP	MAX	MIN	TYP	MAX			
I_{OH}	Output HIGH Current	-1.85 -1.25 -0.9	-2.5								mA	MIN, 25°C MAX	$V_{OUT} = 2.5$ V for $V_{DD} = 5$ V Inputs at 0 or V_{DD} per Function
		-0.62 -0.5 -0.35	-1		-1.85 -1.25 -0.9	-2.5		-5.5 -3.75 -2.7	-7.5		mA	MIN, 25°C MAX	$V_{OUT} = 4.5$ V for $V_{DD} = 5$ V $V_{OUT} = 9.5$ V for $V_{DD} = 10$ V $V_{OUT} = 13.5$ V for $V_{DD} = 15$ V Inputs at 0 or V_{DD} per Function
I_{OL}	Output LOW Current	3.75 3 2.1	6		10 8 5.6	16		30 24 16.8	48		mA	MIN, 25°C MAX	$V_{OUT} = 0.4$ V for $V_{DD} = 5$ V $V_{OUT} = 0.5$ V for $V_{DD} = 10$ V $V_{OUT} = 1.5$ V for $V_{DD} = 15$ V Inputs at 0 or V_{DD} per Function
		3.3 2.6 1.8	5.2								mA	MIN, 25°C MAX	$V_{OUT} = 0.4$ V for $V_{DD} = 4.5$ V Inputs at 0 V or V_{DD} per Function
I_{DD}	Quiescent Power Supply Current			1 30			2 60			4 120	μA	MIN, 25°C MAX	All Inputs at 0 V or V_{DD}

Notes on the following page.

FAIRCHILD CMOS • 4049B • 4050B

SYMBOL	PARAMETER	V_{DD} = 5 V			V_{DD} = 10 V			V_{DD} = 15 V			UNITS	TEMP	TEST CONDITIONS
		MIN	TYP	MAX	MIN	TYP	MAX	MIN	TYP	MAX			
I_{OH}	Output HIGH Current	-1.5 -1.25 -1.0	-2.5								mA mA mA	MIN 25°C MAX	V_{OUT} = 2.5 V for V_{DD} = 5 V Inputs at 0 or V_{DD} per Function
		-0.6 -0.5 -0.4	-1		-1.5 -1.25 -1.0	-2.5		-4.5 -3.75 -3	-7.5		mA mA mA	MIN 25°C MAX	V_{OUT} = 4.5 V for V_{DD} = 5 V V_{OUT} = 9.5 V for V_{DD} = 10 V V_{OUT} = 13.5 V for V_{DD} = 15 V Inputs at 0 or V_{DD} per Function
I_{OL}	Output LOW Current	3.6 3.0 2.5	6		9.6 8 6.6	16		28 24 19	48		mA mA mA	MIN 25°C MAX	V_{OUT} = 0.4 V for V_{DD} = 5 V V_{OUT} = 0.5 V for V_{DD} = 10 V V_{OUT} = 1.5 V for V_{DD} = 15 V Inputs at 0 or V_{DD} per Function
		3.1 2.6 2.1	5.2								mA mA mA	MIN 25°C MAX	V_{OUT} = 0.4 V for V_{DD} = 4.5 V Inputs at 0 V or V_{DD} per Function
I_{DD}	Quiescent Power Supply Current		4 30			8 60				16 120	μA	MIN,25°C MAX	All inputs at 0 V or V_{DD}

SYMBOL	PARAMETER	V_{DD} = 5 V			V_{DD} = 10 V			V_{DD} = 15 V			UNITS	TEST CONDITIONS
		MIN	TYP	MAX	MIN	TYP	MAX	MIN	TYP	MAX		
t_{PLH} t_{PHL}	Propagation Delay		65 50	130 105		30 25	65 50		29 17	52 40	ns	C_L = 50 pF, R_L = 200 kΩ
t_{TLH} t_{THL}	Output Transition Time		73 33	145 65		40 13	80 25		30 9	60 20	ns	Input Transition Times ⩽ 20 ns

NOTES:
1. Additional DC Characteristics are listed in this section under 4000B Series CMOS Family Characteristics.
2. Propagation Delays and Output Transition Times are graphically described in this section under 4000B Series CMOS Family Characteristics.

ANSWERS TO SELECTED PROBLEMS

Chapter One

1-2 (a) 25
     (b) 9.5625
     (c) 1241.6875
1-4 1023
1-5 Nine bits
1-6 (a) four circuits for parallel
     (b) one circuit for serial

Chapter Two

2-1 (a) 22
     (b) 141
     (c) 2313
     (d) 983_{10}
     (e) 191_{10}
2-2 (a) 100101
     (b) 1110
     (a) 10111101
     (d) 11001101
     (e) 100100001001
2-3 65,535
2-4 (a) 483
     (b) 30
     (c) 2047
     (d) 175
     (e) 672
2-5 (a) 73_8
     (b) 564_8
     (c) 1627_8
     (d) 200000_8
     (e) 377
2-6 (a) 111100011
     (b) 011110
     (c) 011111111111

     (d) 010101111
     (e) 001010000100
2-7 (a) 26_8
     (b) 215_8
     (c) 4411_8
     (d) 1727_8
     (e) 277_8
2-8 165, 166, 167, 170, 171, 172, 173, 174, 175, 176, 177, 200
2-9 100100001001
2-10 (a) 146
     (b) 422
     (c) 14333
     (d) 704
     (e) 2047
2-11 (a) 4B
     (b) 13A
     (c) 800
     (d) 6413
     (e) FFF
2-12 (a) 16
     (b) 8D
     (c) 909
     (d) 3D7
     (e) BF
2-13 (a) 10010010
     (b) 000110100110
     (c) 0011011111111101
     (d) 001011000000
     (e) 011111111111
2-14 (a) 65,536 addresses
     (b) 000 to FFF (hex)
2-15 280, 281, 282, 283, 284, 285, 286, 287, 288, 289, 28A, 28B, 28C, 28D, 28E, 28F, 290, 291, 292, 293, 294, 295, 296, 297, 298, 299, 29A, 29B, 29C, 29D, 29E, 29F, 2A0

2-16 (a) 01000111
(b) 100101100010
(c) 000110000111
(d) 0100001001101000100101100100111

2-17 10 bits for binary, 12 bits for BCD

2-18 (a) 9752
(b) 184
(c) 7775
(d) 492

2-19 X = 25/Y

2-20 11011000 (X); 10111101 (=); 10110010 (2); 00110101 (5); 10101111 (/); 01011001 (Y)

2-21 (a) 110110110 (parity bit is leftmost bit)
(b) 000101000
(c) 111110111

2-22 (a) 101110100 (parity bit on the left)
(b) 000111000
(c) 0000101100101
(d) 11001001000000001

2-23 (a) no single-bit error
(b) single-bit error
(c) double error
(d) no single-bit error

2-25 (a) 10110001001
(b) 11111111
(c) 209
(d) 59,943
(e) 4701
(f) 777
(g) 157
(h) 2254
(i) 1961
(j) 15900
(k) 640
(l) 952B
(m) 100001100101
(n) 947
(o) 135_{16}
(p) 5464_8
(q) 1001010
(r) 01011000 (BCD)

Chapter Three

3-1

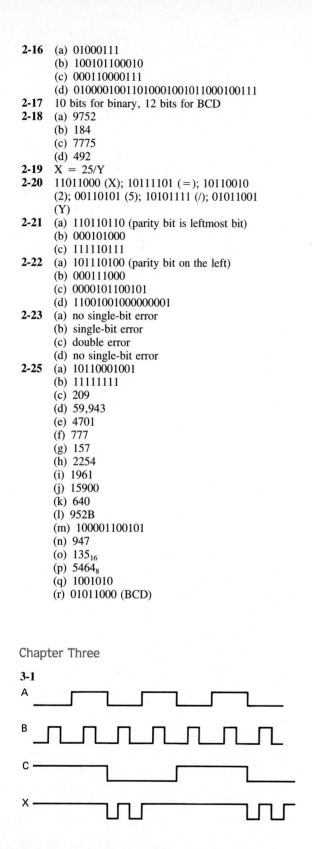

3-2

3-3 X will be a constant HIGH

3-5 31

3-6 (a) x is HIGH only when A, B and C are all HIGH.
(b) x is a constant LOW.
(c) x is HIGH only when B and C are simultaneously HIGH.

3-7 Change the OR gate to an AND gate.

3-8 OUT is always LOW.

3-10 $x = \overline{(\overline{A} + \overline{B})\, BC}$

3-11 $x = \overline{A}\,\overline{B}\,\overline{C} + A\overline{B}\,\overline{C} + \overline{A}\,\overline{B}\,D$

3-12 $x = D \cdot (AB + C) + E$

3-13 x = 1 only when B = C = 1 and A = D = 0.

3-14 (a)

3-15–3-18

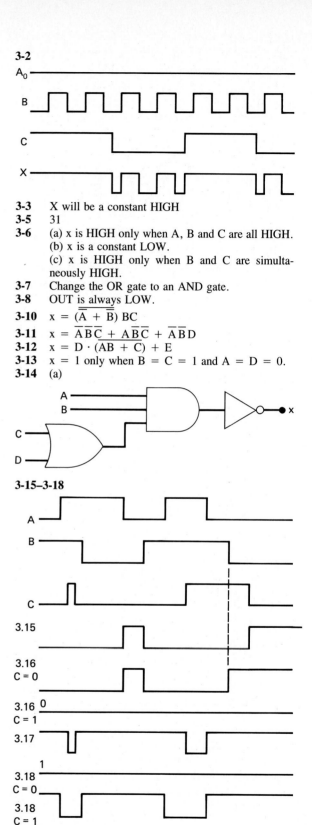

3-19 $x = \overline{(\overline{A} + \overline{B}) \cdot (\overline{BC})}$

3-20 $x = 0$ only when $A = B = 0, C = 1$.

3-21 (c)

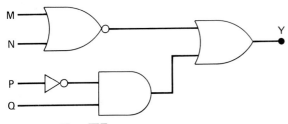

3-23 $M P \overline{N} + \overline{M} \overline{P} N$

3-24 $B \overline{C}$

3-26 (a) $A + \overline{B} + \underline{C}$
(b) $\underline{A} \cdot (\underline{B} + \overline{C})$
(c) $\overline{A} + \overline{B} + CD$
(d) $\overline{A} + B + \overline{C} + \overline{D}$

3-27 $A + B + \overline{C}$

3-33 (a) NOR
(b) AND
(c) NAND

3-34

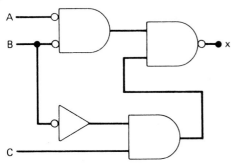

3-35 (a) Z goes HIGH only when $A = B = 0$ and $C = 1$.
(b) Z goes LOW when A or B is HIGH or when C or D is LOW.

3-37 X will go HIGH when $E = 1$, or $D = 0$, or $B = C = 0$, or when $B = 1$ and $A = 0$.

3-39 (a) HIGH
(b) LOW

3-41

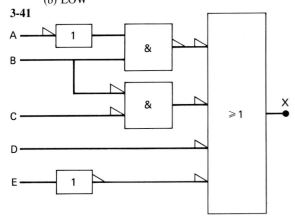

3-42 $Z = \overline{\overline{(\overline{AB} + \overline{C})F} + \overline{C} + \overline{(D + E)}}$

Chapter Four

4-1 (a) $BC + \overline{B}(\overline{C} + A)$
(b) $BC + \overline{B}(\overline{C} + A)$
(c) $\overline{D} + A\overline{B}\overline{C} + \overline{A}\overline{B}C$

4-2 $Q(M + N)$

4-3 $MN + Q$

4-4 One solution: $\overline{x} = \overline{B}C + AB\overline{C}$
another: $x = \overline{A}B + \overline{B}\overline{C} + BC$

4-5 $x = A\overline{B} + A\overline{C} + \overline{B}\overline{C}$

4-6 $x = \overline{A}\overline{B}C + ABC$

4-7 $x = A_3(A_2 + A_1A_0)$

4-8 ALARM $= ID + \overline{I}L$

4-9

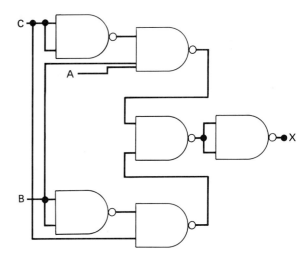

4-12 $x = BC + \overline{B}\overline{C} + AC$; or $BC + \overline{B}\overline{C} + A\overline{B}$

4-13 $y = \overline{D} + \overline{A}\overline{B}C + A\overline{B}\overline{C}$

	$\overline{C}\overline{D}$	$\overline{C}D$	CD	$C\overline{D}$
$\overline{A}\overline{B}$	1		1	1
$\overline{A}B$	1			1
AB	1			1
$A\overline{B}$	1	1		1

4-14 $x = \overline{A}_3A_2 + \overline{A}_3A_1A_0$

4-15 (a) $x = \overline{A}\overline{B} + \overline{A}\overline{C} + \overline{B}D + ABC\overline{D}$

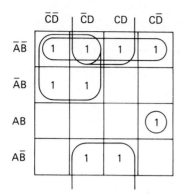

(b) $x = \overline{A}\overline{D} + \overline{B}C + \overline{B}\overline{D}$
(c) $y = \overline{B} + AC$

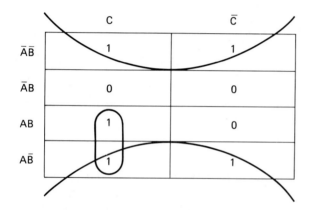

4-16 Best solution: $x = B\overline{C} + AD$
4-17 $x = \overline{S1}\ \overline{S2} + \overline{S1}\ \overline{S3} + \overline{S3}\ \overline{S4} + \overline{S2}\ \overline{S3} + \overline{S2}\ \overline{S4}$
4-18 (a)

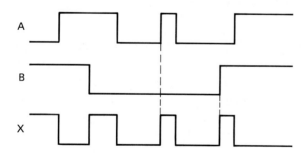

(b) $x = A$
(c) $x = \overline{A}$
4-19 $A = 0, B = C = 1$

4-20 One possibility:

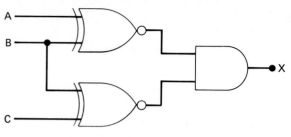

4-21 Four outputs where Z_3 is the MSB
$Z_3 = Y_1 Y_0 X_1 X_0$
$Z_2 = Y_1 X_1 (\overline{Y_0} + \overline{X_0})$
$Z_1 = Y_0 X_1 (\overline{Y_1} + \overline{X_0}) + Y_1 X_0 (\overline{Y_0} + \overline{X_1})$
$Z_0 = Y_0 X_0$
4-22 $x = A_3 A_2 + A_3 A_1$
4-24

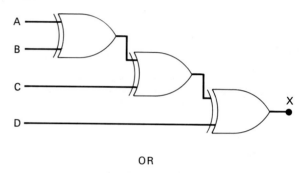

OR

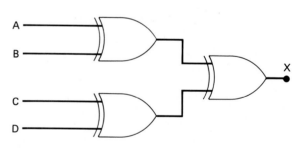

4-25 $x = AB + (\overline{C \oplus D})$
4-26 $(A + B)(C + D)$
4-27 $N/S = \overline{C}\overline{D}(A + B) + AB(\overline{C} + \overline{D})$
$E/W = \overline{N/S}$
4-28 $x = A\overline{B}C$
4-29 $x = A + BCD$
4-30 $x = A + (B \oplus C)$
4-31 $Z = A_0 S + A_1 \overline{S}$
4-32 $Z = X_1 X_0 Y_1 Y_0 + X_1 \overline{X_0} Y_1 \overline{Y_0} + \overline{X_1} X_0 \overline{Y_1} Y_0 + \overline{X_1}\overline{X_0}\overline{Y_1}\overline{Y_0}$
No pairs, quads, or octets.

4-33   (a) indeterminate
      (b) 1.4–1.8V
      (c)

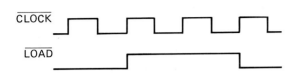

4-37   Possible faults:
      faulty V_{cc} or ground on Z2
         Z2-1 or Z2-2 open internally or externally
         Z2-3 internally open

4-38   YES: C, E, F
      NO: A, B, D, G

4-40   Z2-6 and Z2-11 shorted together

4-42   Most likely faults:
      faulty ground or V_{cc} on Z1
         Z1 plugged in backwards
         Z1 internally damaged

4-43   Possible faults:
      Z2-13 shorted to V_{cc}
      Z2-8 shorted to V_{cc}
      broken connection to Z2-13
      Z2-3, Z2-6, Z2-9, or Z2-10 shorted to ground

Chapter Five

5-1

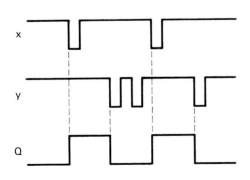

5-2   Same as 5-1

5-3

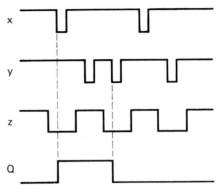

5-6   Z1-4 stuck HIGH
5-7   20ns
5-8

5-9   Q is a 500-Hz squarewave
5-10

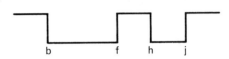

5-12   2-5 kHz
5-13

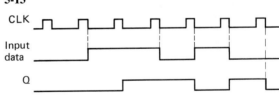

5-14   500 Hz squarewave
5-15   Q stays LOW
5-18

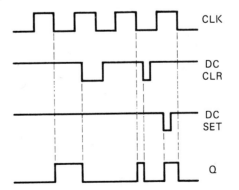

5-19

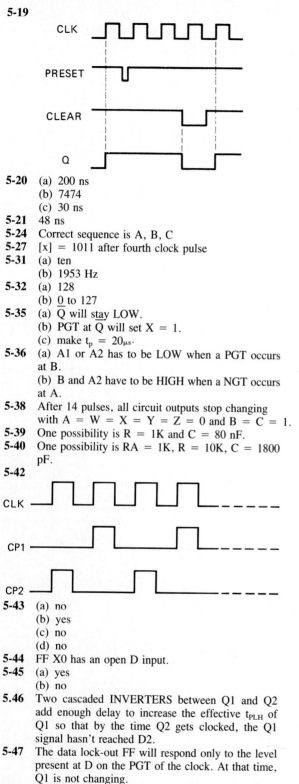

5-20 (a) 200 ns
(b) 7474
(c) 30 ns

5-21 48 ns

5-24 Correct sequence is A, B, C

5-27 [x] = 1011 after fourth clock pulse

5-31 (a) ten
(b) 1953 Hz

5-32 (a) 128
(b) 0 to 127

5-35 (a) Q will stay LOW.
(b) PGT at Q will set X = 1.
(c) make t_p = 20μs.

5-36 (a) A1 or A2 has to be LOW when a PGT occurs at B.
(b) B and A2 have to be HIGH when a NGT occurs at A.

5-38 After 14 pulses, all circuit outputs stop changing with A = W = X = Y = Z = 0 and B = C = 1.

5-39 One possibility is R = 1K and C = 80 nF.

5-40 One possibility is RA = 1K, R = 10K, C = 1800 pF.

5-42

5-43 (a) no
(b) yes
(c) no
(d) no

5-44 FF X0 has an open D input.

5-45 (a) yes
(b) no

5.46 Two cascaded INVERTERS between Q1 and Q2 add enough delay to increase the effective t_{PLH} of Q1 so that by the time Q2 gets clocked, the Q1 signal hasn't reached D2.

5-47 The data lock-out FF will respond only to the level present at D on the PGT of the clock. At that time, Q1 is not changing.

5-49 (a) no
(b) no
(c) yes
(d) no

5-50 First combination: 101
Second combination: 010

5-51 (a) no
(b) no
(c) yes

Chapter Six

6-1 (a) 10101
(b) 10010
(c) 1111.0101
(d) 1.1010
(e) 100111000

6-2 (a) 00100000 (incl. sign bit)
(b) 11110010
(c) 00111111
(d) 10011000
(e) 11111111
(f) 10000000
(g) can't be done
(h) 00000000

6-3 (a) +13
(b) −3
(c) +123
(d) −103
(e) +127
(f) −32
(g) −1
(h) −127

6-4 (a) −2048 to +2047
(b) sixteen bits including sign bit

6-5 -16_{10} to $+15_{10}$

6-6 (a) 164
(b) −92
(c) illegal BCD
(d) $ (e) −36

6-7 0 to 1023; −512 to +511

6-9 (a) 00001111
(b) 11111101
(c) 11111011
(d) 10000000
(e) 00000001
(f) 11011110
(g) 00000000
(h) 00010101

6-11 (a) 100011
(b) 1111001
(c) 100011.00101
(d) .10001111

6-12 (a) 11
(b) 111
(c) 101.11
(d) 1111.0011

6-13 (a) 10010111 (BCD)
(b) 10010101 (BCD)

6-14 (a) 6E24
(b) 100D
6-15 (a) OEFE
(b) 229
(c) 02A6
6-16 16,849 locations
6-18 SUM = A ⊕ B
CARRY = AB
6-20 [A] = 0000
6-21 80ns
6-22 Overflow = $\overline{A_3}\,\overline{B_3}S_3 + A_3B_3\overline{S_3}$
6-25 (a) SUM = 0111
(b) SUM = 1010

6-26

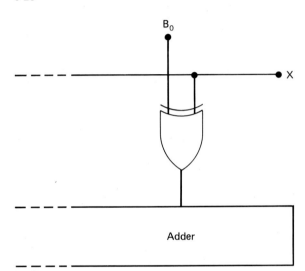

6-27 [S] = 01110; x = 1; [Σ] = 0100
6-28 Yes
6-29 [Σ] = 100001000101 (BCD)
6-30 After fourth clock pulse:
[A] = 00111111; [B] = 01110000
[X] = 0000; [S] = 10101111
6-31 (a) no
(b) yes
(c) no
6-34 (a) yes
(b) no
(c) yes
(d) no

Chapter Seven

7-1 250 kHz, 50%
7-2 Same as 7.1
7-3 10000_2
7-4 Five FFs are required: Q0-Q4 with Q4 as the MSB. Connect Q3 and Q4 outputs to a NAND gate whose output is connected to all CLRs.

7-5

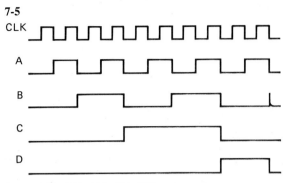

7-7 100, 011, 010, 001, 000 and repeats
7-8 (b) 000, 001, 010, 100, 101, 110 and repeats
7-9

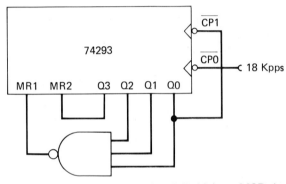

7-10 A MOD-15 (from problem 7-9) driving a MOD-4.
7-11 60 Hz
7-12 1000 and 0000 states never occur.
7-13 (a) 12.5 MHz
(b) 8.33 MHz
7-14 (a) Add two FFs (E and F) to Figure 7-55. Connect AND gates below to appropriate FF inputs.

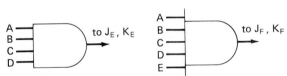

(b) 33 MHz
(c) 16.7 MHz
7-15 0000, 0001, 0010, 0011, 0100, 0101, 0110, 0111, 1000, 1001, and repeats
7-16 (a) Connect Q0 to CP1 and clock signal to CP0; ground MR1, MR2, MS1, and MS2.
(b) Connect clock signal to CPI; connect Q3 to CP0; ground all MR and MS inputs.
7-17 (d) Z will not be cleared and timer cannot be restarted.
7-18 Change the parallel data input to 1010.
7-19 Nine
7-20 (a) ten
(b) clears counter to 0000
(c) sets counter to 1001

7-21 Changes are shown below.

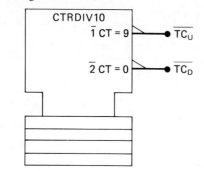

7-24

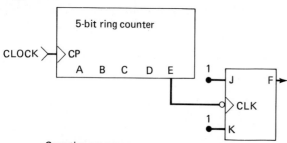

7-26 (a) when the counter goes from 0111 to 1000
7-27 Twelve FFs
7-28 Four; sixteen; thirteen
7-30

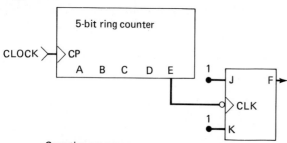

5-bit ring counter

CLOCK >—> CP

A B C D E

Counting sequence

A	B	C	D	E	F
1	0	0	0	0	0
0	1	0	0	0	0
0	0	1	0	0	0
0	0	0	1	0	0
0	0	0	0	1	0
1	0	0	0	0	1
0	1	0	0	0	1
0	0	1	0	0	1
0	0	0	1	0	1
0	0	0	0	1	1

Repeats

7-31 Counting sequence is: 00000, 10000, 11000, 11100, 11110, 11111, 01111, 00111, 00011, 00001, and repeats.
7-32 Frequency at Z is 5Hz.
7-33 (b) 257
7-34 (a) 22
    (b) 450
    (c) 0 or 1
7-41 64 microseconds

7-42 (b) $Q0 \cdot \overline{Q1} + Q0 \cdot Q7$
    (c) $Q3 \cdot \overline{Q6}$

Chapter Eight

8-1 (a) A, B
    (b) A
    (c) A
8-2 The results for the 7432 are:
    P_D (avg) = 41.2 mW
    t_{pd} (avg) = 18.5 ns
    speed-power = 762.2 pJ
8-3 0.9 V; 1.4 V
8-4 (a) 0.5 UL/0.25 UL
    (c) five
8-7 7486 fanout is not exceeded.
8-8 67 ns
8-9 125 ohms if noise margin of 0.4 V is to be maintained
8-14 AB + CD + FG
8-15 R_c (min) = 359 ohms
8-16 (a) 5 V
    (b) R_s = 110 ohms for a LED current of 20 mA
8-18

(a)
x	y	DATA ON BUS
0	0	C
0	1	B
1	0	B
1	1	A

8-19 R_P (max) = 5.77 kilohms
8-20 AND
8-24 V_{DD} = 5 V, f_{max} = 10 kHz
8-25 3.6 V for both states
8-26 Using values at V_{CC} = 6 V, the speed-power product is 1.44 pJ per gate.
8-27

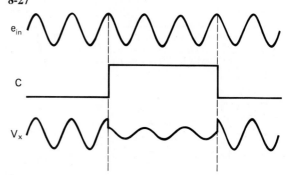

8-28 −1 and −2
8-30 none
8-31 two
8-36 b

Chapter Nine

9-1 six inputs, 64 outputs
9-2 [A] = 110, E_3 = 1, $\overline{E_1}$ = $\overline{E_2}$ = 0

9-3

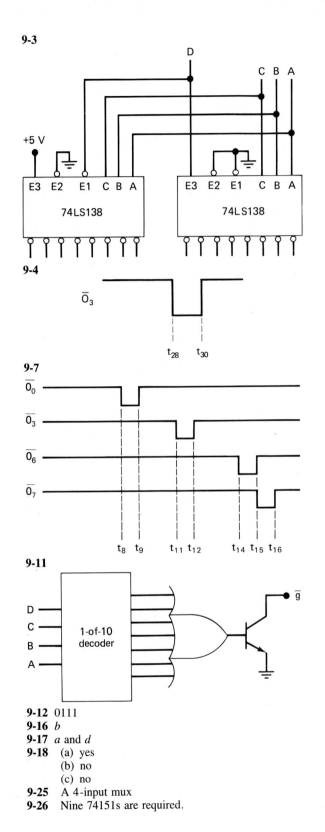

9-4

$\overline{O}_3$

t_{28} t_{30}

9-7

$\overline{O}_0$

$\overline{O}_3$

$\overline{O}_6$

$\overline{O}_7$

t_8 t_9 t_{11} t_{12} t_{14} t_{15} t_{16}

9-11

D —
C —   1-of-10
B —   decoder
A —

$\overline{g}$

9-12 0111
9-16 *b*
9-17 *a* and *d*
9-18 (a) yes
     (b) no
     (c) no
9-25 A 4-input mux
9-26 Nine 74151s are required.

9-27

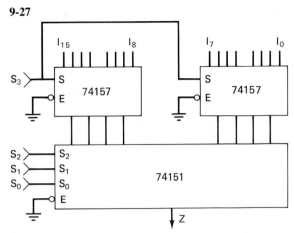

9-28 (b) The total number of connections in the circuit using muxes is 63, not including V_{CC} and GND, and not including the connections to counter clock inputs. The total number for the circuit using separate decoder/drivers is 66.

9-29

⟵ 1 cycle ⟶

9-30

⟵ 1 cycle ⟶

9-31

A	B	C	
0	0	0	$0 \rightarrow I_0$
0	0	1	$0 \rightarrow I_1$
0	1	0	$0 \rightarrow I_2$
0	1	1	$1 \rightarrow I_3$
1	0	0	$0 \rightarrow I_4$
1	0	1	$1 \rightarrow I_5$
1	1	0	$1 \rightarrow I_6$
1	1	1	$1 \rightarrow I_7$

9-32 Connect I_1, I_5, I_9, I_{11}, I_{14}, I_{15} to V_{CC}.
All other inputs connect to ground.

9-34 (b) $Z = \overline{A}\,\overline{B}C + A\overline{B}\,\overline{C} + \overline{A}BC + A\overline{B}C + \overline{A}\,\overline{B}\,\overline{C}D + ABCD$

9-35

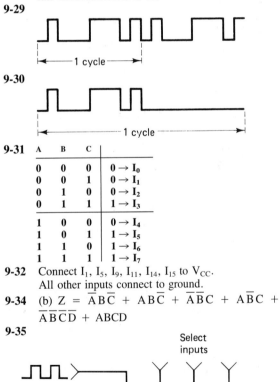

9-38   Five lines

9-40   (a) Sequencing stops after Actuator #3 is activated.

         (b) same as (a)

9-45   (a) no

         (b) yes

9-48 $\overline{OE}_c = 0$, $\overline{IE}_c = 1$; $\overline{OE}_B = \overline{OE}_A = 1$; $\overline{IE}_B = \overline{IE}_A = 0$; apply a clock pulse

9-49   (a) At t_4, each register will have a contents of 1011.

9-50

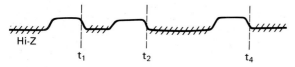

Hi-Z t_1 t_2 t_4

9-51   (b) ON, ON, ON, OFF (left to right).

9-52   (a) At t_3, each register holds 1001.

Chapter Ten

10-1   3.58 V

10-2   LSB = 20 mV

10-3   20mV; approx. 0.4%

10-4   Approx. 5 mV

10-5   eight

10-6   14.3%, 0.286 V

10-8   250.06 RPM

10-9   10 mV, 0.1%, 6.95 V

10.10

BINARY	**BCD**
3.88 mV	**10 mv**
0.392%	**1.01%**

10-11   800 ohms; no

10-12   (a) -0.3125 V, -4.6875 V

         (b) 4.27 kilohms

         (c) $-.0267$

10-13 $V_{out} = 2Vp - p$ for B = 1111

10-14   20 μA; yes

10-15   (a) seven

10-16   242.5 mV is not within specifications.

10-20   (a) 10010111

         (b) 10010111

         (c) 102 μs, 51 μs

10-23   (a) 1.2 mV

         (b) 2.7 mV

10-24   (a) 0111110110

         (b) 0111110111

10-25 $2.834 - 2.844$ V

10-27   Final register contents is $111110_2 = 62_{10} = 6.2$ V.

10-28   80 μs

10-29   100101

10-30   2.96 V

10-31   2.37 V

10-34   1 cycle = 3862 μs

Chapter Eleven

11-1   16,384; 32; 524,288

11-2   16,384

11-3   64K $\times$ 4

11-4   16; 16; 13; 16,384

11-5   (a) Hi-Z

         (b) 11101101

11-6   (a) register 11

         (b) 0100

11-7   (a) 16,384

         (b) four

         (c) two 1-of-128 decoders

11-8   120 ns

11-9   180 ns

11-10   Q13, Q14, Q15

11-11   The following transistors will have open base connections: Q0, Q2, Q5, Q6, Q7, Q9, Q15.

11-13   The waveform at D0 is:

11-15

A5	A4	A3	A2	A1	A0	D2	D1	D0	c3	c2	c1	c0
1	0	0	1	0	1	0	1	1	0	1	1	1
0	0	1	1	1	0	0	0	1	0	1	0	0
1	1	1	1	1	0	1	1	0	0	0	1	0
0	1	1	1	0	0	0	1	0	0	1	1	0
1	1	0	1	0	1	1	0	1	0	0	1	1
0	0	0	1	1	0	0	0	0	0	1	1	0
0	1	0	0	1	1	0	0	1	1	0	0	1
1	1	0	0	0	0	1	0	0	1	0	0	0

11-18   (a) 100 ns

         (b) 30 ns

         (c) ten million

         (d) 20 ns

         (e) 30 ns

         (f) 40 ns

         (g) ten million

11-19   seven

11-20

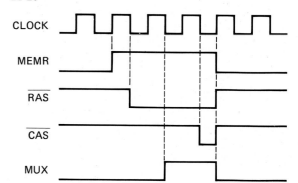

CLOCK

MEMR

$\overline{RAS}$

$\overline{CAS}$

MUX

11-21

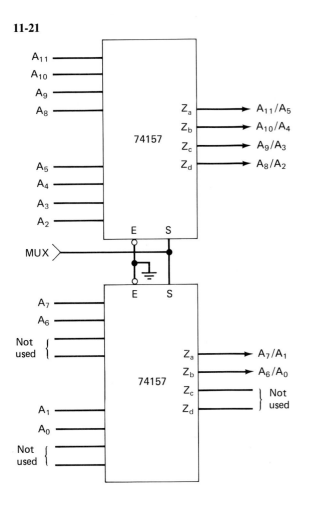

11-27 Add four more PROMs (PROM-4 through PROM-7) to the circuit. Connect their data outputs and address inputs to data and address bus respectively. Connect AB_{13} to C input of decoder, and connect decoder outputs 4-7 to CS inputs of PROMs 4-7 respectively.

11-28 Connect AB_{13}, AB_{14}, and AB_{15} to a 3-input OR gate, and connect the OR gate output to C input of the decoder.

11-29 (a) 32×8
(b) RAMs 2 and 3
(c) 00-0F; 10-1F

11-32 * PEACE

11-33 F000–F3FF; F400–F7FF; F800–FBFF; FC00–FFFF

11-41 $11101010_2 = EA_{16}$

INDEX